Calculus
Preliminary Edition

Calculus
Preliminary Edition

Robert Decker
UNIVERSITY OF HARTFORD

Dale Varberg
HAMLINE UNIVERSITY

 Prentice Hall, Upper Saddle River, New Jersey 07458

Library of Congress Cataloging-in-Publication Data

Decker, Robert, 1954–
    Calculus / Robert Decker, Dale Varberg.—Prelim. ed.
      p. cm.
    Includes bibliographical references and index.
    ISBN 0-13-287640-X
    1. Calculus. I. Varberg, Dale E. II. Title.
  QA303.D315 1996
  515—dc20

98–1291
CIP

Editor-in-Chief: **JEROME GRANT**
Editorial Director: **TIM BOZIK**
Assistant Vice-President of
   Production and Manufacturing: **DAVID W. RICCARDI**
Development Editor: **ALAN MacDONELL**
Production Editors: **RICHARD DeLORENZO/BARBARA MACK**
Managing Editor: **LINDA BEHRENS**
Prepress / Manufacturing Buyer: **ALAN FISCHER**
Manufacturing Manager: **TRUDY PISCIOTTI**
Art Director: **AMY ROSEN**
Interior Designer: **SHEREE GOODMAN**
Cover Designer: **ROLANDO CORUJO**
Editorial Assistant: **JOANNE WENDELKEN**
Supplements Editor: **AUDRA WALSH**
Interior Illustrator: **MONOTYPE COMPOSITION**
Compositor: **NEW ENGLAND TYPOGRAPHIC SERVICE, INC.**

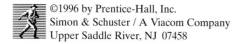 ©1996 by Prentice-Hall, Inc.
Simon & Schuster / A Viacom Company
Upper Saddle River, NJ 07458

Printed in the United States of America

10 9 8 7 6 5 4 3 2 1

ISBN 0-13-287640-X

Prentice-Hall International (UK) Limited, London
Prentice-Hall of Australia Pty. Limited, Sydney
Prentice-Hall Canada Inc., Toronto
Prentice-Hall Hispanoamericana, S.A., Mexico
Prentice-Hall of India Private Limited, New Delhi
Prentice-Hall of Japan, Inc., Tokyo
Simon & Schuster Asia Pte. Ltd., Singapore
Editora Prentice-Hall do Brasil, Ltda., Rio de Janeiro

Annotated Table of Contents

*F*or your convenience, we have annotated this table of contents in order to highlight some specific areas of the text where traditional presentation has been rejuvenated with what we believe to be new and innovative approaches.

The natural log function is introduced as the inverse of the natural exponential function, and its derivative is found numerically at several points.

2

Numerical and Graphical Techniques   89

Graphing calculators and computer algebra systems are discussed and used to approach problems numerically and graphically.

Graphs generated by calculator and computer are discussed in detail, with emphasis on how to interpret output. Examples 2 and 3 involve functions with asymptotes and explore methods for finding an appropriate graph window.

This section explains how to use calculators and computers to generate tables of values and discusses how they can be of use in solving problems. Example 2 explores the interplay between graphs and tables of values, using a model of an oscillating car bumper.

The concept of functions with parameters and families of functions is introduced and explored with the use of graphing technology. In Example 2, curve fitting by trial and error is used to model the cooling of a (human) body, and in the lab at the end of the section to model the motion of a damped pendulum.

Zooming into a function with a calculator or computer is explained and used to solve equations. Differentials are introduced and used to estimate small changes in one variable for a given small change in the other variable. In the lab at the end of the section, zooming is used to estimate the speed of Lab 3's pendulum when it is at the bottom of its swing.

Approximate and exact solutions to equations are discussed in the context of the built-in methods of graphing calculators and computer algebra systems. The bisection method and Newton's method of finding roots are briefly explained.

Linear equations are fit to data by the method of least squares regression, a method built into many graphing calculators and computer algebra systems.

The nonlinear model-fitting capabilities of calculators and computers are explored.

3

Derivatives   151

In this chapter we develop the standard derivative formulas and techniques of calculus.

Some simple derivative formulas are derived from the definition of the derivative. This section also discusses the use of computer algebra to find derivatives.

The power rule, sum, product, and quotient rules are developed.

Strategies for generating graphs of implicit equations by calculator and computer are discussed. Implicit derivatives are calculated and applied to various problem situations.

Differentials are applied to a variety of both functions and implicit equations, and to problems involving estimation and the approximation of solutions to equations. In Example 5 the oscillating car bumper returns as a practical application.

4

Applications of the Derivative   217

The first and second derivatives are used to find maximums, minimums, inflection points, and intervals for which a function is increasing, decreasing, concave up, and concave down. Students learn both exact and numerical/graphical approaches and must determine which approach is appropriate for a particular problem.

Both exact and numerical /graphical (calculator/computer based) methods of finding maximum and minimum values of a function are presented.

Both exact and numerical /graphical methods of finding intervals of monotonicity and concavity of a function are presented. In Example 6, these methods are applied to a cyclical population model.

The second derivative test for maximum and minimum values is presented and, in Example 5, applied to a function with a parameter for which numerical/graphical methods are not appropriate. In the lab at the end of the section, these ideas are applied to the Van der Waals equation relating pressure and volume for non-ideal gases.

We show that there are situations in which calculator/computer graphs are of little help, and for which hand-sketching techniques are useful.

5

The Integral   281

Indefinite integrals and differential equations are presented in the first two sections. The definite integral is approached and studied numerically in the second two sections. These two topics are then tied together with the Fundamental Theorem of Calculus in the remaining sections. Emphasis is placed on choosing between exact and approximate methods of finding definite integrals.

This section covers separable differential equations. In the lab at the end of the section, a model for fluid draining out of a can is developed to fit to real data.

Areas are approximated numerically using sequences of finite (left- and right-endpoint) Riemann sums.

The definite integral is defined and explored numerically using both midpoint Riemann sums and the built-in numerical methods of calculators and computers.

Exact definite integrals are found by hand and with the use of computer algebra.

6

Applications of the Integral   349

The definite integral is applied to numerous situations.

Parametric representation of curves is introduced, including calculator and computer graphing of such curves. In the lab project at the end of the section, a problem involving a guitar string is used to show when it is helpful to be able to evaluate a definite integral exactly, and when a numerical approximation is necessary.

7

Transcendental Functions and Differential Equations   395

The exponential, logarithmic and trigonometric functions are explored in greater detail, as well as the relationship between exponential functions and differential equations.

This section shows that by defining the natural log function using a definite integral, and the natural exponential function as its inverse, it now becomes relatively easy to demonstrate several properties of the log and exponential functions that were previously left unproven. In the lab at the end, a model of falling objects that encounter air resistance is developed and fit to real data taken of a falling balloon.

This section explores the relationship of the exponential function to certain first-order differential equations.

Euler's method of numerically solving differential equations is presented, and the numerical methods built in to some calculators and computer algebra systems are investigated. In the lab at the end of the section, advantages and disadvantages of exact versus numerical solutions are discovered in the context of retirement planning.

8

Techniques of Integration   453

Several methods of finding indefinite and definite integrals are presented and applied to variety of situations.

Simple partial fraction decomposition (both by hand and with computer algebra) is presented. In Example 4, the logistic population model of population growth is developed by solving the appropriate differential equation and is further explored in the lab at the end of the section.

L'Hôpital's Rule is presented.

In the lab at the end of the section, a possible numerical approach to improper integrals is discovered and applied to the problem of estimating the heights of the tallest man and woman in the United States.

9

Infinite Series   513

Infinite sequences and infinite series of constants and functions are presented and applied to a variety of situations. Both exact and numerical methods are presented and compared.

Explicitly and recursively defined sequences are given equal emphasis. The lab at the end introduces a discrete model of population growth, which parallels the continuous model from Lab 15 and leads to the study of chaotic systems.

This section uses order-of-convergence arguments to introduce Newton's method of finding roots and Simpson's method of numerical integration.

Both by-hand and computer algebra methods of finding power series are discussed.

10

Conics, Polar Coordinates and Parametric Curves   587

Parabolas, ellipses and hyperbolas are defined and studied. Polar coordinate and parametric representations of these and other curves are explored.

*This section explores advantages and pitfalls of using
calculators and computers to generate polar graphs.
Calculator/computer graphs are used to study more complicated polar graphs.*

*This section develops area and arclength formulas for
curves given in polar coordinates and then applied to the
motion of Earth in the lab at the end of the section.*

*By-hand as well as calculator/computer-based methods of
graphs of parametric equations are studied.*

Preface

This text is a second-generation reform calculus text. By this, we mean that a reform calculus text is being wedded to the best and necessary aspects of traditional calculus: the full syllabus and a graded range of problem sets. This text covers topics like continuity, sequence and series, parametric equations, etc. It assumes the use of technology as a tool. It does not assume a starting knowledge of how to use technology. The use of technology allows some trimming of techniques of integration. Since this is not a precalculus text, our review material is limited.

Greater Emphasis on Numerical Approaches

In addition to the standard algebraic and graphical techniques of calculus, this book, more than others, incorporates numerical or empirical approaches to problem solving which reflect the way calculus is often used by practicing engineers and scientists. This combination of approaches has been seen as a cornerstone of calculus reform; more important, it brings calculus back into the real world. Furthermore, numerical techniques often help students understand and appreciate the definitions of the derivative and the definite integral.

The interplay of numerical and algebraic techniques plays a large role in the book. While several recent calculus texts discuss numerical and graphical techniques, this book puts a special emphasis on *why* a particular technique works for a particular type of problem. It is often left to the student to choose between numerical and algebraic approaches, and then to explain the reasoning behind the choice. For example, the student must be able to recognize and explain why the equation $x + \sin x = 1$ can be solved numerically but not algebraically for x, and why the equation $x^2 + ax + 5 = 0$ can be solved algebraically but not numerically for x. Similar examples can be given for definite integrals and for the first-order differential equations.

This book also provides an emphasis on *how* calculus works. Like most calculus professors, we expect students to master techniques and a degree of theory. But until students see for themselves how calculus is related to the real world, they can't make it a part of themselves. This book provides frequent and explicit hands-on practice in using calculus for real-life situations.

The Role of Technology

In order to accomplish the goal of introducing numerical and graphical techniques, we have explained in Chapter 2 and at various points throughout the rest of the text how graphing calculators (TI-85, TI-82, Casio 9700, etc.) or computer algebra systems (Mathematica, Maple, Derive, etc.) can be incorporated into the course. Some of the numerical or graphical techniques could also be performed on a spreadsheet program or a computer graphing program, but no specific suggestions are given for those types of software programs.

There are several places in the text where specific technology techniques, key technology sequences, or programs are referenced. Rather than interrupt the text with this material, we have placed it in a set of technology pages in the Solutions Manual. These pages are referenced in the text by a technology icon as shown at the left of this paragraph.

Also note that while Chapter 2 is nominally an introduction to technology, we use this space to help build a strong foundation for the limit and the derivative. Explorations of how the graphs of a graphing calculator work are related carefully to the geometry underlying limits and derivatives.

Lab Projects

In addition to the use of numerical and graphical techniques in the examples and exercises, there are 22 lab projects incorporated into the book. These projects were previously developed by Bob Decker and John Williams and appeared as a separate lab manual under the title "Bringing Calculus to Life". Many of the lab projects involve taking real data (using pendulums, balloons, etc.) and then using the ideas of calculus to learn something new about the problem situation. For most labs, students gather data, graph it, and discuss or interpret it. Many of these projects make good group learning activities. These lab projects have been used extensively at colleges, community colleges and high schools throughout the country, having been presented at several NSF-sponsored summer workshops at the University of Hartford.

Techniques and Philosophy

While numerical techniques are an extremely important part of the book, the number of techniques required is kept to a minimum. Newton's method and the bisection method for solving equations, left-endpoint, right-endpoint and midpoint Riemann sums for definite integrals and Euler's method for first-order differential equations are the techniques explained in detail and used throughout the book. Students are then encouraged to use the numerical methods built in to various graphing calculators and computer algebra systems. Because the more sophisticated built-in methods are usually a modification of these basic methods, when something goes "wrong" with a calculator or computer solution, the students can use their knowledge of the simpler numerical method to understand what the problem might be.

While some of the traditional algebraic methods are de-emphasized, most remain in at least abbreviated form. For example, the methods of substitution and integration by parts for indefinite integrals are fully treated, but trigonometric methods and partial fractions are discussed only for the simpler cases. Similarly, there are fewer techniques and tests for summation of series introduction than in many texts, with an increased emphasis on numerical estimation of series.

We have provided an annotated table of contents to highlight our philosophy and some innovations.

Acknowledgments

I would like to thank the following people: Alan MacDonell for his countless suggestions, changes and editing decisions, John Williams for his help in developing the lab projects, Catherine Hoyser for her encouragement, the folks at the Espresso Bar at Border's Bookstore for providing the caffeine to keep me working on this project, the University of Hartford and its Department of Mathematics for granting me the time to finish this project, and the entire Calculus Reform community of scholars for making this type of book possible.

Preliminaries

0.1 Graphs and Equations

Numbers   Numbers fall into several categories. We will use these categories in discussions throughout this book, so it is important to clarify what is meant by each.

The *natural numbers* consist of the numbers $0, 1, 2, \ldots$. The *integers* consist of the natural numbers along with their negatives; we can represent the integers as $\ldots, -2, -1, 0, 1, 2, \ldots$.

Any number that can be written as a fraction $\frac{n}{m}$ (where n and m are integers and where $m \neq 0$) is called a *rational number*. Any rational number can be turned into a decimal by using long division. When put in decimal form, some rational numbers end after a finite sequence of digits:

$$\frac{1}{4} = 0.25$$

On the other hand, some end in an infinitely repeating sequence of digits:

$$\frac{2}{11} = 0.1818\overline{18}$$

(The bar means that the sequence 18 repeats infinitely many times.) Any such infinitely repeating decimal can also be turned into a fraction.

There also exist decimal numbers that have infinitely many digits, but that don't end in a repeating sequence. These numbers are called *irrational numbers*; they can't be written as the ratio of two integers. Two common irrational numbers are $\sqrt{2}$ and π.

The *real numbers* consist of the rational and irrational numbers together. These are the numbers we will be primarily concerned with in calculus. They are the numbers that make up number lines and coordinate axes; they are also used to make measurements (length, time, area, and so on).

Intervals   We can represent the set of all real numbers between 2 and 10, including both 2 and 10, as $\{x : 2 \leq x \leq 10\}$. We will often shorten this to $\{2 \leq x \leq 10\}$ or just $2 \leq x \leq 10$. Such a set of real numbers is called a *closed interval*; the numbers 2 and 10 are called endpoints. The word "closed" refers to the fact that the endpoints are included in the interval. When the endpoints are not included, as in the interval $\{x : -1 < x < 3\}$, then we call the interval an *open interval*.

Another notation that is sometimes used is $[a, b]$ to represent the closed interval from a to b, and (a, b) to represent the open interval from a to b. Thus, the closed interval $\{x : 2 \leq x \leq 10\}$ could also be written as $[2, 10]$.

Intervals can be mixed—that is, closed at one end and open at the other. For example, the interval $\{x : 1 < x \leq 4\}$ includes 4 but not 1. This interval may also be written as $(1, 4]$. Intervals can also be unbounded (or infinite) at one end. We use the symbol ∞ to represent infinity, so that $\{x : x > 5\}$ and $(5, \infty)$ would both represent all real numbers greater than 5.

Coordinate Systems   Two Frenchmen deserve credit for the idea of a coordinate system. Pierre de Fermat (1601–1665) was a lawyer who made mathematics his hobby. In 1629, he wrote a paper that, in effect, made use of coordinates to describe points and curves. René Descartes (1596–1650) was a philosopher who thought mathematics could unlock the secrets of the universe. He published his *La Geometrie* in 1637. Although this book does emphasize the role of algebra in solving geometric problems, one finds only a hint of coordinates there. By virtue of having the idea first and more explicitly, Fermat ought to get the majority of the credit. History can be a fickle friend; coordinates are known today as Cartesian coordinates, named after Descartes.

Cartesian Coordinates   In the plane, produce two copies of a number line, one horizontal and the other vertical, so that they intersect at the zero points of the two lines. The two lines are called **coordinate axes**; their intersection is labeled 0 and is called the **origin**. By convention, the horizontal line is called the **x-axis** and the vertical line is called the **y-axis**. The positive half of the x-axis points to the right; the positive half of the y-axis points upward. The coordinate axes divide the plane into four regions, called **quadrants**, labeled I, II, III, and IV, as shown in Figure 1.

Each point P in the plane can now be identified by a pair of numbers, called its **Cartesian coordinates**. If vertical and horizontal lines through P intersect the x- and y-axes at a and b, respectively, then P has coordinates (a, b) (see Figure 2). We call (a, b) an **ordered pair** of numbers because it makes a difference which number is first. The first number a is the **x-coordinate** (or abscissa); the second number b is the **y-coordinate** (or ordinate). We will sometimes use the notation $P(a, b)$ to refer to a point with name P and coordinates (a, b). In Figure 3, we have shown the coordinates of several points.

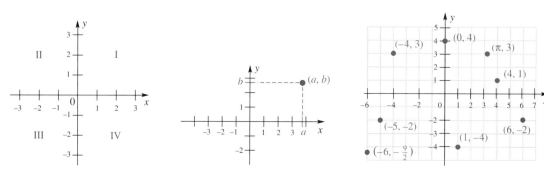

Figure 1   Figure 2   Figure 3

Note: When we write (2, 3) we could either be referring to an ordered pair *or* an open interval of real numbers from 2 to 3. It should be clear from the context of the problem which is meant.

Graphs of Equations
The use of coordinates for points in the plane allows us to describe a curve (a geometric object) by an equation (an algebraic object). The **graph of an equation** in *x* and *y* consists of those points in the plane whose coordinates (*x*, *y*) satisfy the equation—that is, make it a true equality.

The Graphing Procedure
To graph an equation (for example, $y = 2x^3 - x + 19$) we follow a simple three-step procedure:

Step 1 Obtain the coordinates of a few points that satisfy the equation.

Step 2 Plot those points in the plane.

Step 3 Connect the points with a smooth curve.

The best way to do Step 1 is to make a table of values. Assign values to one of the variables, such as *x*, and determine the corresponding values of the other variable, listing the results in tabular form.

EXAMPLE 1: Graph the equation $y = x^2 - 3$.

SOLUTION:

Step 1: Make a table of values. Because the equation is already solved for *y*, we assign values to *x*. See Figure 4.

Step 2: Plot the points from the table. (See Figure 5.) Note that we chose the scales on the *x*- and *y*-axes so that all of the points in the table would show up on the graph. Always look at your table of values when deciding how to choose the scale on your axes.

Step 3: Connect the points with a smooth curve. (See Figure 6.) ◀

x	*y*
-3	6
-2	1
-1	-2
0	-3
1	-2
2	1
3	6

Figure 4

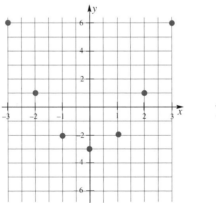

Figure 5

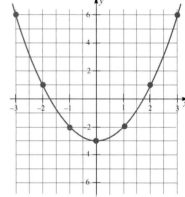

Figure 6

Of course, you need to use common sense and even a little faith when you graph by plotting points and sketching. When you connect the points you have plotted with a smooth curve, you are assuming that the curve behaves nicely between consecutive points, which is faith. That is why you should plot enough points so the outline of the curve seems very clear; the more points you plot, the less faith you will need.

Also, you should recognize that you can seldom display the whole curve. In our example, the "real" curve has infinitely long arms, opening wider and wider. But our graph in Example 1 does show the essential features. This is our goal in graphing: Show enough of the graph so the essential features are visible. Later on, we will use both technology (calculators and/or computers) and the tools of calculus to refine and improve our graphing technique.

Intercepts   The points where the graph of an equation crosses the two coordinate axes play a significant role in many problems. Consider, for example,

$$y = x^3 - 2x^2 - 5x + 6 = (x + 2)(x - 1)(x - 3)$$

Notice that $y = 0$ when $x = -2, 1, 3$. The numbers $-2, 1$, and 3 are called **x-intercepts**; they are the x-values of the points where the graph crosses the x-axis. You can see that factoring the equation helps greatly in locating the x-intercepts.

Similarly, when $x = 0$ we get $y = 6$, and so 6 is called the **y-intercept**. These are the y-values of the points where the graph crosses the y-axis. The graph of this equation is shown in Figure 7.

EXAMPLE 2:   Sketch the graph of $y^2 - x + y - 6 = 0$, showing all intercepts clearly.

SOLUTION: Putting $y = 0$ in the given equation, we get $x = -6$, and so the x-intercept is -6. Putting $x = 0$ in the equation, we find $y^2 + y - 6 = 0$, or $(y + 3)(y - 2) = 0$; the y-intercepts are -3 and 2.

To plot a few points in addition to the intercepts, we create a table of values. In this case, it is easier to solve for x in terms of y and choose values for y first. The table and graph are displayed in Figure 8.   ◄

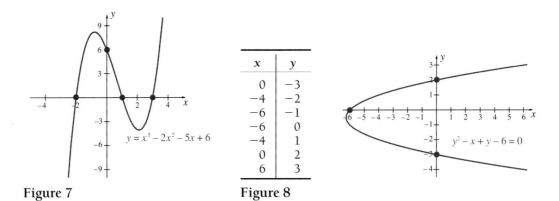

x	y
0	-3
-4	-2
-6	-1
-6	0
-4	1
0	2
6	3

Figure 7   Figure 8

General Quadratic and Cubic Equations   Because quadratic and cubic equations will often be used as examples in later work, we display their typical graphs in Figure 9.

If an equation has the form $y = ax^2 + bx + c$ or $x = ay^2 + by + c$ with $a \neq 0$, it is called a quadratic equation. Its graph will always be a cup-shaped curve called a **parabola**. In the first case ($y = ax^2 + bx + c$), the curve opens up or down, depending on whether $a > 0$ or $a < 0$. In the second case, ($x = ay^2 + by + c$), the curve opens right or left, again depending on whether $a > 0$ or $a < 0$. Note that the equation of Example 2 can be put into the form $x = y^2 + y - 6$. Because $a = 1 > 0$, the parabola opens to the right.

BASIC QUADRATIC AND CUBIC GRAPHS

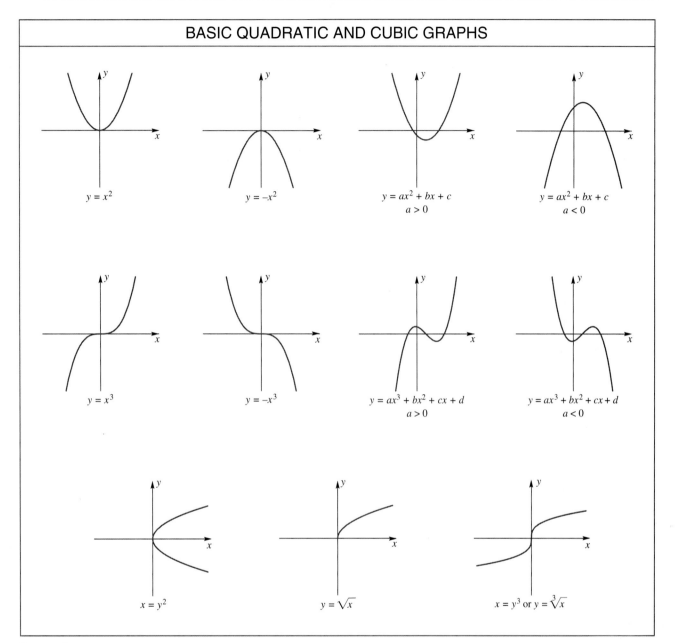

Figure 9

Equations of Circles

The general equation of a circle centered at the points (h, k) with radius r is given by

$$(x - h)^2 + (y - k)^2 = r^2$$

This follows directly from the Pythagorean Theorem. The Pythagorean Theorem says that if a, b, and c are the lengths of the sides of a right triangle, with c the length of the hypotenuse, then $a^2 + b^2 = c^2$. See Figure 10.

On a circle of radius r, centered at (h, k), with (x, y) any point on the circle (as in Figure 11), you can draw a right triangle with (h, k) and (x, y) at opposite ends of the

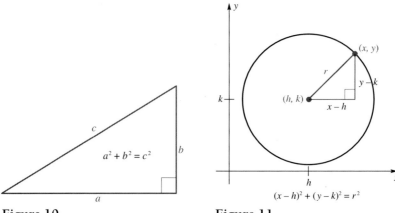

Figure 10   **Figure 11**

hypotenuse. The squared length of the hypotenuse is r^2, and $(x - h)^2$ and $(y - k)^2$ are the squared lengths of the other two sides of the right triangle formed.

Of particular importance, especially in trigonometry, is the **unit circle**, defined as the circle with radius 1, centered at $(0, 0)$. Its equation is $x^2 + y^2 = 1$. A unit circle is shown in Figure 12 along with a table of values.

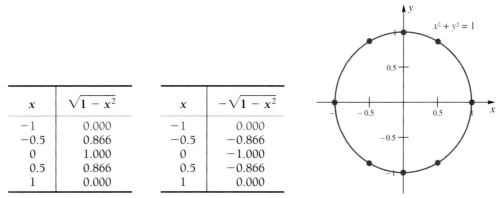

x	$\sqrt{1 - x^2}$
-1	0.000
-0.5	0.866
0	1.000
0.5	0.866
1	0.000

x	$-\sqrt{1 - x^2}$
-1	0.000
-0.5	-0.866
0	-1.000
0.5	-0.866
1	0.000

Figure 12

Notice that in order to generate a table of values we solved the equation $x^2 + y^2 = 1$ for y to get $y = \sqrt{1 - x^2}$ (the upper half of the circle) and $y = -\sqrt{1 - x^2}$ (the lower half of the circle).

The Distance Formula   You should recall from precalculus that the distance d from a point (x_1, y_1) to another point (x_2, y_2) is given by the formula

$$d = \sqrt{(x_1 - x_2)^2 + (y_1 - y_2)^2}$$

This formula also follows directly from the Pythagorean Theorem.

Intersections of Graphs   Occasionally, we need to know the points of intersection of two graphs. These points are found by solving the two equations for the graphs simultaneously.

EXAMPLE 3: Find the points of intersection of the line $y = -2x + 2$ and the parabola $y = 2x^2 - 4x - 2$, and sketch both graphs on the same coordinate plane.

SOLUTION: We must solve the two equations simultaneously. This is easy to do by substituting the expressions for y from the first equation into the second equation, and then solving the resulting equation for x.

$$-2x + 2 = 2x^2 - 4x - 2$$
$$0 = 2x^2 - 2x - 4$$
$$0 = 2(x - 2)(x + 1)$$
$$x = 2, x = -1$$

By substitution into either of the original equations, we find the corresponding values of y to be -2 and 4; the intersection points are therefore $(2, -2)$ and $(-1, 4)$.

The two graphs are shown in Figure 13.   ◄

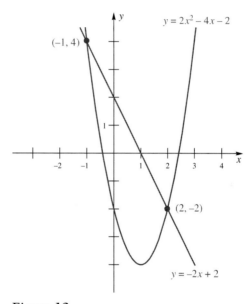

Figure 13

Accuracy and Approximation

When reporting numerical answers to problems, we often give a decimal approximation instead of an exact result. Whether we give an exact answer or an approximation depends on the context of the problem. For example, if the question is, "What is the area of a circle of radius 2?" then we might report the answer as 4π. On the other hand, if the question is, "How many square feet of material are contained in a circular rug of radius 2 feet?" then we would probably approximate and report the answer as 12.6 square feet.

There are two ways we can measure the degree of accuracy in an approximation. We can specify the number of decimal places of accuracy (absolute accuracy) or we can specify the number of digits of accuracy (relative accuracy). For example, a three-*decimal* approximation to π would be 3.142. A three-*digit* approximation to π would be 3.14. Decimal places refer to the numerals *after* the decimal point, whereas digits refer to the total number of numerals reported. The number of decimal places of accuracy depends on the units involved; the number of digits of accuracy does not. For example, 4.134 meters and 413.4 centimeters (two ways to report the same measurement) are both accurate to four digits, but have three and one decimals of accuracy, respectively.

In the table of values in Figure 12 we approximated $\sqrt{0.75}$ as 0.866, a three-digit approximation. A good rule of thumb to use is to report any approximations accurate to three digits, unless there are reasons to do otherwise for a particular problem. You should apply the usual rules of rounding to the last digit reported.

Computer and Calculator Graphing   In Chapter 2 we will investigate the use of computer algebra systems and graphics calculators for producing graphs and tables. In this chapter we will stick to graphs that are simple enough to produce with pencil and paper, and perhaps a scientific calculator.

When a computer or calculator makes a graph, it is just going through the same basic process you go through when you make a graph. A table of values is produced, then the points are plotted and connected. It is important that you thoroughly understand how graphs are produced by hand so that when you start using technology in earnest you can correctly interpret what the computer or calculator is telling you.

We don't want to discourage you, however, from starting to use your computer or calculator graphing tools before Chapter 2. You may want to check your pencil-and-paper results with a computer or calculator graph. It is always good to have more than one way to work a problem, so that each serves as a check on the other. Feel free to take a look at Section 2.1 if you want to get started with computer and calculator graphing early. You will also find a few problems in the exercises in this chapter that require a computer or graphing calculator.

Concepts Review

1. The graph of $y = (x + 2)(x - 1)(x - 4)$ has y-intercept _____ and x-intercepts _____.

2. The graphs of $y = x$ and $y = x^2$ intersect at the points _____ and _____.

3. The graph of $x^2 + y^2 = 4$ is a _____. The upper part of the graph corresponds to the equation $y =$ _____ and the lower part of the graph corresponds to the equation $y =$ _____.

4. The graph of $y = ax^2 + bx + c$ is a _____ if $a = 0$ and a _____ if $a \neq 0$.

5. Describe some limitations of using point-plotting to generate a curve.

Problem Set 0.1

For Problems 1–16 sketch the graph of each equation by creating a table of values and finding all x- and y-intercepts. You may *check* your results with a graphics calculator or computer.

1. $y = -x^2 + 4$

2. $x = -y^2 + 4$

3. $3x^2 + 4y = 0$

4. $y = 2x^2 - x$

5. $x^2 + y^2 = 36$

6. $(x - 2)^2 + y^2 = 4$

7. $4x^2 + 9y^2 = 36$

8. $16x^2 + y^2 = 16$

9. $y = x^3 - 3x$

10. $y = x^3 + 1$

11. $y = \dfrac{1}{x^2 + 1}$

12. $y = \dfrac{x}{x^2 + 1}$

13. $x^3 - y^2 = 0$

14. $x^4 - y^4 = 16$

15. $y = (x - 2)(x + 1)(x + 3)$

16. $y = x(x - 3)(x - 5)$

In Problems 17–24, sketch the graphs of both equations on the same coordinate plane. Find the points of intersection of the two graphs (see Example 3). You will need the quadratic formula in Problems 21–24.

17. $y = -x + 1$
$y = x^2 + 2x + 1$

18. $y = -x + 4$
$y = -x^2 + 2x + 4$

19. $y = -2x + 1$
$y = -x^2 - x + 3$

20. $y = -3x + 15$
$y = 3x^2 - 3x + 12$

21. $y = 1.5x + 3.2$
$y = x^2 - 2.9x$

22. $y = 2.1x + 6.4$
$y = -1.2x^2 + 4.3x$

23. $y = 4x + 3$
$x^2 + y^2 = 4$

24. $y - 3x = 1$
$x^2 + 2x + y^2 = 15$

25. The points $(2, 3)$, $(6, 3)$, $(6, -1)$, and $(2, -1)$ are corners of a square. Find the equations of the inscribed and circumscribed circles.

26. A belt fits tightly around the two circles with equations $(x - 1)^2 + (y + 2)^2 = 16$ and $(x + 9)^2 + (y - 10)^2 = 16$. How long is this belt?

27. Cities at A, B, and C are vertices of a right triangle, with the right angle at vertex B. Also, AB and BC are roads of lengths 214 and 179 miles, respectively. An airplane flies above the route AC, which is not a road. It costs \$3.71 per mile to ship a certain product by truck and \$4.82 per mile by plane. Decide whether it is cheaper to ship the product from A to C by truck or by plane, and find the total cost by the cheaper method.

28. City B is 10 miles downstream from city A and on the opposite side of a river $\frac{1}{2}$ mile wide. Mary Crane will run from city A along the river for 6 miles, then swim diagonally to city B. If she runs at 8 miles per hour and swims at 3 miles per hour, how long will it take her to get from city A to city B? Assume that the rate of the current is negligible.

29. Find the distance between the two points on the graph of $y = 3x^4 - 2x + 1$ with x-coordinates -1 and 1.

30. Find the distance between the points on the curve $y = 3x^2 - 2x + 1$ corresponding to $x = 1$ and $x = \pi$, accurate to four decimal places.

31. Sketch the graph of $y = (1 + x^{3/2})/x$ for $0 < x \le 16$ by making an extensive table of values. *Note*: Be careful near $x = 0$.

32. Rewrite the equation of Problem 31 as $y = 1/x + \sqrt{x}$. Now sketch its graph by separately graphing $y = 1/x$ and $y = \sqrt{x}$ on the same coordinate plane and then adding ordinates (y-coordinates).

33. What information can we deduce about the graph of $y = ax^2 + bx + c$ from the discriminant $d = b^2 - 4ac$? *Hint*: Use the quadratic formula and consider the three cases $d > 0$, $d = 0$, and $d < 0$.

34. Use the process of completing the square to show that the vertex (high or low point) of the parabola $y = ax^2 + bx + c$ has x-coordinate $-b/2a$. Find the corresponding y-coordinate.

35. Construct a proof of the distance formula $d = \sqrt{(x_1 - x_2)^2 + (y_1 - y_2)^2}$. *Hint*: Draw a sketch similar to Figure 11.

Answers to Concepts Review: 1. 8; $-2, 1, 4$ 2. $(0, 0)$; $(1, 1)$ 3. Circle of radius 2 with center at $(0, 0)$; $\sqrt{4 - x^2}$; $-\sqrt{4 - x^2}$.   4. Straight line; parabola   5. The actual curve may not behave as expected between the plotted points; likewise, it may have characteristic features beyond the range of plotted points. In either case, the result may miss important features of the curve unless the points are chosen carefully.

0.2 | Functions

Think of a function as a machine, a calculating machine. (See Figure 1.) It takes a number (the input) and produces a result (the output). Each number put in results in a *single* number as output, but it could happen that several different input values might give the same output value. We can state the definition more formally and introduce some notation at the same time.

Definition

A **function** f is a rule of correspondence that associates with each object x in one set (called the **domain**) a single value $f(x)$ from another set. The set of values so obtained is called the **range** of the function. (See Figure 2.)

The definition puts no restriction on the domain and range sets. The domain might consist of the set of people in your calculus class, the range the set of grades {A, B, C, D, F} that will be given, and the rule of correspondence the procedure your teacher uses in assigning grades.

Note that the definition does not allow more than one output to be matched with any one input. Thus, Figure 3 is *not* the diagram for a function.

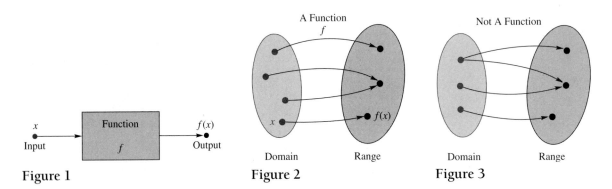

Figure 1

Figure 2

Figure 3

For most functions we will see in calculus, the domain and the range both consist of sets of real numbers. For example, the function g might take a real number and square it, thus producing the real number x^2. In this case, we have a formula that gives the rule of correspondence—namely, $g(x) = x^2$.

Functional Notation   A single letter like f (or g or F) is used to name a function. If x is in the domain, then $f(x)$, read "f of x," is in the range: it defines the value that f assigns to x. Thus, if $f(x) = x^3 - 4$,

$$f(2) = 2^3 - 4 = 4$$

$$f(-1) = (-1)^3 - 4 = -5$$

$$f(a) = a^3 - 4$$

Domain and Range   The rule of correspondence is the heart of a function, but a function is not completely determined until its domain is given. Recall that the *domain* is the set of elements to which the function assigns values. The *range* is the set of values so obtained. For example, if F is the function with rule $F(x) = x^2 + 1$ and if the domain is specified as {-1, 0, 1, 2, 3} (see Figure 4), then the range is {1, 2, 5, 10}. The domain and the rule determine the range.

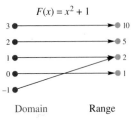

Figure 4

When no domain is specified for a function, we always assume that it is the largest set of real numbers for which the rule for the function makes sense and yields real number values. This is called the **natural domain**.

For example, the natural domain of the function $f(x) = 2x$ is the set of real numbers, because any real number x makes sense in the rule "$2x$." On the other hand, the natural domain for the function $f(x) = \sqrt{x}$ does not include negative numbers, because the square root of a negative number is not a real number.

Often, especially in science, functions are given by equations of the form $y = f(x)$ (for example, $y = x^3 + 3x - 6$). In this case x is called the **independent variable** and y the **dependent variable**. Any element of the domain may be chosen as a value of the independent variable x, but that choice completely determines the corresponding value of the dependent variable. Thus, the value of y depends on the chosen value of x.

Graphs of Functions

When both the domain and range of a function consist of real numbers, we can picture the function by drawing its graph on a coordinate plane. The **graph of a function** f is simply the graph of the equation $y = f(x)$. The difference between function notation and equation notation is that with function notation we think of f as the *name* of the function, whereas we think of y as the *number* that comes out when we put the *number* x into the function *named f*.

EXAMPLE 1: Graph the function $f(x) = x^3$.

SOLUTION: The graph of the function would be the same as the graph of the equation $y = x^3$. In Figure 5 we compare how the plotted points would be represented using function notation, ordered pair notation, and table notation. ◄

$$f(-3) = -27, f(-2) = -8, f(-1) = -1, f(0) = 0,$$
$$f(1) = 1, f(2) = 8, f(3) = 27$$
Function notation

$$(-3, -27), (-2, -8), (-1, -1), (0, 0),$$
$$(1, 1), (2, 8), (3, 27)$$
Ordered pair notation

x	y
-3	-27
-2	-8
-1	-1
0	0
1	1
2	8
3	27

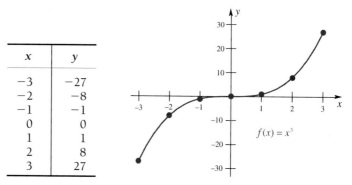

Figure 5
Table notation

Be sure to compare all three forms so that you understand how the notations are related when you are plotting points.

Important: When we write $f(-3) = -27$, note that -3 is the x-coordinate and -27 is the y-coordinate. Sometimes equation notation and function notation are mixed—for example, $y(-3) = -27$ would mean the same thing as $f(-3) = -27$. This tends to confuse the name of the function with the output of the function, but you will come across this notation because many authors find it convenient to use.

Most equations in science use letters other than x and y for the independent and dependent variables. For example, the equation

$$s = 16t^2$$

represents the distance s in feet that an object on the surface of the earth will fall in t seconds if there is no air resistance. Here, t is the independent variable and s is the dependent variable; we say that s is a function of t. When graphing such an equation, we put t on the horizontal axis where we have been putting x and we put s on the vertical axis where we have been putting y. Generally, when an equation is solved for a particular variable (in this case s), then that variable goes on the vertical axis.

The falling-object equation is an example of a **mathematical model**, which we define as follows:

Definition

A mathematical model is a mathematical description of some real-world situation. It is often stated as an equation or set of equations.

EXAMPLE 2: A calculus student has class on the seventh floor of the math building (assuming the ground floor is the first floor). Each floor is about 15 feet higher than the previous floor. If the student drops a water balloon out the window onto the head of her calculus instructor below, about how much time will it take before the balloon hits the instructor? In order to make a direct hit, she must drop the balloon when the instructor is still a few feet from the point directly below the window; estimate this distance if the instructor walks 5 miles per hour.

Use the falling-object equation as a mathematical model for the flight of the balloon. Graph the equation

$$s = 16t^2$$

on the interval $0 \le t \le 3$ and use the graph and/or the table to estimate how long it would take the balloon to fall the distance to the head of the instructor.

SOLUTION: See Figure 6 for the table and graph. The t-scale was chosen to run from 0 to 3 because that was specified as the domain in the problem; the s-scale was chosen to run from 0 to 150 because a 0-to-3 domain results in a 0-to-144 range. (We found these range values by substituting into $s = 16t^2$.) We generally choose a vertical scale that guarantees that the entire range of the function is included (so that all plotted points appear on the graph), and then round to the nearest "nice" number.

t	s
0	0
1	16
2	64
3	144

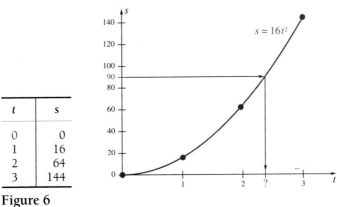

Figure 6

The seventh floor is about 90 feet above the ground (we neglect the height of the instructor). Because $s(2) = 64$ (the balloon falls 64 feet in 2 seconds) and $s(3) = 144$ (the balloon falls 144 feet in 3 seconds), we see that the balloon falls 90 feet somewhere between 2 and 3 seconds.

Using the graph to get a better estimate, we find that 90 feet on the s-axis corresponds to about 2.3 seconds on the t-axis (see Figure 6).

Finally, because $5 \frac{\text{miles}}{\text{hour}}$ converts to feet per second as

$$5 \tfrac{\text{mi}}{\text{hr}} \cdot 5280 \tfrac{\text{ft}}{\text{mi}} \cdot \tfrac{1}{3600} \tfrac{\text{hr}}{\text{sec}} \approx 7.33 \tfrac{\text{ft}}{\text{sec}}$$

we can use this rate to find the distance the instructor travels while the balloon is in the air:

$$2.3 \text{ sec} \cdot 7.33 \tfrac{\text{ft}}{\text{sec}} \approx 16.9 \text{ ft}$$

So the student should drop the balloon when her instructor is about 17 feet from the point directly below the window.

We could, of course, have used algebra to solve for t for a given s value. If we let $s = 90$ and use the equation $s = 16t^2$, we get

$$90 = 16t^2$$

$$t = \sqrt{\frac{90}{16}}$$

$$t \approx 2.37$$

which is not much different from our graphical estimate.

We will find that for many problems, algebraic approaches don't work and we must rely on tables and graphs for our estimates. Also, it is important for students to have several methods of solution for a problem, so that each method can be a check on the others. This is one way to develop self-confidence in mathematics without relying on the answers in the back of the book! ◀

One function that we will find very useful is the absolute value function, which is defined here.

Definition of Absolute Value

The function $f(x) = |x|$, read "the *absolute value* of x," is defined by

$$|x| = \begin{cases} -x & \text{if } x < 0 \\ x & \text{if } x \geq 0 \end{cases}$$

To produce the graph of the absolute value function we graph the equation $y = x$ to the right of the y-axis (for positive values of x and zero), and we graph the equation $y = -x$ to the left of the y-axis (for negative values of x). See Figure 7. When graphing a function that is "pieced together" as this one is, you should make a separate table for each piece. Notice that we have included $x = 0$ in both tables even though it is not actually part of the $y = -x$ piece, because it is an endpoint for that piece.

x	$f(x)$	x	$f(x)$
0	0	0	0
1	1	-1	1
2	2	-2	2
3	3	-3	3

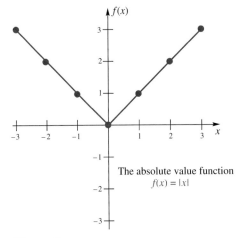

The absolute value function
$f(x) = |x|$

Figure 7

The common sense meaning of absolute value is that it is the function that leaves positive numbers and zero alone and strips the negative sign off the negative numbers. The absolute value of a number is therefore its "size" or *magnitude* without regard to sign. Notice in the definition of absolute value that when $x < 0$, $-x$ is a positive number because "the negative (or *opposite*) of a negative is a positive." For example, from the definition we have

$$|-3| = -(-3) = 3$$

Distance Interpretation of Absolute Value   A more useful interpretation of absolute value is that the absolute value of the difference of two numbers is the distance from one number to the other. For example,

$$|3 - 7| = |7 - 3| = 4$$

shows that the distance from the number 7 to the number 3 is 4 units; it doesn't matter in which order we subtract because the absolute value function makes the output positive. The absolute value of a single number is the distance of that number to zero. For example,

$$|-3| = |-3 - 0| = 3$$

shows that the distance from the number -3 to zero is 3 units. We summarize this interpretation as

$$|x - y| = \text{the distance from } x \text{ to } y$$

Translations   Observing how a function is built up from simpler ones can be a big aid in graphing. We may ask this question: How are the graphs of

$$y = f(x) \qquad y = f(x - 3) \qquad y = f(x) + 2 \qquad y = f(x - 3) + 2$$

related to each other? Consider $f(x) = |x|$ as an example. The corresponding four graphs are displayed in Figure 8.

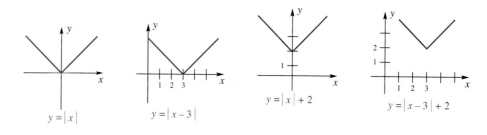

Figure 8

Notice that all four graphs have the same shape; the last three are just *translations* of the first. Replacing x by $x - 3$ translates the graph 3 units to the right; adding 2 translates it upward by 2 units.

What happened with $f(x) = |x|$ is typical. Figure 9 offers an illustration for the function $f(x) = x^3 + x^2$.

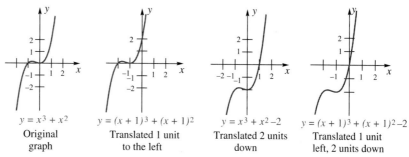

Figure 9

Exactly the same principles apply in the general situation. They are illustrated in Figure 10 with both h and k positive. If $h < 0$, the translation is to the left; if $k < 0$, the translation is downward.

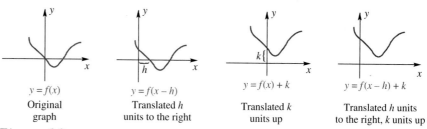

Figure 10

EXAMPLE 3: Sketch the graph of $g(x) = \sqrt{x-3} + 1$ by first graphing $f(x) = \sqrt{x}$ and then making appropriate translations.

SOLUTION: The graph of g (Figure 12) is obtained by translating the graph

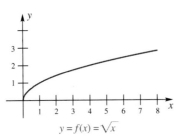

$$y = f(x) = \sqrt{x}$$

Figure 11

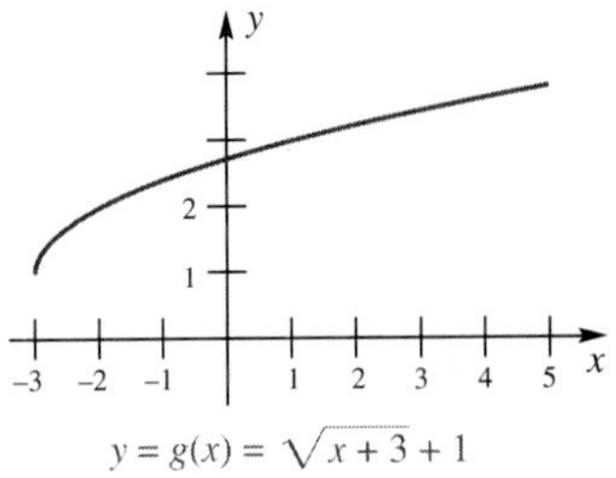

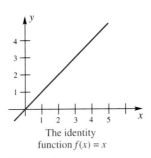

$$y = g(x) = \sqrt{x + 3} + 1$$

Figure 12

Partial Catalog of Functions   A function of the form $f(x) = k$, where k is a constant (real number), is called a **constant function**. Its graph is a horizontal line (see Figure 13). The function $f(x) = x$ is called the **identity function**. Its graph is a line through the origin with equal x- and y-coordinates for each point on the line (see Figure 14). From these simple functions, we can build many of the important functions of calculus.

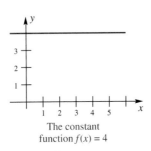

The constant
function $f(x) = 4$

Figure 13

The identity
function $f(x) = x$

Figure 14

Any function that can be obtained from the constant functions and the identity function by use of the properties of addition, subtraction, and multiplication is called a **polynomial function**. This amounts to saying that f is a polynomial function if it is of the form

$$f(x) = a_n x^n + a_{n-1} x^{n-1} + \cdots + a_1 x + a_0$$

where the a's are real numbers and n is a nonnegative integer. If $a_n \neq 0$, then n is the degree of the polynomial function. The function $f(x) = ax + b$ is a first-degree polynomial function, or **linear function**. The function $f(x) = ax^2 + bx + c$ is a second-degree polynomial function, or **quadratic function**. We dealt with a quadratic function in Example 2 when we used the equation $s = 16t^2$ to model a falling water balloon.

Quotients of polynomial functions are called rational functions. Thus, f is a **rational function** if it is of the form

$$f(x) = \frac{a_n x^n + a_{n-1} x^{n-1} + \cdots + a_1 x + a_0}{b_m x^m + b_{m-1} x^{m-1} + \cdots + b_1 x + b_0}$$

An **explicit algebraic function** is one that can be obtained from the constant functions and the identity function by use of the five operations of addition, subtraction, multiplication, division, and root extraction. Examples are

$$f(x) = 3x^{2/5} = 3 \sqrt[5]{x^2} \qquad g(x) = \frac{(x + 2)\sqrt{x}}{x^3 + \sqrt[3]{x^2} - 1}$$

You should recall from your previous work in mathematics that the nth root of a number x, written $\sqrt[n]{x}$, is the number that gives you x when it is multiplied by itself n times. For example,

$$\sqrt[4]{3} \cdot \sqrt[4]{3} \cdot \sqrt[4]{3} \cdot \sqrt[4]{3} = 3$$

We can also write $\sqrt[n]{x^m}$ as $x^{\frac{m}{n}}$; we will discuss exponents more fully in Section 1.4.

The functions listed so far, together with the trigonometric, inverse trigonometric, exponential, and logarithmic functions (to be introduced later), are the basic raw material of calculus.

Concepts Review

1. The set of allowable inputs for a function is called the _____ of the function; the set of outputs that are obtained is called the _____ of the function.

2. If $f(3) = -1$, then _____ would be a point on the graph of $f(x)$.

3. If $f(x) = 3x^2$, then $f(2) =$ _____ and $f(a) =$ _____ .

4. Compared to the graph of $y = f(x)$, the graph of $y = f(x + 2)$ is translated _____ units _____ .

5. What is a mathematical model, and how is it used?

Problem Set 0.2

In Problems 1–14, sketch the graph of the given function. Create a table of values for each function, and label the corresponding points on the graph (see Example 1).

1. $f(x) = -4$

2. $f(x) = 3x$

3. $F(x) = 2x + 1$

4. $F(x) = 3x - \sqrt{2}$

5. $g(x) = 3x^2 + 2x - 1$

6. $g(u) = \dfrac{u^3}{8}$

7. $g(x) = \dfrac{x}{x^2 - 1}$

8. $\phi(z) = \dfrac{2z + 1}{z - 1}$

9. $f(w) = \sqrt{w - 1}$

10. $h(x) = \sqrt{x^2 + 4}$

11. $f(x) = |$

12. $F(t) = -|t + 3|$

13. $g(t) = \begin{cases} 1 & \text{if } t \le 0 \\ t + 1 & \text{if } 0 < t < 2 \\ t^2 - 1 & \text{if } t \ge 2 \end{cases}$

14. $h(x) = \begin{cases} -x^2 + 4 & \text{if } x \le 1 \\ 3x & \text{if } x > 1 \end{cases}$

15. Which of the following determine a function f with formula $y = f(x)$? For those that do, find $f(x)$. *Hint:*

Solve for y in terms of x and note that the definition of function requires a single y for each x.

(a) $x^2 + y^2 = 4$

(b) $xy + y + 3x = 4$

(c) $x = \sqrt{3y + 1}$

(d) $3x = \dfrac{y}{y + 1}$

16. Which of the graphs in Figure 15 are graphs of functions of the form $y = f(x)$? (Is there a single y for each x?)

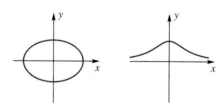

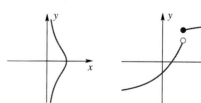

Figure 15

This problem suggests a rule: *For a graph to be the graph of a function $y = f(x)$, each vertical line must meet the graph in at most one point.*

17. A plant has the capacity to produce from 0 to 100 refrigerators per day. The daily overhead for the plant is $2200 and the direct cost (labor and material) of producing one refrigerator is $151. Write a formula for $T(x)$, the total cost of producing x refrigerators in one day, and also the unit cost $u(x)$ (average cost per refrigerator). What are the domains for these functions? Sketch the graph of each function. At what point does the unit cost $u(x)$ become $200? Display your answer on the graph of $u(x)$ as in Example 2.

18. It costs the ABC Company $400 + 5\sqrt{x(x - 4)}$ dollars to make x toy stoves that sell for $6 each.

(a) Find a formula for $P(x)$, the total profit in making x stoves.

(b) Sketch a graph of $P(x)$.

(c) Evaluate $P(200)$ and $P(1000)$, and label the corresponding points on the graph from (b).

(d) How many stoves does ABC have to make to just break even? Display your answer graphically as in Example 2.

19. Find a formula for the amount $E(x)$ by which a number x exceeds its cube. Draw a very accurate graph of $E(x)$ for $0 \le x \le 1$. Use the graph to estimate the positive number that exceeds its cube by the maximum amount.

20. Let p denote the perimeter of an equilateral triangle. Find a formula for $A(p)$, the area of such a triangle. Sketch the graph of $A(p)$ and use the graph to estimate the perimeter p for which the area $A(p)$ is equal to one.

21. The Acme Car Rental Agency charges $24 a day for the rental of a car plus $0.40 per mile.

(a) Write a formula for the total rental expense $E(x)$ for one day, where x is the number of miles driven.

(b) If you rent a car for one day, how many miles can you drive for $120? Find both an exact answer using algebra and an approximate answer using a graph.

22. A right circular cylinder of radius r is inscribed in a sphere of radius $2r$. Find a formula for $V(r)$, the volume of the cylinder, in terms of r. Sketch the graph of $V(r)$ and estimate or calculate the value of r for which $V(r)$ equals one.

23. A one-mile track has parallel sides and equal semicircular ends. Find a formula for the area enclosed by the track, $A(d)$, in terms of the diameter d of the semicircles. What is the natural domain for this function?

24. Let f be a function whose domain is the natural numbers, satisfying $f(1) = 3$, $f(2) = 1$, $f(3) = 4$, $f(4) = 1$, $f(5) = 5$, and $f(6) = 1$. After finding a pattern, give a rule for $f(n)$. What is the range of this function?

25. What is the range of the function f if $f(n)$ is the nth digit in the decimal expansion of $\frac{3}{13}$?

26. A baseball diamond is a square with sides of length 90 feet. Sam Slugger, after hitting a home run, loped around the diamond at 15 feet per second. Let s represent his distance from home plate after t seconds.

(a) Express s as a function of t by means of a four-part formula. Sketch a graph of your function, and estimate the time or times when Sam is 100 feet from home.

(b) Express s as a function of t by means of a three-part formula.

27. Sketch the graph of $f(x) = \sqrt{x - 2} - 3$ by first sketching $g(x) = \sqrt{x}$ and then translating.

28. Sketch the graph of $g(x) = |x + 3| - 4$ by first sketching $h(x) = |x|$ and then translating.

29. Sketch the graph of $f(x) = (x - 2)^2 - 4$ by making use of translations.

30. Sketch the graph of $g(x) = (x + 1)^3 - 3$ by making use of translations.

31. The relationship between the unit price P (in dollars) for a certain product and the demand D (in thousands of units) appears to satisfy

$$P = \sqrt{29 - 3D + D^2}$$

On the other hand, the demand has risen over the t years since 1970 according to $D = 2 + \sqrt{t}$.

(a) Express P as a function of t.

(b) Evaluate P when $t = 15$.

(c) Sketch the graph of P as a function of t and estimate the number of years it takes for the price to reach $6.

32. After being in business x years, a manufacturer of tractors is making $100 + x + 2x^2$ units per year. The sales price (in dollars) per unit has risen according to the formula $P = 500 + 60x$. Write a formula for the manufacturer's yearly revenue $R(x)$ after x years. Sketch a graph of $R(x)$ and estimate or calculate the number of years it takes to reach a revenue of $1,000,000.

33. Starting at noon, plane A flies due north at 400 miles per hour. Starting 1 hour later, plane B flies due east at 300 miles per hour. Neglecting the curvature of the earth and assuming they fly at the same altitude, find a formula for $D(t)$, the distance between the two planes t hours after noon. *Hint*: There will be two formulas for $D(t)$— one if $0 \le t \le 1$, the other if $t > 1$.

34. Find the distance between the planes of Problem 33 at 2:30 P.M.

Use a computer or graphics calculator in Problems 35–40.

35. Let $f(x) = x^2 - 3x$. Using the same axes, draw the graphs of $y = f(x)$, $y = f(x - 0.5) - 0.6$, and $y = f(1.5x)$, all on the domain $[-2, 5]$.

36. Let $f(x) = |x^3|$. Using the same axes, draw the graphs of $y = f(x)$, $y = f(3x)$, and $y = f(3(x - 0.8))$, all on the domain $[-3, 3]$.

37. Let $f(x) = 2\sqrt{x} - 2x + 0.25x^2$. Using the same axes, draw the graphs of $y = f(x)$, $y = f(1.5x)$, and $y = f(x - 1) + 0.5$, all on the domain $[0, 5]$.

38. Let $f(x) = 1/(x^2 + 1)$. Using the same axes, draw the graphs of $y = f(x)$, $y = f(2x)$, and $y = f(x - 2) + 0.6$, all on the domain $[-4, 4]$.

39. A *parameter* is a constant; if a function has a parameter in it, each value of the parameter produces a different graph. In each case below, draw the graph of $y = f(x)$ for the specified values of the parameter k, using the same axes and $-5 \leq x \leq 5$.

(a) $f(x) = |kx|^{0.7}$ for $k = 1, 2, 0.5$, and 0.2.

(b) $f(x) = |x - k|^{0.7}$ for $k = 0, 2, -0.5$, and -3.

(c) $f(x) = |x|^k$ for $k = 0.4, 0.7, 1$, and 1.7.

40. Using the same axes, draw the graph of $f(x) = |k(x - c)|^n$ for the following choices of parameters.

(a) $c = -1, k = 1.4, n = 0.7$

(b) $c = 2, k = 1.4, n = 1.2$

(c) $c = 0, k = 0.9, n = 0.6$

Answers to Concepts Review: 1. Domain; range 2. $(3, -1)$ 3. $12; 3a^2$ 4. 2; to the left 5. A mathematical model is a mathematical description of a real-life situation, sometimes using equations or graphs. It makes it possible to answer questions about the situation being modeled, by manipulating the equations, or by observing the behavior of the graphs.

0.3 The Straight Line and Linear Functions

Of all curves, the straight line is in many ways the simplest. We assume that you have a good intuitive notion of this concept from looking at a taut string or sighting along the edge of a ruler. In any case, let us agree that two points—for example, $A(3, 2)$ and $B(8, 4)$ shown in Figure 1—determine a unique straight line through them. And from now on, we use the word *line* as a synonym for *straight line*.

A line is a geometric object. When it is placed in a coordinate plane, it can be described by a specific equation, just as a circle can. How do we find the equation of a line? To answer, we will need the notion of slope.

The Slope of a Line Consider the line in Figure 1. From point A to point B, there is a **rise** (vertical change) of 2 units and a **run** (horizontal change) of 5 units. We say that the line has a slope of $\frac{2}{5}$. In general (see Figure 2), for a line through $A(x_1, y_1)$ and $B(x_2, y_2)$, where $x_1 \neq x_2$, we define the **slope** m of that line by

$$m = \frac{\text{rise}}{\text{run}} = \frac{\Delta y}{\Delta x} = \frac{y_2 - y_1}{x_2 - x_1}$$

The symbols $\Delta x = x_2 - x_1$ and $\Delta y = y_2 - y_1$ represent the change in x and the change in y in moving from the point $A(x_1, y_1)$ to the point $B(x_2, y_2)$.

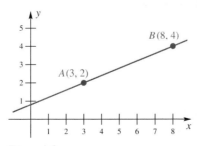

Figure 1

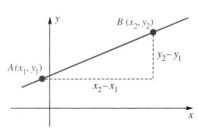

Figure 2

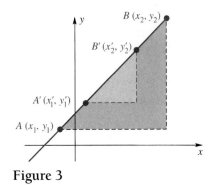

Figure 3

A question should immediately come to mind: A line has many points; does the value we get for the slope depend on which pair of points we use for A and B? The similar triangles in Figure 3 show us that

$$\frac{y'_2 - y'_1}{x'_2 - x'_1} = \frac{y_2 - y_1}{x_2 - x_1}$$

Thus, points A' and B' would do just as well as A and B. It does not even matter whether A is to the left or right of B, because

$$\frac{y_1 - y_2}{x_1 - x_2} = \frac{y_2 - y_1}{x_2 - x_1}$$

The slope of a line, then, is unchanging, no matter which points we use to calculate that slope. All that matters is that we subtract the coordinates in the same order in the numerator and the denominator.

The slope m is a measure of the steepness of a line, as Figure 4 illustrates. Notice that a horizontal line has zero slope, a line that rises to the right has positive slope, and a line that falls to the right has negative slope. The larger the magnitude of the slope, the steeper the line. The concept of slope for a vertical line makes no sense, because it would involve division by zero. Therefore, slope for a vertical line is left undefined.

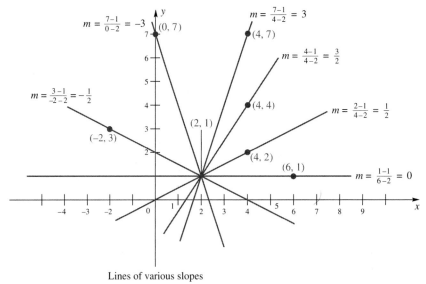

Lines of various slopes

Figure 4

The Point-Slope Form   Consider again the line of our opening discussion; it is reproduced in Figure 5. We know that this line:

1.   passes through $(3, 2)$;
2.   has slope $\frac{2}{5}$.

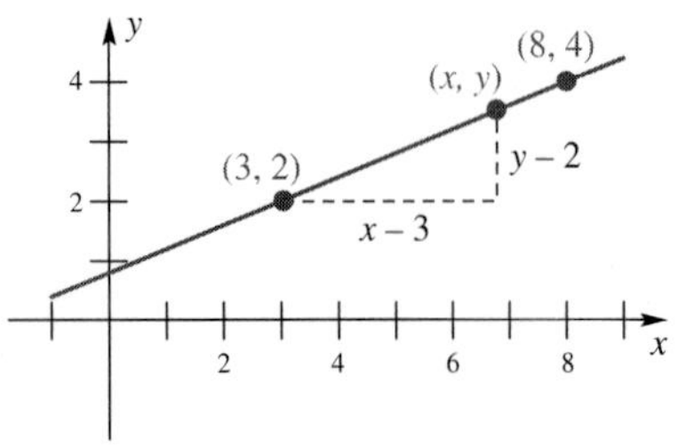

Take any other point on that line, such as one with coordinates (x, y). Because the slope of the line is $\frac{2}{5}$ all along the line, we can use this point and the point $(3, 2)$ to measure slope, and we will always get $\frac{2}{5}$—that is,

$$\frac{y - 2}{x - 3} = \frac{2}{5}$$

Figure 5

Grade and Pitch

The international symbol for the slope of a road (called the *grade*) is shown below. The grade is given as a percentage. A grade of 10% corresponds to a slope of ±0.10.

Carpenters use the term *pitch*. A 9:12 pitch corresponds to a slope of $\frac{9}{12}$.

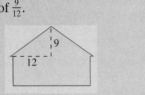

or, after multiplying by $x - 3$,

$$y - 2 = \tfrac{2}{5}(x - 3)$$

Notice that this last equation is satisfied by all points on the line, even by $(3, 2)$. Moreover, none of the points *off* the line can satisfy this equation.

What we have just done in an example can be done in general. The line passing through the (fixed) point (x_1, y_1) with slope m has equation

$$y - y_1 = m(x - x_1)$$

We call this the **point-slope form** of the equation of a line.

Consider once more the line of our example. That line passes through $(8, 4)$ as well as through $(3, 2)$. If we use $(8, 4)$ as (x_1, y_1), we get the equation

$$y - 4 = \tfrac{2}{5}(x - 8)$$

which looks quite different from

$$y - 2 = \tfrac{2}{5}(x - 3)$$

However, both can be simplified to $5y - 2x = 4$; they are equivalent.

EXAMPLE 1: Find an equation of the line through $(-4, 2)$ and $(6, -1)$.

SOLUTION: The slope m is $\frac{-1 - 2}{6 - (-4)} = -\frac{3}{10}$. Thus, using $(-4, 2)$ as the fixed point, we obtain the equation

$$y - 2 = -\tfrac{3}{10}(x + 4) \qquad \blacktriangleleft$$

The Slope-Intercept Form

The equation of a line can be expressed in various forms. Suppose that we are given the slope m for a line and the y-intercept b (that is, the line intersects the y-axis at $(0, b)$), as shown in Figure 6. Choosing $(0, b)$ as (x_1, y_1) and applying the point-slope form, we get

$$y - b = m(x - 0)$$

which we can rewrite as

$$y = mx + b$$

This equation is called the **slope-intercept form.**

EXAMPLE 2: Find the slope and the y-intercept of the line whose equation is given by $3x - 2y + 4 = 0$.

SOLUTION: If we solve for y, we get

$$y = \tfrac{3}{2}x + 2$$

It is the equation of a line with slope $\frac{3}{2}$ and y-intercept 2. Slope-intercept form makes it easy to sketch the graph of a line. The graph of $y = \frac{3}{2}x + 2$ is shown in Figure 7. $\qquad \blacktriangleleft$

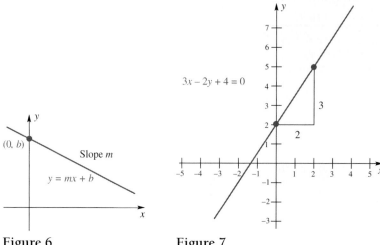

Figure 6   **Figure 7**

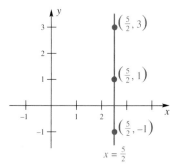

Figure 8

Equation of a Vertical Line   Vertical lines do not fit within the discussion above, because the concept of slope is not defined for them. But they do have equations, very simple ones. The line in Figure 8 has the equation $x = \frac{5}{2}$, because every point on the line satisfies this equation, and every point that satisfies the equation is on the line. The equation of any vertical line can be put in the form

$$x = k$$

where k is a constant. It should be noted that the equation of a horizontal line can be written in the form $y = k$.

The Form $Ax + By + C = 0$ It would be nice to have a form that covered all lines, including vertical lines. Consider, for example, the following **linear equations** (equations that represent lines):

(1) $y - 2 = -4(x + 2)$
(2) $y = 5x - 3$
(3) $x = 5$

These can be rewritten (by taking everything to the left-hand side) as follows:

(1) $4x + y + 6 = 0$
(2) $-5x + y + 3 = 0$
(3) $x + 0y - 5 = 0$

All are of the form

$$Ax + By + C = 0, \quad A \text{ and } B \text{ not both } 0$$

which we call the **general linear equation**. It takes only a moment's thought to see that the equation of any line can be put in this form. Conversely, the graph of the general linear equation is always a line (see Problem 31).

Summary: Equations of Lines

Vertical line: $x = k$
Horizontal line: $y = k$
Point-slope form:

$$y - y_1 = m(x - x_1)$$

Slope-intercept form:

$$y = mx + b$$

General linear equation:

$$Ax + By + C = 0$$

EXAMPLE 3: Write the equation of each line described below in the form $Ax + By + C = 0$.
(a) The line with slope 3 and y-intercept -4.
(b) The line through the point $(1, 2)$ with slope -3.
(c) The line through the points $(2, 3)$ and $(2, 7)$.

SOLUTION:

(a) In slope-intercept form the equation is $y = 3x - 4$. The general form would be $3x - y - 4 = 0$.
(b) In point-slope form the equation would be $y - 2 = -3(x - 1)$. The general form would be $3x + y - 5 = 0$.
(c) The slope would be $\frac{7 - 3}{2 - 2} = \frac{4}{0}$, which is undefined. The line is vertical, so the equation would be $x = 2$. The general form would be $x - 2 = 0$.   ◀

Linear Functions   Any function of the form

$$f(x) = mx + b$$

is called a linear function. From our discussion of linear equations, we recognize that this is essentially the same form as the slope-intercept form of a linear equation. Notice that the graph of a linear *equation* can be a vertical line (the graph of $x = a$ is a vertical line), but the graph of a linear *function* can never be a vertical line ($x = a$ cannot be put into the form $f(x) = mx + b$).

Interpretation of Slope   If the variables x and y are related by the linear equation $y = mx + b$ then the *slope m can be interpreted as the rate of change of x with respect to y.* The slope indicates the number of units that y will change if you change x by one unit. By writing the slope in the form $\frac{\Delta y}{\Delta x}$ it is clear that *the units of slope are units of y over units of x.*

Differentials   By solving the equation $m = \frac{\Delta y}{\Delta x}$ for Δy we get

$$\Delta y = m\Delta x$$

which provides another very useful interpretation of slope. *If you multiply the change in the independent variable by the slope, you get the change in the dependent variable.* We call this the method of *differentials.*

EXAMPLE 4: You have just bought a brand new Miata and plan to drive it from Hartford, Connecticut, to Key West, Florida. The odometer reads 100 miles when you get in the car. You plan to drive around 60 miles per hour. Find or estimate:
(1) The odometer reading after you have driven for 2 hours and 20 minutes.
(2) The time it will take until your odometer reaches 300 miles.

SOLUTION:

The Mathematical Model   If we let t be the time traveled in hours and let y be the odometer reading in miles, the relationship between t and y would be given by

$$y = 60t + 100$$

This is because the rate of travel in miles per hour (60) multiplied by the time in hours (t) would be the distance traveled in miles. We then add to that the initial odometer reading in miles (100) to get the equation above.

Notice that the slope of our equation is

$$m = \frac{\Delta y}{\Delta t} = 60\,\frac{\text{mi}}{\text{hr}}$$

Because the units of y are miles (mi) and the units of t are hours (hr), the units of slope are $\frac{\text{mi}}{\text{hr}}$, which is a measure of speed.

In terms of this linear mathematical model, the questions above become:
(1) Find y when $t = 2\frac{1}{3}$ hours.
(2) Find t when $y = 300$ miles.

Numerical/Graphical Approach   We can build a table of values and sketch a graph based on the table in order to estimate our solutions. Because the graph of a linear equation is a straight line, we really need only to plot two points in order to sketch the graph, but a slightly larger table helps in the estimation process. See Figure 9.

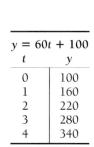

$y = 60t + 100$	
t	y
0	100
1	160
2	220
3	280
4	340

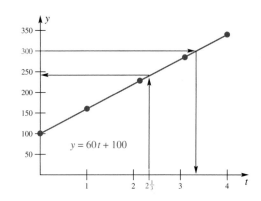

Figure 9

From the table of function values we can estimate the answers to both questions. Because $t = 2\frac{1}{3}$ hours lies between 2 and 3 hours, the corresponding odometer reading lies between 220 and 280 miles. Clearly, if we included more function values we could get a more accurate answer. Similarly, because $y = 300$ miles lies between 280 and 340 miles, the corresponding time value lies between 3 and 4 hours. When estimating answers from tables, we often report our answers as an interval or range of values. The advantage of an interval estimate is that it gives a degree of certainty to the answer, in that we *know* the exact result lies between two given values.

The graph gives a visual representation of the table and a way to *interpolate* or estimate values not in the table. The graph shows that when $t = 2\frac{1}{3}$ we have $y \approx 240$, and that when $y = 300$ we have $t \approx 3.3$. A summary of our results would be
(1) After 2 hours and 20 minutes, the odometer will read between 220 and 280 miles, most likely around 240 miles.
(2) It takes between 3 and 4 hours, most likely around 3.3 hours, for the odometer to reach 300 miles.

Algebraic Approach Number One   We can substitute $2\frac{1}{3}$ directly in for t in the equation $y = 60t + 100$ to get y; converting $2\frac{1}{3}$ to $\frac{7}{3}$ we get

$$y = 60\left(\tfrac{7}{3}\right) + 100 = 240 \text{ miles}$$

Reversing the process we let $y = 300$ and then solve the equation for t. We get

$$300 = 60t + 100$$

$$200 = 60t$$

$$t = \frac{200}{60} = \frac{10}{3} = 3\frac{1}{3} \approx 3.33 \text{ hours}$$

Notice that 240 miles and $3\frac{1}{3}$ hours are exact answers; 3.33 hours is a three-digit approximation. Thus, we summarize our results in this way:

(1) After $2\frac{1}{3}$ hours the odometer will read 240 miles.

(2) It will take $3\frac{1}{3}$ hours for the odometer to reach 300 miles.

Algebraic Approach Number Two   We can use the method of differentials outlined above. The equation we use is $\Delta y = m\Delta t$. To find the odometer reading after $2\frac{1}{3}$ hours we have

$$\Delta y = m\Delta t = \left(60 \tfrac{\text{mi}}{\text{hr}}\right)\left(\tfrac{7}{3} \text{ hr}\right) = 140 \text{ mi}$$

Because Δy is the *change* in the odometer reading (which in this case is the distance traveled), we will need to add the starting odometer reading. We get $100 + 140 = 240$ miles as the final odometer reading.

Similarly, to find how long it will take for the odometer to reach 300 miles, we solve $\Delta y = m\Delta t$ for Δt to get $\Delta t = \frac{1}{m}\Delta y$. Again, the *change* in y from 100 to 300 would be $\Delta y = 300 - 100 = 200$, so

$$\Delta t = \frac{1}{m}\Delta y = \frac{1}{60}\frac{\text{hr}}{\text{mi}} 200\,\text{mi} = \frac{10}{3}\,\text{hr} = 3\frac{1}{3}\,\text{hr}$$

The result, $3\frac{1}{3}$ hr, refers to the *change* in t, which is in this case the same as the total time. Both results are consistent with our first algebraic approach and with our numerical/graphical approach. Keeping track of units helps us remember whether to multiply or divide by the slope.

The Common-Sense Approach   Common sense? In a math book? For some problems, it is possible to use the context in which the problem is set to help solve the problem. For this problem we might reason as follows:

"It takes me about 2 hours and 20 minutes to get from my house in Hartford, Connecticut, to Manhattan, driving about 60 miles per hour. I also know that it is around 130 miles from Hartford to Manhattan, based on signs on the highway. Therefore, because my odometer started at 100 miles, I estimate that my odometer will read about $130 + 100 = 230$ miles after 2 hours and 20 minutes."

Not every problem will have a common-sense solution, but it is important to consider the possibility that one might exist. In this problem we are off a bit, but certainly in the right ballpark. You *are allowed* to use your common sense, even in mathematics!   ◄

Concepts Review

1. The line through (a, b) and (c, d) has slope $m = $ _____ provided $a \neq c$.

2. A horizontal line has slope $m = $ _____, whereas the slope of a vertical line is undefined.

3. The equation of a nonvertical line can always be written in the form _____, whereas the equation of a vertical line is written in the form _____.

4. The units of slope for the line $y = mx + b$ would be _____.

5. Why is it convenient to have a variety of standard forms for linear equations?

Problem Set 0.3

In Problems 1–8, find the slope of the line containing the given two points.

1. (2, 3) and (4, 8)
2. (4, 1) and (8, 2)
3. (−4, 2) and (3, 0)
4. (2, −4) and (0, −6)
5. (3, 0) and (0, 5)
6. (−6, 0) and (0, 6)
7. (−1.732, 5.014) and (4.315, 6.175)
8. $(\pi, \sqrt{3})$ and $(1.642, \sqrt{2})$

In Problems 9–16, find an equation for each line. Then write your answer in the form $Ax + By + C = 0$.

9. Through (2, 3) and slope 4.
10. Through (3, −4) and slope −2.
11. With y-intercept 4 and slope −2.
12. With y-intercept 5 and slope 0.
13. Through (2, 3) and (4, 8).
14. Through (4, 1) and (8, 2).
15. Through (2, −3) and (2, 5).
16. Through (−5, 0) and (−5, 4).

In Problems 17–20, find the slope and y-intercept of each line.

17. $3y = 2x − 4$
18. $2y = 5x + 2$
19. $2x + 3y = 6$
20. $4x + 5y = −20$

Parallel lines have slopes that are the same. Perpendicular lines have slopes that are negative reciprocals of each other (such as 2 and $−\frac{1}{2}$). The next few problems use these facts.

21. Write an equation for the line through (3, −3) that is:

(a) parallel to the line $y = 2x + 5$;

(b) perpendicular to the line $y = 2x + 5$;

(c) parallel to the line $2x + 3y = 6$;

(d) perpendicular to the line $2x + 3y = 6$;

(e) parallel to the line through (−1, 2) and (3, −1);

(f) parallel to the line $x = 8$;

(g) perpendicular to the line $x = 8$;

22. Find the value of k for which the line $4x + ky = 5$:

(a) passes through the point (2, 1);

(b) is parallel to the y-axis;

(c) is parallel to the line $6x − 9y = 10$;

(d) has equal x- and y-intercepts;

(e) is perpendicular to the line $y − 2 = 2(x + 1)$.

23. Does (3, 9) lie above or below the line $y = 3x − 1$?

24. Show that the equation of the line with x-intercept $a \neq 0$ and y-intercept $b \neq 0$ can be written as

$$\frac{x}{a} + \frac{y}{b} = 1$$

25. A bulldozer costs $120,000, and each year it depreciates 8% of its original value. Find a formula for the value V of the bulldozer after t years. When will the value reach $0? Find an estimate using a numerical or graphical approach and an exact result using algebra.

26. The graph of the answer to Problem 25 is a straight line. What is its slope, assuming the t-axis to be horizontal? Interpret the slope. What are the units of slope?

27. Past experience indicates that egg production in Matlin County is growing linearly. In 1980 it was 700,000 cases, and in 1990 it was 820,000 cases. Write a formula for the number N of cases produced n years after 1980 and use it to predict egg production in the year 2000. Also, use the method of differentials to predict when egg production will reach 1,000,000 cases.

28. A piece of equipment purchased today for $80,000 will depreciate linearly to a scrap value of $2000 after 20 years. Write a formula for its value V after n years. What is the value of the equipment after 10 years? When does the value of the equipment reach one-half its original value? Estimate the answers to these two questions using a numerical or graphical approach, and find exact values using algebra.

29. Suppose that the profit P that a company realizes in selling x items of a certain commodity is given by $P = 450x − 2000$ dollars.

(a) Interpret the value of P when $x = 0$.

(b) Find the slope of the graph of the above equation. This slope is called the *marginal profit*. What is its economic interpretation? What are its units?

30. The cost C of producing x items of a certain commodity is given by $C = 0.75x + 200$ dollars. The slope of its graph is called the *marginal cost*. Find it and given an economic interpretation, including a discussion of its units.

31. Show that the graph of $Ax + By + C = 0$ is always a line (provided A and B are not both 0). *Hint:* Consider two cases: (1) $B = 0$ and (2) $B \neq 0$.

32. Find an equation for the line through (2, 3) that has equal *x*- and *y*-intercepts. *Hint*: Use Problem 24.

For the next three problems, review the relationship between the slopes of parallel and perpendicular lines explained just prior to Problem 21.

33. Express the perpendicular distance between the parallel lines $y = mx + b$ and $y = mx + B$ in terms of *m*, *b*, and *B*. *Hint*: The required distance is the same as that between $y = mx$ and $y = mx + B - b$.

34. Show that the line through the midpoints of two sides of a triangle is parallel to the third side. *Hint*: You may assume the triangle has vertices at (0, 0), (*a*, 0), and (*b*, *c*).

35. Show that the line segments joining the midpoints of adjacent sides of any quadrilateral (four-sided polygon) form a parallelogram.

Answers to Concepts Review: 1. $(d - b)/(c - a)$ 2. 0   3. $y = mx + b$; $x = k$ 4. $\frac{\text{units of } y}{\text{units of } x}$ 5. The various forms make it easy to write linear equations when given various kinds of information (slope and intercept, slope and another point, etc.). Likewise, putting the equation of a line into one of the standard forms makes it easy to determine information about the line (slope, *y*-intercept, etc.).

0.4 The Trigonometric Functions

We assume that you have studied trigonometry and are familiar with the definitions of trigonometric functions based on angles and right triangles. We remind you of three of these definitions in Figure 1. Do not forget them. Here, however, we are more interested in the trigonometric functions as based on the unit circle. When considered in this way, their domains are sets of numbers rather than sets of angles.

Let *C* be the unit circle—that is, the circle $x^2 + y^2 = 1$ centered at the origin with radius 1 (see Figure 2). Denote by *A* the point (1, 0) and let *t* be any positive number. Then there is exactly one point $P(x, y)$ on *C* such that the length of arc $A\frown P$, measured in the *counterclockwise* direction from *A* along the unit circle, is *t*. The circumference of *C* is 2π; so if $t > 2\pi$, it will take more than a complete circuit of the unit circle to trace the arc $A\frown P$. If $t = 0$, $P = A$.

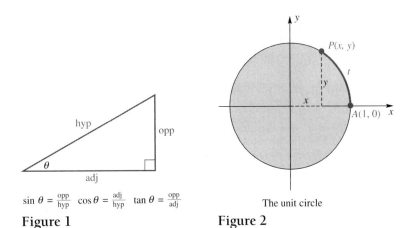

$\sin\theta = \frac{\text{opp}}{\text{hyp}}$ $\cos\theta = \frac{\text{adj}}{\text{hyp}}$ $\tan\theta = \frac{\text{opp}}{\text{adj}}$

The unit circle

Figure 1   **Figure 2**

Similarly if $t < 0$, there is exactly one point $P(x, y)$ on the unit circle such that the length of the arc $A\frown P$, measured *clockwise* on *C*, is $|t|$. Thus, with any real number *t*, we associate a unique point $P(x, y)$. This allows us to make the key definitions of sine and cosine (sin and cos).

Definition
_____

Let t determine the point $P(x, y)$ on a unit circle as indicated above. Then

$$\sin t = y \qquad \cos t = x$$

Basic Properties of Sine and Cosine

Several facts are almost immediately apparent from the definitions just given. First, x and y vary between -1 and 1, so

$$|\sin t| \leq 1 \qquad |\cos t| \leq 1$$

Because t and $t + 2\pi$ determine the point $P(x, y)$,

$$\sin(t + 2\pi) = \sin t \qquad \cos(t + 2\pi) = \cos t$$

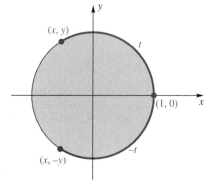

Figure 3

We say that sine and cosine are periodic with period 2π. More generally, a function f is **periodic** if there is a positive number p such that $f(t + p) = f(t)$ for all t in the domain of f. And the smallest such p is called the **period** of f.

The points P corresponding to t and $-t$ are symmetric with respect to the x-axis (Figure 3), and thus their x-coordinates are equal and their y-coordinates differ in sign only. Consequently,

$$\sin(-t) = -\sin t \qquad \cos(-t) = \cos t$$

Even and Odd Functions

Functions $f(x)$ with the property that $f(-x) = -f(x)$ are called **odd functions**. Functions with the property that $f(-x) = f(x)$ are called **even functions**. Thus, from the discussion above, we see that $\sin(t)$ is an odd function and $\cos(t)$ is an even function. The functions $x^1, x^3, x^5, \ldots$ are other examples of odd functions and the functions $x^2, x^4, x^6, \ldots$ are other examples of even functions. One can see where the names "even" and "odd" come from.

The graphs of even and odd functions have certain properties called **symmetries**. Because $f(-x) = f(x)$ for an even function, points that are the same distance from but on opposite sides of the y-axis have the same y-coordinates. Even functions are said to be symmetric with respect to the y-axis. Because $f(-x) = -f(x)$ for an odd function, points that are the same distance from the y-axis—but on opposite sides—have y-coordinates that are negatives of each other. Odd functions are said to be symmetric with respect to the origin.

One consequence of these symmetries is that if a function is either even or odd, you need only to sketch the function to the right of the y-axis (quadrants I and IV), because the other half of the graph is determined by the symmetry. For even functions, just flip the right-hand side of the graph around the y-axis to get the left-hand side. For odd functions, you need to flip twice: once around the y-axis and then again around the x-axis. This is the same as flipping once through the origin $(0, 0)$. See Figure 4, where these symmetries are illustrated with the cosine and sine functions.

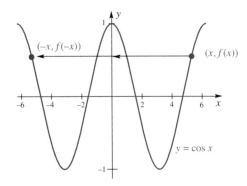

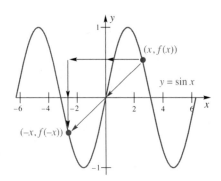

Even functions: Flip points to the right of the *y*-axis an equal distance over the *y*-axis, because $f(-x) = f(x)$.

Odd functions: Flip points to the right of the *y*-axis an equal distance over the *y*-axis, and then an equal distance over the *x*-axis (or once through the origin), because $f(-x) = -f(x)$.

Figure 4

The points P corresponding to t and $\pi/2 - t$ are symmetric with respect to the line $y = x$ (see Figure 5) and thus have their coordinates reversed. This means that

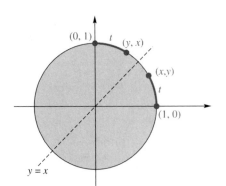

$$\sin\left(\frac{\pi}{2} - t\right) = \cos t \qquad \cos\left(\frac{\pi}{2} - t\right) = \sin t$$

Finally, we mention an important identity connecting the sine and cosine functions.

$$\sin^2 t + \cos^2 t = 1$$

Figure 5

This identity follows from the fact that for any point (x, y) on the unit circle, $y^2 + x^2 = 1$.

Graphs of Sine and Cosine   To graph $y = \sin t$ and $y = \cos t$, we follow our standard procedure (make a table of values, plot the corresponding points, and connect these points with a smooth curve). But how do we make a table of values? Before hand-held scientific calculators were available, one would generally consult a table of values created for this purpose. A brief table for special numbers is shown in Figure 6. With the aid of these tables or computations on a calculator (in radian mode), we can draw the graphs in Figure 7.

t	$\sin t$	$\cos t$
0	0	1
$\pi/6$	1/2	$\sqrt{3}/2$
$\pi/4$	$\sqrt{2}/2$	$\sqrt{2}/2$
$\pi/3$	$\sqrt{3}/2$	1/2
$\pi/2$	1	0
$2\pi/3$	$\sqrt{3}/2$	$-1/2$
$3\pi/4$	$\sqrt{2}/2$	$-\sqrt{2}/2$
$5\pi/6$	1/2	$-\sqrt{3}/2$
π	0	-1

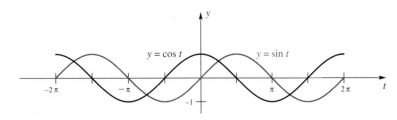

Figure 6

Figure 7

Even a casual observer might notice four things about these graphs:

1. Both sin t and cos t range from -1 to 1.
2. Both graphs repeat themselves on adjacent intervals of length 2π.
3. The graph of $y = \sin t$ is symmetric about the origin, $y = \cos t$ about the y-axis.
4. The graph of $y = \sin t$ is the same as that of $y = \cos t$ but is shifted $\pi/2$ units to the right.

There are no surprises here. These are the graphical interpretations of the basic properties of the sine and cosine function.

Four Other Trigonometric Functions

We could get by with just the sine and cosine, but it is convenient to introduce four additional trigonometric functions: tangent, cotangent, secant, and cosecant.

$$\tan t = \frac{\sin t}{\cos t} \qquad \cot t = \frac{\cos t}{\sin t}$$

$$\sec t = \frac{1}{\cos t} \qquad \csc t = \frac{1}{\sin t}$$

What we know about sine and cosine will automatically give us knowledge about these four new functions.

EXAMPLE 1: Show that tangent is an odd function.

SOLUTION:

$$\tan(-t) = \frac{\sin(-t)}{\cos(-t)} = \frac{-\sin t}{\cos t} = -\tan t$$ ◄

EXAMPLE 2: Verify the following identities.

$$1 + \tan^2 t = \sec^2 t \qquad 1 + \cot^2 t = \csc^2 t$$

SOLUTION:

$$1 + \tan^2 t = 1 + \frac{\sin^2 t}{\cos^2 t} = \frac{\cos^2 t + \sin^2 t}{\cos^2 t} = \frac{1}{\cos^2 t} = \sec^2 t$$

$$1 + \cot^2 t = 1 + \frac{\cos^2 t}{\sin^2 t} = \frac{\sin^2 t + \cos^2 t}{\sin^2 t} = \frac{1}{\sin^2 t} = \csc^2 t$$ ◄

Next we graph the tangent function (Figure 8). Here we are in for two minor surprises. First, we notice that there are t-values for which no y-value is defined ($t = \frac{\pi}{2}$ for example) because cos $t = 0$ at these values of t, which means that (sin t)/(cos t) involves a division by zero. For t-values near these points, the corresponding y-values get either very large and positive or very large and negative. We say that the graph has *asymptotes* at these t-values; asymptotes will be discussed in more detail in Section 1.2. The tangent function has asymptotes at $-3\pi/2$, $-\pi/2$, $\pi/2$, $3\pi/2$, and so on. Second, it appears that the tangent is periodic (which we expected), but with period π (which we might not have expected). You will see the reason for this in Problem 17.

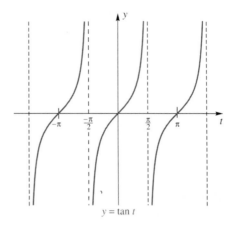

Figure 8

Relation to Angle Trigonometry   Angles are commonly measured either in degrees or in radians. The angle corresponding to complete revolution measures 360°, but only 2π radians. Equivalently, a straight angle measures 180° or π radians, a fact worth remembering.

$$180° = \pi \text{ radians} \approx 3.1415927 \text{ radians}$$

This leads to the common conversions shown in Figure 9 and also to the following facts.

$$1 \text{ radian} \approx 57.29578°$$

$$1° \approx 0.0174533 \text{ radian}$$

The division of a revolution into 360 parts is quite arbitrary (due to the ancient Babylonians, who liked multiples of 60). The division into 2π parts is more fundamental and lies behind the almost universal use of radian measure in calculus. Notice, in particular, that the length s of the arc cut off on a circle of radius r by a central angle of t radians satisfies (see Figure 10)

$$\frac{s}{2\pi r} = \frac{t}{2\pi}$$

That is,

$$s = rt$$

Degrees	Radians
0	0
30	$\pi/6$
45	$\pi/4$
60	$\pi/3$
90	$\pi/2$
120	$2\pi/3$
135	$3\pi/4$
150	$5\pi/6$
180	π

Figure 9   **Figure 10**

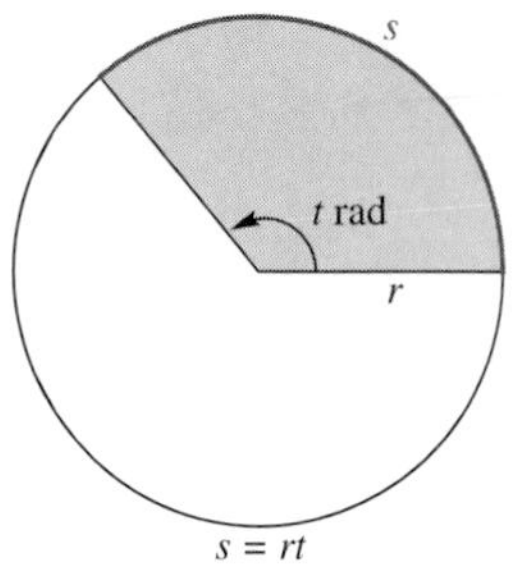

When $r = 1$, this gives $s = t$. In words, the *length of arc on the unit circle cut off by a central angle of t radians is t*. This is correct even if t is negative, provided we interpret length to be negative when measured in the clockwise direction.

EXAMPLE 3: Find the distance traveled by a bicycle with wheels of radius 30 centimeters when the wheels turn through 100 revolutions.

SOLUTION: We use the boxed formula, recognizing that 100 revolutions corresponds to $100 \cdot (2\pi)$ radians.

$$s = (30)(100)(2\pi) = 6000\pi$$

$$\approx 18849.6 \text{ centimeters} \qquad \blacktriangleleft$$

Now we can make the connection between angle trigonometry and unit circle trigonometry. If θ is an angle measuring t radians (Figure 11), then

$$\boxed{\sin \theta = \sin t \qquad \cos \theta = \cos t}$$

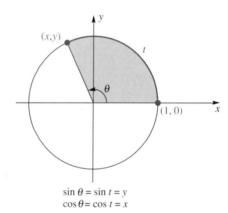

$$\sin \theta = \sin t = y$$
$$\cos \theta = \cos t = x$$

Figure 11

Another View

We have based our discussion of trigonometry on the unit circle. We could as well have used a circle of radius r.

Then

$$\sin \theta = \frac{y}{r}$$

$$\cos \theta = \frac{x}{r}$$

In calculus, when we meet an angle measured in degrees, we almost always change it to radians before doing any calculations. To do this, we use the conversion fraction $\frac{2\pi \text{ radians}}{360 \text{ degrees}}$ or $\frac{\pi}{180}$. For example,

$$\sin 31.6° = \sin\left(31.6 \cdot \frac{\pi}{180} \text{ radian}\right) \approx \sin(0.552 \text{ radian})$$

EXAMPLE 4: Find $\cos 51.8°$.

SOLUTION: We find the result two ways; directly with a scientific calculator in degree mode and again with a scientific calculator in radian mode. First we put our calculator in *degree mode* and get

$$\cos 51.8 \approx 0.618$$

Next we convert 51.8° to radians, put the calculator in *radian mode* and get

$$51.8° = 51.8\left(\frac{\pi}{180}\right) \approx 0.904 \text{ radian}$$

$$\cos 0.904 \approx 0.618 \qquad \blacktriangleleft$$

Generally, you should leave your calculator in radian mode for calculus. *One of the most common errors made using a calculator when dealing with trig functions is having the calculator in degree mode when it should be in radian mode.*

Trigonometric Inverses

We postpone a full discussion of inverse functions, including the inverses of the trig functions, until later. We will, however, need inverses for solving equations involving trig functions.

Definition of Arcsin

If $-1 \leq y \leq 1$, then the solution x to the equation $\sin x = y$ that lies in the interval $-\frac{\pi}{2} \leq x \leq \frac{\pi}{2}$ is given by $x = \arcsin y$. In other words, arcsin y is the angle whose sine is y (provided that the angle is between $-\frac{\pi}{2}$ and $\frac{\pi}{2}$). Another notation for arcsin y is $\sin^{-1}y$; with either notation this should be read as "arc sine y." See Figure 12.

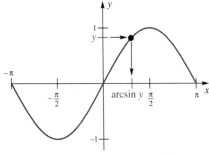

Figure 12

Function Inverses

Let us look again at Figure 6. The sin of the number in the t column is, of course, sin t. The arcsin function goes the other way; *the arcsin of the number in the sin t column is the corresponding value in the t column for t-values between $-\frac{\pi}{2}$ and $\frac{\pi}{2}$.* For example,

$$\sin^{-1}\!\left(\frac{\sqrt{2}}{2}\right) = \frac{\pi}{4} \text{ because } \sin\!\left(\frac{\pi}{4}\right) = \frac{\sqrt{2}}{2}$$

For this reason, the sin and arcsin functions are called *inverses* of each other. Another example of functions that are inverses of each other would be the functions $f(x) = x^2$ for $x \geq 0$ and $g(x) = \sqrt{x}$ for $x \geq 0$ (why?).

It is important to understand that the arcsin function gives you only *one* particular solution of an equation involving the sin function, for which there are actually *infinitely* many possible solutions.

EXAMPLE 5: Find three solutions t of the equation

$$\sin t = 0.5$$

one of which is determined by the $\sin^{-1}$ function.

SOLUTION: Using the $\sin^{-1}$ key on a scientific calculator in radian mode we get

$$t = \sin^{-1}(0.5) \approx 0.5236 \text{ radians}$$

When dealing with radians, it is often helpful to convert your answer to a multiple of π. To convert an answer in "raw" radians to "π" radians, *divide* by π. Thus,

$$0.5236 \text{ radians} = \frac{0.5236}{\pi}\,\pi \text{ radians} \approx 0.1667\,\pi \text{ radians}$$

The number 0.1667 should look familiar; it is a four-decimal approximation of $\frac{1}{6}$. Thus, our answer could also be stated as

$$t = \sin^{-1}(0.5) = \frac{\pi}{6} \text{ radians}$$

This is consistent with our table of function values for the sin function in Figure 6, which shows that $\sin\!\left(\frac{\pi}{6}\right) = \frac{1}{2} = 0.5$. Remember, the arcsin function

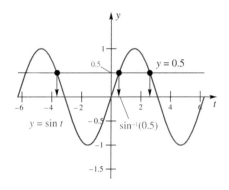

Solutions of sin $t = 0.5$

Figure 13

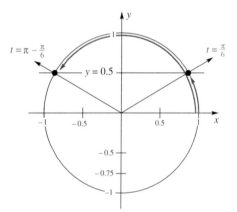

Figure 14

just "goes the other way" in this table of values, taking values in the sin t column back to values in the t column.

What about our other two solutions? One approach is to estimate them from the graph of $y = \sin t$. See Figure 13.

From this graph we estimate that two other solutions of sin $t = 0.5$ are $t \approx 2.5$ radians and $t \approx -3.7$ radians. Converting these to π radians, we get $t \approx \frac{2.5}{\pi}\,\pi \approx 0.8\,\pi$ radians and $t \approx \frac{-3.6}{\pi}\,\pi \approx -1.15\,\pi$ radians.

We can picture these solutions on the unit circle as well. (See Figure 14.) Clearly, $t = \pi - \frac{\pi}{6} = \frac{5\pi}{6} \approx 0.83\pi$ is a solution that corresponds well with one of our estimates. One can see from Figure 14 that another solution would be $t = \frac{5\pi}{6} - 2\pi = -\frac{7\pi}{6} \approx -1.17\pi$, due to the periodic nature of the sine function. Thus $\frac{\pi}{6}$, $\frac{5\pi}{6}$, and $-\frac{7\pi}{6}$ *are all exact solutions* of the equation sin $t = 0.5$, but *only* $\frac{\pi}{6}$ is given by $\sin^{-1}(0.5)$. ◄

Of course, the cosine and tangent functions have inverses as well. The definitions of the arccos ($\cos^{-1}$) and arctan ($\tan^{-1}$) functions are similar to that of the arcsin function. We state the definitions of these below, but we leave problems involving them to the exercises at the end of the section.

Definitions of Arccos and Arctan

Assume that $-1 \le y \le 1$. The solution x to the equation $\cos x = y$ that lies in the interval $0 \le x \le \pi$ is given by $x = $ arccos y. The solution x to the equation $\tan x = y$ that lies in the interval $-\frac{\pi}{2} \le x \le \frac{\pi}{2}$ is given by $x = $ arctan y.

List of Important Identities   We will not take space to verify all of the following identities. We simply assert their truth and suggest that most of them will be needed somewhere in this book.

Trigonometric Identities

Odd-even identities	*Cofunction identities*
$\sin(-x) = -\sin x$	$\sin\left(\dfrac{\pi}{2} - x\right) = \cos x$
$\cos(-x) = \cos x$	$\cos\left(\dfrac{\pi}{2} - x\right) = \sin x$
$\tan(-x) = -\tan x$	$\tan\left(\dfrac{\pi}{2} - x\right) = \cot x$

Pythagorean identities

$$\sin^2 x + \cos^2 x = 1 \qquad 1 + \tan^2 x = \sec^2 x$$

$$1 + \cot^2 x = \csc^2 x$$

Addition identities

$$\sin(x + y) = \sin x \cos y + \cos x \sin y$$

$$\cos(x + y) = \cos x \cos y - \sin x \sin y$$

$$\tan(x + y) = \frac{\tan x + \tan y}{1 - \tan x \tan y}$$

Double-angle identities

$$\sin 2x = 2\sin x \cos x$$

$$\cos 2x = \cos^2 x - \sin^2 x = 2\cos^2 x - 1 = 1 - 2\sin^2 x$$

Half-angle identities

$$\sin^2 x = \frac{1 - \cos 2x}{2} \qquad \cos^2 x = \frac{1 + \cos 2x}{2}$$

Sum identities

$$\sin x + \sin y = 2\sin\left(\frac{x + y}{2}\right)\cos\left(\frac{x - y}{2}\right)$$

$$\cos x + \cos y = 2\cos\left(\frac{x + y}{2}\right)\cos\left(\frac{x - y}{2}\right)$$

Product identities

$$\sin x \sin y = -\tfrac{1}{2}[\cos(x + y) - \cos(x - y)]$$

$$\cos x \cos y = \tfrac{1}{2}[\cos(x + y) + \cos(x - y)]$$

$$\sin x \cos y = \tfrac{1}{2}[\sin(x + y) + \sin(x - y)]$$

Concepts Review

1. The natural real number domain of the sine function is _____; its range is _____.

2. The period of the cosine function is _____; the period of the sine function is _____; the period of the tangent function is _____.

3. Cos θ _____ is always a number between _____ and _____. Cos^{-1} x is always an angle between _____ and _____ radians.

4. Both $\sin(\frac{\pi}{4}) = \frac{\sqrt{2}}{2}$ and $\sin(\frac{3\pi}{4}) = \frac{\sqrt{2}}{2}$; therefore, $\sin^{-1}\left(\frac{\sqrt{2}}{2}\right) =$ _____ radians.

5. How can you tell, by looking at the graph of a function, whether the function is even or odd?

Problem Set 0.4

1. Convert the following degree measures to radians (leave π in your answer).

(a) 240°   (b) −60°   (c) −135°

(d) 540°   (e) 600°   (f) 720°

(g) 33.3°   (h) 471.5°   (i) −391.4°

(j) 14.9°   (k) 4.02°   (l) −1.52°

2. Convert the following radian measures to degrees.

(a) $\dfrac{7\pi}{6}$ (b) $\dfrac{-\pi}{3}$ (c) 8π

(d) $\dfrac{5\pi}{4}$ (e) $\dfrac{3\pi}{2}$ (f) $\dfrac{-11\pi}{12}$

(g) 1.51   (h) −3.1416   (i) 2.31

(j) 34.25   (k) −0.002   (l) 6.28

3.   Calculate

(a) $\dfrac{234.1\sin(1.56)}{\cos(0.34)}$ (b) $\sin^2(2.51) + \sqrt{\cos(0.51)}$

4.   Calculate the following two ways: by using your calculator in degree mode, and again by converting degrees to radians, and then using your calculator in radian mode.

(a) $\dfrac{56.3\tan 34.2°}{\sin 56.1°}$ (b) $\left(\dfrac{\sin 35°}{\sin 26° + \cos 26°}\right)^3$

5.   Evaluate without use of a calculator, then check your results with a calculator.

(a) $\tan\left(\dfrac{\pi}{6}\right)$ (b) $\sec(\pi)$ (c) $\left(\dfrac{3\pi}{4}\right)$

(d) $\csc\left(\dfrac{\pi}{2}\right)$ (e) $\cot\left(\dfrac{\pi}{4}\right)$ (f) $\tan\left(-\dfrac{\pi}{4}\right)$

6.   Evaluate without use of a calculator, then check your results with a calculator.

(a) $\tan\left(\dfrac{\pi}{3}\right)$ (b) $\sec\left(\dfrac{\pi}{3}\right)$ (c) $\cot\left(\dfrac{\pi}{3}\right)$

(d) $\csc\left(\dfrac{\pi}{4}\right)$ (e) $\tan\left(-\dfrac{\pi}{6}\right)$ (f) $\cos\left(-\dfrac{\pi}{3}\right)$

7.   Sketch the graphs of the following on $[-\pi, 2\pi]$. Build a table of values for each; you may *check* with a graphics calculator or computer.

(a) $y = \sin\left(t - \dfrac{\pi}{4}\right)$ (b) $y = 3\sin t$

(c) $y = \sin 2t$ (d) $y = \sec t$

8.   Sketch the graphs of the following on $[-\pi, 2\pi]$. Build a table of values for each; you may *check* with a graphics calculator or computer.

(a) $y = \cos\left(t + \dfrac{\pi}{3}\right)$ (b) $y = 2\cos t$

(c) $y = \cos 3t$ (d) $y = \csc t$

9.   Find all solutions to the given equation on the interval $[-\pi, 2\pi]$ (see Problem 7). Find exact values if possible; otherwise, estimate. Indicate which solution results directly from using an inverse trig function, and illustrate your results using the graphs from Problem 7 as in Example 5.

(a) $\sin\left(t - \dfrac{\pi}{4}\right) = 0.5$ (b) $3\sin t = 1$

(c) $\sin 2t = \dfrac{\sqrt{2}}{2}$ (d) $\sec t = 2$

10.   Find all solutions to the given equation on the interval $[-\pi, 2\pi]$ (see Problem 8). Find exact values if possible; otherwise, estimate. Indicate which solution results directly from using an inverse trig function. Illustrate your results using the graphs from Problem 8 as in Example 5.

(a) $\cos\left(t + \dfrac{\pi}{3}\right) = 0.5$ (b) $2\cos t = \sqrt{2}$

(c) $\cos 3t = 1$ (d) $\csc t = \sqrt{2}$

11.   Verify that the following are identities (see Example 2).

(a) $(1 + \sin z)(1 - \sin z) = \dfrac{1}{\sec^2 z}$

(b) $(\sec t - 1)(\sec t + 1) = \tan^2 t$

12.   Verify that the following are identities.

(a) $\dfrac{\sin u}{\csc u} + \dfrac{\cos u}{\sec u} = 1$

(b) $(1 - \cos^2 x)(1 + \cot^2 x) = 1$

13.   Find the quadrant in which the point $P(x, y)$ will lie for each t below, and thereby determine the sign of $\cos t$. *Hint*: See the unit circle definition of $\cos t$.

(a) $t = 5.97$ (b) $t = 9.34$ (c) $t = -16.1$

14.   Follow the directions of Problem 13 to determine the sign of $\tan t$.

(a) $t = 4.34$ (b) $t = -15$ (c) $t = 21.9$

15.   Which of the following are odd functions? Even functions? Neither?

(a) $\sec t$ (b) $\csc t$

(c) $t \sin t$ (d) $x \cos x$

(e) $\sin^2 x$ (f) $\sin x + \cos x$

16.   Find identities analogous to the addition identities for each expression.

(a) $\sin(x - y)$ (b) $\cos(x - y)$ (c) $\tan(x - y)$

17.   Use the addition identity for the tangent to show that $\tan(t + \pi) = \tan t$ for all values of t in the domain of $\tan t$.

18.   Show that $\cos(x - \pi) = -\cos x$ for all x.

19.   Find the length of arc on a circle of radius 2.5 centimeters cut off by each central angle.

(a) 6 radians   (b) 225°

20.   How far does a wheel of radius 2 feet roll along level ground in making 150 revolutions? (See Example 3.)

21.   Suppose that a tire on a truck has an outer radius of 2.5 feet. How many revolutions per minute does the tire make when the truck is traveling at 60 miles per hour?

22.   A belt passes around two wheels, as shown in Figure 15. How many revolutions per second does the small wheel make when the large wheel makes 21 revolutions per second?

Figure 15

23.   A 50-pound bag of corn is being dragged along the ground by a man whose arm makes an angle of t radians with the ground. The force F (in pounds) required is given by

$$F(t) = \frac{50\mu}{\mu \sin t + \cos t}$$

Here μ is a constant relating to the friction involved. Find F in each case.

(a) $t = \dfrac{\pi}{4}$ (b) $t = 0$

(c) $t = 1$ (d) The angle is 90°.

24.   The *angle of inclination* α of a line is the smallest positive angle from the positive x-axis to the line ($\alpha = 0$ for a horizontal line). Show that the slope m of the line is equal to $\tan \alpha$.

25.   Find the angle of inclination of the following lines (see Problem 24).

(a) $y = \sqrt{3}x - 7$ (b) $\sqrt{3}x + 3y = 6$

26.   Let l_1 and l_2 be two nonvertical lines with slopes m_1 and m_2, respectively. If θ, the angle from l_1 to l_2, is not a right angle, then

$$\tan \theta = \frac{m_2 - m_1}{1 + m_1 m_2}$$

Show this using the fact that $\theta = \theta_2 - \theta_1$ in Figure 16.

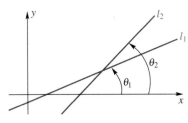

Figure 16

27.   Find the angle (in radians) from the first line to the second (see Problem 26).

(a) $y = 2x, y = 3x$ (b) $y = \dfrac{x}{2}, y = -x$

(c) $2x - 6y = 12, 2x + y = 0$

28.   Derive the formula $A = \frac{1}{2}r^2 t$ for the area of a sector of a circle. Here r is the radius and t is the radian measure of the vertex angle (see Figure 17).

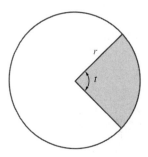

Figure 17

29.   Find the area of the sector of a circle of radius 5 centimeters and vertex angle 2 radians (see Problem 28).

30.   Suppose that the spoke of a mounted wheel of radius 2 feet rotates 10 times a second. Find the area swept out by the spoke in $\frac{1}{40}$ second; in 3 seconds.

31.   A $33\frac{1}{3}$-rpm stereo record has a spiral groove starting 6 inches from the center and ending 3 inches from the center. If it took 18 minutes to play the record, approximately how long is the spiral groove?

32.   A regular polygon of n sides is inscribed in a circle of radius r. Find formulas for the perimeter, P, and area, A, of the polygon in terms of n and r.

33.   An isosceles triangle is topped by a semicircle, as shown in Figure 18. Find a formula for the area A of the whole figure in terms of the side length r and vertex angle t (radians).

Figure 18

34.   From a product identity, we obtain

$$\cos\frac{x}{2}\cos\frac{x}{4} = \frac{1}{2}\left[\cos\frac{3}{4}x + \cos\frac{1}{4}x\right]$$

Find the corresponding sum of cosines for

$$\cos\frac{x}{2}\cos\frac{x}{4}\cos\frac{x}{8}\cos\frac{x}{16}$$

Do you see a generalization?

Use a computer or graphics calculator in Problems 35–38.

35.   Draw the graph of each of the following pairs of functions on $[-\pi, 2\pi]$ using the same axes.

(a) $f(x) = \cos x$ and $g(x) = \cos 2x$

(b) $f(x) = \cos x$ and $g(x) = \cos(x/2)$

(c) $f(x) = \cos x$ and $g(x) = 0.3 \cos x$

(d) $f(x) = \cos x$ and $g(x) = 2 \cos x$

(e) $f(x) = \cos x$ and $g(x) = \cos(x - \pi/4)$

(f) $f(x) = \cos x$ and $g(x) = 0.8 \cos[0.5(x - \pi/4)]$

Draw the graph of $f(x) = A \cos[B(x - c)]$ for various values of the parameters A, B, and c. Can you predict the period, amplitude (one-half the vertical variation), and horizontal shift for such a function from the values of A, B, and c?

36.   Draw the graph of $f(x) = \sin x + 0.25 \cos 4x$ on $[-\pi, 2\pi]$.

37.   Draw the graph of $\sin 2x \cos^2 x$ on $[-\pi, 2\pi]$.

38.   Draw the graph of $\sin 2x - \sin 3x \cos^2 x$ on $[-\pi, 2\pi]$.

Answers to Concepts Review: 1. $(-\infty, \infty)$, $[-1, 1]$ 2. 2π; 2π; π 3. $-1, 1; 0, \pi$ 4. $\frac{\pi}{4}$ 5. If the function is even, it will be symmetric with respect to the y-axis. If the function is odd, it will be symmetric with respect to the origin.

0.5 | Chapter Review

Concepts Test

Respond with true or false to each of the following assertions, and briefly explain your reasoning.

1.   Any number that can be written as a fraction p/q is rational.

2.   It is possible for two closed intervals to have exactly one point in common.

3.   If two open intervals have a point in common, then they have infinitely many points in common.

4.   You always get an accurate graph of an equation by making a table of values, plotting the points in the table, and connecting them with a smooth curve.

5.   The y-intercepts of the graph of an equation are found by setting $y = 0$.

6.   The graph of $x + y + y^2 = 0$ is a parabola opening to the left.

7.   The graph of $(x + 1)^2 + (y + 3)^2 = 4$ is a circle of radius 2 centered at the point $(1, 3)$.

8.   The intersection of the graphs of $x^2 + y^2 = 1$ and $y = x^2$ consists of two points.

9.   The intersection of the graphs of a circle and a parabola consists of two points.

10.   The number 3.333 is a four-decimal approximation to $3\frac{1}{3}$.

11.   If $ab > 0$, then (a, b) lies in either the first or third quadrant.

12.   If $ab = 0$, then (a, b) lies on either the x-axis or the y-axis.

13.   If $\sqrt{(x_2 - x_1)^2 + (y_2 - y_1)^2} = |x_2 - x_1|$, then (x_1, y_1) and (x_2, y_2) lie on the same horizontal line.

14.   The distance between $(a + b, a)$ and $(a - b, a)$ is $|2b|$.

15.   The equation $xy + x^2 = 3y$ determines a function with formula of the form $y = f(x)$.

16.   The equation $xy^2 + x^2 = 3x$ determines a function with formula of the form $y = f(x)$.

17.   The natural domain of

$$f(x) = \sqrt{\frac{x}{4 - x}}$$

is the interval $[0, 4)$.

18.   The range of $f(x) = x^2 - 6$ is the interval $[-6, \infty)$.

19.   If $x < 0$, then $\sqrt{x^2} = -x$.

20.   The falling-object equation $s = 16t^2$ is a mathematical model that will predict the exact number of feet s a real object will fall in t seconds when dropped from rest.

21.   If we have an equation relating two variables x and y, and the value of y is given, we must use algebra to find or estimate the corresponding value of x.

22.   If the range of a function consists of just one number, then its domain also consists of just one number.

23.   If the domain of a function contains at least two numbers, then the range also contains at least two numbers.

24.   If f and g have the same domain, then f/g also has that domain.

25.   If the graph of $y = f(x)$ has an x-intercept at $x = a$, then the graph of $y = f(x + h)$ has an x-intercept at $x = a - h$.

26.   If (a, b) is on a line with slope $\frac{3}{4}$, then $(a + 4, b + 3)$ is also on that line.

27.   The equation of any line can be written in point-slope form.

28.   If two nonvertical lines are parallel, they have the same slope.

29.   It is possible for two lines to have positive slopes and be mutually perpendicular.

30. $(3x - 2y + 4) + m(2x + 6y - 2) = 0$ is the equation of a straight line for each real number m.

31.   For the linear equation $y = 3x - 5$, if y changes by the amount $\Delta y = 12$ units, then x changes by 4 units.

32.   If it is possible to use both a numerical/graphical approach and an algebraic approach to solve a problem, one or the other should be used, but not both (because the answers may be different).

33.   The cotangent is an odd function.

34.   The function $f(x) = (2x^3 + x)/(x^2 + 1)$ is odd.

35.   The natural domain of the tangent function is the set of all real numbers.

36.   If $\cos s = \cos t$, then $s = t$.

37.   The solution to the equation $\cos t = 0.3$ is given by $\cos^{-1} 0.3$.

38.   There are infinitely many solutions to the equation $\cos t = 0.3$.

Sample Test Problems

1.   Find an irrational number between $\frac{1}{2}$ and $\frac{13}{25}$.

2.   Find the distance from $(3, -6)$ to the point on the curve $y = x^3 + 5$ with x-coordinate 1.

3.   Find the equation of the circle with diameter AB if $A = (2, 0)$ and $B = (10, 4)$.

In Problems 4–7, sketch the graph of the equation.

4. $3y - 4x = 6$

5. $x^2 + y^2 = 3$

6. $y = \dfrac{2x}{x^2 + 2}$

7. $x = y^2 - 3$

8.   Find the points of intersection of the graphs of $y = x^2 - 2x + 4$ and $y - x = 4$.

9.   For $f(x) = 1/(x + 1) - 1/x$, find each value (if possible).

(a) $f(1)$ (b) $f(-\frac{1}{2})$ (c) $f(-1)$

10.   For $g(x) = (x + 1)/x$, find and simplify.

(a) $g(2)$ (b) $g(\frac{1}{2})$ (c) $g(\frac{1}{10})$

11.   Describe the natural domains of each function.

(a) $f(x) = \dfrac{x}{x^2 + 1}$ (b) $g(x) = \sqrt{4 - x^2}$

(c) $h(x) = \dfrac{\sqrt{1 + x^2}}{|2x + 3|}$

12.   Sketch the graphs of each of the following functions.

(a) $f(x) = x^2 - 1$ (b) $g(x) = \dfrac{x}{x^2 + 1}$

13.   An open box is made by cutting squares of side x inches from the four corners of a sheet of cardboard 24 inches by 32 inches and then turning up the sides. Express the volume $V(x)$ in terms of x. Estimate the value of x for which the volume is 1000 cubic inches.

14.   Sketch the graphs of each of the following, making use of translations.

(a) $y = \frac{1}{4}x^2$ (b) $y = \frac{1}{4}(x + 2)^2$

(c) $y = -1 + \frac{1}{4}(x + 2)^2$

15.   Write the equation of the line through $(-2, 1)$ that:

(a) goes through $(7, 3)$;

(b) is parallel to $3x - 2y = 5$;

(c) is perpendicular to $y = 4$;

(d) has y-intercept 3.

16.   Show that $(2, -1)$, $(5, 3)$, and $(11, 11)$ are on the same line.

17.   A person is walking 3 miles per hour. She starts her trip 4 miles from her house, and walks toward her house. Set up an equation relating the distance to her house and time. How long does it take her to get to the point where she is 2.5 miles from her house? Use the method of differentials.

18.   Calculate each of the following without using a calculator or tables.

(a) $\sin(570°)$ (b) $\cos\left(\dfrac{9\pi}{2}\right)$

(c) $\sin^2(5) + \cos^2(5)$ (d) $\cos\left(\dfrac{-13\pi}{6}\right)$

19.   If $\sin t = 0.8$ and $\cos t < 0$, find each value.

(a) $\sin(-t)$ (b) $\cos t$

(c) $\sin 2t$ (d) $\tan t$

(e) $\cos\left(\dfrac{\pi}{2} - t\right)$ (f) $\sin(\pi + t)$

20.   Write $\sin 3t$ in terms of $\sin t$. *Hint*: $3t = 2t + t$.

21.   A fly sits on the rim of a wheel spinning at the rate of 20 revolutions per minute. If the radius of the wheel is 9 inches, how far does the fly travel in 1 second?

22.   Find all solutions to the given equation on the interval $[0, 3\pi]$. Find exact values if possible; otherwise, estimate. Indicate which solution results directly from using an inverse trig function. Illustrate your results graphically.

(a) $3 \sin t = 2$ (b) $2 \sin t = 3$ (c) $2 \tan t = 3$

CHAPTER 1

Calculus: A First Look

1.1 Introduction to Limits: Part I

The topics discussed so far are part of what is called *precalculus*. They provide the foundation for calculus, but they are not calculus. Now we are ready for an important new idea, the notion of *limit*. It is this idea that distinguishes calculus from other branches of mathematics. In fact, we might define calculus as *the study of limits*.

Of course, the word *limit* is used often in everyday language, as when one says, "I can't eat any more. That's my limit." That usage has something to do with calculus, but not very much.

An Intuitive Understanding   Consider the function determined by the formula

$$f(x) = \frac{x^3 - 1}{x - 1}$$

Note that it is not defined at $x = 1$, because at this point $f(x)$ has the form $\frac{0}{0}$, which is meaningless. We can, however, still ask what is happening to $f(x)$ as x approaches 1. More precisely, is $f(x)$ approaching some specific number as x approaches 1? To get at this question, we have done two things. We have calculated some values of $f(x)$ for x near 1, and we have sketched the graph of $y = f(x)$ (Figure 1).

All the information we have assembled seems to point to the same conclusion: $f(x)$ approaches 3 as x approaches 1. In mathematical symbols, we write

$$\lim_{x \to 1} \frac{x^3 - 1}{x - 1} = 3$$

This is read "the limit as x approaches 1 of $(x^3 - 1)/(x - 1)$ is 3."

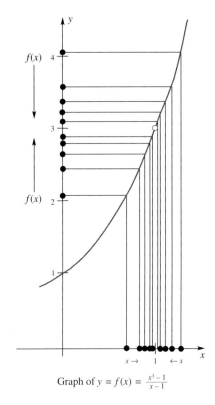

Graph of $y = f(x) = \frac{x^3 - 1}{x - 1}$

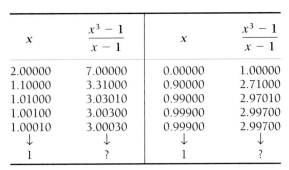

x	$\dfrac{x^3 - 1}{x - 1}$	x	$\dfrac{x^3 - 1}{x - 1}$
2.00000	7.00000	0.00000	1.00000
1.10000	3.31000	0.90000	2.71000
1.01000	3.03010	0.99000	2.97010
1.00100	3.00300	0.99900	2.99700
1.00010	3.00030	0.99900	2.99700
↓	↓	↓	↓
1	?	1	?

Table of values

Figure 1

Being good algebraists (and therefore knowing how to factor the difference of cubes), we can provide more and better evidence:

$$\lim_{x \to 1} \frac{x^3 - 1}{x - 1} = \lim_{x \to 1} \frac{(x - 1)(x^2 + x + 1)}{x - 1}$$

$$= \lim_{x \to 1} (x^2 + x + 1) = 1^2 + 1 + 1 = 3$$

Note that $(x - 1)/(x - 1) = 1$ as long as $x \neq 1$. This justifies the second step: we do not have to worry about division by zero, because x never quite reaches 1. The third step should seem reasonable; a rigorous justification is left for an analysis course.

To be sure we are on the right track, we need to have a clearly understood meaning for the word *limit*. Here is our first attempt at a definition.

> **Definition**
>
> **(Intuitive meaning of limit.)** To say that $\lim\limits_{x \to c} f(x) = L$ means that as x gets closer and closer to c, but is different from c, then $f(x)$ gets closer and closer to L. Another notation that is sometimes used is $f(x) \to L$ as $x \to c$, read "$f(x)$ approaches L as x approaches c."

Notice that we may not require anything to be true right at c. The function f may not even be defined at c; it was not defined in the example $f(x) = (x^3 - 1)/(x - 1)$ just considered. The notion of limit is associated with the behavior of a function *near c*, not *at c*.

Accuracy of Estimated Limits

When a limit can be found using algebraic means, as when we factored $f(x) = (x^3 - 1)/(x - 1)$, the result is exact. When a limit is determined using a table of values, the result is generally not exact (though it could coincide with the exact limit). From the tables in Figure 1 we can conclude that the limit is 3.00 accurate to two decimal places (and hence three digits). This follows from the fact that the last two entries in the y column agree to two decimal places as x approaches 1, using either values greater than 1 (first table) or values less than 1 (second table). When estimating limits using tables of values, we will use the same convention we established in the tables in Figure 1.

Convention for Estimating Limits Numerically

To estimate $\lim_{x \to a} f(x)$, choose a sequence of x-values that approach a in powers of $\frac{1}{10}$ when feasible. Specifically, use the sequence of values $x = a + 1, a + 0.1, a + 0.01, \ldots$ and the sequence $x = a - 1, a - 0.1, a - 0.01, \ldots$. Then the accuracy of the limit can be estimated by comparing consecutive values in the $f(x)$ column. *When two or more consecutive values agree to n decimal places, assume n decimal place accuracy.*

Note: This convention for estimating accuracy works well for most "nice" functions. However, functions can be constructed for which this method fails, as we will see below. *Numerically estimated limits are not exact; only algebraic methods yield exact results.*

Examples

Our first example may seem trivial, but it nonetheless demonstrates important ideas.

EXAMPLE 1: Find $\lim_{x \to 3} (10x - 5)$.

SOLUTION: When x is near 3, $10x - 5$ is near $10 \cdot 3 - 5 = 25$. We write

$$\lim_{x \to 3} (10x - 5) = 25$$

The limit of 25 is exact. It is still useful to look at a table of values. See Figure 2. ◄

x	$10x - 5$	x	$10x - 5$
4	35	2	15
3.1	26	2.9	24
3.01	25.1	2.99	24.9
3.001	25.01	2.999	24.99
3.0001	25.001	2.9999	24.999

Figure 2

The numerical evidence is clearly consistent with the exact result of the limit at 3; as x gets closer and closer to 3, $10x - 5$ gets closer and closer to 25. In addition, we can see why we chose the "powers of $\frac{1}{10}$" convention for numerically estimating limits.

For each additional decimal place of accuracy in the x column (as an approximation to the limiting value 3), we get one additional decimal place of accuracy in the $10x - 5$ column (as an approximation to the limiting value 25). Note that we consider 2.999 a two-decimal approximation to 3, because if we round 2.999 to two decimal places we get 3.00.

We can see why each decimal of accuracy in the input results in another decimal of accuracy in the output by reasoning as follows. Multiplication of a number by 10 just moves the decimal point over one place (resulting in one less decimal place), and subtraction of 5 has no effect on the number of decimal places. Thus, the number of decimal places of each number in the $10x - 5$ column is always one less than the number of decimal places of each number in the x column. Because the number of decimals of *accuracy* in the x column (as an approximation to the number 3) increases by 1 for each new entry (by our chosen convention), the number of decimals of *accuracy* in the $10x - 5$ column (as an approximation to the number 25) must increase by 1 for each new entry as well. The effect is essentially the same for any linear function (why?).

Of course, not all functions are linear. The good news is that most of the functions that we will be dealing with are "locally linear," meaning that if we look at a function up close, it looks linear near any point in its domain. Thus, the powers of $\frac{1}{10}$ convention works well for just about any function we will be interested in; generally, for each additional decimal place of accuracy in the x column, we get at least one additional place of accuracy in the $f(x)$ column, when x is sufficiently close to the limit point c.

Next we show how you can use algebra in certain cases to simplify the process of finding a limit.

EXAMPLE 2: Find $\lim\limits_{x \to 3} \dfrac{x^2 - x - 6}{x - 3}$.

SOLUTION: Note that $(x^2 - x - 6)/(x - 3)$ is not defined at $x = 3$, but that is all right. To get an idea of what is happening as x approaches 3, we could use a calculator to evaluate the given expression at 4, 3.1, 3.01, 3.001, and so on. But *if possible* it is better to use a little algebra to simplify the problem. In this case the algebra is easy, so we use that approach.

$$\lim_{x \to 3} \frac{x^2 - x - 6}{x - 3} = \lim_{x \to 3} \frac{(x - 3)(x + 2)}{x - 3} = \lim_{x \to 3} (x + 2) = 3 + 2 = 5$$

The cancellation of $x - 3$ in the second step is legitimate because the definition of limit ignores the behavior right at $x = 3$. Thus, we did not divide by 0. ◄

EXAMPLE 3: Find $\lim\limits_{x \to 1} \dfrac{x - 1}{\sqrt{x} - 1}$.

SOLUTION: This time we use both a numerical approach with a table of function values, and an algebraic approach; each is a check on the other. In Figure 3 the numerical approach shows that the limit is 2.000, accurate to three decimal places. The calculation below shows that the exact result is in fact 2.

$$\lim_{x \to 1} \frac{x - 1}{\sqrt{x} - 1} = \lim_{x \to 1} \frac{(\sqrt{x} - 1)(\sqrt{x} + 1)}{\sqrt{x} - 1} = \lim_{x \to 1} (\sqrt{x} + 1) = \sqrt{1} + 1 = 2 \blacktriangleleft$$

x	$\dfrac{x - 1}{\sqrt{x} - 1}$	x	$\dfrac{x - 1}{\sqrt{x} - 1}$
1.1	2.04880885	0.9	1.9486833
0.01	2.00498756	0.99	1.99498744
0.001	2.00049988	0.999	1.99949987
0.0001	2.00005	0.9999	1.99995

Figure 3

EXAMPLE 4: Find $\lim\limits_{x\to 0} \dfrac{\sin x}{x}$.

SOLUTION: No algebraic trick will simplify our task; certainly we cannot cancel the x's. We use a numerical approach. Use your own calculator (radian mode) to check the values in the table shown in Figure 4. Our conclusion based on numerical estimation is that

$$\lim_{x\to 0} \frac{\sin x}{x} = 1.000000$$

accurate to six decimal places (with rounding). We will give a rigorous demonstration in Section 3.4 that the limit is in fact exactly 1. ◀

x	$\dfrac{\sin x}{x}$	x	$\dfrac{\sin x}{x}$
1	0.84147098	−1	0.84147098
0.1	0.99833417	−0.1	0.99833417
0.01	0.99998333	−0.01	0.99998333
0.001	0.99999983	−0.001	0.99999983

Figure 4

Some Warning Flags

Things may not be quite as simple as they appear. Calculators may mislead us; so may our own intuition. The examples that follow suggest some possible pitfalls.

EXAMPLE 5: **(Numerical estimates may fool you.)** Find $\lim\limits_{x\to 0} \sin\left(\frac{\pi}{x}\right)$.

SOLUTION: Following the procedure used earlier, we construct the table of values labeled Table 1 in Figure 5. It seems clear from this table that as the x-values approach 0, the values of $\sin\left(\frac{\pi}{x}\right)$ approach 0 as well, accurate to at least six decimal places. However, Table 2 in Figure 5 leads to a completely different conclusion. For this table, even though the x-values are very close to the x-values used in Table 1, the limit appears to be 1.000000, accurate to six decimal places. ◀

x	$\sin(\pi/x)$	x	$\sin(\pi/x)$
1.00000000	0.000000	1.000000000	0.000000
0.10000000	0.000000	0.095000000	0.996584
0.01000000	0.000000	0.009950000	0.999969
0.00100000	0.000000	0.000999500	1.000000
0.00010000	0.000000	0.000099995	1.000000
Table 1		Table 2	

Figure 5

This example points out the problem with choosing one particular sequence of x-values when numerically estimating a limit. The "powers of one-tenth" convention we recommend is not infallible. A computer-generated graph of $y = \sin\left(\frac{\pi}{x}\right)$ on the interval $0 \le x \le 0.2$ is shown in Figure 6, along with a point from each of the tables in Figure 5. As x approaches 0,

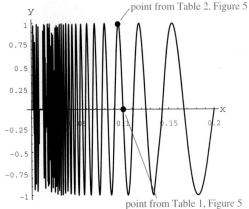

point from Table 2, Figure 5

point from Table 1, Figure 5

Figure 6

y oscillates between -1 and 1, without approaching any particular number. Between $x = 0$ and about $x = 0.05$ there are so many oscillations that the graph becomes a blur. *Thus the limit does not in fact exist.*

In Table 1 of Figure 5, the *x*-values all happen to correspond to points where the graph of the function crosses the *x*-axis, so that the *y*-values are all zero. In Table 2 of Figure 5, the *x*-values correspond approximately to the tops of the humps of the graph of the function, so the *y*-values are all about equal to 1. Because the limit does not exist, both estimates are wrong!

Numerical methods are extremely important in the practical world of science and engineering. Most real-world problems are too complicated to solve exactly, so that numerical methods must be used. This example points out that it is possible to find a function for which a particular numerical method fails, and it shows why the method fails. *Use numerical estimates when you cannot get exact answers, but always check your results in as many ways as possible.*

EXAMPLE 6: **(No limit at a jump.)** Define the **unit step function** as

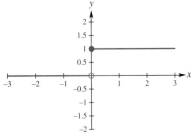

$$U(x) = \begin{cases} 0 & x < 0 \\ 1 & x \geq 0 \end{cases}$$

The graph of $U(x)$ is shown in Figure 7. Find $\lim_{x \to 0} U(x)$.

SOLUTION: For all numbers *x* less than 0 but near 0, $U(x) = 0$, but for all numbers *x* greater than 0 but near 0, $U(x) = 1$. Is $U(x)$ near a single number *L* when *x* is near 0? No. No matter what number we propose for *L*, there will be *x*'s very close to 0 on one side or the other, where $U(x)$ differs from *L* by at least $\frac{1}{2}$. Our conclusion is that $\lim_{x \to 0} U(x)$ does not exist. If you check back, you will see that we have not claimed that every limit we can write must exist.   ◄

The unit step function $U(x)$.

Figure 7

One-Sided Limits   When a function takes a jump (as does $U(x)$ at $x = 0$, in Example 6), then the limit does not exist at the jump points. For such functions, it is natural to introduce one-sided limits. Let the symbol $x \to c^+$ mean that *x* approaches *c* from the right, and let $x \to c^-$ mean that *x* approaches *c* from the left.

> **Definition**
> ___
> **(Right- and left-hand limits.)** To say that $\lim_{x \to c^+} f(x) = L$ means that as *x* gets closer and closer to *c*, but is to the right of *c*, then $f(x)$ gets closer and closer to *L*. Similarly, to say that $\lim_{x \to c^-} f(x) = L$ means that when *x* is close to but on the left of *c*, then $f(x)$ is close to *L*.

Thus, while $\lim_{x \to 0} U(x)$ does not exist, it is correct to write (see Figure 7)

$$\lim_{x \to 0^-} U(x) = 0 \text{ and } \lim_{x \to 0^+} U(x) = 1$$

We believe that you will find the following theorem quite reasonable.

Theorem A

$$\lim_{x \to c} f(x) = L \text{ if and only if } \lim_{x \to c^-} f(x) = L \text{ and } \lim_{x \to c^+} f(x) = L$$

Figure 8 should give additional insight.

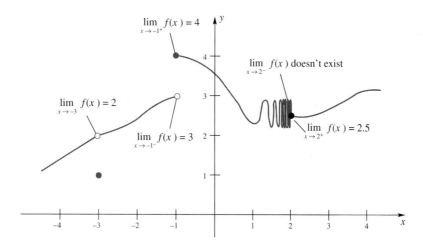

Figure 8

The next example of a limit defines a very important constant in mathematics, known simply as *e*.

EXAMPLE 7: Find $\lim_{h \to 0} (1 + h)^{1/h}$.

SOLUTION: Here is another example where algebra does not help us to simplify first. Thus, we form tables to estimate the limit. From the tables shown in Figure 9 we estimate that

$$\lim_{h \to 0^-} (1 + h)^{1/h} \approx 2.7183 \text{ and } \lim_{h \to 0^+} (1 + h)^{1/h} \approx 2.7183$$

so that by Theorem A we also have

$$\lim_{h \to 0} (1 + h)^{1/h} \approx 2.7183$$

accurate to four decimal places.

◄

The Words "If and Only If"

If we let A and B represent mathematical statements, then "A if and only if B" is short-hand for the two sentences "If A is true then B is also true" and "If B is true then A is also true." To prove an "if and only if" statement you must assume that A is true and show that B follows *and* you must assume that B is true and show that A follows.

h	$(1 + h)^{1/h}$	h	$(1 + h)^{1/h}$
0.1	2.593742	−0.1	2.867972
0.01	2.704814	−0.01	2.731999
0.001	2.716924	−0.001	2.719642
0.0001	2.718146	−0.0001	2.718418
0.00001	2.718268	−0.00001	2.718295
0.000001	2.718280	−0.000001	2.718283

Figure 9

Definition of the Constant e

We define e as

$$e = \lim_{h \to 0}(1 + h)^{1/h}$$

which according to the last example has the approximate value 2.7183.

Concepts Review

1. $\lim_{x \to c} f(x) = L$ means that $f(x)$ gets near to _____ when x gets sufficiently near to (but is different from) _____.

2. Let $f(x) = (x^2 - 9)/(x - 3)$ and note that $f(3)$ is undefined. Nevertheless, $\lim_{x \to 3} f(x) =$ _____.

3. $\lim_{x \to c^+} f(x) = L$ means that $f(x)$ gets near to _____ when x approaches a from the _____.

4. If both $\lim_{x \to c^-} f(x) = M$ and $\lim_{x \to c^+} f(x) = M$, then _____.

Problem Set 1.1

In Problems 1–6, find the indicated limit by inspection. Check your results by building a table of values as in Example 1; use the "powers of $\frac{1}{10}$" convention.

1. $\lim_{x \to 3}(2x - 8)$

2. $\lim_{x \to 3}\left(\dfrac{2}{x} + 1\right)$

3. $\lim_{x \to -2}(x^2 - 3x + 1)$

4. $\lim_{x \to 4}\dfrac{\sqrt{9 + x^2}}{x - 3}$

5. $\lim_{x \to \sqrt{3}}\dfrac{\sqrt{12 - x^2}}{x^4}$

6. $\lim_{x \to 1}\dfrac{5x - x^2}{x^2 + 2x - 4}$

In Problems 7–16, find the indicated limit. Build a table of values using the "powers of $\frac{1}{10}$" convention to estimate the limit numerically. Also find the exact limit, if feasible, using algebra (see Examples 2 and 3).

7. $\lim_{x \to 1}\dfrac{x^2 + 3x - 4}{x - 1}$

8. $\lim_{x \to 3}\dfrac{x^2 - 2x - 3}{x - 3}$

9. $\lim_{x \to -3}\dfrac{2x^2 + 5x - 3}{x + 3}$

10. $\lim_{x \to 0}\dfrac{x^3 - 16x}{x^2 + 4x}$

11. $\lim_{x \to 9}\dfrac{x - 9}{\sqrt{x} - 3}$

12. $\lim_{x \to 2}\dfrac{x^3 - 8}{x - 2}$

13. $\lim_{x \to 2}\dfrac{x^2 + x - 6}{x + 2}$

14. $\lim_{x \to -3}\dfrac{x^2 + 4x - 4}{x - 3}$

15. $\lim_{t \to 2}\dfrac{t^2 - 5t + 6}{t^2 - t - 2}$

16. $\lim_{u \to 1}\dfrac{u^2 + 6u - 7}{u^2 - 1}$

In Problems 17–26, use a calculator to calculate values of the given function near the limit point c. Organize these values in a table using the "powers of $\frac{1}{10}$" convention and use the results to estimate the required limit or to conclude that it does not exist. Be sure to put your calculator in radian mode for problems that involve trig functions.

17. $\lim_{x \to 0}\dfrac{\tan x}{2x}$

18. $\lim_{x \to 0}\dfrac{1 + \cos x}{x}$

19. $\lim_{x \to 0}\dfrac{x - \sin x}{x^3}$

20. $\lim_{x \to 0}\dfrac{1 - \cos x}{x^2}$

21. $\lim_{t \to 0}\dfrac{\sin t}{t^2}$

22. $\lim_{t \to 0}\dfrac{\sin^2 t}{t}$

23. $\lim\limits_{x\to\pi} \dfrac{1 + \cos x}{\sin 2x}$

24. $\lim\limits_{x\to 0} \dfrac{\tan x - \sin x}{x^3}$

25. $\lim\limits_{x\to 1} \dfrac{\sin 3x}{x - 1}$

26. $\lim\limits_{u\to\pi/2} \dfrac{\tan u}{u}$

For Problems 27–38, use any method to estimate the limit accurate to four decimal places, or conclude that the limit does not exist. If you can also find an exact value for the limit, do so.

27. $\lim\limits_{x\to 0} \sqrt{x}$

28. $\lim\limits_{x\to 0^+} x^x$

29. $\lim\limits_{x\to 0} \sqrt{|x|}$

30. $\lim\limits_{x\to 0} |x|^x$

31. $\lim\limits_{h\to 0} (1+2h)^{1/h}$

32. $\lim\limits_{h\to 0} (1+h)^{2/h}$

33. $\lim\limits_{x\to 0} \cos(1/x)$

34. $\lim\limits_{x\to 0} x\cos(1/x)$

35. $\lim\limits_{x\to 1} \dfrac{x^3 - 1}{\sqrt{2x + 2} - 2}$

36. $\lim\limits_{x\to 0} \dfrac{x \sin 2x}{\sin(x^2)}$

37. $\lim\limits_{x\to 2^-} \dfrac{x^2 - x - 2}{|x - 2|}$

38. $\lim\limits_{x\to 1^+} \dfrac{2}{1 + 2^{1/(x-1)}}$

39. For the function f graphed in Figure 10, find the indicated limit or function value, or state that it does not exist.

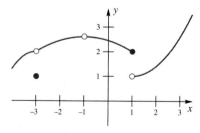

Figure 10

(a) $\lim\limits_{x\to -3} f(x)$ (b) $f(-3)$ (c) $f(-1)$
(d) $\lim\limits_{x\to -1} f(x)$ (e) $f(1)$ (f) $\lim\limits_{x\to 1} f(x)$
(g) $\lim\limits_{x\to 1^-} f(x)$ (h) $\lim\limits_{x\to 1^+} f(x)$

40. Follow the directions of Problem 39 for the function f graphed in Figure 11.

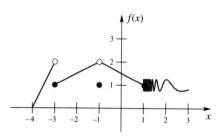

Figure 11

41. Sketch the graph of
$$f(x) = \begin{cases} x^2 & \text{if } x \le 0 \\ x & \text{if } 0 < x < 1 \\ 1 + x & \text{if } x \ge 1 \end{cases}$$
Then find each of the following or state that it does not exist.

(a) $\lim\limits_{x\to 0} f(x)$ (b) $f(1)$
(c) $\lim\limits_{x\to 1} f(x)$ (d) $\lim\limits_{x\to 1^-} f(x)$

42. Sketch the graph of
$$g(x) = \begin{cases} -x + 1 & \text{if } x < 1 \\ x - 1 & \text{if } 1 < x < 2 \\ 5 - x^2 & \text{if } x \ge 2 \end{cases}$$
Then find each of the following or state that it does not exist.

(a) $\lim\limits_{x\to 1} g(x)$ (b) $g(1)$
(c) $\lim\limits_{x\to 2} g(x)$ (d) $\lim\limits_{x\to 2^+} g(x)$

43. Sketch the graph of $f(x) = U(x) - U(x - 1)$ where $U(x)$ is the unit step function defined in Example 6. Then find each of the following or state that it does not exist.

(a) $f(1)$ (b) $\lim\limits_{x\to 1} f(x)$
(c) $\lim\limits_{x\to 1^-} f(x)$ (d) $\lim\limits_{x\to 1^+} f(x)$

44. Follow the directions of Problem 43 for $f(x) = \dfrac{x}{|x|}$. How does this function differ from the unit step function $U(x)$?

(a) $f(0)$ (b) $\lim\limits_{x\to 0} f(x)$
(c) $\lim\limits_{x\to 0^-} f(x)$ (d) $\lim\limits_{x\to 0^+} f(x)$

45. Find $\lim\limits_{x\to 1} \dfrac{x^2 - 1}{|x - 1|}$ or state that it does not exist.

46. Evaluate $\lim\limits_{x\to 0} \dfrac{\sqrt{x + 3} - \sqrt{3}}{x}$. *Hint:* Rationalize the numerator.

47. Let
$$f(x) = \begin{cases} x & \text{if } x \text{ is rational} \\ -x & \text{if } x \text{ is irrational} \end{cases}$$

Find each value, if possible.

(a) $\lim_{x \to 1} f(x)$ (b) $\lim_{x \to 0} f(x)$

48.   Sketch, as best you can, the graph of a function f that satisfies all the following conditions.

(a) Its domain is the interval $[0, 4]$.

(b) $f(0) = f(1) = f(2) = f(3) = f(4) = 1$

(c) $\lim_{x \to 1} f(x) = 2$ (d) $\lim_{x \to 2} f(x) = 1$

(e) $\lim_{x \to 3^-} f(x) = 2$ (f) $\lim_{x \to 3^+} f(x) = 1$

49.   Let

$$f(x) = \begin{cases} x^2 & \text{if } x \text{ is rational} \\ x^4 & \text{if } x \text{ is irrational} \end{cases}$$

For what value of a does $\lim_{x \to a} f(x)$ exist?

50.   The function $f(x) = x^2$ had been carefully graphed, but during the night a mysterious visitor changed the values of f at a million different places. Did this affect the value of $\lim_{x \to a} f(x)$ at any a? Explain.

51.   Find each of the following limits or state that it does not exist.

(a) $\lim_{x \to 1} \dfrac{|x - 1|}{x - 1}$ (b) $\lim_{x \to 1^-} \dfrac{|x - 1|}{x - 1}$

(c) $\lim_{x \to 1^-} \dfrac{x^2 - |x - 1| - 1}{|x - 1|}$ (d) $\lim_{x \to 1^-} \left[\dfrac{1}{x - 1} - \dfrac{1}{|x - 1|} \right]$

Answers to Concepts Review: 1. L; c 2. 6   3. L; right
4. $\lim_{x \to c} f(x) = M$

LAB 1: EXPLORING LIMITS

Mathematical Background

In calculus, many problems do not have exact solutions like $2, -5.34, \sqrt{3} + 4$, or 2π. So instead of finding an exact solution, one looks for a sequence of approximations that come closer and closer to the exact answer. For instance π does not have a finite decimal representation but members of the sequence $3, 3.1, 3.14, 3.142$, and 3.1416 give better and better approximations for the value of π.

With each approximation, one should include an estimate of the size of the error in the approximation. One common method is to specify the number of decimal places that agree with the exact answer. We say an approximation has n decimal place accuracy if the approximation differs from the exact answer by less than $0.5 \times 10^{-n} = 0.00\ldots05$, where there are n zeros between the decimal and the 5. Another way to write this condition is

$$|\text{exact solution} - \text{approximate solution}| \le 0.5 \times 10^{-n}$$

The problem is you usually do not know the exact solution. A good rule of thumb to estimate the number of decimal places of accuracy is to compare two successive approximations; if the first n decimal places agree then assume n decimal place accuracy.

Sometimes the total number of *digits* of accuracy rather than the number of decimal places of accuracy is desired; use the same rule of thumb as above. For example, for the sequence of approximations $31.2583, 32.9835, 33.3956, 33.3829, 33.3815$ we would conclude that we have two decimal places of accuracy or four digits of accuracy after the fifth term in the sequence.

This process of successive approximations can be described using the language of limits. When we write that

$$\lim_{x \to a} f(x) = L$$

we mean that for any sequence of successive approximations $a_1, a_2, a_3, \ldots$ that gets closer and closer to a the sequence of approximations $f(a_1), f(a_2), f(a_3), \ldots$ gets closer and closer to L.

Lab Introduction

The limits investigated below are "calculus classics" in that they appear in nearly every calculus textbook; some are results that will be seen again. Keep in mind that we will be approximating these limits experimentally; thus, the results will not have been proven.

Calculator Experiment

(1) Estimate $\lim_{x \to 0} \frac{\sin x}{x}$. We need a sequence of x-values that get closer to 0 from the positive side; use the sequence, 1, 0.1, 0.01, 0.001, ... Notice that each x-value is one-tenth of the previous one. For each x calculate the corresponding y using $y = \frac{\sin x}{x}$. Form a table of x-versus y-values; choose enough x-values so that y is accurate to eight decimal places (that is, the y-value repeats in the eighth decimal place). Now repeat using x-values that get closer to 0 from the negative side. Once you think you know what the limit is, go back and write down how many additional decimal places of accuracy you get for the limit (that is, the y-value) for each additional decimal place of accuracy for the x-value.

(2) Estimate $\lim_{x \to 0} \frac{1 - \cos x}{x}$. Proceed as in part (1).

(3) Estimate $\lim_{h \to 0} (1 + h)^{1/h}$. Proceed as in part (1).

(4) Estimate $\lim_{x \to 0} \sin\left(\frac{1}{x}\right)$. Proceed as in part (1).

(5) Estimate $\lim_{h \to 0} \frac{a^h - 1}{h}$ for $a = 2$ and 3. POroceed as in part (1).

(6) Find the value of a that makes the limit in (5) equal to 1.

Discussion

1. Carefully compare the rates of convergence for parts (1), (2), and (3). Which converged fastest to its limit? Is there a relationship between the rate of convergence and the shape of the graph near the limiting value? (You will need to sketch a graph of each function near the limiting value using a computer or graphing calculator.)

2. What was the problem in part (4)? Can you explain what happened? Think about what happens to $\frac{1}{x}$ when x gets small, and use what you know about the sin function.

3. Explain how you found the value of a in part (6). Is there a similarity to your answer to (3)? See if you can explain this connection. *Hint*: Choose a small value for h and then solve the approximate equation $\frac{a^h - 1}{h} \approx 1$ for a (by hand). *Note*: Parts (3) and (6) represent two possible ways of defining the constant e.

1.2 Introduction to Limits: Part II

Infinity The concept of infinity has been a topic of interest to both philosophers and mathematicians for a long time. In mathematics, we use the symbol ∞ to represent infinity, although *we do not consider ∞ to be a real number*. If you go further in mathematics and take a course in analysis (more or less the theory of calculus), you will find that it is OK to treat ∞ as a special type of number—but not now!

If infinity is not a number, then just what is it? We use "approaches infinity" in the context of limits to mean "gets larger and larger." To be a little more specific, if we say "x approaches infinity," which we can write in symbols as "$x \to \infty$," we mean that x gets larger than any fixed number, no matter how large. We can define "negative infinity" in a similar way; if we write "$x \to -\infty$," we mean that x gets larger and larger in magnitude (absolute value), but in the negative direction.

It's convenient to think of ∞ as "way to the right" on the number line and $-\infty$ as "way to the left" on the number line, though neither represents an actual location on the number line. See Figure 1.

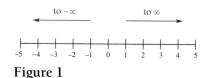

Figure 1

Limits Involving Infinity Taking the above definitions of positive and negative infinity, we can now make sense of limits that involve infinity. We interpret

$$\lim_{x \to \infty} f(x) = L$$

to mean that as x gets larger and larger, $f(x)$ gets closer and closer to L. Similarly, we interpret

$$\lim_{x \to -\infty} f(x) = L$$

to mean that as x gets larger and larger in magnitude, but in the negative direction, $f(x)$ gets closer and closer to L.

We can also interpret limits where the y-value approaches positive or negative infinity. For example,

$$\lim_{x \to a^+} f(x) = \infty$$

means that as x gets closer and closer to a from the right, $f(x)$ gets larger and larger. You can use similar interpretations for $-\infty$ or for left-hand limits as x approaches a.

EXAMPLE 1: Find

(a) $\displaystyle \lim_{x \to \infty} \frac{2}{x + 3}$

(b) $\displaystyle \lim_{x \to -3^+} \frac{2}{x + 3}$

SOLUTION:

(a) We want to determine what happens to $y = \frac{2}{x+3}$ as x gets larger and larger. In Figure 2 we build a table of function values, letting x get larger and larger by powers of 10, which suggests that $\lim_{x \to \infty} \frac{2}{x+3} = 0.00$ accurate to two decimal places.

(b) We want to determine what happens to $y = \frac{2}{x+3}$ as x gets closer and closer to -3 from the right. In Figure 3 we build a table of function values, which suggests that $\lim_{x \to -3^+} \frac{2}{x+3} = +\infty$. ◄

x	$y = \dfrac{2}{x + 3}$
10	0.15384615
100	0.01941748
1000	0.00199402
10000	0.00019994

Figure 2

x	$y = \dfrac{2}{x + 3}$
-2.000	2
-2.900	20
-2.990	200
-2.999	2000

Figure 3

Horizontal Asymptotes

Limits involving infinity show up graphically as asymptotes. If $\lim\limits_{x\to\infty} f(x) = L$, then we say that the graph of $f(x)$ has a *horizontal asymptote* at $y = L$ as x approaches infinity. The larger x gets, the closer the graph of $f(x)$ gets to the asymptote $y = L$. Asymptotes behave in a similar way for limits as x approaches negative infinity.

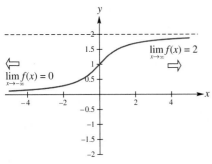

Figure 4

In Figure 4 we show the graph of a function with horizontal asymptotes at $y = 2$ as $x \to \infty$ and $y = 0$ as $x \to -\infty$. In the language of limits, $\lim\limits_{x\to\infty} f(x) = 2$ and $\lim\limits_{x\to-\infty} f(x) = 0$.

Vertical Asymptotes

If $\lim\limits_{x\to a^+} f(x) = \pm\infty$ or $\lim\limits_{x\to a^-} f(x) = \pm\infty$, we say that the graph of $f(x)$ has a *vertical asymptote* at $x = a$. As the graph of $f(x)$ gets close to the line $x = a$, the y-coordinate of the graph gets larger and larger in magnitude (in either the positive or the negative direction).

In Figure 5 we show the graph of a function with vertical asymptotes at $x = 1$ and $x = 3$. In the language of limits, we say that $\lim\limits_{x\to1^-} f(x) = -\infty$ and $\lim\limits_{x\to1^+} f(x) = +\infty$; $\lim\limits_{x\to3^-} f(x) = +\infty$ and $\lim\limits_{x\to3^+} f(x) = -\infty$.

Functions can have both horizontal and vertical asymptotes. The function whose graph is shown in Figure 6 has a horizontal asymptote at $y = 2$ as $x \to \infty$, a horizontal asymptote at $y = -2$ as $x \to -\infty$, and a vertical asymptote at $x = 4$. In the language of limits, $\lim\limits_{x\to\infty} f(x) = 2$, $\lim\limits_{x\to-\infty} f(x) = -2$, $\lim\limits_{x\to4^+} f(x) = -\infty$, and $\lim\limits_{x\to4^-} f(x) = \infty$.

The functions in Figures 4, 5, and 6 were not given by formulas. How do you determine where asymptotes are located if you are given an algebraic formula for the function? The following limit results provide much of the basis for determining vertical and horizontal asymptotes of functions.

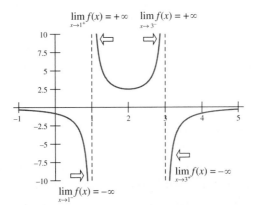

Figure 5

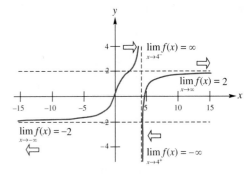

The graph of a function $f(x)$ with horizontal asymptotes at $y = 2$ and $y = -2$ and a vertical asymptote at $x = 4$.

Figure 6

Important Limits Let k and C be positive real constants. Then

$$\lim_{x\to 0^+} \frac{C}{x^k} = \infty \text{ and } \lim_{x\to 0^-} \frac{C}{x^k} = \pm\infty, \text{ provided these limits exist}$$

Thus, for example, $\lim_{x\to 0^+} \frac{4}{x^2} = \infty$. These limits follow from the fact that as the denominator of a fraction gets smaller (while the numerator remains constant) the fraction gets larger. For example, we have $\frac{2}{0.01} = 200$ and $\frac{2}{0.001} = 2000$.

$$\lim_{x\to \infty} \frac{C}{x^k} = 0 \text{ and } \lim_{x\to -\infty} \frac{C}{x^k} = 0, \text{ provided these limits exist}$$

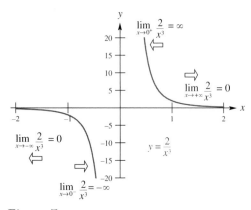

$$\lim_{x\to 0^+} \frac{2}{x^3} = \infty$$
$$\lim_{x\to +\infty} \frac{2}{x^3} = 0$$
$$\lim_{x\to -\infty} \frac{2}{x^3} = 0$$
$$y = \frac{2}{x^3}$$
$$\lim_{x\to 0^-} \frac{2}{x^3} = -\infty$$

Figure 7

Thus, for example, $\lim_{x\to\infty} \frac{5}{\sqrt{x}} = \lim_{x\to\infty} \frac{5}{x^{1/2}} = 0$. These limits follow from the fact that as the denominator of a fraction gets larger (while the numerator remains constant) the fraction gets smaller. For example, we have $\frac{1}{100} = 0.01$ and $\frac{1}{1000} = 0.001$.

We can illustrate these ideas by looking at the graph of $y = \frac{2}{x^3}$ shown in Figure 7. There is a vertical asymptote at $x = 0$ because the denominator is zero there. There is a horizontal asymptote at $y = 0$ because $\frac{2}{x^3} \to 0$ as $x\to\infty$ and as $x\to -\infty$. When we replace the term $\frac{2}{x^3}$ with zero in the equation $y = \frac{2}{x^3}$, we are left with $y = 0$, which is the equation of the asymptote.

Keeping these ideas in mind, we state the following techniques for finding asymptotes.

Techniques for Finding Asymptotes The following techniques will reveal the location of some, though not necessarily all, of the asymptotes of a function.

Vertical asymptotes

1. Set the denominator or denominators of the function that involve x equal to zero and solve for x.

2. Determine the right- and left-hand limits at the vertical asymptote by building tables of values as x approaches the asymptote from the left and from the right.

Horizontal asymptotes

1. For functions that consist of a quotient of two polynomials, divide both numerator and denominator by the largest power of x.

2. Determine which fractions within the function approach zero as x gets large and eliminate them; whatever is left represents the equation of the asymptote. We can use the same reasoning we used under the heading "Important Limits" above.

3. Build tables of function values letting $x \to \infty$ and letting $x \to -\infty$. You can use a "powers of 10" convention as in Example 1 by choosing $x = 10, 100, 1000, \ldots$ when letting $x \to \infty$ and choosing $x = -10, -100, -1000, \ldots$ when letting $x \to -\infty$.

Warning: If the numerator $f(x)$ of a fraction $\frac{f(x)}{x}$ does not remain constant, that fraction does not necessarily approach zero as x gets large. A table of function values is especially important in determining the limit in such a case.

***EXAMPLE* 2:** Find all asymptotes of the function $f(x) = 1 + \frac{3}{2-x}$ and determine the corresponding limits. Plot a few additional points and sketch the graph.

SOLUTION: The function $f(x)$ has one denominator, which is equal to zero when $x = 2$, so we suspect a vertical asymptote there. Thus, we need tables of values to find out what happens as $x \to 2^+$ and as $x \to 2^-$ (remember that $x \to 2^-$ means "x approaches 2 from the left").

We also see that as x gets large in magnitude, the denominator of $\frac{3}{2-x}$ gets large in magnitude, so that the value of the fraction approaches zero. As x gets large, then, $1 + \frac{3}{2-x}$ approaches $1 + 0 = 1$. We have a horizontal asymptote at $y = 1$ because only the first term of the function $f(x) = 1 + \frac{3}{2-x}$ is left when the fractional part goes to zero. We check our results by building tables of function values as $x \to \infty$ and as $x \to -\infty$.

Figure 8 shows all tables, as well as a table with a few additional points and a sketch of the graph.

Our conclusion based on both algebraic and numerical evidence is that there is a horizontal asymptote at $y = 1$ as $x \to \infty$ and as $x \to -\infty$. Also, there is a vertical asymptote at $x = 2$, where $y \to -\infty$ as $x \to 2^+$ and $y \to \infty$ as $x \to 2^-$.   ◄

x	$f(x) = 1 + \dfrac{3}{2-x}$
10	0.625
100	0.969387755
1000	0.996993988

$$\lim_{x \to \infty} f(x) \approx 1.0$$

x	$f(x) = 1 + \dfrac{3}{2-x}$
2.1	-29
2.01	-299
2.001	-2999

$$\lim_{x \to 2^+} f(x) = -\infty$$

x	$f(x) = 1 + \dfrac{3}{2-x}$
-10	1.25
-100	1.029411765
-1000	1.002994012

$$\lim_{x \to -\infty} f(x) \approx 1.0$$

x	$f(x) = 1 + \dfrac{3}{2-x}$
1.9	31
1.99	301
1.999	3001

$$\lim_{x \to 2^-} f(x) = \infty$$

Estimates of limits and asymptotes.

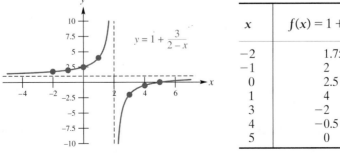

Figure 8

x	$f(x) = 1 + \dfrac{3}{2-x}$
-2	1.75
-1	2
0	2.5
1	4
3	-2
4	-0.5
5	0

Additional points.

EXAMPLE 3: Find $\lim\limits_{x\to\infty} \dfrac{2 - 3x + x^2}{7 + 4x - 5x^2}$.

SOLUTION: We divide numerator and denominator by the largest power of x, which is x^2. Thus,

$$\lim_{x\to\infty} \frac{2 - 3x + x^2}{7 + 4x - 5x^2} = \lim_{x\to\infty} \frac{2/x^2 - 3/x + 1}{7/x^2 + 4/x - 5}$$

$$= \frac{0 - 0 + 1}{0 + 0 - 5} = -\frac{1}{5}$$

using the important limits listed previously.   ◄

Continuity   Roughly speaking, we say that a function is *continuous* if its graph doesn't have any "breaks" or "jumps" in it. To be a little more specific, we say that a function is continuous at a point $x = c$ if the graph doesn't have a break at $x = c$. Thus, a function could be continuous at some points and not continuous (discontinuous) at others (see Figure 9). For example, the unit step function $U(x)$ from Example 6 of Section 1.1 is continuous for all real numbers x except for $x = 0$ (see Figure 7 of Section 1.1). Generally when we say that a function is continuous, we mean that it is continuous everywhere it is defined.

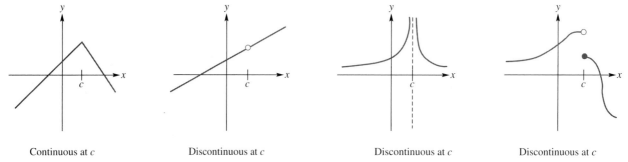

Continuous at c Discontinuous at c Discontinuous at c Discontinuous at c

Figure 9

Talking about breaks and jumps gives a good intuitive understanding of continuity, but sometimes we need a more precise definition. We can use the language of limits to define continuity.

Definition of Continuity

A function $f(x)$ is *continuous* at a point $x = c$ if $\lim\limits_{x\to c} f(x) = f(c)$. If $f(x)$ is not continuous at a point, it is called *discontinuous* there. *Note*: If $\lim\limits_{x\to c} f(x) = \pm\infty$ (that is, there is a vertical asymptote at $x = c$), then $f(x)$ is discontinuous at $x = c$.

In plain English this says that on the graph of $f(x)$, as the x-coordinate approaches c from either the right or left, the y-coordinate approaches $f(c)$. This is how we translate "no jump at $x = c$" into more precise language.

EXAMPLE 4: Consider the life of a battery that is being used to power a flashlight. At time $t = 0$ the flashlight is turned on and the battery delivers its maximum voltage. The voltage then decreases while the battery is being used. Let time (t) be measured in hours and voltage (v) in volts. Suppose that the curve describing the amount of voltage supplied to the flashlight as a function of time can be approximated by the equation $v = f(t)$, where

$$f(t) = \begin{cases} \frac{5}{1+t} & t \geq 0 \\ 0 & t < 0 \end{cases}$$

Sketch the graph of voltage versus time. Discuss continuity and any asymptotes for this function, both in mathematical terms and in terms of the context of the problem.

SOLUTION: The graph of $v = \frac{5}{1+t}$ is shown in Figure 10. It can be found by the techniques of Example 2 (and checked with a graphing calculator or computer).

The graph of $v = f(t)$ is shown in Figure 11. We graph $v = \frac{5}{1+t}$ to the right of $t = 0$ and graph $v = 0$ to the left of $t = 0$. ◄

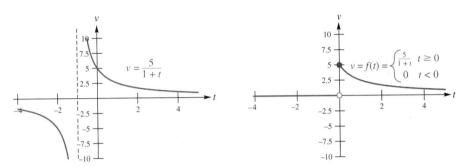

Figure 10   Figure 11

The function $f(t)$ is continuous everywhere except at $t = 0$, where there is a jump of 5 units. We have $\lim_{t \to 0^-} f(t) = 0$, $\lim_{t \to 0^+} f(t) = 5$, and $f(0) = 5$, which shows that $f(t)$ is not continuous at $t = 0$ using the definition of continuity given above. This means that the battery delivers zero voltage until the flashlight is turned on, at which point there is a discontinuous jump to 5 volts.

There is a horizontal asymptote at $v = 0$ as $t \to \infty$. This corresponds to the fact that the flashlight is draining the battery so that the voltage approaches zero as t approaches infinity.

Example 4 is typical of how discontinuous functions are used in the sciences. Often there is a discontinuity at the beginning of some phenomenon or experiment, when an action is taken such as applying an external force (a voltage in Example 4) or moving an object.

All of the elementary functions we will be using regularly in this book, such as polynomials and trig, logarithmic, and exponential functions (introduced later in this chapter), have the property that they *are continuous everywhere they are defined*. That leaves us wondering what kinds of functions are not continuous. Often, discontinuous functions result from piecing together elementary functions, the way the function $f(t)$ from Example 4 was pieced together from the two functions $v = \frac{5}{1+t}$ and $v = 0$.

Concepts Review

1. To say that $x \to \infty$ means that _____ ; to say that $\lim_{x \to \infty} f(x) = L$ means that _____ . And to say that $\lim_{x \to a^+} f(x) = \infty$ means that _____ . Give your answers in informal language.

2. If $\lim_{x \to \infty} f(x) = 6$, then the line _____ is a _____ asymptote of the graph of $y = f(x)$. If $\lim_{x \to 6^+} f(x) = \infty$,

then the line _____ is a _____ asymptote of the graph of $y = f(x)$.

3. A function f is continuous at c if _____ $= f(c)$.

4. The unit step function $U(x)$ from Example 6 of Section 1.1 is discontinuous at the point $x = $ _____ because _____ .

Problem Set 1.2

1. Give a common-sense interpretation for each of the following limits. Illustrate each with a sketch.

(a) $\lim_{x \to a^-} f(x) = \infty$ (b) $\lim_{x \to a^+} f(x) = -\infty$ (c) $\lim_{x \to a^-} f(x) = -\infty$

2. Give a common-sense interpretation for each of the following limits. Illustrate each with a sketch.

(a) $\lim_{x \to \infty} f(x) = \infty$ (b) $\lim_{x \to \infty} f(x) = -\infty$ (c) $\lim_{x \to -\infty} f(x) = \infty$

In Problems 3–8, estimate the limits by building a table of values as in Example 1.

3. $\lim_{x \to \infty} \dfrac{3 - 2x}{x + 5}$

4. $\lim_{x \to 1^+} \dfrac{5x + 1}{x - 1}$

5. $\lim_{x \to 1^-} \dfrac{2x + 7}{x^2 - x}$

6. $\lim_{x \to -\infty} \dfrac{2x^2 - x + 5}{5x^2 + 6x - 1}$

7. $\lim_{x \to 0^-} \dfrac{1 + \cos x}{\sin x}$

8. $\lim_{x \to \infty} \dfrac{\sin x}{x}$

In Problems 9–20, find all asymptotes of the given function and determine the corresponding limits. Plot a few additional points and sketch the graph (see Example 2).

9. $f(x) = \dfrac{3}{x + 1}$

10. $f(x) = \dfrac{3}{(x + 1)^2}$

11. $f(x) = \dfrac{2x}{x - 3}$

12. $f(x) = \dfrac{3}{9 - x^2}$

13. $g(x) = \dfrac{14}{2x^2 + 7}$

14. $g(x) = \dfrac{2x}{\sqrt{x^2 + 5}}$

15. $f(x) = \sqrt{\dfrac{2x^2 - 3x}{5x^2 + 1}}$

16. $f(x) = \dfrac{3x^2 + x + 1}{2x^2 - 1}$

17. $H(x) = \dfrac{2x + 1}{2x^2 - x - 3}$

18. $H(x) = \dfrac{2x}{x^2 - 6x + 9}$

19. $h(x) = \dfrac{2x^2 - 18}{x - 3}$

20. $h(x) = \dfrac{x^2 - x - 6}{x - 3}$

In Problems 21–46, find each limit accurate to three decimal places. Also attempt to find the exact value of the limit (see Examples 1 and 3).

21. $\lim_{x \to \infty} \dfrac{2x}{x + 4}$

22. $\lim_{x \to \infty} \dfrac{2}{x + 4}$

23. $\lim_{t \to 3^+} \dfrac{3 + t}{3 - t}$

24. $\lim_{t \to 3^-} \dfrac{3 + t}{3 - t}$

25. $\lim_{x \to -\infty} \dfrac{x^2 - 2x + 5}{8x^2 - 6x}$

26. $\lim_{x \to \infty} \dfrac{2x^3 - 3x^2 + 1}{5x^3 - 4x + 7}$

27. $\lim_{x \to 2^+} \dfrac{x^2 + 2x - 8}{x^2 - 4}$

28. $\lim_{x \to 2^+} \dfrac{3}{x^2 - 4}$

29. $\lim_{x \to \infty} \dfrac{(3x - 2)(2x + 4)}{(2x + 1)(x + 2)}$

30. $\lim_{x \to -\infty} \dfrac{3x^3 - 4x + 1}{(x^2 + 1)(x^2 - 1)}$

31. $\lim_{x \to (3/2)^-} \dfrac{4x + 1}{2x - 3}$

32. $\lim_{x \to 3^-} \dfrac{x^2}{x^2 - 9}$

33. $\lim_{x \to \infty} \dfrac{3x\sqrt{x} + 3x + 1}{x^2 - x + 11}$

34. $\lim_{x \to \infty} \sqrt[3]{\dfrac{1 + 8x^2}{x^2 + 4}}$

35. $\lim_{x \to \infty} \dfrac{\cos(x - 3)}{x - 3}$

36. $\lim_{x \to (\pi/2)^+} \dfrac{\cos x}{x - \pi/2}$

37. $\lim_{x \to 0^-} \dfrac{1 + \cos x}{\sin x}$

38. $\lim_{x \to \infty} \dfrac{\sin x}{x}$

39. $\displaystyle\lim_{x\to\infty}\left(1+\frac{1}{x}\right)^{10}$

40. $\displaystyle\lim_{x\to\infty}\left(1+\frac{1}{x}\right)^{x}$

41. $\displaystyle\lim_{x\to0^{+}}\left(1+\sqrt{x}\right)^{1/\sqrt{x}}$

42. $\displaystyle\lim_{x\to0^{+}}\left(1+\sqrt{x}\right)^{1/x}$

43. $\displaystyle\lim_{x\to\infty}\left(1+\frac{1}{x}\right)^{x^{2}}$

44. $\displaystyle\lim_{x\to\infty}\left(1+\frac{1}{x}\right)^{\sin x}$

45. $\displaystyle\lim_{x\to\infty}\frac{2x+1}{\sqrt{x^{2}+3}}$

Hint: To find the exact limit, divide the numerator and the denominator by x. Note that for $x>0$,
$\sqrt{x^{2}+3}/x = \sqrt{x^{2}+3/x^{2}}$.

46. $\displaystyle\lim_{x\to\infty}\left(\sqrt{2x^{2}+3}-\sqrt{2x^{2}-5}\right)$

Hint: To find the exact limit, multiply and divide by $\sqrt{2x^{2}+3}+\sqrt{2x^{2}-5}$.

47. Find $\displaystyle\lim_{x\to\infty}\frac{a_{0}x^{n}+a_{1}x^{n-1}+\cdots+a_{n-1}x+a_{n}}{b_{0}x^{n}+b_{1}x^{n-1}+\cdots+b_{n-1}x+b_{n}}$

where $a_{0}\neq0$, $b_{0}\neq0$, and n is a natural number.

Sketch a graph of each function in Problems 48–55. Find all asymptotes and points of discontinuity for each function.

48. $f(x)=4x^{2}-2x+12$

49. $f(x)=\dfrac{8}{x-2}$

50. $g(x)=\dfrac{3x^{2}}{x-2}$

51. $g(x)=\sqrt{x-1}$

52. $f(x)=\begin{cases} x+3 & \text{if } x<2 \\ x^{2}+1 & \text{if } x\geq2 \end{cases}$

53. $f(x)=\begin{cases} -3x+4 & \text{if } x\leq2 \\ -2 & \text{if } x>2 \end{cases}$

54. $f(x)=\begin{cases} x & \text{if } x<0 \\ x^{2} & \text{if } 0\leq x\leq1 \\ 2-x & \text{if } x>1 \end{cases}$

55. $g(x)=\begin{cases} x^{2} & \text{if } x<0 \\ -x & \text{if } 0\leq x\leq1 \\ x & \text{if } x>1 \end{cases}$

56. Find the values of a and b so that the following function is continuous everywhere.

$$f(x)=\begin{cases} x+1 & \text{if } x<0 \\ ax+b & \text{if } 0\leq x\leq1 \\ 3x & \text{if } x\geq1 \end{cases}$$

57. A thin equilateral triangular block of side length 1 unit has its face in the vertical xy-plane with a vertex V at the origin. Under the influence of gravity, it will rotate about V until a side hits the x-axis floor (Figure 12). Let x denote the initial x-coordinate of the midpoint M of the side opposite V and let $f(x)$ denote the final x-coordinate of this point. Assume the block balances when M is directly above V.

(a) Determine the domain and range of f.

(b) Where on this domain is f discontinuous?

(c) Sketch a graph of $f(x)$.

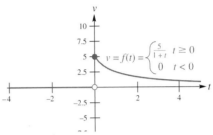

Figure 12

58. Figure 13 contains the a graph of voltage (in volts) supplied by a battery to a laptop computer as a function of time. The computer was turned on at time -1 hour (-60 minutes) and turned off at time 0. This computer has a battery-saving feature built into it, which switches the processor into low voltage mode if no keystrokes are made for a few seconds (called processor resting). This results in the "spikes" in the graph.

(a) Discuss continuity for the graph of voltage versus time.

(b) What would the graph of voltage versus time look like on the interval $-2\leq t\leq0$ (where t is time measured in hours)? Discuss continuity for this case.

(c) If the computer were left on longer, what do you think the graph of voltage versus time would look like? Discuss continuity and asymptotes.

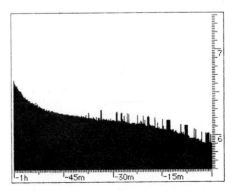

Figure 13

59.   Sketch the graph of a function f that satisfies all the following conditions.

(a) Its domain is $[-2, 2]$.

(b) $f(-2) = f(-1) = f(1) = f(2) = 1$.

(c) It is discontinuous at -1 and 1.

60.   The line $y = ax + b$ is an **oblique asymptote** to the graph of $y = f(x)$ if either $\lim\limits_{x \to \infty} [f(x) - (ax + b)] = 0$ or $\lim\limits_{x \to -\infty} [f(x) - (ax + b)] = 0$. Find the oblique asymptote for

$$f(x) = \frac{2x^4 + 3x^3 - 2x - 4}{x^3 - 1}$$

Hint: Begin by dividing the denominator into the numerator.

61.   Find the oblique asymptote for

$$f(x) = \frac{3x^3 + 4x^2 - x + 1}{x^2 - 1}$$

62.   We have given meaning to $\lim\limits_{x \to A} f(x)$ for $A = a$, a^-, a^+, $-\infty$, ∞. Moreover, in each case, this limit may be L (finite), $-\infty$, ∞, or may fail to exist in any sense. Make a table illustrating each of the 20 possible cases.

Answers to Concepts Review: 1. x increases without bound; $f(x)$ gets close to L as x increases without bound; $f(x)$ increases without bound as x approaches a from the right. 2. $y = 6$; horizontal; $x = 6$; vertical. 3. $\lim\limits_{x \to c} f(x)$ 4. 0; $\lim\limits_{x \to 0^+} U(x) = 1$ but $\lim\limits_{x \to 0^-} U(x) = 0$.

1.3 | The Derivative: Two Problems with One Theme

Our first problem is very old; it dates back to the great Greek scientist Archimedes (287?–212 B.C.). We refer to the problem of the *tangent line*.

Our second problem is newer. It grew out of attempts by Kepler (1571–1630), Galileo (1564–1642), Newton (1642–1727), and others to describe the speed of a moving body. It is the problem of *instantaneous velocity*.

These two problems, one geometric and the other mechanical, appear to be quite unrelated. In this case, appearances are deceptive. The two problems are identical twins.

The Tangent Line   Euclid's notion of a tangent as a line touching a curve at just one point is all right for circles (Figure 1) but completely unsatisfactory for most other curves (Figure 2). The idea of a tangent to a curve at P as the line that best approximates the curve near P is better, but it is still too vague for mathematical precision. The concept of limit provides a way of getting the best description.

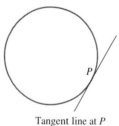

Tangent line at P

Figure 1

Tangent line at P

Figure 2

Let P be a fixed point on a curve and let Q be a nearby movable point on that curve. Consider the line through P and Q, called a **secant line**. The **tangent line** at P is the limiting position (if it exists) of the secant line as Q approaches P along the curve (Figure 3). Suppose that the curve is the graph of the equation $y = f(x)$, as in Figure 4. Then P has coordinates $(c, f(c))$, a nearby point Q has coordinates $(c + h, f(c + h))$, and the secant line through P and Q has slope m_{sec} given by

$$m_{\text{sec}} = \frac{f(c + h) - f(c)}{h}$$

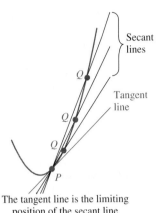

The tangent line is the limiting position of the secant line

Figure 3

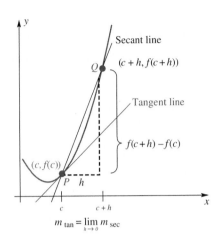

$$m_{\tan} = \lim_{h \to 0} m_{\sec}$$

Figure 4

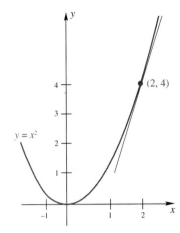

Figure 5

Because the tangent line is the limiting position of the secants as $h \to 0$, its slope $m_{\tan}$ will be the limit of the secant slopes $m_{\sec}$ as $h \to 0$. Consequently, the tangent line to the curve $y = f(x)$ at the point $P(c, f(c))$ is that line through P with slope $m_{\tan}$ satisfying

$$m_{\tan} = \lim_{h \to 0} m_{\sec} = \lim_{h \to 0} \frac{f(c + h) - f(c)}{h}$$

Note that this definition does not hold true for vertical tangents. Why not?

EXAMPLE 1: Find the slope of the tangent line to the curve $y = f(x) = x^2$ at the point $(2, 4)$. Also find the slope of the secant line through the points $(2, 4)$ and $(2.1, 2.1^2) = (2.1, 4.41)$.

SOLUTION: The line whose slope we are seeking is shown in Figure 5. Clearly it has a large positive slope.

$$m_{\tan} = \lim_{h \to 0} \frac{f(2 + h) - f(2)}{h}$$

$$= \lim_{h \to 0} \frac{(2 + h)^2 - 2^2}{h} \qquad (because\ f(x) = x^2)$$

$$= \lim_{h \to 0} \frac{4 + 4h + h^2 - 4}{h}$$

$$= \lim_{h \to 0} \frac{\cancel{h}(4 + h)}{\cancel{h}}$$

$$= 4$$

The slope of the secant through the points $(2, 4)$ and $(2.1, 4.41)$ would be

$$\frac{4.41 - 4}{2.1 - 2} = \frac{0.41}{0.1} = 4.1$$

The slope of the secant and the slope of the tangent are almost the same because the two points that define the secant are close together. In terms of the definition of the tangent line $m_{\text{tan}} = \lim\limits_{h \to 0} m_{\text{sec}} = \lim\limits_{h \to 0} \dfrac{f(c + h) - f(c)}{h}$, we have chosen $h = 0.1$, a number fairly close to zero. ◄

The Derivative   We can think of the slope of the tangent line to the graph of a function at a point as the instantaneous slope (or just slope for short) of the curve at that point. The slope of a function $f(x)$ at a point c is called the *derivative* of $f(x)$ at c and is denoted $f'(c)$. Thus, $f'(c)$ and m_{tan} are just two ways of writing the same thing. We repeat the definition in the new notation:

$$f'(c) = \lim\limits_{h \to 0} \dfrac{f(c + h) - f(c)}{h}$$

Note: We usually think of x as a variable (it can have many possible values) and c as a *particular* value of x. Don't be thrown if it seems that c and x are being used somewhat interchangeably in what follows.

EXAMPLE 2:   Find the slope of the tangent line to the curve $y = f(x) = -x^2 + 2x + 2$ at the points with x-coordinates $-1, \frac{1}{2}, 2$, and 3. In the notation of derivatives, this means finding $f'(-1), f'\left(\frac{1}{2}\right), f'(2)$, and $f'(3)$.

SOLUTION: Rather than make four separate calculations, it is more efficient to calculate the slope at the point with x-coordinate c and then obtain the four desired answers by substituting for c.

$$m_{\text{tan}} = f'(c) = \lim\limits_{h \to 0} \dfrac{f(c + h) - f(c)}{h}$$

$$= \lim\limits_{h \to 0} \dfrac{-(c + h)^2 + 2(c + h) + 2 - (-c^2 + 2c + 2)}{h}$$

$$= \lim\limits_{h \to 0} \dfrac{-c^2 - 2ch - h^2 + 2c + 2h + 2 + c^2 - 2c - 2}{h}$$

$$= \lim\limits_{h \to 0} \dfrac{\cancel{h}(-2c - h + 2)}{\cancel{h}}$$

$$= -2c + 2$$

The four desired slopes (obtained by letting $c = -1, \frac{1}{2}, 2, 3$) are $4, 1, -2$, and -4. These answers do appear to be consistent with the graph in Figure 6. ◄

In Example 2 we found our first derivative formula. We saw that if $f(x) = -x^2 + 2x + 2$, then $f'(c) = -2c + 2$. If we replace the c with an x we get $f'(x) = -2x + 2$. As we saw in the last example, when such derivative formulas can be found, they can be quite helpful in solving problems. In Chapter 3 we will develop quite a few derivative formulas. For now, in cases where a derivative formula can't easily be found, we can still resort to numerical techniques for estimating derivatives.

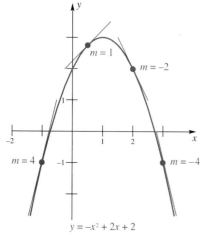

$y = -x^2 + 2x + 2$

Figure 6

EXAMPLE 3: Find the derivative of $f(x) = \sin x$ at the points $x = 0, \frac{\pi}{2}, \pi$. Illustrate your results by sketching the graph of the function $f(x)$ and using slope marks (as in Figure 6) to indicate the derivatives at the x-values given.

SOLUTION: We want to find

$$f'(c) = \lim_{h \to 0} \frac{\sin(c + h) - \sin(c)}{h}$$

for $c = 0, \frac{\pi}{2}, \pi$. Letting $c = 0$, we get

$$f'(0) = \lim_{h \to 0} \frac{\sin(0 + h) - \sin(0)}{h} = \lim_{h \to 0} \frac{\sin(h)}{h} \qquad (because \sin 0 = 0)$$

This time there are no algebraic simplifications we can use. Instead, we must estimate this limit numerically using a table, as we did with some of the problems in the last section. In fact, we already estimated this limit in Example 4 of Section 1.1. The letter x was used as the variable instead of h, but the result is still the same; we get $f'(0) \approx 1$. The tables corresponding to the values $c = \frac{\pi}{2}$ and $c = \pi$ are shown in Figure 7. We conclude that $f'\left(\frac{\pi}{2}\right) \approx 0$ and $f'(\pi) \approx -1$.

Two comments are in order regarding the tables shown in Figure 7. In the table used to estimate $f'\left(\frac{\pi}{2}\right)$, we see numbers like $-4.996\text{E-}08$. This is the notation calculators and

h	$\dfrac{\sin\left(\dfrac{\pi}{2} + h\right) - \sin\left(\dfrac{\pi}{2}\right)}{h}$
1	-0.459697694
0.1	-0.049958347
0.01	-0.004999958
0.001	-0.0005
0.0001	$-5\text{E-}05$
0.00001	$-5\text{E-}06$
0.000001	$-5.00044\text{E-}07$
0.0000001	$-4.996\text{E-}08$
0.00000001	0

Table to estimate $f'\left(\frac{\pi}{2}\right)$

h	$\dfrac{\sin(\pi + h) - \sin(\pi)}{h}$
1	-0.841470985
0.1	-0.998334166
0.01	-0.999983333
0.001	-0.999999833
0.0001	-0.999999998
0.00001	-1
0.000001	-1
0.0000001	-0.999999998
0.00000001	-0.999999994

Figure 7   Table to estimate $f'(\pi)$.

computers usually use to represent scientific notation. In this case, the E-08 represents multiplication by 10^{-8}, so that

$$-4.996\text{E-08} = -4.996 \times 10^{-8} = -0.00000004996$$

We left the table in computer notation, because that is the form you are likely to encounter when using technology. When you write up your results for homework exercises or lab projects, you should normally use standard mathematical notation rather than computer notation.

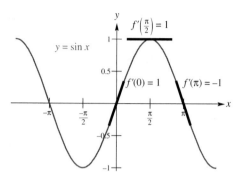

The second comment refers to the table representing the case $c = \pi$. It appears that the numbers in the column labeled $\dfrac{\sin(\pi + h) - \sin(\pi)}{h}$ approach the number -1 at first and then drift away from -1 as h gets even closer to 0. This is an example of numerical round-off, which occurs with all calculators and computer programs. We must often use our judgment when interpreting tables created by technology.

Our final sketch with slope marks is shown in Figure 8. Notice that a slope of 1 *does not* generally represent a line at a 45° angle with the x-axis; only when the scales on the x- and y-axes are the same is this true. ◄

Figure 8

Average Velocity and Instantaneous Velocity

If we drive an automobile from one town to another 80 miles away in 2 hours, our average velocity is 40 miles per hour. That is, *average velocity* is the distance from the first position to the second position divided by the elapsed time.

But during our trip the speedometer reading was often different from 40. At the start, it registered 0; at times it rose as high as 57; at the end it fell back to 0 again. Just what does the speedometer measure? Certainly it does not indicate average velocity.

Consider the more precise example of an object P falling in a vacuum. Experiment shows that if it starts from rest, P falls $16.09t^2$ feet in t seconds; we use the approximation $16t^2$ in what follows. Thus, it falls 16 feet in the first second and 64 feet during the first 2 seconds (Figure 9); clearly it falls faster and faster as time goes on.

During the second second (that is, in the time interval from $t = 1$ to $t = 2$), P fell $(64 - 16)$ feet. Its average velocity during that second was

$$v_{\text{ave}} \frac{64 - 16}{2 - 1} = 48 \text{ feet per second}$$

During the time interval from $t = 1$ to $t = 1.5$, it fell $16(1.5)^2 - 16 = 20$ feet. Its average velocity was

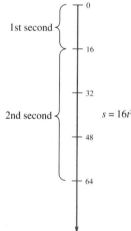

Figure 9

$$v_{\text{ave}} = \frac{16(1.5)^2 - 16}{1.5 - 1} = \frac{20}{0.5} = 40 \text{ feet per second}$$

Similarly, on the time intervals $t = 1$ to $t = 1.1$ and $t = 1$ to $t = 1.01$, we calculate the respective average velocities to be

$$v_{\text{ave}} = \frac{16(1.1)^2 - 16}{1.1 - 1} = \frac{3.36}{0.1} = 33.6 \text{ feet per second}$$

$$v_{\text{ave}} = \frac{16(1.01)^2 - 16}{1.01 - 1} = \frac{0.3216}{0.01} = 32.16 \text{ feet per second}$$

What we have done is to calculate the average velocity over shorter and shorter time intervals, each starting at $t = 1$. The shorter the time interval, the better our approximation of the *instantaneous velocity* at the instant $t = 1$. Looking at the numbers 48, 40, 33.6, and 32.16, you might guess 32 feet per second to be the instantaneous velocity.

But let us be more precise. Suppose that an object P moves along a coordinate line so that its position at time t is given by $s = f(t)$. At time c the object is at $f(c)$; at the nearby time $c + h$, it is at $f(c + h)$ (see Figure 10). Thus, the **average velocity** on this interval is

$$v_{\text{ave}} = \frac{f(c + h) - f(c)}{h}$$

And now we define the **instantaneous velocity** v at c by

$$v = \lim_{h \to 0} v_{\text{ave}} = \lim_{h \to 0} \frac{f(c + h) - f(c)}{h}$$

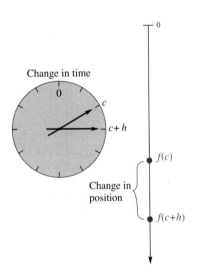

Change in time

Change in position

Figure 10

In the case where $f(t) = 16t^2$ (free fall in a vacuum on the earth), the instantaneous velocity at $t = 1$ is

$$v = \lim_{h \to 0} \frac{f(1 + h) - f(1)}{h}$$

$$= \lim_{h \to 0} \frac{16(1 + h)^2 - 16}{h}$$

$$= \lim_{h \to 0} \frac{16 + 32h + 16h^2 - 16}{h}$$

$$= \lim_{h \to 0} (32 + 16h) = 32$$

This confirms our earlier guess. In the problem set we ask you to do this problem again with the more precise function $f(t) = 16.09t^2$ to see what happens.

Now you can see why we called the *slope of the tangent line* (the derivative) and *instantaneous velocity* identical twins. Look at the two boxed formulas in this section. They give different names to the same mathematical concept.

EXAMPLE 4: Find the instantaneous velocity of a falling object, starting from rest, at $t = 3.8$ seconds and at $t = 5.4$ seconds.

SOLUTION: We calculate the instantaneous velocity at $t = c$ seconds. Because $f(t) = 16t^2$,

$$v = \lim_{h \to 0} \frac{f(c + h) - f(c)}{h}$$

$$= \lim_{h \to 0} \frac{16(c + h)^2 - 16c^2}{h}$$

$$= \lim_{h \to 0} \frac{16c^2 + 32ch + 16h^2 - 16c^2}{h}$$

$$= \lim_{h \to 0} (32c + 16h) = 32c$$

Thus, the instantaneous velocity at 3.8 seconds is $32(3.8) = 121.6$ feet per second; at 5.4 seconds, it is $32(5.4) = 172.8$ feet per second.   ◄

EXAMPLE 5: How long will it take the falling object of Example 4 to reach an instantaneous velocity of 112 feet per second?

SOLUTION: We learned in Example 4 that the instantaneous velocity after c seconds is $32c$. Thus, we must solve the equation $32c = 112$. The solution is $c = \frac{112}{32} = 3.5$ seconds. ◀

EXAMPLE 6: A particle moves along a coordinate line; s, its directed distance in centimeters from the origin at the end of t seconds, is given by $s = f(t) = \sqrt{5t + 1}$. Find the instantaneous velocity of the particle at the end of 3 seconds.

SOLUTION:

$$v = \lim_{h \to 0} \frac{f(3 + h) - f(3)}{h}$$

$$= \lim_{h \to 0} \frac{\sqrt{5(3 + h) + 1} - \sqrt{5(3) + 1}}{h}$$

$$= \lim_{h \to 0} \frac{\sqrt{16 + 5h} - 4}{h}$$

This limit is somewhat more difficult to find exactly than the one in Example 5. We will complete the problem two ways: one using some algebra to find the exact limit, and the other using a table of values to get an estimate of the limit. It is always good to do a problem more than one way, so that each method is a check on the other. Also, to the beginning calculus student it is not easy to recognize when an exact solution is possible and when it is not; if you can't see how to get an exact result, try a numerical estimate first.

The numerical estimate of the limit is shown in Figure 11. It appears that the limit to three decimal places is 0.625. Note that we really estimated only the right-hand limit; we leave it to the reader to show that the result is the same for the left-hand limit (that is, for negative h values).

h	$\dfrac{\sqrt{16 + 5h} - 4}{h}$
0.1	0.62019
0.01	0.62451
0.001	0.62495
0.0001	0.62500

Figure 11

To evaluate the limit exactly, we rationalize the numerator (by multiplying numerator and denominator by $\sqrt{16 + 5h} + 4$). We obtain

$$v = \lim_{h \to 0} \left(\frac{\sqrt{16 + 5h} - 4}{h} \cdot \frac{\sqrt{16 + 5h} + 4}{\sqrt{16 + 5h} + 4} \right)$$

$$= \lim_{h \to 0} \frac{16 + 5h - 16}{h(\sqrt{16 + 5h} + 4)}$$

$$= \lim_{h \to 0} \frac{5h}{h\sqrt{16 + 5h} + 4} = \frac{5}{8}$$

We conclude that the instantaneous velocity at the end of 3 seconds is $\frac{5}{8}$ centimeter per second. This agrees with our numerical estimate from above.   ◄

Velocity or Speed?
For the time being, we will use the terms *velocity* and *speed* interchangeably. Later, in Section 3.5, we will distinguish between these two words.

Rates of Change   Velocity is only one of many rates of change that will be important in this course; it is the rate of change of distance with respect to time. Other rates of change that will interest us are density for a wire (the rate of change of mass with respect to distance), marginal revenue (the rate of change of revenue with respect to the number of items produced), and current (the rate of change of electrical charge with respect to time). These rates and many more are discussed in the problem set. In each case, we must distinguish between an *average* rate of change on an interval and an *instantaneous* rate of change at a point. The phrase *rate of change* without an adjective will mean instantaneous rate of change.

Concepts Review

1. The line that most closely approximates a curve near the point P is the _____ through that point.

2. More precisely, the tangent line to a curve at P is the limiting position of the _____ line through P and Q as Q approaches P along the curve.

3. The slope $m_{\tan}$ of the tangent line to the curve $y = f(x)$ at $(c, f(c))$ is given by $m_{\tan} = \lim\limits_{h \to 0}$ _____.

4. The instantaneous velocity of a point P (moving along a line) at a time c is the limit of the _____ on the time interval c to $c + h$ as h approaches zero.

Problem Set 1.3

In Problems 1 and 2, a tangent line to a curve is drawn. Estimate its slope (slope = rise/run). Be careful to note the difference in scales on the two axes.

1.

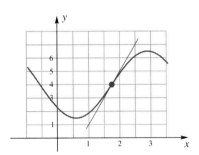

2.

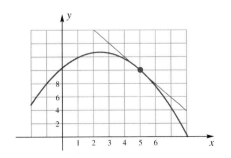

In Problems 3–6, draw the tangent line to the curve through the indicated point and estimate its slope.

3.

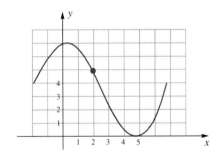

4.

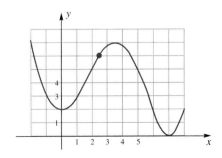

5.

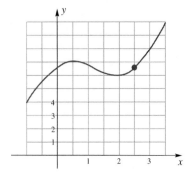

6.

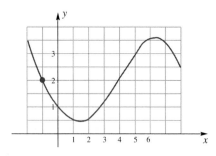

7. Consider $y = 4 - x^2$.

(a) Sketch its graph as carefully as you can.

(b) Draw the tangent line at $(3, -5)$.

(c) Estimate the slope of this tangent line.

(d) Calculate the slope of the secant line through $(3, -5)$ and $(3.01, 4 - 3.01^2)$.

(e) Find by the limit process (see Example 1) the exact slope of the tangent at $(3, -5)$.

(f) Discuss the relationship between your answers to (d) and (e) (see Example 1).

8. Consider $y = x^3 + 1$.

(a) Sketch its graph.

(b) Draw the tangent line at $(2, 9)$.

(c) Estimate the slope of this tangent line.

(d) Calculate the slope of the secant line through $(2, 9)$ and $(1.999, 1.999^3 + 1)$.

(e) Use the limit process to find the exact slope of the tangent at $(2, 9)$.

(f) Discuss the relationship between your answers to (d) and (e).

9. Find the slope of the tangent line to the curve $y = x^2 - 3x + 2$ at the points where $x = -2, 1.5, 2, 5$ (see Example 2).

10. Find the slope of the tangent line to the curve $y = x^3 - 2x$ at the points where $x = -3, -1.5, 0, 3$.

11. Approximate the derivative of $f(x) = \cos x$ at the points $x = 0, \frac{\pi}{2}, \pi$ numerically. Illustrate your results by sketching the graph of the function $f(x)$ and using slope marks to indicate the derivatives at the x-values given (see Example 3).

12. Approximate the derivative of $f(x) = \tan x$ at points $x = 0, \frac{\pi}{4}, \pi$ numerically. Illustrate your results by sketching the graph of the function $f(x)$ and using slope marks to indicate the derivatives at the x-values given.

13. Approximate the derivative of $f(x) = \frac{x}{x+1}$ at the points $x = 0, 1, 2$ numerically. Illustrate your results by sketching the graph of the function $f(x)$ and using slope marks to indicate the derivatives at the x-values given.

14. Approximate the derivative of the function

$$f(x) = \begin{cases} \frac{\sin x}{x} & x \neq 0 \\ 1 & x = 0 \end{cases}$$

at the points $x = 0, \frac{\pi}{2}, \pi$ numerically. Illustrate your results by sketching the graph of the function $f(x)$ and using slope marks to indicate the derivatives at the x-values given.

15. Assume that a falling body will fall $16t^2$ feet in t seconds.

(a) How far will it fall between $t = 3$ and $t = 4$?

(b) What is its average velocity on the interval $3 \le t \le 4$?

(c) What is its average velocity on the interval $3 \le t \le 3.02$?

(d) Find its instantaneous velocity at $t = 3$ (see Example 4).

(e) Discuss the relationship between your answers to (b) and (d) in terms of the definition of instantaneous velocity.

16. An object travels along a line so that its position s is $s = 2t^2 + 2$ meters after t seconds.

(a) What is its average velocity on the interval $2 \le t \le 3$?

(b) What is its average velocity on the interval $2 \le t \le 2.001$?

(c) What is its average velocity on the interval $2 \le t \le 2 + h$?

(d) Find its instantaneous velocity at $t = 2$.

(e) Discuss the relationship between your answers to (b) and (d) in terms of the definition of instantaneous velocity.

17. For the falling body in Problem 15, find

(a) its instantaneous velocity at $t = c, c > 0$.

(b) the time when its velocity reaches 100 feet per second. (See Examples 4 and 5.)

18. For the object in Problem 16, find

(a) its instantaneous velocity at $t = c, c > 0$.

(b) the time when its velocity reaches 50 meters per second.

19. Suppose that the position of an object along a line is $\sqrt{t}$ feet after t seconds.

(a) Find its instantaneous velocity at $t = c, c > 0$.

(b) When will it reach a velocity of $\frac{1}{6}$ foot per second? (See Examples 4 and 5.)

20. If a particle moves along a coordinate line so that its directed distance from the origin after t seconds is $(-t^2 + 4t)$ feet, when did the particle come to a momentary stop? (That is, when did its instantaneous velocity become zero?)

For Problems 21–30 find exact results if you can, and if not, find numerical approximations. State whether your result is exact or approximate, and if approximate, the number of digits or decimals of accuracy. (See Example 6.)

21. A certain bacterial culture is growing so that it has a mass of $10 - \frac{9}{t+1}$ grams after t hours.

(a) How much did it grow during the interval $2 \leq t \leq 2.01$?

(b) What was its average growth rate during the interval $2 \leq t \leq 2.01$?

(c) What was its instantaneous growth rate at $t = 2$?

22. A business is prospering in such a way that its total (accumulated) profit after t years is $1000t^3$ dollars.

(a) How much did the business make during the third year (that is, between $t = 2$ and $t = 3$)?

(b) What was its average rate of profit during the first half of the third year (that is, between $t = 2$ and $t = 2.5$)?

(c) What was its instantaneous rate of profit at $t = 2$?

23. A wire of length 8 centimeters is such that the mass between its left end and a point x centimeters to the right is x^3 grams (Figure 12).

(a) What is the average density of the middle 2-centimeter segment of this wire? *Note*: Average density equals mass/length.

(b) What is the actual density at the point 3 centimeters from the left end?

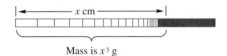

Figure 12

24. Suppose that the revenue from producing x pounds of a product is given by $R(x) = 0.5x - 0.002x^2$. Find the instantaneous rate of change of revenue when $x = 10$; when $x = 100$. (The instantaneous rate of change of revenue with respect to the amount produced is called the *marginal revenue*.)

25. The weight in grams of a malignant tumor at time t is $W(t) = 0.2t^2 + 0.09t^3$, where t is measured in weeks. Find the rate of growth of the tumor when $t = 10$.

26. A city is hit by an Asian flu epidemic. Officials estimate that t days after the beginning of the epi-

demic, the number of persons sick with the flu is given by $p(t) = 120t^2 - 2t^3$, when $0 \leq t \leq 40$. At what rate is the flu spreading at time $t = 10; t = 20; t = 40$?

27. The graph in Figure 13 shows the amount of water in a city water tank during one 24-hour day when no water was pumped into the tank. What was the average rate of water usage during the day? How fast was water being used at 8 A.M.?

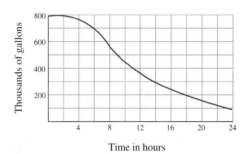

Figure 13

28. The rate of change of electric charge with respect to time is called **current**. Suppose that $\frac{1}{3}t^3 + t$ coulombs of charge flow through a wire in t seconds. Find the current in amperes (coulombs per second) after 3 seconds. When will a 20-amp fuse in the line blow?

29. The radius of a circular oil spill is growing at a constant rate of 2 kilometers per day. At what rate is the area of the spill growing 3 days after it began?

30. Find the rate of change of the area of a circle with respect to its circumference when the circumference is 6 centimeters.

Use a computer or graphing calculator to do Problems 31–32.

31. Draw the graph of $y = f(x) = x^3 - 2x^2 + 1$. Then find the slope of the tangent line at (a) -1, (b) 0, (c) 1, (d) 3.2.

32. Draw the graph of $y = f(x) = \sin x \sin^2 2x$. Then find the slope of the tangent line at (a) $\pi/3$, (b) 2.8, (c) π, (d) 4.2.

33. If a point moves along a line so that its distance s (in feet) from 0 is given by $s = t + t\cos^2 t$ at time t seconds, find its instantaneous velocity at $t = 3$.

34. If a point moves along a line so that its distance s (in meters) from 0 is given by $s = (t + 1)^3/(t + 2)$ at time t minutes, find its instantaneous velocity at $t = 1.6$.

35. Assume that an object falls a distance of $16.09t^2$ feet in t seconds. Find the instantaneous velocity at time $t = 1$.

Answers to Concepts Review: 1. Tangent line   2. Secant
3. $[f(c + h) - f(c)]/h$ 4. Average velocity

1.4 | Exponential Functions

Functions like x^2 and x^3 are examples of power functions. They are also polynomial functions. A power function has a variable as the base and a constant as the exponent. If we reverse these we get exponential functions; 2^x and 3^x are examples. Exponential functions are very useful in describing growth, such as growth of populations or of money in bank accounts or other types of investments.

EXAMPLE 1: An account yields interest at a rate of 8% per year. Interest is paid off once per year and then reinvested. Given that $1000 is initially invested, develop a way to predict how much is in the account after n years, where n is an integer. Calculate and then graph your predictions for 20 years. When will the account reach $4000?

SOLUTION: We let A represent the amount of money in the account and let n represent the number of years after the initial investment. Because A is really a function of n, we will write $A(n)$ to represent the amount of money in the account after n years.

The amount in the account after 1 year would be

$$A(1) = 1000 + (0.08)(1000) = (1.08)(1000)$$

and the amount after 2 years would be

$$A(2) = (1.08)(1000) + (0.08)[(1.08)(1000)]$$
$$= (1.08)(1.08)(1000) = (1.08)^2(1000)$$

Repeating this process, we find that after n years, the amount of money in the account would be

$$A(n) = (1.08)^n(1000)$$

The predictions of this function for 20 years, and the corresponding graph, are seen in Figure 1.

After 18 years the account is nearly $4000, but it does not actually reach that amount until the 19th year.   ◄

The function $A(n)$ from the previous example is an exponential function, because the variable n is in the exponent of the function. Because of the way the problem was set up, this function makes sense only for the natural numbers—that is, for $n = 1, 2, 3, \ldots$. In other words, the domain of the function $A(n)$ is the set of natural numbers. If our account paid interest twice each year, then we would want to use values of n such as 3.5 (though we would need a different exponential function in this case—the function $(1.08)^n(1000)$ works only for annual compounding). We want to extend the domain of exponential functions to include all real numbers.

Clearly, *any real number is the limit of a sequence of finite rational numbers.* Why clearly? Because any real number can be expressed as a decimal (possibly with an infinite number of digits); just choose the first digit, then the first two digits, then the first three digits, and so on. For example, a sequence of finite rational numbers that approaches $\pi = 3.141592654\ldots$ would be $3, 3.1, 3.14, 3.141, \ldots$. This approach can be used to approximate either irrational numbers or infinitely repeating rational numbers.

n	$1000\,(1.08)^n$
0	1000.00
1	1008.00
2	1166.40
3	1259.71
4	1360.49
5	1469.33
6	1586.87
7	1713.82
8	1850.93
9	1999.00
10	2158.92
11	2331.64
12	2518.17
13	2719.62
14	2937.19
15	3172.17
16	3425.94
17	3700.02
18	3996.02
19	4315.70
20	4660.96

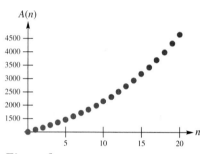

Figure 1

Definitions of Exponents

We need to define what we mean by various types of exponents, such as positive and negative integer exponents, rational (fractional) exponents, and irrational exponents. Let a be a positive real number for each definition below.

1. $a^n = \underbrace{a \cdot a \cdot a \cdot \; \cdots \; \cdot a}_{n \text{ times}}$ if n is a positive integer.

2. $a^{-n} = \dfrac{1}{a^n}$ if n is a positive integer (so that $-n$ is a negative integer).

3. $a^{n/m} = \sqrt[m]{a^n} = \left(\sqrt[m]{a}\right)^n$ if n and m are positive integers (so that n/m is a rational number).

4. $a^0 = 1$.

5. $a^x = \lim\limits_{r \to x} a^r$ where x is an irrational number, and the r-values are finite rational numbers (which can be chosen as described above).

EXAMPLE 2: Find 2^3, 2^{-3}, $8^{2/3}$, 5^0, and 3^{π}.

SOLUTION:

1. $2^3 = 2 \cdot 2 \cdot 2 = 8$

2. $2^{-3} = \frac{1}{2^3} = \frac{1}{8}$

3. $8^{2/3} = \left(\sqrt[3]{8}\right)^2 = 2^2 = 4$

4. $5^0 = 1$

5. $3^{\pi} =$ the limit of $3^3,\ 3^{3.1},\ 3^{3.14},\ 3^{3.141}, \ldots$
    $=$ the limit of $3^3, \sqrt[10]{3^{31}}, \sqrt[100]{3^{314}}, \sqrt[1000]{3^{3141}}, \ldots$
    $\approx$ the limit of $27, 30.1353, 31.4891, 31.5237, \ldots$
    ≈ 31.5 (accurate to 3 digits)   ◄

Note: We would not normally compute 3^π as in part 5. Most scientific calculators have a π key and a key for exponents, so that we could key in 3^π directly. The point is this: Calculators can handle only finite rational numbers, so when a calculator evaluates 3^π, it is internally going through a process similar to the one we went through in part 5. And by outlining the process here, we are showing how one can *define* exponentiation for real number exponents, based on the previous definitions for integer and rational exponents.

Rules of Exponents

Rules of Exponents   The following rules for exponents should be familiar from your previous study of mathematics. In each of the following rules, a is a positive real number and x and y are real numbers.

$$
\begin{aligned}
&\textbf{1. } a^x \cdot a^y = a^{x+y} \\
&\textbf{2. } (a^x)^y = a^{xy} \\
&\textbf{3. } a^{-x} = \frac{1}{a^x}
\end{aligned}
$$

Rules 1 and 2 are clearly true for whole numbers. (Why?) Proving these rules for real exponents is harder, and we leave the proofs to the real analysis books.

These rules of exponents help to explain why we defined fractional exponents as we did. For example, because

$$x^{\frac{1}{3}} \cdot x^{\frac{1}{3}} \cdot x^{\frac{1}{3}} = x^{\frac{2}{3}} \cdot x^{\frac{1}{3}} = x^1 = x$$

by rule 1 above, it makes sense to define $x^{\frac{1}{3}}$ as the cube root of x—that is, the number you multiply by itself three times to get x. As we extend the concept of exponentiation from whole numbers to negative numbers to rational numbers, we want to preserve the same rules we use for whole number exponents.

Definition of Exponential Functions

A function of the form

$$f(x) = a^x$$

where a is any positive real number and x is any real number is called an exponential function. *Note*: Using the rules of exponents from above, we have

$$a^{-x} = (a^{-1})^x = \left(\frac{1}{a}\right)^x$$

so that, for example, 2^{-x} and $\left(\frac{1}{2}\right)^x$ are the same function. We generally use the form with $a > 1$ and the negative sign in the exponent.

We have already seen an exponential function in Example 1. Now we can deal with more general exponential functions, where the exponent can be any real number.

EXAMPLE 3: Graph the exponential functions 2^x and 2^{-x} by building a table of values for $x = -2, -1, 0, 1, 2$.

SOLUTION: The table and graphs are in Figure 2.   ◄

x	2^x	2^{-x}
-2	0.25	4
-1	0.5	2
0	1	1
1	2	0.5
2	4	0.25

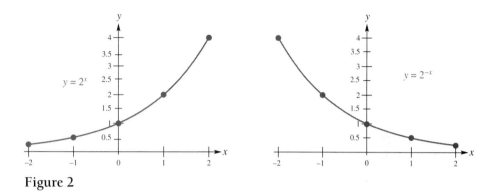

Figure 2

The two graphs in Example 3 have slopes that are typical of *all* exponential functions: If the base a of the function is a real number greater than 1, then the functions a^x and a^{-x} have the same shapes as 2^x and 2^{-x}; the only difference is the scale on the y-axis.

Notice that we have "filled in" the graph between the plotted points, because we have now defined exponential functions for all real numbers x. Compare these graphs to the graph in Figure 1 (from Example 1) where we were not "allowed" to fill in between the plotted points. The definitions for rational and real exponents given above are consistent, so that when we connect the plotted points to make a smooth curve, we do in fact get the correct graph.

An Important Limit   One of the most important properties of exponential functions is that they grow large *very* quickly—in particular, any exponential function of the form a^x, where $a > 1$, eventually becomes larger than any power function of the form x^n, where n is a positive integer. In Figure 3 we illustrate this graphically with the functions 2^x, x^2, and x^5. As we zoom out, we see that the function 2^x eventually catches up with and passes both x^2 and x^5. We will postpone the proof of this general property until later in the book.

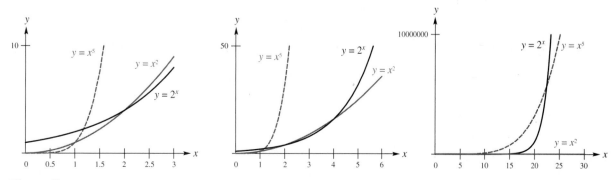

Figure 3

An important consequence of this growth property of exponential functions is the following limit.

$$\text{If } a > 1 \text{ and } n \text{ is a positive integer, then}$$
$$\lim_{x \to \infty} \frac{x^n}{a^x} = 0$$

In plain English, this says that exponential functions approach infinity faster than power functions do as x gets larger and larger.

EXAMPLE 4: Find $\lim_{x \to \infty} (1 + x - x^2)2^{-x}$.

SOLUTION: We have

$$(1 + x - x^2)2^{-x} = (1 + x - x^2)\frac{1}{2^x}$$

$$= \frac{1}{2^x} + \frac{x}{2^x} - \frac{x^2}{2^x}$$

Because each term gets very small when x gets large (by the general limit stated above), the sum also gets small and so we get

$$\lim_{x \to \infty} (1 + x - x^2)2^{-x} = 0$$

As a check, we can build a table of values as in Section 1.2. See Figure 4; we see that the numerical evidence is consistent with our theoretical result. Note the use of scientific notation, and how quickly the function approaches zero as x approaches infinity. ◄

x	$(1 + x - x^2)2^{-x}$
10	-0.0869141
100	$-7.809\text{E-}27$
1000	$-9.32\text{E-}296$
10000	0

Figure 4

Calculators and Computers

When calculating values for the exponential function $f(x) = e^x$ with a computer or calculator, we could replace e with an approximate value such as 2.7183 and then calculate 2.7183^x. However, scientific and graphics calculators, and many computer programs that perform numerical calculations (spreadsheets, computer algebra systems) can calculate values for the exponential function directly.

Mathematical computer programs often use the notation $\exp(x)$ for the exponential function (also commonly used in mathematics writing). Thus, for example, to calculate e^2 we might type something like $\exp(2)$ or $\text{Exp}[2]$ into the program. On calculators, look for an e^x or exp key; it is often obtained by pressing $\boxed{\text{2nd}}$, $\boxed{\text{SHIFT}}$, or $\boxed{\text{INV}}$ and then the $\boxed{\text{LN}}$ key.

The Natural Exponential Function

We define the natural exponential function as

$$f(x) = e^x$$

where e is defined by $e = \lim_{h \to 0} (1 + h)^{1/h} \approx 2.7183$ (as at the end of Section 1.1).

An obvious question is "Just what is so natural about using the base e?" The reason for using e as the base of the exponential function is related to the concept of derivative, or instantaneous slope, developed in the last section.

Theorem A

On the graph of $y = e^x$, the slope at the point $x = c$ is the same as the y-coordinate at $x = c$. In the language of derivatives, this means that if $f(x) = e^x$, then $f'(c) = e^c$.

Sketch of Proof.   Because $e = \lim\limits_{h \to 0} (1 + h)^{\frac{1}{h}}$, we know that for small values of h we have $e \approx (1 + h)^{\frac{1}{h}}$. Using a little algebra we get

$$e \approx (1 + h)^{\frac{1}{h}}$$
$$e^h \approx ((1 + h)^{\frac{1}{h}})^h = 1 + h$$
$$e^h - 1 \approx h$$
$$\frac{e^h - 1}{h} \approx 1$$

if h is close to 0.

Now, to calculate the slope of e^x at c, we write $f'(c) = \lim\limits_{h \to 0} \dfrac{f(c + h) - f(c)}{h}$. We have

$$\frac{f(c + h) - f(c)}{h} = \frac{e^{c+h} - e^c}{h}$$
$$= \frac{e^c \cdot e^h - e^c}{h}$$
$$= e^c \frac{e^h - 1}{h}$$
$$\approx e^c \cdot 1$$

if h is close to 0, as we showed above. This suggests that

$$f'(c) = \lim\limits_{h \to 0} \frac{f(c + h) - f(c)}{h} = e^c$$

which is what we wanted to show.   ◄

The exponential function $y = e^x$ is unique in that it is the *only* exponential function for which the y-coordinate and the slope are the same at each point on the graph. This is why e is a "natural" base to use for exponential functions. See Figure 5.

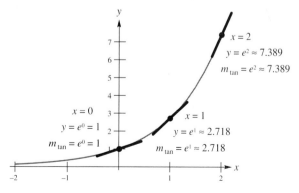

The graph of $y = e^x$; at the point, the slope and the y-coordinate are the same.

Figure 5

EXAMPLE 5: Let $f(x) = e^{2x}$. Find a formula for $f'(c)$. Estimate both the y-coordinate and the slope of $f(x)$ at $x = -1, 0, 1$. Discuss the relationship between the y-coordinate and the slope of $f(x)$, and display this relationship on a sketch similar to Figure 5.

SOLUTION: Using the definition of the derivative $f'(c) = \lim\limits_{h \to 0} \dfrac{f(c + h) - f(c)}{h}$, we have

$$\frac{f(c + h) - f(c)}{h} = \frac{e^{2(c+h)} - e^{2c}}{h}$$

$$= \frac{e^{2c + 2h} - e^{2c}}{h}$$

$$= \frac{e^{2c} \cdot e^{2h} - e^{2c}}{h}$$

$$= e^{2c} \frac{(e^{2h} - 1)}{h}$$

so that

$$f'(c) = \lim_{h \to 0} e^{2c} \frac{(e^{2h} - 1)}{h}$$

When h is small we can estimate the value of the expression $\dfrac{(e^{2h} - 1)}{h}$ using a table of values. See Figure 6. We find that $\dfrac{(e^{2h} - 1)}{h} \approx 2.000$ as h approaches zero, so that

$$f'(c) = 2.000 \cdot e^{2c}$$

Thus,

$$f'(-1) \approx 2.000 \cdot e^{2(-1)} \approx 0.2707$$

$$f'(0) \approx 2.000 \cdot e^{2(0)} = 2.000$$

$$f'(1) \approx 2.000 \cdot e^{2(1)} \approx 14.78$$

In plain English, the slope of the curve $y = e^{2x}$ is twice the y-coordinate at any point on the curve. We display these results graphically in Figure 7. In the problems you are asked to show that the equation $f'(c) = 2e^{2c}$ is exact. ◀

h	$\dfrac{e^{2h} - 1}{h}$
1	7.38906
0.1	2.21403
0.01	2.02013
0.001	2.00200
0.0001	2.00020

Figure 6

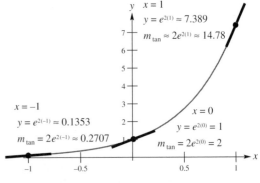

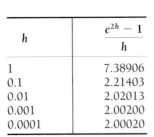

The slope of the curve $y = e^{2x}$ is twice the y-coordinate at any point on the curve.

Figure 7

Concepts Review

1. We define negative exponents with the equation $a^{-n} = $ _____. We define fractional exponents with the equation $a^{n/m} = $ _____ = _____ .

2. Two important properties of exponents are that $a^x \cdot a^y = $ _____ and that $(a^x)^y = $ _____ .

3. $\lim\limits_{x \to \infty} \dfrac{x^{10}}{2^x} = $ _____ .

4. The y-coordinate and the slope of the exponential function $y = e^x$ at any point x are _____ .

Problem Set 1.4

1. Reduce each expression below to its simplest integer or fractional form (no exponents). See Example 2, parts 1–4.

(a) 10^3 (b) 10^{-3} (c) 10^0

(d) $8^{\frac{2}{3}}$ (e) $8^{-\frac{2}{3}}$ (f) 8^0

2. Reduce each expression below to its simplest integer or fractional form (no exponents).

(a) 2^5 (b) 2^{-5} (c) 2^0

(d) $\left(\dfrac{1}{16}\right)^{\frac{3}{4}}$ (e) $\left(\dfrac{1}{16}\right)^{-\frac{3}{4}}$ (f) $\left(\dfrac{1}{16}\right)^0$

3. Graph the exponential functions 10^x and 10^{-x} by building a table of values for $x = -2, -1, 0, 1, 2$.

4. Graph the exponential functions $(2/3)^x$ and $(2/3)^{-x}$ by building a table of values for $x = -2, -1, 0, 1, 2$. Explain any differences between these graphs and those of the previous problem.

In Problems 5–8, sketch and describe the graph of the given function by building a table of values.

5. $f(x) = x^2 e^{-x}$

6. $f(x) = e^x - x$

7. $f(x) = e^{-x^2}$

8. $f(x) = e^x + e^{-x}$

In Problems 9 and 10, find each limit exactly and then check your results with a table of values as in Example 4.

9. (a) $\lim\limits_{x \to \infty} 1 + \dfrac{x}{3^x}$ (b) $\lim\limits_{x \to \infty} 10^{-x}(x^2 + 1)$

(c) $\lim\limits_{x \to \infty} x^{-1}(x^2 + 1)$ (d) $\lim\limits_{x \to \infty} x\, e^{-x} + 5$

10. (a) $\lim\limits_{x \to \infty} 5 - 5^{-x} x^{10}$ (b) $\lim\limits_{x \to \infty} \dfrac{10 - 3^{-x} x^3}{5 + 2^{-x} x^2}$

(c) $\lim\limits_{x \to \infty} x^{-1}(x^2 + 1)$ (d) $\lim\limits_{x \to \infty} x^{100} e^{-x} + 5$

For each function $f(x)$ in Problems 11–14, find a formula for $f'(c)$. Estimate both the y-coordinate and the slope of $f(x)$ at $x = -1, 0, 1$. Discuss the relationship between the y-coordinate and the slope of $f(x)$, and display this relationship on a sketch similar to Figure 5. (See Example 5.)

11. $f(x) = e^{3x}$

12. $f(x) = e^{-3x}$

13. $f(x) = 2^x$

14. $f(x) = 2^{-x}$

15. A bank account yields interest at a rate of 3% per year. Interest is paid off once per year and then reinvested. Given that \$500 is initially invested, calculate and graph the amount of money in the bank for 10 years. When will the account reach \$600?

16. The population of a small country is growing at the constant rate of 5% per year; that is, each year the population is exactly 5% larger than the previous year. Develop a model to predict the size of the population as a function of the year n. If there are currently 400,000 inhabitants of the country, predict and graph the population as a function of n for the next 10 years. Should we plot points only at the integer values of n as in Example 1, or should we "fill in" the graph as in Example 3?

17. Explain why $a < b$ implies that $e^{-a} > e^{-b}$.

18. Stirling's Formula says that for large n, we can approximate $n! = 1 \cdot 2 \cdot 3 \cdots n$ by

$$n! \approx \sqrt{2\pi n}\left(\frac{n}{e}\right)^n$$

(a) Calculate 10! exactly and then approximately using the above formula.

(b) Calculate 60! exactly and then approximately using the above formula.

19. It will be shown later (Section 10.1) that for small x,

$$e^x \approx 1 + x + \frac{x^2}{2!} + \frac{x^3}{3!} + \frac{x^4}{4!}$$

$$= \left\{\left[\left(\frac{x}{4} + 1\right)\frac{x}{3} + 1\right]\frac{x}{2} + 1\right\}x + 1$$

Use this result to calculate $e^{0.3}$ and compare your result with what you get by calculating it directly.

20. If customers arrive at a check-out counter at the average rate of k per minute, then (see books on probability theory) the probability that exactly n customers will arrive in a period of x minutes is given by the formula

$$P_n(x) = \frac{(kx)^n e^{-kx}}{n!}$$

Find the probability that exactly eight customers will arrive during a 30-minute period if the average rate for this check-out counter is one customer every 4 minutes.

21. Estimate $\lim\limits_{n \to \infty} \dfrac{e^{1/n} + e^{2/n} + \cdots + e^{n/n}}{n}$ using a table of values.

Answers to Concepts Review: 1. $\dfrac{1}{a^n}$; $\sqrt[m]{a^n}$; $\left(\sqrt[m]{a}\right)^n$

2. a^{x+y}; a^{xy} 3. 0   4. the same

1.5 Logarithms and the Logarithmic Function

Logarithms   We need logarithms as a means of solving equations that involve exponentials and as a way of converting between exponential functions of different bases. Once logarithms have been defined, we can study the logarithmic function as a useful function in its own right.

> ### Definition of Logarithms
>
> We define the natural base logarithm of a positive real number y as the solution x to the equation $y = e^x$. The natural base logarithm (or natural log) of y is written $\ln y$. Thus, for example, $\ln 12$ would be defined as the solution to the equation $12 = e^x$.
>
> In other words, if we want to find the x-value for a given y-value on the graph of $y = e^x$, just take the natural log of y. See Figure 1.

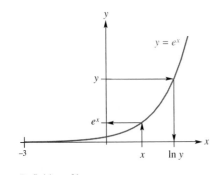

Definition of $\ln y$

Figure 1

It should be clear from Figure 1 that the domain of the natural logarithm function is the positive real numbers, because e^x is positive for all x. Thus, just as the square root of a negative number is not defined (as a real number), the natural log of a negative number or zero is not defined (as a real number). In mathematical slang, *you can't take the log of a negative number or zero*.

You may notice that the natural log function is defined with respect to the natural exponential function the same way that the arcsin function is defined with respect to the sin function. In other words, $\ln x$ and e^x are inverses of each other. That is why they generally appear on the same key of a calculator, one being obtained using the "shift" or "2nd" key first.

The definition of the natural logarithm leads to the following rules for solving equations.

Solving Equations with Logarithms and Exponentials

> **1.** To solve $y = e^x$ for x we get $x = \ln y$ (where $y > 0$).
> **2.** To solve $y = \ln x$ for x we get $x = e^y$ (where $y > 0$).

EXAMPLE 1: Suppose that the population P of a small city is growing according to the equation

$$P = 10e^{0.03t}$$

where t is time measured in years, and P is measured in thousands of people.
(a) What will the population be after 5 years?
(b) When will the population reach 11,000?

SOLUTION:
(a) We let $t = 5$ in the equation $P = 10e^{0.03t}$ and use a calculator to get

$$P = 10e^{0.03(5)} \approx 11.6$$

which is about 11,600 people.

(b) Let $P = 11$ in the equation $P = 10e^{0.03t}$ and solve for t. Using algebra and a calculator we get

$$11 = 10e^{0.03t}$$

$$\frac{11}{10} = e^{0.03t}$$

$$0.03t = \ln\left(\frac{11}{10}\right) \quad \textit{(Rule 1 from above)}$$

$$t = \frac{\ln\left(\frac{11}{10}\right)}{0.03} \approx 3.18$$

Thus, it takes a little over 3 years for the population of the city to reach 11,000. ◄

Another way of looking at the relationship between logs and exponentials is this: If you take the natural log of a number and then apply the natural exponential function to the result, you get back the number you started with (and similarly if you apply the exponential function first and then the log function). This is just another way of saying that the natural log function and the natural exponential function are inverses of each other. Symbolically, we can express this relationship as follows.

Inverse Properties of the Natural Log and Natural Exponential Functions

1. $e^{\ln x} = x$ for positive x.
2. $\ln(e^x) = x$ for all x.

Because of these properties, we can write any exponential function as a natural exponential function. This is the way it works. If you start with an exponential function such as $y = 2^x$, you can just rewrite the base as $2 = e^{\ln 2}$ (rule 1 above), so that the function becomes $y = (e^{\ln 2})^x = e^{(\ln 2)x} \approx e^{0.693x}$ (rule 2 above). We can summarize this process as follows.

CONVERTING BASES WITH EXPONENTIAL FUNCTIONS Any exponential function $y = a^x$ can be written as a natural exponential function of the form $y = e^{kx}$, where $k = \ln a$. If $0 < a < 1$, then k is negative (a decreasing exponential), and if $a > 1$, then k is positive (an increasing exponential).

EXAMPLE 2: Return to Example 1 of Section 1.4. In that example, an investment returned 8% annually on an initial investment of $1000. Interest was paid once per year and reinvested. The formula that gave the value of the account $A(n)$ after n years was $A(n) = (1.08)^n(1000)$.

Suppose that we want to devise a way to pay interest to an investor who closes an account before the interest period ends. If, for instance, the account is closed after 4.5 years, the value of the account should be consistent with the values at 4 and 5 years. We want to "fill in" the spaces in the graph of the function $A(n)$ in Figure 1 of Section 1.4.

Develop a *natural* exponential function that gives you the value of the account after t years, where t can be any real number. The function should give the same values as $A(n)$ for integer values of t. Sketch the graph of your function and find the amount of money in the account after 4.5 years and the time it would take the account to reach a total value of $3000.

SOLUTION: If we let $A(t)$ represent the amount of money in the account after t years, we can just use the same function we used for integer values of t, that is $A(t) = (1.08)^t(1000)$. The graph of this function fills in the spaces of the graph of $A(n) = (1.08)^n(1000)$, where n is a positive integer.

Next we convert this exponential function to a natural exponential function using the procedure outlined previously. We have

$$1.08^t = (e^{\ln 1.08})^t$$
$$\approx (e^{0.07696})^t$$
$$= e^{0.07696t}$$

so that we get

$$A(t) = 1000e^{0.07696t}$$

for our function (with k accurate to four digits). Of course, we could use the function $A(t) = (1.08)^t(1000)$ just as well (in fact it's a bit more accurate because there is no rounding); we are just showing that all exponential functions can be expressed in a "standard" way.

Note: Recall that this function was originally developed in Section 1.4 to describe an 8% yearly interest rate (an interest rate is an example of a growth rate). The constant k in this problem turned out to be about 0.077 or 7.7%. If the account is compounded *instantaneously* (every instant) at 7.7% we get the same result as 8% compounded yearly. We can interpret the constant k in an exponential function e^{kx} as the instantaneous percentage growth rate of the function. We will develop this idea more fully in Chapter 7.

The graph of this function is shown in Figure 2; just "connect the dots" of the graph of $A(n) = (1.08)^n(1000)$ in Figure 1 of Section 1.4.

To find the amount in the account after 4.5 years we calculate

$$A(4.5) = 1000e^{0.07696(4.5)} \approx 1414$$

so that there would be about $1414 in the account after 4.5 years.

To find the time it would take the account to reach a value of $3000 we set $A(t) = 3000$ and solve for t. We get

$$1000e^{0.07696t} = 3000$$

$$e^{0.07696t} = \frac{3000}{1000} = 3$$

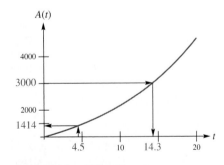

Graph of $A(t) = 1000e^{0.07696t}$

Figure 2

$$0.07696t = \ln 3$$

$$t = \frac{\ln 3}{0.07696} \approx 14.3$$

so that it would take a little more than 14 years for the account to reach a value of $3000 with this method of paying interest. Figure 2 shows the two numerical results on the graph of $A(t)$. ◄

An Important Limit Revisited

In Section 1.4 we stated that exponential functions approach infinity more quickly than power functions. Because any exponential function can be written as a natural exponential function, we restate this limit property using natural exponentials. If $k > 0$ and n is a positive integer, then

$$\lim_{x \to \infty} \frac{x^n}{e^{kx}} = 0$$

The Natural Logarithm Function

Up to this point we have been interested in logs because they help with problems that involve exponential functions. The function $\ln x$ is of interest in its own right also. A table of values and a graph of $\ln x$ are shown in Figure 3.

x	$\ln x$
0.125	−2.079
0.250	−1.386
0.500	−0.693
1.000	0.000
2.000	0.693
3.000	1.099
4.000	1.386
5.000	1.609

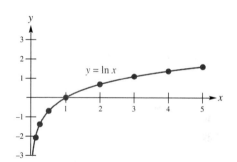

Figure 3

Of particular interest is the point $(1, 0)$ on the graph of $\ln x$, which tells us that $\ln 1 = 0$. This comes directly from the definition of the natural logarithm; if we solve for the zero in the equation $e^0 = 1$, we get $\ln 1 = 0$. Also, notice that we have chosen some points for our table close to $x = 0$. This is because the natural log function has a vertical asymptote at $x = 0$. In particular, $\ln 0$ is *not defined*. Look back at Figure 1 to see why.

Properties of Logarithms

There are some important properties of logarithms that follow directly from the properties of exponents from the last section and the relationship between logs and exponential functions.

If x and y are positive real numbers, then:

1. $\ln xy = \ln x + \ln y$

2. $\ln \frac{x}{y} = \ln x - \ln y$

Furthermore, if x is a positive real number and y is any real number, then:

3. $\ln(x^y) = y \ln x$

Proof   If x and y are positive real numbers, then $x = e^{\ln x}$ and $y = e^{\ln y}$. Therefore,

$$xy = e^{\ln x}e^{\ln y} = e^{\ln x + \ln y}$$

Now just take the natural log of both the first and last expression above to get

$$\ln xy = \ln(e^{\ln x + \ln y}) = \ln x + \ln y$$

which is what we wanted to show for part 1. The steps in this proof used only elementary algebra and the properties of logs and exponents from this section and the last one; you should go through and determine which property was used at each step.   ◀

See the problem set for proofs of parts 2 and 3.

EXAMPLE 3:   Find the derivative (slope) of the function $\ln x$ at the points $x = 0.5$ and $x = 3$. Sketch the graph of $y = \ln x$ and indicate on it the derivatives you found using slope marks.

SOLUTION: To find the derivative of $f(x) = \ln x$ at $x = 0.5$ we need to find

$$
\begin{aligned}
f'(0.5) &= \lim_{h \to 0} \frac{f(0.5 + h) - f(0.5)}{h} \\[2mm]
&= \lim_{h \to 0} \frac{\ln(0.5 + h) - \ln(0.5)}{h} \\[2mm]
&= \lim_{h \to 0} \frac{\ln\left(\dfrac{0.5 + h}{0.5}\right)}{h} \\[2mm]
&= \lim_{h \to 0} \frac{\ln(1 + 2h)}{h} \\[2mm]
&= \lim_{h \to 0} \frac{1}{h}\ln(1 + 2h) \\[2mm]
&= \lim_{h \to 0} \ln(1 + 2h)^{\frac{1}{h}}
\end{aligned}
$$

Each step above uses either elementary algebra or one of the properties of logarithms stated previously. Similarly, for $x = 3$ we get

$$f'(3) = \lim_{h \to 0} \ln(1 + \tfrac{1}{3}h)^{\frac{1}{h}}$$

Numerical Solution   We can estimate these limits numerically with tables of values. (See Figure 4.) To four digits we estimate the derivatives to be $f'(0.5) \approx 2.000$ and $f'(3) \approx 0.3333$.

h	$\ln(1 + 2h)^{\frac{1}{h}}$	h	$\ln(1 + \frac{1}{3}h)^{\frac{1}{h}}$
0.1	1.82321557	0.1	0.32789823
0.01	1.98026273	0.01	0.33277901
0.001	1.99800266	0.001	0.33327779
0.0001	1.99980003	0.0001	0.33332778
0.00001	1.99998000	0.00001	0.33333278

Figure 4

Exact Solution   It turns out that we can also find these limits exactly. It requires a standard mathematical technique that consists of substituting one variable for another. Notice that

$$f'(0.5) = \lim_{h \to 0} \ln(1 + 2h)^{\frac{1}{h}}$$

looks a lot like the definition of the constant e from Section 1.5. That definition was

$$e = \lim_{h \to 0} (1 + h)^{\frac{1}{h}}$$

The idea is to realize that as h gets small, $2h$ also gets small, so that we can make the substitution $k = 2h$ and then take the limit as k goes to zero. To deal with the h in the exponent we write $\frac{1}{h} = \frac{2}{2h}$ so that with the replacement $k = 2h$ we get

$$\ln(1 + 2h)^{\frac{1}{h}} = \ln(1 + 2h)^{\frac{2}{2h}} = \ln(1 + k)^{\frac{2}{k}}$$

$$= \ln\left((1 + k)^{\frac{1}{k}}\right)^2$$

where one of the rules of exponents from Section 1.8 was used on the last step. (Which rule was used?) If we now let k approach zero, $(1 + k)^{\frac{1}{k}}$ approaches e, so that

$$f'(0.5) = \lim_{h \to 0} \ln(1 + 2h)^{\frac{1}{h}} = \ln e^2 = 2$$

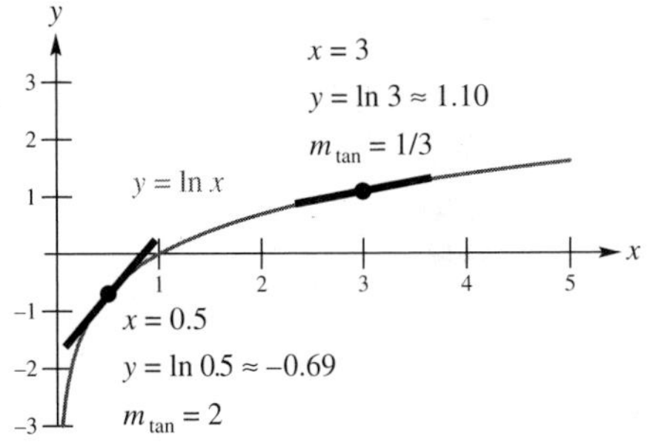

We are allowed to take the limit "inside" the natural log function because this function is continuous (the proof of which we leave to the analysis books). This result checks with our numerical estimate of the limit. In the problems at the end of the section you are asked to show that the exact value of $f'(3)$ is $\frac{1}{3}$. In Figure 5 we show the graph of $y = \ln x$ with slope marks at $x = 0.5$ and $x = 3$.   ◄

Figure 5

Logarithms with Other Bases
We can define logs with bases other than e in the same way we defined the natural log. Let a be any positive real number. Then for positive y we define $\log_a y$, read "the log of y, base a," as the solution x to the equation $y = a^x$. Notice that this means that $\ln y$ and $\log_e y$ represent exactly the same thing; we use the ln notation (for "log natural") because it is shorter.

All of the properties of logarithms that were stated for natural logs are true for logarithms with any other base. For example, the inverse property of logs and exponentials would mean that $\log_a a^x = x$ and $a^{\log_a x} = x$.

The only logarithm in common use besides the natural logarithm is the logarithm base 10. This is because our number system uses a base of 10. Many calculators have an "ln" key and a "log" key; the "log" key is for base 10 logs. Notice that for base 10 logs we would have

$$\log_{10} 10 = \log_{10} 10^1 = 1$$

$$\log_{10} 100 = \log_{10} 10^2 = 2$$

$$\log_{10} 1000 = \log_{10} 10^3 = 3$$

from the inverse properties. This provides a nice common-sense interpretation of logs. The base 10 log of a positive integer n is approximately the number of digits in n. This gives you a feel for log functions; they grow *very* slowly. For example, the number consisting of a 1 followed by 100 zeros (called a googol in the popular literature) would only have a base 10 log of 100.

Converting Between Logarithms of Different Bases   Just as we could convert between exponential functions of different bases, we can convert between logs of different bases. The only conversion equation we will need is

$$\log_a x = \frac{\ln x}{\ln a}$$

for any positive x and positive a. We can use this equation to convert logarithms of any other base to natural logarithms. Thus, we will rarely encounter logs other than natural logs in this book. The proof of the above equation is in the problem set.

EXAMPLE 4: Make a table of values for the function $f(x) = \log_2 x$ and sketch its graph. The logarithm base 2 is used occasionally in computer science, because computers use a base 2 (or binary) number system.

SOLUTION: Use the natural log (ln) key on your calculator and the conversion equation

$$\log_2 x = \frac{\ln x}{\ln 2}$$

to generate the table. For example, $\log_2 (5) = \dfrac{\ln 5}{\ln 2} \approx 2.322$. We show our table and graph in Figure 6. Remember to use some values of x close to zero, as we did with the natural log function. Question: Why did the values of $\log_2 x$ come out "nicely" in the table except for $x = 3$ and $x = 5$?   ◄

x	$\log_2 x$
0.125	-3.000
0.250	-2.000
0.500	-1.000
1.000	0.000
2.000	1.000
3.000	1.585
4.000	2.000
5.000	2.322

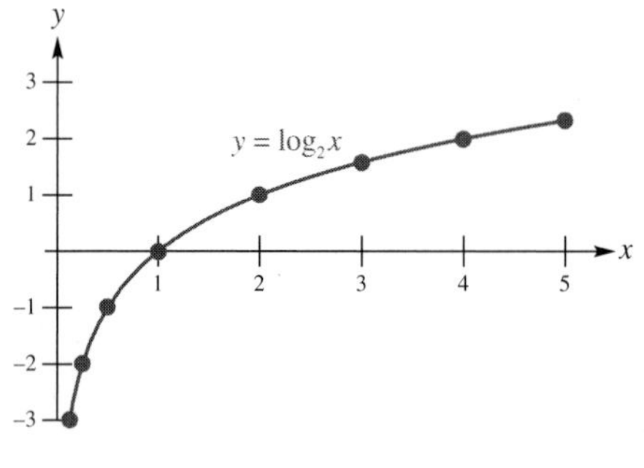

Figure 6

Concepts Review

1.   To solve $y = e^x$ for x we get $x =$ ___ (where $y > 0$). To solve $y = \ln x$ for x we get $x =$ ___ (where $x > 0$).

2. $e^{\ln x} =$ ___ for positive x and $\ln(e^x) =$ ___ for all x.

3.   Two important properties of logarithms are that $\ln xy =$ ___ + ___ and that $\ln x^y =$ ___ · ___.

4.   We can convert logarithms of any base to natural logarithms with the equation $\log_a x =$ ___/___.

Problem Set 1.5

1.   For each function, find the y-values corresponding to the x-values $x = -4, -2, 0, 2, 4$ and sketch a graph based on these points. Then use logarithms to find the x-value corresponding to the y-value $y = 2$, or explain why no such x-value exists.

(a) $y = 2e^x$

(b) $y = e^{2x}$

(c) $y = 1 + e^x$

(d) $y = 3 + e^x$

(e) $y = 3e^{-x}$

(f) $y = e^{-3x}$

(g) $y = 3 - e^{2x}$

(h) $y = 1 - e^{-2x}$

2. For each function, find the y-values corresponding to the x-values $x = 0, 2, 4, 6, 8$ and sketch a graph based on these points. Then use logarithms to find the x-value corresponding to the y-value $y = 5$, or explain why no such x-value exists.

(a) $y = \dfrac{10}{1 + e^x}$

(b) $y = \dfrac{20}{1 + e^x}$

(c) $y = \dfrac{10}{1 + 9e^{-x}}$

(d) $y = \dfrac{5}{1 + 4e^{-x}}$

(e) $y = \dfrac{2}{1 + e^{-x}}$

(f) $y = \dfrac{6}{1 + e^{-x}}$

3. Rewrite each exponential function as a natural exponential function.

(a) $y = 10^x$

(b) $y = 3^x$

(c) $y = 2^{-x}$

(d) $y = 5^{-x}$

(e) $y = 4 \cdot 5^x$

(f) $y = 5 \cdot 2^{-x}$

(g) $y = 2^{-3x}$

(h) $y = -4^{5x}$ (remember order of operations)

4. Rewrite each natural exponential function as an exponential function of the form $y = a \cdot b^x$. Use a three-decimal approximation to the values of a and b.

(a) $y = e^{2x}$

(b) $y = 3e^{2x}$

(c) $y = 3e^{-x}$

(d) $y = e^{-5x}$

5. The value of an investment (y, in thousands dollars) is growing according to the equation $y = 2e^{0.07t}$, where t is time (in years).

(a) What is the initial value of the investment ($t = 0$)?

(b) What is the value of the investment after 5 years?

(c) How long will it take for the value of the investment to reach $5000?

6. An investment gains 7% in value each year. $2000 is initially invested.

(a) Write an equation of the form $y = a \cdot b^t$ that describes the growth of the investment, where y is the value of the investment in thousands of dollars and t is time measured in years.

(b) Convert the exponential function in (a) to a natural exponential function.

(c) What is the value of the investment after 5 years?

(d) How long will it take for the value of the investment to reach $5000?

(e) Compare the results of this problem to those of the previous problem. Does the investment in Problem 5 gain more or less than 7% each year?

7. Use the approximations $\ln 2 = 0.693$ and $\ln 3 = 1.099$ together with the properties of logarithms to calculate approximations to each of the following. For example, $\ln 6 = \ln(2 \cdot 3) = \ln 2 + \ln 3 = 0.693 + 1.099 = 1.792$.

(a) $\ln 6$

(b) $\ln 1.5$

(c) $\ln 81$

(d) $\ln \sqrt{2}$

(e) $\ln\left(\frac{1}{36}\right)$

(f) $\ln 48$

8. Use your calculator to make the computations in Problem 7 directly.

In Problems 9–16, solve for x. *Hint*: $\log_a b = c \Leftrightarrow a^c = b$.

9. $\log_3 9 = x$

10. $\log_4 x = 2$

11. $\log_9 x = \frac{3}{2}$

12. $\log_x 81 = 4$

13. $2 \log_{10}\left(\dfrac{x}{3}\right) = 1$

14. $\log_4\left(\dfrac{1}{x}\right) = 3$

15. $\log_2(x + 1) - \log_2 x = 2$

16. $\log_6(x + 1) + \log_6 x = 1$

In Problems 17–20, use natural logarithms to solve each of the exponential equations. *Hint*: To solve $3^x = 11$, take $\ln$ of both sides, obtaining $x \ln 3 = \ln 11$; then $x = (\ln 11)/(\ln 3) \approx 2.1827$.

17. $2^x = 19$

18. $5^x = 12$

19. $4^{3x-1} = 5$

20. $12^{1/(x-1)} = 3$

In Problems 21–28, simplify the given expression.

21. $e^{2 \ln x}$

22. $e^{-\ln x}$

23. $\ln e^{\sin x}$

24. $\ln e^{-x+2}$

25. $\ln(x^2 e^{-2x})$

26. $e^{x + \ln x}$

27. $e^{\ln 2 + \ln x}$

28. $e^{\ln x - 2 \ln y}$

In Problems 29–32, use the properties of logarithms to write each of the following expressions as a logarithm of a single quantity.

29. $2 \ln(x + 1) - \ln x$

30. $\frac{1}{2} \ln(x - 9) + \frac{1}{2} \ln x$

31. $\ln(x - 2) - \ln(x + 2) + 2 \ln x$

32. $\ln(x^2 - 9) - 2 \ln(x - 3) - \ln(x + 3)$

In Problems 33–36, make use of the known graph of $y = \ln x$ to sketch the graphs of the equations.

33. $y = \ln|x|$

34. $y = \ln \sqrt{x}$

35. $y = \ln\left(\dfrac{1}{x}\right)$

36. $y = \ln(x - 2)$

37. Sketch the graph of $y = \ln \cos x + \ln \sec x$ on $(-\pi/2, \pi/2)$, but think before you begin.

Use your calculator and $\log_a x = (\ln x)/(\ln a)$ to calculate each of the logarithms in Problems 38–41.

38. $\log_5 13$

39. $\log_6 (0.12)$

40. $\log_{11}(8.16)^{\frac{1}{5}}$

41. $\log_{10}(91.2)^3$

42. Complete Example 3 by showing that if $y = \ln x$, then $f'(3) = \frac{1}{3}$ (exactly).

43. For each function below, find the derivative (slope) of the function at the points $x = 0.5$ and $x = 3$. If you can find exact results do so; otherwise find numerical approximations. Sketch the graph of the function and indicate on it the derivatives you found using slope marks.

(a) $y = \ln 2x$ (b) $y = \ln(2 + x)$

(c) $y = \ln(x - 1)$ (d) $y = 2 \ln x$

44. For each function below, find the derivative (slope) of the function at the points $x = 0.5$ and $x = 3$. If you can find exact results do so; otherwise find numerical approximations. Sketch the graph of the function and indicate on it the derivatives you found using slope marks. You may find the equation $\log_a x = \frac{\ln x}{\ln a}$ helpful.

(a) $y = \log_{10} x$ (b) $y = \log_{10}(2 + x)$

(c) $y = \log_5 2x$ (d) $y = 2 \log_5 x$

45. Show that $\log_a x = \frac{\ln x}{\ln a}$ for $a > 0$ and $x > 0$. *Hint:* Start with the equation $a^{\log_a x} = x$ and take the natural log of both sides.

46. How are $\log_{1/2} x$ and $\log_2 x$ related?

47. Sketch the graphs of $\log_{1/3} x$ and $\log_3 x$ using the same coordinate axes.

48. Explain why $\lim_{x \to 0} \ln \frac{\sin x}{x} = 0$.

49. Use the fact that $\ln 4 > 1$ to show that $\ln 4^m > m$ for $m > 0$. Conclude that $\ln x$ can be made as large as desired by choosing x sufficiently large. What does this imply about $\lim_{x \to \infty} \ln x$?

50. Use the fact that $\ln x = -\ln(1/x)$ and Problem 49 to show that $\lim_{x \to \infty} \ln x = -\infty$.

51. A famous theorem (the Prime Number Theorem) says that the number of primes less than n for large n is approximately $n/(\ln n)$. About how many primes are there less than 1,000,000?

52. The magnitude M of an earthquake on the Richter scale is

$$M = 0.67 \log_{10}(0.37E) + 1.46$$

where E is the energy of the earthquake in kilowatt-hours. Find the energy of an earthquake of magnitude 7; of magnitude 8.

53. The loudness of sound is measured in decibels in honor of Alexander Graham Bell (1847–1922), inventor of the telephone. If the variation in pressure is P pounds per square inch, then the loudness L in decibels is

$$L = 20 \log_{10}(121.3P)$$

Find the variation in pressure caused by a rock band at 115 decibels.

54. In the equally tempered scale to which keyed instruments have been tuned since the days of J. S. Bach (1685–1750), the frequencies of successive notes C, C#, D, D#, E, F, F#, G, G#, A, A#, B, C form a geometric sequence (progression), with C having twice the frequency of C. Thus, the ratios of consecutive notes are always the same (as with an exponential function). What is the ratio r between the frequencies of successive notes? If the frequency of A is 440, find the frequency of C.

55. Prove that $\log_2 3$ is irrational. *Hint:* Assume that $\log_2 3 = \frac{m}{n}$ for integers m and n and derive a contradiction.

56. You are suspicious that the xy-data you have collected lie on either an exponential curve, $y = A \cdot b^x$, or a power curve, $y = A \cdot x^b$. To check, you plot $\ln y$ against x and also $\ln y$ against $\ln x$. Explain how this will help you come to a conclusion.

57. Draw the graphs of $y = x^3$ and $y = 3^x$ using the same axes and find all their intersection points.

Answers to Concepts Review: 1. $\ln y$; e^y 2. x; x 3. $\ln x$; $\ln y$; y; $\ln x$ 4. $\ln x$; $\ln a$

1.6 | Chapter Review

Concepts Review

Respond with true or false to each of the following assertions. Be prepared to justify your answer.

1. If $\lim_{x \to c} f(x) = L$, then $f(c) = L$.

2. If $f(c)$ is not defined, then $\lim_{x \to c} f(x)$ does not exist.

3. The coordinates of the hole in the graph of $y = \dfrac{x^2 - 25}{x - 5}$ are (5, 10).

4. If $p(x)$ is a polynomial, then $\lim_{x \to c} p(x) = p(c)$.

5. If $\lim_{x \to c^+} f(x) = \lim_{x \to c^-} f(x)$, then f is continuous at

$x = c.$

6. If $f(x) \neq g(x)$ for all x, then $\lim\limits_{x \to c} f(x) \neq \lim\limits_{x \to c} g(x)$.

7. If $f(x) < 10$ for all x and $\lim\limits_{x \to 2} f(x)$ exists, then $\lim\limits_{x \to 2} f(x) < 10$.

8. If $\lim\limits_{x \to a} f(x) = b$, then $\lim\limits_{x \to a} |f(x)| = |b|$.

9. Estimating a limit numerically using a table of values is sometimes necessary.

10. It is always a good idea to find a limit both numerically and exactly using algebra if both methods are possible.

11. Numerical methods of estimating limits using tables are always very accurate.

12. We use the symbol ∞ to denote a number that is bigger than any other number.

13. The graph of $y = \dfrac{x^2 - x - 6}{x - 3} = \dfrac{(x + 2)(x - 3)}{x - 3}$ has a vertical asymptote at $x = 3$.

14. The graph of $y = \dfrac{x^2 + 1}{1 - x^2}$ has a horizontal asymptote of $y = -1$.

15. The tangent line to a curve at a point cannot cross the curve at that point.

16. The slope of the tangent line to the curve $y = x^4$ is different at every point of the curve.

17. If the tangent line to the graph of $y = f(x)$ is horizontal at $x = c$, then $f'(c) = 0$.

18. For any moving object, the average velocity between $t = 1$ and $t = 3$ is different from the instantaneous velocity at $t = 2$.

19. For any moving object, the average velocity between $t = 1$ and $t = 3$ is the same as the instantaneous velocity at $t = 2$.

20. Any function of the form a^x where a is a positive number gets larger and larger as x gets larger and larger.

21. The function $\left(\dfrac{1}{2}\right)^x$ has a horizontal asymptote as x approaches infinity.

22. The function xe^{-x} approaches infinity as x approaches infinity.

23. The slope of the function $f(x) = e^x$ at each point on its graph is the same as the y-coordinate at each point.

24. The slope of the function $f(x) = e^{-x}$ at each point on its graph is the same as the y-coordinate at each point.

25. The only solution to the equation $2e^x = 4$ is given by $x = \ln 2$.

26. Any function of the form a^x where a is positive can be written as a natural exponential function of the form e^{kx}.

27. The graphs of $y = \log_2 x$ and $\dfrac{\ln x}{\ln 2}$ are the same.

28. $\lim\limits_{x \to \infty} \dfrac{e^{2x}}{x^{50}} = \infty$.

Sample Test Problems

In Problems 1–10, find or estimate the indicated limit or state that it doesn't exist. Show your work, and indicate whether you used a numerical method (table of values) or an exact method.

1. $\lim\limits_{u \to 1} \dfrac{u^2 - 1}{u + 1}$

2. $\lim\limits_{u \to 1} \dfrac{u^2 - 1}{u - 1}$

3. $\lim\limits_{u \to 1} \dfrac{u + 1}{u^2 - 1}$

4. $\lim\limits_{x \to 2} \dfrac{1 - 2/x}{x^2 - 4}$

5. $\lim\limits_{z \to 2} \dfrac{z^2 - 4}{z^2 + z - 6}$

6. $\lim\limits_{x \to 0} \dfrac{\tan x}{\sin 2x}$

7. $\lim\limits_{y \to 1} \dfrac{y^3 - 1}{y^2 - 1}$

8. $\lim\limits_{x \to 4} \dfrac{x - 4}{\sqrt{x} - 2}$

9. $\lim\limits_{x \to 0} \dfrac{\cos x}{x}$

10. $\lim\limits_{x \to 0^-} \dfrac{|x|}{x}$

11. Determine each limit (possibly ∞ or $-\infty$) or state that it does not have a limit.

(a) $\lim\limits_{x \to \infty} \dfrac{3x^2 - 2x + 7}{2x^2 + 5x + 9}$

(b) $\lim\limits_{x \to \infty} \dfrac{3x + 9}{\sqrt{2x^2 + 1}}$

(c) $\lim\limits_{x \to -\infty} \dfrac{\sin x}{x}$

(d) $\lim\limits_{x \to -\infty} \cos x$

(e) $\lim\limits_{x \to 3^+} \dfrac{x + 3}{x^2 - 9}$

(f) $\lim\limits_{x \to 2} \dfrac{x + 2}{x^2 - 4}$

(g) $\lim\limits_{x \to 1^-} \dfrac{|x - 1|}{x - 1}$

(h) $\lim\limits_{x \to \infty} \cos\left(\dfrac{1}{x}\right)$

12.   Let $f(x) = \begin{cases} x^3 & \text{if } x < -1 \\ x & \text{if } -1 < x < 1 \\ 1 - x & \text{if } x \geq 1 \end{cases}$

Find each value.

(a) $f(1)$

(b) $\lim\limits_{x \to 1^+} f(x)$

(c) $\lim\limits_{x \to 1^-} f(x)$

(d) $\lim\limits_{x \to -1} f(x)$

13.   Refer to f of Problem 12. (a) What are the values of x at which f is discontinuous? (b) How should f be defined at $x = -1$ to make it continuous there?

14.   Sketch the graph of the following function.

$$h(x) = \begin{cases} x^2 & \text{if } 0 \leq x \leq 2 \\ 6 - x & \text{if } x > 2 \end{cases}$$

15.   Sketch the graph of a function f that satisfies all of the following conditions.

(a) Its domain is $[0, 6]$.

(b) $f(0) = f(2) = f(4) = f(6) = 2$.

(c) f is continuous except at $x = 2$.

(d) $\lim\limits_{x \to 2^-} f(x) = 1$ and $\lim\limits_{x \to 5^+} f(x) = 3$.

16.   Let

$$f(x) = \begin{cases} -1 & \text{if } x < 0 \\ ax + b & \text{if } 0 < x < 1 \\ 1 & \text{if } x \geq 1 \end{cases}$$

Determine a and b so that f is continuous everywhere.

17.   Use $f'(c) = \lim\limits_{x \to 1} [f(c + h) - f(c)]/h$ to find $f'(2)$ for each of the following. Use both a numerical method and an exact method where possible.

(a) $f(x) = x^2 - 5x$

(b) $f(x) = \dfrac{1}{x - 3}$

(c) $f(x) = \sqrt{9 - x}$

(d) $f(x) = \ln x$

18.

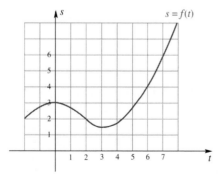

Figure 1

Use the sketch in Figure 1 to approximate each of the following.

(a) $f'(2)$

(b) $f'(6)$

(c) v_{ave} on $[3, 7]$

19.   Sketch the graph of $f(x) = xe^{-x}$ on the interval $-1 \leq x \leq 4$ by building a table of values and plotting points. Find or numerically estimate $f'(2)$ and show what it represents on the graph of $f(x)$.

20.   Find each limit exactly and then check your results with a table of values.

(a) $\lim\limits_{x \to \infty} 10 + \dfrac{x^{10}}{2^x}$

(b) $\lim\limits_{x \to \infty} e^{-x}(x^2 + \sin x)$

21.   For each function $f(x)$, find a formula for $f'(c)$. Estimate both the y-coordinate and the slope of $f(x)$ at $x = -1, 0, 1$. Discuss the relationship between the y-coordinate and the slope of $f(x)$, and display this relationship on a sketch.

(a) $f(x) = e^{-x}$

(b) $f(x) = 10^{-x}$

22.   Rewrite each exponential function as a natural exponential function.

(a) $y = 3 \cdot 10^x$

(b) $y = 2 \cdot 3^{-x}$

23.   Use logarithms to solve each of the exponential equations.

(a) $e^x = 19$

(b) $3e^x = 12$

(c) $e^{3x-1} = 5$

(d) $12^x = 3$

24.   For each function below, find the derivative (slope) of the function at the point $x = 1$. If you can find exact results do so; otherwise find numerical approximations. Sketch the graph of the function and indicate on it the derivative you found using slope marks.

(a) $y = -2 \ln x$

(b) $y = 5 \ln(3+x)$

(c) $y = 3 \log_2 x$

CHAPTER 2

Numerical and Graphical Techniques

2.1 Calculator and Computer Graphs

Computer algebra systems and graphing calculators produce mathematical graphs in basically the same way. A brief overview of how graphs are produced will help a great deal in determining "what went wrong" when you don't get what you expect on your calculator or computer screen.

Pixels   The screen of the computer or calculator consists of a rectangular patch of small dots called *pixels*, each of which can either be lit or dark (each pixel is like a tiny light bulb). Thus, when you look at your screen and see numbers, letters, and graphs, what you are actually looking at are the pixels that have been lit. If you look closely, you can see the individual pixels. In Figure 1, the pixels that are lit appear to form the numeral two.

In order to generate the graph of a function on an interval, you type in (or "input") the formula for the function and establish the endpoints of the interval; a table of values is formed internally. The table of values is precisely the same type of table you made in order to graph functions by hand in the previous chapter. As you would by hand, the computer or calculator determines the *y*-values by substituting the *x*-values into the formula for the function.

The graph is produced by lighting the pixels that correspond to the points in the internal table of values, just as you plotted points in the previous section. Normally, the calculator or computer then fills in the graph with additional pixels in order to give the appearance of a smooth curve. Again, this is essentially the same process you went through when graphing by hand.

If you remember that *the computer or calculator is going through the same process that you use when making a graph*, only much faster, you will be able to handle problems much more easily when they come up.

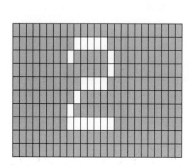

x	y
−2.00	4.00
−1.50	2.25
−1.00	1.00
−0.50	0.25
0.00	0.00
0.50	0.25
1.00	1.00
1.50	2.25
2.00	4.00

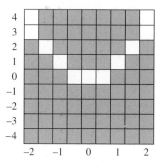

Figure 1   **Figure 2**

EXAMPLE 1: Suppose that we have a screen that is only nine pixels by nine pixels, and suppose that the coordinates of the pixels are equally spaced and range from −2 to 2 in the x-direction and from −4 to 4 in the y-direction. Sketch what the graph of $y = x^2$ would look like on this screen. Assume that there is one entry in the table for each pixel in the x-direction (as is the case with graphing calculators).

SOLUTION: See Figure 2. Because there are nine pixels in the x-direction and because x ranges from −2 to 2, the x-values in the table will be $x = -2.0, -1.5, -1.0, -0.5, 0, 0.5, 1.0, 1.5, 2.0$. The y-values are determined by the equation $y = x^2$. Note that when the points are plotted, the y-coordinates of the pixels are not the same as the y-coordinates of the points in the table. For example, the point $(0.5, 0.25)$ is plotted at the pixel with coordinates $(0.5, 0.0)$ because that is the nearest pixel to the point. Also note that we have filled in the graph with lit pixels at $(-2, 3)$ and $(2, 3)$ in order to give the appearance of a continuous curve. We could have lit the pixels at $(-1.5, 3)$ and $(1.5, 3)$; the choice is somewhat arbitrary.   ◄

The Graph Window   The way your computer or calculator graph looks depends on how you define the *graph window*. When you make a graph by forming a table and then plotting points as in Chapter 1, you still had to determine a scale and a range of values for the x- and y-axes of your coordinate system. The same is true of computer or calculator graphs; to generate a graph you must specify a range of values for the x- and y-axes as well as a formula for the function.

CALCULATOR GRAPHS   On a graphing calculator the range key accesses the range screen (some calculators use the term *window* instead of range). On the range screen you set the minimum and maximum values that appear on both the x- and y-axes. Don't confuse this terminology with the range of a function, which refers only to the y-values of the function.

The minimum and maximum x-values (usually called *xmin* and *xmax*) determine which x's are used in the table of values; a calculator with 95 pixels in the x direction would generate a table with *xmin* as the smallest x-value and *xmax* as the largest x-value, with 93 equally spaced x-values in between. The distance between x-values, which we will call Δx, is given by $\Delta x = (xmax - xmin)/94$ because with 95 pixels there are 94 intervals between pixels. The function being graphed determines the corresponding y-values.

When we say "set the graph window to $-2 \le x \le 2$ and $-1 \le y \le 1$," we mean set the values of $xmin$ and $xmax$ to -2 and 2 and set the values of $ymin$ and $ymax$ to -1 and 1. You also can set the x- and y-*scales* with the range key; on a graphing calculator this establishes the distance between tick marks on the x- and y-axes. Changing the $xscale$ and $yscale$ settings does *not* affect the shape of the graph that you generate but can make the graph easier to interpret.

Graphing calculators fill in the spaces between points from the internal table with pixels that approximate straight lines. Thus, a calculator graph is essentially what you would get by hand if you plotted a lot of points and then connected the points with straight line segments. The final graph does not look like a collection of straight line segments, but rather, like a smooth curve, because the segments are so short and close together.

THE TRACE FEATURE   You can view the table that is formed internally when you produce a graph by using the *Trace* feature of the calculator. A small cross or cursor follows the curve, and the x- and y-coordinates of the plotted points are displayed at the bottom of the screen. See Figure 3.

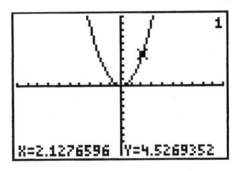

The Trace feature. The curve shown here is $y = x^2$, with graph window $-10 \le x \le 10$, $-10 \le y \le 10$ on a 95-by-63-pixel calculator.

Figure 3

95 Pixels	96 Pixels
-10	-10
-9.787234	-9.789474
-9.574468	-9.578947
-9.361702	-9.368421
-9.148936	-9.157895
-8.936170	-8.947368

The x-values from the Trace for two calculators with the same x-range setting of $-10 \le x \le 10$. One has 95 pixels in the x-direction and the other has 96.

Figure 4

Only the coordinates of the data points in the table of values are kept in memory when a function is graphed. Thus, when you trace a curve, the cursor "jumps" from point to point on the graph, appearing to skip over some points. This is because the cursor goes only to the points that are in the internal table of values, and not to the additional points that were used to fill in the graph. In Example 1, the points at $(-2, 3)$ and $(2, 3)$ would be skipped over when tracing. In order to get coordinates of points between the points in the table, you must produce a new graph with narrower range settings.

In Figure 4 we show the first few x-values obtained using the Trace feature for two calculators both set with $xmin = -10$ and $xmax = 10$ (any function and y-range can be used because we are only looking at the x-values). The first calculator has 95 pixels in the x-direction (as does the TI-82 and the Casio fx-7700) and the second has 96 pixels in the x-direction. As you can see, one additional pixel makes a small difference in the internal table of values, and hence in the values reported by the Trace.

COMPUTER ALGEBRA GRAPHS Computer algebra systems handle range settings in a variety of ways. More often than not, you give the *x*-range for the function (that is, you specify the domain of the function for that plot) and the computer determines a *y*-range that seems suitable. If you don't like the *y*-range the computer picked (that is, the *default y*-range), you can override it with one of your own choice. With other computer algebra systems (older versions of Derive, for example), you may be asked to choose the center and scale of the graph instead of an *x*- or *y*-range.

With most computer algebra systems, the table, and hence the graph produced, is *device independent*. This means that the *x*-values used in the table are not tied to particular pixels, or to a particular number of pixels, so that the appearance of the graph is pretty much the same no matter what computer or printer is used. Sophisticated graphing programs generally sample the graph more frequently (that is, compute more points for the table) in places where the graph is curving rapidly, and sample less frequently where the graph is straighter. Thus, a typical internal table of values would not have equally spaced *x*-values, as a graphing calculator does.

LOCATING POINTS Unlike graphing calculators, computer algebra systems generally don't have a Trace feature. Most, however, allow you to get the coordinates of any pixel on the graph screen by using the mouse or the arrow keys. Keep in mind that because this is not tracing, the indicated ordered pairs represent points *near* the graph of the function, but not necessarily *on* the graph.

EXAMPLE 2: Graph the function $f(x) = \frac{10}{x-2} + e^x$ on the interval $-10 \leq x \leq 10$. Generate three graphs, using the *y*-range settings of $-10 \leq y \leq 10$, $-100 \leq y \leq 100$, and $-1000 \leq y \leq 1000$. Additionally, if you are using a computer algebra system or graphing calculator that can generate a default *y*-range, use that setting also. Which graph best displays the important features of this function? *Note*: Another common notation for e^x is $\exp(x)$; this notation is used in many computer programs.

SOLUTION:

Graphing Calculators

1. Store the function in the first function memory.

2. Set the range to the first specified setting of $-10 \leq x \leq 10$ and $-10 \leq y \leq 10$. Figure 5 shows the range screen of a typical calculator. Make appropriate changes for the range settings $-100 \leq y \leq 100$ and $-1000 \leq y \leq 1000$.

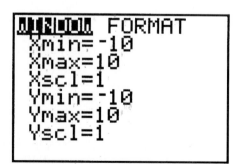

Figure 5

3.   Give the command to graph the function stored in memory.

The results are shown in Figure 6.

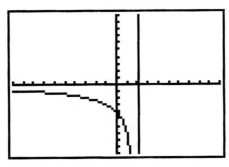

Graph window $-10 \leq x \leq 10$, $-10 \leq y \leq 10$.

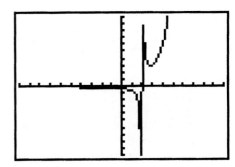

Graph window $-10 \leq x \leq 10$, $-100 \leq y \leq 100$.

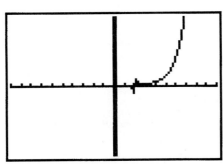

Graph window $-10 \leq x \leq 10$, $-1000 \leq y \leq 1000$.

Figure 6

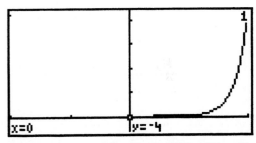

Default y-range on $-10 \leq x \leq 10$ using TI-85 and Zoom Fit.
(This option is not available on all graphing calculators.)

Graphing calculator graphs of the function $f(x) = \frac{10}{x-2} + e^x$

Computer Algebra Systems   With menu-driven computer algebra systems (Derive, for example), to graph a function you enter the formula for the function, specify the window, and give the graph command (similar to graphing calculators). With command-based computer algebra systems like Mathematica or Maple, you normally specify the function in the same line as the command—and generally the window size as well. The line may look something like this:

$$\text{Plot}[10/(x - 2) + E^{\wedge}x, \{x, -10, 10\}, \text{PlotRange} \rightarrow \{-10, 10\}]$$

The results are shown in Figure 7.

Discussion   The computer and calculator graphs are quite similar, except that the computer graphs have a higher resolution (that is, more and smaller pixels), and hence greater accuracy. Because the term $x - 2$ appears in the denominator of the first term of the function, we expect to see an asymptote at $x = 2$. The second term is exponential; because this term gets large very fast, we expect to see rapid growth of the function as x gets large (the right side of the graph).

The y-range setting of $-100 \leq y \leq 100$ does the best job of showing the function's behavior near the asymptote at $x = 2$. The y-range setting of

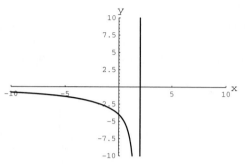

Graph window −10 ≤ x ≤ 10, −10 ≤ y ≤ 10.

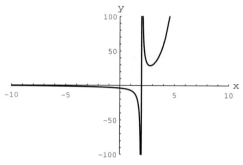

Graph window −10 ≤ x ≤ 10, −100 ≤ y ≤ 100.

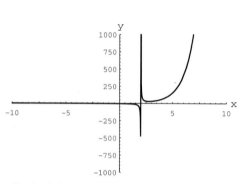

Graph window −10 ≤ x ≤ 10, −1000 ≤ y ≤ 1000.

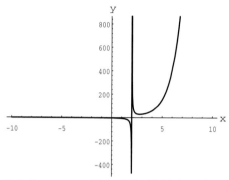

Default y-range on −10 ≤ x ≤ 10 with Mathematica. The Mathematica command is **Plot[10/(x−2)+E^x, {x,−10,10}]**

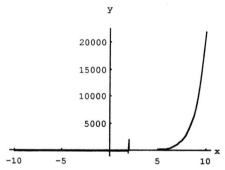

Default y-range on −10 ≤ x ≤ 10 with Maple. The Maple command is **plot(10/(x−2)+exp(x),x=−10..10);**

Computer algebra graphs of the function $f(x) = \frac{10}{x-2} + e^x$

Figure 7

−1000 ≤ y ≤ 1000 does a good job of representing the exponential growth for large x, but the asymptote shows up just as a blip (here the computer graph is somewhat better than the calculator graph). The y-range setting of −10 ≤ y ≤ 10 is not large enough to show much of the function to the right of the asymptote, so is not very useful. Therefore the answer to the question, "Which graph best displays the important features of the function?" is that it depends what feature you want to focus on.

Default y-Ranges   The TI-85 calculator (Figure 6) and the computer algebra system Maple (Figure 7) both choose a default y-range so as to include all of the points in the internal table of values. For the function we are investigating, the largest y-value occurs at the right endpoint $x = 10$, where we have $y = \frac{10}{10-2} + e^{10} \approx 22{,}028$. Even though the function approaches infinity as x approaches 2, no values in the internal table near $x = 2$ are nearly as large as the value at the right endpoint (the table "skips over" $x = 2$), so that the asymptote is reduced almost to invisibility. The function looks almost completely exponential.

Mathematica attempts to guess where an asymptote occurs by looking for rapid growth in the function, and it chooses a y-range that is not "too" large. For our example, we get a graph very similar to the one with y-range setting $-100 \le y \le 100$, emphasizing the asymptote at $x = 2$ over the exponential growth beyond $x = 6$ or so (Figure 7). The default y-range may not always be the best choice for viewing a given function; it depends on what aspect of the function you want to emphasize.   ◀

Graphing and Asymptotes

Sometimes a vertical line will appear where an asymptote should be (see Figures 6 and 7), and other times nothing will appear. Generally, if the x-coordinate of one of the points in the internal table of values happens to fall on (or close enough to) an asymptote, then no vertical line appears. The computer or calculator cannot plot a point where the function is undefined.

If the asymptote is "skipped over" in the internal table of values, then a point just to the left of the asymptote (with, say, a large negative y-coordinate) is connected to a point just to the right of the asymptote (with, say, a large positive y-coordinate), and a nearly vertical line is the result. Because computers and calculators have a difficult time dealing with asymptotes, it's important to know how to look for asymptotes algebraically, as explained in Section 1.2. See Figure 8.

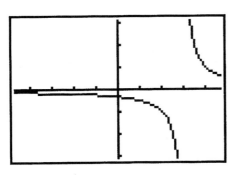

Graph of $y = \frac{1}{x-3}$ with window $-4.7 \le x \le 4.7$, $-3.1 \le y \le 3.1$ on a 95-by-63-pixel calculator.

Graph of $y = \frac{1}{x-3}$ with window $-5 \le x \le 5$, $-3.1 \le y \le 3.1$ on a 95-by-63-pixel calculator.

Figure 8

EXAMPLE **3:** Find a graph window for the function

$$h(x) = \frac{1}{2\sqrt{2\pi}} e^{-\frac{1}{2}\left(\frac{x-70}{2}\right)^2}$$

that displays the important features of the graph. Describe your graph.

SOLUTION: When a "standard" graph window of $-10 \le x \le 10$, $-10 \le y \le 10$ is used we get the picture in Figure 9, taken from a TI-85 calculator. With the Trace feature invoked, we see that at $x = 0$ we have $y = 1.9702 \cdot 10^{-267}$. Thus, the y-coordinate is so close to zero that the graph of the function coincides with the x-axis and it appears that there is no graph at all.

Such situations are typical for functions taken from the "real" world. It is important to look at the algebraic form of the function to get clues for a good graph window. For this function you should notice the quantity $x - 70$ in the exponent; recall that this represents a translation of 70 units to the right. Thus, we should look for the graph near $x = 70$. In Figure 10 we show the graph with the window $65 \le x \le 75$, $-10 \le y \le 10$, which is our original window translated 70 units to the right.

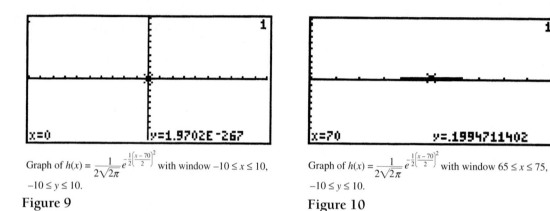

Graph of $h(x) = \dfrac{1}{2\sqrt{2\pi}} e^{-\frac{1}{2}\left(\frac{x-70}{2}\right)^2}$ with window $-10 \le x \le 10$,
$-10 \le y \le 10$.
Figure 9

Graph of $h(x) = \dfrac{1}{2\sqrt{2\pi}} e^{-\frac{1}{2}\left(\frac{x-70}{2}\right)^2}$ with window $65 \le x \le 75$,
$-10 \le y \le 10$.
Figure 10

We see some of the graph, but not much. However, the Trace shows that the y-coordinate is now $y = 0.1994711402$ when $x = 70$, which gives us an idea of how to set the y-range. In Figure 11 we show the graph with the window $65 \le x \le 75$, $0 \le y \le 0.2$, which gives a pretty good picture.

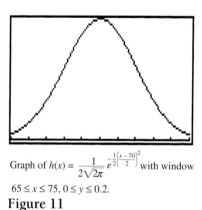

Graph of $h(x) = \dfrac{1}{2\sqrt{2\pi}} e^{-\frac{1}{2}\left(\frac{x-70}{2}\right)^2}$ with window
$65 \le x \le 75$, $0 \le y \le 0.2$.
Figure 11

Even with a computer algebra system or graphing calculator that chooses a default y-range for you, you need to be able to pick a reasonable x-range first, which can only be done by looking at the algebraic form of the function.

The graph of the function turns out have the well-known bell shape, with a maximum y-value at $x = 70$, and approaches the x-axis asymptotically as x gets farther from $x = 70$.

Note: This is not a "phony" made-up function, but one that is used in probability theory, and that you will work with in a Lab later in this book. ◀

<hr>

Concepts Review

1. The screen of the computer or calculator consists of a rectangular patch of small dots called _____ , each of which can either be lit or dark.

2. A calculator with an array of 41 pixels (x-direction) by 21 pixels (y-direction) is set with a graph window of $-5 \leq x \leq 5$, $-5 \leq y \leq 5$. The x-coordinate of the left-most pixels would be -5, and the next pixel to the right would have x-coordinate _____ .

3. The default y-range of a computer algebra system will (always, never, sometimes) determine the best y-range setting for the graph of a function.

4. To view the important features of the graph of $y = \frac{100}{x - 20}$, you would want to include $x =$ _____ in the graph window.

Problem Set 2.1

1. Suppose we have a screen that is five pixels by five pixels, and suppose the coordinates of the pixels range from -2 to 2 in the x-direction and from -8 to 8 in the y-direction. Sketch what the graph of each of the following equations might look like on this screen. State any assumptions you make when you "fill in" between points in the table. Assume there is one entry in the table for each pixel in the x-direction (as is the case for graphing calculators). See Example 1.

(a) $y = x^2$ (b) $y = x^3$

(c) $y = e^x$ (d) $y = \sin x$

2. Suppose we have a screen that is nine pixels across and seven pixels up and down, and suppose the coordinates of the pixels range from 0 to 2 in the x-direction and from 0 to 8 in the y-direction. Sketch what the graph of each of the following equations might look like on this screen. State any assumptions you make when you "fill in" between points in the table. Assume there is one entry in the table for each pixel in the x-direction (as is the case for graphing calculators).

(a) $y = x^2$ (b) $y = x^3$

(c) $y = e^x$ (d) $y = \sin x$

3. Graph each of the following functions on the interval $-10 \leq x \leq 10$. Generate three graphs, using the y-range settings of $-0.1 \leq y \leq 0.1$, $-1 \leq y \leq 1$, and $-10 \leq y \leq 10$. Additionally, if you are using a computer algebra system or graphing calculator that can generate a default y-range, use that setting also. Make accurate hand sketches of each case. Which graph best displays the important features of each function? Briefly discuss why you made the choice you did.

(a) $f(x) = 5 \sin x$ (b) $f(x) = \frac{5}{x}$

(c) $f(x) = x^2$ (d) $f(x) = 0.01x + 0.001x^2$

4. Graph each of the following functions on the intervals $-1 \leq x \leq 1$, $-10 \leq x \leq 10$, and $-100 \leq x \leq 100$. Choose an appropriate y-range setting for each case. Make accurate hand sketches of each case. Which graph best displays the important features of each function? Briefly discuss why you made the choice you did.

(a) $g(x) = \cos(0.1x)$ (b) $g(x) = \frac{x}{3} - x^3$

(c) $g(x) = \frac{1}{x - 20}$ (d) $g(x) = e^{-x^2}$

5. Find a graph window for the function

$$C = 50 \sin(2\pi t) + 5t + 80$$

that shows the important features of the graph. Briefly describe the features that you find, and how you found them.

6. Find a graph window for the function

$$P(V) = \frac{RT}{V - b} - \frac{a}{V^2}$$

that shows the important features of the graph. Use the values $R = 81.80241$, $a = 3{,}583{,}858.8$, $b = 42.7$, and $T = 280$. Briefly describe the features that you find, and how you found them. This function with these values will appear later in this book in a Lab project.

For graphing calculator users:

7. Determine a window setting for your calculator for which the coordinates of the pixels correspond to consecutive integers.

<hr>

Answers to Concepts Review: 1. Pixels   2. -4.75
3. Sometimes   4. 20

LAB 2: CALCULATOR AND COMPUTER GRAPHS

Lab Introduction

In the first part of the lab below, we investigate a few important "building block" functions. You will use your calculator or computer to generate the graphs of these functions, then you will sketch them on paper. These functions are so important that you should know their graphs by heart. In the second part we find that by combining these functions and adding some parameters (constants), we can build many new functions. In several of the labs that follow, we collect data and try to come up with a function that fits the data; the functions in this lab will come in handy.

Note: Parameters will be discussed more fully in Section 2.3.

Graphing Experiments

Use a calculator or computer to generate the graphs indicated below. The x-interval is given for each function; you must choose a suitable y-interval for each.

1. (a) Graph the functions x and $\frac{1}{x}$ on the interval $-4 \le x \le 4$. Sketch their graphs on one set of axes, and label them.

(b) Graph the functions x^2, x^3, x^4, and x^5 on the interval $-2 \le x \le 2$. Sketch their graphs on one set of axes, and label them.

(c) Graph $\sin x$ and $\cos x$ on the interval $-2\pi \le x \le 2\pi$. Sketch their graphs on one set of axes, and label them.

(d) Graph e^x and e^{-x} on the interval $-2 \le x \le 2$. Sketch their graphs on one set of axes, and label them.

(e) Graph $\ln x$ on the interval $-2 \le x \le 2$ and sketch its graph on a set of axes.

2. (a) Graph $\cos(wx)$ for $w = 0.5, 1, 2$ together on the interval $-2\pi \le x \le 2\pi$. Sketch this family of curves.

(b) Graph $a \cos x$ for $a = 0.5, 1, 2$ together on the interval $-2\pi \le x \le 2\pi$. Sketch this family of curves.

(c) Graph e^{-kx} for $k = 0.5, 1, 2$ together on the interval $0 \le x \le 2$. Sketch this family of curves.

(d) Graph $x \cos (x)$ on the interval $-20 \le x \le 20$. Sketch this graph together with the graphs of x and $-x$ on the same set of axes.

(e) Create a function that looks like the one below. Use what you discovered in parts (a)–(d).

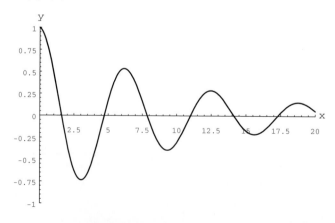

Discussion

1. Give a brief verbal description of each of the functions in (1) above. Use plain English; you don't need to use the "correct" technical jargon. For example, you could discuss whether a function increases or decreases (from left to right), whether it oscillates or not, and where it seems to be headed as it leaves the screen on the left and right sides.

2. What patterns do you see in part 1(b)? Discuss even versus odd exponents. Explain any patterns you found.

3. Describe how each of the parameters (constants) in parts 2(a)–2(c) affects the shape of each graph.

4. Discuss the principle involved in parts 2(d) and 2(e). Can you state a general form of this principle?

5. Discuss the difference between a parameter (constant) and a variable. How would this difference show up if you were graphing these functions by hand?

2.2 Calculator and Computer Tables

In Chapter 1 we generated graphs by first forming a table of values, then plotting the points in the table. With computers and graphing calculators we bypass the table stage; we just specify the formula for the function and the x- and y-ranges to generate a graph. Still, in many cases it is useful to look at a table. In particular, if you are having trouble adjusting the y-range to get a good view of your graph, a quick look at a table often helps.

Calculator Tables   Some calculators have a built-in table feature (the TI-82, for example). For calculators without this feature, we outline two methods for generating tables.

METHOD ONE Use the Trace to access the internal table of values created when you produce a graph. To build a table with a particular starting x-value (x_0) and a particular change in x from one table entry to the next (Δx), set your x-range as follows:

1. Set *xmin* to x_0.
2. Set *xmax* to $[x_0 + \Delta x \cdot (pixels - 1)]$ where *pixels* is the number of pixels in the x-direction for your calculator.

In the accompanying manual we provide a program called Table, which automates this process.

METHOD TWO Build the table point by point by storing the formula for the function, then specifying each x-value and letting the calculator calculate the y-value. An advantage of this method is that you can build tables for which Δx is not constant, such as the tables in Chapter 1 that were used to estimate limits.

Some calculators (for example, the TI-85 and TI-82) have a built-in evaluation command. For other calculators, you need to store an x-value and ask for the resulting function value (usually listed as f_1 or Y_1). In either case, each table entry will take five to eight keystrokes. In the accompanying manual we provide a program called Evaluate, which automates this process.

Computer Algebra Tables   Most computer algebra systems have a built-in command for generating tables. The output is a set of ordered pairs or a table, depending on the software.

EXAMPLE 1: Build a table of values for the function $f(x) = 5xe^{-x} - 1$ with $x = 0, 0.5, 1, 1.5,$ . . . 5. Use the table to give an interval estimate of the x-value that corresponds to the maximum (largest) y-value on the interval $0 \le x \le 5$. Also give interval estimates of any x-values for which $f(x) = 0$.

SOLUTION: Using one of the methods outlined above, we get the table shown in Figure 1. We see that for the values given in the table, the maximum y-value occurs at the point where $x = 1.0$. Of course, there could be a larger y-value just about anywhere on the interval $0 \le x \le 5$, at points between the ones given. However, for a typical function, we would expect the maximum y-value to occur between $x = 0.5$ and $x = 1.5$ (why?). Thus, our best interval estimate of where the maximum y-value occurs would be $0.5 \le x \le 1.5$.

We can see that there must be at least one point where $f(x) = 0$ between $x = 0$ and $x = 0.5$, because $f(x)$ is continuous and the y-value of the function changes sign between these points. Similarly, another zero of the function must occur between $x = 2.5$ and $x = 3.0$. Thus, our best interval estimates of the points where $f(x) = 0$ would be $0 \le x \le 0.5$ and $2.5 \le x \le 3.0$.   ◄

$-x$	$5xe^{-x} - 1$
0.0	−1.0000
0.5	0.5163
1.0	0.8394
1.5	0.6735
2.0	0.3534
2.5	0.0261
3.0	−0.2532
3.5	−0.4715
4.0	−0.6337
4.5	−0.7500

Figure 1

EXAMPLE 2: In order to evaluate the quality of the rear shock absorbers of a car, the following test was devised. The rear bumper is pushed down 5 inches from its rest position and then released. The displacement of the bumper from its rest position (that is, the number of inches moved) is measured as a function of time. The bumper bounces up and then rebounds back toward the point at which it was released. (See Figure 2.)

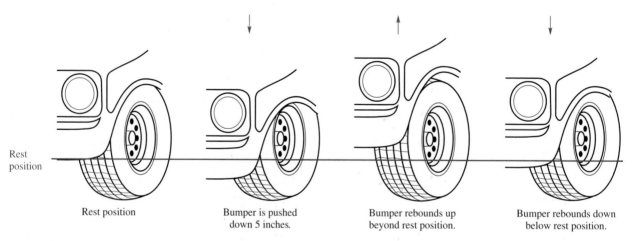

Rest position

Rest position

Bumper is pushed down 5 inches.

Bumper rebounds up beyond rest position.

Bumper rebounds down below rest position.

Figure 2

The distance that the bumper moves below the rest position on the rebound is then used as a measure of the effectiveness of the shock. Dividing this rebound distance by the initial displacement of 5 inches gives a rebound ratio; for example, if the rebound distance is 0.5 inches (below rest position) then we have a 10% rebound ratio. A ratio below 1% is considered acceptable.

For one particular brand of shock absorber, engineers have determined that the distance from rest position is approximated by the function

$$y = 5e^{-10t}(\cos(10t) + \sin(10t))$$

where t is time measured in seconds and y is distance of the bumper from rest position measured in inches. Use graphs and/or tables of function values to find the rebound ratio. Is this brand of shock acceptable by our 1% criterion?

SOLUTION: We need to look at the context of the problem in order to see how to set the graph window. Experience tells us that when we push a car bumper down a few inches and release it, all of the motion takes place within a second or two. Thus, we might look at the function on the interval $0 \leq t \leq 1$ to start. Because the bumper is pushed down 5 inches, we could set the y-range to be $-7 \leq y \leq 7$ to start.

Before diving in with a computer or calculator graph, let's take a moment to think about what we *expect* a graph of y versus t to look like. That way, if our calculator graph is different from what we expect, we will know we need to examine things more closely. From the context of the problem, we know the bumper starts 5 inches below rest position, then moves above rest position, then moves back down again below rest position. Thus, we expect *oscillations*; the graph should look something like Figure 3. The y-coordinate of the point shown in the figure is the rebound distance we are looking for. This point is called a *local minimum* of the function. ◄

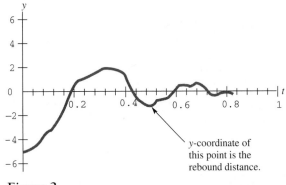

Figure 3

Graphing Calculators Even though the independent variable in the problem is labeled t, we must change it to x in order to enter it into most graphing calculators. Set the graph window to $0 \leq x \leq 1$, $-7 \leq y \leq 7$ and give the command to graph. (You should choose appropriate values for the x- and y-scales to make the graph easy to interpret).

Computer Algebra Systems If we wanted nothing more than a graph of the function, we could give the plot command with the function definition as part of the command, as we did before. In this case, however, we may need to create a table as well, to help us adjust range values. To save time, we can define a function $f(t)$ first, and then give the graph and table commands both in terms of $f(t)$.

The graphs for a typical graphing calculator and for a typical computer algebra system are shown in Figure 4.

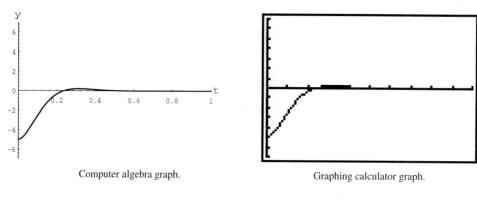

Computer algebra graph.   Graphing calculator graph.

Graph of $y = -5e^{-10t}(\cos(10t) + \sin(10t))$ using the window
$0 \le x \le 1$ and $-7 \le y \le 7$.

Figure 4

$y = -5e^{-10t}(\cos(10t) + \sin(10t))$	
t	y
0	-5.00000000000
0.1	-2.54162993000
0.2	-0.33370337407
0.3	0.21131436311
0.4	0.12916610868
0.5	0.02274940084
0.6	-0.00843711207
0.7	-0.00643282050
0.8	-0.00141541447
0.9	0.00030791473
1	0.00031396154

Figure 5

Our graph is not exactly what we expected to see. What happened? This is where a table of values can be very helpful. For quick information, we can produce a table of values with t running from 0 to 1 in increments of $\Delta t = 0.1$.

The numbers in the resulting table are shown in Figure 5. The number of digits reported will depend on the system you are using.

The table shows us what the graph could not. The y-values do indeed follow the pattern that we expected when we made the sketch in Figure 3. The bumper starts at 5 inches below rest position ($y = -5$), then moves above rest position (positive y-values), then back down below rest position (negative y-values) and so on. *The problem is one of scale.*

Consider one oscillation to correspond to the movement of the bumper from one maximum y-value to the next minimum y-value, or vice versa. Roughly, the size of each oscillation is one *order of magnitude* (one power of ten) less than the previous one. The y-scale we have chosen is adequate to show the initial displacement of 5 inches, but we can barely see the first oscillation above $y = 0$, and we cannot see any of the other oscillations. This is because the nearest pixel to the plotted point is on the x-axis for these oscillations.

Using the table as our guide to setting the y-range, we show two more views of the graph in Figure 6. In the first view, the y-range is set to $-0.4 \le y \le 0.4$ so that we can see the first oscillation above $y = 0$. In the second view the y-range is set to $-0.02 \le y \le 0.02$ so that we can see the next oscillation. Notice that as we focus on smaller oscillations, the larger oscillations are no longer in view and appear as almost vertical lines (computer and calculator graphs are sufficiently similar that we present only one graph for each window).

We can now finish the problem by estimating the rebound distance, which is the y-coordinate of the local minimum shown in the second graph in Figure 6. We find that the rebound distance is $y = -0.009$ inches, accurate to three decimal places. Calculator users can Trace to estimate this value. Computer users can either read the coordinates from the free-floating cursor, or can build a new table of values between 0.6 and 0.7 with increment $\Delta t = 0.01$.

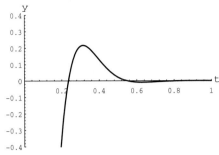

Graph of $y = -5e^{-10t}(\cos(10t) + \sin(10t))$ using the window $0 \le t \le 1$ and $-0.4 \le y \le 0.4$.

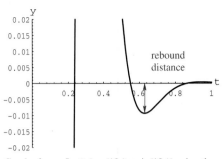

Graph of $y = -5e^{-10t}(\cos(10t) + \sin(10t))$ using the window $0 \le t \le 1$ and $-0.02 \le y \le 0.02$.

Figure 6

As a percentage of the initial displacement, we have a rebound of about

$$\frac{0.009}{5} = 0.0018 \approx 0.2\%$$

which is well within the acceptable 1% level.   ◄

Make sure that you understand the relationship between the graphs in Figures 4 and 6 and the table in Figure 5 in the last example. Finding the "right" graph window is not a trivial matter. As we saw, the ideal window depends on what part of the graph you want to focus on. A graph can display information about a function over one or two orders of magnitude, but no more. A table can display information about a function over many orders of magnitude. *If a graph is not coming out as you expect, look at a table.*

EXAMPLE 3: Monthly heating oil costs for a typical three-bedroom home in Hartland, Indiana, are approximated by the equation

$$C = 50 \sin(2\pi t) + 5t + 80$$

where C is the cost in dollars of heating a home for the month corresponding to the time t. The time t is measured in years starting with January 1990 (that is, $t = 0$ corresponds to January 1990) and each $\frac{1}{12}$ unit corresponds to a month. Thus, $t = \frac{1}{12}$ is February 1990, $t = \frac{2}{12}$ is March 1990, and so on.

Create both a graph and a table of values for the heating cost equation for $0 \le t \le 5$. Describe the function and explain its main features in terms of the context of the problem. Estimate the year and month when costs will first reach $150.

SOLUTION:

Numerical/Graphical Approach   To see where the function reaches 150 for the first time, we plot the cost equation and the equation $C = 150$ as two functions graphed on one screen. We also create a table of function values from $t = 0$ to $t = 5$ using an increment of $\Delta t = \frac{1}{12}$, because $\frac{1}{12}$ year is 1 month.

The results are shown in Figure 7. By producing the table first, we get a good idea of how to set the range. From the problem we know to set the t-range to $0 \le t \le 5$. From the table we see that a reasonable choice for the C-range would be $30 \le C \le 160$.

There are two important features to the graph. There is a general upward trend (coming from the $5t$ term), and there are oscillations (coming from the $50 \sin(2\pi t)$ term). Because the function represents monthly heating costs, the oscillations result from higher costs in the winter than in the summer. The

t	C	t	C
0	80	2.58333	67.9167
0.08333	105.417	2.66667	50.0321
0.16667	124.135	2.75	43.75
0.25	131.25	2.83333	50.8654
0.33333	124.968	2.91667	69.5833
0.41667	107.083	3	95
0.5	82.5	3.08333	120.417
0.58333	57.9167	3.16667	139.135
0.66667	40.0321	3.25	146.25
0.75	33.75	3.33333	139.968
0.83333	40.8654	3.41667	122.083
0.91667	59.5833	3.5	97.5
1	85	3.58333	72.9167
1.08333	110.417	3.66667	55.0321
1.16667	129.135	3.75	48.75
1.25	136.25	3.83333	55.8654
1.33333	129.968	3.91667	74.5833
1.41667	112.083	4	100
1.5	87.5	4.08333	125.417
1.58333	62.9167	4.16667	144.135
1.66667	45.0321	4.25	151.25
1.75	38.75	4.33333	144.968
1.83333	45.8654	4.41667	127.083
1.91667	64.5833	4.5	102.5
2	90	4.58333	77.9167
2.08333	115.417	4.66667	60.0321
2.16667	134.135	4.75	53.75
2.25	141.25	4.83333	60.8654
2.33333	134.968	4.91667	79.5833
2.41667	117.083	5	105
2.5	92.5		

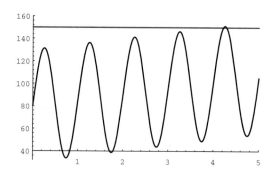

Figure 7

general upward trend probably results from inflation.

The graph shows that the heating cost reaches $150 approximately when $t = 4.2$. The table gives us more accurate results; the first time when the cost goes over $150 is at $t = 4.25$. This corresponds to April of 1994.

Algebraic/Exact Solution   None. The equation we want to solve is

$$150 = 50 \sin(2\pi t) + 5t + 80$$

We could combine the two constant terms, but after that we are stuck. The 50 $\sin(2\pi t)$ term and the $5t$ term cannot be combined. Therefore, there is no way to solve algebraically for t. A numerical/graphical approach is the *only* approach to this problem. ◄

Notice that the t variable in the previous example is actually *discrete*. This means that there really are no t values between the ones listed in the table. The graph makes the t variable look continuous—that is, it fills in the gaps between points. Approximating a discrete variable with a continuous one is a common practice in mathematics, and a useful one, as long as we remember that the variable is defined only at specific points.

Concepts Review

1. If a computer or calculator graph does not appear as we think it should, it is sometimes helpful to look at a _____ to determine how to set the graph window.

2. In a table of function values of a continuous function $f(x)$, the solutions to the equation $f(x) = 0$ lie between points where the output of the function changes from _____ to _____.

3. Assume the table of values shown was determined by the equation $y = f(x)$, where $f(x)$ is continuous.

x	0	1	2	3
y	-0.5	0.25	1	0.5

We can tell that a solution to $f(x) = 0$ lies between $x =$ _____ and $x =$ _____.

4. For the table in 3 above, we suspect that the maximum (largest) y-value lies between $x =$ _____ and $x =$ _____.

Problem Set 2.2

1. Build a table of values for each function $f(x)$ below, with $x = 0, 0.5, 1, 1.5, \ldots, 5$. Use the table to give an interval estimate of the x-value that corresponds to the maximum (largest) y-value on the interval $0 \leq x \leq 5$. Also give interval estimates of any x-values for which $f(x) = 0$.

(a) $f(x) = 2e^{-x}\cos x + 1$

(b) $f(x) = 2x^2 e^{-x} + 1$

(c) $f(x) = 1 + 47x - 12x^2 + x^3$

(d) $f(x) = \dfrac{3\ln(x+1)}{x+1} - 1$

2. Build a table of values for each function $g(x)$ below, with $x = -5, -4, -3, \ldots, 5$. Use the table to give an interval estimate of the x-value that corresponds to the maximum (largest) y-value on the interval $-5 \leq x \leq 5$. Also give interval estimates of any x-values for which $g(x) = 0$.

(a) $g(x) = e^{-x^2} - x$

(b) $g(x) = x + 3\cos(0.5x)$

(c) $g(x) = 1279 - 251x - 10x^2 + 10x^3 - x^4 + x^5$

(d) $g(x) = x - 2\sin x$

3. Estimate the *second* rebound ratio for the function $y = -5e^{-10t}(\cos(10t) + \sin(10t))$ from Example 2.

That is, use a table of function values to estimate the distance the bumper travels below the rest position $y = 0$ on the second rebound, and divide by the initial displacement of 5 inches. Use a table of at least 10 values with increment $\Delta t = 0.01$. *Hint:* Use the table from Example 2 to determine which x-values to use.

4. Use a graph and/or a table of values for the heating cost equation $C = 50\sin(2\pi t) + 5t + 80$ of Example 3 to estimate the year and month when costs will first reach $160.

5. Form a table of values for the function

$$P(V) = \frac{RT}{V - b} - \frac{a}{V^2}$$

on the interval $80 \leq V \leq 220$ in increments of 10. Use the values $R = 81.80421$, $a = 3{,}583{,}858.8$, $b = 42.7$, and $T = 280$. Estimate any peaks (maximum y-values) or valleys (minimum y-values), and briefly describe the features that you find. Compare your findings to those of Problem 6, Section 2.1.

Answers to Concepts Review: 1. Table 2. One sign; the other 3. 0; 1 4. 1; 3

2.3 | Parameters

Most of the equations we have been dealing with so far have contained just two letters, which we interpret as the independent variable and the dependent variable. Equations that occur in science and mathematics, even when there are only two variables involved, sometimes contain more than just two letters; the other letters are often called constants or *parameters*.

In the equation $y = ax^2$, if we consider x to be the independent variable and y to be the dependent variable, then a would be considered a parameter. From your previous work in mathematics you probably know that the graph of $y = ax^2$ is a parabola with vertex at $x = 0$ and that a determines the shape of the parabola. In particular, the smaller the magnitude of a, the wider the parabola. Also, when a is positive the parabola opens upward, and when a is negative the parabola opens downward. See Figure 1.

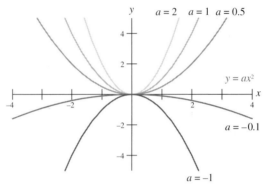

The family of curves defined by $y = ax^2$. In this family, a is a parameter.

Figure 1

Parameters are quantities that remain constant for a particular table of function values or graph. If a parameter changes we get a new graph. Thus, an equation with one or more parameters is said to define a *family* of curves.

Graphing with Parameters   A function or equation with parameters can't be graphed until values are specified for the parameters. Without a graphing calculator or computer this would be very tedious; one would have to calculate an extensive table of values for each parameter value. We outline below general procedures that can be used with calculators or computer algebra systems.

PROCEDURE 1 Type in the function just once using a letter for the parameter, then assign various numerical values to the letter and graph the function for each value. For the family of functions $y = ax^2$ shown in Figure 1, you could define or store the function as ax^2 and then assign the values $-1, -0.1, 0.5, 1, 2$ in turn to a and graph for each value.

PROCEDURE 2 Type in the function several times using a different numerical value for the parameter each time. For the family of functions in Figure 1 you could type in the functions $-1x^2, -0.1x^2, 0.5x^2, 1x^2, 2x^2$ separately and then graph them together.

EXAMPLE **1:** The mathematical model $y = Ae^{kt}$ is often used to describe population growth under certain conditions. This model works particularly well if there are no constraints to growth, such as lack of food or overcrowding. Populations can consist of people, animals, plants, or microscopic organisms such as bacteria. Because $y = A$ when $t = 0$, the parameter A represents the *initial population*.

Suppose that we are modeling the growth of a small town whose current population is 100; thus we know that $A = 100$. Investigate the equation $y = 100e^{kt}$ by graphing it for various values of k. Start with $k = 0.2$ and graph on a domain of $0 \leq t \leq 5$, with t measured in years. Then vary k a bit, using at least one larger and one smaller value. Sketch all of your graphs together on one set of axes. Explain in words the effect that k has on the graph, and interpret your results in terms of population growth also.

SOLUTION: Use either Procedure 1 or 2 from above, whichever you find easier on your calculator or computer. The sketches are shown in Figure 2.

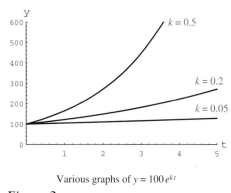

Various graphs of $y = 100\,e^{kt}$

Figure 2

We chose to graph the function using $k = 0.05$, $k = 0.2$, $k = 0.5$, and a y-range of $0 \leq y \leq 600$. There are no "correct" values for k and the y-range to use; these values were chosen with some trial and error, in an attempt to separate the different curves but still keep them all in the picture.

One can see from the sketch that for all values of k, y increases as t increases, and that the rate of growth—that is, the slope or derivative—also increases as t increases. The effect of the parameter k is that the larger the value of k, the more rapidly the function increases. In terms of population, this means that populations with large k values would represent rapidly growing populations, and ones with small k values would represent slowly growing populations. Thus, k could be called the "growth constant" for this model. See Section 1.4 for another explanation of why k can be interpreted as a growth constant. ◄

Parameters and Data

A mathematical model with one or more parameters in it can be used to fit a curve to data. If the model is appropriate for the data, one can simply vary the parameter or parameters until a good fit is attained. What is a "good fit?" In this section we will consider a good fit to be one where the data points fall on or almost on the curve when the data and curve are graphed together. In Section 2.6 we will approach curve fitting a bit more quantitatively; for now we will be satisfied with an intuitive idea of a good fit.

EXAMPLE 2: Positive exponential functions (that is, functions of the form e^{kx}, where k is positive) are often used to model growth as in Example 1. Similarly, negative exponentials (where k is negative) are often used to model decay. One common application is the cooling of a hot object.

A coroner arrives at the scene of a murder; the temperature of the victim is 80°F. The coroner would like to know how long the victim has been dead. She assumes that the temperature of the victim prior to death was 98°F, and that eventually the temperature of the body will reach 70°F, as that is the temperature of the room. She also has some experimental data on the cooling of a human body similar in height and weight to that of the victim (see Figure 3). Develop a mathematical model for the victim's body temperature and use it to estimate the time of death of the victim.

Time	Temperature
0	98.0°F
4	89.5
8	81.3
12	77.6
16	76.5
20	72.1
24	71.3

Temperature of a human body after time of death.

Figure 3

SOLUTION: Let y represent temperature (°F) and let t represent time (hours). We know that a negative exponential function $y = e^{-kx}$ with k positive has a horizontal asymptote at $y = 0$ as $x \to \infty$. We want a function that starts at 98 and approaches 70 as $t \to \infty$. A function of the form

$$y = 70 + 28e^{-kt}$$

has these properties. At time zero, e^{-kt} equals 1 so that y equals 98. As t gets large, e^{-kt} goes to zero and eliminates the second term, so we get a horizontal asymptote at $y = 70$, just as we want. We now must determine the value of k that fits the data.

We need to plot the data and the equation $y = 70 + 28e^{-kt}$ together and adjust k as in Example 1 until we get a curve that approximates the data points.

Graphing Calculators Most graphing calculators have a data-entry capability that allows you to plot points based on a list or table of data you have entered. Once you have plotted the data from Figure 3, draw the graph of the equation with various values of k. See the solution manual's technology pages for details.

Computer Algebra Systems Define the data as a list, or vector, of ordered pairs. Then plot the ordered pairs and the graph of the function in a combined display, varying k until you have a good fit. See the technology pages for details.

After some experimentation with k, we settled on a value of $k = 0.1$ for the best fit. See Figure 4, where we plotted the data with the function for the parameter values $k = 0.05, 0.1, 0.2$. We see that larger values of k correspond to bodies that cool more quickly and smaller values of k correspond to bodies that cool more slowly. Our complete model is therefore $y = 70 + 28e^{-0.1t}$.

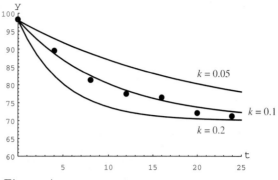

Figure 4

We can now use the model we developed to estimate the time of death. We let $y = 80$ in the equation $y = 70 + 28e^{-0.1t}$, and then solve for t. Of course, we could estimate the answer from the graph, but the algebraic approach is straightforward in this case. We have

$$80 = 70 + 28e^{-0.1t}$$

$$28e^{-0.1t} = 10$$

$$e^{-0.1t} = \frac{10}{28}$$

$$-0.1t = \ln\left(\frac{10}{28}\right)$$

$$t = -10\ln\left(\frac{10}{28}\right) \approx 10.3$$

so we estimate that the murder occurred a little over 10 hours ago. ◄

Trigonometric Families of Curves

The family of curves given by

$$y = A \sin(2\pi w(x - d))$$

is among the most useful and important in all of applied mathematics. It is used to describe all kinds of oscillating or vibrating systems, from airplane wings to atoms. In the exercises you will show that it doesn't matter whether one uses the sin or cos function, because they yield the same family of curves.

A is called the *amplitude*, and it represents the maximum distance from the x-axis to the curve (either above or below). The parameter w is called the *frequency*, and represents the number of full cycles of the sine curve per unit in the x direction. The parameter d is called the *phase shift*, and it represents the number of units the basic sine curve is shifted to the right in order to produce the new curve. The *period*, for which we will use the letter T, is defined as $T = \frac{1}{w}$, and it represents the length of one full cycle of the curve. *Note*: Some authors do not include the factor 2π in the definition of the family of curves; the relationship between period and frequency must then include this factor.

In Figure 5 we illustrate each of these parameters. For the curve shown, it should be clear that $A = 3$, $w = 2$, $d = 0.1$, and $T = 0.5$.

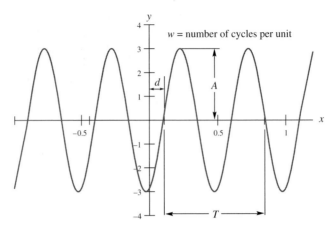

Graph of $y = A \sin(2\pi w(x - d))$ with $A = 3$, $w = 2$, $d = 0.1$, and with period $T = \frac{1}{w} = \frac{1}{2}$.

Figure 5

EXAMPLE 3: The trigonometric functions are good for describing natural phenomena that repeat a pattern over time. One pattern that tends to repeat over time is the sequence of average monthly temperature readings taken from a fixed location. The farther one gets from the equator the more pronounced this effect becomes, with warm temperatures in the summer and cold temperatures in the winter.

In Figure 6, we show the actual average monthly temperature in Hartford, Connecticut (averages taken over a 30-year period), displayed in table form and in graph form. We use x to represent the month, with $x = 1$ corresponding to January, $x = 2$ corresponding to February, and so on. The temperature, y, is in degrees Fahrenheit. Use a model based on the equation $y = A \sin(2\pi w(x - d))$ to fit a curve to the data. *Hint:* You will need to add something to this model. Describe your method for fitting the model, and discuss how good the fit is.

Month	x	Temperature y
January	1	25 °F
February	2	28
March	3	37
April	4	49
May	5	59
June	6	69
July	7	73
August	8	71
September	9	63
October	10	52
November	11	42
December	12	29

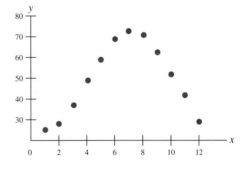

Figure 6

SOLUTION: We see from Figure 5 that the graph of $y = A \sin(2\pi w(x - d))$ is centered around the x-axis ($y = 0$), but from Figure 6 we see that the data are not. The trick is to draw a horizontal line through the "middle" of the data, and then simply read the values of A, d, and T directly from the graph by comparing with Figure 5.

One quick way to estimate the middle temperature is to average the high and the low; we get $\frac{25 + 73}{2} = 49$. The graph of the data with a line at $y = 49$ is shown in Figure 7.

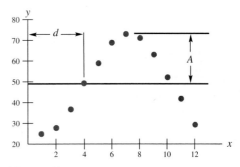

Figure 7

By comparing Figure 7 to Figure 5, we see that A is about 24 (the distance from the "middle" line at $y = 49$ to the maximum temperature of 73 is $73 - 49 = 24$), and d is about 4. Because one cycle of temperature takes 12 months, $T = 12$, so we have $w = \frac{1}{T} = \frac{1}{12}$ (corresponding to $\frac{1}{12}$ cycles per month). Finally, we must raise the graph of $y = A \sin(2\pi w(x - d))$ up 49 units by adding 49 to the right-hand side; thus we come up with the model

$$y = 24 \sin((2\pi/12)\,(x - 4)) + 49$$

The graph of our model along with the data points is shown in Figure 8. The fit is quite good, although we might be able to improve the fit somewhat with minor adjustments of the parameters.   ◄

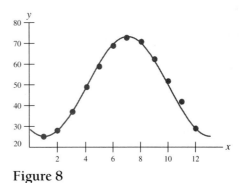

Figure 8

Concepts Review

1. If a letter in an equation represents a constant for a set of functions, the letter is called a _____, and the graphs of the set of functions are a _____ of curves.

2. Exponential functions of the form e^{kx} are often used to model _____ if k is positive. If k is negative, they can be used to model _____.

3. If a set of data points falls on or close to a curve, then the curve is said to be a _____ for the data.

4. The maximum distance from the x-axis to a sine curve is the _____ of the curve, and the length of one full cycle of the curve is the _____.

Problem Set 2.3

1. Investigate each equation below by graphing it for the given values of the parameter. Choose an appropriate y-range for the given x-range. Sketch all of your graphs together on one set of axes and label each graph. Explain in words the effect that the parameter has on the graph.

(a) $y = x^3 - ax$ for $a = -2, -1, 0, 1, 2$. Use $-2 \le x \le 2$.

(b) $y = xe^{-ax}$ for $a = 0.1, 0.2, 0.3$. Use $-5 \le x \le 30$.

(c) $y = x^n e^{-x}$ for $n = 1, 2, 3, 4$. Use $-1 \le x \le 10$.

2. Investigate each equation below by graphing it for the given values of the parameter. Choose an appropriate y-range for the given x-range. Sketch all of your graphs together on one set of axes and label each graph. Explain in words the effect that the parameter has on the graph.

(a) $y = x^2 + ax$ for $a = -2, -1, 0, 1, 2$. Use $-2 \le x \le 2$.

(b) $y = \dfrac{20}{1 + ae^{-0.2x}}$ for $a = 1, 2, 5, 9$. Use $-10 \le x \le 50$.

(c) $y = \dfrac{20}{1 + 9e^{-ax}}$ for $a = 0.1, 0.2, 0.3$. Use $-10 \le x \le 50$.

3. Use the population growth model $y = Ae^{kt}$ of Example 1 to fit a curve to each of the following data sets. Graph data and function together as in Example 2.

(a)

Year	Population
0	100
1	104
2	108
3	113
4	117
5	122

(b)

Year	Population
0	500,000
1	610,701
2	745,912
3	911,059
4	1,112,770
5	1,359,141

4. The population model of Example 1 can also be used to model declining populations. We generally write the model in the form $y = Ae^{-kt}$ in this case. Fit a curve to each of the following data sets representing declining populations. Graph data and function together as in Example 2.

(a)

Year	Population
0	100
1	74
2	55
3	41
4	30
5	22

(b)

Year	Population
0	500,000
1	490,099
2	480,395
3	470,882
4	461,558
5	452,419

5. A model of the form $y = a + be^{-kt}$ can be used to describe the warming or cooling of an object, where t represents time and y represents temperature (as in Example 2). Investigate the effect of each of the parameters a, b, and k on the graph of this equation. Start with the values $a = 70$, $b = 28$, and $k = 0.1$ (the values settled on in Example 2). Then vary each parameter one at a time to determine its effect on the graph. Describe in words the effect that each parameter has, both in terms of the graph and as a model of a warming or cooling object.

6. A laptop computer has been left in a car overnight and is at $40°F$. It is brought into a room at $72°F$. The time versus temperature data is given in the following table, with time given in hours.

Time	Temperature
0	40
1	53
2	61
3	65
4	67
5	70
6	71
7	71

Month	x	Temperature y
January	1	40
February	2	43
March	3	46
April	4	49
May	5	55
June	6	61
July	7	65
August	8	66
September	9	61
October	10	53
November	11	45
December	12	41

Use the approach of Example 2 and the information you gained from Problem 5 to find a mathematical model that fits this data.

7. The average monthly temperature in Seattle, Washington, is given in the following table. Use the method of Example 3 to develop and fit a trigonometric model to this data. Describe your method for fitting the model, and discuss how good the fit is.

Answers to Concepts Review: 1. Parameter; family 2. Growth; decay   3. Good fit   4. Amplitude; period

LAB 3:   THE DAMPED HARMONIC OSCILLATOR

Mathematical Background

Vibrations (oscillations) are everywhere. Some vibrations are desirable and some are not. A microphone's vibrations are used to record the vibrations of a guitar string, which can then be reproduced by the vibration of a speaker. Vibrations of an airplane's engines cause the wings to vibrate; if the vibrations get too large the wing can fall off. Thus, in many cases it is necessary to "damp out" or stop the vibrations. The shock absorbers of a car are examples of dampers; without them the springs of a car would cause the car to vibrate too much when it hits a bump.

The simplest vibrating or oscillating system is called a *harmonic oscillator*. If damping is added (friction is a form of damping) then we have a damped harmonic oscillator. In the real world, all systems have some friction and hence some damping. There are two simple ways to build a damped harmonic oscillator: A mass at the end of a spring, and a mass at the end of a rope or string (a pendulum). By first studying the simplest form of oscillation we learn how to approach more complicated forms of oscillation.

Lab Introduction

In this lab we are going to model the motion of a damped harmonic oscillator. We have chosen to model the motion of a pendulum because it is easy to set up and take measurements on. In this case the damping is caused by the friction in the rope and air resistance. The simplest mathematical model that one finds for damped oscillatory motion (for instance, in a differential equations book) is the equation.

$$y = ae^{-kt} \cos(wt + d)$$

where y is the displacement (in inches) of the mass from its rest position, t is the elapsed time (in seconds), and a, k, w, and d are parameters. We will simplify this equation somewhat by assuming $d = 0$ (this will still give you a reasonable model as long there is not too much damping). The $\cos(wt)$ part gives the oscillatory (back and forth) motion of the pendulum, while the e^{-kt} part gives the damping [see part 2(e) of Lab 2].

Your goal is to find the values of these parameters that best fit the data taken in class (or given in the appendix if you choose not to do the physical experiment) by using computer or calculator graphs. In later labs we will use this mathematical model to answer questions about the position and velocity of the pendulum that would be hard to answer experimentally.

Data

The goal is to set up a pendulum in such a way that you can measure both the distance and the time when the maximum distance from the rest position is reached. We describe one possible experimental setup below, which we use to collect the data given in the appendix. We strongly urge, however, that you actually do this experiment, in groups or as a class, and collect your own data.

Suspend a mass from a heavy rope (about 6 feet long) that is attached to a point above the blackboard. The mass should be free to swing next to the blackboard. In chalk, draw the arc that the mass traces out on the board. With the mass in the rest position, mark the point on the arc that corresponds to a point on the mass (this represents $y = 0$), and then mark the points on the arc that correspond to $y = 1$, $y = 2$, ..., $y = 10$. Pull the mass so that the point is at the 10-inch mark. Let go of the mass and start the stopwatch at the same time. Count the number of times the mass returns; on the second return mark on the board the farthest point reached by the mass and at the same time stop the stopwatch. Record both the time and the distance. Repeat this a second time and average the results (if time permits, repeat more often). Now, repeat the whole process for the fourth return of the mass (again, if time permits, repeat for the sixth return also). Thus, you will collect at least three data points altogether, including the beginning point. You can then fill in a table such as the one below.

	Trial 1		Trial 2		Trial 3	
Return	Time	Distance	Time	Distance	Time	Distance
0	0	10	0	10	0	10
2	....	....	....	....	....	....
4	....	....	....	....	....	....

Finally, time how long it takes the pendulum to stop; a reasonable stopping criterion might be to determine when the largest positive oscillation is less than 0.5 inch. This time value will be used in Lab 4 to check how well the model makes predictions.

Experiment

1. Use a graphing calculator or computer algebra system to graph the function $ae^{-kt}\cos(wt)$. You will have to choose values for a, k, and w. To start with: try $a = 10$, $k = 0.1$, and $w = 1$. Graph the function on the interval $0 \le t \le 10$.

2. Discover the effect that each parameter has on the shape of the graph by varying one parameter at a time and then graphing. First keep k and w constant and try $a = 5, 20$; sketch these graphs together with the one from part 1. Then going back to the part 1 values for a and w try $k = 0.5, 0.2$ and sketch these graphs together with the one from

part 1. Finally, use the part 1 values for a and k and try $w = 0.5, 2$ and sketch these graphs together with the one from part 1. Thus, you should have three sets of sketches, each with three graphs.

3.   Using what you learned from part 2, find parameter values that give a good fit for the data. To do this, *plot the data points and the function* for each choice of parameter values so that you can see when you have a good fit. Make a sketch of your final graph with your final equation written on it *and the data points drawn in.*

Discussion

1.   What effect did varying each parameter have on the shape of your graph? Relate your answer both to the graph and to the pendulum itself. Explain in plain English. You should use the sketches you made in step 2 of the Experiment section to support your answers.

2. Describe the strategy you used for fitting the graph to the data.

3. Were you able to get a good fit for all the data? If not, why not? Do you think that this equation provides a good model for pendulum motion?

2.4 | **Zooming and Local Linearity**

Zooming   One of the most useful capabilities of computer or calculator graphing is the ability to **zoom** in or out on a graph. By zooming *in* on a part of the graph, that part is magnified as if you were getting closer to the graph or using a microscope. By zooming *out* on a graph, you see more of the graph, as if you were moving away from it (though of course you see less detail as a result). Zooming is actually just a quick way to adjust the graph window.

GRAPHING CALCULATORS   Most graphing calculators have at least two ways to zoom. The **box zoom** is useful for zooming in on a specific part of the graph of the function. First you draw a rectangle around the part of the graph you want to magnify. Then the boundaries of the rectangle become the minimum and maximum x- and y-values of the range settings after the zoom.

The other common method of zooming can be used to zoom in or to zoom out; we will call it the **point zoom**. With this method, you put the cursor at the point you want for the center of the screen *after* the zoom, then you zoom by a factor that you choose. For example, suppose you start with the graph window $-8 \leq x \leq 8$, $-4 \leq y \leq 4$—that is, a 16×8 screen centered at $(0, 0)$. If you now put your cursor at $(1, 0)$ and then zoom in by a factor of 4, the new range settings will be approximately $-1 \leq x \leq 3$ and $-1 \leq y \leq 1$: a 4×2 screen centered at $(1, 0)$. If you had zoomed out by a factor of 4, the new range settings would be approximately $-31 \leq x \leq 33$ and $-16 \leq y \leq 16$, or a 64×32 screen centered at $(1, 0)$.

COMPUTER ALGEBRA SYSTEMS   Users of command-based computer algebra systems (such as Mathematica or Maple) can zoom in or out on a graph most easily by changing the x-range of the graph window, and using the default y-range. These systems generally do not have a specific zoom feature. Menu-based computer graphing programs, like Derive, often do have a Zoom option similar to the point zoom already described in the Graphing Calculators section.

One important use of zooming is in solving equations by graphing and zooming.

EXAMPLE 1: Find all solutions to the equation $x^3 - 3x + 1 = 0$ accurate to two decimal places by graphing $y = x^3 - 3x + 1$ and zooming on each point where $y = 0$—that is, each x-intercept. These points are also called *roots* of the function $f(x) = x^3 - 3x + 1$. Start with the graph window $-5 \leq x \leq 5$, $-5 \leq y \leq 5$.

SOLUTION: The graph of the equation $f(x) = x^3 - 3x + 1$ with the indicated graph window is shown in Figure 1. One can see from this graph that there are three solutions to the equation $x^3 - 3x + 1 = 0$, because there are three points where the graph of $f(x)$ crosses the x-axis. (You may recall that the Fundamental Theorem of Algebra guarantees that there are no more than three solutions.) We can estimate from the graph that the largest root occurs near $x = 1.5$.

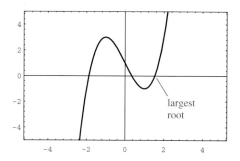

Graph of $f(x) = x^3 - 3x + 1$. Graph window is $-5 \leq x \leq 5$, $-5 \leq y \leq 5$.

Figure 1

Graphing Calculators   We illustrate the use of the box zoom by zooming on the largest root.

1.   Store the function $f(x) = x^3 - 3x + 1$ in the first function memory, and graph the function using the window $-5 \leq x \leq 5$, $-5 \leq y \leq 5$, as in Figure 1.

2.   Use the **box zoom** to draw a box around the largest root. Put the bottom left corner of the box on the curve, below and to the left of the root; put the top right corner of the box on the curve above and to the right of the root.

3.   Give the command to zoom.

4.   Continue zooming until two-decimal-place accuracy is achieved by repeating steps 2 and 3.

The results are shown in Figure 2. After the first zoom the new graph window is $1.170 \leq x \leq 1.702$, $-0.806 \leq y \leq 0.806$. Thus we know after one zoom that the largest root of $f(x)$ lies in the interval $1.170 \leq x \leq 1.702$. After two zooms we know that the root lies in the interval $1.498 \leq x \leq 1.549$, and after three zooms we know that the root lies in the interval $1.528 \leq x \leq 1.534$. The Trace can be used to find an even smaller interval, if that is desired.

Your pictures will vary somewhat, depending on how you draw your rectangles. By drawing a fairly small rectangle, it is easy to get a magnification factor of 10 or more for each zoom. *Hint*: Try to keep two opposite corners of the zoom box *on* the graph itself for best results.

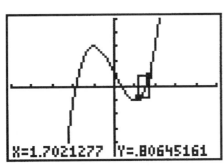

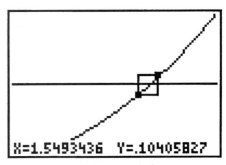

Graph of $f(x) = x^3 - 3x + 1$ before zoom. Graph window is $-5 \le x \le 5$, $-5 \le y \le 5$.

Graph of $f(x) = x^3 - 3x + 1$ after first zoom. Graph window is $1.170 \le x \le 1.702$, $-0.806 \le y \le 0.806$. (Your window may differ, depending on what size box you chose.)

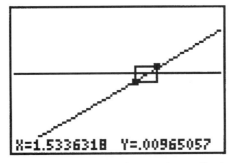

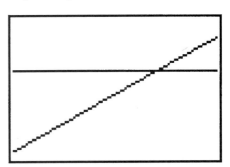

Graph of $f(x) = x^3 - 3x + 1$ after second zoom. Graph window is $1.498 \le x \le 1.549$, $-0.130 \le y \le 0.104$.

Graph of $f(x) = x^3 - 3x + 1$ after third zoom. Graph window is $1.528 \le x \le 1.534$, $-0.017 \le y \le 0.010$.

Figure 2

Computer Algebra Systems   For command-based systems (such as Mathematica or Maple):

1. Define the given function as $f(x)$ and graph the function using the indicated graph window, as in Figure 1.

2. Determine a new, smaller x-range, which contains the largest root. This can be done either by using the scale on the graph itself, or by using the cross in the graph window (if there is one). From Figure 1 we see that the interval $1 \le x \le 2$ should contain the root, so we produce a new graph on that interval using the default y-range.

3. Repeat step 2 until two-decimal-place accuracy is achieved.

The results for a command-based computer algebra system are shown in Figure 3 (except for the original graph window, shown in Figure 1).

For menu-based systems (such as Derive):

The approach is similar to that for graphing calculators. *Derive hint*: Make sure to Center before zooming.

Discussion   We can determine the accuracy of our estimate of the root at each stage of the zoom. When a root is **bracketed** or contained in an interval, and when the endpoints of the interval agree to n digits, then we have n-digit-accuracy (and similarly for decimal places of accuracy).

For example, in Figure 2, after two zooms on the calculator the x-range of the graph window is $1.498 \le x \le 1.549$. The endpoints both round to 1.5, so

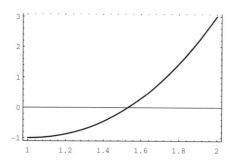

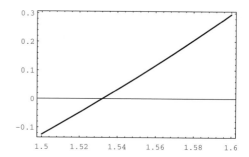

Graph of $f(x) = x^3 - 3x + 1$ after first zoom, using x-range $1 \le x \le 2$, and default y-range.

Graph of $f(x) = x^3 - 3x + 1$ after second zoom, using x-range $1.5 \le x \le 1.6$, and default y-range.

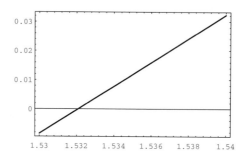

Graph of $f(x) = x^3 - 3x + 1$ after third zoom, using x-range $1.53 \le x \le 1.54$, and default y-range.

Figure 3

we have one-decimal place accuracy. After three zooms the x-range is $1.528 \le x \le 1.534$; thus the root occurs at $x = 1.53$, accurate to two decimal places.

We get the same result from the computer algebra systems. The last graph in Figure 3 shows that the root lies in the interval $1.53 \le x \le 1.54$. Though the endpoints don't quite agree to two decimals, we still get a two-decimal answer of $x = 1.53$ because the function crosses the x-axis between the left endpoint $x = 1.53$ and the midpoint of the interval at $x = 1.535$.

We can now zoom on the other two roots using the same techniques outlined for the largest root. We get $x = -1.88$ and $x = 0.35$.   ◄

Zooming by a factor of 10 (or more) is convenient, because our number system is based on powers of 10. Thus, after each zoom by a power of 10, you get one more decimal place of accuracy in your estimate of a root. The methods outlined in the previous example allow you to zoom each time by at least a factor of 10. *Warning*: If your zoom factor is too large, you are likely to lose sight of the function.

Local Linearity   You may have noticed something interesting in the series of zooms shown in Figures 2 and 3. *As we zoomed in on the root, the graph of the function looked more and more like a straight line.* This is a reflection of something we saw back in Section 1.3. Recall that we defined the derivative at a point $x = c$ as

$$f'(c) = \lim_{h \to 0} \frac{f(c + h) - f(c)}{h}$$

The numerator of $\frac{f(c + h) - f(c)}{h}$ represents the change in y and the denominator the change in x, from a point on the graph of $f(x)$ at $x = c$ to a point at $x = c + h$. If we let $\Delta x = h$ and let $\Delta y = f(c + h) - f(c)$, then the definition becomes

$$f'(c) = \lim_{\Delta x \to 0} \frac{\Delta y}{\Delta x}$$

If this limit exists, it means that when Δx is sufficiently small, the slope $\frac{\Delta y}{\Delta x}$ is about the same regardless of the size of Δx. This implies that the graph of $y = f(x)$ is approximately straight for small Δx, because only straight lines have the property that the slope is the same for different pairs of points.

The straight line that appears when we zoom in on a curve at a point where the derivative exists is a very good approximation to the tangent line that we encountered previously. In other words, *up close, the graph of the curve $y = f(x)$ looks like the graph of the tangent line*. This is essentially what we mean when we say that functions are *locally linear* at points where the derivative exists.

DIFFERENTIALS If we know the derivative $f'(c)$ at a point $x = c$ on the graph of $y = f(x)$, we can use local linearity to estimate either Δy or Δx, if the other is known. When Δx is small, we have the approximate equation

$$f'(c) \approx \frac{\Delta y}{\Delta x}$$

which can be solved for Δy to get

$$\Delta y \approx f'(c)\Delta x$$

When Δx is very small, Δx and $f'(c)\Delta x$ are referred to as *differentials*. Notice that the equations above are only approximate equations for most curves; they are exact only if $f(x)$ is already a linear function (in which case the tangent line is the same as the graph of the function itself).

We can picture these equations by using a graph of $y = f(x)$. (See Figure 4.) The vertical change along the function $f(x)$ is represented by Δy and the change along the tangent line is represented by the differential $f'(c)\Delta x$. Use of the equations above is also called *linear approximation*.

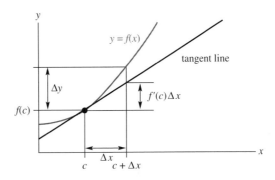

As Δx gets smaller, Δy gets smaller, and Δy and $f'(c)\Delta x$ get closer in value.

Figure 4

EXAMPLE 2:   Let $y = f(x) = x + \sin x$.

1.   Estimate $f'(1)$ to three decimals.

2.   Use differentials to estimate the change in y when x changes from 1 to 1.05 and from 1 to 1.5. Calculate the same quantities directly using the formula for $f(x)$.

3.   When $x = 1$, $y = f(1) = 1 + \sin 1 \approx 1.841$. Use differentials to estimate the change in x when y changes from 1.841 to 2. Can you calculate this change directly?

4.   Discuss your answers from 2 and 3. Use a zoomed-in graph to explain why differentials give good approximations in some cases and not in others.

SOLUTION:

1.   We use a table of values to estimate $f'(1)$ in Figure 5. The result is $f'(1) \approx 1.540$.

2.   The change in x from 1 to 1.05 would be $\Delta x = 1.05 - 1 = 0.05$. Using the equation $\Delta y \approx f'(1)\,\Delta x$ we get

$$\Delta y \approx f'(1)\,\Delta x \approx (1.540)(0.05) = 0.077$$

Similarly, for a change in x from 1 to 1.5, $\Delta x = 0.5$, so

$$\Delta y \approx f'(1)\,\Delta x \approx (1.540)(0.5) = 0.770$$

From the formula for $f(x)$ we have

$$\Delta y = f(1.05) - f(1) = (1.05 + \sin(1.05)) - (1 + \sin(1)) \approx 0.076$$

in the first case and

$$\Delta y = f(1.5) - f(1) = (1.5 + \sin(1.5)) - (1 + \sin(1)) \approx 0.656$$

in the second case. Differentials gives a good approximation in the first case (within 0.001) but not as good in the second case (off by more than 0.1).

3.   The change in y from 1.841 to 2 would be $\Delta y \approx 2 - 1.841 = 0.159$. Rearranging $\Delta y \approx f'(1)\,\Delta x$ to get $\Delta x \approx \frac{\Delta y}{f'(1)}$, we have

$$x = \frac{y}{f(1)} = \frac{0.159}{1.540} \approx 0.103$$

To calculate Δx directly, we need to find x when $y = 2$. Thus, we need to solve the equation

$$x + \sin(x) = 2$$

This equation cannot be solved exactly, because the terms on the left side of the equation cannot be combined. We could, however, solve the equation by graphing and zooming, as in Example 1. We leave this for the exercises at the end of the section.

4.   In Figure 6, we graphed $f(x)$ and the tangent line at $x = 1$. We used a graph window of $1 \le x \le 1.5$, $1.8 \le y \le 2.6$ to show why the differentials approached worked well for some cases and not for others. The point shown on the graph of $y = x + \sin(x)$ at $x = 1.05$ lies almost exactly on the tangent line as well as on the graph of the function. Clearly, linear approximation (differentials) works well in this case. At $x = 1.5$, the function and the tangent line are much farther apart, so approximation with differentials is not as accurate here.

The point shown at $y = 2$ gives us reason to believe that the approximation in part 3 is fairly accurate, even though we could not calculate Δx directly in this case.   ◄

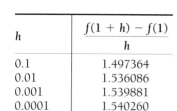

h	$\dfrac{f(1 + h) - f(1)}{h}$
0.1	1.497364
0.01	1.536086
0.001	1.539881
0.0001	1.540260

Estimate of $f'(1)$ for $f(x) = x + \sin x$.

Figure 5

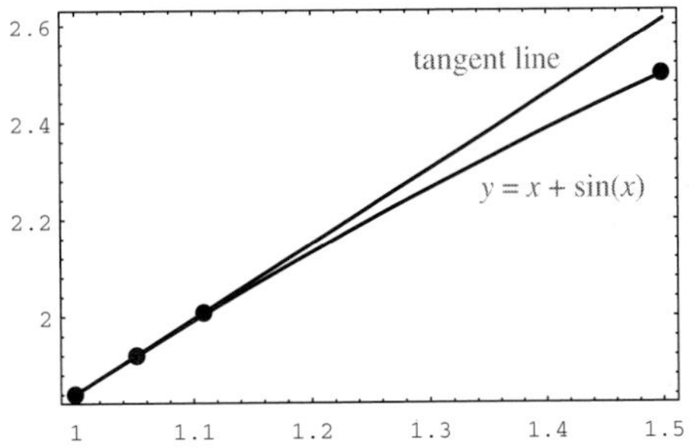

Graph of $y = x + \sin(x)$ and tangent line at $x = 1$ using
graph window $1 \leq x \leq 1.5$, $1.8 \leq y \leq 2.7$. Plotted points
are at $x = 1$, $x = 1.05$, $x = 1.5$, and $y = 2$.

Figure 6

Concepts Review

1. If you want to see a calculator graph more closely, you should _____ on the graph; if you want to see more of the graph, you should _____ .

2. If you establish that a root lies between 3.246 and 3.252, then you know that the root occurs at _____ , accurate to _____ decimal places.

3. When examined up close, many curves taken on the appearance of a straight line. This characteristic of curves is called _____ , and the "straight line" is a close approximation to the _____ to the curve.

Problem Set 2.4

1. Use the method of graphing and zooming (see Example 1) to find all solutions to the given equations on the interval $0 \leq x \leq 5$ accurate to two decimal places. Compare your results to those of Problem 1, Section 2.2.

(a) $2e^{-x} \cos x + 1 = 0$

(b) $2x^2 e^{-x} + 1 = 0$

(c) $1 + 47x + 12x^2 + x^3 = 0$

(d) $\dfrac{3\ln(x + 1)}{x + 1} - 1 = 0$

2. Use the method of graphing and zooming (see Example 1) to find all solutions to the given equations on the interval $-5 \leq x \leq 5$ accurate to two decimal places. Compare your results to those of Problem 2, Section 2.2.

(a) $e^{-x^2} - x = 0$

(b) $x + 3 \cos(0.5x) = 0$

(c) $1279 - 251x - 10x^2 + 10x^3 - x^4 + x^5 = 0$

(d) $x - 2 \sin x = 0$

3. Solve the equation $x + \sin x = 2$ by graphing and zooming. First put the equation in the form $x + \sin x - 2 = 0$. Use your answer to complete part 3 of Example 2 (see the solution to Example 2).

4. Estimate $f'(1)$ accurate to three decimals for each function $f(x)$. Sketch each function on the interval

$0 \leq x \leq 3$ and indicate the derivative you found on the sketch using a slope mark (as in Section 1.3).

(a) $f(x) = x^2$

(b) $f(x) = \sqrt{x}$

(c) $f(x) = e^{2x}$

(d) $f(x) = e^{-2x}$

5. Estimate $g'(2)$ accurate to three decimals for each function $g(x)$. Sketch each function on the interval $0 \leq x \leq 4$ and indicate the derivative you found on the sketch using a slope mark (as in Section 1.3).

(a) $g(x) = \dfrac{e^x}{x + 1}$

(b) $g(x) = \dfrac{x}{x + 1}$

(c) $g(x) = x \ln x$

(d) $g(x) = x \sin x$

6. For each function in Problem 4, use differentials and your results from Problem 4 to estimate the change in y when x changes from 1 to 1.05 and from 1 to 1.5. Calculate the same quantities directly using the formula for $f(x)$. Use your graphs from Problem 4 to explain why the approximation is a good one or a poor one.

7. For each function in Problem 5, use differentials and your results from Problem 5 to estimate the change in y when x changes from 2 to 2.05 and from 2 to 3. Calculate the same quantities directly using the formula for $f(x)$. Use your graphs from Problem 5 to explain why the approximation is a good one or a poor one.

8. For each function in Problem 4, find $f(1)$. Then use differentials to estimate the change in x when y increases from $f(1)$ to $f(1) + 1$. Do you think your estimate is a good one? Explain why or why not.

9. For each function in Problem 5, find $g(2)$. Then use differentials to estimate the change in x when y increases from $g(2)$ to $g(2) + 1$. Do you think your estimate is a good one? Explain why or why not.

Answers to Concepts Review: 1. Zoom in; zoom out
2. 3.25; two   3. Local linearity; tangent line

LAB 4: ROOTS AND SLOPES: WHEN AND HOW FAST?

Mathematical Background

We continue the process of using successive approximations to get closer and closer to an exact answer, which we started in Lab 1. We will approximate two types of quantities in the lab below: solutions to equations (also called roots) and slopes of curves. Both can be approximated by zooming in on a point of the graph.

To estimate the solutions of an equation of the form $f(x) = g(x)$, we proceed as follows. Graph both $y = f(x)$ and $y = g(x)$; the points of intersection of the two curves are the solutions of the equation. By zooming in on one of the points of intersection one finds a sequence of successive approximations to the true solution. Alternatively, it is always possible to rearrange the equation to the form $h(x) = 0$ by subtracting off the right-hand side; then one is looking for points where the function $h(x)$ crosses the x-axis. These points are called *roots* of $h(x)$.

To estimate the slope of a curve at a point, we calculate $\frac{\Delta y}{\Delta x}$ after each zoom to get a sequence of successive approximations to the slope of the tangent (derivative) at the point into which we are zooming.

Lab Introduction

In this lab, you are going to examine the function along with the parameters that you found in Lab 3. Your goal is to estimate:

1. The maximum speed of the pendulum.
2. The time it will take the pendulum to stop.

It will be helpful to make some reasonable modeling assumptions in order to answer these questions. First, we assume that the pendulum is moving fastest as it passes through the rest position $y = 0$ for the first time. This assumption will be tested in a later lab. Second, we will consider that the pendulum is stopped when its largest positive oscillation (positive y-value) is less than one-twentieth of its initial y-value (that is, less than 0.5 inch). This is the same assumption that we made when taking the actual data in Lab 3.

To estimate 1 above, you need to find the *first* positive solution (in terms of t) of the equation

$$ae^{-kt}\cos(wt) = 0 \tag{1}$$

and to estimate the velocity (derivative) at this t-value. Remember that for motion problems, velocity is just the slope of the curve. Thus, as we described above, we can answer these questions by using the computer as we might use a microscope: enlarge a portion of the graph (zoom in) in order to read the answer from the graph itself. The more we magnify, the better the approximation to both the solution to equation (1) and the velocity (derivative) at that point.

To answer question 2, we want the *last* solution (in terms of t) of the equation

$$ae^{-kt}\cos(wt) = 0.5 \qquad (2)$$

Again, we can solve this equation by graphing and zooming.

Experiment

1. Let t_F represent the first time the pendulum crosses the rest position, that is, the value of t that solves equation (1). To find t_F, graph the pendulum function with the parameters you obtained in Lab 3. Start with the viewing window $0 \le t \le 2$ and $-10 \le y \le 10$. Then zoom in on the first point where $y = 0$—that is, the first point where the graph crosses the t-axis. Zoom by approximately a factor of 10 each time. *At each stage of the zoom*, estimate t_F, and pick another t-value contained in the window of the graph. Find the y-coordinates for each of the two points and then estimate the velocity (slope, derivative) using the formula $\frac{\Delta y}{\Delta t}$. (*Note to computer users*: Unless you have a Trace feature on your computer, don't use the mouse and the graph to get the y-coordinate for the given t-coordinate; use the function and the t-coordinate to get the y-coordinate.) Thus, you get a sequence of approximations to both t_F and the velocity at t_F. Zoom until you have four-decimal-place accuracy for the root t_F. Make a sketch of the zoomed-in graph that shows the root to four decimal places.

2. Let t_L represent the last t-value that is a solution to equation (2). Rewrite equation (2) as

$$ae^{-kt}\cos(wt) - 0.5 = 0$$

and then graph the function $f(t) = ae^{-kt}\cos(wt) - 0.5$. The last place where $f(t)$ touches the t-axis is t_L. To get the idea of where t_L lies, you will need to look much farther out on the t-axis than in part 1. *Suggestion*: Keep the width of your interval about the same, but keep increasing the endpoints. For example, you might look at the graph on the interval $20 \le t \le 25$, then $40 \le t \le 45$, and so on, until you find the last place where $f(t)$ touches the t-axis. Then zoom in as in part 1. Again, find an approximation to the root t_L with four-decimal-place accuracy (you need not find the velocity this time). Make a sketch of the zoomed-in graph.

Discussion

1. Discuss the relationship between solutions to equations and graphs. Why is zooming necessary?

2. Solve equation (1) exactly. Equation (2) cannot be solved exactly; discuss why.

3. How can you be sure you have four-decimal-place accuracy for the roots in parts 1 and 2 above?

4. How much accuracy did you get for the velocity in part 1? How do you know?

5. How do you know that you found the *last* positive solution to equation (2)? Include sketches to justify your answer.

6. Summarize your results and relate them to the pendulum itself. How good was the estimate of the time it took the pendulum to stop? What are some possible sources of error?

2.5   Solving Equations

In Section 2.4 we saw that zooming could be used to find approximate solutions (numerical solutions) to equations. Though this method is quite general and can be used to obtain any desired degree of accuracy, it is somewhat tedious to use. Several graphing calculators have built-in methods for solving equations numerically, and those that don't can be programmed to do so. Computer algebra systems can find either numerical or exact solutions to equations (when they exist).

Root Finding   A **root** of a function $f(x)$ is an x-value that gives an output of zero—that is, a solution to the equation $f(x) = 0$. For this reason, it is sometimes called a *zero* of the function. Any equation can be solved by finding the root or roots of an appropriate function. For example, any solution of the equation $x = 10 - x^2$ is also a solution of the equation $x^2 + x - 10 = 0$ and therefore a root of the function $f(x) = x^2 + x - 10$.

Bisection   With the bisection method of finding roots, one starts with an x-interval that contains the root (perhaps using a graph or table). Then one divides this interval into two equal subintervals and determines which of the two contains the root, by looking at the value of the function at the endpoints and at the midpoint of the original interval.

    The key idea is that *if a continuous function changes sign on an interval* (either positive to negative or negative to positive) *then there must be a root on that interval*. This result is called the Intermediate Value Theorem, and it is usually proven in an analysis course.

    By repeating this procedure, one gets a sequence of smaller and smaller intervals, each half the size of the previous one, and each containing the root. This method is easily implemented on a computer or programmable calculator.

EXAMPLE 1: Find all solutions to the equation

$$x^3 = x + 1$$

accurate to one decimal.

SOLUTION: One way of picturing solutions to this equation is to graph both $y = x^3$ and $y = x + 1$ and look for points of intersection. See Figure 1; from this graph it is clear that there is exactly one solution, and that the solution lies between $x = 1$ and $x = 2$. To apply a root-finding method, we need to rearrange the equation to get

$$x^3 - x - 1 = 0$$

and then let $f(x) = x^3 - x - 1$. Then the roots of $f(x)$ are the solutions to the original equation.

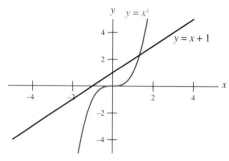

Figure 1

Our first interval containing the root is $1 \leq x \leq 2$, as determined above. The midpoint is at $x = \frac{1+2}{2} = 1.5$. We have $f(1) = -1$, $f(1.5) = 0.875$, and $f(2) = 5$. The numerical values of the function at the endpoints are not important; we need only the signs. We can represent the values of the function on this interval as $- + +$, meaning that the function is negative at the left endpoint, positive at the midpoint and positive at the right endpoint. Therefore the root must lie in the left subinterval $1 \leq x \leq 1.5$, because the function (which is continuous) changes from negative to positive there.

We now summarize the rest of the procedure below. Because we are interested in one-decimal accuracy, we round results below to two decimals. You should check these results on your calculator.

Interval	f(midpoint)	Signs	Root Lies in
$1 \leq x \leq 2$	$f(1.5) = 0.875$	$- + +$	Left half
$1 \leq x \leq 1.5$	$f(1.25) \approx -0.30$	$- - +$	Right half
$1.25 \leq x \leq 1.5$	$f(1.38) \approx 0.25$	$- + +$	Left half
$1.25 \leq x \leq 1.38$	$f(1.32) \approx -0.02$	$- - +$	Right half
$1.32 \leq x \leq 1.38$	$f(1.35) \approx 0.11$	$- + +$	Left half
$1.32 \leq x \leq 1.35$			

We don't need to go any further, because all of the points in the last interval $1.32 \leq x \leq 1.35$ can be rounded to 1.3 to one decimal place. Thus, our one-decimal-place solution is $x = 1.3$.   ◄

It took quite a lot of computation to get just one decimal of accuracy. In fact, each additional decimal of accuracy takes more than three steps. (Three steps reduces the interval by 1/8, and to get another decimal we need to reduce by 1/10.) By zooming and tracing, or by using a table of values, we could have gotten this result much more quickly. The reason for going through this exercise is the fact that many calculators and computer programs use this method for numerically solving equations. Knowing what is going on behind a "solve" command can help, when things don't seem to work out right.

Newton's Method   We don't really have the tools available yet to include a full treatment of Newton's method. Still, it will be helpful to describe the method geometrically, because some computer algebra systems and graphing calculators have this method built into them.

Suppose that a function $f(x)$ has a root at the point $\bar{x}$—that is, $f(\bar{x}) = 0$. Let x_1 be a point near $\bar{x}$. Draw the tangent to the graph of $y = f(x)$ at the point on the graph where $x = x_1$. The point where the tangent line crosses the x-axis (call it x_2) is an approximation to $\bar{x}$, the point where the graph of $y = f(x)$ crosses the x-axis. We now repeat the process, drawing the tangent to the curve at $x = x_2$, and letting $x = x_3$ be the intersection of this second tangent line with the x-axis. In this way, we get a sequence of points $x_1, x_2, x_3, \ldots$, which should get closer and closer to (converge to) the root $\bar{x}$. See Figure 2.

The reason the method works is based on the idea of local linearity as explained in the previous section. Near the root $\bar{x}$, the graph of $y = f(x)$ will be approximately linear. Therefore, if the initial guess x_1 is close enough to the root $\bar{x}$, the tangent to the curve at x_1 should approximate the curve $y = f(x)$; thus, the x-intercept of the tangent (x_2) should approximate the x-intercept of the curve ($\bar{x}$). Repeating the process should result in an even better approximation, because the curve looks more and more like the tangent line as you zoom in.

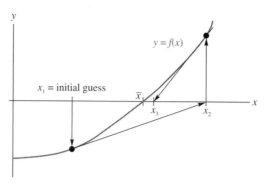

Newton's method. Sequence $x_1, x_2, x_3, \ldots$ approaches root $\bar{x}$.

Figure 2

Recall from Section 1.7 that in order to find the slope of the tangent line at a point, we need to find the derivative of the function at that point. Thus, in order to use Newton's method on a particular function we need to be able to find derivatives easily. We postpone the details and limitations of Newton's method until after Chapter 3.

Built-In Numerical Root Finding and Equation Solving

Bisection and Newton's method are examples of a *numerical* methods of finding roots. The root is found to a desired degree of accuracy (a certain number of decimal places or digits), but the result is usually not an *exact* root. Computer algebra systems and many graphing calculators have built-in numerical methods of finding roots. With any numerical root finder, you must specify either a starting interval (as with the bisection method) or a single initial guess (as with Newton's method). If an interval is required, the interval must contain a root. With the second method, the initial guess should be as close as possible to the root. Normally, a graph or table of values is needed to estimate the starting interval or starting point.

For graphing calculators that do not have numerical root finding or equation solving built into them, the Bisection method can easily be programmed.

EXAMPLE 2: Find all solutions to the equation

$$x^2 = 2^x$$

to the default accuracy of your calculator or computer. Look for any exact solutions as well.

SOLUTION: If we graph $y = x^2$ and $y = 2^x$ separately on one screen, we get the first picture in Figure 3 (we used trial and error to adjust the graph window to get a good view). It appears that the graphs of the functions cross at three points, but the two functions are too close to distinguish one from the other very easily.

We can rewrite the equation that we are solving as

$$x^2 - 2^x = 0$$

so that the roots of $f(x) = x^2 - 2^x$ are solutions to the original equation. In the second picture in Figure 3 we graph $f(x)$; the roots (x-intercepts) are easy to distinguish. From the printed scale on the graph (or using the Trace with a graphing calculator) we can establish that the interval $-2.5 \le x \le 0$ contains the leftmost root, $0 \le x \le 2.5$ contains the middle root, and $2.5 \le x \le 5$ contains the rightmost root. If we are using a numerical root finder that needs an

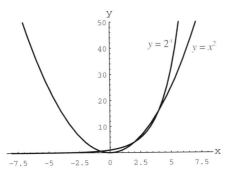

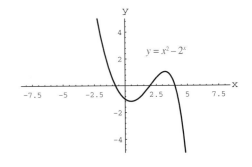

Figure 3

initial guess, $x = -1$, $x = 2$, and $x = 4$ would be reasonable. (If greater accuracy is needed, we could zoom in.)

We now run a numerical root finder using either starting intervals or initial guesses (depending on the calculator or computer algebra system) as described in the solution manual. The results are $x = -0.76666$, $x = 2$, and $x = 4$.

When round numbers, such as 2 or 4, result from a numerical method, one wonders whether or not these are exact results. We can substitute these values into the equation we are solving to see if they satisfy the equation. Because $2^2 = 2^2$ and $2^4 = 4^2$, we see that $x = 2$ and $x = 4$ are indeed solutions to the equation $x^2 = 2x$. Sometimes solutions to an equation can be found by just using some common sense. ◀

Exact Equation Solving   Finding exact solutions to equations is much more difficult in general than finding approximate solutions. You are familiar from your previous work in mathematics with methods for solving linear and quadratic equations (see the box on page 128). For polynomial equations of degree 5 or higher, there is no general method of solution. It is often impossible to solve equations involving trig, log, or exponential functions, especially when the variable appears in more than one location.

Computer algebra systems provide a Solve command, which attempts to find exact solutions to equations. The command has many limitations, however. Even for equations that have solutions, the command may find only one of several, or it may find none at all. Still, the command is worth trying, because an exact solution may provide more (or different) information than a numerical one.

Exact results are often given in nonnumerical form. To get a decimal approximation to an exact result, one usually needs to give an additional command.

EXAMPLE 3: Use the Solve command (exact equation solving) to find solutions to the following equations. Convert the exact results to numerical form. Discuss the results. Were all solutions found? Would numerical solutions provide the same information? What approach would you take if you were to do the algebra by hand?

(a) $2\sin^2 x + 3\sin x = 2$

(b) $x^2 = 2^x$ (the equation from Example 2)

SOLUTION: These are the results:

(a) Exact result is $x = \frac{\pi}{6}$. Numerical approximation is $x \approx 0.5236$.

(b) No solutions found.

Solving Cubic Equations

You are all familiar with the quadratic formula $x = \dfrac{-b \pm \sqrt{b^2 - 4ac}}{2a}$, which provides the two solutions to the equation $ax^2 + bx + c = 0$. What about higher degree
polynomial equations? Is there a formula for solving the general cubic equation
$ax^3 + bx^2 + cx + d = 0$?

The general cubic equation has either 1, 2, or 3 real solutions. There are formulas
for each of the three solutions; one is always a real number, and the other two can be
either real or complex. (Recall that the quadratic formula can also give complex results.) We show the formula for the real solution to $ax^3 + bx^2 + cx + d = 0$ below.

$$x = -\frac{b}{3a} - \frac{\sqrt[3]{2}(-b^2 + 3ac)}{3a\{-2b^3 + 9abc - 27a^2d + [4(-b^2 + 3ac)^3 + (-2b^3 + 9abc - 27a^2d)^2]^{\frac{1}{2}}\}^{\frac{1}{3}}}$$

$$+ \frac{\{-2b^3 + 9abc - 27a^2d + [4(-b^2 + 3ac)^3 + (-2b^3 + 9abc - 27a^2d)^2]^{\frac{1}{2}}\}^{\frac{1}{3}}}{3a\sqrt[3]{2}}$$

At one time in your mathematics career you probably thought the quadratic formula
was hard to remember!

If you have a computer algebra system, you can generate this formula and the
formulas for the other two solutions yourself. Just apply the Solve command to the
equation $ax^3 + bx^2 + cx + d = 0$. The general fourth-degree polynomial equation
$ax^4 + bx^3 + cx^2 + dx + e = 0$ can also be solved exactly, and the solutions are quite
a bit more complicated than the one shown above—try it!

Note: Some computer algebra systems give another solution to equation (a)
whose numerical approximation is $x = -1.571 + 1.317i$, a complex number,
not a real number. If you get solutions involving $i = \sqrt{-1}$, disregard them at
this point.

Discussion

(a) Only one of infinitely many real solutions was found. Because trig
functions are periodic, solutions to equations involving such functions often
have many solutions. As discussed in Section 1.4, arcsin(0.5) is only one of infinitely many solutions to $\sin x = 0.5$. In Figure 4 we show the graph of $f(x) =
2 \sin^2 x + 3 \sin x - 2$; the roots of this function are the solutions to equa-
tion (a). We can see the solution we found at $x \approx 0.5236$, as well as several
others.

An advantage of the exact solution is that the answer involves π, an
important number in mathematics. This is not apparent from a numerical
solution.

The approach to take if you do the algebra by hand is to rewrite the equation as

$$2 \sin^2 x + 3 \sin x - 2 = 0$$

and then factor the left-hand side to get

$$(2 \sin x - 1)(\sin x + 2) = 0$$

It is now clear that one solution comes from setting $2 \sin x - 1 = 0$ and solving
for $\sin x$ to get $\sin x = 0.5$. From the periodicity of the sine function we can get

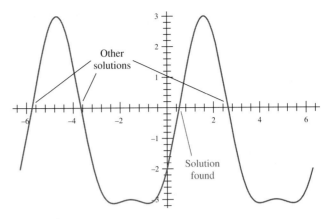

Figure 4

all the solutions pictured in Figure 4. There are no real solutions to $\sin x = -2$ because the output of the sine function is always between -1 and 1. (This is where the complex solutions came from.)

Is there an advantage to going through the algebra by hand? Of course! The more approaches you can take to a problem, the more reliable your solution is. You also get a deeper understanding of where the solution comes from. On the other hand, you will almost surely come across problems where your algebraic skills will not be developed enough to come up with a solution, even when one is possible. In these cases, finding exact solutions by computer can be a great help.

(b) The computer cannot find any exact solutions, even though we know from Example 2 that $x = 2$ and $x = 4$ are exact solutions. In this case, the numerical approach was of more help in finding exact solutions than the exact approach!

There is no algebraic procedure that helps us to solve this equation. We can't factor $x^2 - 2^x$, and we can't combine these terms in any helpful way. We will come across many equations that have no algebraic solution, where a numerical approach is the only possibility.   ◄

The next example requires numerical equation solving, as well as several of the techniques covered earlier, such as numerical derivative finding, zooming, and differentials.

EXAMPLE 4: A small Caribbean country is undergoing an exponential growth in its local population. In addition, tourism contributes a seasonal component to the total number of inhabitants of the country.

Using available data, a government minister has developed a rough mathematical model for the total number of inhabitants of the country (in Section 2.6 we will see how to develop models from data). Letting $P(t)$ represent population (in thousands) at time t (in years), the model is

$$P(t) = 50\, e^{0.03t} + 2 \sin(2\pi t) + 3$$

The first term in this equation represents the growth of the local population (see Example 2 of Section 1.4), and the next two terms represent the seasonal (periodic) nature of the tourist population. Use the model to answer the following questions. Assume that $t = 0$ represents the current time.

Note: We will assume that the population $P(t)$ is a continuous variable, even though populations actually increase or decrease in discrete units. Here it is convenient to approximate a discrete variable with a continuous one.

1.   What is the current total population?
2.   What is the current rate of growth of the total population?
3.   What will the total population be 1 month from now?
4.   What will the total population be 3 years from now?
5.   How long will it take for the total population to reach 54,000?
6.   How long will it take for the total population to reach 60,000?
7.   Of all the predictions made, which do you think is the most reliable? Why?

Use any method that is appropriate. Consider the use of differentials.

SOLUTION: Our first step is to look at a graph. We use the context of the problem to set the graph window. Because we are interested in projecting the population out to 3 years, we are using a t-range of $0 \le t \le 4$. At $t = 0$ we have $P(0) = 50e^{0.03(0)} + 2 \sin(2\pi(0)) + 3 = 53$. We also want to know when the population will reach 60,000, so we have a P-range of $50 \le P \le 62$. In Figure 5 we show the graph of $P(t)$, along with points plotted on the graph at $t = 0$, $t = \frac{1}{12}$, and $t = 3$ (for questions 1, 3, and 4), and with horizontal lines at $P = 54$ and at $P = 60$ (for questions 5 and 6).

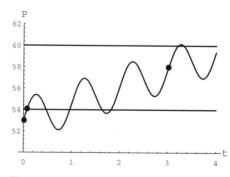

Figure 5

We can now answer the questions posed.

1. We have already found that $P(0) = 53$, so the current population is 53,000.

2. The current rate of growth of population is the value of the derivative at 0:

$$P'(0) = \lim_{h \to 0} \frac{P(0 + h) - P(0)}{h}$$

$$= \lim_{h \to 0} \frac{50e^{0.03h} + 2 \sin(2\pi h) + 3 - 53}{h}$$

In Figure 6 we calculate a table of values to estimate this quantity. The result to three decimals is 14.066 thousand people per year—that is, 14,066 people per year.

h	$\dfrac{50e^{0.03h} + 2\,\sin(2\pi h) + 3 - 53}{h}$
1	1.52272670
0.1	13.25795730
0.01	14.05832893
0.001	14.06631043
0.0001	14.06637204

Figure 6

3. One month is 1/12 year, so we want to find $P(\frac{1}{12})$. Calculating directly, we get

$$P(\tfrac{1}{12}) = 50e^{0.03(\frac{1}{12})} + 2\,\sin(2\pi(\tfrac{1}{12})) + 3$$

$$\approx 54.125$$

or 54,125 people. Population models cannot be expected to be accurate down to the last person. An estimated population of about 54,000 is more realistic.

Differentials can be used to predict the population in one month, because this represents a small change in t. How do we know this is small enough? Look at a graph! In Figure 7 we zoom on the initial portion of the graph from Figure 5 (graph window $0 \le t \le 0.1$, $53 \le P \le 54.4$). The point plotted at $t = \frac{1}{12} \approx 0.083$ shows that a linear approximation should work well, because the graph is nearly a straight line. We have

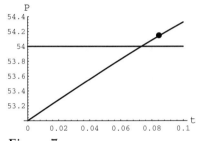

Figure 7

$$\Delta P \approx P'(0)\,\Delta t \approx 14.066\,\frac{\text{thousand}}{\text{year}} \cdot \tfrac{1}{12}\,\text{year} \approx 1.172\,\text{thousand}$$

so that our estimate of population would be $54 + 1.172 = 54.172$ thousand people. This is not much different from the value calculated directly.

4. In 3 years the predicted population would be

$$P(3) = 50e^{0.03(3)} + 2\,\sin(2\pi(3)) + 3$$

$$\approx 57.709$$

or 57,709 people. From the graph in Figure 5 it is clear that a linear approximation using differentials and $P'(0)$ would *not* be a good idea. (Why?)

5. To find when the population reaches 54 thousand we want to solve $P(t) = 54$ or

$$50e^{0.03t} + 2\,\sin(2\pi t) + 3 = 54$$

We can use a numerical root-finder as outlined in Example 2 to solve this, and if we have a computer algebra system, we can try for an exact solution, as in Example 3.

Attempts at an exact solution fail (try it!), and we can see why. The exponential and trig terms cannot be combined to get the t's together.

For a numerical solution, we rewrite the equation as

$$50e^{0.03t} + 2\,\sin(2\pi t) - 51 = 0$$

and use our root-finder on the function $f(t) = 50e^{0.03t} + 2\,\sin(2\pi t) - 51$. From Figure 5 we see that there are several points where the population reaches 54 thousand. We are interested in the first such point. From Figure 7 we can see that a starting interval of $0 \leq t \leq 0.1$ or an initial guess of $t = 0.08$ should work well. In fact, an initial guess of $t = 0$ should be close enough because of the near-linearity of the function in this region. We get

$$t \approx 0.0734 \text{ year}$$

which works out to a little less than a month.

From Figure 7 we see that because of linearity, differentials should also work here. Using $\Delta t \approx \frac{1}{P'(0)} \Delta P$ and using $\Delta P = 54{,}000 - 53{,}000 = 1000$, we get

$$\Delta t \approx \frac{1}{14.066 \frac{\text{thousand}}{\text{year}}} \cdot 1 \text{ thousand} \approx 0.0711 \text{ year}$$

which is quite close to our first estimate, and very easy to find.

6. We want to solve the equation $P(t) = 60$ or

$$50e^{0.03t} + 2 \sin(2\pi t) + 3 = 60$$

From Figure 5 we see that this first occurs between 3 and 4 years. There are, however, two points where the population hits 60 thousand, and we want the first one. We need to zoom on this part of the graph to get a starting interval or initial guess. In Figure 8 we show the function with the graph window $3.1 \leq t \leq 3.4$, $59.6 \leq P \leq 60.2$. We can now estimate a starting interval of $3.15 \leq t \leq 3.25$, or an initial guess of $t = 3.2$. We get

$$t \approx 3.212 \text{ years}$$

Common sense tells us that three decimals of accuracy is too much to expect from a rough population model; this would pinpoint the time to better than the exact day! 3.2 years is more realistic, which gives the time to within about a month.

$P'(0)$ is of no help in estimating this value, as $t = 0$ is not close enough to $t = 3.2$. To use differentials, we would need the derivative at a point closer to the solution we are looking for.

7. Whether you use the model to predict directly or use the current rate of growth (differentials), you will find that short-term predictions are always more reliable than long-term ones. For this example, the predictions in 3 and 5 have the shortest terms and are therefore the most reliable.

With the longer-term predictions, the main problem is with the reliability of the model itself, not with the mathematics. The solution we found in part 6) using a numerical root-finder *is* accurate to the number of decimals printed, *for the equation given*. However, if the equation is not a good model of population over the long term, then the prediction is not likely to be accurate. ◄

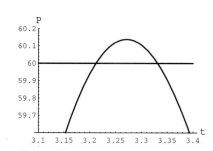

Figure 8

Concepts Review

1. Given any function f, an x-value for which $f(x) = 0$ is called a _____ of the function.

2. The _____ method of approximating roots involves finding an interval that contains the root; dividing the interval in half and finding which half contains the root; and then repeating the process several times.

3. For each case, decide which would probably be more appropriate—a numerical solution or an exact solution.

(a) a problem involving π

(b) a fourth-degree polynomial equation

(c) a population-growth model

Problem Set 2.5

1. Find all solutions to the given equation on the given interval, accurate to one decimal place, using the bisection method as in Example 1.

(a) $x^4 = x + 1$ on $0 \le x \le 2$

(b) $x^4 = x + 1$ on $-2 \le x \le 2$

(c) $\tan x = x$ on $\frac{\pi}{2} \le x \le \frac{3\pi}{2}$

(d) $e^{-x} = x$ on $0 \le x \le 2$

2. Find all solutions to the given equation on the given interval, accurate to one decimal place, using the bisection method as in Example 1.

(a) $2e^{-x} \cos 4x = 1$ on $0 \le x \le 0.5$

(b) $2e^{-x} \cos 4x = 1$ on $0 \le x \le 1$

(c) $x + 2 \sin x = 10$ on $10 \le x \le 15$

(d) $x + 2 \sin x = 10$ on $0 \le x \le 15$

(e) $x + 2 \sin x = 10$ on $-\infty < x < \infty$

3. Use your built-in equation-solver or root-finder to find all solutions to each equation on the given interval to the default accuracy of your calculator or computer. Use a graph to get starting values or starting intervals. Look for any exact solutions as well. See Example 2.

(a) $x^3 = 3^x$ on $-5 \le x \le 5$

(b) $x^4 = 4^x$ on $-5 \le x \le 5$

(c) $x^4 = x + 1$ on $-2 \le x \le 2$. Compare to 1(b).

(d) $e^{-x} = x$ on $0 \le x \le 2$. Compare to 1(d).

4. Use your built-in equation-solver or root-finder to find all solutions to each equation on the given interval to the default accuracy of your calculator or computer. Use a graph to get starting values or starting intervals. Look for any exact solutions as well. See Example 2.

(a) $x^4 = 3^x$ on $-5 \le x \le 5$

(b) $x^4 = 3^x$ on $-10 \le x \le 10$

(c) $2e^{-x} \cos 4x = 1$ on $0 \le x \le 1$. Compare to 2(b).

(d) $x + 2 \sin x = 10$ on $0 \le x \le 15$. Compare to 2(d).

Problems 5 and 6 are for those with access to a computer algebra system.

5. Use the Solve command of a computer algebra system (exact equation solving) to find solutions to the following equations. Convert the exact results to numerical form. Discuss the results. Were all solutions found? Would numerical solutions give the same information? What approach would you take if you were to do the algebra by hand? See Example 3.

(a) $e^{2x} - 3e^x = -2$

(b) $x^3 = 3^x$

6. Use the Solve command of a computer algebra system (exact equation solving) to find solutions to the following equations. Convert the exact results to numerical form. Discuss the results. Were all solutions found? Would numerical solutions give the same information? What approach would you take if you were to do the algebra by hand? See Example 3.

(a) $2 \sin x = x$

(b) $e^{-x} \cos x = 0$

7. Suppose that tourism is dropping in the country described in Example 4. To account for this, the "tourism" factor $2 \sin(2\pi t)$ must decay as time increases; the new model becomes $P(t) = 50e^{0.03t} + 2e^{-t} \sin(2\pi t) + 3$. As in Example 4, $t = 0$ represents the current time.

(a) What is the current total population?

(b) What is the current rate of growth of the total population?

(c) What will the total population be 1 month from now?

(d) What will the total population be 3 years from now?

(e) How long will it take for the total population to reach 54,000?

(f) How long will it take for the total population to reach 60,000?

(g) Of all the predictions made, which do you think is the most reliable? Why?

Use any method that is appropriate. Consider the use of differentials.

Answers to Concepts Review: 1. Root (or zero) 2. Bisection 3. (a) Exact (b) Numerical (c) Numerical

---

2.6 Linear Curve Fitting

Problem The issue of global warming has provoked heated debate for more than a decade. Many scientists believe that the earth's temperature is increasing due to human influence such as pollution from automobile exhaust. Others think that these scientists are misinterpreting past temperature data.

Determine a method of predicting the average yearly temperature in the future in the hypothetical city of Hartland. Determine whether or not temperature is increasing, and if so, how fast.

Data   Figure 1 shows the average yearly temperature data taken from Hartland over a 21-year period.

Problems like the one above are the type that people in the real world are interested in solving, because they have real consequences. Real-world problems are generally not clearly stated in mathematical terms; one needs to develop a mathematical model first.

Note: The problem of determining whether or not global warming is a real phenomenon is much more complicated than is indicated here. For example, temperature may be increasing in one location but decreasing in another. Also, a rise in temperature over a hundred years may not be anomalous in the context of a few thousand years. Still, understanding how linear models are used to make predictions is fundamental to any understanding of global weather forecasting.

EXAMPLE 1: Find a linear equation that is a good model for the temperature data of Figure 1. Find the rate at which temperature is increasing and use the model to predict when average temperature will reach 52 °F.

SOLUTION: We could fit a linear model to the data using trial and error, as we did in Section 2.3. Instead, we will take the first and last data points and find the equation of the line through these points. We let t represent the year and let y represent the average temperature for that year. Then the first and last data points from Figure 1 would be (1970, 46.97) and (1990, 47.75).

From the formula for slope we get

$$m = \frac{\Delta y}{\Delta t} = \frac{(47.75 - 46.97)}{(1990 - 1970)} = 0.039$$

This gives us the linear equation

$$y = 0.039t + b$$

where the y-intercept b is not yet determined. To find b we can substitute the first data point into this equation. We get

$$46.97 = 0.039(1970) + b$$

which we can solve for b to get $b = -29.86$. Our mathematical model of temperature becomes

$$y = 0.039t - 29.86.$$

In Figure 2 we graph this equation together with the original data points from Figure 1. Note that we have chosen the y-scale to range from 46.9 to 47.9, because that is the approximate range in temperatures we are dealing with.

Year	Temperature
1970	46.97 °F
1971	47.07
1972	47.04
1973	47.13
1974	47.17
1975	47.23
1976	47.24
1977	47.31
1978	47.29
1979	47.40
1980	47.38
1981	47.39
1982	47.48
1983	47.52
1984	47.56
1985	47.55
1986	47.66
1987	47.72
1988	47.69
1989	47.80
1990	47.75

Figure 1

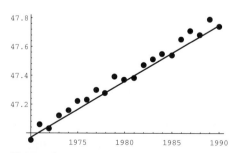

Figure 2

We see that the model comes reasonably close to the data points. One drawback is that most of the data points lie above the equation of the line $y = 0.039t - 29.86$.

Because the slope of the line is 0.039, we conclude that temperature is increasing at a rate of about 0.039 °F per year. The units of slope are units of the dependent variable y over units of the independent variable t.

To predict what year the temperature will reach 52 °F, we use an algebraic approach: we let $y = 52$ and solve for t. We get

$$52 = 0.039t - 29.86$$

$$t \approx 2099$$

rounded to the nearest year.   ◀

The Least Squares Regression Line   Linear models are the most commonly used models for fitting curves to data. This is partly because linear equations are easy to handle mathematically, and partly because many data sets of interest are approximately linear. Because linear models are so common, we need a reliable, standard method of fitting straight lines to approximately linear data sets. Most graphing calculators and computer algebra systems have built into them the *least squares* method of fitting lines to data.

Conceptually, the method is quite simple. For a given set of data points, and a given linear equation, one can measure the vertical distance of each data point from the graph of the linear equation. (See Figure 3.) These vertical distances are called deviations. One can then square these deviations and add them up. Now just pick the straight line that minimizes the sum of the squares of the deviations; this line is called the least squares regression line.

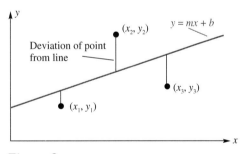

Figure 3

Let's make this a bit more precise. Suppose we have three data points as in Figure 3, labeled (x_1, y_1), (x_2, y_2), (x_3, y_3). The equation of a straight line is $y = mx + b$. The squared deviation of the first point from the line would be $(mx_1 + b - y_1)^2$ and similarly for the other points. The sum of the squared deviations is usually written

$$\sum_{i=1}^{3} (mx_i + b - y_i)^2 = (mx_1 + b - y_1)^2 + (mx_2 + b - y_2)^2 + (mx_3 + b - y_3)^2$$

The symbol $\sum$ (the uppercase sigma of the Greek alphabet) is called a summation symbol and should be read "the sum of." The subscript and superscript on the $\sum$ indicate the first and last values of the parameter in the summation expression; i is called the summation index. Of course, any number of data points can be used; in general, for n data points we would have $\displaystyle\sum_{i=1}^{n}(mx_i + b - y_i)^2$ or just $\displaystyle\sum(mx_i + b - y_i)^2$ for short. (When the summation index i is left off, assume the sum is over all relevant values of i.)

We now just have to choose an m and b that will make $\displaystyle\sum(mx_i + b - y_i)^2$ as small as possible. This could be done using trial and error for each set of data points, but that would be very tedious. Fortunately, we can easily solve the problem in general by using ideas from multivariable calculus. We summarize the result in the next theorem.

Theorem A (The Least Squares Regression Line)

For any set of n points $(x_1, y_1), (x_2, y_2), \ldots, (x_n, y_n)$ the straight line $y = mx + b$ that minimizes the sum of the squares of the deviations of the points from the line, given by $\displaystyle\sum(mx_i + b - y_i)^2$, is determined by the following formulas for m and b.

$$m = \frac{n\sum x_i y_i - \sum x_i \sum y_i}{n\sum x_i^2 - \left(\sum x_i\right)^2}$$

$$b = \frac{\sum x_i^2 \sum y_i - \sum x_i \sum x_i y_i}{n\sum x_i^2 - \left(\sum x_i\right)^2}$$

EXAMPLE 2: Find the least squares regression line for the data points $(1, 1)$, $(2, 3)$ $(3, 4)$. Sketch a graph of the data points and the equation of the line together.

SOLUTION: Calculating all of the quantities needed to find m and b is done most easily in table form. See the table in Figure 4.

The last row in the table is labeled $\sum$ to indicate that it represents the sum of each column. From the table one finds that $\sum x_i = 6$, $\sum y_i = 8$, $\sum x_i^2 = 14$, and $\sum x_i y_i = 19$. For m and b we get

$$m = \frac{(3)(19) - (6)(8)}{(3)(14) - (6)^2} = 1.5$$

$$b = \frac{(14)(8) - (6)(19)}{(3)(14) - (6)^2} = -\frac{1}{3} \approx -0.333$$

	x_i	y_i	x_i^2	$x_i y_i$
	1	1	1	1
	2	3	4	6
	3	4	9	12
Σ	6	8	14	19

Figure 4

Thus, the equation of the least squares line is $y = 1.5x - \frac{1}{3}$. In Figure 5 we graph the data points and the least squares line together.   ◀

The fit is quite good here; we can see that the regression line does a good job of approximating the data points. However, the procedure required to find the regression line is

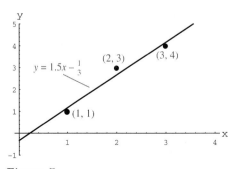

Figure 5

a bit tedious. Clearly, if one had to deal with a large number of data points, this method would take quite a long time to carry out. Fortunately, technology comes to the rescue here. Both graphing calculators and computer algebra systems can be used to find the regression line by simply inputting the data points and giving a single command to find m and b. *Note: In the statistics literature, the regression line is often written as $y = a + bx$ rather than as $y = mx + b$.* The next example shows how this can be done.

EXAMPLE 3: Find the least squares regression line for the temperature data given in Figure 1 using either your calculator or computer. Plot the data and the least squares line together. As in Example 1, find the rate at which temperature is increasing and estimate when the average temperature will reach 52 °F. Compare the models developed in this example and in Example 1.

SOLUTION: The procedures for calculators and computer algebra systems are briefly outlined below.

Graphing Calculators:

1. Enter the data into the two variable statistical data parts of the calculator.
2. Give the linear regression command. This will give you the slope (b) and the y-intercept (a) of the regression equation $y = a + bx$.
3. Graph the data.
4. Graph the regression line.

Computer Algebra Systems

1. Define the data as a list or vector.
2. Give the command to find the least squares regression line.
3. Graph the data and the regression line together.

The regression equation that results from using a calculator or computer is $y = -31.08 + 0.03963x$ to four digits in both the y-intercept and the slope. The graph of the regression line with the data is given in Figure 6.

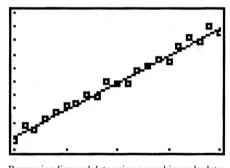

Regression line and data using a graphing calculator.

Figure 6

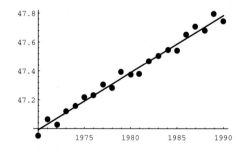

Regression line and data using a computer algebra system.

To find when the temperature will reach 52 °F we set $y = 52$ and solve for x:

$$52 = -31.08 + 0.03963x$$

$$x \approx 2096$$

Thus, we find that, accurate to two digits, the temperature is increasing at a rate of 0.040 °F per year, and that temperature will reach 52 °F in about 2096.

This forecast is based on the assumption that current trends continue into the future. Because we are making a prediction that is quite far from the data points at which we started, this assumption is, at the least, quite suspect. ◄

In comparing this model to the one developed in Example 1 where we fit a line to the first and last data points, we find that the estimated slope of the temperature versus time graph is almost the same to two digits. The estimated time it will take to reach 52 °F differs in the two models by only 3 years. In this problem it is not particularly important which model we choose, because the information we get out of them is so similar. For most problems the least squares regression line gives the "best" linear model; even for this problem we can see that the fit is better for the least squares line by comparing the two graphically. The line in Figure 6 comes closer to more of the data points than the line in Figure 2.

Remarks on Modeling

1. Using a model to make predictions outside the range of the model's data points is not always reliable. Anytime you read about or hear about a prediction that is prefaced by the remark "If current trends continue, . . ." then the prediction is probably being made outside the range of data points. This type of prediction is called extrapolation. Though not totally reliable, *extrapolations* are often the only information we have to go by.

2. We have seen two approaches to fitting a curve (in this case a straight line) to data points. One approach is to take selected data points and use them to determine the parameters, as in Example 1. The other approach is to find the "best fitting" line with the aid of a calculator or computer. Both of these methods are important and can be generalized to work with nonlinear models, as we shall see in the next section. Another approach is simply to use trial and error to vary the parameters (the slope and y-intercept for linear models), and graph the data points and model together until a good visual fit is achieved, as in Section 2.3. Lab 3 contains a good example of this approach. Finding the right model for a given set of data points is often more of an art than a science.

Concepts Review

1. The least squares regression line minimizes the _____ of the deviations of the data points from the line.

2. The symbol Σ is called a _____ and should be read "_____."

3. An extrapolation is a prediction based on given data. The prediction lies where in relation to the data points?

Problem Set 2.6

1. For each data set, find a linear model for the data by choosing two points and finding the equation of the line that passes through the two points as in Example 1. Sketch the data together with the graph of the line.

(a)

x	y
1	10
2	8
3	5
4	5

(b)

x	y
10	0
20	30
30	40
40	80

2. For each data set, find a linear model for the data by choosing two points and finding the equation of the line that passes through the two points, as in Example 1. Sketch the data together with the graph of the line. Discuss any differences between these results and those of Problem 1.

(a)

x	y
−2	1
−1	2
0	3
1	4
2	5

(b)

x	y
−2	5
−1	4
0	3
1	2
2	1

3. Find the least squares regression line for each data set in Problem 1 using Theorem A (see Example 2). Check your results using the built-in capability of your computer or calculator to find regression lines (see Example 3). Sketch the data together with the regression line. Compare these models to the ones from Problem 1.

4. Find the least squares regression line for each data set in Problem 2 using Theorem A (see Example 2). Check your results using the built-in capability of your computer or calculator to find regression lines (see Example 3). Sketch the data together with the regression line. Compare these models to the ones from Problem 2.

5. Gross domestic product (GDP) is the total output of goods and services produced by labor and property located in the United States. It measures the size of the domestic economy. In the table below, GDP is measured in billions of dollars.

Year	GDP
1984	3777.2
1985	4038.7
1986	4268.6
1987	4539.9
1988	4900.4
1989	5250.8
1990	5546.1
1991	5722.9
1992	6038.5
1993	6377.9

U. S. GDP in billions of dollars, from *The American Almanac*, The Reference Press, 1994.

(a) Develop a linear model for GDP as a function of year by fitting a linear equation to the first and last data points as in Example 1. Interpret the slope of the line in the context of the problem.

(b) Develop a linear model for GDP as a function of year by finding the least squares regression line as in Example 3. Interpret the slope of the line in the context of the problem.

(c) Predict GDP for 1994, 1995, and 1996 using the models developed in parts (a) and (b).

(d) Predict the year that the U.S. GDP will reach $10,000 billion ($10 trillion). Use whichever model you think is best. Do you think this prediction will be accurate? Why or why not?

6. The Dow Jones Industrial Index is a common index used to measure the size of the U.S. stock market. It takes into account 30 industrial stocks.

Year	Dow
1984	1178.5
1985	1328.2
1986	1792.8
1987	2276.0
1988	2060.8
1989	2508.9
1990	2678.9
1991	2929.3
1992	3284.3
1993	3754.1

Dow Jones Industrial Index, from *The American Almanac*, The Reference Press, 1994.

(a) Develop a linear model for the Dow Jones Industrial Index as a function of year by fitting a linear equation to the first and last data points as in Example 1. Interpret the slope of the line in the context of the problem.

(b) Develop a linear model for the Dow Jones Industrial Index as a function of year by finding the least squares regression line as in Example 3. Interpret the slope of the line in the context of the problem.

(c) Predict the Dow Jones Industrial Index for 1994, 1995, and 1996 using the models developed in parts (a) and (b).

(d) Predict the year when the Dow Jones Industrial Index will reach 4000. Use whichever model you think is best. If the year when the Dow should hit 4000 by your prediction has already passed, go to the library to find out if the prediction was accurate.

7. Why might businesses be interested in making the types of predictions you were asked to make in Problems 5 and 6? Briefly discuss both strengths and weaknesses of using linear models to make business predictions.

Answers to Concepts Review: 1. Sum of the squares 2. Summation symbol; the sum of 3. Outside the range of the data points

2.7 | Nonlinear Curve Fitting

In Section 2.3 we looked at functions that involve parameters, and we were introduced to the idea of fitting a curve to data by adjusting parameters (trial and error). In Section 2.6 we looked at a general method of fitting linear equations to data that nearly follow a straight line, without resorting to trial and error (the least squares line). Are there methods similar to least squares that work for data that do not seem to follow a straight line? In some cases, the answer is yes, and we take up these procedures in this section.

The Exponential Family   We have already encountered the family of curves given by

$$y = Ae^{kx}$$

in Section 2.3, Example 1 (where we used t in place of x). This family of curves is often used to describe population growth or decay. The quantities A and k are considered parameters, with x the independent variable and y the dependent variable.

When $x = 0$, $y = Ae^{k(0)} = A$, so that A represents the initial y value. For this reason, the symbol y_0 (read "y nought") is often used in place of A. We will consider only the case where A is positive; you can explore the case for A negative on your own.

Warning: y is a variable, whereas y_0 is a constant, the particular value of y at $x = 0$.

The parameter k represents the growth rate. If k is positive, the function increases (growing population), and if k is negative, the function decreases (decaying population). The larger the magnitude of k, the more rapidly the function either grows or dies out. In Figure 1, we show the effects of both A and k on the shape of the curve. In the exercises at the end of the section you are asked to verify these pictures with your own calculator or computer graphs.

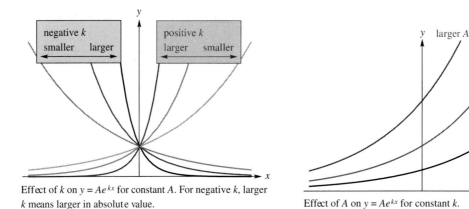

Effect of k on $y = Ae^{kx}$ for constant A. For negative k, larger k means larger in absolute value.

Effect of A on $y = Ae^{kx}$ for constant k.

Figure 1

Curve Fitting with the Exponential Family   If we have data points that seem to follow one of the patterns shown in Figure 1, then we can try fitting an exponential function to the data. By using logarithms, we can employ the regression-line technique from Section 2.6.

Here's how it works. We start with the exponential function equation $y = Ae^{kx}$ and then take the natural log of both sides (actually, any base logarithm would work). Using the properties of logarithms from Section 1.5 we get

$$\ln y = \ln(Ae^{kx})$$

$$\ln y = \ln A + \ln e^{kx}$$

$$\ln y = \ln A + kx$$

This shows that $\ln y$ is a linear function of x.

Suppose now that we have some data points $(x_1, y_1), (x_2, y_2), \ldots, (x_n, y_n)$ and we want to determine if we can fit some member of the exponential family to these data points. We can use the following procedure:

1. Take the natural logarithms of the y-coordinates of the data points—that is, $\ln y_1$, $\ln y_2$, $\ln y_3, \ldots$, or, in general, $\ln y_i$. *Note*: The y-coordinates must be positive, because the logarithm of a negative number or zero is not defined.

2. Plot $\ln y_i$ (y-axis) versus x_i (x-axis).

3. If the plot in 2 looks approximately linear, perform a linear regression as in Section 2.6, using the original x_i's as the x-coordinates and the $\ln y_i$'s as the y-coordinates.

4. The slope of the regression line is k, the growth constant, and the y-intercept is $\ln A$, the natural log of the initial y value. To get A, apply the natural exponential function to the y-intercept of the regression equation (because $e^{\ln A} = A$).

Many graphing calculators will give you the values of the exponential parameters directly, without having to take logs of the y's. Some calculators use an exponential model of the form $y = Ab^x$ rather than $y = Ae^{kx}$. If we rewrite the standard form as $y = A(e^k)^x$, we see that $b = e^k$ and hence $k = \ln b$. Even when the parameters are given directly, it is still a good idea to plot $\ln y_i$ versus x_i in order to determine whether or not an exponential model is appropriate. If this plot is not approximately linear, a different choice of model may be in order. *Note*: The exponential model can only be used if the y-values of the data points are positive, as explained above. If they are not, you will get an error message from your calculator.

Linear and Exponential Models
How does one know when a linear model or an exponential model (or neither) is appropriate for a given set of data? One method, described above, is to use graphs. If a plot of x versus y is approximately linear, then a linear model is a good choice. If a plot of x versus $\ln y$ is approximately linear, then an exponential model is a good choice.

Another approach is to use tables. If a table of function values is made for a linear function, and the change in x (that is, $\Delta x = x_{i+1} - x_i$) is constant, then the change in y ($\Delta y = y_{i+1} - y_i$) will also be constant. This follows from the fact that the slope between any two points on the graph of a linear function is always the same.

For exponential functions, in a table of values with Δx constant, the ratio $\dfrac{y_{i+1}}{y_i}$ will be constant. We can see why by writing the exponential family as $y = A(e^k)^x$. Our assumption is that $x_{i+1} = x_i + \Delta x$ (with Δx constant), so that

$$\frac{y_{i+1}}{y_i} = \frac{A(e^k)^{x_{i+1}}}{A(e^k)^{x_i}} = \frac{(e^k)^{\Delta x + x_i}}{(e^k)^{x_i}} = (e^k)^{\Delta x} = \text{constant}$$

This is easy to see with a simple example. For the exponential equation $y = 2^x$, if we choose $x = 0, 1, 2, 3$ then we get $y = 1, 2, 4, 8$. Each new y-value is twice the previous one, so the ratios of consecutive y-values will always be 2.

To summarize, for a data set $(x_1, y_1), (x_2, y_2), \ldots, (x_n, y_n)$, where $\Delta x = x_{i+1} - x_i$ is constant:

1. If $\Delta y = y_{i+1} - y_i$ is approximately constant, then a linear model is a good choice.

2. If $\dfrac{y_{i+1}}{y_i}$ is approximately constant, then an exponential model is a good choice.

EXAMPLE 1: The exponential model of population growth works best for populations that are undergoing rapid growth, where there are few or no limits to growth (such as limited space or food). Let's look at the United States population from 1790 to 1860, which in many ways fits these conditions.

Figure 2 shows population data (rounded to the nearest thousand), for the United States from 1790 to 1860, in ten-year increments.

1.   Fit a curve to the data in Figure 2. Use either a linear or exponential model, depending on which you think fits the data better. Explain how you decided on which model to use.

2.   Discuss how good your fit is. Use both a graph and a table to compare the population values indicated by the model to the actual population values.

3.   Predict the population in 1870, and compare to the actual population in 1870 (obtained from a library). Discuss the result of this prediction given in the data.

Year	Population
1790	3,929,000
1800	5,308,000
1810	7,240,000
1820	9,639,000
1830	12,861,000
1840	17,063,000
1850	23,192,000
1860	31,443,000

Figure 2

SOLUTION: In Figure 3 we show the original data, along with additional columns for $\ln y_i$, $\Delta y = y_{i+1} - y_i$, and $\dfrac{y_{i+1}}{y_i}$, which will help in determining which model fits the data better. The units of the original data have been changed to make them a bit easier to work with; the time units represent number of years after 1790 ($x = 0$ is 1790), and the population units are millions of people. You can calculate the values for the additional columns using a scientific or graphing calculator. (Spreadsheets also do a good job calculating these kinds of tables.) Because we have population data every 10 years, $\Delta x = x_{i+1} - x_i = 10$.

Year x_i	Population y_i	$\ln y_i$	$\Delta y = y_{i+1} - y_i$	$\dfrac{y_{i+1}}{y_i}$
0	3.929	1.36838494	1.379	1.35097989
10	5.308	1.66921512	1.932	1.3639789
20	7.24	1.97962121	2.399	1.33135359
30	9.639	2.26581737	3.222	1.33426704
40	12.861	2.55419948	4.202	1.3267242
50	17.063	2.83691238	6.129	1.35919827
60	23.192	3.14380739	8.251	1.35576923
70	31.443	3.44817638		

Figure 3

In Figure 4, we show plots of y_i versus x_i and $\ln y_i$ versus x_i.

From the graphs in Figure 4 we conclude that an exponential model is a good candidate. The plot of y_i versus x_i is clearly not linear; it seems to have the typical exponential shape. The plot of $\ln y_i$ versus x_i confirms this: Because this plot does appear linear, an exponential model should work well.

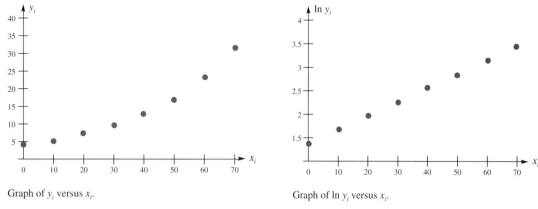

Graph of y_i versus x_i.   Graph of ln y_i versus x_i.

Figure 4

From the table in Figure 3 we get similar information. The numbers in the column representing Δy are clearly not constant. This means that the rate at which the population is growing (the slope) increases, rather remaining constant as would be the case for linear data. The numbers in the column representing $\dfrac{y_{i+1}}{y_i}$ are, however, nearly constant, which is indicative of exponential growth.

We now proceed to find the values of A and k in the exponential model $y = Ae^{kx}$ that best fit the data. This can be done directly from the original data on most graphing calculators; with computer algebra systems, one fits a linear model to the points $(x_i, \ln y_i)$ as explained above.

The results of fitting an exponential model to the data are $A \approx 3.956$ and $k \approx 0.02951$. For graphing calculators that use an exponential model of the form $y = Ab^x$ (TI calculators, for example), the reported values are $A \approx 3.956$ and $b \approx 1.02995$; we then get $k = \ln 1.002995 \approx 0.02951$. For computer algebra systems where one fits a linear model to the data points $(x_i, \ln y_i)$, the results are a slope of 0.02951 and a y-intercept of 1.37528. Thus, we get $k \approx 0.02951$ and $A = e^{1.37528} \approx 3.956$. Our model becomes $y = 3.956e^{0.02951x}$.

In Figure 5 we show a graph of the data, together with the fitted curve. In Figure 6, we show the predicted values using the equation $y = 3.956e^{0.02951\,x}$ along with the actual data in table form.

From Figure 5 we find that the fit of the curve to the data is outstanding. On graphing calculators that display data points as single pixels (such as the

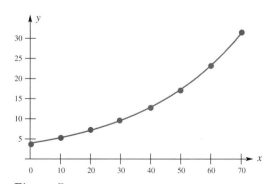

Figure 5

Year x_i	Population y_i	Predicted Population $3.956\,e^{0.02951\,x_i}$
0	3.929	3.956
10	5.308	5.314
20	7.24	7.138
30	9.639	9.588
40	12.861	12.879
50	17.063	17.300
60	23.192	23.239
70	31.443	31.216

Figure 6

Ti-85 and Casio 7700) the fit is so good that the curve coincides with all of the data points, covering them up (which is why it is best to plot the data first). From Figure 6, we see that the predicted values are accurate to the nearest million people, and for most data points, to the nearest hundred thousand (remember that the units are millions of people).

The predicted population for 1870 ($x = 80$) would be

$$y = 3.956e^{0.02951(80)} \approx 41.931 \text{ million.}$$

The actual population from an almanac is 38.558 million. This prediction is quite a bit further off than the predicted values of the other data points. Why? Because this point was not included when fitting the model, it is not surprising that there is some error. Predicting outside the original range of data points is always less reliable than predicting within the range of data points.

There is another explanation for this discrepancy, though: the Civil War! Before 1860, the growth of the United States population was relatively unaffected by outside factors, but between 1860 and 1865 the Civil War disrupted this trend, for a variety of reasons. For one thing, a half-million soldiers died during the war, though this number does not entirely make up the discrepancy. The actual population in 1870 was less than predicted based on the previous years; this is to be expected after a war. ◄

Other Types of Nonlinear Models   There are too many possible nonlinear models to use when trying to fit a curve to data to describe each in detail in a calculus book. We briefly describe a few below, which are common and which are built into many graphing calculators.

POWER MODEL A model of the form

$$y = ax^b$$

can be used for situations where it is suspected that one variable is proportional to the other variable raised to some power. Generally one is most interested in finding the value of the exponent b. Because $\ln y = \ln a + b \ln x$, one can fit a linear model to the data points ($\ln x_i$, $\ln y_i$). The slope is the exponent b, and the constant a is obtained by applying the natural exponential function to the y-intercept (as with the exponential model). If a graph of the points ($\ln x_i$, $\ln y_i$) is linear, then this model will work. *Note:* Both the x and y values of the data points must be positive to use this model. Even though (0, 0) is assumed to be a data point, it should *not* be included in the list of data (the logarithm of a negative number or zero is not defined).

POLYNOMIAL MODELS The simplest polynomial model is the quadratic model, which is of the form

$$y = ax^2 + bx + c$$

This is a three-parameter model (our other models have been two-parameter models). The cubic and quartic models would be of the form

$$y = ax^3 + bx^2 + cx + d \qquad \text{(cubic)}$$

and

$$y = ax^4 + bx^3 + cx^2 + dx + e \qquad \text{(quartic)}$$

There are no restrictions on the data points when using these models (recall that for the exponential model the y_i's had to be positive, and for the power model the x_i's and the y_i's both had to be positive).

EXAMPLE 2: In Figure 7 we show a table representing the average (mean) distance to the sun and the time it takes for one revolution around the sun (called the period of the planet) for each of the nine planets of our solar system. Let x represent the mean distance to the sun in astronomical units (one astronomical unit is the distance from Earth to the sun), and let y represent the period of the planet in Earth years.

Find a model of the form $y = f(x)$ that fits the data as well as possible. Compare the data to the predictions of the model and discuss the results.

Planet	Distance	Period
Mercury	0.387	0.24
Venus	0.732	0.62
Earth	1.000	1.00
Mars	1.524	1.88
Jupiter	5.203	11.86
Saturn	9.555	29.46
Uranus	19.218	84.01
Neptune	30.110	164.79
Pluto	39.810	247.68

Figure 7

SOLUTION: First we plot the data; see Figure 8. When trying to determine which model to try, it is sometimes helpful to think about what happens when one or the other variable is zero. From the plot of the data it is clear that as the planets get closer to the sun, it takes less time to go around the sun. It appears that as x (distance) approaches zero, y (period) also approaches zero. This means that the power model is a good candidate; anytime $(0, 0)$ is a theoretical data point, the power model can be considered.

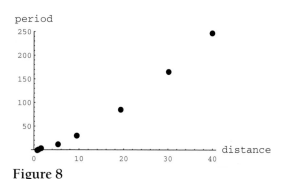

Figure 8

The exponential model is not a possibility, because the assumption with that model is that $y = A \neq 0$ when $x = 0$. This would mean that a planet at zero distance from the sun would take a finite (nonzero) time to go around it, a rather strange idea. A linear model could be tried, but that possibility is included in the power model in this instance; if the value of b in the model $y = ax^b$ turns out to be equal to 1, then we have a linear relationship.

To test whether a power model will work, we plot the data points ($\ln x_i$, $\ln y_i$) and check to see if it looks as if the points lie on a straight line. In Figure 9 we show the table of values and a graph of these data points.

Planet	ln(distance)	ln(period)
Mercury	−0.95	−1.42
Venus	−0.31	−0.49
Earth	0.00	0.00
Mars	0.42	0.63
Jupiter	1.65	2.47
Saturn	2.26	3.38
Uranus	2.96	4.43
Neptune	3.40	5.10
Pluto	3.68	5.51

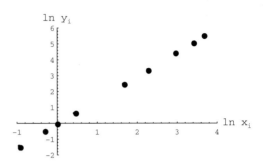

Figure 9

The data points ($\ln x_i$, $\ln y_i$) do indeed seem to fall along a straight line. With a graphing calculator, we can now use the original data from Figure 7, and choose to fit a power model. The results are $b = 1.4993$ and $a = 0.996954$ for a model of the form $y = ax^b$.

With a computer algebra system, we use the data points ($\ln x_i$, $\ln y_i$) from Figure 9 and fit a linear model. The results are a slope of 1.4993 and a y-intercept of −0.00305081. As we saw in the power model section above, this gives a value for a of $e^{-0.00305081} = 0.996954$, consistent with the calculator approach.

Rounding to three digits, our model becomes

$$y = 0.997x^{1.5}$$

This equation is a statement of Kepler's third law of planetary motion. Kepler stated it as "The square of the period of a planet is proportional to the cube of the mean orbital distance." This can be seen by squaring both sides of the equation we just found. The proportionality constant reflects our choice of units. Because the distance to the sun and the period for Earth are both equal to 1 by definition, the proportionality constant should be about 1.

Figure 10 compares the model to the data in graphical and tabular form.

Planet	Distance x	Period y	Predicted Period $y = 0.997\,x^{1.5}$
Mercury	0.387	0.24	0.24
Venus	0.732	0.62	0.62
Earth	1.000	1.00	1.00
Mars	1.524	1.88	1.88
Jupiter	5.203	11.86	11.83
Saturn	9.555	29.46	29.45
Uranus	19.218	84.01	84.00
Neptune	30.110	164.79	164.73
Pluto	39.810	247.68	250.43

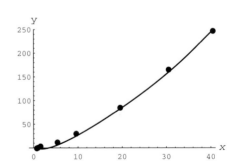

Figure 10

The table and graph both show a good fit. In this case one can see the advantage of using a table. Several of the data points are bunched so close together that it is hard to see how good the fit is for these points from the graph. ◄

Concepts Review

1. If $y = f(x)$ is an exponential function of x, then ln y is a _____ function of x.

2. In an exponential function, the ratio $\frac{y_{i+1}}{y_i}$ is _____ .

3. If a power model is the appropriate model for a set of data, then one variable is proportional to some power of the other variable. Express this as an equation using parameters.

4. How many parameters are necessary to create a quadratic model for curve fitting?

Problem Set 2.7

1. *Proposition:* When making short-term predictions in the growth of a population, a linear model may be sufficient, but for long-term predictions an exponential model is probably needed.

Discuss the validity of the above proposition. In your discussion include the concepts of local linearity and differentials.

2. For each data set below, determine whether a linear or exponential model would be more appropriate, and then fit such a model to the data. Sketch both the data and fitted curve together.

(a)

x	y
0	1
1	2
2	4
3	7
4	17
5	31

(b)

x	y
0	1
1	2
2	4
3	6
4	7
5	10

3. Below is listed the world population (in millions) for the years 1985 to 1992.

Year	Population
1985	4850
1986	4936
1987	5024
1988	5112
1989	5202
1990	5294
1991	5384
1992	5478

World population in millions, from *The American Almanac*, The Reference Press, 1994.

(a) Fit a curve to the world population data. Use either a linear or exponential model, depending on which you think best fits the data. Explain how you decided which model to use.

(b) Discuss how good your fit from (a) is. Use both a graph and a table to compare the population values predicted from the model to the actual population values.

(c) Predict the world population in 1993 and in 2000. Find the actual world population for 1993 from a reference source and compare to your prediction. Discuss your prediction for the year 2000; how accurate do you think it might be?

(d) Predict when the world population will reach 10 billion (1000 million). How accurate do you think this prediction might be?

4. *Guess the equation* Each set of data below was generated directly by a relatively simple equation. Use the methods of this chapter to find the equation used. Sketch the data and a graph of the equation together. Briefly describe how you decided on an equation for each data set.

(a)

x	y
1	6.80
2	5.60
3	4.40
4	3.20
5	2.00

(b)

x	y
1	6.67
2	5.56
3	4.63
4	3.86
5	3.22

(c)

x	y
1	6.50
2	4.60
3	3.75
4	3.25
5	2.91

Answers to Concepts Review: 1. Linear 2. A constant 3. $y = ax^b$ 4. Three

2.8 | Chapter Review

Concepts Test

Respond with true or false to each of the following assertions, and provide an explanation of your response.

1. For graphing calculators and computer algebra systems that have the ability to choose a y-range for your graphs (a default y-range), the machine's choice will be the best one.

2. A graph always provides more information about a function than a table of values does.

3. In order to correctly interpret a computer or calculator graph with vertical asymptotes, you should try to determine the location of those asymptotes algebraically before graphing.

4. If no graph appears when using a computer or calculator, it means that you have made an error in entering the function into the computer or calculator.

5. The choice of graph window has very little effect on the appearance of the graph of a function.

6. The graph of $y = \sin x$ could appear as a straight line on a computer or calculator.

7. There are so many things that can go wrong when using a computer or calculator to graph a function that you are probably better off smashing your machine and just doing everything by hand.

8. With a table of function values one can observe the behavior of a function over many orders of magnitude, but with a graph one can only observe about two orders of magnitude.

9. One can estimate roots of a function using either a graph or a table of values.

10. A parameter is constant for a given graph, but can vary from one graph to another.

11. Equations with parameters rarely occur in real-world situations.

12. The best mathematical model for a given set of data points is the equation whose graph hits the most points.

13. Models of growth phenomena often involve exponential functions and models of repeating phenomena often involve trig functions.

14. The family of curves determined by the equation $y = ax^2$ where x is the independent variable and a is a parameter consists of mostly parabolas.

15. The family of curves determined by the equation $y = ax^2$ where a is the independent variable and x is a parameter consists of mostly parabolas.

16. In theory, any equation with one unknown quantity can be solved numerically be graphing and zoom-

ing (though in practice you are limited by the number of digits stored by your computer or calculator).

17. In theory, any equation with one unknown quantity can be solved exactly using algebra (though in practice your are limited by your own algebraic ability).

18. If $f'(c)$ exists, then $f(c) + f'(c)\Delta x$ should be a good estimate of $f(c + \Delta x)$ if Δx is small enough.

19. Up close (after several zooms) the graph of a function near a point $x = c$ and the graph of the tangent line at $x = c$ are nearly the same.

20. All numerical methods of solving equations require an initial guess or an initial interval, which can be obtained from a graph or table of values.

21. If an exact solution to an equation exists, then a computer algebra system will find it.

22. Numerical methods can be used to solve equations with two unknown quantities for one of the unknowns in terms of the other.

23. A linear model $y = mx + b$ can be fit to any data set using least squares regression.

24. A power model $y = ax^b$ can be fit to any data set by taking logarithms of the x- and y-coordinates of the data points, and then using least squares regression.

25. An exponential model $y = Ae^{kx}$ can be fit to a data set by taking logarithms of the y-coordinates of the data points and then using least squares regression, but only if the y-coordinates are positive.

26. In deciding between a linear or exponential model for a given set of data, it is impossible to tell which model will give the better fit without actually calculating both models.

Sample Test Problems

1. Suppose that we have a screen that is nine pixels by five pixels, and suppose that the coordinates of the pixels range from -2 to 2 in the x-direction and from -8 to 8 in the y-direction. Sketch what the graph of each of the following equations might look like on this screen. State any assumptions you make when you "fill in" between points in the table. Assume there is one entry in the table for each pixel in the x-direction (as is the case for graphing calculators).

(a) $y = \ln x$ (b) $y = \dfrac{1}{x}$

2. Find a graph window for the function

$$y = e^{-(x-20)^2} + \frac{1}{x + 30}$$

that shows the important features of the graph. Briefly describe the features that you find, and how you found them.

3. Build a table of values for each $f(x)$ below, with $x = -2.5, -2.0, \ldots, 2.5$. Use the table to give an interval estimate of the x-value that corresponds to the maximum (largest) y-value on the interval $-2.5 \le x \le 2.5$. Also give interval estimates of any x-values for which $f(x) = 0$.

(a) $f(x) = 1 + x + x^3 - x^4$

(b) $f(x) = \sin(2x) + x + 1$

4. Investigate each equation below by graphing it for the given values of the parameter. Choose an appropriate y-range for the given x-range. Sketch all of your graphs together on one set of axes and label each graph. Explain in words the effect that parameter has on the graph.

(a) $y = x^4 - ax^2$ for $a = -2, -1, 0, 1, 2$. Use $-2 \le x \le 2$.

(b) $y = \dfrac{1}{\sqrt{2\pi a^2}} e^{-\frac{x^2}{2a^2}}$ for $a = 1, 2, 3, 4$. Use $-10 \le x \le 10$.

5. Fit a curve to each data set below. First plot the data, and then use whatever techniques from Sections 2.3, 2.6, and 2.7 seem appropriate.

(a)

x	y
0	30.29
1	43.26
2	47.47
3	49.18
4	50.08

(b)

x	y
0	0.00
1	3.02
2	4.46
3	5.78
4	6.84

(c)

x	y
0	7.47
1	10.95
2	13.91
3	16.89
4	19.93

(d)

x	y
0	-0.29
1	6.75
2	9.57
3	7.43
4	0.34
5	-7.15
6	-9.66
7	-7.13
8	-0.09

6. Use the method of graphing and zooming to find all solutions to the given equation on the interval $-1 \le x \le 4$ accurate to two decimal places.

(a) $x^5 + x + 1 = 0$

(b) $x + 2\sin(2x) + 2 = 0$

7. Do Problem 6 again using the bisection method. Show all work.

8. Do Problem 6 again using either a program or the built-in numerical root finding capability of a calculator or computer. Give the starting value or interval that you used, and explain how you came up these values.

9. For the function $f(x) = x^2 e^{-x}$:

(a) Estimate $f'(1)$ accurate to three decimals.

(b) Use differentials to estimate the change in y when x changes from 1 to 1.05 and from 1 to 2. Calculate the same quantities directly using the formula for $f(x)$. Use a graph to explain why the approximation is a good one or a poor one.

(c) Find $f(1)$ and use differentials to estimate the change in x when y increases from $f(1)$ to $f(1) + 1$. Do you think your estimate is a good one? Explain why or why not.

10. Use the Solve command of a computer algebra system (exact equation solving) to find solutions to the following equations. Convert the exact results to numerical form. Discuss the results. Were all solutions found? Would numerical solutions give the same information? What approach would you take if you were to do the algebra by hand?

(a) $\dfrac{e^x + 2}{e^x} = e^x + e^{-x}$ (b) $\sin x = \cos x + 1$

11. Develop a mathematical model for the population data given below for the years 1980 through 1993. Predict the population for the years 1994, 1995, and 1996.

x	y
1980	1000
1981	1145
1982	1159
1983	1091
1984	1105
1985	1250
1986	1395
1987	1409
1988	1341
1989	1355
1990	1500
1991	1645
1992	1659
1993	1591

Derivatives

3.1 The Derivative as a Function

In Section 1.3 we saw that *slope of the tangent line* and *instantaneous velocity* are manifestations of the same basic idea. Rate of growth of an organism (biology), marginal profit (economics), density of a wire (physics), and dissolution rates (chemistry) are other versions of the same basic concept. Good mathematical sense suggests that we study this concept independently of these specialized vocabularies and diverse applications. That is why we chose the neutral name *derivative*.

Up to this point we have emphasized the concept of the derivative of a function at a point $x = c$. Because we can find the slope, or derivative, of a function at many points, we can also think of the derivative as a function. For a given input x, the output of the derivative function $f'(x)$ is the slope of the function f at the point x. The definition of $f'(x)$ is the same definition as we gave for $f'(c)$ in Section 1.3, except that we replace c with x to generalize the rule.

Definition

The **derivative** of a function f is another function f' (read "eff prime") whose value at any number x is

$$f'(x) = \lim_{h \to 0} \frac{f(x + h) - f(x)}{h}$$

provided that this limit exists.

If this limit does exist, we can say that f is **differentiable** at x. Finding a derivative is called **differentiation**; the part of calculus associated with the derivative is called **differential calculus**.

Finding Derivatives   We illustrate with several examples.

EXAMPLE 1: Let $f(x) = 3x + 60$. Find $f'(x)$. Sketch graphs of both $f(x)$ and $f'(x)$ and explain the relationship between them.

SOLUTION:

$$f'(x) = \lim_{h \to 0} \frac{f(x + h) - f(x)}{h} = \lim_{h \to 0} \frac{[3(x + h) + 60] - [3(x) + 60]}{h}$$

$$= \lim_{h \to 0} \frac{3h}{h} = \lim_{h \to 0} 3 = 3$$

The graphs of $f(x)$ and $f'(x)$ are shown in Figure 1. Because $f(x) = 3x + 60$ is a linear function, its slope is the same at any point x. Because the slope is always 3, the graph of the derivative is a horizontal line at $y = 3$.

Note: If we write $y = f(x)$, then we can refer to the units of y as the *output units* of the function. The derivative function $f'(x)$ will have output units that are different from the ones the original function $f(x)$ has. As a result, a good choice of the y-scale for the graph of $f(x)$ may not be a good choice for $f'(x)$. Look at the different y-scales used in Figure 1.   ◄

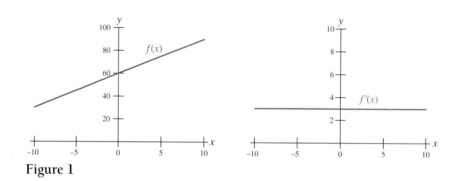

Figure 1

EXAMPLE 2: If $f(x) = 1/x$, find $f'(x)$. Use this formula to find $f'(2)$. Also, numerically estimate $f'(2)$ accurate to three decimal places, using a table as in Section 1.3.

SOLUTION:

$$f'(x) = \lim_{h \to 0} \frac{f(x + h) - f(x)}{h} = \lim_{h \to 0} \frac{\dfrac{1}{x + h} - \dfrac{1}{x}}{h}$$

$$= \lim_{h \to 0} \left[\frac{x - (x + h)}{(x + h)x} \cdot \frac{1}{h} \right] = \lim_{h \to 0} \left[\frac{-h}{(x + h)x} \cdot \frac{1}{h} \right]$$

$$= \lim_{h \to 0} \frac{-1}{(x + h)x} = \frac{-1}{x^2}$$

Thus, f' is the function given by $f(x) = -1/x^2$. Its domain is all real numbers except $x = 0$. Thus, we find that

$$f'(2) = -\frac{1}{2^2} = -\frac{1}{4}$$

To estimate $f'(2)$ numerically, we need to go back to the definition of derivative with $x = 2$. Because

$$f'(2) = \lim_{h \to 0} \frac{f(2 + h) - f(2)}{h} = \lim_{h \to 0} \frac{\frac{1}{2 + h} - \frac{1}{2}}{h}$$

we form a table of values for $\frac{\frac{1}{2+h} - \frac{1}{2}}{h}$ using h's that approach 0. See Figure 2. The last two entries in the table are both -0.250 when rounded to three decimals. Thus, our numerical estimate agrees with our exact calculation.

h	$\dfrac{\frac{1}{2+h} - \frac{1}{2}}{h}$
1	-0.1666667
0.1	-0.2380952
0.01	-0.2487562
0.001	-0.2498751
0.0001	-0.2499875

Figure 2   ◀

EXAMPLE 3: If $G(x) = ax^2 + bx + c$, find $G'(x)$.

SOLUTION:

$$G'(x) = \lim_{h \to 0} \frac{G(x + h) - G(x)}{h}$$

$$= \lim_{h \to 0} \frac{(a(x + h)^2 + b(x + h) + c) - (ax^2 + bx + c)}{h}$$

$$= \lim_{h \to 0} \left[\frac{ax^2 + 2axh + ah^2 + bx + bh + c - ax^2 - bx - c}{h} \right]$$

$$= \lim_{h \to 0} \left[\frac{2axh + ah^2 + bh}{h} \right] = \lim_{h \to 0} \left[\frac{(2ax + ah + b)\cancel{h}}{\cancel{h}} \right]$$

$$= 2ax + a(0) + b = 2ax + b \qquad ◀$$

Notice that in the previous example, we have found a derivative formula for an entire family of curves with the three parameters a, b, and c. The goal of the next section is to find very general derivative formulas that can be applied to large groups of functions, rather than just single functions. In the next example we use the derivative formula developed in Example 3.

EXAMPLE 4: Let $h(x) = 3x^2 - 5x + 3$.

(a) Find $h'(x)$ and use it to find $h'(2)$, $h'(-1)$, and $h'\left(\frac{5}{6}\right)$.

(b) Sketch graphs of both $h(x)$ and $h'(x)$, and on the graphs indicate the three values you found in part (a).

SOLUTION: We could go back to the definition of derivative as we did for the previous examples, or we could use the derivative formula from Example 3, which is much quicker and easier. With $a = 3$, $b = -5$, and $c = 2$, we have

$$h'(x) = 2(3)x - 5 = 6x - 5$$

Therefore, $h'(2) = 6(2) - 5 = 7$, $h'(-1) = 6(-1) - 5 = -11$, and $h'\left(\frac{5}{6}\right) = 6\left(\frac{5}{6}\right) - 5 = 0$.

On the graph of $h(x)$, these values represent the *slopes* at the points $x = 2$, $x = -1$, and $x = \frac{5}{6}$. On the graph of $h'(x)$, these values are the *y-coordinates* at $x = 2$, $x = -1$, and $x = \frac{5}{6}$. See Figure 3.   ◄

Form is All that Matters

We use $f'(x)$ in place of $f'(c)$ in this chapter, and $G(x)$ in place of $f(x)$ in Example 3. The symbols we use are matters of taste only. "Mathematicians do not study objects, but relations between objects; thus, they are free to replace some object by others so long as the relations remain unchanged. Content to them is irrelevant: they are interested in form only."

Henri Poincaré

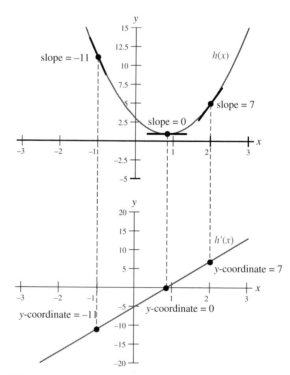

Figure 3

EXAMPLE 5: Each of the following is a derivative, but of what function and what point?

(a) $\lim\limits_{h \to 0} \dfrac{(4 + h)^2 - 16}{h}$ (b) $\lim\limits_{h \to 0} \dfrac{\dfrac{2}{3 + h} - \dfrac{2}{3}}{h}$

SOLUTION: (a) This is the derivative of $f(x) = x^2$ at $x = 4$.

(b) This is the derivative of $f(x) = 2/x$ at $x = 3$.   ◄

Derivatives by Computer Algebra

Those who have access to a computer algebra system can find derivative formulas directly. Of course, this method does not help you understand where the formulas come from, but it can be very useful in checking results and in finding derivative formulas for cases where your algebra skills fail you.

There are two ways to find derivatives with a computer algebra system. One is to use the Derivative command directly. The other is to set up a difference quotient based on the definition of derivative, and then use the Limit command. All commonly used computer algebra systems have both of these commands built into them.

EXAMPLE 6: If $f(x) = x^3$, $g(x) = x^4$, and $h(x) = x^5$, find $f'(x)$, $g'(x)$, and $h'(x)$ using both the Limit command and the Derivative command of a computer algebra system. Make a conjecture of what the derivative of x^{10} would be.

SOLUTION: To use the Limit command we must first set up the correct difference quotient. For the first function $f(x) = x^3$ we have

$$f'(x) = \lim_{h \to 0} \frac{f(x + h) - f(x)}{h}$$

$$= \lim_{h \to 0} \frac{(x + h)^3 - x^3}{h}$$

When we use the Limit command to let h approach zero in the expression $((x + h)^3 - x^3)/h$, the result is $3x^2$.

To use the Derivative command, we need only to specify the function whose derivative we want, and which letter represents the variable in the function (as opposed to any possible parameters). When we apply the Derivative command to the expression x^3, the result is $3x^2$, which checks with our first approach.

The results for all three functions are

$$\begin{array}{ll} f(x) = x^3 & f'(x) = 3x^2 \\ g(x) = x^4 & g'(x) = 4x^3 \\ h(x) = x^5 & h'(x) = 5x^4 \end{array}$$

Based on these results it appears that the derivative of x^{10} should be $10x^9$ (why?). In the next section we will find that this is indeed the case, and we will examine a proof of this result. Finding these derivatives directly from the definition is certainly possible, but it's quite tedious without computer algebra.   ◄

Using Computer Algebra to Check Results

It may seem that it would be easy to check whether the result from using a computer to find a derivative matches the result of doing the problem by hand. This is not always the case. Different computer algebra systems have different conventions for how they display a result.

We suggest the following method for checking results.

(1) Type in your by-hand result and check it to make sure you have entered it correctly (the computer will "pretty print" what you type in to make it easy to check).

(2) Generate the computer solution.

(3)   Subtract (1) from (2) and Simplify (all computer algebra systems have a Simplify command). If your by-hand result and the computer solution are equal, the difference should, of course, equal zero.

Differentiability Implies Continuity

If a curve has a tangent line at a point, then that curve cannot take a jump or wiggle too badly at that point. The precise formulation of this fact is an important theorem.

Theorem A

If $f'(c)$ exists, then f is continuous at c.

Proof   We need to show that $\lim_{x \to c} f(x) = f(c)$. We begin by writing $f(x)$ in a fancy way. (You should simplify this expression to prove to yourself that the left- and right-hand sides are equal.)

$$f(x) = f(c) + \frac{f(x) - f(c)}{x - c} \cdot (x - c), \quad x \neq c$$

Therefore,

$$\lim_{x \to c} f(x) = \lim_{x \to c} \left[f(c) + \frac{f(x) - f(c)}{x - c} \cdot (x - c) \right]$$

Because $\lim_{x \to c} f(c) = f(c)$, $\lim_{x \to c} \frac{f(x) - f(c)}{x - c} = f'(c)$, and $\lim_{x \to c}(x - c) = 0$, when we let x approach c in each expression we get

$$\lim_{x \to c} f(x) = f(c)$$

which proves the theorem.   ◄

The converse of this theorem is false. If a function f is continuous at c, it does not follow that f has a derivative at c. This is easily seen by considering $f(x) = |x|$ at the origin (Figure 4). This function is certainly continuous at zero. However, it does not have a derivative there, as we now show. Note that

$$\frac{f(0 + h) - f(0)}{h} = \frac{|0 + h| - |0|}{h} = \frac{|h|}{h}$$

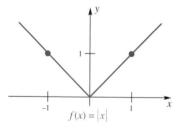

$f(x) = |x|$

Figure 4

Thus,

$$\lim_{h \to 0^+} \frac{f(0 + h) - f(0)}{h} = \lim_{h \to 0^+} \frac{|h|}{h} = \lim_{h \to 0^+} \frac{h}{h} = 1$$

whereas

$$\lim_{h \to 0^-} \frac{f(0 + h) - f(0)}{h} = \lim_{h \to 0^-} \frac{|h|}{h} = \lim_{h \to 0^-} \frac{-h}{h} = -1$$

Because the right- and left-hand limits are different,

$$f'(0) = \lim_{h \to 0} \frac{f(0 + h) - f(0)}{h}$$

does not exist.

A similar argument shows that at any point where the graph of a function has a sharp corner, it is continuous but not differentiable. The graph in Figure 5 indicates a number of ways for a function to be non-differentiable at a point.

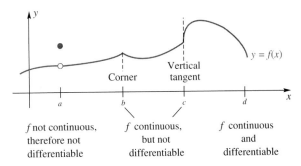

Figure 5

We claim in Figure 5 that the derivative does not exist at the point c where the tangent line is vertical. This is because

$$\frac{f(c + h) - f(c)}{h}$$

increases without bound as $h \to 0$. It corresponds to the fact that the slope of a vertical line is undefined.

Surprise!!!

The converse to Theorem A is spectacularly false. It came as a great surprise to mathematicians when they discovered functions that were continuous everywhere but differentiable nowhere. The first three steps in the construction of such a function are shown below. Continue the process ad infinitum. In the limit, you will obtain a continuous nondifferentiable function. If this interests you, ask your teacher for more information or look at any book with the title *Real Analysis*.

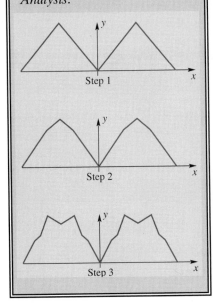

Concepts Review

1. The derivative of f at x is given by $f'(x) = \lim_{h \to 0}$ _____.

2. The derivative of $f(x) = 10 - 2x$ would be given by $f'(x) =$ _____ because the graph of $f(x) = 10 - 2x$ is a _____.

3. If f is differentiable at c, then f is _____ at c. The converse is false, as is shown by the example $f(x) =$ _____.

4. $\lim_{h \to 0} \dfrac{2(c + h)^2 - 2c^2}{h}$ is the derivative of $f(x) =$ _____ at $x =$ _____.

Problem Set 3.1

Note: In any of the problems below where you need to find a derivative, you should *check* your results with the Derivative command of a computer algebra system if one is available (see Example 6).

In Problems 1–4, use the definition

$$f'(x) = \lim_{h \to 0} \frac{f(x + h) - f(x)}{h}$$

to find the indicated derivative two ways. First find an exact result, then find a three-decimal-place numerical approximation using a table to estimate the limit as in Section 1.3. (See Examples 1 and 2.)

1. $f'(3)$ if $f(x) = x^2 - x$

2. $f'(-2)$ if $f(x) = x^3$

3. $f'(-1)$ if $f(x) = x^3 + 2x^2$

4. $f'(4)$ if $f(x) = \dfrac{3}{x + 1}$

In Problems 5–22, use $f'(x) = \lim_{h \to 0}[f(x + h) - f(x)]/h$ or any other derivative formula *that you have proven* (including the result from Example 3) to find the derivative at x. Sketch both $f(x)$ and $f'(x)$ (when possible) and indicate on each the values of $f'(-1)$ and $f'(2)$. (See Examples 3 and 4.)

5. $f(x) = 5x - 4$

6. $f(x) = ax + b$

7. $f(x) = 8x^2 - 1$

8. $f(x) = x^2 + 3x + 4$

9. $f(x) = ax^3 + bx^2 + cx + d$

10. $f(x) = 2x^3$

11. $f(x) = x^3 - 2x$

12. $f(x) = x^4$

13. $g(x) = \dfrac{3}{5x}$

14. $g(x) = \dfrac{2}{x + 6}$

15. $F(x) = \dfrac{6}{x^2 + 1}$

16. $F(x) = \dfrac{x - 1}{x + 1}$

17. $G(x) = \dfrac{2x - 1}{x - 4}$

18. $G(x) = \dfrac{2x}{x^2 - x}$

19. $g(x) = \sqrt{3x}$

20. $g(x) = \dfrac{1}{\sqrt{3x}}$

21. $H(x) = \dfrac{3}{\sqrt{x - 2}}$

22. $H(x) = \sqrt{x^2 + 4}$

In Problems 23–26, find $f'(1)$ any way you can (but using only the definition of derivative or derivative formulas you have proven). Try for an exact result, but failing that, find a numerical approximation (both is even better).

23. $f(x) = e^x$ *Hint*: Look at Section 1.4.

24. $f(x) = \sin x$

25. $f(x) = \ln x$ *Hint*: Look at Section 1.5.

26. $f(x) = x + \cos x$

In Problems 27–30, the given limit is the derivative, but of what function and at what point? (See Example 5.)

27. $\lim\limits_{h \to 0} \dfrac{2(5 + h)^3 - 2(5)^3}{h}$

28. $\lim\limits_{h \to 0} \dfrac{(3 + h)^2 + 2(3 + h) - 15}{h}$

29. $\lim\limits_{h \to 0} \dfrac{\cos(x + h) - \cos(x)}{h}$

30. $\lim\limits_{h \to 0} \dfrac{\tan(t + h) - \tan(t)}{h}$

For Problems 31–34, use a computer algebra system to find the derivative of the function, using both the Limit command and the Derivative command. (See Example 6).

31. $f(x) = 5x^7 - 3x^2 + x + 1$

32. $f(x) = e^{2x}$

33. $f(x) = e^{x^2}$

34. $f(x) = \sin x$

35. From Figure 6, estimate $f'(0), f'(2), f'(5)$, and $f'(7)$.

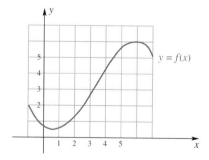

Figure 6

36. From Figure 7, estimate $g'(-1)$, $g'(1)$, $g'(4)$, and $g'(6)$.

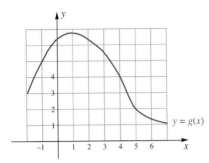

Figure 7

37. Sketch the graph of $y = f'(x)$ on $-1 < x < 7$ for the function f of Problem 35.

38. Sketch the graph of $y = g'(x)$ on $-1 < x < 7$ for the function g of Problem 36.

39. Consider the function $y = f(x)$, whose graph is sketched in Figure 8.

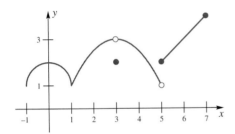

Figure 8

(a) Estimate $f(2)$, $f'(2)$, $f(0.5)$, $f'(0.5)$.

(b) Estimate the average rate of change in f on the interval $0.5 \leq x \leq 2.5$.

(c) Where on the interval $-1 < x < 7$ does the $\lim_{u \to x} f(u)$ fail to exist?

(d) Where on the interval $-1 < x < 7$ does f fail to be continuous?

(e) Where on the interval $-1 < x < 7$ does f fail to have a derivative?

(f) Where on the interval $-1 < x < 7$ is $f'(x) = 0$?

(g) Where on the interval $-1 < x < 7$ is $f'(x) = 1$?

40. Let $f(x) = \begin{cases} mx + b & \text{if } x < 2 \\ x^2 & \text{if } x \geq 2 \end{cases}$

Determine m and b so that f is differentiable everywhere. *Hint*: Match up the derivatives of the two parts of the function at $x = 2$.

41. Draw the graphs of $f(x) = x^3 - 4x^2 + 3$ and its derivative $f'(x)$ on the interval $[-2, 5]$ using the same axes.

(a) Where on this interval is $f'(x) < 0$?

(b) Where on this interval is $f(x)$ decreasing as x increases?

(c) Make a conjecture. Experiment with other intervals and other functions to confirm this conjecture.

Answers to Concepts Review: 1. $[f(x + h) - f(x)]/h$
2. -2; straight line with slope -2 3. Continuous; $|x|$
4. $2x^2$; c

LAB 5: THE DERIVATIVE AS A FUNCTION

Mathematical Background

The definition of the derivative of a function can be written

$$f'(x) = \lim_{\Delta x \to 0} \frac{f(x + \Delta x) - f(x)}{\Delta x}$$

where $f(x + \Delta x) - f(x)$ is called the change in y and is sometimes written Δy. For simplicity, we often replace the symbol Δx with h; thus the definition becomes

$$f'(x) = \lim_{h \to 0} \frac{f(x + h) - f(x)}{h}$$

The derivative of a function can also be written $\dfrac{d}{dx} f(x)$; the symbol $\dfrac{d}{dx}$ should be read "the derivative of."

Lab Introduction

To look at these functions graphically, we are going to graph the difference quotient,

$$\frac{f(x + h) - f(x)}{h}$$

for smaller and smaller values of h. This will approximate the graph of the derivative of the function f. Again, we want to stress that this is an approximation and not a proof.

Experiment

(1) For the function, $f(x) = \sin x$:

    (a) Graph the function using the window $-2\pi \leq x \leq 2\pi, -2 \leq y \leq 2$.

    (b) Graph the difference quotient $\dfrac{f(x + h) - f(x)}{h}$ on the same graph. Start with $h = 1$ and decrease h by a factor of $1/10$ until you get the same graph twice.

    (c) Sketch the graph of the function together with your final difference quotient graph (that is, the graph of its derivative).

    (d) Use your knowledge of graphs of simple functions (you may want to refer to Lab 2) to make a conjecture about the formula for the derivative of the function. Test your conjecture by graphing your conjectured function and the final difference quotient on the same set of axes.

(2) Repeat (1) for the functions $\ln(x), x^2$, and $\cos(x)$. You may need to change the graph window in order to see all that is happening. If you are using a graphing calculator, remember to set it to radian mode for the trigonometric functions.

(3) (a) Repeat (1b) for 2^x and 3^x.

    (b) Find a so that $\dfrac{d}{dx}(a^x) = a^x$. To do this, vary a until the graphs of a^x and $\dfrac{a^{x+h} - a^x}{h}$ coincide for a small value of h.

Discussion

1. How small did you have to make h in order to get the picture to settle down (that is, repeat) for each function in (1) and (2) above? Was it the same for all the functions?

2. How did you determine a formula for the derivative of the given functions?

3. Look back at Lab 1 and compare the a value found in part (3) above with the a value found in part (6) of Lab 1. Show why they are the same by manipulating the definition of the derivative of a^x:

$$\frac{d}{dx}a^x = \lim_{h \to 0} \frac{a^{x+h} - a^x}{h}$$

Hint: The condition for finding a in this lab is that the above limit should equal a^x; set up this equation and factor a^x out of both sides.

3.2 Rules for Finding Derivatives

The process of finding the derivative of a function directly from the definition of the derivative—that is, by setting up the difference quotient

$$\frac{f(x + h) - f(x)}{h}$$

and evaluating its limit—can be time-consuming and tedious. We are going to develop tools that will allow us to shorten this lengthy process—that will, in fact, allow us to find derivatives of the most complicated-looking functions almost instantly.

Recall that the derivative of a function f is another function f'. For example, if $f(x) = x^2$ is the formula for f, then $f'(x) = 2x$ is the formula for f'. Taking the derivative of f (differentiating f) means operating on f to produce f'. We often use the symbol $\frac{d}{dx}$ to indicate this operation (Figure 1), and so $\frac{d}{dx}$ is called an *operator*. Thus, we write $\frac{d}{dx} f = f'$, $\frac{d}{dx} f(x) = f'(x)$, or (in the example just mentioned) $\frac{d}{dx}(x^2) = 2x$. All the theorems below are stated both in functional notation and in the operator $\frac{d}{dx}$ notation.

Other Letters The x in the operator $\frac{d}{dx}$ specifies that x is considered to be the variable in the function f. If the variable were t, then one would use the operator $\frac{d}{dt}$, as in, for example, $\frac{d}{dt}(t^2) = 2t$.

The Constant and Power Rules The graph of the constant function $f(x) = k$ is a horizontal line (Figure 2) which, therefore, has slope zero everywhere. This is one way to understand our first theorem.

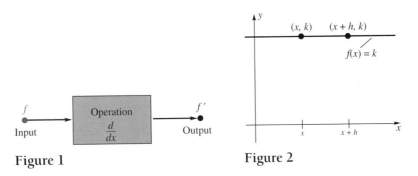

Figure 1

Figure 2

Theorem A

(Constant Function Rule). If $f(x) = k$, where k is a constant, then for any x, $f'(x) = 0$—that is,

$$\frac{d}{dx} k = 0$$

Proof

$$f'(x) = \lim_{h \to 0} \frac{f(x + h) - f(x)}{h} = \lim_{h \to 0} \frac{k - k}{h} = \lim_{h \to 0} 0 = 0 \qquad \blacktriangleleft$$

The graph of $f(x) = x$ is a line through the origin with slope 1 (Figure 3); so we should expect the derivative of this function to be 1 for all x.

Theorem B

(Identity Function Rule). If $f(x) = x$, then $f'(x) = 1$—that is,

$$\frac{d}{dx}(x) = 1$$

Proof

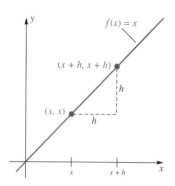

$$f'(x) = \lim_{h \to 0} \frac{f(x+h) - f(x)}{h} = \lim_{h \to 0} \frac{x+h-x}{h} = \lim_{h \to 0} \frac{h}{h} = 1 \qquad \blacktriangleleft$$

Before stating our next theorem, we recall something from algebra: how to raise a binomial to a power.

$$(a+b)^2 = a^2 + 2ab + b^2$$

$$(a+b)^3 = a^3 + 3a^2b + 3ab^2 + b^3$$

$$(a+b)^4 = a^4 + 4a^3b + 6a^2b^2 + 4ab^3 + b^4$$

$$\vdots$$

$$(a+b)^n = a^n + na^{n-1}b + \frac{n(n-1)}{2}a^{n-2}b^2 + \cdots + nab^{n-1} + b^n$$

Figure 3

Theorem C

(Power Rule). If $f(x) = x^n$, where n is a positive integer, then $f'(x) = nx^{n-1}$—that is,

$$\frac{d}{dx}(x^n) = nx^{n-1}$$

Proof

$$f'(x) = \lim_{h \to 0} \frac{f(x+h) - f(x)}{h} = \lim_{h \to 0} \frac{(x+h)^n - x^n}{h}$$

$$= \lim_{h \to 0} \frac{x^n + nx^{n-1}h + \dfrac{n(n-1)}{2}x^{n-2}h^2 + \cdots + nxh^{n-1} + h^n - x^n}{h}$$

$$= \lim_{h \to 0} \frac{h\left[nx^{n-1} + \dfrac{n(n-1)}{2}x^{n-2}h + \cdots + nxh^{n-2} + h^{n-1} \right]}{h}$$

Within the brackets, all terms except the first have h as a factor, and so each of these terms has limit zero as h approaches zero. Thus,

$$f'(x) = nx^{n-1} \qquad \blacktriangleleft$$

As illustrations of Theorem C, note that

$$\frac{d}{dx}(x^3) = 3x^2, \frac{d}{dx}(x^9) = 9x^8, \frac{d}{dx}(x^{100}) = 100x^{99}$$

Proofs of Derivative Rules In the proofs that follow, and in the proofs that you are asked to do in the problem set, we will be using some properties of limits that have not formally been stated yet. We state the following theorem, which contains most of the limit results that we will need.

Theorem D

(The Main Limit Theorem).
Let n be a positive integer, k be a constant, and f and g be functions that have limits at c. Then

1. $\lim_{x \to c} k = k$;

2. $\lim_{x \to c} x = c$;

3. $\lim_{x \to c} k f(x) = k \lim_{x \to c} f(x)$;

4. $\lim_{x \to c} [f(x) + g(x)] = \lim_{x \to c} f(x) + \lim_{x \to c} g(x)$;

5. $\lim_{x \to c} [f(x) - g(x)] = \lim_{x \to c} f(x) - \lim_{x \to c} g(x)$;

6. $\lim_{x \to c} [f(x) \cdot g(x)] = \lim_{x \to c} f(x) \cdot \lim_{x \to c} g(x)$;

7. $\lim_{x \to c} \dfrac{f(x)}{g(x)} = \dfrac{\lim_{x \to c} f(x)}{\lim_{x \to c} g(x)}$, provided $\lim_{x \to c} g(x) \neq 0$;

8. $\lim_{x \to c} [f(x)]^n = \left[\lim_{x \to c} f(x) \right]^n$;

9. $\lim_{x \to c} \sqrt[n]{f(x)} = \sqrt[n]{\lim_{x \to c} f(x)}$, provided $\lim_{x \to c} f(x) > 0$

   when n is even.

10. $\lim_{x \to c} g(f(x)) = g\left[\lim_{x \to c} f(x) \right]$ provided that g is continuous on its domain and that $\lim_{x \to c} f(x)$
    is in the domain of g.

The properties stated in the Main Limit Theorem generally follow from common sense and have been used that way in previous sections. For example, property 6 says that if $f(x)$ gets close to L and $g(x)$ gets close to M as x gets close to c, then $f(x)g(x)$ should get close to LM as x gets close to c. Proofs of these properties can be found in an analysis book.

$\dfrac{d}{dx}$ **Is a Linear Operator** The operator $\frac{d}{dx}$ behaves very well when applied to constant multiples of functions or to sums of functions. This fact will be demonstrated by the following theorems.

Theorem E

(Sum Rule). If f and g are differentiable, then

$$\frac{d}{dx} [f(x) + g(x)] = \frac{d}{dx} f(x) + \frac{d}{dx} g(x)$$

In words, this says that *the derivative of a sum is the sum of the derivatives.*

Proof

$$\frac{d}{dx}[f(x) + g(x)] = \lim_{h \to 0} \frac{[f(x + h) + g(x + h)] - [f(x) + g(x)]}{h}$$

$$= \lim_{h \to 0} \left[\frac{f(x + h) - f(x)}{h} + \frac{g(x + h) - g(x)}{h} \right]$$

$$= \lim_{h \to 0} \frac{f(x + h) - f(x)}{h} + \lim_{h \to 0} \frac{g(x + h) - g(x)}{h}$$

$$= \frac{d}{dx}f(x) + \frac{d}{dx}g(x)$$

The next-to-the-last step is justified by the Main Limit Theorem (Part 4).   ◄

An example that illustrates this is

$$\frac{d}{dx}(x^3 + 1) = 3x^2 + 0 = 3x^2$$

Theorem F

(Constant Multiple Rule). If k is a constant and f is a differentiable function, then

$$\frac{d}{dx}[k \cdot f(x)] = k \cdot \frac{d}{dx}f(x)$$

In words, this says, that *a constant multiplier k can be passed across the operator* $\frac{d}{dx}$.

Proof   Let $F(x) = k \cdot f(x)$ and apply the definition of derivative. In the problem set you are asked to complete the proof.   ◄

Examples that illustrate this are

$$\frac{d}{dx}(-7x^3) = -7\frac{d}{dx}(x^3) = -7 \cdot 3x^2 = -21x^2$$

and

$$\frac{d}{dx}\left(\frac{4}{3}x^9\right) = \frac{4}{3}\frac{d}{dx}(x^9) = \frac{4}{3} \cdot 9x^8 = 12x^8$$

Linear Operator   The fundamental meaning of the word *linear* as used in mathematics is that given in this section. An operator L is linear if it satisfies the two key conditions:

$$L(ku) = kL(u)$$

$$L(u + v) = L(u) + L(v)$$

Linear operators play the central role in the *linear algebra* course that many readers of this book will eventually take. Unfortunately, functions of the form $f(x) = mx + b$ are called linear functions (because of their connection with lines) even though they are not linear in the operator sense. To see this, note that

$$f(kx) = mkx + b$$

whereas

$$kf(x) = k(mx + b)$$

Thus, $f(kx) \neq kf(x)$ unless b happens to be zero or k happens to be 1.

Any operator L with the properties stated in Theorems E and F is called *linear*; that is, L is a **linear operator** if:

1. $L(kf) = kL(f), k$ a constant;
2. $L(f + g) = L(f) + L(g)$.

Linear operators will appear again in this book; $\frac{d}{dx}$ is a particularly important example. A linear operator always satisfies the difference rule $L(f - g) = L(f) - L(g)$ stated next for $\frac{d}{dx}$.

Theorem G

(Difference Rule). If f and g are differentiable functions, then

$$\frac{d}{dx}[f(x) - g(x)] = \frac{d}{dx}f(x) - \frac{d}{dx}g(x)$$

Proof   See the problem set at the end of this section.   ◄

EXAMPLE 1: Find the derivatives of $5x^2 + 7x - 6$ and $4x^6 - 3x^5 - 10x^2 + 5x + 16$.

   SOLUTION:

$$\frac{d}{dx}(5x^2 + 7x - 6) = \frac{d}{dx}(5x^2 + 7x) - \frac{d}{dx}(6) \qquad \text{(Theorem G)}$$

$$= \frac{d}{dx}(5x^2) + \frac{d}{dx}(7x) - \frac{d}{dx}(6) \qquad \text{(Theorem F)}$$

$$= 5\frac{d}{dx}(x^2) + 7\frac{d}{dx}(x) - \frac{d}{dx}(6) \qquad \text{(Theorem E)}$$

$$= 5 \cdot 2x + 7 \cdot 1 + 0 \qquad \text{(Theorems C, B, A)}$$

$$= 10x + 7$$

To find the next derivative, we note that the theorems on sums and differences extend to any finite number of terms. Thus,

$$\frac{d}{dx}(4x^6 - 3x^5 - 10x^2 + 5x + 16)$$

$$= \frac{d}{dx}(4x^6) - \frac{d}{dx}(3x^5) - \frac{d}{dx}(10x^2) + \frac{d}{dx}(5x) + \frac{d}{dx}(16)$$

$$= 4\frac{d}{dx}(x^6) - 3\frac{d}{dx}(x^5) - 10\frac{d}{dx}(x^2) + 5\frac{d}{dx}(x) + \frac{d}{dx}(16)$$

$$= 4(6x^5) - 3(5x^4) - 10(2x) + 5(1) + 0$$

$$= 24x^5 - 15x^4 - 20x + 5 \qquad ◄$$

The method of Example 1 allows us to find the derivative of any polynomial. If you know the Power Rule and do what comes naturally, you are almost sure to get the right result. If you can write the answer without any intermediate steps, that is fine.

Product and Quotient Rules   Now we are in for a surprise. The derivative of a product of functions is *not* equal to the product of the derivatives of the functions.

Theorem H

(Product Rule). If f and g are differentiable functions, then

$$\frac{d}{dx}[f(x)g(x)] = f(x)\frac{d}{dx}g(x) + g(x)\frac{d}{dx}f(x)$$

The Product Rule should be memorized in words as follows: *The derivative of a product of two functions is the first times the derivative of the second plus the second times the derivative of the first.*

Proof   Let $F(x) = f(x)g(x)$. Then

$$F'(x) = \lim_{h \to 0} \frac{F(x + h) - F(x)}{h}$$

$$= \lim_{h \to 0} \frac{f(x + h)g(x + h) - f(x)g(x)}{h}$$

$$= \lim_{h \to 0} \frac{f(x + h)g(x + h) - f(x + h)g(x) + f(x + h)g(x) - f(x)g(x)}{h}$$

$$= \lim_{h \to 0} \left[f(x + h) \cdot \frac{g(x + h) - g(x)}{h} + g(x) \cdot \frac{f(x + h) - f(x)}{h} \right]$$

$$= \lim_{h \to 0} f(x + h) \cdot \lim_{h \to 0} \frac{g(x + h) - g(x)}{h} + g(x) \cdot \lim_{h \to 0} \frac{f(x + h) - f(x)}{h}$$

$$= f(x)g'(x) + g(x)f'(x)$$

The derivation just given relied first on the trick of adding and subtracting the same thing, namely, $f(x + h)g(x)$. Second, at the very end, we used the fact that

$$\lim_{h \to 0} f(x + h) = f(x)$$

This is just an application of the theorem that says that differentiability at a point implies continuity there.   ◄

EXAMPLE 2: Find the derivative of $(3x^2 - 5)(2x^4 - x)$ by the Product Rule. Check your answer by doing the problem a different way.

SOLUTION:

$$\frac{d}{dx}[(3x^2 - 5)(2x^4 - x)] = (3x^2 - 5)\frac{d}{dx}(2x^4 - x) + (2x^4 - x)\frac{d}{dx}(3x^2 - 5)$$

$$= (3x^2 - 5)(8x^3 - 1) + (2x^4 - x)(6x)$$

$$= 24x^5 - 3x^2 - 40x^3 + 5 + 12x^5 - 6x^2$$

$$= 36x^5 - 40x^3 - 9x^2 + 5$$

To check, we first multiply and then take the derivative.

$$(3x^2 - 5)(2x^4 - x) = 6x^6 - 10x^4 - 3x^3 + 5x$$

Thus,

$$\frac{d}{dx}[(3x^2 - 5)(2x^4 - x)] = \frac{d}{dx}(6x^6) - \frac{d}{dx}(10x^4) - \frac{d}{dx}(3x^3) + \frac{d}{dx}(5x)$$

$$= 36x^5 - 40x^3 - 9x^2 + 5 \qquad \blacktriangleleft$$

Theorem I

(Quotient Rule). Let f and g be differentiable functions with $g(x) \neq 0$. Then

$$\frac{d}{dx}\left(\frac{f(x)}{g(x)}\right) = \frac{g(x)\dfrac{d}{dx}f(x) - f(x)\dfrac{d}{dx}g(x)}{g^2(x)}$$

We strongly urge you to memorize this in words, as follows: *The derivative of a quotient is equal to the denominator times the derivative of the numerator minus the numerator times the derivative of the denominator, all divided by the square of the denominator.*

Proof   Let $F(x) = f(x)/g(x)$. Then

$$F'(x) = \lim_{h \to 0} \frac{F(x + h) - F(x)}{h}$$

$$= \lim_{h \to 0} \frac{\dfrac{f(x + h)}{g(x + h)} - \dfrac{f(x)}{g(x)}}{h}$$

$$= \lim_{h \to 0} \frac{g(x)f(x + h) - f(x)g(x + h)}{h} \cdot \frac{1}{g(x)g(x + h)}$$

$$= \lim_{h \to 0} \left[\frac{g(x)f(x + h) - g(x)f(x) + f(x)g(x) - f(x)g(x + h)}{h} \right.$$

$$\left. \cdot \frac{1}{g(x)g(x + h)} \right]$$

$$= \lim_{h \to 0} \left\{ \left[g(x)\frac{f(x + h) - f(x)}{h} - f(x)\frac{g(x + h) - g(x)}{h} \right] \right.$$

$$\left. \cdot \frac{1}{g(x)g(x + h)} \right\}$$

$$= [g(x)f'(x) - f(x)g'(x)]\frac{1}{g(x)g(x)} \qquad \blacktriangleleft$$

EXAMPLE 3: Find the derivative of $\dfrac{(3x - 5)}{(x^2 + 7)}$.

SOLUTION:

$$\frac{d}{dx}\left[\frac{3x - 5}{x^2 + 7}\right] = \frac{(x^2 + 7)\dfrac{d}{dx}(3x - 5) - (3x - 5)\dfrac{d}{dx}(x^2 + 7)}{(x^2 + 7)^2}$$

$$= \frac{(x^2 + 7)(3) - (3x - 5)(2x)}{(x^2 + 7)^2}$$

$$= \frac{-3x^2 + 10x + 21}{(x^2 + 7)^2} \qquad \blacktriangleleft$$

EXAMPLE 4: Find $\dfrac{d}{dx}y$ if $y = \dfrac{2}{x^4 + 1} + \dfrac{3}{x}$.

SOLUTION:

$$\frac{d}{dx}y = \frac{d}{dx}\left(\frac{2}{x^4 + 1}\right) + \frac{d}{dx}\left(\frac{3}{x}\right)$$

$$= \frac{(x^4 + 1)\dfrac{d}{dx}(2) - 2\dfrac{d}{dx}(x^4 + 1)}{(x^4 + 1)^2} + \frac{x\dfrac{d}{dx}(3) - 3\dfrac{d}{dx}(x)}{x^2}$$

$$= \frac{(x^4 + 1)(0) - (2)(4x^3)}{(x^4 + 1)^2} + \frac{(x)(0) - (3)(1)}{x^2}$$

$$= \frac{-8x^3}{(x^4 + 1)^2} - \frac{3}{x^2} \qquad \blacktriangleleft$$

EXAMPLE 5: Show that the Power Rule holds for negative integral exponents; that is,

$$\boxed{\frac{d}{dx}(x^{-n}) = -nx^{-n-1}}$$

SOLUTION:

$$\frac{d}{dx}(x^{-n}) = \frac{d}{dx}\left(\frac{1}{x^n}\right) = \frac{x^n \cdot 0 - 1 \cdot nx^{n-1}}{x^{2n}} = \frac{-nx^{n-1}}{x^{2n}} = -nx^{-n-1} \qquad \blacktriangleleft$$

We saw as part of Example 4 that $\frac{d}{dx}(3/x) = -3/x^2$. Now we have another way to see the same thing.

$$\frac{d}{dx}\left(\frac{3}{x}\right) = \frac{d}{dx}(3x^{-1}) = 3\frac{d}{dx}(x^{-1}) = 3(-1)x^{-2} = -\frac{3}{x^2}$$

The General Power Rule

We now formally state that the power rule, which we have proved for positive and negative integers, is in fact true for all real numbers.

$$\boxed{\text{If } r \text{ is a real number, then } \frac{d}{dx}x^r = r\,x^{r-1}.}$$

For the case where r is a rational number, we can prove this rule using implicit differentiation, which will be introduced in Section 3.6. In the problems at the end of that section you will be asked to prove the rule for the rational case. We will prove the case for irrational r in Section 7.3.

EXAMPLE 6: If $f(x) = 3\sqrt{x} + \dfrac{5}{\sqrt{x}} - x^\pi$, find $f'(x)$.

SOLUTION: First we must convert $f(x)$ to a form where we can use the General Power Rule. We have

$$f(x) = 3x^{\frac{1}{2}} + 5x^{-\frac{1}{2}} - x^\pi$$

so that

$$f'(x) = 3 \cdot \frac{1}{2}x^{-\frac{1}{2}} + 5 \cdot \left(-\frac{1}{2}\right)x^{-\frac{3}{2}} - \pi x^{\pi - 1}$$

$$= \frac{3}{2}x^{-\frac{1}{2}} - \frac{5}{2}x^{-\frac{3}{2}} - \pi x^{\pi - 1} \qquad \blacktriangleleft$$

Concepts Review

1. The derivative of a product of two functions is the first times _____ plus the _____ times the derivative of the first. In symbols, this would be $\frac{d}{dx}[f(x) \cdot g(x)] = $ _____.

2. The derivative of a quotient is the _____ times the derivative of the numerator minus the numerator times the derivative of the _____, all divided by the _____. In symbols, $\frac{d}{dx}[f(x)/g(x)] = $ _____.

3. The second term (the term involving h) in the expansion of $(x + h)^n$ is _____. It is a fact that leads to the formula $\frac{d}{dx}[x^n] = $ _____.

4. L is called a linear operator if $L(kf) = $ _____ and $L(f + g) = $ _____. The derivative operator denoted by _____ is such an operator.

Problem Set 3.2

In Problems 1–44, find $\frac{d}{dx}y$ using the rules of this section. Check your results with the Derivative command of a computer algebra system if one is available.

1. $y = 2x^3$

2. $y = 3x^4$

3. $y = \pi x^2$

4. $y = \sqrt{2}x^5$

5. $y = -3x^{-3}$

6. $y = 4x^{-2}$

7. $y = \dfrac{-2}{x^4}$

8. $y = \dfrac{2}{3x^6}$

9. $y = -8\sqrt{x}$

10. $y = \dfrac{3}{5\sqrt{x}}$

11. $y = -x^3 + 2x$

12. $y = 2x^4 - 3x$

13. $y = -x^4 + 3x^2 - 6x + 1$

14. $y = 11x^4 - 3x + 19$

15. $y = 5\sqrt{x} - 3\sqrt[3]{x} + 11\sqrt[3]{x^2} - 9$

16. $y = 3x^{\frac{7}{3}} - 9x^{\frac{3}{7}} + 21$

17. $y = 3x^{-5} + 2x^{-3}$

18. $y = 2x^{-6} + x^{-1}$

19. $y = \dfrac{2}{\sqrt{x}} - \dfrac{1}{x^2}$

20. $y = \dfrac{3}{x^{\frac{3}{2}}} - \dfrac{1}{x^{\frac{2}{3}}}$

21. $y = \dfrac{1}{2x} + 2x$

22. $y = \dfrac{2}{3x} - \dfrac{2}{3}$

23. $y = x(x^2 + 1)$

24. $y = 3x(x^3 - 1)$

25. $y = (2x + 1)^2$

26. $y = (-3x + 2)^2$

27. $y = (x^2 + 2)(x^3 + 1)$

28. $y = (x^4 - 1)(x^2 + 1)$

29. $y = (\sqrt{x} + 17)(\sqrt[3]{x} + 1)$

30. $y = (x^{\frac{4}{3}} + 2x)(x^{\frac{3}{4}} + 1)$

31. $y = (5x^2 - 7)(3x^2 - 2x + 1)$

32. $y = (3x^2 + 2x)(x^4 - 3x + 1)$

33. $y = \dfrac{1}{3x^2 + 1}$

34. $y = \dfrac{2}{5x^2 - 1}$

35. $y = \dfrac{1}{4x^2 - 3x + 9}$

36. $y = \dfrac{4}{2x^3 - 3x}$

37. $y = \dfrac{x - 1}{x + 1}$

38. $y = \dfrac{2x - 1}{x - 1}$

39. $y = \dfrac{2x^2 - 1}{3\sqrt{x} + 5}$

40. $y = \dfrac{5\sqrt[3]{x} - 4}{3x^2 + 1}$

41. $y = \dfrac{2x^2 - 3x + 1}{2x + 1}$

42. $y = \dfrac{5x^2 + 2x - 6}{3x - 1}$

43. $y = \dfrac{x^2 - x + 1}{x^2 + 1}$

44. $y = \dfrac{x^2 - 2x + 5}{x^2 + 2x - 3}$

45. If $f(0) = 4$, $f'(0) = -1$, $g(0) = -3$, and $g'(0) = 5$, find (a) $(f \cdot g)'(0)$; (b) $(f + g)'(0)$; (c) $(f/g)'(0)$.

46. If $f(3) = 7$, $f'(3) = 2$, $g(3) = 6$, and $g'(3) = -10$, find (a) $(f \cdot g)'(3)$; (b) $(f + g)'(3)$; (c) $(g/f)'(3)$.

47. Use the Product Rule to demonstrate that
$$\frac{d}{dx}[f(x)]^2 = 2 \cdot f(x) \cdot \frac{d}{dx} f(x).$$

48. Develop a rule for $\dfrac{d}{dx}[f(x)g(x)h(x)]$.

49. Find the equation of the tangent line to $y = 3x^2 - 6x + 1$ at $(1, -2)$.

50. Find the equation of the tangent line to $y = 1/(x^2 + 1)$ at $(1, \frac{1}{2})$.

51. Find all the points on the graph of $y = x^3 - x^2$ where the tangent line is horizontal.

52. Find all the points on the graph of $y = \dfrac{1}{3}x^3 + x^2 - x$ where the tangent line has slope 1.

53. The height s in feet of a ball above ground at t seconds is given by $s = -16t^2 + 40t + 100$.

(a) What is its instantaneous velocity at $t = 2$?

(b) What is its instantaneous velocity at $t = 0$?

54. A ball rolls down a long inclined plane so that its distance s from its starting point after t seconds is $s = 4.5t^2 + 2t$ feet. When will its instantaneous velocity be 30 feet per second?

55. A space traveler is moving from left to right along the curve $y = x^2$. When she shuts off the engines, she will go off along the tangent line at the point where she is at that time. At what point should she shut off the engines in order to reach the point $(4, 15)$?

56. A fly is crawling from left to right along the top of the curve $y = 7 - x^2$ (Figure 4). A spider waits at the point $(4, 0)$. Find the distance between the two insects when they first see each other.

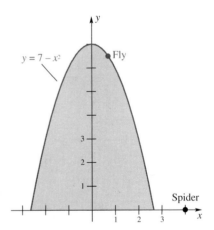

Figure 4

57. The radius of a spherical watermelon is growing at a constant rate of 2 centimeters per week. The thickness of the rind is always one-tenth of the radius. How fast is the volume of the rind growing at the end of the fifth week? Assume the radius is initially 0.

58. Finish the proof of Theorem F.

59. Prove Theorem G.

Answers to Concepts Review: 1. The derivative of the second; second; $f(x)g'(x) + g(x)f'(x)$ 2. Denominator; denominator; square of the denominator; $[g(x)f'(x) - f(x)g'(x)]/g^2(x)$ 3. $nx^{n-1}h$; nx^{n-1} 4. $kL(f)$; $L(f) + L(g)$; $\frac{d}{dx}$

3.3 | Derivatives of Sines, Cosines, Logs, and Exponentials

The trigonometric functions sin x and cos x, and the inverse pair of functions e^x and ln x are called *transcendental* functions, because they cannot be reduced to algebraic operations such as sums, products, and roots. We have seen that the transcendental functions are important for describing natural phenomena such as population growth and vibrations. We find the derivatives of these functions in this section.

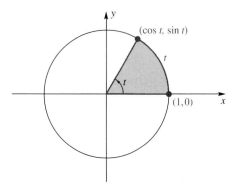

Figure 1

Figure 1 reminds us of the definition of the sine and cosine functions. In what follows, t should be thought of as a number measuring the length of an arc on the unit circle or, equivalently, as the number of radians in the corresponding angle. Thus, $f(t) = \sin t$ and $g(t) = \cos t$ are functions for which both domain and range are a set of real numbers.

The Derivatives of Sin and Cos   We choose to use x rather than t as our basic variable. To find $\frac{d}{dx}(\sin x)$, we use the definition of derivative and use the addition identity for $\sin(x + h)$.

$$\frac{d}{dx}(\sin x) = \lim_{h \to 0} \frac{\sin(x + h) - \sin x}{h}$$

$$= \lim_{h \to 0} \frac{\sin x \cos h + \cos x \sin h - \sin x}{h}$$

$$= \lim_{h \to 0} \left(-\sin x \frac{1 - \cos h}{h} + \cos x \frac{\sin h}{h} \right)$$

$$= (-\sin x)\left[\lim_{h \to 0} \frac{1 - \cos h}{h} \right] + (\cos x)\left[\lim_{h \to 0} \frac{\sin h}{h} \right]$$

To finish our derivation, we have two limits to evaluate. A calculator provided the table in Figure 2 (the first limit was encountered in Example 4 of Section 1.1). It suggests, and later in this section we prove, that

$$\lim_{h \to 0} \frac{1 - \cos h}{h} = 0 \qquad \lim_{h \to 0} \frac{\sin h}{h} = 1$$

Thus,

$$\frac{d}{dx}(\sin x) = (-\sin x) \cdot 0 + (\cos x) \cdot 1 = \cos x$$

h	$\dfrac{\sin h}{h}$	h	$\dfrac{\sin h}{h}$
1	0.841470985	-1	0.841470985
0.1	0.998334166	-0.1	0.998334166
0.01	0.999983333	-0.01	0.999983333
0.001	0.999999833	-0.001	0.999999833

h	$\dfrac{1-\cos h}{h}$	h	$\dfrac{1-\cos h}{h}$
1	0.459697694	-1	-0.459697694
0.1	0.049958347	-0.1	-0.049958347
0.01	0.004999958	-0.01	-0.004999958
0.001	0.000500000	-0.001	-0.000500000

Figure 2

Similarly,

$$\frac{d}{dx}(\cos x) = \lim_{h\to 0}\frac{\cos(x+h) - \cos x}{h}$$

$$= \lim_{h\to 0}\frac{\cos x \cos h - \sin x \sin h - \cos x}{h}$$

$$= \lim_{h\to 0}\left(-\cos x\,\frac{1-\cos h}{h} - \sin x\,\frac{\sin h}{h}\right)$$

$$= (-\cos x)\cdot 0 - (\sin x)\cdot 1$$

$$= -\sin x$$

We summarize these results in an important theorem.

Could You Have Guessed?

The solid curve below is the graph of $y = \sin x$. Note that the slope is 1 at 0, 0 at $\pi/2$, -1 at π, and so on. When we graph the slope function (the derivative), we obtain the dashed curve. Could you have guessed that $\frac{d}{dx}\sin x = \cos x$?

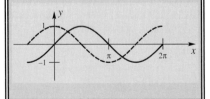

Theorem A

The functions $f(x) = \sin x$ and $g(x) = \cos x$ are both differentiable. In fact,

$$\frac{d}{dx}(\sin x) = \cos x \qquad \frac{d}{dx}(\cos x) = -\sin x$$

EXAMPLE 1: Find $\dfrac{d}{dx}(3\sin x - 2\cos x)$.

SOLUTION:

$$\frac{d}{dx}(3\sin x - 2\cos x) = 3\frac{d}{dx}(\sin x) - 2\frac{d}{dx}(\cos x)$$

$$= 3\cos x + 2\sin x \qquad \blacktriangleleft$$

EXAMPLE 2: Find $\dfrac{d}{dx}(\tan x)$. Find the equation of the line tangent to $\tan x$ at $x = \dfrac{\pi}{4} \approx 0.785$ and sketch the graph of $y = \tan x$ along with the tangent line indicated.

SOLUTION:

$$\frac{d}{dx}(\tan x) = \frac{d}{dx}\left(\frac{\sin x}{\cos x}\right)$$

$$= \frac{\cos x \dfrac{d}{dx}(\sin x) - \sin x \dfrac{d}{dx}(\cos x)}{\cos^2 x}$$

$$= \frac{\cos x \cos x + \sin x \sin x}{\cos^2 x}$$

$$= \frac{1}{\cos^2 x} = \sec^2 x$$

At $x = \dfrac{\pi}{4}$ we get a y-coordinate of

$$\tan\left(\frac{\pi}{4}\right) = \frac{\sin\left(\dfrac{\pi}{4}\right)}{\cos\left(\dfrac{\pi}{4}\right)} = \frac{\dfrac{\sqrt{2}}{2}}{\dfrac{\sqrt{2}}{2}} = 1$$

and a slope of

$$\sec^2\left(\frac{\pi}{4}\right) = \frac{1}{\cos^2\left(\dfrac{\pi}{4}\right)} = \frac{1}{\left(\dfrac{\sqrt{2}}{2}\right)^2} = 2$$

Thus, using point-slope form of a line $y - y_1 = m(x - x_1)$, we get

$$y - 1 = 2\left(x - \frac{\pi}{4}\right)$$

which to three decimal places is equivalent to $y = 2x - 0.571$. A sketch is shown in Figure 3.   ◀

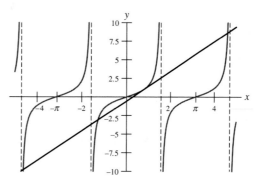

Graph of $y = \tan x$ and tangent line at $x = \frac{\pi}{4} \approx 0.785$

Figure 3

Derivatives of the Natural Logarithm and Natural Exponential Functions   In Section 1.4, Theorem A, we showed that if $f(x) = e^x$, then $f'(c) = e^c$. The proof given for this theorem was not complete; in Chapter 7 we will give a more complete proof of this result. Using our current notation we would write this as $f'(x) = e^x$. (Recall that we thought of c as a specific fixed point on the curve, whereas we think of x as a variable that can take on any value in the domain of the function.)

For the derivative of the natural logarithm function $f(x) = \ln x$ we have

$$\frac{d}{dx}(\ln x) = \lim_{h \to 0} \frac{\ln(x + h) - \ln x}{h}$$

$$= \lim_{h \to 0} \frac{1}{h} \ln\left(\frac{x + h}{x}\right)$$

$$= \lim_{h \to 0} \left(\ln\left(\frac{x + h}{x}\right)^{\frac{1}{h}}\right)$$

$$= \ln\left(\lim_{h \to 0} \left(1 + \frac{h}{x}\right)^{\frac{1}{h}}\right)$$

In the problem set you are asked to justify the steps in this proof. Now, let $k = \dfrac{h}{x}$ so that $\dfrac{1}{h} = \dfrac{1}{k} \cdot \dfrac{1}{x}$. For a fixed value of x, it should seem reasonable that $k \to 0$ when $h \to 0$. Thus, we can replace $\dfrac{h}{x}$ with k, $\dfrac{1}{h}$ with $\dfrac{1}{k} \cdot \dfrac{1}{x}$, and $h \to 0$ with $k \to 0$ to get

$$\ln\left(\lim_{h \to 0} \left(1 + \frac{h}{x}\right)^{\frac{1}{h}}\right)$$

$$= \ln\left(\lim_{k \to 0} (1 + k)^{\frac{1}{k} \cdot \frac{1}{x}}\right) = \ln\left(\left(\lim_{k \to 0} (1 + k)^{\frac{1}{k}}\right)^{\frac{1}{x}}\right)$$

$$= \ln\left(e^{\frac{1}{x}}\right) = \frac{1}{x}$$

using the definition $e = \lim_{k \to 0} (1 + k)^{\frac{1}{k}}$.

We summarize these results in the next theorem.

Theorem B

The functions $f(x) = e^x$ and $g(x) = \ln x$ are both differentiable. In fact,

$$\frac{d}{dx}(e^x) = e^x \qquad \frac{d}{dx}(\ln x) = \frac{1}{x}$$

EXAMPLE 3: Find $\dfrac{d}{dx}(x \ln x)$.

SOLUTION: We have

$$\frac{d}{dx}(x \ln x) = x \cdot \frac{d}{dx} \ln x + \ln x \cdot \frac{d}{dx} x$$

$$= x \cdot \frac{1}{x} + \ln x \cdot 1 = 1 + \ln x$$

using the product rule. ◄

EXAMPLE 4: Suppose that an equation of the form $y = f(t)$ is used to describe the vertical displacement of a car bumper from its normal position after hitting a speed bump (we investigated a similar situation in Example 2 of Section 2.2). Let y represent the displacement from normal position in feet and let t represent time after impact in seconds.

If $f(t) = e^{-t} \sin t$, find $f'(t)$ and $f'(1.5)$. Sketch a graph of $f(t)$ and indicate on it the derivative at $t = 1.5$ using a slope mark (see Section 1.3). Explain the meaning of this number in the context of the car bumper.

SOLUTION: We don't have a derivative formula for e^{-t} yet, but we can rewrite it as $\dfrac{1}{e^t}$ and use the quotient rule. We have

$$f'(t) = \frac{d}{dt} e^{-t} \sin t$$

$$= \frac{d}{dt} \frac{\sin t}{e^t}$$

$$= \frac{e^t \cdot \dfrac{d}{dt} \sin t - \sin t \cdot \dfrac{d}{dt} e^t}{(e^t)^2}$$

$$= \frac{e^t \cdot \cos t - \sin t \cdot e^t}{e^{2t}}$$

$$= \frac{\cos t - \sin t}{e^t}$$

Therefore, $f'(1.5) = \dfrac{\cos 1.5 - \sin 1.5}{e^{1.5}} \approx -0.207$. In Figure 4 we sketch a graph of $f(t)$ and indicate our result with a slope mark.

This means that the bumper has reached its maximum distance above the ground and is now moving back down at a rate of 0.207 feet per second. ◄

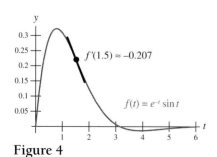

$f'(1.5) \approx -0.207$

$f(t) = e^{-t} \sin t$

Figure 4

Generalized Derivative Rules
As they stand, functions like $\sin x$ and e^x are not very useful for real-world modeling problems. We have seen, however, that the families of functions given by $A \sin(2\pi w(x - d))$ and Ae^{kt} can be extremely useful (see Section 2.3).

In the next section we will encounter the Chain Rule, which will allow us to find derivatives of functions like the ones above. The Chain Rule is quite a bit more powerful than what we need right now, though, so we present the following theorem, which will take care of the special case where we replace the argument of a function (such as the x in sin x) with a linear function.

Theorem C

Assume that a and b are constants. The following derivative rules generalize the rules from Theorems A and B:

(1) $\dfrac{d}{dx}\sin(ax + b) = a\cos(ax + b)$

(2) $\dfrac{d}{dx}\cos(ax + b) = -a\sin(ax + b)$

(3) $\dfrac{d}{dx}e^{ax} = ae^{ax}$

Proof   We prove (1) here; the others are similar. If we let $f(x) = \sin(ax + b)$ then

$$f'(x) = \lim_{h \to 0} \frac{\sin(a(x + h) + b) - \sin(ax + b)}{h}$$

$$= \lim_{h \to 0} a\,\frac{\sin(ax + b + ah) - \sin(ax + b)}{h}$$

Now, let $k = ah$ and therefore $\dfrac{1}{h} = \dfrac{a}{k}$; also let $u = ax + b$. Because $h \to 0$ and a is constant we must also have $k \to 0$, and so

$$f'(x) = \lim_{h \to 0} \frac{\sin(u + k) - \sin(u)}{k}$$

$$= a\cos(u) = a\cos(ax + b) \qquad \blacktriangleleft$$

Note: The reason that rule (3) was written as it was, rather than as $\dfrac{d}{dx}e^{ax + b} = ae^{ax + b}$, is that because $e^{ax + b} = e^{ax}e^{b}$, we do not find expressions of the form $e^{ax + b}$ very often in practice.

EXAMPLE 5: Find the derivatives of $5\cos(2\pi x)$ and $100e^{0.05t}$.

> **SOLUTION:** We combine the constant multiple rule (Theorem F of Section 3.2) with the previous theorem to get

$$\frac{d}{dx}5\cos(2\pi x) = 5\frac{d}{dx}\cos(2\pi x)$$

$$= -5 \cdot 2\pi \sin(2\pi x)$$

$$= -10\pi \sin(2\pi x)$$

and

$$\frac{d}{dx}100e^{0.05t} = 100\frac{d}{dx}e^{0.05t}$$

$$= 100(0.05)e^{0.05t}$$

$$= 5e^{0.05t}$$ ◄

EXAMPLE 6: In Section 2.3, Example 2, we developed the model $y = 70 + 28e^{-0.1t}$ to describe the cooling of a human body after death. The variable t represented time after death in hours, and the variable y represented temperature in degrees Fahrenheit. Find the derivative $\frac{d}{dt}y$ at the time $t = 2$ and explain what this number represents in the context of the problem.

SOLUTION: Using Theorems E and F from Section 3.2 and Theorem C of this section we get

$$\frac{d}{dt}y = \frac{d}{dt}(70 + 28e^{-0.1t})$$

$$= \frac{d}{dt}70 + \frac{d}{dt}28e^{-0.1t}$$

$$= 0 + 28\frac{d}{dt}e^{-0.1t}$$

$$= 28(-0.1)e^{-0.1t}$$

$$= -2.8e^{-0.1t}$$

At $t = 2$ the derivative would be $-2.8e^{-0.1(2)} \approx -2.29$. The units of the derivative are degrees Fahrenheit per hour. Thus, two hours after death, the body was cooling at a rate of -2.29 °F/hr. ◄

Proof of Two Limit Statements   The derivations of the derivative formulas for sin and cos depend on the two limit statements

$$\lim_{t \to 0} \frac{\sin t}{t} = 1 \qquad \lim_{t \to 0} \frac{1 - \cos t}{t} = 0$$

They require proofs.

Proof   Consider the diagram in Figure 5. Note that as $t \to 0$, the point P moves toward $(1, 0)$ and, therefore, that

$$\lim_{t \to 0} \cos t = 1 \qquad \lim_{t \to 0} \sin t = 0$$

Next, for $-\pi/2 < t < \pi/2$, $t \neq 0$, draw the vertical line segment BP and the circular arc BC, as shown in Figure 6. (If $t < 0$, the shaded region will be reflected across the x-axis.) Clearly,

$$\text{Area(sector } OBC) \leq \text{area}(\Delta OBP) \leq \text{area(sector } OAP)$$

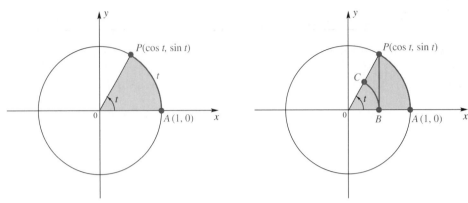

Figure 5   Figure 6

From the formulas $\frac{1}{2}bh$ (area of a triangle) and $\frac{1}{2}r^2|t|$ (area of a circular sector), we obtain

$$\frac{1}{2}(\cos t)^2|t| \le \frac{1}{2}\cos t|\sin t| \le \frac{1}{2}(1)^2|t|$$

or, after multiplying by 2 and dividing by the positive number $|t|\cos t$ and noting that $(\sin t)/t$ is positive,

$$\cos t \le \frac{\sin t}{t} \le \frac{1}{\cos t}$$

We now let t approach 0; because $\cos 0 = 1$, both ends of the above inequality approach 1. This means that the middle term must also approach 1 (a result called the Squeeze Theorem in analysis books) so that

$$\lim_{t \to 0} \frac{\sin t}{t} = 1$$

the first of our claimed results.

The second result follows easily from the first.

$$\lim_{t \to 0} \frac{1 - \cos t}{t} = \lim_{t \to 0} \frac{1 - \cos t}{t} \cdot \frac{1 + \cos t}{1 + \cos t} = \lim_{t \to 0} \frac{1 - \cos^2 t}{t(1 + \cos t)}$$

$$= \lim_{t \to 0} \frac{\sin^2 t}{t(1 + \cos t)} = \lim_{t \to 0} \frac{\sin t}{t} \cdot \frac{\displaystyle\lim_{t \to 0} \sin t}{\displaystyle\lim_{t \to 0}(1 + \cos t)}$$

$$= 1 \cdot \frac{0}{2} = 0 \qquad \blacktriangleleft$$

Concepts Review

1. The derivative formulas for $\sin x$ and $\cos x$ are $\frac{d}{dx}(\sin x) = $ _____ and $\frac{d}{dx}(\cos x) = $ _____ .

2. At $x = \pi/3$, $\frac{d}{dx}(\sin x)$ has the value _____ . Thus, the equation of the tangent line to $y = \sin x$ at $x = \pi/3$ is _____ .

3. The derivative formulas for e^x and $\ln x$ are $\frac{d}{dx}e^x = $ _____ and $\frac{d}{dx}\ln x = $ _____ .

4. The derivative of $\sin(\pi x + 5)$ with respect to x is _____ .

Problem Set 3.3

In Problems 1–14, find $\frac{d}{dx}y$.

1. $y = 3 \sin x - 5 \cos x$

2. $y = \sin x \cos x$

3. $y = e^x - \ln x$

4. $y = e^x + e^{-x}$

5. $y = \cot x = \dfrac{\cos x}{\sin x}$

6. $y = \sec x = \dfrac{1}{\cos x}$

7. $y = x^2 e^x$

8. $y = x^2 e^{-x}$

9. $y = 3 \sin(2x + 1)$

10. $y = x \sin(3x)$

11. $y = e^{2x} \cos \pi x$

12. $y = 2 \ln x \cos 2\pi x$

13. $y = \dfrac{\ln x}{e^{0.1x} + \cos x}$

14. $y = \dfrac{\tan 2x}{x + 3e^x}$

In Problems 15–18, find the equation of the line tangent to the given curve at the given point. Sketch both the curve and the tangent line together.

15. $y = 3 \cos(\pi x)$ at $x = 1$

16. $y = 100 e^{0.05t}$ at $t = 5$

17. $y = \dfrac{\cos x}{x}$ at $x = 1$

18. $y = \dfrac{x^2 + 1}{\sin x}$ at $x = 1$

In Problems 19–22, find the indicated derivative at the indicated point. Sketch the curve and represent the derivative you found with a slope mark.

19. $f'(2)$ for $f(t) = 10 e^{-0.1t} \cos(3t)$

20. $g'(1)$ for $g(t) = 2t + 5 \sin(2\pi t)$

21. $h'(50)$ for $h(t) = \dfrac{100}{9 e^{-0.1t} + 1}$

22. $f'(4)$ for $f(x) = \dfrac{\sin x}{x}$

23. In Example 1 of Section 2.7 we developed the model $y = 3.956 e^{0.02951x}$ of the U.S. population for the years 1790 to 1860. In this model, t represents the number of years after 1790 and y represents the U.S. population in millions. Find $\frac{d}{dt}y$ at $t = 10$ and at $t = 60$. Explain what these numbers represent in the context of the problem. Sketch a graph and illustrate your findings on this graph.

24. In Example 3 of Section 2.3 we developed the model $y = 24 \sin((2\pi/12)(x - 4)) + 49$ for the average monthly temperature in Hartford, Connecticut. In this model, x represents the number of the month, starting with $x = 1$ corresponding to January, and y represents average temperature for that month. Find $\frac{d}{dx}y$ at $x = 3$ and explain what this number represents in the context of the problem. Sketch a graph and illustrate your finding on this graph.

25. Show that the curves $y = \sqrt{2} \sin x$ and $y = \sqrt{2} \cos x$ intersect at right angles at a certain point with $0 < x < \pi/2$. Recall that two lines intersect at right angles if their slopes are negative reciprocals of each other.

26. At time t seconds, the center of a bobbing cork is $2 \sin\left(\frac{\pi}{2}t\right)$ centimeters above (or below) water level. What is the velocity of the cork at $t = 0, 1, 2$?

27. Use the definition of derivative to show that $\frac{d}{dx}(\sin x^2) = 2x \cos x^2$.

28. From area$(\triangle OBP) \le$ area(sector $AOP) \le$ [area $(\triangle OBP)$ + area$(ABPQ)$] in Figure 7, show that

$$\cos t \le \frac{t}{\sin t} \le 2 - \cos t$$

and thus obtain another proof that $\lim_{t \to 0}(\sin t)/t = 1$.

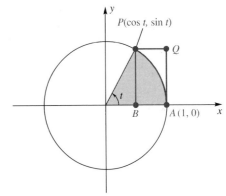

Figure 7

29. In Figure 8, let D be the area of the triangle ABP and E the area of the shaded region.

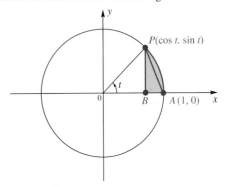

Figure 8

(a) Guess the value of $\lim_{t \to 0^+}(D/E)$ by looking at the figure.

(b) Find a formula for D/E in terms of t.

(c) Use a calculator to get an accurate estimate of $\lim_{t \to 0^+}(D/E)$.

30. An isosceles triangle is topped by a semicircle, as shown in Figure 9. Let D be the area of triangle AOB and E be the area of the shaded region. Find a formula for D/E in terms of t and then calculate $\lim_{t \to 0^+} \dfrac{D}{E}$.

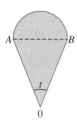

Figure 9

31. Let $f(x) = x \sin x$.

(a) Draw the graphs of $f(x)$ and $f'(x)$ on $[\pi, 6\pi]$.

(b) How many solutions does $f(x) = 0$ have on $[\pi, 6\pi]$? How many solutions does $f'(x) = 0$ have on this interval?

(c) What is wrong with the following conjecture? If f and f' are both continuous and differentiable on $[a, b]$, if $f(a) = f(b) = 0$, and if $f(x) = 0$ has exactly n solutions on $[a, b]$, then $f'(x) = 0$ has exactly $n - 1$ solutions on $[a, b]$.

(d) Determine the maximum value of $|f(x) - f'(x)|$ on $[\pi, 6\pi]$.

32. Let $f(x) = \cos^3 x - 1.25 \cos^2 x + 0.225$. Find $f'(x_0)$ at that point x_0 in $[\pi/2, \pi]$, where $f(x_0) = 0$.

Answers to Concepts Review: 1. $\cos x, -\sin x$ 2. $\frac{1}{2}$; $y - \sqrt{3}/2 = \frac{1}{2}(x - \pi/3)$ 3. $e^x, \frac{1}{x}$ 4. $\pi \cos(\pi + 5)$

3.4 | The Chain Rule

Imagine trying to find the derivative of

$$F(x) = (2x^2 - 4x + 1)^{60}$$

You would first have to multiply together 60 quadratic factors of $2x^2 - 4x + 1$ and then differentiate the resulting polynomial of degree 120.

Fortunately, there is a better way. After you have learned the Chain Rule, you will be able to write the answer

$$F'(x) = 60(2x^2 - 4x + 1)^{59}(4x - 4)$$

as fast as you can move your pencil. In fact, the Chain Rule is so important that you will seldom again differentiate any function without using it.

Function Composition   Two functions can be combined to form a third function by a process called *function composition*. If we write $u = g(x)$, then x is the input to function g and u is the output. If we then write $y = f(u)$, u is now the input to function f and y is the output. Using the principle that we can replace a quantity with an equal quantity, we can combine the two statements $y = f(u)$ and $u = g(x)$ into one to get $y = f(g(x))$.

Definition

If f and g are functions, then $f \circ g$ is called the composition of f with g and is defined by $(f \circ g)(x) = f(g(x))$.

EXAMPLE 1: Find $(f \circ g)(x)$; and $(g \circ f)(x)$, where $f(x) = \sin x$ and $g(x) = 2x$. Are these two compositions the same function?

SOLUTION: We have

$$(f \circ g)(x) = f(g(x))$$
$$= f(2x)$$
$$= \sin(2x)$$

and

$$(g \circ f)(x) = g(f(x))$$
$$= g(\sin x)$$
$$= 2\sin(x)$$

The functions $\sin(2x)$ and $2\sin x$ are not the same; try graphing them both if you are not sure! ◄

Leibniz Notation If we think of $\frac{d}{dx}$ as a fraction (which it is not), then we could rewrite $\frac{d}{dx}y$ as $\frac{dy}{dx}$. This notation for the derivative was invented by Gottfried Wilhelm Leibniz (1646–1716), one of the two principal founders of calculus. (The other was Isaac Newton.)

If we want to substitute a particular x-value into a derivative formula using Leibniz notation, we draw a vertical line after the derivative symbol with the substitution indicated at the bottom of the vertical line. Thus, for example, if $y = x^2$, then $\frac{dy}{dx} = 2x$; at the point $x = 3$ we get $\frac{dy}{dx}\Big|_{x=3} = 2(3) = 6$. This is equivalent to writing $f(x) = x^2$, $f'(x) = 2x$, $f'(3) = 2(3) = 6$.

This notation is consistent with the definition of the derivative. If we let $\Delta x = h$ represent the change in x from x to $x + h$, and let $\Delta y = f(x + h) - f(x)$ represent the change in y from $f(x)$ to $f(x + h)$, then our definition of derivative

$$f'(x) \lim_{h \to 0} \frac{f(x + h) - f(x)}{h}$$

becomes

$$\frac{dy}{dx} = \lim_{\Delta x \to 0} \frac{\Delta y}{\Delta x}$$

In this form it is clear that the derivative represents a slope, or a rate of change.

The symbol $\frac{dy}{dx}$ should be read "the derivative of y with respect to x"; it measures how fast y is changing with respect to x. The x indicates that x is being treated as the basic variable. Thus, if $y = s^2x^3$, we may write

$$\frac{dy}{dx} = 3s^2x^2 \quad \text{and} \quad \frac{dy}{ds} = 2sx^3$$

In the first case, s is treated as a constant and x is the basic variable; in the second case, x is constant and s is the basic variable.

More important is the following example. Suppose $y = u^{60}$ and $u = 2x^2 - 4x + 1$. Then, $\frac{dy}{du} = 60u^{59}$ and $\frac{du}{dx} = 4x - 4$. But notice that when we substitute $u = 2x^2 - 4x + 1$ into $y = u^{60}$, we obtain

$$y = (2x^2 - 4x + 1)^{60}$$

so it makes sense to ask for $\frac{dy}{dx}$. What is $\frac{dy}{dx}$ and how is it related to $\frac{dy}{du}$ and $\frac{du}{dx}$? More generally, how do you differentiate a composite function?

Differentiating a Composite Function   If David can type twice as fast as Mary, and Mary can type three times as fast as Jack, then David can type $2 \cdot 3 = 6$ times as fast as Jack. The two rates are multiplied.

Suppose that

$$y = f(u) \quad \text{and} \quad u = g(x)$$

determine the composite function $y = f(g(x))$. Because a derivative indicates a rate of change, we can say that

$$y \text{ changes } \frac{dy}{du} \text{ times as fast as } u$$

$$u \text{ changes } \frac{du}{dx} \text{ times as fast as } x$$

It seems reasonable to conclude that

$$y \text{ changes } \frac{dy}{du} \cdot \frac{du}{dx} \text{ times as fast as } x$$

This is in fact true and we will suggest a formal proof at the end of this section. The result is called the **Chain Rule**.

Theorem A

(Chain Rule). Let $y = f(u)$ and $u = g(x)$ determine the composite function $y = f(g(x)) = (f \circ g)(x)$. If g is differentiable at x and f is differentiable at $u = g(x)$, then $f \circ g$ is differentiable at x and

$$(f \circ g)'(x) = f'(g(x))g'(x)$$

That is,

$$\frac{dy}{dx} = \frac{dy}{du} \cdot \frac{du}{dx}$$

In this form, the chain rule just looks like cancellation of fractions. Keep in mind that derivatives are not in fact fractions, but limits of fractional quantities.

Applications of the Chain Rule   We begin with the example $(2x^2 - 4x + 1)^{60}$ introduced at the beginning of this section.

EXAMPLE 2: If $y = (2x^2 - 4x + 1)^{60}$, find $\frac{dy}{dx}$.

SOLUTION: We think of this as

$$y = u^{60} \quad \text{and} \quad u = 2x^2 - 4x + 1$$

Thus,

$$\frac{dy}{dx} = \frac{dy}{du} \cdot \frac{du}{dx}$$

$$= (60u^{59})(4x - x)$$

$$= 60(2x^2 - 4x + 1)^{59}(4x - x) \qquad \blacktriangleleft$$

EXAMPLE 3: In Example 1 of Section 2.7 we derived the equation $y = 3.956e^{0.02951x}$ as a good model of the U.S. population from 1790 to 1860, with y representing the population in millions, and x representing the number of years after 1790.

Find y and $\frac{dy}{dx}$ at $x = 10$ and explain in plain English what these numbers represent.

SOLUTION: To find $\frac{dy}{dx}$, we can think of it this way:

$$y = 3.956e^u \quad \text{and} \quad u = 0.02951x$$

Thus,

$$\frac{dy}{dx} = \frac{dy}{du} \cdot \frac{du}{dx}$$

$$= (3.956e^u)(0.02951)$$

$$= 0.117e^{0.02951x}$$

At $x = 10$ we have $y = 3.956e^{0.02951(10)} \approx 5.314$ and $\frac{dy}{dx} = 0.117e^{0.02951(10)} \approx 0.157$. This means that the population in 1800 (ten years after 1790) was about 5,314,000 people, and that the population was increasing at a rate of about 157,000 people per year. Remember that the units for the derivative are units of y per unit of x. $\qquad \blacktriangleleft$

Note: We could have found the derivative $\frac{dy}{dx}$ using the methods of the previous section. You should now see that Theorem C of Section 3.3 is just a special case of the chain rule, as was claimed at the time.

EXAMPLE 4: If $y = \sin(x^3 - 3x)$, find $\frac{dy}{dx}$.

SOLUTION: We may write

$$y = \sin u \quad \text{and} \quad u = x^3 - 3x$$

Hence,

$$\frac{dy}{dx} = \frac{dy}{du} \cdot \frac{du}{dx}$$

$$= (\cos u) \cdot (3x^2 - 3)$$

$$= [\cos(x^3 - 3x)] \cdot (3x^2 - 3)$$

$$= (3x^2 - 3)\cos(x^3 - 3x) \qquad \blacktriangleleft$$

Notation

The Chain Rule says that

$$\frac{d}{dx} \sin x^3 = \cos x^3 \cdot 3x^2$$

but the answer written this way is ambiguous. Does $3x^2$ multiply x^3 or $\cos x^3$? We mean, of course, the latter. Therefore, we should either introduce parentheses, as in

$$(\cos x^3)(3x^2)$$

or, better, write the answer as

$$3x^2\cos x^3$$

Note that we give our final answer in Example 4 this way. It is a practice we normally follow.

Here is another important notational point. When we write $\cos x^3$, we mean $\cos(x^3)$. If we intend $(\cos x)^3$, we write $\cos^3 x$.

EXAMPLE 5: Find $\dfrac{d}{dt}\left(\dfrac{t^3 - 2t + 1}{t^4 + 3}\right)^{13}$.

SOLUTION: Think of this as finding $\frac{dy}{dt}$, where

$$y = u^{13} \quad \text{and} \quad u = \frac{t^3 - 2t + 1}{t^4 + 3}$$

Then, the Chain Rule followed by the Quotient Rule gives

$$\frac{dy}{dt} = \frac{dy}{du} \cdot \frac{du}{dt}$$

$$= 13u^{12}\,\frac{(t^4 + 3)(3t^2 - 2) - (t^3 - 2t + 1)(4t^3)}{(t^4 + 3)^2}$$

$$= 13\left(\frac{t^3 - 2t + 1}{t^4 + 3}\right)^{12} \cdot \frac{-t^6 + 6t^4 - 4t^3 + 9t^2 - 6}{(t^4 + 3)^2}$$ ◄

Soon you will learn to make a mental introduction of the middle variable without actually writing it. Thus, an expert immediately writes:

$$\frac{d}{dx}\left(\ln(x^2 + 1)\right) = \frac{1}{x^2 + 1} \cdot 2x = \frac{2x}{x^2 + 1}$$

or

$$\frac{d}{dx}(x^3 + \sin x)^6 = 6(x^3 + \sin x)^5(3x^2 + \cos x)$$

or

$$\frac{d}{dt}\left(\frac{t}{\cos t}\right)^4 = 4\left(\frac{t}{\cos t}\right)^3 \cdot \frac{\cos t - t(-\sin t)}{\cos^2 t}$$

$$= \frac{4t^3(\cos t + t \sin t)}{\cos^5 t}$$

EXAMPLE 6: Find $\frac{d}{dx}[\sin^3(4x)]$.

SOLUTION:

$$\frac{d}{dx}[\sin(4x)]^3 = 3\sin^2(4x)\frac{d}{dx}\sin(4x)$$

$$= 3\sin^2(4x)\cos(4x)\frac{d}{dx}(4x)$$

$$= 3\sin^2(4x)\cos(4x) \cdot 4$$

$$= 12\sin^2(4x)\cos(4x)$$

> **The Last First**
>
> Here is an informal rule that may help you in using the derivative rules.
>
> *The last step in calculation corresponds to the first step in differentiation.*
>
> For instance, in calculating $\sin^3(4x)$ for a particular value of x, the last step is to cube $\sin(4x)$, so the first rule to use in differentiating $\sin^3(4x)$ is that for u^3. Similarly, the last step in calculating
>
> $$\frac{x^2 - 1}{x^2 + 1}$$
>
> is to take the quotient, so the first rule to use in differentiating is the Quotient Rule.

Notice that we use the operator $\frac{d}{dx}$ as a kind of "placeholder" so that we only use one derivative rule at a time. Above we used the Chain Rule twice; this is where the "chain" in the Chain Rule comes from.   ◄

EXAMPLE 7: Find $\frac{d}{dx}\sin(\cos x^2)$.

SOLUTION:

$$\frac{d}{dx}\sin(\cos x^2) = \cos(\cos x^2) \cdot \frac{d}{dx}\cos x^2$$

$$= \cos(\cos x^2) \cdot (-\sin x^2) \cdot \frac{d}{dx}x^2$$

$$= \cos(\cos x^2) \cdot (-\sin x^2) \cdot 2x$$

$$= -2x(\sin x^2)\cos(\cos x^2) \qquad \blacktriangleleft$$

Partial Proof of the Chain Rule

Proof   We suppose that $y = f(u)$ and $u = g(x)$, that g is differentiable at x, and that f is differentiable at $u = g(x)$. When x is given an increment Δx, there are corresponding increments in u and y given by

$$\Delta u = g(x + \Delta x) - g(x)$$

$$\Delta y = f(g(x + \Delta x)) - f(g(x))$$

$$= f(u + \Delta u) - f(u)$$

Thus,

$$\frac{dy}{dx} = \lim_{\Delta x \to 0}\frac{\Delta y}{\Delta x} = \lim_{\Delta x \to 0}\frac{\Delta y}{\Delta u}\frac{\Delta u}{\Delta x}$$

$$= \lim_{\Delta x \to 0}\frac{\Delta y}{\Delta u} \cdot \lim_{\Delta x \to 0}\frac{\Delta u}{\Delta x}$$

Because g is differentiable at x, it is continuous there, and so $\Delta x \to 0$ forces $\Delta u \to 0$. Hence,

$$\frac{dy}{dx} = \lim_{\Delta u \to 0}\frac{\Delta y}{\Delta u} \cdot \lim_{\Delta x \to 0}\frac{\Delta u}{\Delta x} = \frac{dy}{du} \cdot \frac{du}{dx} \qquad \blacktriangleleft$$

This proof is very slick, but unfortunately it contains a subtle flaw. There are functions $u = g(x)$ that have the property $\Delta u = 0$ for some points in every neighborhood of x. (The constant function $g(x) = k$ is a good example.) This means the division by Δu at our first step might be undefined. There is no simple way to get around this difficulty, though the Chain Rule is valid even in this case. A complete proof of the Chain Rule can be found in analysis books.

Concepts Review

1.   If $y = f(u)$, where $u = g(t)$, then $\frac{dy}{dt} = \frac{dy}{du} \cdot$ _____.

2.   If $w = G(v)$, where $v = H(s)$, then $\frac{dw}{ds} =$ _____ $\cdot \frac{dv}{ds}$.

3.   If $y = e^{x^2}$, then $\frac{dy}{dx} =$ _____.

4.   If $y = (\sin x^2)(2x + 1)^3$, then $\frac{dy}{dx} = \sin(x^2) \cdot$ _____ $+ (2x + 1)^3$ _____.

Problem Set 3.4

In Problems 1–18, find $\frac{dy}{dx}$ by explicitly defining an intermediate variable u and using the Chain Rule in the form $\frac{dy}{dx} = \frac{dy}{du} \cdot \frac{du}{dx}$. See Examples 2, 3, 4, and 5.

1. $y = (2 - 9x)^{15}$

2. $y = (4x + 7)^{23}$

3. $y = (5x^2 + 2x - 8)^5$

4. $y = (3x^3 - 11x)^7$

5. $y = (x^3 - 3e^x + 4 \ln x)^9$

6. $y = (2x^4 - 12\sqrt{x} + e^{2x})^{10}$

7. $y = \sqrt{3x^4 + x - 8}$

8. $y = \dfrac{1}{\sqrt{3x^4 + x - 8}}$

9. $y = \ln(4x^3 - 3x^2 + 11x - 1)$

10. $y = e^{(4x^3 + 11x)}$

11. $y = \sin(3x^2 + 11x)$

12. $y = \cos(4x^5 - 11x)$

13. $y = \sin^3 x$

14. $y = \cos^5 x$

15. $y = \left(\dfrac{e^x}{x + 4}\right)^4$

16. $y = \left(\dfrac{\ln x}{2x + 5}\right)^6$

17. $y = \sin\left(\dfrac{3x - 1}{2x + 5}\right)$

18. $y = \cos\left(\dfrac{x^2 - 1}{x + 4}\right)$

In Problems 19–26, find $\frac{dy}{dx}$.

19. $y = (4x - 7)^2(2x + 3)$

20. $y = (5x + 6)^2(x - 13)^3$

21. $y = (2x - 1)^3 e^{5x}$

22. $y = (3x^2 + 5)^2 e^{x^3}$

23. $y = \dfrac{(x + 1)^2}{e^{3x - 4}}$

24. $y = \dfrac{\ln(x + 1)}{(x^2 + 4)^2}$

25. $y = \dfrac{(\ln x)^2}{2x^2 - 5}$

26. $y = \dfrac{(x^2 - 1)^3}{\cos^2 x}$

In Problems 27–42, find the indicated derivative. Use the method of Examples 6 and 7 (don't explicitly define an intermediate variable u).

27. $\dfrac{d}{dt}\left(\dfrac{3t - 2}{t + 5}\right)^3$

28. $\dfrac{d}{ds}\left(\dfrac{s^2 - 9}{s + 4}\right)$

29. $\dfrac{d}{dt}\left(\dfrac{(3t - 2)^3}{e^{5t}}\right)$

30. $\dfrac{d}{ds}\left(\dfrac{5 \ln s}{s + 4}\right)^3$

31. $\dfrac{d}{d\theta}(\sin^3 \theta)$

32. $\dfrac{d}{d\theta}(\cos^4 \theta)$

33. $\dfrac{d}{dx}\left(\dfrac{\sin x}{\ln 2x}\right)^3$

34. $\dfrac{d}{dt} 4e^{\sin(t^2 + 1)}$

35. $\dfrac{d}{dx}[\sin^4(x^2 + 3x)]$

36. $\dfrac{d}{dt}[\cos^5(4t - 19)]$

37. $\dfrac{d}{dt}[\sin^3(e^t)]$

38. $\dfrac{d}{du}\left[\ln\dfrac{(u + 1)}{(u - 1)}\right]^4$

39. $\dfrac{d}{d\theta}[\cos(\sin \theta^2)]$

40. $\dfrac{d}{dx}[x \sin^2(2x)]$

41. $\dfrac{d}{dx} e^{\cos(\sin 2x)}$

42. $\dfrac{d}{dt} \cos[\ln(\cos t)]$

In Problems 43–46, evaluate the indicated derivative. Sketch a graph of the function and on it show the indicated derivative with a slope mark.

43. $f'(1)$ if $f(x) = e^{-x^2}$

44. $G'(1)$ if $G(t) = (t^2 + 9)^3(t^2 - 2)^4$

45. $F'(1)$ if $F(t) = \sin(t^2 + 3t + 1)$

46. $g'(1)$ if $g(s) = 3 \ln(s^2 + 1)$

47. A one-kilogram object dropped from 100 meters is falling toward the ground according to the equation $y = -32t + 32(1 - e^{-t}) + 100$ (we will investigate models of this type in Lab 12). Here, y represents distance from the ground in meters and t represents time in seconds after the object is dropped. The kinetic energy (E) of a moving object is given by $E = \frac{1}{2}mv^2$, where m is the mass of the object in kilograms and v is the velocity of the object in meters per second.

(a) Find the velocity of the object 2 seconds after it is dropped.

(b) Find E and dE/dt 2 seconds after the object is dropped. Explain what each of these numbers represents in the context of the problem.

48. Find the equation of the tangent line to $y = (x^2 + 1)^3(x^4 + 1)^2$ at $(1, 32)$.

49. A point P is moving in the plane so that its coordinates after t seconds are $(4 \cos 2t, 7 \sin 2t)$, measured in feet.

(a) Show that P is following an elliptical path. *Hint*: Show $\left(\frac{x}{4}\right)^2 + \left(\frac{y}{7}\right)^2 = 1$, which is an equation of an ellipse.

(b) Obtain an expression for L, its distance from the origin at time t. *Hint*: Use the Pythagorean Theorem.

(c) How fast is P moving away from the origin at $t = \frac{\pi}{8}$?

50. A wheel centered at the origin and of radius 10 centimeters is rotating counterclockwise at a rate of 4 revolutions per second. A point P on the rim is at $(10, 0)$ at $t = 0$.

(a) What are the coordinates of P at time t seconds?

(b) At what rate is P rising (or falling) at time $t = 1$?

51. Consider the wheel-piston device in Figure 1. The wheel has radius 1 foot and rotates counterclockwise at 2 radians per second. The connecting rod is 5 feet long. The point P is at $(1, 0)$ at time $t = 0$.

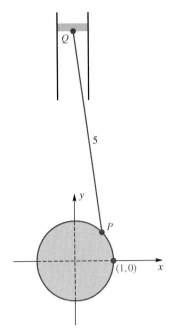

Figure 1

(a) Find the coordinates of P at time t.

(b) Find the y-coordinate of Q at time t (the x-coordinate is always zero).

(c) Find the velocity of Q at time t.

52. Show that $\dfrac{d}{dx}|x| = |x|/x$, $x \neq 0$. *Hint*: Write $|x| = \sqrt{x^2}$ and use the Chain Rule with $u = x^2$.

53. Apply the result in Problem 52 to find each derivative.

(a) $\dfrac{d}{dx}|x^2 - 1|$

(b) $\dfrac{d}{dx}|\sin x|$

54. Let $f(0) = 0$ and $f'(0) = 2$. Find the derivative of $f(f(f(f(x))))$ at $x = 0$.

55. Let $f(x) = \sin(\sin(\sin(\sin(x))))$ on $[-3\pi, 3\pi]$.

(a) Estimate the largest value of $f(x)$.

(b) Estimate the largest value of $|f'(x)|$.

Answers to Concepts Review: 1. du/dt 2. dw/dv 3. $2xe^{x^2}$ 4. $6(2x + 1)^2$; $2x \cos(x^2)$

3.5 Higher Derivatives

The operation of differentiation takes a function f and produces a new function f'. If we now differentiate f', we produce still another function, denoted by f'' (read "eff double prime") and called the **second derivative** of f. It, in turn, may be differentiated, thereby producing

f''', which is called the **third derivative** of f. With the **fourth derivative** and beyond we use a number in parentheses, as with $f^{(4)}$, instead of multiple primes. Thus, for example, let

$$f(x) = 2x^3 - 4x^2 + 7x - 8$$

Then

$$f'(x) = 6x^2 - 8x + 7$$
$$f''(x) = 12x - 8$$
$$f'''(x) = 12$$
$$f^{(4)}(x) = 0$$

Because the derivative of the zero function is zero, all derivatives of f higher than the third derivative will be zero.

We have introduced two notations for the derivative (now also called the *first derivative*) of $y = f(x)$. They are

$$f'(x) \quad \text{and} \quad \frac{dy}{dx}$$

called, respectively, the *prime notation* and the *Leibniz notation*. There is a variation of the prime notation—namely, y'—that we will also use occasionally. All of these notations have extensions for higher derivatives, as shown in the chart below. Note especially the Leibniz notation, which—though complicated—seemed most appropriate to Leibniz. What, thought he, is more natural than to write

$$\frac{d}{dx}\left(\frac{dy}{dx}\right) \quad \text{as} \quad \frac{d^2y}{dx^2}$$

Notations for Derivatives of $y = f(x)$

Derivative	f' Notation	y' Notation	Leibniz Notation
First	$f'(x)$	y'	$\dfrac{dy}{dx}$
Second	$f''(x)$	y''	$\dfrac{d^2y}{dx^2}$
Third	$f'''(x)$	y'''	$\dfrac{d^3y}{dx^3}$
Fourth	$f^{(4)}(x)$	$y^{(4)}$	$\dfrac{d^4y}{dx^4}$
Fifth	$f^{(5)}(x)$	$y^{(5)}$	$\dfrac{d^5y}{dx^5}$
Sixth	$f^{(6)}(x)$	$y^{(6)}$	$\dfrac{d^6y}{dx^6}$
$\vdots$	$\vdots$	$\vdots$	$\vdots$
nth	$f^{(n)}(x)$	$y^{(n)}$	$\dfrac{d^ny}{dx^n}$

Operator Notation We use the symbol $\frac{d}{dx}$ to mean "the derivative of"; it must be followed by some function. The symbol by itself is meaningless. We have seen that this notation is closely related to Leibniz notation, in that $\frac{d}{dx}y$ and $\frac{dy}{dx}$ both mean the derivative of y with respect to x. This idea extends to higher derivatives, in that the symbol $\frac{d^2}{dx^2}$ means "the second derivative of" whatever function follows. Thus, for example, $\frac{d^2}{dx^2}\sin x = \frac{d}{dx}\cos x = -\sin x$.

EXAMPLE 1:

(1) If $y = \sin 2x$, find d^3y/dx^3, d^4y/dx^4, and $d^{12}y/dx^{12}$.

(2) If $y = e^{3x}$, find d^3y/dx^3, and d^ny/dx^n.

SOLUTION:

(1)   **(2)**

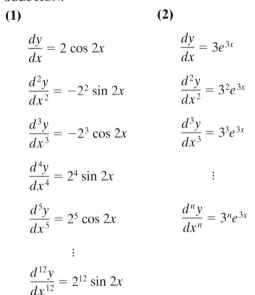

$$\frac{dy}{dx} = 2\cos 2x \qquad\qquad \frac{dy}{dx} = 3e^{3x}$$

$$\frac{d^2y}{dx^2} = -2^2\sin 2x \qquad\qquad \frac{d^2y}{dx^2} = 3^2e^{3x}$$

$$\frac{d^3y}{dx^3} = -2^3\cos 2x \qquad\qquad \frac{d^3y}{dx^3} = 3^3e^{3x}$$

$$\frac{d^4y}{dx^4} = 2^4\sin 2x \qquad\qquad \vdots$$

$$\frac{d^5y}{dx^5} = 2^5\cos 2x \qquad\qquad \frac{d^ny}{dx^n} = 3^ne^{3x}$$

$$\vdots$$

$$\frac{d^{12}y}{dx^{12}} = 2^{12}\sin 2x$$

Velocity and Acceleration In Section 1.3, we used the notation of instantaneous velocity to motivate the definition of the derivative. Let's review this notion by means of an example. Also, from now on, we will use the single word *velocity* in place of the more cumbersome phrase *instantaneous velocity*.

EXAMPLE 2: An object moves along the coordinate line in such a way that its position s satisfies $s = 2t^2 - 12t + 8$, where s is measured in centimeters and t in seconds with $t \geq 0$. Determine the velocity of the object when $t = 1$ and when $t = 6$. When is the velocity 0? When is it positive? Briefly describe the motion of the object in plain English.

SOLUTION: If we use the symbol $v(t)$ for the velocity at time t, then

$$v(t) = \frac{ds}{dt} = 4t - 12$$

Thus,

$$v(1) = 4(1) - 12 = -8 \text{ centimeters per second}$$

$$v(6) = 4(6) - 12 = 12 \text{ centimeters per second}$$

The velocity is 0 when $4t - 12 = 0$, that is, when $t = 3$. The velocity is positive when $4t - 12 > 0$, or when $t > 3$. All of this is shown schematically in Figure 1.

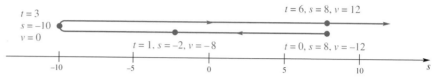

Figure 1

The object is, of course, moving along the s-axis, not on the curved path above it. But the path shows what is happening to the object. Between $t = 0$ and $t = 3$, the velocity is negative; the object is moving to the left (backing up). By the time $t = 3$, it has "slowed" to a zero velocity, and then starts moving to the right as its velocity becomes positive. Thus, negative velocity corresponds to moving in the direction of decreasing s; positive velocity corresponds to moving in the direction of increasing s. ◄

There is a technical distinction between the words *velocity* and *speed*. Velocity has a sign associated with it; it may be positive or negative (or zero). **Speed** is defined to be the absolute value of the velocity. Thus, in the example above, the speed at $t = 1$ is $|-8| = 8$ centimeters per second. The meter in most cars is a speedometer; it always gives nonnegative values.

Now we want to give a physical interpretation of the second derivative d^2s/dt^2. It is, of course, just the first derivative of the velocity. Thus, it measures the rate of change of velocity with respect to time, which has the name **acceleration**. If it is denoted by a, then

$$a = \frac{dv}{dt} = \frac{d^2s}{dt^2}$$

In Example 2, $s = 2t^2 - 12t + 8$. Thus,

$$v = \frac{ds}{dt} = 4t - 12$$

$$a = \frac{d^2s}{dt^2} = 4$$

This means that the velocity is increasing at a constant rate of 4 centimeters per second every second, which we write as 4 centimeters per second per second.

EXAMPLE 3: A point moves along a horizontal coordinate line in such a way that its position at time t is specified by

$$s = t^3 - 12t^2 + 36t - 30$$

Here s is measured in feet and t in seconds.

(a) When is the velocity 0?

(b) When is the velocity positive?
(c) When is the point moving backward (that is, to the left)?
(d) When is the acceleration positive?

SOLUTION:

(a) $v = \frac{ds}{dt} = 3t^2 - 24t + 36 = 3(t - 2)(t - 6)$. Thus, $v = 0$ at $t = 2$ and $t = 6$.

(b) The graph of the velocity $v = 3(t - 2)(t - 6)$ is shown in Figure 2. From the graph it is clear that the function is positive when $t < 2$ or $t > 6$.

(c) The point is moving left when $v < 0$—that is, when $3(t - 2)(t - 6) < 0$. From Figure 2 we see that this inequality has the interval $2 < t < 6$ as its solution.

(d) $a = \frac{dv}{dt} = 6t - 24 = 6(t - 4)$. The graph of the acceleration function is shown in Figure 3. We see that $a > 0$ when $t > 4$. The motion of the point is shown schematically in Figure 4. ◄

Measuring Time

If $t = 0$ corresponds to the present moment, then $t < 0$ corresponds to the past and $t > 0$ to the future. In many problems, it will be obvious that we are concerned only with the future. However, because Example 3 does not specify, it seems reasonable to allow t to have negative as well as positive values.

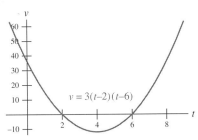

Figure 2

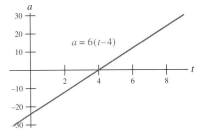

Figure 3

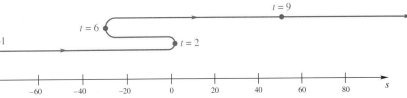

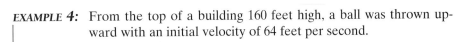

Figure 4

Falling-Body Problems

If an object is thrown straight up (or down) from an initial height s_0 feet with an initial velocity of v_0 feet per second and if s is its height above the ground in feet after t seconds, then

$$s = -16t^2 + v_0 t + s_0$$

This assumes that the experiment takes place near sea level and that the air resistance can be neglected. The diagram in Figure 5 portrays the situation we have in mind.

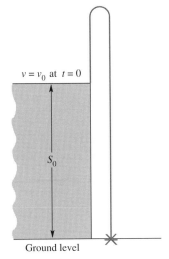

$v = v_0$ at $t = 0$

s_0

Ground level

Figure 5

EXAMPLE 4: From the top of a building 160 feet high, a ball was thrown upward with an initial velocity of 64 feet per second.

(a) When did it reach maximum height?
(b) What was its maximum height?
(c) When did it hit the ground?

(d) With what speed did it hit the ground?

(e) What was its acceleration at $t = 2$?

SOLUTION: Let $t = 0$ correspond to the instant when the ball was thrown. Then $s_0 = 160$ and $v_0 = 64$, so

$$s = -16t^2 + 64t + 160$$

$$v = \frac{ds}{dt} = -32t + 64$$

$$a = \frac{dv}{dt} = -32$$

> ### The Book of Nature
>
> "The great book of Nature lies ever open before our eyes and the true philosophy is written in it. . . . But we cannot read it unless we have first learned the language and the characters in which it is written. . . . It is written in mathematical language and the characters are triangles, circles, and other geometrical figures."
>
> Galileo Galilei

(a) The ball reached its maximum height at the time its velocity was 0—that is, when $-32t + 64 = 0$, or at $t = 2$ seconds.

(b) At $t = 2$, $s = -16(2)^2 + 64(2) + 160 = 224$ feet.

(c) The ball hit the ground when $s = 0$—that is, when

$$-16t^2 + 64t + 160 = 0$$

If we divide by -16 and use the quadratic formula, we obtain

$$t^2 - 4t - 10 = 0$$

$$t = \frac{4\sqrt{16 + 40}}{2} = \frac{42\sqrt{14}}{2} = 2\sqrt{14}$$

Only the positive answer makes sense. Thus, the ball hit the ground at $t = 2 + \sqrt{14} \approx 5.74$ seconds. *Note*: We could also have solved this equation using a root-finder, as in Section 2.4.

(d) At $t = 2 + \sqrt{14}$, $v = -32(2 + \sqrt{14}) + 64 \approx -119.73$. Thus, the ball hit the ground at a speed of 119.73 feet per second.

(e) The acceleration is always -32 feet per second per second. This is the acceleration of gravity near sea level. ◄

Derivatives and Mathematical Modeling

Galileo may have been right in claiming that the book of nature is written in mathematical language. Certainly, the scientific enterprise seems largely an effort to prove him correct. One of the basic elements of mathematical modeling is translating word descriptions into mathematical language. Doing this, especially in connection with derivatives, or rates of change, will become increasingly important as we go on. Here are some simple illustrations.

Word Description		*Mathematical Model*
Water is leaking from a cylindrical tank at a rate proportional to the depth of the water.		If V denotes the volume of the water at time t, then $\frac{dV}{dt} = -kh$.
A wheel is spinning at a constant rate of 6 revolutions per minute—that is, at $6(2\pi)$ radians per minute.		$\frac{d\theta}{dt} = 6(2\pi)$

Word Description		***Mathematical Model***
The density (in grams per centimeter) of a wire at a point is twice its distance from the left end.		If m denotes the mass of the left x centimeters of the wire, then $\dfrac{dm}{dx} = 2x$.
The height of a tree continues to increase but at a slower and slower rate.		$\dfrac{dh}{dt} > 0,\ \dfrac{d^2h}{dt^2} < 0$

The use of mathematical language is not limited to the physical sciences; it is also appropriate in the social sciences, especially in economics.

EXAMPLE 5: KTS News reported in May 1980 that unemployment was continuing to increase at an increasing rate. On the other hand, the price of food was increasing, but at a slower rate than before. Interpret these statements in mathematical language.

SOLUTION: Let $u = f(t)$ denote the number of people unemployed at time t. Though u actually jumps by unit amounts, we follow standard practice in representing u by a nice smooth curve, as in Figure 6. To say unemployment is increasing is to say $du/dt > 0$; to say that it is increasing at an increasing rate is to say $d^2u/dt^2 > 0$.

Similarly, if $p = g(t)$ represents the price of food (for instance, the typical cost of one day's groceries for one person) at time t, then $dp/dt > 0$. But because the rate of increase is becoming smaller, $d^2p/dt^2 < 0$; see Figure 7.

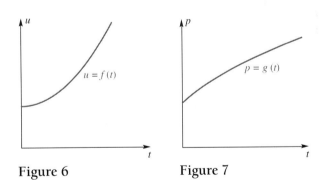

Figure 6   Figure 7

Concepts Review

1. If $y = f(x)$, then the third derivative of y with respect to x can be denoted by any of the following symbols: _____ .

2. If $s = f(t)$ denotes the position of a particle on a coordinate line at time t, then its velocity is given by _____ , its speed is given by _____ , and its acceleration is given by _____ .

3. Assume that an object is thrown straight up (positive direction), so that its height s at time t is given by $s = f(t)$. The object reaches its maximum height when $ds/dt =$ _____ , after which ds/dt _____ .

4. If the amount W of water in a tank at time t is increasing but at a slower rate, then dW/dt is _____ and d^2W/dt^2 is _____ .

Problem Set 3.5

In Problems 1–8, find d^3y/dx^3.

1. $y = x^3 + 3x^2 - 2x - 8$
2. $y = 2x^5 - x^4$
3. $y = (2x + 5)^4$
4. $y = (3x - 2)^5$
5. $y = \sin(3x)$
6. $y = \cos(x^2)$
7. $y = \ln(x - 3)$
8. $y = e^{x^2}$

In Problems 9–16, find $f''(2)$.

9. $f(x) = 2x^3 - 7$
10. $f(x) = 5x^3 + 1$
11. $f(t) = \dfrac{1}{t}$
12. $f(u) = \dfrac{1}{2u - 5}$
13. $f(x) = xe^x$
14. $f(x) = x \ln x$
15. $f(x) = \sin^2(\pi x)$
16. $f(x) = x \cos(\pi x)$

17. Let $n! = n(n - 1)(n - 2) \cdots 3 \cdot 2 \cdot 1$. Thus, $4! = 4 \cdot 3 \cdot 2 \cdot 1 = 24$ and $5! = 5 \cdot 4 \cdot 3 \cdot 2 \cdot 1$. We give $n!$ the name **n factorial**. Show that $\frac{d^n}{dx^n}(x^n) = n!$.

18. Using the factorial symbol of Problem 17, find the formula for

$$\frac{d^n}{dx^n}(a_nx^n + a_{n-1}x^{n-1} + \cdots + a_1x + a_0)$$

19. Without doing any calculating, find each derivative.

(a) $\dfrac{d^4}{dx^4}(3x^3 + 2x - 19)$

(b) $\dfrac{d^{12}}{dx^{12}}(100x^{11} - 79x^{10})$

(c) $\dfrac{d^{11}}{dx^{11}}(x^2 - 3)^5$

20. (a) Find a formula for $\dfrac{d^n}{dx^n}(1/x)$.

(b) Find a formula for $\dfrac{d^n}{dx^n} \ln x$.

21. If $f(x) = x^3 + 3x^2 - 45x - 6$, find the value of f'' at each zero of f'—that is, at each point c where $f'(c) = 0$.

22. Suppose that $g(t) = at^2 + bt + c$ and $g(1) = 5$, $g'(1) = 3$, and $g''(1) = -4$. Find a, b, and c.

In Problems 23–28, an object is moving along a horizontal coordinate line according to the formula $s = f(t)$, where s, the directed distance from the origin, is in feet and t is in seconds. In each case, answer the following questions (see Examples 2 and 3).

(a) What are $v(t)$ and $a(t)$, the velocity and acceleration at time t?

(b) When is the object moving right?

(c) When is it moving left?

(d) When is the acceleration negative?

(e) Draw a schematic diagram, showing the motion of the object.

23. $s = 12t - 2t^2$
24. $s = t^3 - 6t^2$
25. $s = e^t - 5t$
26. $s = e^t - 5t^2$
27. $s = t^2 + \dfrac{16}{t}, t > 0$
28. $s = t + \dfrac{4}{t}, t > 0$

29. If $s = \left(\frac{1}{2}\right)t^4 - 5t^3 + 12t^2$, find the velocity of the moving object when its acceleration is zero.

30. If $s = \frac{1}{10}(t^4 - 14t^3 + 60t^2)$, find the velocity of the moving object when its acceleration is zero.

31. Two particles move along a coordinate line. At the end of t seconds their directed distances from the origin, in feet, are given by $s_1 = 4t - 3t^2$ and $s_2 = t^2 - 2t$, respectively.

(a) When do they have the same velocity?

(b) When do they have the same speed? (The speed of a particle is the absolute value of its velocity.)

(c) When do they have the same position?

32. The positions of two particles, P_1 and P_2, on a coordinate line at the end of t seconds are given by $s_1 = 3t^3 - 12t^2 + 18t + 5$ and $s_2 = -t^3 + 9t^2 - 12t$, respectively. When do the two particles have the same velocity?

33. An object thrown directly upward is at a height $s = -16t^2 + 48t + 256$ feet after t seconds (see Example 4).

(a) What was its initial velocity?

(b) When did it reach maximum height?

(c) What was its maximum height?

(d) When did it hit the ground?

(e) With what speed did it hit the ground?

34. An object thrown directly upward from the ground level with an initial velocity of 48 feet per second is approximately $s = 48t - 16t^2$ feet high at the end of t seconds.

(a) What is the maximum height attained?

(b) How fast is it moving, and in which direction, at the end of 1 second?

(c) How long does it take to return to its original position?

35. A projectile is fired directly upward from the ground with an initial velocity of v_0 feet per second. Its height in t seconds is given by $s = v_0t - 16t^2$ feet. What must its initial velocity be for the projectile to reach a maximum height of 1 mile?

36. An object thrown directly downward from the top of a cliff with an initial velocity of v_0 feet per second falls approximately $s = v_0t + 16t^2$ feet in t seconds. If it strikes the ocean below in 3 seconds with a velocity of 140 feet per second, how high is the cliff?

37. A point moves along a horizontal line in such a way that its position at time t is specified by $s = t^3 - 3t^2 - 24t - 6$. Here, s is measured in centimeters and t in seconds. When is the point slowing down, that is, when is its *speed* decreasing?

38. Convince yourself that a point moving along a line is slowing down when its velocity and acceleration have opposite signs (see Problem 37).

39. Translate each of the following into the language of first, second, and third derivatives of distance with respect to time.

(a) The speed of that car is proportional to the distance it has traveled.

(b) That car is speeding up.

(c) I didn't say that car was slowing down; I said its rate of increase in speed was slowing down.

(d) That car's velocity is increasing 10 miles per hour every minute.

(e) That car is slowing very gently to a stop.

(f) That car always travels the same distance in equal time intervals.

40. Translate each of the following into the language of derivatives.

(a) Water is evaporating from that tank at a constant rate.

(b) Water is being poured into that tank at 3 gallons per minute but is also leaking out at $\frac{1}{2}$ gallon per minute.

(c) Since water is being poured into that conical tank at a constant rate, the water level is rising at a slower and slower rate.

(d) Inflation held steady this year but is expected to rise more and more rapidly in the years ahead.

(e) At present the price of oil is dropping but this trend is expected to slow and then reverse directions in 2 years.

(f) David's temperature is still rising but the penicillin seems to be taking effect.

41. Translate each of the following statements into mathematical language as in Example 5.

(a) The cost of a car continues to increase at a faster and faster rate.

(b) During the last 2 years, the United States has continued to cut its consumption of oil, but at a slower and slower rate.

(c) World population continues to grow, but at a slower and slower rate.

(d) That car is going faster and faster at a constant rate.

(e) The angle the Leaning Tower of Pisa makes with the vertical is increasing more rapidly.

(f) Upper Midwest firm's profit growth slows.

(g) The XYZ Company has been losing money but will soon turn this situation around.

42. Translate each statement from the following newspaper column into a statement about derivatives.

(a) In the United States, the ratio R of government debt to national income remained unchanged at around 28% up to 1981, but (b) then it began to increase more and more sharply, reaching 36% during 1983. (c) The IMF released a table showing that the speed increase of R was greater in the United States than in Japan.

43. Leibniz obtained a general formula for $\frac{d^n}{dx^n}(uv)$, where u and v are both functions of x. See if you can find it. *Hint*: Begin by considering the cases $n = 1$, $n = 2$, and $n = 3$.

44. Use the formula of Problem 43 to find $\frac{d^4}{dx^4}(x^4 \sin x)$.

45. Let $f(x) = x[\sin x - \cos(x/2)]$.

(a) Draw the graphs of $f(x)$, $f'(x)$, $f''(x)$, and $f'''(x)$ on $[0, 6]$ using the same axes.

(b) Evaluate $f'''(2.13)$.

46. Repeat Problem 45 for $f(x) = (x + 1)/(x^2 + 2)$.

Answers to Concepts Review: 1. $f'''(x)$; y'''; d^3y/dx^3
2. ds/dt; $|ds/dt|$; d^2s/dt^2 3. 0; <0 4. Positive; negative

3.6 | Implicit Differentiation and Related Rates

With a little work, you should be able to check that the graph of

$$y^3 + 7y = x^3$$

looks something like that shown in Figure 1. One approach to graphing such an equation is to build a table of values; for a given x-value we could use numerical root-finding to calculate the possible y-values. Also, some computer algebra systems can graph equations that are not solved in terms of y. Certainly the point $(2, 1)$ is on the graph, and there appears to be a well-defined tangent line at that point. How shall we find the slope of this tangent? Just calculate $\frac{dy}{dx}$ at this point, you may answer. But that is the rub; we do not know how to find $\frac{dy}{dx}$ in this situation.

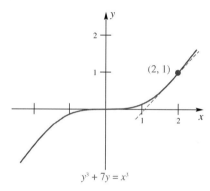

$$y^3 + 7y = x^3$$

Figure 1

The new element in this problem is that we are given an equation that is not explicitly solved for y. We will call such an equation an *implicit* equation for y. Try as we will, we cannot seem to solve it for y. Is it possible to find $\frac{dy}{dx}$ in circumstances like these? Yes. Differentiate both sides of the equation

$$y^3 + 7y = x^3$$

with respect to x and equate the results. In doing this, we think of y as a function of x (we can *think* of y as a function of x without knowing how to find it explicitly). And because y is a function of x, we know that $3y^2$ is a composite function that will require the Chain Rule. Thus, after using the Chain Rule on the first term, we get

$$3y^2 \cdot \frac{dy}{dx} + 7\frac{dy}{dx} = 3x^2$$

The latter can be solved for $\frac{dy}{dx}$ as follows.

$$\frac{dy}{dx}(3y^2 + 7) = 3x^2$$

$$\frac{dy}{dx} = \frac{3x^2}{3y^2 + 7}$$

Note that our expression for $\frac{dy}{dx}$ involves both x and y, a fact that is often a nuisance. But if we wish only to find a slope at a point where we know both coordinates, no difficulty exists. At $(2, 1)$,

$$\frac{dy}{dx} = \frac{3(2)^2}{3(1)^2 + 7} = \frac{12}{10} = \frac{6}{5}$$

The slope is $\frac{6}{5}$.

Note that we just found $\frac{dy}{dx}$ without first solving the given implicit equation for y explicitly in terms of x. This method is called **implicit differentiation**.

For some implicit equations, we may be able to solve for y, and for some others we may not be able to solve for y. When we can solve for y, we can take the derivative using the methods of the previous sections. The next example show that our new method yields the same result as our previous methods.

EXAMPLE 1: Sketch the graph of the equation $x^2 + y^2 = 4$. Find $\frac{dy}{dx}$ and use it to find the equation of the tangent at $(1, \sqrt{3}) \approx (1, 1.732)$. Sketch the tangent on your graph of $x^2 + y^2 = 4$.

SOLUTION: First we graph the equation $x^2 + y^2 = 4$. Because this is an implicit equation for y, we cannot graph the equation directly on a calculator or computer as we did before. We must either first solve for y, or else use a computer algebra system that can graph implicit equations (such as Mathematica or Maple).

Graphing $x^2 + y^2 = 4$: Method 1   We solve for y and get the two solutions $y = \sqrt{4 - x^2}$ and $y = -\sqrt{4 - x^2}$. We must graph *both* to get the entire graph of the equation. In Figure 2 we show a graphing calculator graph of these two equations. From our work in Chapter 1 we should have expected the graph of a circle centered at the origin, with radius 2. The upper part is the graph of $y = \sqrt{4 - x^2}$, and the lower part is the graph of $y = -\sqrt{4 - x^2}$.

Notice that the upper and lower parts are not quite connected; this reflects the difficulty that computing devices have when graphing curves that have a nearly vertical slope. When you make a hand sketch of the result, make sure that you connect these parts; you must use common sense to compensate for the lack of intelligence of a computer or calculator.

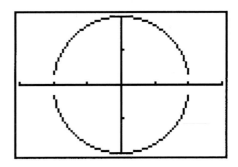

Calculator graph of $x^2 + y^2 = 4$. Upper part is $y = \sqrt{4 - x^2}$; lower part is $y = -\sqrt{4 - x^2}$.

Figure 2

Note: In order for the graph of the circle to appear circular, the scales or ranges for your graphing device must be set so that one unit on the *x*-axis is the same length as one unit on the *y*-axis.

Graphing $x^2 + y^2 = 4$: Method 2   If you have available a computer algebra system that can graph implicit equations directly, such as Mathematica or Maple, just specify the equation and the graph window and give the command to graph. Figure 3 shows the result of using implicit function graphing in Mathematica.

The advantage of this method is that one can graph any implicit equation, even when it is impossible to solve for *y*. For such equations, it is quite difficult to generate a graph any other way.

Next we find the derivative by two methods, which parallel the two methods of graphing.

Finding $\frac{dy}{dx}$: Method 1   We solve the given equation explicitly for *y* to get $y = \sqrt{4 - x^2}$ and $y = -\sqrt{4 - x^2}$. The point $(1, \sqrt{3})$ lies on the upper part of the graph, so we differentiate as follows:

$$\frac{dy}{dx} = \frac{d}{dx}(4 - x^2)^{\frac{1}{2}} = \frac{1}{2}(4 - x^2)^{-\frac{1}{2}} \cdot \frac{d}{dx}(4 - x^2)$$

$$= \frac{1}{2}(4 - x^2)^{-\frac{1}{2}} \cdot (-2x) = -\frac{x}{\sqrt{4 - x^2}}$$

Finding $\frac{dy}{dx}$: Method 2 (Implicit Differentiation).   We equate the derivatives of the two sides of

$$x^2 + y^2 = 4$$

treating *y* as an unknown function.

We obtain

$$2x + 2y\frac{dy}{dx} = 0$$

$$\frac{dy}{dx} = \frac{-2x}{2y} = -\frac{x}{y}$$

Though this answer at first looks different from the result of Method 1, the two derivatives are equivalent. To see this, substitute $y = \sqrt{4 - x^2}$ in the expression for $\frac{dy}{dx}$ just obtained.

$$\frac{dy}{dx} = -\frac{x}{y} = -\frac{x}{\sqrt{4 - x^2}}$$

At the point $(1, \sqrt{3})$ we get $\frac{dy}{dx} = -\frac{x}{y} = -\frac{1}{\sqrt{3}} \approx -0.577$. The equation of the tangent line is $y - \sqrt{3} = -\frac{1}{\sqrt{3}}(x - 1)$, or $y = -0.577x + 2.309$ to three decimal places. The graph of $x^2 + y^2 = 4$ is shown in Figure 4, along with the graph of the tangent line at $(1, \sqrt{3})$.   ◀

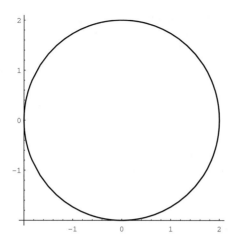

Graph of $x^2 + y^2 = 4$ using implicit graphing command.

Figure 3   **Figure 4**

More Examples

In the examples that follow, we assume that the given equation determines one or more differentiable functions whose derivatives can be found by implicit differentiation. It may not be possible to solve for y, as it was in Example 1. Note that in each case, we begin by taking the derivative of each side of the given equation with respect to the appropriate variable. When we come across the dependent variable we treat it as an unknown function and use the Chain Rule.

EXAMPLE 2: Find dy/dt if $t^3 + t^2y - 10y^4 = 0$.

SOLUTION:

$$\frac{d}{dt}(t^3 + t^2y - 10y^4) = \frac{d}{dt}(0)$$

$$\frac{d}{dt}t^3 + \frac{d}{dt}(t^2y) - \frac{d}{dt}10y^4 = 0$$

$$3t^2 + \left(t^2\frac{dy}{dt} + y(2t)\right) - 40y^3\frac{dy}{dt} = 0$$

$$t^2\frac{dy}{dt} - 40y^3\frac{dy}{dt} = -3t^2 - 2ty$$

$$\frac{dy}{dt}(t^2 - 40y^3) = -3t^2 - 2ty$$

$$\frac{dy}{dt} = \frac{3t^2 + 2ty}{40y^3 - t^2}$$

Notice that the term t^2y is a product of the function t^2 and the function y, which we think of as an unknown function of t. Thus, the derivative of this term is $t^2\frac{dy}{dt} + y(2t)$ using the product rule, as shown in the third step above.   ◄

EXAMPLE 3: Find the equation of the tangent line to the curve

$$y^3 - xy^2 + \cos xy = 2$$

at the point $(0, 1)$. Graph both the curve and the tangent line.

SOLUTION: For simplicity, let us use the notation y' for $\frac{dy}{dx}$. When we differentiate both sides and equate the results, we obtain

$$3y^2y' - x(2yy') - y^2 - (\sin xy)(xy' + y) = 0$$
$$y'(3y^2 - 2xy - x\sin xy) = y^2 + y\sin xy$$
$$y' = \frac{y^2 + y\sin xy}{3y^2 - 2xy - x\sin xy}$$

At $(0, 1)$, $y' = \frac{1}{3}$. Thus, the equation of the tangent line at $(0, 1)$ is

$$y - 1 = \frac{1}{3}(x - 0)$$

The equation $y^3 - xy^2 + \cos xy = 2$ cannot be solved for y (or for x). To graph it, we resort to computer software that can graph implicit equations. A Mathematica graph is shown in Figure 5.   ◄

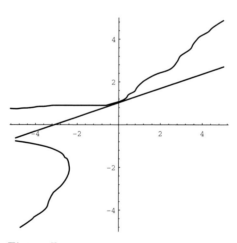

Figure 5

Related Rates   If a variable y depends on time t, then its derivative dy/dt is called a **time rate of change**. Of course, if y measures distance, then this time rate of change is also called velocity. We are interested in a wide variety of time rates: the rate at which water is flowing into a bucket, the rate at which the area of an oil spill is growing, the rate at which the value of a piece of real estate is increasing, and so on. If y is given explicitly in terms of t, the problem is simple; we just differentiate and then evaluate the derivative at the required time.

It may be that, in place of knowing y explicitly in terms of t, we know a relationship that connects y and another variable x and that we also know something about dx/dt. We may still be able to find dy/dt, because dy/dt and dx/dt are **related rates**. This will usually require implicit differentiation.

A Simple Example

In preparation for outlining a systematic procedure for solving related rates problems, we discuss the following example.

EXAMPLE 4: Water is pouring into a conical cistern at the rate of 8 cubic feet per minute. If the height of the cistern is 12 feet and the radius of its circular opening is 6 feet, how fast is the water level rising when the water is 4 feet deep?

SOLUTION: Denote the depth of the water in the cistern at any time t by h and let r be the corresponding radius of the surface of the water (see Figure 6).

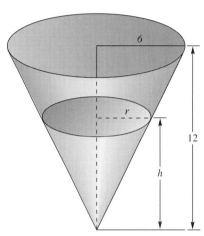

Figure 6

Similar Triangles

Two triangles are similar if their corresponding angles are congruent.

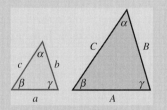

From geometry, we learn that ratios of corresponding sides of similar triangles are equal. For example,

$$\frac{b}{a} = \frac{B}{A}$$

This fact, used in Example 5, will be needed often in the problem set.

We are *given* that the volume, V, of water in the cistern is increasing at the rate of 8 cubic feet per minute; that is, $dV/dt = 8$. We *want to know* how fast the water is rising—that is, dh/dt—at the instant when $h = 4$.

We need to find an equation relating V and h. The formula for the volume of water in the cistern, $V = \frac{1}{3}\pi r^2 h$, contains the unwanted variable r, unwanted because we do not know its rate dr/dt. However, by similar triangles (see the marginal box), we have $r/h = 6/12$, so $r = h/2$. Substituting this into $V = \frac{1}{3}\pi r^2 h$, we get

$$V = \frac{\pi h^3}{12}$$

a relation holding for all $t > 0$.

Now we differentiate implicitly, keeping in mind that h depends on t. Now we obtain

$$\frac{dV}{dt} = \frac{3\pi h^2}{12}\frac{dh}{dt}$$

or

$$\frac{dV}{dt} = \frac{\pi h^2}{4}\frac{dh}{dt}$$

At this point, and not earlier, we consider the situation when $h = 4$. Substituting $h = 4$ and $dV/dt = 8$, we obtain

$$8 = \frac{\pi(4)^2}{4} \frac{dh}{dt}$$

from which

$$\frac{dh}{dt} = \frac{2}{\pi} \approx 0.637$$

When the depth of the water is 4 feet, the water level is rising at 0.637 foot per minute.   ◄

A Systematic Procedure

Example 4 suggests the following method for solving a related rates problem.

Step 1   Let t denote the elapsed time. Draw a diagram that is valid for all $t > 0$. Label those quantities whose values do not change as t increases, indicating their given constant values. Assign letters to the quantities that vary with t, and label the appropriate parts of the figure with these variables.

Step 2   State what is given about the variables and what information is wanted about them. This information will be in the form of the derivatives with respect to t.

Step 3   Write an equation relating the variables that is valid at all times $t > 0$, not just at some particular instant.

Step 4   Differentiate the equation found in Step 3 implicitly with respect to t. The resulting equation, containing the derivatives with respect to t, is true for all $t > 0$.

Step 5   At this point, and no earlier, substitute into the equation found in Step 4 all data that are valid *at the particular instant* for which the answer to the problem is required. Solve for the desired variables.

EXAMPLE 5:   A plane flying north at 640 miles per hour passes over a certain town at noon, and a second plane going east at 600 miles per hour is directly over the same town 15 minutes later. If the planes are flying at the same altitude, how fast will they be separating at 1:15 P.M.?

SOLUTION:

Step 1   Let t denote the number of hours after 12:15 P.M. Figure 7 shows the situation for all $t > 0$. The distance in miles from the town to the northbound plane when $t = 0$ (12:15 P.M.) is labeled with the constant $\frac{640}{4} = 160$. For any $t > 0$, we let y denote the distance in miles flown by the northbound plane (after 12:15 P.M.), x the distance flown by the eastbound plane, and s the distance between planes.

Step 2   We are given that for all $t > 0$, $dy/dt = 640$ and $dx/dt = 600$. We want to know ds/dt at $t = 1$—that is, at 1:15 P.M.

Step 3   By the Pythagorean Theorem

$$s^2 = x^2 + (y + 160)^2$$

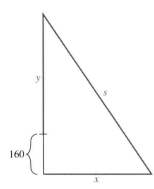

Figure 7

Step 4   Differentiating implicitly with respect to t and using the Chain Rule, we have

$$2s\frac{ds}{dt} = 2x\frac{dx}{dt} + 2(y + 160)\frac{dy}{dt}$$

or

$$s\frac{ds}{dt} = x\frac{dx}{dt} + (y + 160)\frac{dy}{dt}$$

Step 5   For all $t > 0$, $dx/dt = 600$ and $dy/dt = 640$, while at the particular instant $t = 1$, we have $x = 600$, $y = 640$, and $s = \sqrt{(600)^2 + (640 + 160)^2} = 1000$. When we substitute these data into the equation of Step 4, we obtain

$$1000\frac{ds}{dt} = (600)(600) + (640 + 160)(640)$$

from which

$$\frac{ds}{dt} = 872$$

At 1:15 P.M., the planes are separating at 872 miles per hour.

Now let's see if our answer makes sense. Look at Figure 7 again. Clearly, s is increasing faster than either x or y is increasing, so ds/dt exceeds 640. On the other hand, s is surely increasing more slowly than the sum of x and y; that is, $ds/dt < 600 + 640$. Our answer, $ds/dt = 872$, is reasonable.   ◄

Concepts Review

1.   The implicit relation $yx^3 - 3y = 9$ can be solved explicitly for y, yielding $y =$ _____.

2.   Implicit differentiation of $y^3 + x^3 = 2x$ with respect to x yields _____ $+ 3x^2 = 2$.

3.   To ask how fast u is changing with respect to time t after 2 hours is to ask the value of _____ at _____.

4.   An airplane flew directly over an observer, moving away at a constant airspeed of 400 miles per hour. The distance between the observer and the plane grew at an increasing rate eventually approaching a rate of _____.

Problem Set 3.6

Assuming that each equation in Problems 1–6 defines a differentiable function of x, find dy/dx by implicit differentiation.

1. $x^2 - y^2 = 9$

2. $4x^2 + 9y^2 = 36$

3. $xy = 4$

4. $b^2x^2 + a^2y^2 = a^2b^2$, where a and b are constants.

5. $6x - \sqrt{2xy} + xy^3 = y^2$

6. $\cos(xy) = y^2 + 2x$

In Problems 7–10, find the equation of the tangent line at the indicated point (see Examples 1, 3). Sketch both the curve and the tangent, if possible. If you can't graph the curve, explain why.

7. $x^2 - y^4 = 9$; $(5, 2)$

8. $xy^2 = 8$; $(2, 2)$

9. $\sin(xy) = y$; $(\pi/2, 1)$

10. $y + \cos(xy^2) + 3x^2 = 4$; $(1, 0)$

In Problems 11 and 12, find the indicated derivative(s).

11.   If $s^2t + t^3 = 1$, find ds/dt and dt/ds.

12. If $y = \sin(x^2) + 2x^3$, find dx/dy.

13. A particle of mass m moves along the x-axis so that its position x and velocity $v = dx/dt$ satisfy

$$m(v^2 - v_0^2) = k(x_0^2 - x^2)$$

where v_0, x_0, and k are constants. Show by implicit differentiation that

$$m\frac{dv}{dt} = -kx$$

whenever $v \neq 0$.

14. Find any points on the curve $x^2y - xy^2 = 2$ where the tangent line is vertical, that is, where $dx/dy = 0$.

The following are related-rates problems; use the four-step procedure outlined in this section.

15. Each edge of a variable cube is increasing at the rate of 3 inches per second. How fast is the volume of the cube increasing when the edge is 10 inches long?

16. Assuming that a soap bubble retains its spherical shape as it expands, how fast is its radius increasing when its radius is 2 inches, if the air is blown into it at the rate of 4 cubic inches a second?

17. An airplane, flying horizontally at an altitude of 1 mile, passes directly over an observer. If the constant speed of the plane is 240 miles per hour, how fast is its distance from the observer increasing 30 seconds later? *Hint*: Use Figure 8 and note that in 30 seconds (1/120 hour), the plane goes 2 miles.

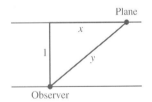

Figure 8

18. A student is using a straw to drink from a conical paper cup, whose axis is vertical, at the rate of 3 cubic centimeters a second. If the height of the cup is 10 centimeters and the diameter of its opening is 6 centimeters, how fast is the level of the liquid falling when the depth of the liquid is 5 centimeters?

19. An airplane, flying west at 400 miles per hour, goes over a certain town at 11:30 A.M., and a second plane at the same altitude, flying south at 500 miles per hour, goes over the town at noon. How fast are they separating at 1:00 P.M.? *Hint*: See Example 3.

20. A man on a dock is pulling in a rope fastened to the bow of a small boat. If the man's hands are 12 feet higher than the point where the rope is attached to the

boat and if he is retrieving the rope at the rate of 3 feet per second, how fast is the boat approaching the dock when 20 feet of rope are still out?

21. A 20-foot ladder is leaning against a wall. If the bottom of the ladder is pulled along the level pavement directly away from the wall at 2 feet per second, how fast is the top of the ladder moving down the wall when the foot of the ladder is 4 feet from the wall?

22. Oil from a ruptured tanker spreads in a circular pattern. If the radius of the circle increases at the constant rate of 1.5 feet per second, how fast is the enclosed area increasing at the end of 2 hours?

23. Sand is pouring from a pipe at the rate of 16 cubic feet per second. If the falling sand forms a conical pile on the ground whose altitude is always $\frac{1}{4}$ the diameter of the base, how fast is the altitude increasing when the pile is 4 feet high? *Hint*: Refer to Figure 9 and use the fact that $V = (\frac{1}{3})\pi r^2 h$.

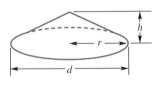

Figure 9

24. A child is flying a kite. If the kite is 90 feet above the child's hand level and the wind is blowing it on a horizontal course at 5 feet per second, how fast is the child paying out cord when 150 feet of cord is out? (Assume the cord forms a line—actually an unrealistic assumption.)

25. A swimming pool is 40 feet long, 20 feet wide, 8 feet deep at the deep end, and 3 feet deep at the shallow end; the bottom is rectangular (see Figure 10). If the pool is filled by pumping water into it at the rate of 40 cubic feet per minute, how fast is the water level rising when it is 3 feet deep at the deep end?

Figure 10

26. A particle P is moving along the graph of $y = \sqrt{x^2 - 4}$, $x \geq 2$, so that the x-coordinate of P is increasing at the rate of 5 units per second. How fast is the y-coordinate of P increasing when $x = 3$?

27. A metal disk expands during heating. If its radius increases at the rate of 0.02 inch per second, how fast is the area of one of its faces increasing when its radius is 8.1 inches?

28.   Two ships sail from the same island port, one going north at 24 knots (24 nautical miles per hour) and the other east at 30 knots. The northbound ship departed at 9:00 A.M. and the eastbound ship left at 11:00 A.M. How fast is the distance between them increasing at 2:00 P.M.? *Hint*: Let $t = 0$ at 11:00.

29.   A light in a lighthouse 1 kilometer offshore from a straight shoreline is rotating at 2 revolutions per minute. How fast is the beam moving along the shoreline when it passes the point $\frac{1}{2}$ kilometer from the point opposite the lighthouse?

30.   Water is pumped at a uniform rate of 2 liters (1 liter = 1000 cubic centimeters) per minute into a tank shaped like a frustum of a right circular cone. The tank has altitude 80 centimeters and lower and upper radii of 20 and 40 centimeters, respectively (Figure 11). How fast is the water level rising when the depth of the water is 30 centimeters? *Note*: The volume, V, of a frustum of a right circular cone of altitude h and lower and upper radii a and b is $V = \frac{1}{3}\pi h \cdot (a^2 + ab + b^2)$.

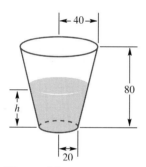

Figure 11

31.   Water is leaking out the bottom of a hemispherical tank of radius 8 feet at a rate of 2 cubic feet per hour. The tank was full at a certain time. How fast is the water level changing when its height h is 3 feet? *Note*: The volume of a segment of height h in a hemisphere of radius r is $\pi h^2[r - (\frac{h}{3})]$. (See Figure 12.)

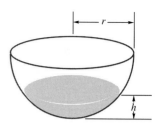

Figure 12

32.   A right circular cylinder with a piston at one end is filled with gas. Its volume is continually changing because of the movement of the piston. If the temperature of the gas is kept constant, then—by Boyle's Law—$PV = k$, where P is pressure (pounds per square inch), V is the volume (cubic inches), and k is a constant. The pressure was monitored by a recording device over one 10-minute period. The results are shown in Figure 13. Approximately how fast was the volume changing at $t = 6.5$ if its volume was 300 cubic inches at that instant?

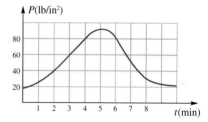

Figure 13

33.   Prove the Power Rule $\frac{d}{dx}x^r = rx^{r-1}$ for the case where r is rational, that is, $r = \frac{p}{q}$ where p and q are integers. *Hint*: Let $y = x^{p/q}$, then take both sides to the power q to get $y^q = x^p$. Then use implicit differentiation to find $\frac{dy}{dx}$.

Answers to Concepts Review: 1. $y = 9/(x^3 - 3)$
2. $3y^2 \frac{dy}{dx}$ 3. du/dt; $t = 2$ 4. 400 mi/h

3.7   Differentials and Approximations

We have been using Leibniz notation dy/dx for the derivative of y with respect to x. Up to now, we have treated dy/dx as a single symbol and have not tried to give separate meanings to dy and dx, as we will do next.

Differential Defined   Here is the formal definition of *differential*.

Definition

(Differential). Let $y = f(x)$ be differentiable at x and suppose that dx, the differential of the independent variable x, denotes an arbitrary increment of x. The corresponding differential dy of the independent variable y is defined by

$$dy = f'(x)dx$$

EXAMPLE 1: Find dy if (a) $y = x^3 - 3x + 1$, (b) $y = \sqrt{x^2 + 3x}$, (c) $y = \sin(x^4 - 3x^2 + 11)$.

SOLUTION: If we know how to calculate derivatives, we know how to calculate differentials. We simply calculate the derivative and multiply it by dx.

(a) $dy = (3x^2 - 3)dx$

(b) $dy = \dfrac{1}{2}(x^2 + 3x)^{-\frac{1}{2}}(2x + 3)dx = \dfrac{2x + 3}{2\sqrt{x^2 + 3x}}\,dx$

(c) $dy = \cos(x^4 - 3x^2 + 11) \cdot (4x^3 - 6x)dx$ ◀

Be careful to distinguish between derivatives and differentials. They are not the same. When you write $\frac{d}{dx}y$ or dy/dx, you are using a symbol for derivative; when you write dy, you are denoting a differential. Do not be sloppy and write dy when you mean to label a derivative.

Approximations

In Section 2.4 we introduced the idea of linear approximation through the equation

$$\Delta y \approx f'(c)\Delta x$$

where Δx represents the change in x from a point c to a nearby point $c + \Delta x$ and where $\Delta y = f(c + \Delta x) - f(c)$ represents the change in y along the graph of $y = f(x)$, and where $f'(c)\Delta x$ represents the change in y along the tangent line at $x = c$.

If we replace c with x and Δx with dx, then we see that $dy = f'(x)dx$ represents the *change in y along the tangent*, and our approximation equation becomes

$$\Delta y \approx dy$$

Thus, we can use the formula for dy to approximate small changes in a function $f(x)$. Another useful form of the approximation equation is

$$f(x + \Delta x) \approx f(x) + f'(x)\Delta x$$

In Figure 4 of Section 2.4 we displayed these quantities in graph form; in Figure 1 we reproduce this figure using the new notation. Looking at the figure, it is easy to see that $\Delta y \approx dy$ for *small* values of Δx.

Notation Note: The symbols Δx, dx, and h are all used to represent a small change in x and are used somewhat interchangeably. (Thus, $dx = \Delta x$ but $dy \neq \Delta y$). We could therefore write the last equation as

$$f(x + h) \approx f(x) + f'(x)h$$

We used differentials and linear approximation before, but we did not have the ability then to find derivative formulas easily. Now that we do, we can exploit the power of differentials to a larger extent.

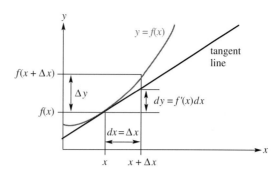

When Δx is small, $\Delta y \approx dy$ and hence
$f(x + \Delta x) \approx f(x) + f'(x)\Delta x$.

Figure 1

EXAMPLE 2: Use differentials to approximate the increase in the surface area of a soap bubble when its radius increases from 3 inches to 3.025 inches.

SOLUTION: The area of a spherical soap bubble is given by $A = 4\pi r^2$. We may approximate the exact change, ΔA, by the differential dA, where

$$dA = 8\pi r\,dr$$

At $r = 3$ and $dr = \Delta r = 0.025$,

$$dA = 8\pi(3)(0.025) \approx 1.885 \text{ square inches} \qquad \blacktriangleleft$$

Estimating Errors   Here is a typical problem in science. A researcher measures a certain variable x to have a value x_0 with possible error of size $\pm\Delta x$. The value x_0 is then used to calculate a value y_0 for a function y that depends on x. The value y_0 is contaminated by the error in x, but how badly? The standard procedure is to estimate this error by means of differentials.

EXAMPLE 3: The side of a cube is measured as 11.4 centimeters with a possible error of ±0.05 centimeter. Evaluate the volume of the cube and give an estimate for the error in this value.

SOLUTION: The volume V of a cube of side x is $V = x^3$. Thus $dV = 3x^2dx$. If $x = 11.4$ and $dx = 0.05$, then $V = (11.4)^3 \approx 1482$ and

$$dV = 3(11.4)^2(0.05) \approx 19$$

Thus, we might report the volume of the cube as 1482 ± 19 cubic centimeters. $\qquad \blacktriangleleft$

We can use the method of differentials with implicit equations, which were studied in the last section. The next example explores this idea, and also employs numerical equation solving as a check on the estimate obtained using differentials.

EXAMPLE 4: **(Implicit Equations)**
For the implicit equation $y^3 - xy + x^3 = 1$, when $x = 0$ we get $y^3 = 1$, and therefore $y = 1$. Use differentials to estimate y when $x = 0.1$. Also, find a five-digit approximation for the value of y when $x = 0.1$ by using the equation $y^3 - xy + x^3 = 1$ and numerical root-finding. Compare the two answers. If possible, sketch a graph of the given equation and use it to explain the difference between (or similarity of) your two answers.

SOLUTION: In order to find dy, we first need to find dy/dx by using implicit differentiation. We get

$$\frac{d}{dx}(y^3 - xy + x^3) = \frac{d}{dx} 1$$

$$3y^2\frac{dy}{dx} - \left(1 \cdot y + x \cdot \frac{dy}{dx}\right) + 3x^2 = 0$$

$$3y^2\frac{dy}{dx} - x\frac{dy}{dx} = y - 3x^2$$

$$\frac{dy}{dx} = \frac{y - 3x^2}{3y^2 - x}$$

In the language of differentials, we solve for dy to get $dy = \frac{y - 3x^2}{3y^2 - x}dx$.

Thus, at the point $(0, 1)$ we get $dy = \frac{1 - 3(0)^2}{3(1)^2 - 0}dx = \frac{1}{3}dx$. Because we want y when x changes from 0 to 0.1, we have $dx = \Delta x = 0.1$; this gives us $dy = \frac{1}{3}(0.1) = \frac{1}{30} \approx 0.03333$. This is the *change* in y, so the new value of y would be about $1 + 0.03333 = 1.03333$.

In order to find y directly from the equation, we will need to solve $y^3 - (0.1)y + (0.1)^3 = 1$ using a numerical root-finder. We first rewrite the equation as $y^3 - 0.1y - 0.999 = 0$ and proceed as explained in Section 2.4. If an initial guess is required we could use $y = 1$ because we don't expect y to change very much for a small change in x; if an initial interval is required we could use $0.9 \le y \le 1.1$ by the same reasoning. Our result is 1.032999053, which to five digits is 1.0330 (remember that on a graphing calculator you must change the y's to x's). Our differentials estimate rounded to five digits is 1.0333, so we are off only in the ten-thousandths place.

Can the equation $y^3 - xy + x^3 = 1$ be solved for y in terms of x? In theory, yes. There is a formula for solving cubic equations, just as there is the more familiar one for solving quadratics (see the box in Section 2.5). However, the formula is so complicated, it is not of much use; therefore, we resort to using software that can graph implicit equations.

Figure 2 shows a computer-generated graph of this equation along with the tangent line $y = \frac{1}{3}x + 1$ and points plotted on the graph at $x = 0$ and $x = 0.1$. Notice that near $x = 1$ there is more than one y-value for a given x-value; this means that we could not, even in theory, have solved *uniquely* for y in terms of x. ◀

EXAMPLE 5: In Example 1 of Section 2.2 we introduced the equation

$$y = -5e^{-10t}(\cos(10t) + \sin(10t))$$

as a model for the vertical displacement from rest position of a car bumper that has been pushed down 5 inches and released. Time is represented by t, in seconds, and vertical displacement is represented by y, in inches.

(1) Find the time when the bumper first reaches the rest position ($y = 0$).

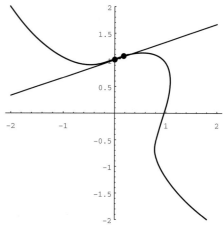

Figure 2

(2) Find the velocity of the bumper when it first reaches the rest position.

(3) Use differentials to estimate the time it will take to move 0.1 inch after reaching rest position for the first time, and illustrate your answer on a graph of $y = -5e^{-10t}(\cos(10t) + \sin(10t))$.

SOLUTION:

(1) We need to find the *first* solution to $-5e^{-10t}(\cos(10t) + \sin(10t)) = 0$ for $t > 0$. A graph will help identify where this point is. In Figure 3 we produce a graph with the window $0 \le t \le 1$, $-5 \le y \le 5$.

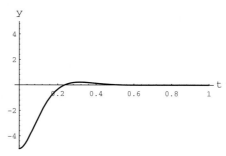

Graph of $y = -5e^{-10t}(\cos(10t) + \sin(10t))$ with graph window $0 \le t \le 1$, $-5 \le y \le 5$.

Figure 3

From the graph we see that we want a root of $f(t) = -5e^{-10t}(\cos(10t) + \sin(10t))$ near $t = 0.2$. We should consider three possible approaches:

(a) Find an approximate solution using a numerical root-finder (Section 2.5).

(b) Attempt an exact solution using the Solve command of a computer algebra system (Section 2.5).

(c) Attempt an exact solution using algebra.

Using approach (a), we can apply our numerical root-finder to the expression $-5e^{-10t}(\cos(10t) + \sin(10t))$ with an initial guess of $t = 0.2$ or an initial interval of $0.1 \leq t \leq 0.3$, based on the sketch in Figure 3. The result is a root of $t = 0.235619$.

An algebraic solution is possible in this case. To solve $-5e^{-10t}(\cos(10t) + \sin(10t)) = 0$ we set each factor equal to zero. The equation $-5e^{-10t} = 0$ has no real solutions, because the function e^{-10t} can never equal zero (why?). When we set the second factor equal to zero we get

$$\cos(10t) + \sin(10t) = 0$$

$$\frac{\cos(10t)}{\cos(10t)} + \frac{\sin(10t)}{\cos(10t)} = 0$$

$$\tan(10t) = -1$$

If we now use the arctan function to solve for t we get

$$t = \frac{1}{10} \arctan(-1) = \frac{1}{10}\left(-\frac{\pi}{4}\right) = -\frac{\pi}{40} \approx -0.0785398$$

which is *not* the solution we are looking for. Recall that inverse trig functions only give *one* of many possible solutions. We could, of course, use the fact that the tangent function repeats every π units to find the solution we are interested in. Numerical methods have the advantage that it is often easy to get the particular solution to an equation that you want, when there are several possible ones. *Note*: Computer algebra systems that are able to solve this equation exactly will find the same solution we found by hand.

(2) For a function that measures distance as a function of time, the velocity is given by the first derivative. With $f(t) = -5e^{-10t}(\cos(10t) + \sin(10t))$, we have

$$f'(t) = \frac{d}{dt}(-5e^{-10t}(\cos(10t) + \sin(10t)))$$

$$= \left[\frac{d}{dt}(-5e^{-10t})\right] \cdot (\cos(10t) + \sin(10t))$$

$$+ (-5e^{-10t}) \cdot \frac{d}{dt}(\cos(10t) + \sin(10t))$$

$$= 50e^{-10t}(\cos(10t) + \sin(10t))$$

$$- 5e^{-10t}(-10\sin(10t) + 10\cos(10t))$$

$$= 100e^{-10t}\sin(10t)$$

If you have a computer algebra system available, you should definitely check this result using the derivative command. There comes a point in the level of complexity of a problem where it is quicker and more accurate to do the algebra by computer than by hand. You must make this type of decision for yourself. We warn you, however, that you will need to retain the ability to take simpler derivatives quickly by hand in the chapters that follow. In particular, the basic derivative rules of this chapter *must* be memorized and easily recalled.

We now substitute $t = 0.235619$ from part (1) into the derivative above to get a velocity of

$$f'(0.235619) = 100e^{-10(0.235619)}\sin(10(0.235619))$$

$$\approx 6.70203 \text{ inches per second.}$$

(3) To estimate the time it takes to travel the next 0.1 inches, we use differentials in the form of the equation

$$dy = f'(t)dt$$

where $f'(t) = 100e^{-10t}\sin(10t)$ from part (2). We have $dy = 0.1$ and $t = 0.235619$, so $f'(t)$ becomes $f'(0.235619) \approx 6.70203$, as shown above. Thus

$$0.1 = 6.70203dt$$

$$dt = \frac{0.1}{6.70203} \approx 0.014921$$

so it will take about 0.015 seconds for the bumper to move the next 0.1 inches.

To get an idea of how good an estimate we got, we can look at a zoomed-in graph of $f(t)$ and the tangent line at $t = 0.235619$. See Figure 4. The actual time it takes to move 0.1 inch is a bit greater than the estimated time using differentials. From the sketch we see that the actual time is about $0.254 - 0.236 = 0.018$.

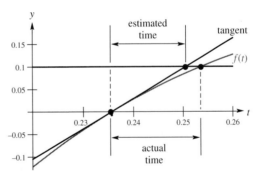

Graph of $y = -5e^{-10t}(\cos(10t) + \sin(10t))$ on $0.22 \le t \le 0.26$, with estimated time using differentials versus actual time to move from $y = 0$ to $y = 0.1$.

Figure 4

Summary of Results   The equation $y = -5e^{-10t}(\cos(10t) + \sin(10t))$ is used as a model of vertical position (y) of a car bumper as a function of time (t). Using this model we conclude the following. The bumper is pushed down 5 inches and released. It takes about 0.236 second for the bumper to return to its original position. Using differentials we estimate that it will take another 0.015 second to move 0.1 inch beyond rest position. From Figure 4 we see that the actual time it takes to move another 0.1 inch is about 0.018. The error in the estimate is due to the fact that the tangent line departs somewhat from the graph of the above equation in the region from $y = 0$ to $y = 0.1$.   ◄

Concepts Review

1. Let $y = f(x)$. The differential of y in terms of dx is defined by $dy = $ _____ .

2. Consider the curve $y = f(x)$ and suppose x is given an increment Δx. The corresponding change in y on the curve is denoted by _____ , whereas the corresponding change in y on the tangent line is denoted by _____ .

3. We can expect $dy \approx \Delta y$ to be a good approximation provided _____ .

4. On the curve $y = \sqrt{x}$, we should expect dy to be close to Δy but always _____ than Δy. On the curve $y = x^2$, $x \geq 0$, we should expect dy to be _____ than Δy.

Problem Set 3.7

In Problems 1–6, find a formula for dy.

1. $y = 2x^2 - 3x + 5$

2. $y = 7x^3 - 3x^2 + 4$

3. $y = (3 + 2x^3)^{-4}$

4. $y = \dfrac{13x}{5x^2 + 2}$

5. $y = \sqrt{4x^5 + 2x^4 - 5}$

6. $y = (6x^8 - 11x^5 + x^2)^{-\frac{2}{3}}$

7. If $s = \sqrt[5]{(t^2 - 3)^3}$, find ds.

8. If $F(x) = (5x^2 + 1)^2(x - 7)^5$, find dF.

9. Let $y = f(x) = x^3$. Find the value of dy in each case.

(a) $x = 0.5, dx = 1$ (b) $x = -1, dx = 0.75$

Make a careful drawing of the graph of f for $-1.5 \leq x \leq 1.5$ and the tangents to the curve at $x = 0.5$ and $x = -1$; on this drawing label dy and dx for each of the given sets of data in (a) and (b). Briefly discuss how good the approximation $\Delta y \approx dy$ is, based on your sketch.

10. Let $y = 1/x$. Find the value of dy in each case.

(a) $x = 1, dx = 0.5$ (b) $x = -2, dx = 0.75$

Make a large-scale drawing, as in Problem 9, for $-3 \leq x < 0$ and $0 < x \leq 3$. Briefly discuss how good the approximation $\Delta y \approx dy$ is, based on your sketch.

11. For the data of Problem 9, find the actual changes in y, namely, Δy.

12. For the data of Problem 10, find the changes in y, namely, Δy.

13. If $y = x^2 - 3$, find the values of Δy and dy in each case.

(a) $x = 2$ and $dx = \Delta x = 0.5$

(b) $x = 3$ and $dx = \Delta x = -0.12$

14. If $y = x^4 + 2x$, find the values of Δy and dy in each case.

(a) $x = 2$ and $dx = \Delta x = 1$

(b) $x = 2$ and $dx = \Delta x = 0.005$

15. For the implicit equation $x^4 + y^4 = 2$, when $x = 1$ we get $y^4 = 1$; thus, $y = 1$ or $y = -1$. Use differentials to estimate two values of y when $x = 0.9$. Also, find a five-digit approximation for the values of y when $x = 0.9$ by using the equation $x^4 + y^4 = 2$. Compare the two answers. If possible, sketch a graph of the given equation and use it to explain the difference between (or similarity of) your two answers.

16. For the implicit equation $x + y + e^x + e^y = 2$, when $x = 0$ we get $y + e^y = 1$; solving by inspection we get $y = 0$. Use differentials to estimate y when $x = 0.1$. Also, find a five-digit approximation for the value of y when $x = 0.1$ by using the equation $x + y + e^x + e^y = 2$ and numerical root-finding. Compare the two answers. If possible, sketch a graph of the given equation and use it to explain the difference between (or similarity of) your two answers.

17. Continue Example 5 using the equation

$$y = -5e^{-10t}(\cos(10t) + \sin(10t))$$

as a model for the vertical displacement from rest position of a car bumper that has been pushed down 5 inches and released. Time is represented by t, in seconds, and vertical displacement is represented by y, in inches.

(a) Find the time when the bumper reaches the rest position ($y = 0$) for the *second* time.

(b) Find the velocity of the bumper when it reaches the rest position for the second time.

(c) Use differentials to estimate the time it will take to move 0.1 inch after reaching rest position for the second time, and illustrate your answer on a graph of $y = -5e^{-10t}(\cos(10t) + \sin(10t))$.

18. In Example 4 of Section 2.5 we used the model

$$P(t) = 50e^{0.03t} + 2\sin(2\pi t) + 3$$

to describe the growth in the population ($P(t)$, in thousands) of a country as a function of time (t, in years), where $t = 0$ corresponds to the current time.

(a) According to the model, what will the population be in one year?

(b) What will the rate of growth of the population be in one year?

(c) Use differentials to estimate the population 13 months from now based on your answers to parts (a) and (b).

(d) Use differentials to estimate when the population will first reach 56,000, based on your answers to parts (a) and (b).

(e) Estimate when the population will first reach 56,000 by using the model $P(t) = 50e^{0.03t} + 2 \sin(2\pi t) + 3$ directly.

(f) Compare your answers to parts (d) and (e) and explain any differences. Use a sketch in your explanation.

19. Approximate the volume of material in a spherical shell of inner radius 5 centimeters and outer radius 5.125 centimeters. (See Example 3.)

20. All six sides of a cubical metal box are 0.25 inch thick, and the volume of the interior of the box is 40 cubic inches. Use differentials to find the approximate volume of metal used to make the box.

21. The outside diameter of a thin spherical shell is 12 feet. If the shell is 0.3 inch thick, use differentials to approximate the volume of the interior region of the shell.

22. The interior of an open cylindrical tank is 12 feet in diameter and 8 feet deep. The bottom is copper and the sides are steel. Use differentials to find approximately how many gallons of waterproofing paint are needed to apply a 0.05-inch coat to the steel part of the inside of the tank (1 gallon ≈ 231 cubic inches).

23. Assuming that the equator is a circle whose radius is approximately 4000 miles, how much longer than the equator would a rope stretched around the earth be if each point on the rope were 2 feet above the equator? Use differentials.

24. The period of a simple pendulum of length L feet is given by $T = 2\pi\sqrt{L/g}$ seconds. We assume that g, the acceleration due to gravity on (or very near) the surface of the earth, is 32 feet per second per second. If the pendulum is that of a clock that keeps good time when $L = 4$ feet, how much time will the clock gain in 24 hours if the length of the pendulum is decreased to 3.97 feet?

25. The diameter of a sphere is measured as 20 ± 0.1 centimeter. Calculate the volume with an estimate for the error. (See Example 3.)

26. A cylindrical roller is exactly 12 inches long and its diameter is measured as 6 ± 0.005 inch. Calculate its volume with an estimate for the error.

27. A tank has the shape of a cylinder with hemispherical ends. If the cylindrical part is 100 centimeters long and has a radius of 10 centimeters, about how much paint is required to coat the outside of a tank to a thickness of 1 millimeter?

28. Einstein's Special Theory of Relativity says that mass m is related to velocity v by the formula

$$m = \frac{m_0}{\sqrt{1 - v^2/c^2}} = m_0\left(1 - \frac{v^2}{c^2}\right)^{-\frac{1}{2}}$$

Here, m_0 is the rest mass and c is the velocity of light. Use differentials to determine the percent increase in mass of an object when its velocity is increased from $0.9c$ to $0.92c$.

Answers to Concepts Review: 1. $f'(x)\,dx$ 2. Δy; dy 3. Δx is small 4. Larger, smaller

3.8 | Chapter Review

Concepts Test

Respond with true or false to each of the following assertions. Be prepared to justify your answer.

1. For a given x-value, the y-coordinate on the curve $y = f(x)$ is the same as slope of the curve $y = f'(x)$.

2. The exact value of $\lim\limits_{h\to 0}\dfrac{(1 + h)^\pi - 1}{h}$ is π.

3. It is possible for the velocity of an object to be increasing while its speed is decreasing.

4. If f is continuous at c, then $f'(c)$ exists.

5. If $f'(x) = g'(x)$ for all x, then $f(x) = g(x)$ for all x.

6. If $y = e^5$, then $\dfrac{dy}{dx} = e^5$.

7. If $f'(c)$ exists, then f is continuous at c.

8. The graph of $y = \sqrt[3]{x}$ has a tangent line at $x = 0$ and yet $\dfrac{dy}{dx}$ does not exist there.

9. The derivative of a product is the product of the derivatives.

10. If the acceleration of an object is negative, then its velocity is decreasing.

11. If x^3 is a factor of the differentiable function $f(x)$, then x^2 is a factor of its derivative.

12. The equation of the line tangent to the graph of $y = x^3$ at $(1, 1)$ is $y - 1 = 3x^2(x - 1)$.

13. If $y = f(x)g(x)$, then $\dfrac{d^2y}{dx^2} = f(x)g''(x) + g(x)f''(x)$.

14. If $y = (x^3 + x)^8$, then $\dfrac{d^{25}y}{dx^{25}} = 0$.

15. The derivative of a polynomial is a polynomial.

16. The function $f(x) = e^x$ is the only function for which $\dfrac{d}{dx} f(x) = f(x)$.

17. If $f'(c) = g'(c) = 0$ and $h(x) = f(x)g(x)$, then $h'(c) = 0$.

18. $\dfrac{d}{dx}\ln(e^x) = 1$.

19. $\dfrac{d}{dx} \ln(\sin x) = (\cos x)\ln(\sin x)$.

20. If $h(x) = f(g(x))$ where both f and g are differentiable, then $g'(c) = 0$ implies $h'(c) = 0$.

21. If $f'(2) = g'(2) = g(2) = 2$, then $(f \circ g)'(2) = 4$.

22. If f is differentiable and increasing and if $dx = \Delta x > 0$, then $\Delta y > dy$.

23. If the radius of a sphere is increasing at 3 feet per second, then its volume is increasing at 27 cubic feet per second.

24. $\dfrac{d^{n+4}}{dx^{n+4}}(\sin x) = \dfrac{d^n}{dx^n} (\sin x)$ for every positive integer n.

25. $\displaystyle\lim_{x \to 0} \dfrac{\tan x}{3x} = \dfrac{1}{3}$.

26. If $s = 5t^3 + 6t - 300$ gives the position of an object on a horizontal coordinate line at time t, then that object is always moving to the right (the direction of increasing s).

27. If air is being pumped into a spherical rubber balloon at a constant rate of 3 cubic inches per second, then the radius will increase but at a slower and slower rate.

28. If water is being pumped into a spherical tank of fixed radius at a rate of 3 gallons per second, the height of the water in the tank will increase more and more rapidly as the tank nears being full.

29. If an error Δr is made in measuring the radius of a sphere, the corresponding error in the calculated volume will be approximately $S \cdot \Delta r$ where S is the surface area of the sphere.

30. If $y = x^5$, then $dy \geq 0$.

Sample Test Problems

1. Use $f'(x) = \displaystyle\lim_{h \to 0}[f(x + h) - f(x)]/h$ to prove that $\dfrac{d}{dx}x^n = nx^{n-1}$.

2. Let $h(x) = e^{\sin x}$.

(a) Find $h'(x)$ and use it to find $h'(0)$, $h'\left(\dfrac{\pi}{2}\right)$, and $h'(2)$.

(b) Sketch graphs of both $h(x)$ and $h'(x)$ on the interval $0 \leq x \leq \pi$, and on the graphs indicate the three values you found in part (a).

In Problems 3 and 4, the given limit is a derivative, but of what function f and at what point?

3. $\displaystyle\lim_{h \to 0} \dfrac{\dfrac{1}{2 + h} - \dfrac{1}{2}}{h}$

4. $\displaystyle\lim_{\Delta x \to 0} \dfrac{\tan(\pi/4 + \Delta x) - 1}{\Delta x}$

In Problems 5–14, find each derivative by using the rules we have developed.

5. $\dfrac{d}{dx}(x^3 - 3x^2 + 2e^x)$

6. $\dfrac{d}{dx}\left(\dfrac{3x - 2\ln x}{x^2 + 1}\right)$

7. $\dfrac{d^2}{dx^2}e^{3x^2 + 2}$

8. $\dfrac{d}{dt}(t\sqrt{2t + 6})$

9. $\dfrac{d}{dx} 5\ln(x^2 + 4)$

10. $\dfrac{d}{dx}[\sin(t^2) - \sin^2(t)]$

11. $\dfrac{d}{dx} (\cos^3 5x)$

12. $\dfrac{d}{dt}[\sin^2(\cos 4t)]$

13. $f'(2)$ if $f(x) = (x^2 - 1)^2(3x^3 - 4x)$

14. $g''(0)$ if $g(x) = \sin 3x + \sin^2 3x$

15. A spherical balloon is expanding from the sun's heat. Find the rate of change of the balloon with respect to its radius when the radius is 5 meters.

16. Use the differentials to approximate the change in volume of the balloon of Problem 15 when its radius increased from 5 to 5.1 meters.

17. If the volume of the balloon of Problem 15 is increasing at a constant rate of 10 cubic meters per hour, how fast is its radius increasing when the radius is 5 meters?

18. A trough 12 feet long has a cross section in the form of an isosceles triangle 4 feet deep and 6 feet across at the top. If water is filling the trough at the rate of 9 cubic feet per minute, how fast is the water level rising when the water is 3 feet deep?

19. An object is projected directly upward from the ground with an initial velocity of 128 feet per second. Its height s at the end of t seconds is approximately $s = 128t - 16t^2$ feet.

(a) When does it reach its maximum height and what is this height?

(b) When does it hit the ground and with what velocity?

20. An object moves on a horizontal coordinate line. Its directed distance s from the origin at the end of t seconds is $s = te^{-t}$.

(a) When is the object moving to the left?

(b) What is its acceleration when its velocity is zero?

(c) When is its acceleration positive?

21. Find $\dfrac{d^{20}y}{dx^{20}}$ in each case.

(a) $y = 13x^{19} - 2x^{12} - 6x^5 + 18$

(b) $y = \ln x$

(c) $y = 3e^{2x}$

22. Find dy/dx in each case.

(a) $x^3 + y^3 = x^3y^3$

(b) $x \sin x(xy) = x^2 + 1$

(c) $x + e^y = y$

23. Show that the tangents to the curves $y^2 = 4x^3$ and $2x^2 + 3y^2 = 14$ at $(1, 2)$ are perpendicular to each other. *Hint*: Use implicit differentiation.

24. Let $y = \sin(\pi x) + x^2$. If x changes from 2 to 2.01, approximately how much does y change?

25. Suppose that $f(2) = 3, f'(2) = 4, f''(2) = -1$, $g(2) = 2$, and $g'(2) = 5$. Find each value.

(a) $\dfrac{d}{dx}[f^2(x) + g^3(x)]$ at $x = 2$

(b) $\dfrac{d}{dx}[f(x)g(x)]$ at $x = 2$

(c) $\dfrac{d}{dx}[f(g(x))]$ at $x = 2$

26. A 13-foot ladder is leaning against a vertical wall. If the bottom of the ladder is being pulled along the ground at a constant rate of 2 feet per second, how fast is the top end of the ladder moving down the wall when it is 5 feet above the ground?

27. An airplane is climbing at a $15°$ angle to the horizon. How fast is it gaining altitude if its speed is 400 miles per hour?

28. Given that $\dfrac{d}{dx}|x| = |x|/x, \ x \neq 0$, find a formula for $\dfrac{d}{dx}|\sin x|$.

CHAPTER 4

Applications of the Derivative

4.1 Maximums and Minimums

Often in life we are faced with the problem of finding the *best* way to do something. For example, a farmer wants to choose the mix of crops that is likely to produce the largest profit. A doctor wishes to select the smallest dosage of a drug that will cure a certain disease. A manufacturer would like to minimize the cost of distributing its products. Sometimes a problem of this type can be formulated so it involves maximizing or minimizing a function over a specified set. If so, the methods of calculus provide a powerful tool for solving the problem.

Suppose we are given a function f and a domain S, as in Figure 1. Our first job is to decide whether f even has a maximum value or a minimum value on S. Assuming that such values exist, we want to know

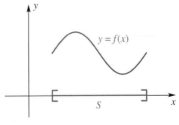

Figure 1

where on S they occur. Finally, we wish to determine the maximum and the minimum values. Analyzing these three tasks is the principal goal of this section.

We begin by introducing a precise vocabulary.

Definition

Let S, the domain of f, contain the point c. We say that:
(i) $f(c)$ is the **maximum value** of f on S if $f(c) \geq f(x)$ for all x in S;
(ii) $f(c)$ is the **minimum value** of f on S if $f(c) \leq f(x)$ for all x in S;
(iii) $f(c)$ is an **extreme value** of f on S if it is either the maximum value or the minimum value.

In plain English, this just says that the maximum value of a function $f(x)$ (or max for short) is the largest possible y-coordinate on the graph of $y = f(x)$. Similarly, the minimum value (or min) is the smallest possible y-coordinate on the graph of $y = f(x)$. *Warning*: By smallest, we mean farthest to the left on the number line, so that -4 is "smaller" than -1.

EXAMPLE 1: Find the maximum value of $f(x) = x - x^3$ on the interval $[0, 1]$.

SOLUTION: In Figure 2 we show a sketch of the graph of $f(x)$ on the interval $[0, 1]$ and a table of values for the function on this interval with increment $\Delta x = 0.1$. The points in the table are plotted on the graph.

x	$x - x^3$
0.0	0.000
0.1	0.099
0.2	0.192
0.3	0.273
0.4	0.336
0.5	0.375
0.6	0.384 maximum?
0.7	0.357
0.8	0.288
0.9	0.171
1.0	0.000

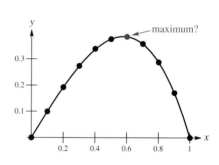

Figure 2

From the table, it appears that a maximum of 0.384 occurs at $x = 0.6$. But how do we know that there is not a larger y-value somewhere on the graph of $y = f(x)$? In fact, from the graph, it appears that there is no point plotted exactly at the top of the hump. We can be relatively confident (though not certain) that the maximum lies between $x = 0.5$ and $x = 0.7$, and by using a table with a smaller increment Δx we could get more accuracy.

With a graphing calculator, we could get the same information by using the Trace feature. Trace values near the top of the curve on the calculator screen. If you can find three consecutive Trace points for which the middle point has the highest y-coordinate, then you can be fairly sure that a maximum lies somewhere between the outer two points.

In Figure 3 we show the process just described. We estimate that the maximum y-value is about 0.385, which occurs somewhere between $x = 0.564$ and

$x = 0.585$. Note that the fact that eight digits are printed out for the x- and y-coordinates does *not* imply that we have found the location of the maximum accurate to eight digits.

By zooming, we can get increased accuracy in our estimate of the maximum, but we would never be sure of having an exact result. Remember that the Trace is just displaying an internal table of values, and that the exact maximum could lie between two of these values.   ◀

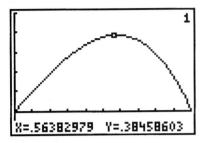
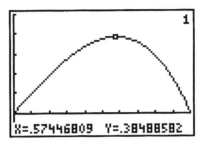
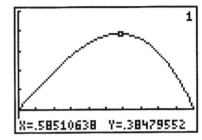

Figure 3

The point of the rest of this section is to show that it is possible to use calculus to find the *exact* maximum and minimum of a function under certain conditions. It turns out that instead of looking at 10 or 100 or more points to find maximums and minimums, we usually need to look only at a few key points, called critical points.

Notation Note

Recall from Section 1.1 that $[a, b]$ and $\{x: a \le x \le b\}$ both represent the closed interval of real numbers between a and b, where "closed" means the endpoints are included. Also, (a, b) and $\{x: a < x < b\}$ both represent the open interval from a to b, where "open" means the endpoints are not included. An interval can be open at one end and closed at the other; the interval $(1, 4]$ would include 4 but not 1.

The Existence Question   *Does f have a maximum (or minimum) value on S?* The answer depends first of all on the set S. Consider $f(x) = 1/x$ on $S = (0, \infty)$; it has neither a maximum value nor a minimum value (Figure 4). On the other hand, the same function on $S = [1, 3]$ has a maximum value of $f(1) = 1$ and a minimum value of $f(3) = \frac{1}{3}$. On $S = (1, 3]$, f has no maximum value, but has a minimum value of $f(3) = \frac{1}{3}$.

Why is it that f has no maximum value $S = (1, 3]$? The y-values get larger as you look left along the graph of the function. But because the interval is open, there is no single left-most point in the domain; so there can't be a largest, or maximum, y-value.

The answer also depends on the type of function. Consider the discontinuous function g (Figure 5) defined by

$$g(x) = \begin{cases} x & \text{if } 1 \le x < 2 \\ x - 2 & \text{if } 2 \le x \le 3 \end{cases}$$

On $S = [1, 3]$, g has no maximum value (it gets very close to 2 but never reaches it). However, g has the minimum value $g(2) = 0$.

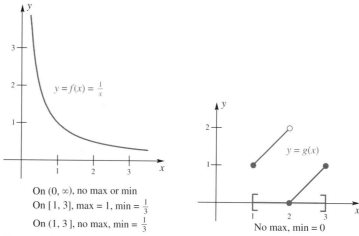

On $(0, \infty)$, no max or min
On $[1, 3]$, max $= 1$, min $= \frac{1}{3}$
On $(1, 3\,]$, no max, min $= \frac{1}{3}$

Figure 4

No max, min $= 0$

Figure 5

There is a useful theorem that answers the existence question for some of the problems that come up in practice. Though it is intuitively obvious, a rigorous proof is quite difficult; we leave that for more advanced textbooks.

Theorem A

(Max-Min Existence Theorem). If f is continuous on a closed interval $[a, b]$, then f reaches both a maximum value and a minimum value there.

Note the key words: f is required to be *continuous* and the set S is required to be a *closed interval*.

Where Do Extreme Values Occur?

Usually a function that we want to maximize or minimize will have an interval I as its domain. Some intervals contain their endpoints; some do not. For instance, $I = [a, b]$ contains both its endpoints; $[a, b)$ contains only its left endpoint; (a, b) contains neither endpoint. *Extreme values of functions defined on closed intervals often occur at endpoints* (see Figure 6).

If c is a point at which $f'(c) = 0$, we call c a **stationary point.** The name derives from the fact that at a stationary point, the graph of f levels off, because the tangent line is horizontal. *Extreme values often occur at stationary points* (see Figure 7).

Finally, if c is an interior point of I where f' fails to exist, we call c a **singular point**. It is a point where the graph of f has a sharp corner, a vertical tangent, or perhaps takes a jump (or near which it wiggles very badly). *Extreme values can occur at singular points* (see Figure 8), though in practical problems this is quite rare.

These three kinds of points (endpoints, stationary points, and singular points) are the key points of max-min theory. If any point in the domain of a function f is one of these three types, its value is called a **critical number** of f.

EXAMPLE 2: Find the critical numbers of $f(x) = -2x^3 + 3x^2$ on $\left[-\frac{1}{2}, 2\right]$.

    SOLUTION: The endpoints are $-\frac{1}{2}$ and 2. To find the stationary points, we solve $f'(x) = -6x^2 + 6x = 0$ for x, obtaining 0 and 1. There are no singular points. Thus, the critical numbers are $-\frac{1}{2}, 0, 1, 2$.   ◄

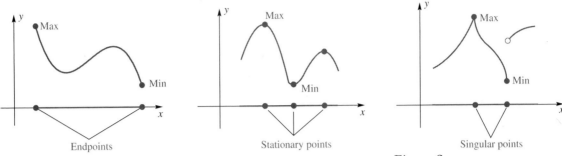

Figure 6   Figure 7   Figure 8

Theorem B

(Critical Number Theorem). Let f be defined on an interval I containing c. If $f(c)$ is an extreme value, then c must be a critical number—that is,

(i) an endpoint of I;

(ii) a stationary point of f (the derivative $f'(c) = 0$); or

(iii) a singular point of f (the derivative $f'(c)$ does not exist).

Outline of Proof Consider first the case where $f(c)$ is the maximum value of f on I, and suppose that c is neither an endpoint nor a singular point. It will be enough to show that c is then a stationary point (why?).

Now because $f(c)$ is the maximum value, $f(x) \leq f(c)$ for all x in I; that is,

$$f(x) - f(c) \leq 0$$

Thus, if $x < c$, so that $x - c < 0$, then

(1)
$$\frac{f(x) - f(c)}{x - c} \geq 0$$

and if $x > c$, then

(2)
$$\frac{f(x) - f(c)}{x - c} \leq 0$$

Because we assumed that c is not a singular point, $f'(c)$ exists. Consequently, when we let $x \to c^-$ in (1) and $x \to c^+$ in (2), we obtain, respectively, $f'(c) \geq 0$ and $f'(c) \leq 0$. We conclude that $f'(c) = 0$, as desired. We used the fact that the inequality $\leq$ is preserved under the operation of taking limits. This fact is proven in any standard analysis book.

The case where $f(c)$ is the minimum value is handled similarly.

What Are the Extreme Values?

In view of Theorems A and B, we can now state a very simple procedure for finding the maximum value or minimum value of a continuous function f on a *closed interval I*.

Step 1 Find the critical numbers of f on I.

Step 2   Evaluate f at each of the critical numbers. The largest of these values of f is the maximum value; the smallest is the minimum value.

EXAMPLE 3: Find the maximum and minimum values of

$$f(x) = -2x^3 + 3x^2$$

on $\left[-\frac{1}{2}, 2\right]$.

SOLUTION: In Example 2, we identified $-\frac{1}{2}, 0, 1, 2$ as the critical numbers. Now $f\left(-\frac{1}{2}\right) = 1, f(0) = 0, f(1) = 1$, and $f(2) = -4$. Thus, the maximum value is 1 (reached at both $-\frac{1}{2}$ and 1) and the minimum value is -4 (reached at 2). The graph of f is shown in Figure 9. All the relevant information is summarized in table form in Figure 10.

Notice that we needed to put only four points into our table in order to determine the exact maximum and minimum. Contrast this with Example 1, where we generated a table of 10 values and still did not find the exact maximum.   ◄

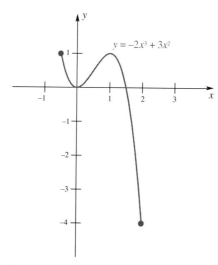

$y = -2x^3 + 3x^2$

Figure 9

x	$-2x^3 + 3x^2$	
-0.5	1	max
0	0	
1	1	max
2	-4	min

Maximum of 1 occurs at $x = -0.5$ and at $x = 1$. Minimum of -4 occurs at $x = 2$.

Figure 10

EXAMPLE 4: The function $F(x) = x^{\frac{2}{3}}$ is continuous everywhere. Find its maximum and minimum values on $[-1, 2]$.

SOLUTION: $F'(x) = \frac{2}{3}x^{-\frac{1}{3}}$, which is never 0. However, $F'(0)$ does not exist, so 0 is a critical number, as are the endpoints -1 and 2. Now $F(-1) = 1$, $F(0) = 0$, and $F(2) = \sqrt[3]{4} \approx 1.59$. Thus, the maximum value is $\sqrt[3]{4}$; the minimum value is 0. The graph is shown in Figure 11. A table of approximate values is shown in Figure 12.   ◄

What happens when it is difficult or impossible to solve the equation $f'(x) = 0$? We must then resort to a numerical solution, as we demonstrate in the next example.

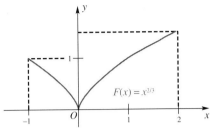

Figure 11

x	$F(x)$	
-1	1	
0	0	min
2	1.587	max

Figure 12

EXAMPLE 5: Find the maximum and minimum values of $f(x) = \dfrac{\sin x}{x}$ on the interval $\pi \le x \le 2\pi$.

SOLUTION: We have

$$f'(x) = \frac{d}{dx}\frac{\sin x}{x} = \frac{\cos x \cdot x - \sin x \cdot 1}{x^2}$$

$$= \frac{x\cos x - \sin x}{x^2}$$

using the quotient rule. To solve $f'(x) = 0$ we set the numerator equal to zero, since a fraction is zero only if the numerator is zero. However, the equation

$$x\cos x - \sin x = 0$$

cannot be solved algebraically. It can be transformed into the equation $\tan x = x$ (show this!), but we still cannot isolate x.

We need a numerical solution. There are two ways to proceed. We can use numerical root finding as explained in Section 2.5 to solve the equation $f'(x) = 0$. Some calculators also have the ability to numerically find maximums and minimums directly from the original function.

With either method, we need an initial estimate of the critical numbers. We can get these estimates from a graph of $f(x)$ or from a graph of $f'(x)$. We show both in Figure 13.

Notice that the graph of $f(x)$ shows that we have a minimum at about $x = 4.5$; this corresponds to a zero (x-intercept) on the graph of $f'(x)$ at the same point.

Approach 1:   Apply Numerical Root Finding to $f'(x)$.
Enter the derivative $\frac{x\cos x - \sin x}{x^2}$—*not* the original function $f(x)$—into your calculator or computer. (With computer algebra, you could enter the original function first and then use the Derivative command.) Now use the root-finding command as explained in Section 2.5, with an appropriate initial guess or initial interval. Graphing the derivative first to get the initial guess or interval is usually a good idea (remember, you are looking for an x-intercept). From the graph in Figure 13, a possible initial guess could be $x = 4.5$, and a possible initial interval could be $[4, 5]$. The result is a critical number of $x \approx 4.49341$, accurate to six digits.

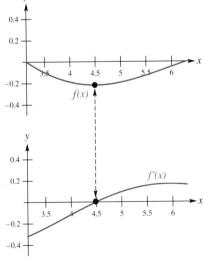

Figure 13

Approach 2: *Apply Numerical Max-Min Finding to* $f(x)$.
Enter the original function $\frac{\sin x}{x}$ into your calculator and graph it. Because it is clear from the graph that the only nonendpoint extreme value on the interval is a minimum near $x = 4.5$, use the numerical minimum command. (The maximum command is similar, but we don't need it for this problem.) After giving the appropriate command, we find that a minimum occurs at $x \approx 4.49341$, consistent with our first approach. *Note:* Computer algebra systems also often can find maximums and minimums numerically, but because they can generate derivatives easily, you can stick to Approach 1 explained above.

We now finish the problem by evaluating the critical numbers as we did in the previous examples. We need to include the endpoints π and 2π, as well as the critical number 4.49341 we found above. The results are displayed in Figure 14 in table form. Our conclusion is that the maximum is 0, which occurs at both endpoints $x = \pi$ and $x = 2\pi$, and the minimum is -0.21723, which occurs at $x \approx 4.49341$. When you use a numerical method that is built into a calculator or computer, the results are generally accurate to the number of digits printed out, though exceptions can occur. ◄

x	$\dfrac{\sin x}{x}$	
$\pi \approx 3.14159$	0	max
4.49341	-0.21723	min
$2\pi \approx 6.28319$	0	max

Figure 14

Practical Problems

By a practical problem, we mean a problem that might arise in everyday life. Such problems rarely have singular points; in fact, for them, maximum and minimum values usually occur at stationary points, though you should check endpoints as well. Here are two typical examples.

EXAMPLE 6: A rectangular box is to be made from a piece of cardboard 24 inches long and 9 inches wide by cutting out identical squares from the four corners and turning up the sides, as in Figure 15. Find the dimensions of the box of maximum volume. What is this volume?

SOLUTION: Let x be the width of the square to be cut out and V the volume of the resulting box. Then

$$V = x(9 - 2x)(24 - 2x) = 216x - 66x^2 + 4x^3$$

We know that x cannot be less than 0 nor more than 4.5. Thus, our problem is to maximize V on $[0, 4.5]$. The stationary points are found by setting dV/dx equal to 0 and solving the resulting equation:

$$\frac{dV}{dx} = 216 - 132x + 12x^2 = 12(18 - 11x + x^2) = 12(9 - x)(2 - x) = 0$$

This gives $x = 2$ or $x = 9$, but 9 is not in the interval $[0, 4.5]$. We see that there are only three critical numbers, namely, 0, 2, and 4.5. At the endpoints 0 and 4.5, $V = 0$; at 2, $V = 200$. (At 0 and 4.5, where $V = 0$, we may even make a case that there *is* no box, because it has only length or width.) We conclude that the box has a maximum volume of 200 cubic inches if $x = 2$—that is, if the box is 20 inches long, 5 inches wide, and 2 inches deep. The results are displayed in table form in Figure 16. ◄

Our next example illustrates a problem faced by a firm that delivers its products by truck. As the speed at which a truck is driven increases, the operating cost (gasoline, oil, and so on) increases, whereas the driver cost goes down. What is the most economical speed at which a truck should be driven?

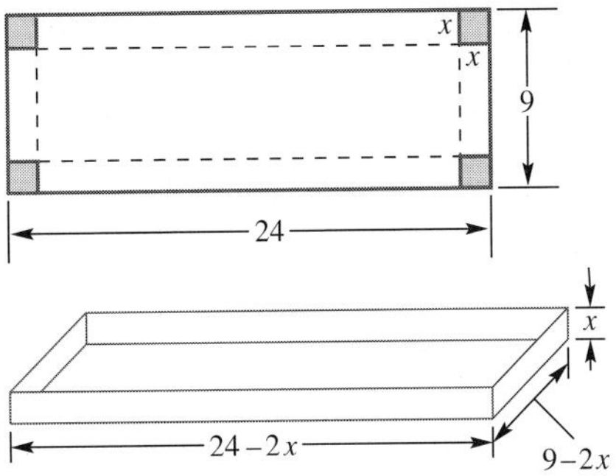

Figure 15

x	V	
0	0	min
2	200	max
4.5	0	min

Figure 16

EXAMPLE 7: The operating cost for a certain truck is estimated to be $(30 + v/2)$ cents per mile when it is driven at a speed of v miles per hour. The driver is paid \$14 per hour. What speed will minimize the cost of making a delivery to a city k miles away? Assume that the law restricts the speed to $40 \leq v \leq 60$.

SOLUTION: Let C be the total cost in cents of driving the truck k miles. Then

$$C = \text{driver cost} + \text{operating cost}$$

$$= \frac{k}{v}(1400) + k\left(30 + \frac{v}{2}\right) = 1400kv^{-1} + \left(\frac{k}{2}\right)v + 30k$$

Thus,

$$\frac{dC}{dv} = -1400kv^{-2} + \frac{k}{2} + 0$$

Setting dC/dv equal to 0 yields

$$\frac{1400k}{v^2} = \frac{k}{2}$$

$$v^2 = 2800$$

$$v \approx 53$$

A speed of 53 miles per hour would appear to be optimum, but we must evaluate C at the three critical numbers 40, 53, and 60 to be sure.

$$\text{At } v = 40, \quad C = k\left(\frac{1400}{40}\right) + k(20 + 30) = 85k$$

$$\text{At } v = 53, \quad C = k\left(\frac{1400}{53}\right) + k\left(\frac{53}{2} + 30\right) = 82.9k$$

$$\text{At } v = 60, \quad C = k\left(\frac{1400}{60}\right) + k(30 + 30) = 83.3k$$

We conclude that a speed of 53 miles per hour is best. Notice that we did not need to know the value of k to reach our conclusion.   ◄

We will have more to say about applied maximum and minimum problems in Section 4.3.

Concepts Review

1. A _____ function on a _____ interval will always have both a maximum value and a minimum value on that interval.

2. The term _____ value denotes either a maximum or a minimum value.

3. A function can attain an extreme value only at a critical number. Critical numbers are of three types: _____ , _____ , and _____ .

4. A stationary point for f is a number c such that _____ ; a singular point for f is a number c such that _____ .

Problem Set 4.1

In Problems 1 and 2, estimate the maximum and minimum values of the function on the given interval using a table of values or the Trace command of a graphing calculator. (See Example 1.)

1. $f(x) = x^2 + \sin x; I = [-1, 1]$

2. $f(x) = x^2 + e^x; I = [-1, 1]$

In Problems 3–16, identify the critical numbers and find the maximum value and minimum value. (See Examples 2 through 4.) If exact values for the critical numbers and extreme values can't be found, find approximations to them (see Examples 1 and 5), and state the degree of accuracy of the approximations.

3. $f(x) = -x^2 + 4x - 1; I = [0, 3]$

4. $h(t) = 4t^3 + 3t^2 - 6t + 1; I = [-2, 1]$

5. $f(x) = x^4 - 2x^2 + 4x; I = [-2, 2]$

6. $f(x) = x^5 - 5x^2 + 5x; I = [0, 2]$

7. $g(x) = \dfrac{1}{1 + x^2}; I = [-2, 1]$

8. $g(x) = \dfrac{1}{1 + x^2}; I = (-\infty, \infty)$

9. $f(x) = \dfrac{x^2}{x^4 + 1}; I = [-1, 4]$

10. $g(x) = x^{2/5}; I = [-1, 32]$

11. $f(x) = |x - 2|; I = [1, 5]$

12. $f(x) = |5 - 3x|; I = [0, 3]$

13. $F(t) = \sin t - \cos t; I = [0, \pi]$

14. $f(x) = x \sin x; I = [0, \pi]$

15. $f(t) = x^2 e^{-x}; I = [-1, 3]$

16. $f(t) = x^2 + e^{-x}; I = [-1, 3]$

17. Find two nonnegative numbers whose sum is 10 and whose product is a maximum. *Hint:* If x is one number, $10 - x$ is the other.

18. What number exceeds its square by the maximum amount? Begin by convincing yourself that this number is on the interval $[0, 1]$.

19. Joan has 200 feet of fence with which she plans to enclose a rectangular yard for her dog. If she wishes the area to be a maximum, what should the dimensions be?

20. Show that for a rectangle of given perimeter K, the one of maximum area is a square.

21. Find the volume of the largest open box that can be made from a piece of cardboard 24 inches square by cutting equal squares from the corners and turning up the sides. (See Example 4.)

22. A piece of wire 16 inches long is cut into two pieces; one piece is bent to form a square and the other is bent to form a circle. Where should the cut be made in order that the sum of the areas of the square and the circle be a minimum? A maximum? (Allow the possibility of no cut.)

23. Farmer Brown has 80 feet of fence with which he plans to enclose a rectangular pen along one side of his 100-foot barn, as shown in Figure 17 (the side along the barn needs no fence). What are the dimensions of the pen that has maximum area?

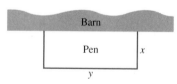

Figure 17

24. Farmer Brown of Problem 23 decides to make three identical pens with his 80 feet of fence, as shown in Figure 18. What dimensions for the total enclosure make the area of the pens as large as possible?

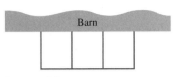

Figure 18

25. Suppose that Farmer Brown of Problem 23 has 180 feet of fence and wants the pen to adjoin to the whole side of the barn, as shown in Figure 19. What should the dimensions be for maximum area? Note that $0 \leq x \leq 40$ in this case.

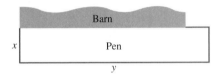

Figure 19

26. Suppose that Farmer Brown of Problem 23 decides to use his 80 feet of fence to make a rectangular pen to fit a 20-foot by 40-foot corner, as shown in Figure 20 (all of the corner must be used and does not require fence). What dimensions give the pen a maximum area? *Hint*: Begin by deciding on the allowable values for x.

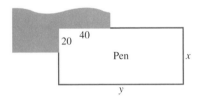

Figure 20

27. A rectangle has two corners on the x-axis and the other two on the parabola $y = 12 - x^2$, with $y \geq 0$. What are the dimensions of the rectangle of this type (Figure 21) with maximum area?

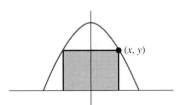

Figure 21

28. A rectangle is to be inscribed in a semicircle of radius r, as shown in Figure 22. What are the dimensions of the rectangle if its area is to be maximized?

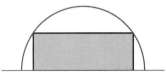

Figure 22

29. A metal rain gutter is to have 3-inch sides and a 3-inch horizontal bottom, the sides making an equal angle θ with the bottom (Figure 23). What should θ be in order to maximize the carrying capacity of the gutter? *Note*: $0 \leq \theta \leq \pi/2$.

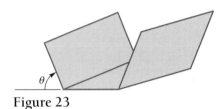

Figure 23

30. A huge conical tank is to be made from a circular piece of sheet metal of radius 10 meters by cutting out a sector with vertex angle θ and then welding the straight edges of the remaining piece together (Figure 24). Find θ so that the resulting cone has the largest possible volume.

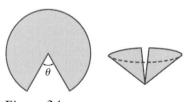

Figure 24

31. The operating cost of a certain truck is $25 + x/4$ cents per mile if the truck travels at x miles per hour. In addition, the driver gets $12 per hour. What is the most economical speed at which to operate the truck on a 400-mile run if the highway speed is required to be between 40 and 55 miles per hour?

32.   Redo Problem 31 assuming the operating cost is $40 + 0.05x^{\frac{3}{2}}$ cents per mile.

33.   Find the points P and Q on the curve $y = x^2/4$, $0 \le x \le 2\sqrt{3}$, which are closest to and farthest from the point $(0, 4)$. *Hint:* The algebra is simpler if you consider the square of the required distance rather than the distance itself.

34.   A humidifier uses a rotating disk of radius r, which is partially submerged in water. The most evaporation occurs when the exposed wetted region (shown shaded in Figure 25) is maximized. Show that this happens when h (the distance from the center to the water) is equal to $r/\sqrt{1 + \pi^2}$.

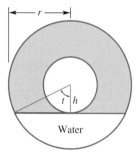

Figure 25

35.   A covered box is to be made from a rectangular sheet of cardboard measuring 5 feet by 8 feet. This is done by cutting out the shaded regions of Figure 26 and then folding on the dotted lines. What are the dimensions x, y, and z that maximize the volume?

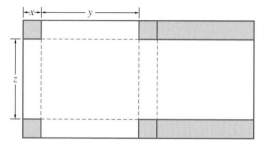

Figure 26

36.   Identify the critical numbers and find the extreme values on $[-1, 5]$ for each function.

(a) $f(x) = x^3 - 6x^2 + x + 2$
(b) $g(x) = |f(x)|$

37.   Follow the instructions of Problem 36 for $f(x) = \cos x + x \sin x + 2$.

Answers to Concepts Review: 1. Continuous, closed
2. Extreme   3. Endpoints, stationary points, singular points   4. $f'(c) = 0; f'(c)$ does not exist.

LAB 6: INCREASING, DECREASING, AND THE DERIVATIVE

Mathematical Background

A function is called *increasing* if it slopes upward (positive slope) reading the graph from left to right. Mathematically, we can write, "if $x_1 > x_2$, then $f(x_1) > f(x_2)$." Similarly, a function is called *decreasing* if it slopes downward (negative slope) reading from left to right; this means that if $x_1 > x_2$, then $f(x_1) < f(x_2)$. Generally, a function will be increasing for some regions and decreasing for others. For instance, the function $y = x^2$ is decreasing for $x < 0$ and increasing for $x > 0$.

A function has a *maximum* at a point if the y-coordinate is smaller for all nearby points; a function has a *minimum* at a point if the y-coordinate is larger for all nearby points. In other words, the tops of the mountains are maximums and the bottoms of the valleys are minimums.

Lab Introduction

In this lab, we will investigate the relationship between the graph of a function and the graph of its derivative. Just as a function can be graphed, so can its derivative; the important point to realize here is that the y-coordinate of a point on the graph of the derivative represents the *slope* of the original function at the same x-coordinate.

Experiment

1. Graph the function $f(x) = x^5 - 2x^4 - 5x^3 + 8x^2 - 7x - 5$ on the interval $-4 \leq x \leq 4$. Choose the y-limits so that you can see the maximums and minimums. Zoom to find the maximums and minimums accurate to four decimal places. Write down the intervals of x-values for which the function is increasing and the intervals for which the function is decreasing.

2. Find the derivative of the function in part 1 and then graph it. Find the x-values for which *the derivative* is zero to four decimal places—that is, solve the equation $f'(x) = 0$ by inspecting the graph and then using a numerical method of solution. Write down the x-intervals for which $f'(x)$ is positive (above the x-axis). Write down the x-intervals for which $f'(x)$ is negative (below the x-axis). Make a sketch of the graphs of the function and its derivative together on the same set of axes.

3. Graph the function $ae^{-kt}\cos(wt)$ and its derivative $-ae^{-kt}(w \sin(wt) + k \cos(wt))$, together on the interval $0 \leq t \leq 4$. Either use the values of a, k, w you found in Lab 3, or use $a = 10, k = 0.05, w = 3$ if you did not do Lab 3. Make a sketch of the two graphs together. *Use what you learned in parts 1 and 2* to determine all maximums and minimums of the original function on the t-interval given to four decimal places, by solving an appropriate equation.

4. Determine whether the derivative (velocity) is at a maximum when the original function (position) is zero. How is this different from what you found in part 3? *Be careful*: you must zoom in quite close.

Discussion

1. Explain in words the relationship between the graph of the original function $f(x)$ and the graph of the derivative $f'(x)$ that you found in parts 1 and 2. Give an explanation *why* this should be the case for graphs in general. Remember that $f'(x)$ gives you a formula for the *slope* of $f(x)$.

2. List both the t- and y-coordinates for all maximums and minimums that you found for the pendulum function, and explain how you found them. Explain what these points represent in terms of the original pendulum experiment. How do you know that you got four decimal places of accuracy?

3. Explain how you went about solving part 4 above. Describe what your answer to 4 means in terms of the pendulum. Is this surprising or not?

4.2 Monotonicity and Concavity

Consider the graph in Figure 1. No one will be surprised when we say that f is decreasing on the left of c and increasing on the right of c. But to make sure that we agree on terminology, we give precise definitions.

Definition

Let f be defined on an interval I (open, closed, or neither). We say that

(i) f is **increasing** on I if for every pair of numbers x_1 and x_2 in I,

$$x_1 < x_2 \Rightarrow f(x_1) < f(x_2)$$

(ii) f is **decreasing** on I if for every pair of numbers x_1 and x_2 in I,

$$x_1 < x_2 \Rightarrow f(x_1) > f(x_2)$$

(iii) f is **strictly monotonic** on I if it is either always increasing on I or always decreasing on I.

How shall we decide where a function is increasing? We could generate a computer or calculator graph and look at it, but remember that computer and calculator graphs are drawn by plotting and connecting points, as is done when creating graphs by hand. Who can be sure that the graph does not wiggle between the plotted points, even when there are many points? We need a better procedure.

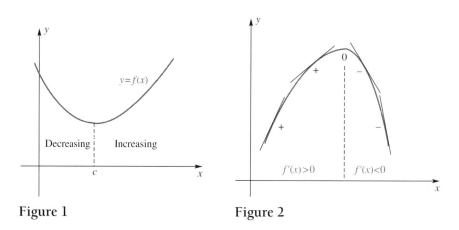

Figure 1   Figure 2

The First Derivative and Monotonicity   Recall that the first derivative $f'(x)$ gives us the slope of the tangent line to the graph of f at the point x. Then, if $f'(x) > 0$, the tangent line is rising to the right (see Figure 2). Similarly, if $f'(x) < 0$, the tangent line is falling to the right. These facts make the following theorem intuitively clear. We postpone a rigorous proof until Section 4.7.

Theorem A

(Monotonicity Theorem).

Let f be continuous on an interval I and differentiable at every interior point of I.
(i) If $f'(x) > 0$ for all x interior to I, then f is increasing on I.
(ii) If $f'(x) < 0$ for all x interior to I, then f is decreasing on I.

This theorem usually allows us to determine precisely where a differentiable function increases and where it decreases. It is a matter of solving two inequalities.

EXAMPLE 1: If $f(x) = 2x^3 - 3x^2 - 12x + 7$, find where f is increasing and where it is decreasing. Sketch a graph of $f(x)$.

SOLUTION: We begin by finding the derivative of f.

$$f'(x) = 6x^2 - 6x - 12 = 6(x + 1)(x - 2)$$

We need to determine where

$$6(x + 1)(x - 2) > 0$$

and also where

$$6(x + 1)(x - 2) < 0$$

The points -1 and 2 are the zeros of $f'(x)$, so they split the x-axis into three intervals: $(-\infty, -1)$, $(-1, 2)$, and $(2, \infty)$. When a function changes sign from positive to negative, or from negative to positive, there must be a point in between where the function is zero (a consequence of the Intermediate Value Theorem). This implies that on each of the three intervals, the function defined by $f'(x) = 6(x + 1)(x - 2)$, and therefore the slope of f, must be entirely positive or entirely negative. Thus, we need only test a single point from each interval to determine whether the function $f'(x)$ is positive or negative there.

Using test points -2, 0, and 3, we get $f'(-2) = +24$ (positive), $f'(0) = -12$ (negative), and $f'(3) = +24$ (positive). We conclude that $f'(x) > 0$ on the first and last of these intervals and that $f'(x) < 0$ on the middle interval (Figure 3).

A graph of $f'(x)$ generated with a computer or calculator gives us the same information; where the graph is above the x-axis, $f'(x)$ is positive, and where it is below the x-axis, $f'(x)$ is negative. See Figure 4. In the marginal box on solving inequalities we outline the procedure that we have just used.

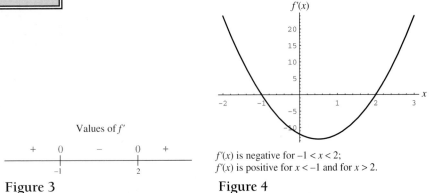

$f'(x)$ is negative for $-1 < x < 2$;
$f'(x)$ is positive for $x < -1$ and for $x > 2$.

Values of f'

$$+ \quad 0 \quad - \quad 0 \quad +$$
$$\underset{-1}{\rule{0pt}{0pt}} \qquad \underset{2}{\rule{0pt}{0pt}}$$

Figure 3

Figure 4

Thus, by Theorem A, f is increasing on $(-\infty, -1]$ and $[2, \infty)$; it is decreasing on $[-1, 2]$. Note that the theorem allows us to include the endpoints on these intervals, even though $f'(x) = 0$ at those points. The graph of f is shown in Figure 5. ◄

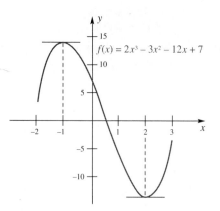

$f(x) = 2x^3 - 3x^2 - 12x + 7$

Figure 5

EXAMPLE 2: Determine where $g(x) = x/(1 + x^2)$ is increasing and where it is decreasing. Sketch the graphs of both $g(x)$ and $g'(x)$ and discuss the relationship between the two.

SOLUTION:

$$g'(x) = \frac{(1 + x^2) - x(2x)}{(1 + x^2)} = \frac{1 - x^2}{(1 + x^2)^2} = \frac{(1 - x)(1 + x)}{(1 + x^2)}$$

Because the denominator is always positive, $g'(x)$ has the same sign as $(1 - x)(1 + x)$. The split points, -1 and 1, determine the three intervals $(-\infty, -1), (-1, 1)$, and $(1, \infty)$. When we test them, we find that $g'(x) < 0$ on the first and last of these intervals and that $g'(x) > 0$ on the middle one (Figure 6).

The graphs of both $g(x)$ and $g'(x)$ are shown in Figure 7. By comparing the graphs of $g(x)$ and $g'(x)$ one can see that $g(x)$ is increasing where $g'(x) > 0$, and $g(x)$ is decreasing where $g'(x) < 0$. We conclude from Theorem A that g is decreasing on $(-\infty, -1]$ and $[1, \infty)$, and that it is increasing on $[-1, 1]$. ◄

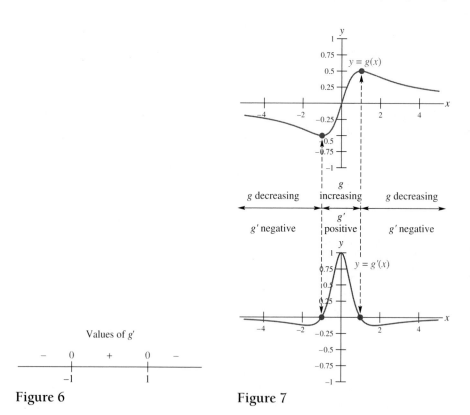

g decreasing   g increasing   g decreasing

g' negative   g' positive   g' negative

$y = g'(x)$

Values of g'

$-$ 0 $+$ 0 $-$

-1 1

Figure 6   **Figure 7**

The Second Derivative and Concavity A function may be increasing and still have a very wiggly graph (Figure 8). To analyze wiggles, we need to study how the tangent line turns as we move from left to right along the graph. If the tangent line turns steadily in the counterclockwise direction, we say that the graph is *concave up*; if the tangent line turns in the clockwise direction, the graph is *concave down*. Both definitions are better stated in terms of functions and their derivatives.

Definition

Let f be differentiable on an open interval I. We say that f (as well as its graph) is **concave up** on I if f' is increasing on I and we say that f is **concave down** on I if f' is decreasing on I.

The diagrams in Figure 9 will help to clarify these notions. Note that a curve that is concave *up* is shaped like a *cup*.

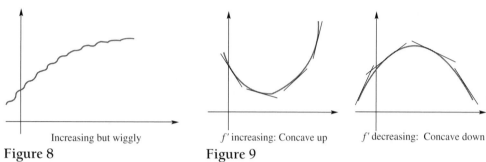

Increasing but wiggly

Figure 8

f' increasing: Concave up f' decreasing: Concave down

Figure 9

In view of Theorem A, we have a simple criterion for deciding where a curve is concave up and where it is concave down. We simply keep in mind that the second derivative of f is the first derivative of f'. Thus, f' is increasing if f'' is positive; it is decreasing if f'' is negative.

Theorem B

(Concavity Theorem). Let f be twice differentiable on the open interval I.
(i) If $f''(x) > 0$ for all x in I, then f is concave up on I.
(ii) If $f''(x) < 0$ for all x in I, then f is concave down on I.

For most functions, this theorem reduces the problem of determining concavity to the problem of solving inequalities.

EXAMPLE 3: Where is $f(x) = \frac{1}{3}x^3 - x^2 - 3x + 7$ increasing, decreasing, concave up, and concave down? Sketch the graphs of f, f', and f'' and discuss the relationships between them.

SOLUTION:

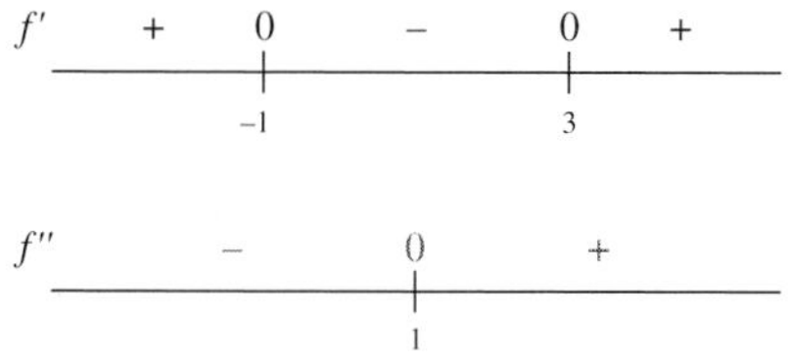

$$f'(x) = x^2 - 2x - 3 = (x + 1)(x - 3)$$
$$f''(x) = 2x - 2 = 2(x - 1)$$

By solving the inequality $(x + 1)(x - 3) > 0$ and its opposite we conclude that f is increasing on $(-\infty, -1]$ and $[3, \infty)$ and decreasing on $[-1, 3]$ (Figure 10). Similarly, solving $2(x - 1) > 0$ and $2(x - 1) < 0$ shows that f is concave up on $(1, \infty)$, concave down on $(-\infty, 1)$.

f' +   0   −   0   +

             −1   3

f'' −   0   +

             1

Figure 10

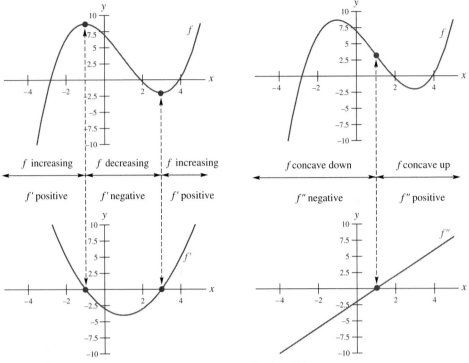

f increasing f decreasing f increasing

f' positive f' negative f' positive

f concave down f concave up

f'' negative f'' positive

Figure 11   Figure 12

The graphs of f and f' are shown together in Figure 11, and graphs of f and f'' are shown together in Figure 12. The x-intervals where the graph of f is increasing are the same as the x-intervals where the graph of f' is positive; similarly, f is decreasing where f' is negative. The x-intervals where the graph of f is concave up are the same as the x-intervals where the graph of f'' is positive, and similarly, f is concave down where f'' is negative. ◄

EXAMPLE 4: Where is $g(x) = x/(1 + x^2)$ concave up and where is it concave down? Sketch the graph of g.

SOLUTION: We began our study of this function in Example 2. There we learned that g is decreasing on $(-\infty, -1]$ and $[1, \infty)$ and increasing on $[-1, 1]$. To analyze concavity, we calculate g''.

$$g'(x) = \frac{1 - x^2}{(1 + x^2)^2}$$

$$g''(x) = \frac{(1 + x^2)^2(-2x) - (1 - x^2)(2)(1 + x^2)(2x)}{(1 + x^2)^4}$$

$$= \frac{(1 + x^2)[(1 + x^2)(-2x) - (1 - x^2)(4x)]}{(1 + x^2)^4}$$

$$= \frac{2x^3 - 6x}{(1 + x^2)^3} = \frac{2x(x^2 - 3)}{(1 + x^2)^3}$$

Because the denominator is always positive, we need only solve $x(x^2 - 3) > 0$ and its opposite. The split points are $-\sqrt{3} \approx -1.732$; 0; and $\sqrt{3} \approx 1.732$. These three split points determine four intervals. After testing them, either by graphing g'' or by substituting values into g'' (Figure 13), we conclude that g is concave up on $(-\sqrt{3}, 0) \approx (-1.73, 0)$ and $(\sqrt{3}, \infty) \approx (1.73, \infty)$ and that it is concave down on $(-\infty, -\sqrt{3}) \approx (-\infty, -1.73)$ and $(0, \sqrt{3}) \approx (0, 1.73)$.

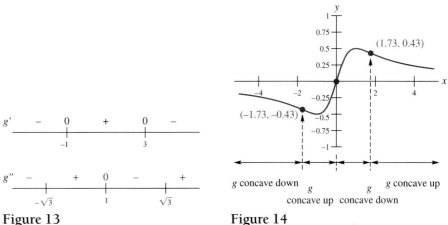

Figure 13   **Figure 14**

In Figure 14 we show the graph of g. Notice that g is an odd function because $g(-x) = \frac{-x}{1 + (-x)^2} = -\frac{x}{1 + x^2}$; thus, the graph is symmetric with respect to the origin. ◄

Inflection Points Let f be continuous at c. We call $(c, f(c))$ an **inflection point** of the graph of f if f is concave up on one side of c and concave down on the other side. The graph in Figure 15 indicates a number of possibilities.

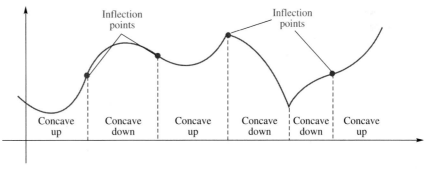

Figure 15

As you might guess, *points where $f''(x) = 0$ or where $f''(x)$ does not exist are the candidates for points of inflection*. We use the word *candidate* deliberately. Just as a candidate for political office may fail to be elected, so—for example—may a point where $f''(x) = 0$ fail to be a point of inflection. Consider $f(x) = x^4$, which has the graph shown in Figure 16. It is true that $f''(0) = 0$; yet the origin is not a point of inflection. However, in searching for inflection points, we begin by identifying those points where $f''(x) = 0$ (and where $f''(x)$ does not exist). Then we check to see if they really are inflection points.

Look back at the graph in Example 4. You will see that it has three inflection points, plotted on the graph in Figure 14. They are $(-\sqrt{3}, -\sqrt{3}/4) \approx (-1.73, -0.43)$, $(0, 0)$, and $(\sqrt{3}, \sqrt{3}/4) \approx (1.73, 0.43)$.

EXAMPLE 5: Find all inflection points of the graph of $f(x) = \frac{1}{6}x^3 - 2x$.

SOLUTION:

$$f'(x) = \frac{1}{2}x^2 - 2x$$

$$f''(x) = x$$

There is only one candidate for an inflection point, namely, the point where $f''(x) = 0$. This occurs at the origin, $(0, 0)$. That $(0, 0)$ is an inflection point follows from the fact that $f''(x) < 0$ for $x < 0$ and $f''(x) > 0$ for $x > 0$. Thus, the concavity changes direction at $(0, 0)$. The graph is shown in Figure 17.   ◀

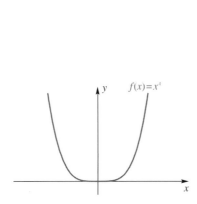

Figure 16   Figure 17

EXAMPLE 6: In Section 2.4 we investigated a model of population growth for the inhabitants of a small Caribbean country. We let $P(t)$ represent population (in thousands) at time t (in years); the model was

$$P(t) = 50e^{0.03t} + 2\sin(2\pi t) + 3$$

Looking at the graph on a one-year period, determine the intervals where the graph of population versus time is increasing, where it is decreasing, where it is concave up, and where it is concave down, and find any inflection points. Discuss your findings in the context of the problem situation.

SOLUTION: Because we are interested in a one-year period, we graph the function on the interval $0 \leq t \leq 1$. See Figure 18.

From the graph it is apparent that there are points where $P'(t) = 0$ near $t = 0.3$ and near $t = 0.7$. It also appears that there is an inflection point (and hence $P''(t) = 0$) near $t = 0.5$, and there could be another inflection point either near $t = 0$ or near $t = 1$ (why?). The derivatives are

$$P'(t) = 50(0.03)e^{0.03t} + 2(2\pi)\cos(2\pi t)$$

$$= 1.5e^{0.03t} + 4\pi\cos(2\pi t)$$

and

$$P''(t) = 1.5(0.03)e^{0.03t} - 4\pi(2\pi)\sin(2\pi t)$$

$$= 0.045e^{0.03t} - 8\pi^2\sin(2\pi t)$$

Increasing/Decreasing: Setting $P'(t) = 0$ we get

$$1.5e^{0.03t} + 4\pi \cos(2\pi t) = 0$$

which cannot be solved algebraically. We graph $P'(t)$ to get initial guesses for numerical root-finding; see Figure 19. Using starting values like $t = 0.3$ and $t = 0.7$ or appropriate starting intervals, such as $[0.2, 0.4]$ and $[0.6, 0.8]$, we get the solutions $t = 0.269198$ and $t = 0.730533$. These solutions could also be obtained directly from $P(t)$ using numerical max-min finding, as explained in the last section.

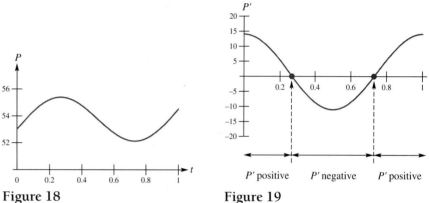

Figure 18   **Figure 19**

This gives us the three intervals $[0, 0.269198]$, $[0.269198, 0.730533]$, and $[0.730533, 1]$, because we are concerned only with the first year. By testing points from each interval in $P'(t)$, or simply by looking at the graphs of $P(t)$ and $P'(t)$, we conclude that the function is increasing on $[0, 0.269198]$ and $[0.730533, 1]$ and decreasing on $[0.269198, 0.730533]$.

Concave Up/Concave Down/Inflection Points: Setting $P''(t) = 0$ we get

$$0.045e^{0.03t} - 8\pi^2 \sin(2\pi t) = 0$$

which is again not solvable by algebraic means. We graph $P''(t)$ to get starting values for a numerical solution; see Figure 20. Using the starting points $t = 0$, $t = 0.5$, and $t = 1$ (or appropriate starting intervals) we get, respectively, $t = 0.000090$, $t = 0.499908$, and $t = 1.000093$. The last point is not in the interval

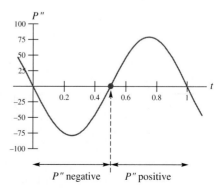

Figure 20

$0 \leq t \leq 1$ so we end up with the intervals $[0, 0.000090]$, $[0.000090, 0.499908]$, and $[0.499908, 1]$.

Some calculators may also be able to find inflection points numerically directly from the original function $P(t)$. As with numerical max-min finding, one graphs the original function, gives the command to find inflection points, then traces to a point near the inflection point to determine an initial guess. The calculator then determines a numerical approximation to the inflection point.

It is easy to see from the graph of $P''(t)$ that the second derivative is positive on $[0.499908, 1]$ and negative on $[0.000090, 0.499908]$. The interval $[0, 0.000090]$ is really too small in Figure 20 to be useful—so small, in fact, that we can't show it on the graph. We could zoom in on this part of the graph, but in this case it is easier just to test the point $t = 0$ in $P''(t)$; we get

$$P''(0) = 0.045e^{0.03(0)} - 8\pi^2\sin(2\pi(0)) = 0.045$$

This tells us that the second derivative is positive on $[0, 0.000090]$. Notice that we can get away with testing an endpoint in this case, because the point $t = 0$ did not arise from setting $P''(t) = 0$.

Our results are that $P(t)$ is concave up on $[0.499908, 1]$ and on $[0, 0.000090]$, and $P(t)$ is concave down on $[0.000090, 0.499908]$. By substituting $t = 0.000090$ and $t = 0.499908$ into the original function $P(t)$ we get 53.0013 and 53.7567, respectively. Thus we have inflection points at $(0.000090, 53.0013)$ and $(0.499908, 53.7567)$.

Interpretation: To interpret these results in terms of the problem situation, it is useful to determine which t-values correspond to which months. Assuming that each month is $\frac{1}{12}$ of a year (an approximation, because some months are longer than others) we can form a conversion chart by converting the fractions $\frac{1}{12}, \frac{2}{12}, \frac{3}{12}, \ldots,$ to decimals. See Figure 21.

Figure 21

We see that the point $t = 0.269198$ corresponds to early April, the point $t = 0.730533$ corresponds to late September, and the point $t = 0.499908$ corresponds to late June. The interval from $t = 0$ to $t = 0.000090$ in early January is so small that we will neglect it in our discussion. (It corresponds to the first 47 minutes of the year!)

From the beginning of January up to early April, the population is increasing ($P' > 0$) but the rate of increase is slowing down ($P'' < 0$). From early April through late June, the population decreases ($P' < 0$) with the rate of decrease speeding up ($P'' < 0$). From late June through late September the population is still decreasing ($P' < 0$) but the rate of decrease slows down ($P'' > 0$). Finally, from late September through the end of the year, the population increases ($P' > 0$) with the rate of increase speeding up ($P'' > 0$).

These results could possibly be explained in terms of tourism. For this country, tourists start arriving in late September, with the population increasing at an accelerating rate. From January through early April the tourist population is still building up, but the rate of increase is now going down. From early April through late June tourists flee the island at an increasing rate, and from late June through late September people are still leaving, but the rate at which they are leaving slows down.

Perhaps the most confusing part of the above interpretation is when both P' and P'' are negative. Recall that when P'' is negative, the tangent is turning in the clockwise direction; thus, if P' is also negative, the tangent slopes down, and by turning clockwise it will have a steeper and steeper downward slope. ◀

Concepts Review

1. If $f'(x) > 0$ everywhere, then f is _____ everywhere; if $f''(x) > 0$ everywhere, then f is _____ everywhere.

2. If _____ and _____ on an open interval I, then f is both decreasing and concave down on I.

3. A point on the graph of a continuous function where the concavity changes direction is called _____ .

4. In trying to locate the inflection points for the graph of a function f, we should look at numbers c, where either _____ or _____ .

Problem Set 4.2

In Problems 1–10, find where the given function is increasing and where it is decreasing. You may choose between numerical and exact methods, but justify your choice. Sketch the graphs of both the function and its derivative and discuss the relationship between the two.

1. $f(x) = x^2 - 4x + 2$

2. $f(x) = 2x - x^2$

3. $F(x) = x^3 - 1$

4. $F(x) = 2x^3 + 9x^2 - 13$

5. $g(t) = t^4 + 4t^3 + 12t$

6. $g(t) = e^{2t} - 4 \cos t$

7. $f(x) = 2x^5 - 15x^4 + 30x^3 - 6x$

8. $f(x) = \dfrac{2 - x}{x^2}$

9. $H(t) = \sin^2 2t, \ 0 \le t \le \pi$

10. $H(t) = \cos t + \sin t, \ 0 \le t \le 2\pi$

In Problems 11–18, determine where the given function is concave up and where it is concave down. Also find all inflection points. You may choose between numerical and exact methods, but justify your choice. Sketch the graphs of both the function and its second derivative and discuss the relationship between the two.

11. $f(x) = (x - 3)^2$

12. $f(x) = 4 - x^2$

13. $F(x) = x^3 - 12x + \ln x$

14. $F(x) = (x - 3)^3 + 4e^x$

15. $g(x) = 3x^2 - \dfrac{1}{x^2}$

16. $g(x) = x^4 - 6x^3 - 24x^2 + x + 2$

17. $g(x) = 2x^6 + 15x^4 + 90x^2 + 120x - 4$

18. $g(x) = 2x^2 + \cos^2 x$

In Problems 19–28, determine where the graph of the given function is increasing, decreasing, concave up, and concave down. You may choose between numerical and exact methods, but justify your choice. Sketch the graphs of the function, its first derivative and its second derivative and discuss the relationships between them. (See Example 3.)

19. $f(x) = -x^3 - 3x - 1$

20. $g(x) = x^3 - 2x^2 + x + 1$

21. $g(x) = 3x^4 - 4x^3 + 2$

22. $F(x) = x^6 - 3x^4$

23. $G(x) = 3x^5 - 5x^3 + 1$

24. $H(x) = \dfrac{x^2}{x^2 + 1}$

25. $f(x) = \sqrt{\sin x}$ on $[0, \pi]$

26. $g(x) = x\sqrt{x - 2}$

27. $f(x) = x^{2/3}(1 - x)$

28. $g(x) = 8x^{1/3} + x^{4/3}$

29. In Example 4 of Section 3.3 we used the equation $y = e^{-x}\sin x$ to describe the motion of a car bumper after it hits a speed bump. Recall that y represents the displacement from normal position in feet and x represents time after impact in seconds.

(a) Sketch the graph of displacement versus time on the interval $0 \le t \le 6$.

(b) Determine where the graph in (a) is increasing and where it is decreasing. Sketch the graph of $\frac{dy}{dt}$ and explain its relationship to the graph in (a) and its meaning in the context of the problem.

(c) Determine where the graph in (a) is concave up and concave down. Sketch the graph of $\frac{d^2y}{dt^2}$ and explain its relationship to the graph in (a) and its meaning in the context of the problem.

30.   Suppose that instead of the model of population growth given in Example 6 we used the model

$$P(t) = 50e^{0.11t} + \sin(2\pi t) + 3$$

Determine the intervals where the graph of population versus time is increasing, where it is decreasing, where it is concave up, where it is concave down, and find any inflection points for a two-year period. Discuss your findings in the context of the problem situation.

In Problems 31–34, on $[0, 6]$ sketch the graph of a continuous function f that satisfies all of the stated conditions.

31. $f(0) = 3; f(3) = 0; f(6) = 4;$
$f'(x) < 0$ on $(0, 3); f'(x) > 0$ on $(3, 6);$
$f''(x) > 0$ on $(0, 5); f''(x) < 0$ on $(5, 6).$

32. $f(0) = 3; f(2) = 2; f(6) = 0;$
$f'(x) < 0$ on $(0, 2) \cup (2, 6); f'(2) = 0;$
$f''(x) < 0$ on $(0, 1) \cup (2, 6); f''(x) > 0$ on $(1, 2).$

33. $f(0) = f(4) = 1; f(2) = 2; f(6) = 0;$
$f'(x) > 0$ on $(0, 2); f'(x) < 0$ on $(2, 4) \cup (4, 6);$
$f'(2) = f'(4) = 0; f''(x) > 0$ on $(0, 1) \cup (3, 4);$
$f''(x) < 0$ on $(1, 3) \cup (4, 6).$

34. $f(0) = f(3) = 3; f(2) = 4; f(4) = 2; f(6) = 0;$
$f'(x) > 0$ on $(0, 2); f'(x) < 0$ on $(2, 4) \cup (4, 5);$
$f'(2) = f'(4) = 0; f'(x) = -1$ on $(5, 6);$
$f''(x) < 0$ on $(0, 3) \cup (4, 5); f''(x) > 0$ on $(3, 4).$

35.   Prove that a quadratic function has no points of inflection.

36.   Prove that a cubic function has exactly one point of inflection.

37.   Use the Monotonicity Theorem to prove each statement if $0 < x < y.$

(a) $x^2 < y^2$ (b) $\sqrt{x} < \sqrt{y}$ (c) $\frac{1}{x} > \frac{1}{y}$

38.   What conditions on a, b, and c will make $f(x) = ax^3 + bx^2 + cx + d$ always increasing?

39.   Determine a and b so that $f(x) = a\sqrt{x} + \sqrt{b}/\sqrt{x}$ has the point $(4, 13)$ as an inflection point.

40.   The general cubic function $f(x)$ has three zeros, $r_1, r_2,$ and $r_3.$ Show that its inflection point has an x-coordinate of $(r_1 + r_2 + r_3)/3.$ *Hint*: Keep in mind that $f(x) = a(x - r_1)(x - r_2)(x - r_3).$

41.   Suppose that $f'(x) > 0$ and $g'(x) > 0$ for all $x.$ What simple additional conditions (if any) are needed to guarantee that:

(a) $f(x) + g(x)$ is increasing for all x?
(b) $f(x) \cdot g(x)$ is increasing for all x?
(c) $f(g(x))$ is increasing for all x?

42.   Suppose that $f''(x) > 0$ and $g''(x) > 0$ for all $x.$ What simple additional conditions (if any) are needed to guarantee that:

(a) $f(x) + g(x)$ is concave up for all x?
(b) $f(x) \cdot g(x)$ is concave up for all x?
(c) $f(g(x))$ is concave up for all x?

43.   Let $f(x) = \sin x + \cos(x/2)$ on the interval $I = (-2, 7).$

(a) Draw the graph of f on $I.$
(b) Use this graph to estimate where $f'(x) < 0$ on $I.$
(c) Use this graph to estimate where $f''(x) < 0$ on $I.$
(d) Draw the graph of f' to confirm your answer to (b).
(e) Draw the graph of f'' to confirm your answer to (c).

44.   Repeat Problem 43 for $f(x) = x \cos^2(x/3)$ on $(0, 10).$

45.   Let $f'(x) = x^3 - 5x^2 + 2$ on $I = [-2, 4].$ Where on I is f increasing?

46.   Let $f''(x) = x^4 - 5x^3 + 4x^2 + 4$ on $I = [-2, 3].$ Where on I is f concave downward?

Answers to Concepts Review: 1. Increasing; concave up
2. $f'(x) < 0; f''(x) < 0$ 3. An inflection point
4. $f''(c) = 0, f''(c)$ does not exist.

4.3   Local Maximums and Minimums

We recall from Section 4.1 that the maximum value (if it exists) of a function f on a set S is the largest value f attains on the whole set $S.$ It is sometimes referred to as the **global maximum value**, or the *absolute maximum value* of $f.$ Thus, for the function f with domain $S = [a, b]$ whose graph is sketched in Figure 1, $f(a)$ is the global maximum value. But what about $f(c)$? It may not be king of the country, but at least it is chief of its own locality. We call it a **local maximum value**, or a *relative maximum value*. Of course, a global maximum value is automatically a local maximum value. Figure 2 illustrates a number of possibilities. Note that the global maximum value (if it exists) is simply the largest of the local maximum values. Similarly, the global minimum value is the smallest of the local minimum values.

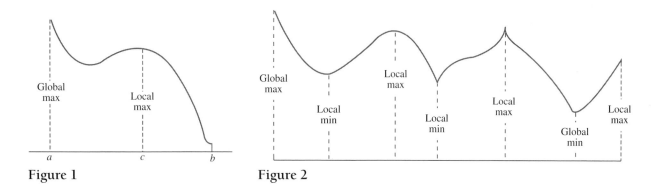

Figure 1   Figure 2

Here is the formal definition of local maximums and local minimums. Recall that the symbol ∩ denotes the intersection (common part) of two sets.

Definition

Let S, the domain of f, contain the point c. We say that:

(i) $f(c)$ is a **local maximum value** of f if there is an interval (a, b) containing c such that $f(c)$ is the maximum value of f on $(a, b) \cap S$;

(ii) $f(c)$ is a **local minimum value** of f if there is an interval (a, b) containing c such that $f(c)$ is the minimum value of f on $(a, b) \cap S$;

(iii) $f(c)$ is a **local extreme value** of f if it is either a local maximum or a local minimum value.

Where Do Local Extreme Values Occur?   The Critical Number Theorem (Theorem 4.1B) holds as stated, with the phrase *extreme value* replaced by *local extreme value*; the proof is essentially the same. Thus, the critical numbers (endpoints, stationary points, and singular points) are the candidates for points where local extrema may occur. We say *candidates* because we are not claiming that there must be a local extremum at every critical point on the graph. The left graph in Figure 3 makes this clear. However, if the deriva-

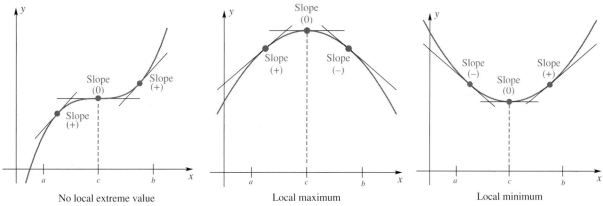

No local extreme value   Local maximum   Local minimum

Figure 3

tive is positive on one side of the critical point and negative on the other, then we have a local extremum, as shown in the middle and right graphs of Figure 3.

Theorem A

(First Derivative Test for Local Extrema). Let f be continuous on an open interval (a, b) that contains a critical number c.

(i) If $f'(x) > 0$ for all x on (a, c) and $f'(x) < 0$ for all x in (c, b), then $f(c)$ is a local maximum value of f.

(ii) If $f'(x) < 0$ for all x on (a, c) and $f'(x) > 0$ for all x in (c, b), then $f(c)$ is a local minimum value of f.

(iii) If $f'(x)$ has the same sign on both sides of c, then $f(c)$ is not a local extreme value of f.

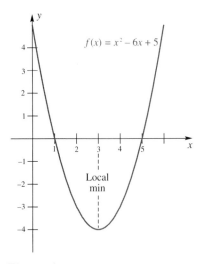

Figure 4

Proof of (i) Because $f'(x) > 0$ for all x on (a, c), f is increasing on $(a, c]$ by the Monotonicity Theorem. Again, because $f'(x) < 0$ for all x on (c, b), f is decreasing on $[c, b)$. Thus, $f(x) < f(c)$ for all x on (a, b), except, of course, at $x = c$. We conclude that $f(c)$ is a local maximum.

The proofs of (ii) and (iii) are similar. ◄

EXAMPLE 1: Find the local extreme values of the function $f(x) = x^2 - 6x + 5$ on $(-\infty, \infty)$.

SOLUTION: The polynomial function f is continuous everywhere, and its derivative, $f'(x) = 2x - 6$, exists for all x. Thus, the only critical number for f is the single solution of $f'(x) = 0$, namely, $x = 3$.

Because $f'(x) = 2(x - 3) < 0$ for $x < 3$, f is decreasing on $(-\infty, 3]$; and because $2(x - 3) > 0$ for $x > 3$, f is increasing on $[3, \infty)$. Therefore, by the First Derivative Test, $f(3) = -4$ is a local minimum value of f. Because 3 is the only critical number, there are no other extreme values. The graph of f is shown in Figure 4. Note that $f(3)$ is actually the global minimum value in this case. ◄

EXAMPLE 2: Find the local extreme values of $f(x) = \frac{1}{3}x^3 - x^2 - 3x + 4$ on $(-\infty, \infty)$.

SOLUTION: Because $f'(x) = x^2 - 2x - 3 = (x + 1)(x - 3)$, the only critical numbers of f are -1 and 3. When we use the test values -2, 0, and 4, we learn that $(x + 1)(x - 3) > 0$ on $(-\infty, -1)$ and $(3, \infty)$ and $(x + 1)(x - 3) < 0$ on $(-1, 3)$. We could get the same information from a computer or calculator graph of $f'(x)$ in addition to (or instead of) testing points; see Figure 5. By the First Derivative Test, we conclude that $f(-1) = \frac{17}{3}$ is a local maximum value and that $f(3) = -5$ is a local minimum value (Figure 6). ◄

EXAMPLE 3: Find the local extreme values of $f(x) = (\sin x)^{2/3}$ on $(-\pi/6, 2\pi/3) \approx (-0.524, 2.094)$.

SOLUTION:

$$f'(x) = \frac{2 \cos x}{3(\sin x)^{1/3}}, \qquad x \neq 0$$

The points 0 and $\pi/2 \approx 1.571$ are critical points, because $f'(0)$ does not exist and $f'(\pi/2) = 0$. Now $f'(x) < 0$ on $(-\pi/6, 0)$ and on $(\pi/2, 2\pi/3)$, while

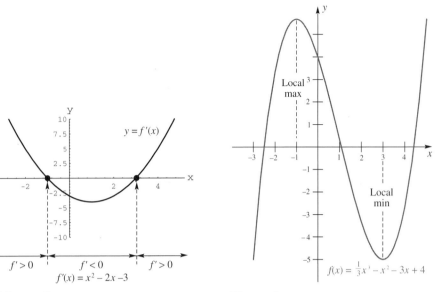

Figure 5

Figure 6

$f'(x) > 0$ on $(0, \pi/2)$; see the graph of f' in Figure 7. By the First Derivative Test, we conclude that $f(0) = 0$ is a local minimum value and that $f(\pi/2) = 1$ is a local maximum value. The graph of f is shown in Figure 8. ◄

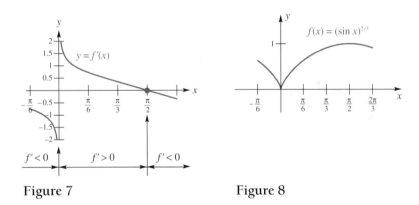

Figure 7

Figure 8

The Second Derivative Test   There is another test for local maximums and minimums that is sometimes easier to apply than the First Derivative Test. It involves evaluating the second derivative at the stationary points. It does not apply to singular points.

Theorem B

(Second Derivative Test for Local Extrema). Let f' and f'' exist at every point in an open interval (a, b) containing c, and suppose $f'(c) = 0$.
(i)   If $f''(c) < 0$, $f(c)$ is a local maximum value of f.
(ii)   If $f''(c) > 0$, $f(c)$ is a local minimum value of f.

Proof of (i)   It is tempting to say that because $f''(c) < 0$, f is concave downward near c and to claim that this proves (i). However, to be sure that f is concave downward in a neighborhood of c, we need $f''(x) < 0$ in that neighborhood (not just at c), and nothing in our hypothesis guarantees that. We take a different tack.

By definition and hypothesis,

$$f''(c) = \lim_{x \to c} \frac{f'(x) - f'(c)}{x - c} = \lim_{x \to c} \frac{f'(x) - 0}{x - c} < 0$$

so we can conclude that there is a (possibly small) interval (α, β) around c where

$$\frac{f'(x)}{x - c} < 0, \qquad x \neq c$$

But this inequality implies that $f'(x) > 0$ for $\alpha < x < c$ (because $x - c < 0$ there) and $f'(x) < 0$ for $c < x < \beta$ (because $x - c > 0$ there). Thus, by the First Derivative Test, $f(c)$ is a local maximum value.

The proof of (ii) is similar. ◄

EXAMPLE 4: For $f(x) = \frac{1}{3}x^3 - x^2 - 3x + 4$, use the Second Derivative Test to identify local extremes.

SOLUTION: This is the function of Example 2.

$$f'(x) = x^2 - 2x - 3 = (x + 1)(x - 3)$$
$$f''(x) = 2x - 2$$

The critical numbers are -1 and 3 (because $f'(-1) = f'(3) = 0$). Because $f''(-1) = -4$ and $f''(3) = 4$, we conclude by the Second Derivative Test that $f(-1) = \frac{17}{3} \approx 5.667$ is a local maximum value and that $f(3) = -5$ is a local minimum value. See Figure 6 for a graph. ◄

EXAMPLE 5: For the function $f(x) = xe^{-ax}$ where $a > 0$, use the Second Derivative Test to identify local extrema. Describe what happens to any local extrema as a changes.

SOLUTION: With the previous examples, it may have seemed like a lot of work to use the First or Second Derivative Tests to identify local extreme values, when a quick look at a graph gives the answer. For the current problem, a is an unspecified parameter, so we can't graph this function directly with a computer or calculator. The power of the Second Derivative Test becomes clear in this case.

We have

$$f'(x) = 1 \cdot e^{-ax} + x \cdot (-a)e^{-ax}$$
$$= e^{-ax} - axe^{-ax}$$

and

$$f''(x) = -ae^{-ax} - (ae^{-ax} + ax(-a)e^{-ax})$$
$$= -2ae^{-ax} + a^2xe^{-ax}$$

For critical numbers we set $f'(x) = 0$ and get

$$e^{-ax} - axe^{-ax} = 0$$
$$e^{-ax}(1 - ax) = 0$$
$$1 - ax = 0$$
$$x = \frac{1}{a}$$

because e^{-ax} can't equal zero (why?). We now test this critical number in the second derivative. We have

$$f''\left(\tfrac{1}{a}\right) = -2ae^{-a\left(\tfrac{1}{a}\right)} + a^2\left(\tfrac{1}{a}\right)e^{-a\left(\tfrac{1}{a}\right)}$$

$$= -2ae^{-1} + ae^{-1} = -ae^{-1}$$

$$= -\frac{a}{e} < 0$$

because we assumed that a was positive. Thus we conclude that the critical number $x = \tfrac{1}{a}$ must indicate a local maximum. The important point here is that this must be true *regardless of the value of the parameter a.*

The value of the function $f(x)$ at $x = \tfrac{1}{a}$ would be $f(x) = \tfrac{1}{a}e^{-a\tfrac{1}{a}} = \tfrac{1}{a}e^{-1}$. We can conclude that $f(x)$ has a local maximum of $\tfrac{1}{a}e^{-1}$ at $x = \tfrac{1}{a}$ for any positive value of a. We can picture a piece of the graph of $f(x)$ near $x = \tfrac{1}{a}$; see Figure 9. ◀

Unfortunately, the Second Derivative Test sometimes fails, because $f''(x)$ may be 0 at a stationary point. For both $f(x) = x^3$ and $f(x) = x^4$, $f'(0) = 0$ and $f''(0) = 0$ (see Figure 10). The first does not have a local maximum or minimum value at 0; the second has a local minimum there. This shows that if $f''(0) = 0$ at a stationary point, we are unable to draw a conclusion about maximums or minimums without more information.

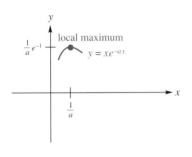

Figure 9

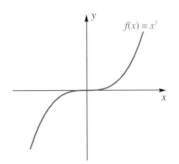

Figure 10a

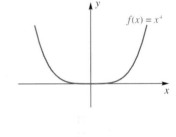

Figure 10b

Concepts Review

1. If f is continuous at c, $f'(x) > 0$ near to c on the left, and $f'(x) < 0$ near to c on the right, then $f(c)$ is a local _____ value for f.

2. If $f'(x) = (x + 2)(x - 1)$, then $f(-2)$ is a local _____ value for f and $f(1)$ is a local _____ value for f.

3. If $f'(x) = (x + 2)(x - 1)^2$, then $f(-2)$ is _____ for f and $f(1)$ is _____ for f.

4. If $f'(c) = 0$ and $f''(c) < 0$, we expect to find a local _____ for f at c.

Problem Set 4.3

In Problems 1–6, identify the critical numbers. Then use (a) the First Derivative Test, and (if possible) (b) the Second Derivative Test to decide which of the critical numbers gives a local maximum value and which gives a local minimum value.

1. $f(x) = x^3 - 3x^2 + 2$

2. $f(x) = x^3 - 3x + 4$

3. $f(x) = \dfrac{1}{2}x - \sin x, 0 < x < 2\pi$

4. $f(x) = \cos^2 x, -\pi/2 < x < 3\pi/2$

5. $g(x) = x^2e^{-x}$

6. $g(x) = x \ln x$

In Problems 7–16, find the critical numbers and use the test you prefer to decide which give a local maximum value and which give a local minimum value. What are these local maximum and minimum values? You may use exact or approximate (numerical) methods, but explain your reason for using the method you choose.

7. $f(x) = \frac{1}{4}x^3 - 3x - 1$

8. $g(x) = x^4 - 2x^2 + 3$

9. $h(x) = (x - a)^4 - 4x$, a an unknown parameter

10. $f(x) = x^3 - 3ax$, a an unknown parameter

11. $g(t) = t^2 + e^{-2t}$

12. $h(t) = t^4 + 2t^2 + 4t$

13. $f(x) = x + \frac{1}{x}, x \neq 0$

14. $g(x) = \frac{x^2}{\sqrt{x^2 + 1}}$

15. $f(t) = \frac{\sin t}{2 + \cos t}, 0 < t < 2\pi$

16. $g(t) = |\cos t|, 0 < t < 2\pi$

17. Find the (global) maximum and minimum values of $F(x) = 6\sqrt{x} - 3x$ on $[0, 9]$.

18. Do Problem 17 on the interval $[0, \infty)$.

19. Find (if possible) the maximum and minimum values of

$$f(x) = \frac{64}{\sin x} + \frac{27}{\cos x}$$

on $(0, \pi/2)$.

20. Find (if possible) the maximum and minimum values of

$$f(x) = x^2 + \frac{1}{x^2}$$

on $(0, \infty)$.

21. Find the minimum value of

$$g(x) = x^2 + \frac{16x^2}{(8 - x)^2}, x > 8$$

22. Consider $f(x) = Ax^2 + Bx + C$ with $A > 0$. Show that $f(x) \geq 0$ for all x if and only if $B^2 - 4AC \leq 0$.

23. If $f'(x) = 2(x + 2)(x + 1)^2(x - 2)^4(x - 3)^3$, what values of x make $f(x)$ a local maximum? A local minimum?

24. What conclusions can you draw about f from the information that $f'(c) = f''(c) = 0$ and $f'''(c) > 0$?

25. Let f be a continuous function and let f' have the graph shown in Figure 11. Try to sketch a graph for f and answer the following questions.

(a) Where is f increasing? Decreasing?

(b) Where is f concave up? Concave down?

(c) Where does f attain a local maximum? A local minimum?

(d) Where are there inflection points for f?

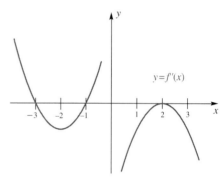

Figure 11

26. Repeat Problem 25 for Figure 12.

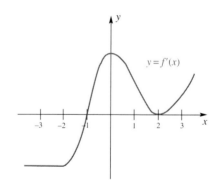

Figure 12

27. Identify all local and global extrema for $f(x) = x^5 - 5x^3 + 4$ on $[-2, 2.5]$.

28. Identify all local and global extrema for $f(x) = \sin x - \cos(x/2)$ on $[-2\pi, 2\pi]$.

29. Suppose that $g'(x) = x^5 - 5x^3 + 4$. Approximately where on $(-2, 2.5)$ does g attain local maximum values? Local minimum values?

30. Suppose that $g'(x) = \sin x - \cos(x/2)$. Approximately where on $(-2\pi, 2\pi)$ does g attain local maximum values? Local minimum values?

Answers to Concepts Review: 1. Maximum
2. Maximum; minimum   3. Local maximum value; neither a maximum nor minimum value   4. Maximum

LAB 7: IDEAL GASES AND REAL GASES

Mathematical Background

The Ideal Gas Law states that for a fixed amount of (ideal) gas

$$PV = RT \tag{1}$$

where P is pressure, V is volume, T is temperature, and R is a constant. Thus, if T is held constant, then when pressure is increased volume is decreased (think of squeezing a basketball).

An equation that is a better model for real gases (like air) is Van der Waals's equation

$$\left(P + \frac{a}{V^2}\right)(V - b) = RT \tag{2}$$

For $a = b = 0$ we get back the Ideal Gas Law. The term b represents the effect arising from the finite (nonzero) size of the molecules, and the term $\frac{a}{V^2}$ represents the effect of the attractive forces between molecules.

If we keep the temperature constant, we can think of P as a function of V; in this case the graph of V versus $P(V)$ is called an isothermal. If we change T to another value and then keep it constant again, we get another isothermal. In this sense, T is a parameter, and various isothermals are a family of functions.

Lab Introduction

For each isothermal we can find the derivatives $\frac{dP}{dV}$ and $\frac{d^2P}{dV^2}$. Our goal is to understand the isothermals and what they say about real and ideal gases.

In equation (1), if we solve for P we obtain

$$P(V) = \frac{RT}{V}$$

For this model, we also have

$$\frac{dP}{dV} = -\frac{RT}{V^2} \text{ and } \frac{d^2P}{dV^2} = \frac{2RT}{V^3}$$

In equation (2), if we solve for P we obtain

$$P(V) = \frac{RT}{V - b} - \frac{a}{V^2}$$

For this model, we also have

$$\frac{dP}{dV} = -\frac{RT}{(V - b)^2} + \frac{2a}{V^3} \text{ and } \frac{d^2P}{dV^2} = \frac{2RT}{(V - b)^3} - \frac{6a}{V^4}$$

Data

For a given number of carbon dioxide molecules, the approximate values of R, a, and b are 81.80421, 3583858.8, and 42.7, respectively. Here the units on P are atmospheres, on V cubic centimeters, and on T degrees Kelvin. We will use these values for the rest of the lab.

Calculator Experiment

1. Graph the function $P(V)$ on the interval $0 \leq V \leq 600$ using the Ideal Gas Law for the values $T = 200, 300, 320$ (a good choice for the P range is $-200 \leq P \leq 200$). Make sketches of each of these graphs. These are called the isothermal curves. Now repeat using Van der Waals's equation. Take a few moments to study and discuss these curves before going on to the next step.

2. For each T value used in 1, determine where the curve $P(V)$ is increasing, decreasing, concave up, and concave down by setting the first and second derivatives equal to zero and solving the resulting equation. You may want to graph the derivatives first to estimate the roots. Label all maximums, minimums, and inflection points on the graphs from 1. Do this for both the Ideal Gas Law and Van der Waals's equation. Identify anything else interesting that is going on for either model (such as asymptotes).

3. According to the theory of thermodynamics, there should be one isothermal, called the critical isothermal, that has a horizontal inflection point (that is, a point for which both the first and second derivatives are zero). Which of the two models given above has a critical isothermal? For this model, find the T value that gives the critical isothermal. Also find the P and V values where the horizontal inflection point occurs. Sketch the critical isothermal and label the horizontal inflection point.

Discussion

1. For which T values does the Van der Waals model have maximums and minimums? Do the isothermals that have maximums and minimums seem physically realistic? Discuss what is happening to pressure as volume is decreased for these curves.

2. The critical isothermal and the corresponding critical temperature is important because for temperatures above the critical temperature the substance can exist as either a gas or a liquid or a combination of both. Can a gas that obeys the ideal gas law have a critical isothermal? Discuss what this implies about ideal gases.

3. Briefly discuss the limitations of each of the two models studied. What is physically unrealistic for each model?

4.4 | More Max-Min Problems

The problems we studied in Section 4.1 usually assumed that the set on which we wanted to maximize or minimize a function was a *closed* interval. However, the intervals that arise in practice are not always closed; they are sometimes open, sometimes open at one end and closed at the other, and sometimes infinite at one or both ends. We can still handle these problems if we correctly apply the theory developed in Section 4.3. Keep in mind that maximum (minimum) with no qualifying adjective means global maximum (minimum).

Extrema on Open Intervals   We give two examples to illustrate appropriate procedures for intervals that are open at one or both endpoints.

EXAMPLE 1: Find (if possible) the minimum and maximum values of $f(x) = x^4 - 4x$ on $(-\infty, \infty)$.

SOLUTION:

$$f'(x) = 4x^3 - 4 = 4(x^3 - 1) = 4(x - 1)(x^2 + x + 1)$$

Because $x^2 + x + 1 = 0$ has no real solutions (quadratic formula), there is only one critical number, namely, $x = 1$. For $x < 1$, $f'(x) < 0$, whereas for $x > 1$, $f'(x) > 0$. We conclude that $f(1) = -3$ is a local minimum value for f; and because f is decreasing on the left of 1 and increasing on the right of 1, it must actually be the minimum value of f.

The facts stated above imply that f cannot have a maximum value. The graph of f is shown in Figure 1. ◀

A computer or graphing calculator can do a pretty good job of graphing a continuous function on a closed interval. The x-range of the graph window is just set to the closed interval given. No computing device can graph a function on an infinite interval, however. For the function in Example 1, how can we be *sure* that the graph will not eventually turn down and reach a lower point than the one at $(1, -3)$? No matter how large we set the x-range, we can never be totally certain of our result using a calculator or computer alone.

The theory developed in this chapter, on the other hand, shows that we can be *absolutely certain* that the point $(1, -3)$ is the lowest point on the graph, no matter what interval we choose to graph it on. This is part of the power of mathematics.

EXAMPLE 2: Find (if possible) the maximum and minimum values of $g(x) = x/(x^3 + 2)$ on $[0, \infty)$.

SOLUTION:

$$g'(x) = \frac{x^3 + 2 - x(3x^2)}{(x^3 + 2)^2} = \frac{2 - 2x^3}{(x^3 + 2)^2} = \frac{2(1 - x)(1 + x + x^2)}{(x^3 + 2)^2}$$

On $[0, \infty)$, there are two critical numbers: the endpoint 0 and the stationary point 1. For $0 < x < 1$, $g'(x) > 0$, while for $x > 1$, $g'(x) < 0$. See the graph of g' in Figure 2. Thus, $g(1) = \frac{1}{3}$ is the maximum value of g on $[0, \infty)$.

If g has a minimum value, it must occur at the other critical number, namely $x = 0$. We know that $g(0) = 0$ and $g(x) > 0$ for $x > 0$, so $g(0) = 0$ is the minimum value of g on $[0, \infty)$. The graph of g is shown in Figure 2 with the graph of g'.

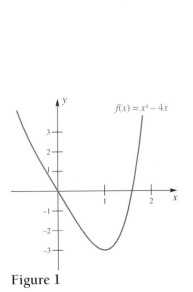

Figure 1

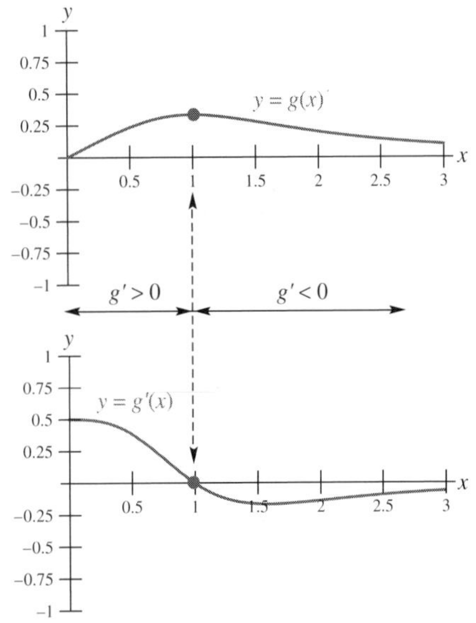

Figure 2

Practical Problems   Each of the examples below is different; yet there are common elements in the procedures that we use to solve them. At the end of the section, we will suggest a set of steps to use in solving any max-min problem.

EXAMPLE 3: A handbill is to contain 50 square inches of printed matter, with 4-inch margins at top and bottom and 2-inch margins on each side. What dimensions for the handbill would use the least paper?

SOLUTION: Let x be the width and y the height of the handbill (see Figure 3). Its area is

$$A = xy$$

We wish to minimize A.

As it stands, A is expressed in terms of two variables, a situation we do not know how to handle. However, we will find an equation connecting x and y so that one of these variables can be eliminated in the expression for A. The dimensions of the printed matter are $x - 4$ and $y - 8$ and its area is 50 square inches; so $(x - 4)(y - 8) = 50$. When we solve this equation for y, we obtain

$$y = \frac{50}{x - 4} + 8$$

Substituting this expression for y in $A = xy$ gives A in terms of x:

$$A = \frac{50x}{x - 4} + 8x$$

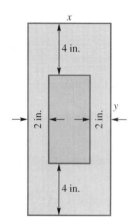

Figure 3

Because there must be two 2-inch side margins, the allowable values for x are $4 < x < \infty$; we want to minimize A on the open interval $(4, \infty)$. We examine the first derivative:

$$\frac{dA}{dx} = \frac{(x - 4)50 - 50x}{(x - 4)^2} + 8 = \frac{8x^2 - 64x - 72}{(x - 4)^2} = \frac{8(x + 1)(x - 9)}{(x - 4)^2}$$

The derivative is undefined at $x = 4$, but that is outside the open interval $(4, \infty)$. The only critical numbers are obtained by solving $dA/dx = 0$; this yields $x = 9$ and $x = -1$. We reject $x = -1$ because it is not in the interval $(4, \infty)$. By substituting $x = 9$ into the formula for A we get $A = 162$. The graph of A as a function of x shows that our critical number is in fact a minimum; see Figure 4. Alternatively, we could have used the First or Second Derivative Test to show that we have a minimum.

By finding critical numbers *before* graphing, we get a better idea of how to set the graph window. In this case the x-range of the graph should start at $x = 4$ and include $x = 9$, and the A-range should include 162.

Note that the graph plots the relationship between x and A. To find y, we substitute x into the equation connecting x and y. Letting $x = 9$ makes $y = 18$. So the dimensions for the handbill that will use the least amount of paper are 9 inches by 18 inches. The amount of paper used would be 162 square inches.

EXAMPLE 4: A rectangular beam is to be cut from a log with circular cross section. If the strength of the beam is proportional to the product of its width and the square of its depth, find the dimensions of the cross section that give the strongest beam.

SOLUTION: Denote the diameter of the log by a (constant) and the width and depth of the beam by w and d, respectively (Figure 5). We wish to maximize S, the strength of the beam.

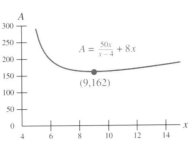

Figure 4

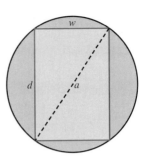

Figure 5

From the condition of the problem,

$$S = kwd^2$$

where k is a constant of proportionality. The strength S depends on the two variables w and d, but there is a simple relationship between them,

$$d^2 + w^2 = a^2$$

When we solve this equation for d^2 and substitute in the formula for S, we obtain S in terms of the single variable w:

$$S = kw(a^2 - w^2) = ka^2w - kw^3$$

Common Sense
It would be hard to make any preliminary estimates in Example 3. However, common sense tells us that the handbill's height should be larger than its width. Why? Because we should capitalize on the narrower margins along the sides.

We consider the allowable values for w to be $0 < w < a$, an open interval.

To find the critical numbers, we calculate dS/dw, set it equal to 0, and solve for w.

$$\frac{dS}{dw} = ka^2 - 3kw^2 = k(a^2 - 3w^2)$$

$$k(a^2 - 3w^2) = 0$$

$$w = \frac{a}{\sqrt{3}}$$

Because $a/\sqrt{3}$ is the only critical number in $(0, a)$, it is likely that it gives the maximum S. We can use the Second Derivative Test to confirm this. We have

$$\frac{d^2S}{dw^2} = \frac{d}{dw}k(a^2 - 3w^2)$$

$$= k(0 - 6w) = -6kw$$

so that at $w = a/\sqrt{3}$ we have $\left.\frac{d^2S}{dw^2}\right|_{w = a/\sqrt{3}} = \frac{-6ka}{\sqrt{3}} < 0$, which implies that we indeed have a maximum.

By substituting $w = a/\sqrt{3}$ in $d^2 + w^2 = a^2$, we learn that $d = \sqrt{2}a/\sqrt{3}$. The desired dimensions are $w = a/\sqrt{3}$ and $d = \sqrt{2}a/\sqrt{3}$. Note that $d = \sqrt{2}w$. ◄

For this problem, it was not possible to graph S as a function of w because of the parameter a. Notice that the solution depends on a. For problems that involve a parameter, numerical approaches such as graphing or numerical root-finding are generally of little help.

On the other hand, we have seen several problems that had no algebraic solution and could be approached only numerically. It is important that you learn to recognize when to use a numerical approach (generally involving technology) and when to use an algebraic approach (which might or might not involve computer algebra).

EXAMPLE 5: Find the dimensions of the right circular cylinder of greatest volume that can be inscribed in a given right circular cone.

SOLUTION: Let a be the altitude and b the radius of the base of the given cone (both constants). Denote by h, r, and V the altitude, radius, and volume, respectively, of an inscribed cylinder (see Figure 6).

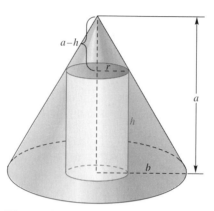

Figure 6

The volume of the cylinder is

$$V = \pi r^2 h$$

From similar triangles,

$$\frac{a - h}{r} = \frac{a}{b}$$

which gives

$$h = a - \frac{a}{b}r$$

When we substitute this expression for h into the formula for V, we obtain

$$V = \pi r^2 \left(a - \frac{a}{b}r\right) = \pi a r^2 - \pi \frac{a}{b}r^3$$

We wish to maximize V for r in the interval $[0, b]$. Because the interval is closed, we can use the methods of Section 4.1. (Someone is sure to argue—and with good reason—that the appropriate interval is $(0, b)$. Actually, the answer is the same either way, though we have to use the First or Second Derivative Test if we do the problem using $(0, b)$ as the domain.)

Now

$$\frac{dV}{dr} = 2\pi a r - 3\pi \frac{a}{b}r^2 = \pi a r\left(2 - \frac{3}{b}r\right)$$

This yields the stationary point $r = 2b/3$, giving us three critical numbers on $[0, b]$ to consider: 0, $2b/3$, and b. One look at the picture shows that $r = 0$ and $r = b$ both give a volume of 0. Thus, $r = 2b/3$ has to give the maximum volume. When we substitute this value for r into the equation connecting r and h, we find that $h = a/3$.   ◀

Theoretical versus Experimental Models

In the previous examples, we developed an equation for some quantity that we wanted to maximize or minimize by using reason and the information given in the problem. The equation we ended up with in each case was a *theoretical* mathematical model of the problem situation. We say that the model was developed from first principles.

We saw in Chapter 2 that it is also possible to develop mathematical models from data; we could call these *experimental* mathematical models. One still must choose a model to fit to the data (linear, exponential, and so on), but the model is often chosen from the scientific literature rather than developed from first principles. In the next example, we encounter the use of an experimental model in solving a max-min problem.

EXAMPLE 6: We have previously used the exponential model $y = y_0 e^{kt}$ to describe the growth of certain populations. This model does not work well for populations that are constrained by limited space or food. Under these conditions, the population will stop increasing and reach a constant long-term level.

For such situations population biologists often use the *logistic* population model. It is given by the equation

$$y = \frac{my_0}{(m - y_0)e^{-kt} + y_0}$$

where y is the population, t is time, and y_0 is the initial population. The parameter m represents the long-term population; it is often called the *carrying capacity* of the environment. We will assume that the initial population y_0 is less than the long-term population m, though the model can be used even when this is not the case.

Even though the equation looks somewhat complicated, we can understand it as follows. When $t = 0$, the exponential term e^{-kt} equals 1, so the denominator equals m, resulting in the equation $y = y_0$. In other words, at the beginning of the "experiment," $y = y_0$. On the other hand, letting $t \to \infty$, the exponential term e^{-kt} goes to 0, leaving the denominator equal to y_0 and yielding $y = m$. Thus, as t goes from 0 to ∞, y starts at y_0 and levels off at m, which is what we want the model to do. Finally, we know that the larger the value of k, the faster the term e^{-kt} goes from 1 to 0, and therefore the faster the population goes from y_0 to m. A typical logistic curve is shown in Figure 7.

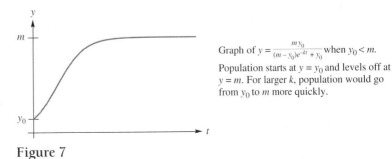

Graph of $y = \frac{my_0}{(m - y_0)e^{-kt} + y_0}$ when $y_0 < m$.

Population starts at $y = y_0$ and levels off at $y = m$. For larger k, population would go from y_0 to m more quickly.

Figure 7

A colony of bacteria grown under laboratory conditions is known to follow the logistic model. The colony starts out at 100 organisms. After 1 day there are 200 organisms, and after a very long time there are 1000 organisms. At what point is the colony growing the fastest?

SOLUTION: The starting population is given as $y_0 = 100$ and the long-term population is given as $m = 1000$. To determine the value of k, we use a data point to determine a parameter, as in Example 1 of Section 2.6. We know that $y = 200$ when $t = 1$; substituting these into the model we get

$$200 = \frac{(1000)(100)}{(1000 - 100)e^{-k(1)} + 100}$$

and solving for k we get

$$900e^{-k(1)} + 100 = 500$$

$$900e^{-k} = 400$$

$$e^{-k} = \frac{4}{9}$$

$$k = -\ln\left(\frac{4}{9}\right) \approx 0.81093$$

Our model is now $y = \dfrac{100{,}000}{900e^{-0.811t} + 100}$. The graph is shown in Figure 8. The known data points at $(0, 100)$ and $(1, 200)$ are shown.

We want to find the time at which the population is increasing the fastest; this would be where the rate of change, or derivative of the population function is the greatest. Thus, we are looking for a maximum of the *derivative* rather than a maximum of the original function. This would also correspond to an inflection point of the population function (why?). From Figure 8, we can see that an inflection point occurs somewhere on the graph of $y = \dfrac{100{,}000}{900e^{-0.811t} + 100}$ between $t = 1$ and $t = 4$.

Because it is hard to see inflection points graphically, we can find the derivative $\dfrac{dy}{dt}$, graph it, and look for a maximum.

$$\frac{dy}{dt} = \frac{d}{dt}\frac{my_0}{(m - y_0)e^{-kt} + y_0}$$

$$= \frac{0 - my_0(m - y_0)e^{-kt}(-k)}{\left((m - y_0)e^{-kt} + y_0\right)^2}$$

$$= \frac{my_0 k(m - y_0)e^{-kt}}{\left((m - y_0)e^{-kt} + y_0\right)^2}$$

$$\approx \frac{73{,}000{,}000e^{-0.811t}}{(900e^{-0.811t} + 100)^2}$$

Notice that we took the derivative before replacing the parameters with their numerical values; this is often conceptually simpler, and can avoid some numerical round off error. The graph of $\dfrac{dy}{dt}$ is shown in Figure 9.

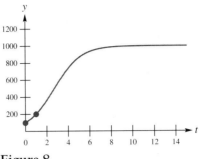

Figure 8   **Figure 9**

There is clearly a global maximum for the derivative $\frac{dy}{dt}$ at $t \approx 2.5$. To find the maximum, we must set the derivative of $\frac{dy}{dt}$, which is $\frac{d^2y}{dt^2}$, equal to zero. Finding $\frac{d^2y}{dt^2}$ from $\frac{dy}{dt}$ is a bit tedious, so we resort to computer algebra. The computer algebra system Derive gives the result

$$\frac{d^2y}{dt^2} = -\frac{k^2 m y_0 e^{kt}(m - y_0)(y_0 e^{kt} - m + y_0)}{(y_0 e^{kt} + m - y_0)^3}$$

Warning: Other computer algebra systems may give results that look different, but they are algebraically equivalent (you are asked to verify this derivative by hand in the exercises at the end of this section).

Setting this expression equal to zero, there is only one nonconstant term in the numerator that could equal zero, so we get

$$y_0 e^{kt} - m + y_0 = 0$$

$$e^{kt} = \frac{m - y_0}{y_0}$$

$$t = \frac{1}{k}\ln\left(\frac{m - y_0}{y_0}\right)$$

$$t \approx 2.71$$

after substituting in the numerical values for m, y_0, and k. Substituting this value into the formula for $\frac{dy}{dt}$ we get

$$\frac{dy}{dt} \approx \frac{73{,}000{,}000 e^{-0.811(2.71)}}{(900 e^{-0.811(2.71)} + 100)^2} \approx 203 \text{ organisms per day}$$

Alternatively, we could avoid calculating the derivative $\frac{d^2y}{dt^2}$ by finding the maximum of $\frac{dy}{dt} \approx \frac{73{,}000{,}000 e^{-0.811t}}{(900 e^{-0.811t} + 100)^2}$ numerically (see Example 5 of Section 4.1), or by finding the inflection point of the original function $y = \frac{10{,}000}{900 e^{-0.811t} + 100}$ numerically (see Example 6 of Section 4.2). We leave it to the student to decide which method to use; the result is, of course, the same as given above.

Summary of Results   We found that the bacteria population is increasing the fastest about 2.71 days after the start of the experiment. At that time, the population is increasing at a rate of about 203 organisms per day.   ◄

A Summary of the Method   Based on the examples above, we suggest a step-by-step method to use in applied max-min problems. Do not follow it slavishly; common sense may sometimes suggest an alternative approach or omission of some steps.

Step 1   Draw a picture, if appropriate, for the problem and assign variables to key quantities.

Step 2   Write a formula for the quantity Q to be maximized (minimized) in terms of these variables, or develop a formula by curve-fitting if data are given.

Step 3   Use the conditions of the problem to eliminate all but one of these variables and thereby express Q as a function of a single variable, such as x.

Step 4   Determine the set of possible values for x, usually an interval.

Step 5   Find the critical numbers (endpoints, stationary points, singular points). Most often, the key critical numbers are the stationary points where $dQ/dx = 0$.

Step 6   Use the theory of this chapter and/or a graph of Q as a function of x to decide which critical number gives the maximum (minimum). When graphing, make sure the graph window includes any critical numbers found in Step 5.

Concepts Review

1.   If a rectangle of area 100 has length x and width y, then the allowable values for x are _____ .

2.   The perimeter P of the rectangle of Question 1 expressed in terms of x is given by $P =$ _____ .

3.   If the rectangle of Question 1 is partitioned down the middle in both directions, the total length of fence L required to enclose and partition it can be expressed in terms of x by $L = P =$ _____ .

4.   If f is continuous on $(0, \infty)$ and if $f'(x) < 0$ for $x < c$ and $f'(x) > 0$ for $x > c$, then f has a _____ at $x = c$. It has no _____ on $(0, \infty)$.

Problem Set 4.4

1.   Find two numbers whose product is -12 and the sum of whose squares is a minimum.

2.   For what number does the fourth root exceed twice the number by the largest amount?

3.   Find the points on the parabola $x = 2y^2$ that are closest to the point $(-10, 0)$. *Hint*: Minimize the square of the distance between (x, y) and $(-10, 0)$.

4.   Find the points on the hyperbola $x^2/4 - y^2 = 1$ that are closest to the point $(5, 0)$.

5.   A farmer wishes to fence off two identical adjoining rectangular pens, each with 900 square feet of area, as shown in Figure 10. What are x and y so that the least amount of fence is required?

Figure 10

6.   Suppose that the outer boundary of the pens of Problem 5 requires heavy fence that costs $2 per foot, but that the middle partition requires fence costing only $1 per foot. What dimensions x and y will produce the least expensive fence?

7.   Suppose the farmer of Problem 5 chose to make three adjoining pens each of 900 square feet, as shown in Figure 11. What are x and y so that the least amount of fence is required?

Figure 11

8. Study the solutions to Problems 5 and 7 to see if you can make a conjecture about the ratio of the amount of fence required in the x-direction to the amount required in the y-direction in all problems of this type. Then, if you are ambitious, try to prove your conjecture.

9. A cistern with a square base is to be constructed to hold 12,000 cubic feet of water. If the metal top costs twice as much per square foot as the concrete side and base, what are the most economic dimensions for the cistern?

10. An open box with a capacity of 36,000 cubic inches is needed. If the box must be twice as long as it is wide, what dimensions would require the least amount of material?

11. If the strength of a rectangular beam is proportional to the product of its width and the square of its depth, find the dimensions of the strongest beam that can be cut from a log whose cross section has the form of the ellipse $9x^2 + 8y^2 = 72$. See Example 4.

12. If $y = \dfrac{my_0}{(m - y_0)e^{-kt} + y_0}$, show that

$$\frac{d^2y}{dt^2} = \frac{k^2 my_0 e^{kt}(m - y_0)(y_0 e^{kt} - m + y_0)}{(y_0 e^{kt} + m - y_0)^3}$$

(See Example 5).

13. A colony of bacteria grown under laboratory conditions is known to follow the logistic model given by

$$y = \frac{my_0}{(m - y_0)e^{-kt} + y_0}$$

where y is the population, t is time, and y_0 is the initial population. (See Example 5.) The colony starts out at 100 organisms. After 1 day there are 200 organisms, and after two days there are 350 organisms. At what point was the colony growing the fastest? How many organisms will there be in the long run? Use sketches to illustrate your solution. *Hint:* You will have to solve two equations in the two unknowns m and k. To do this, solve one equation for m and substitute the result into the other equation. Then solve the resulting equation for k (you may want to do this numerically, or use computer algebra).

14. In Example 4 of Section 3.3 we used the equation $y = e^{-t} \sin t$ to describe the motion of a car bumper after it hits a speed bump, where y represented the displacement from normal position in feet and t represented time after impact in sounds.

We can generalize this model using parameters to $y = Ae^{-kt} \sin wt$. Suppose that for a particular bumper we know from experiment that it oscillates one time per second. Then $w = 2\pi$ (recall Section 2.3). From the data given in the table below, find the values of A and k. Then determine the maximum displacement of the bumper from the normal position. Use sketches to illustrate your solution.

t	0.2	0.4
y	0.2	0.3

15. The illumination at a point is inversely proportional to the square of the distance of the point from the light source and directly proportional to the intensity of the light source. If two light sources are s feet apart and their intensities are I_1 and I_2, respectively, at what point between them will the sum of their illuminations be a minimum?

16. A powerhouse is located on one bank of a straight river that is w feet wide. A factory is situated on the opposite bank of the river, L feet downstream from the point A directly opposite the powerhouse. What is the most economical path for a cable connecting the powerhouse to the factory if it costs a dollars per foot to lay the cable under water and b dollars on land ($a > b$)?

17. At 7:00 A.M. one ship was 60 miles due east from a second ship. If the first ship sailed west at 20 miles per hour and the second ship sailed southeast at 30 miles per hour, when were they closest together?

18. Find the equation of the line that is tangent to the ellipse $b^2 x^2 + a^2 y^2 = a^2 b^2$ in the first quadrant and that forms with the coordinate axes the triangle with smallest possible area (a and b are positive constants).

19. Find the greatest volume that a right circular cylinder can have if it is inscribed in a sphere of radius r.

20. Show that the rectangle with maximum perimeter that can be inscribed in a circle is a square.

21. What are the relative dimensions of the right circular cylinder with greatest curved surface area that can be inscribed in a given sphere?

22. A right circular cone is to be inscribed in another right circular cone of given volume, with the same axis and with the vertex of the inner cone touching the base of the outer cone. What must be the ratio of their altitudes for the inscribed cone to have maximum volume?

23. A wire of length 100 centimeters is cut into two pieces; one is bent to form a square, the other an equilateral triangle. Where should the cut be made if (a) the sum of the two areas is to be a minimum; (b) a maximum? (Allow the possibility of no cut.)

24. A closed box in the form of a rectangular parallelepiped with a square base is to have a given volume. If the material used in the bottom costs 20% more per square inch than the material in the sides, and the material in the top costs 50% more per square inch than that of the sides, find the most economical proportions for the box.

25.   An observatory is to be in the form of a right circular cylinder surmounted by a hemispherical dome. If the hemispherical dome costs twice as much per square foot as the cylindrical wall, what are the most economical proportions for a given volume?

26.   A weight connected to a spring moves along the x-axis so that its x-coordinate at time t is

$$x = \sin 2t + \sqrt{3} \cos 2t$$

What is the farthest the weight gets from the origin?

27.   A flower bed will be in the shape of a sector of a circle (a pie-shaped region) of radius r and vertex angle θ. Find r and θ if its area is a constant A and the perimeter is a minimum.

28.   A fence h feet high runs parallel to a tall building and w feet from it (Figure 12). Find the length of the shortest ladder that will reach from the ground across the top of the fence to the wall of the building.

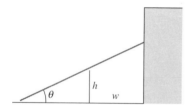

Figure 12

29.   (Snell's Law). Fermat's Principle in optics says that light travels from point A to point B along the path that requires least time. Suppose that light travels in one medium at velocity c_1 and in a second medium at velocity c_2. If A is in medium 1 and B in medium 2 and the x-axis separates the two media, as shown in Figure 13, show that

$$\frac{\sin \theta_1}{c_1} = \frac{\sin \theta_2}{c_2}$$

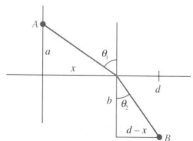

Figure 13

30.   Light from A is reflected to B by a plane mirror. Use Fermat's Principle (Problem 29) to show that the angle of incidence is equal to the angle of reflection.

31.   One end of a 27-foot ladder rests on the ground and the other end rests on the top of an 8-foot wall. As the bottom end is pushed along the ground, the top end extends beyond the wall. Find the maximum horizontal overhang of the top end.

32.   I have enough pure silver to coat 1 square meter of surface area. I plan to coat a sphere and a cube. What dimensions should they be if the total volume of the silvered solids is to be a maximum? A minimum? (Allow the possibility of all the silver going onto one solid.)

33.   One corner of a long narrow strip of paper is folded over so it just touches the opposite side, as shown in Figure 14. With parts labeled as indicated, determine x in order to:

(a) maximize the area of triangle A;

(b) minimize the area of triangle B;

(c) minimize the length z

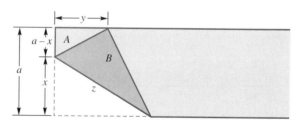

Figure 14

34.   Determine θ so that the area of the symmetric cross shown in Figure 15 is maximized. Then find this maximum area.

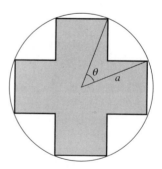

Figure 15

35.   A clock has hour and minute hands of lengths h and m, respectively, with $h \le m$. We wish to study this clock at times between 12:00 and 12:30. Let θ, ϕ, and L be as in Figure 16 and note that θ increases at a constant rate.

By the Law of Cosines, we can establish that $L = L(\theta) = (h^2 + m^2 - 2hm \cos \theta)^{1/2}$ and so

$$L'(\theta) = hm(h^2 + m^2 - 2hm \cos \theta)^{1/2} \sin \theta$$

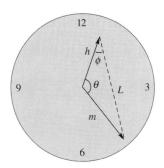

Figure 16

(a) For $h = 3$ and $m = 5$, determine L', L, and ϕ at the instant when L' is largest.

(b) Rework part (a) when $h = 5$ and $m = 13$.

(c) Based on parts (a) and (b), make conjectures about the values of L', L, and ϕ at the instant when the tips of the hands are separating most rapidly.

(d) Try to prove your conjectures.

Answers to Concepts Review: 1. $0 < x < 100$
2. $2x + 200/x$ 3. $3x + 300/x$ 4. Global minimum; global maximum

4.5 | Applications from Economics

Discrete versus Continuous

Most problems in the social sciences are properly viewed as discrete in nature. Moreover, the digital computer is a fast, accurate tool for handling discrete quantities. A natural question arises: Why not study discrete problems using discrete tools rather than by first modeling them with continuous curves? In answer to this question, most colleges now offer courses in discrete mathematics. However, because of its beauty and power, calculus continues to enjoy popularity as a tool for analyzing social science problems, as well as science problems.

Each discipline has its own language. This is certainly true of economics, which has a highly developed special vocabulary. Once we learn this vocabulary, we will discover that many of the problems of economics are just ordinary calculus problems dressed in new clothes.

Consider a typical company, the ABC Company. For simplicity, assume that ABC produces and markets a single product; it might be television sets, car batteries, or bars of soap. If it sells x units of the product in a fixed period of time (for example, a year), it will be able to charge a **price**, $p(x)$, for each unit. We indicate that p depends on x because if ABC increases its output, it will probably need to reduce the price per unit in order to sell its total output. The **total revenue** that ABC can expect is given by $R(x) = xp(x)$, the number of units times the price per unit.

To produce and market x units, ABC will have a total cost, $C(x)$. This is normally the sum of a **fixed cost** (office utilities, real estate taxes, and so on) plus a **variable cost**, which depends directly on the number of units produced.

The key concept for a company is the **total profit**, $P(x)$. It is just the difference between revenue and cost—that is,

$$P(x) = R(x) - C(x) = xp(x) - C(x)$$

Generally, a company seeks to maximize its total profit.

Note a feature that tends to distinguish problems in economics from those in the physical sciences. In most cases, the product will be in discrete units (you can't make or sell 0.23 television sets or π car batteries). Thus, the functions $R(x)$, $C(x)$, and $P(x)$ are usually defined only for $x = 0, 1, 2, \ldots$, and consequently, their graphs consist of discrete points (Figure 1). In order to make the tools of calculus available, we connect these points with a smooth curve (Figure 2), thereby pretending that R, C, and P are nice differentiable functions. This illustrates an aspect of mathematical modeling that is almost always necessary, especially in economics. To model a real-world problem, we must make simplifying assumptions. This means that the answers we get only approximate the answers we seek—one of the reasons economics is a less than perfect science.

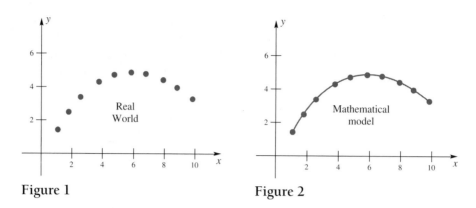

Figure 1　　　　　　　　　　Figure 2

A related problem for an economist is how to obtain formulas for the functions $C(x)$ and $p(x)$. In a simple case, $C(x)$ might have the form

$$C(x) = 10{,}000 + 50x$$

If so, \$10,000 is the *fixed cost* and \$50x is the *variable cost*, based on a \$50 direct cost for each unit produced. Perhaps a more typical situation is

$$\hat{C}(x) = 10{,}000 + 45x + 100\sqrt{x}$$

Note that in this case the *average variable cost* per unit is

$$\frac{45x + 100\sqrt{x}}{x} = 45 + \frac{100}{\sqrt{x}}$$

an amount that decreases as x increases (efficiency of size). The cost functions $C(x)$ and $\hat{C}(x)$ are graphed together in Figure 3.

Selecting appropriate functions to model cost and price is a nontrivial task. Occasionally, they can be inferred from basic assumptions. In other cases, a careful study of the history of the firm will suggest reasonable choices. Sometimes, we must simply make intelligent guesses.

Use of the Word *Marginal*

Suppose that ABC knows its cost function $C(x)$ and that it has tentatively planned to produce 2000 units this year. President Hornblower would like to determine the additional cost per unit if ABC increased production slightly. Would it, for example, be less than the additional revenue per unit? If so, it would make good economic sense to increase production.

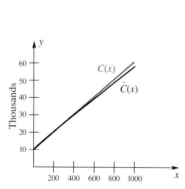

Figure 3

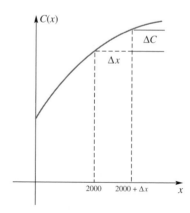

Figure 4

Economics Vocabulary

Because economics tends to be a study of discrete phenomena, your economics professor may define marginal cost at x as the cost of producing one additional unit—that is, as

$$C(x + 1) - C(x)$$

In the mathematical model, this number will be very close in value to dC/dx, and because the latter is a principal concept in calculus, we choose to take it as the definition of marginal cost. Similar statements hold for marginal revenue and marginal profit.

If the cost function is the one shown in Figure 4, President Hornblower is asking for the value of $\Delta C/\Delta x$ when $\Delta x = 1$. But we expect that this will be very close to the value of

$$\lim_{\Delta x \to 0} \frac{\Delta C}{\Delta x}$$

when $x = 2000$. This is called the **marginal cost**. We mathematicians recognize it as dC/dx, the derivative of C with respect to x.

In a similar vein, we define **marginal price** as dp/dx, **marginal revenue** as dR/dx, and **marginal profit** as dP/dx.

Examples We now illustrate how to solve a wide variety of economics problems.

EXAMPLE 1: Suppose $C(x) = 8300 + 3.25x + 40\sqrt[3]{x}$ dollars. Find the average cost per unit and the marginal cost and then evaluate them when $x = 1000$.

SOLUTION:

$$\text{Average cost: } \frac{C(x)}{x} = \frac{8300 + 3.25x + 40\sqrt[3]{x}}{x}$$

$$\text{Marginal cost: } \frac{dC}{dx} = 3.25 + \frac{40}{3}x^{-2/3}$$

At $x = 1000$, these have the values 11.95 and 3.38, respectively. This means that it costs, on the average, \$11.95 per unit to produce the first 1000 units; to produce one additional unit beyond 1000 costs only about \$3.38. ◄

EXAMPLE 2: A company estimates that it can sell 1000 units per week if it sets the unit price at \$3.00, but that its weekly sales will rise by 100 units for each 10¢ decrease in price. If x is the number of units sold each week ($x \geq 1000$), find:
(a) the price function, $p(x)$;
(b) the number of units and the corresponding price that will maximize weekly revenue;
(c) the maximum weekly revenue.

SOLUTION:

(a) We are given that

$$x = 1000 + \frac{3.00 - p(x)}{0.10}(100)$$

or, equivalently,

$$p(x) = 3.00 - (0.10)\frac{(x - 1000)}{100} = 4 - 0.001x$$

(b) $$R(x) = xp(x) = 4x - 0.001x^2$$

$$\frac{dR}{dx} = 4 - 0.002x$$

The only critical numbers are the endpoint 1000 and the stationary point 2000, obtained by setting $dR/dx = 0$. The First Derivative Test ($R'(x) > 0$ for $1000 \leq$

$x < 2000$ and $R'(x) < 0$ for $x > 2000$) shows that $x = 2000$ gives the maximum revenue. This corresponds to a unit price $p(2000) = \$2.00$.

(c) The maximum weekly revenue is $R(2000) = \$4000$.   ◀

EXAMPLE 3: In manufacturing and selling x units of a certain commodity, the price function p and the cost function C (in dollars) are given by

$$p(x) = 5.00 - 0.002x$$

$$C(x) = 3.00 + 1.10x$$

Find expressions for the marginal revenue, marginal cost, and marginal profit; determine the production level that will produce the maximum total profit.

SOLUTION:

$$R(x) = xp(x) = 5.00x - 0.002x^2$$

$$P(x) = R(x) - C(x) = -3.00 + 3.90x - 0.002x^2$$

Thus, we have the following derivatives.

$$\text{Marginal revenue: } \frac{dR}{dx} = 5 - 0.004x$$

$$\text{Marginal cost: } \frac{dC}{dx} = 1.1$$

$$\text{Marginal profit: } \frac{dP}{dx} = \frac{dR}{dx} - \frac{dC}{dx} = 3.9 - 0.004x$$

To maximize profit, we set $dP/dx = 0$ and solve. This gives $x = 975$ as the only critical number to consider. It does provide a maximum, as may be checked by the First Derivative Test. The maximum profit is $P(975) = \$1898.25$.   ◀

Note that at $x = 975$, both the marginal revenue and the marginal cost are $\$1.10$. In general, a company should expect to be at a maximum profit level when the cost of producing an additional unit equals the revenue from that unit. If the cost of the next unit were less than the revenue of the next unit, the company would want to produce that unit; if the marginal cost were greater than the marginal revenue the company would want to produce one less unit.

The statement just made assumes that the cost function and the revenue function are nice, differentiable functions and that endpoints are not significant. In some situations, the cost function may take large jumps, as when a new employee or a new piece of equipment is added; also, a manufacturing plant may have a maximum capacity, thereby introducing an important endpoint. We illustrate these possibilities in the next two examples.

EXAMPLE 4: The XYZ Company manufactures wicker chairs. With its present machines, it has a maximum yearly output of 500 units. If it makes x chairs, it can set a price of $p(x) = 200 - 0.15x$ dollars each and will have a total yearly cost of $C(x) = 4000 + 6x - (0.001)x^2$ dollars. What production level maximizes the total yearly profit?

SOLUTION:

$$R(x) = xp(x) = x(200 - 0.15x) = 200x - 0.15x^2$$

and so

$$P(x) = R(x) - C(x)$$
$$= 200x - 0.15x^2 - (4000 + 6x - 0.001x^2)$$
$$= -4000 + 194x - 0.149x^2$$

Thus,

$$\frac{dP}{dx} = 194 - 0.298x$$

which yields the stationary point 651. However, 651 is not in the interval [0, 500], so the only critical numbers to check are the two endpoints, 0 and 500. If the maximum is at 0, the company had better go out of the wicker-chair business fast. It is not. The maximum occurs at 500, and the maximum profit is $P(500) = \$55,750$. ◄

EXAMPLE 5: With the addition of a new machine, the XYZ Company of Example 4 could boost its yearly production of chairs to 750. However, its cost function $C(x)$ will then take the form

$$C(x) = \begin{cases} 4000 + 6x - (0.001)x^2 & \text{if } 0 \le x \le 500 \\ 4000 + 6x - (0.003)x^2 & \text{if } 500 \le x \le 750 \end{cases}$$

What production level maximizes total yearly profit under these circumstances?

SOLUTION: The new cost function results in the new profit function

$$P(x) = \begin{cases} -4000 + 194x - 0.149x^2 & \text{if } 0 \le x \le 500 \\ -6000 + 194x - 0.147x^2 & \text{if } 500 \le x \le 750 \end{cases}$$

In the previous example we saw that there are no stationary points on the interval $0 \le x \le 500$. On the interval $500 \le x \le 750$,

$$\frac{dP}{dx} = 194 - 0.294x$$

which gives the stationary point 660. There are four critical numbers, namely, 0, 500, 660, 750. The corresponding values of P are -4000, 55,750, 58,007, 56,813. We conclude that a production level of 660 units gives the maximum profit. The graph in Figure 5 clarifies this example. ◄

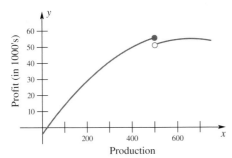

Figure 5

---

1. To use calculus to model economics problems that are usually discrete in nature, we must transform these discrete problems into _____ ones.

2. Total revenue is $R(x) = xp(x)$, where x denotes _____ and $p(x)$ denotes _____ .

3. Total cost $C(x)$ usually consists of two kinds

of cost: _____ cost; consisting of the cost of utilities, real estate taxes, and so on, and _____ cost, which depends on the number of units produced.

4. In economics, dR/dx is called _____ and dC/dx is called _____ .

Problem Set 4.5

1. The fixed monthly cost for operating a manufacturing plant that makes zeebars is $8000 and there are direct costs of $110 for each unit produced. Write an expression for $C(x)$, the total cost of making x zeebars in a month.

2. The manufacturer of Problem 1 estimates that 100 units per month can be sold if the unit price is $250, and that sales will increase by 20 units for each $10 decrease in price. Write an expression for the price $p(x)$ and the revenue $R(x)$ if x units are sold in 1 month, $x \geq 100$. (See Example 2.)

3. Use the information in Problems 1 and 2 to write an expression for the total monthly profit $P(x)$, $x \geq 100$.

4. Sketch the graph of $P(x)$ of Problem 3, and from it estimate the value of x that maximizes P. You may want to use the Trace command of a graphing calculator or the Table feature of a computer algebra system. Then find this x exactly by the methods of calculus.

5. The total cost of producing and selling x units of a certain commodity per month is $C(x) = 1200 + (3.25)x - (0.002)x^2$. If the production level is 1800 units per month, find the average cost, $C(x)/x$, of each unit and the marginal cost.

6. The total cost of producing and selling x units of a certain commodity per week is $C(x) = 1100 + x^2/1200$. Find the average cost, $C(x)/x$, of each unit and the marginal cost at a production level of 900 units per week.

7. The total cost of producing and marketing x units of a certain commodity is given by

$$C(x) = \frac{80{,}000 - 400x^2 + x^3}{40{,}000}$$

For what number x is the average cost a minimum?

8. The total cost of producing and selling $100x$ units of a particular commodity per week is

$$C(x) = 1000 + 33x - 9x^2 + x^3$$

Find (a) the level of production at which the marginal cost is a minimum, and (b) the minimum marginal cost.

9. A price function, p, is defined by

$$p(x) = 20 + 4x - \frac{x^2}{3}$$

where $x \geq 0$ is the number of units.

(a) Find the total revenue function and the marginal revenue function.

(b) On what interval is the total revenue increasing?

(c) For what number x is the marginal revenue a maximum?

10. For the price function defined by $p(x) = (182 - x/36)^{1/2}$, find the number of units x_1 that makes the total revenue a maximum and state the maximum possible revenue. What is the marginal revenue when the optimum number of units, x_1, are sold?

11. For the price function given by

$$p(x) = 800/(x + 3) - 3$$

find the number of units x_1 that makes the total revenue a maximum and state the maximum possible revenue. What is the marginal revenue when the optimum number of units, x_1, is sold?

12. A river boat company offers a Fourth of July excursion to a fraternal organization with the understanding that there will be at least 400 passengers. The price of each ticket will be $12.00, and the company agrees to refund $0.20 to every passenger for each 10 passengers in excess of 400. Write an expression for the price function $p(x)$ and find the number x_1 of passengers that makes the total revenue a maximum.

13. A merchant finds that he can sell 4000 yards of a particular fabric each month if he prices it at $6.00 per yard, and that his monthly sales will increase by 250 yards for each $0.15 reduction in the price per yard. Write an expression for $p(x)$ and find the price per yard that would bring maximum revenue.

14. A manufacturer estimates that she can sell 500 articles per week if her unit price is $20.00, and that her weekly sales will rise by 50 with each $0.50 reduction in price. The cost of producing and selling x articles a week is $C(x) = 4200 + 5.10x + 0.0001x^2$. Find each of the following.

(a) The price function.

(b) The level of weekly production for maximum profit.

(c) The price per article at the optimum level of production.

(d) The marginal price at that level of production.

15. The monthly overhead of a manufacturer of a certain commodity is $6000, and the cost of material is $1.00 per unit. If not more than 4500 units are manufactured per month, labor cost is $0.40 per unit; but for each unit over 4500, the manufacturer must pay time-and-a-half for labor. The manufacturer can sell 4000 units per month at $7.00 per unit and estimates that monthly sales will rise by 100 for each $0.10 reduction in price. Find (a) the total cost function, (b) the price function, and (c) the number of units that should be produced each month for maximum profit.

16. The ZEE Company makes zingos, which it markets at a price $p(x) = 10 - 0.001x$ dollars, where x is the number produced each month. Its total monthly cost is $C(x) = 200 + 4x - 0.01x^2$. At peak production it can make 300 units. What is its maximum monthly profit and what level of production gives this profit?

17. If the company of Problem 16 expands its facilities so that it can produce up to 450 units each month, its monthly cost function takes the form $C(x) = 800 + 3x - 0.01x^2$ for $300 < x \le 450$. Find the production level that maximizes monthly profit and evaluate this profit. Sketch the graph of the monthly profit function $P(x)$ on $0 \le x \le 450$. (See Example 5.)

18. Let us suppose that a manufacturer has m employees, who produce a total of x units of product per week. These are sold at a price $p = p(x)$. Then the total weekly revenue $R(x) = x \cdot p$ can be thought of as depending on m. The derivative dR/dm is called the *marginal revenue product*. It is (approximately) the change in revenue when the manufacturer adds one employee. Show that

$$\frac{dR}{dm} = \frac{dx}{dm}\left(p + x\frac{dp}{dx}\right)$$

Hint: Use the Product Rule and then the Chain Rule.

19. Refer to Problem 18. A manufacturer has determined that in one week, m employees can produce

$x = 5m^2/\sqrt{m^2 + 13}$ units, which the manufacturer can then sell at a price $p = 10x - (0.1)x^2$ dollars. Determine the marginal revenue product when $m = 6$.

20. To be successful, a retail store must control its inventory. Overstocking results in excessive interest costs, extra warehouse rental, and the danger of obsolescence. Too small an inventory involves more paperwork in reordering, extra delivery charges, and the greater likelihood of running out of stock. Carvers Appliance Outlet estimates that it costs $20 to hold a microwave oven in stock for a year. To reorder a lot of ovens costs $200 plus $3 for each oven. What lot size will result in the smallest inventory cost? Assume that Carver sells 1000 ovens per year and that ordering lots of size x means that, on the average, $x/2$ ovens will be in stock.

21. Suppose that a store expects to sell N units of a particular item this year, that it costs A dollars to keep one unit in stock for a year, and that to reorder a lot of size x costs $(F + Bx)$ dollars. Show that $x = \sqrt{2FN/A}$ is the lot size that minimizes inventory cost. See Problem 20.

22. If expected sales of an item quadruple, what will happen to the ideal lot size? See Problem 21.

Answers to Concepts Review: 1. Continuous 2. Number of units sold; price per unit   3. Fixed; variable   4. Marginal revenue; marginal cost

4.6 | Graphing with Parameters

Up to this point we have depended on plotting points and/or using technology (calculators or computers) to graph functions. We have seen that calculus provides a powerful tool for analyzing the fine structure of a graph, especially in identifying those points where the character of the graph changes. We can locate maximum points, local minimum points, and inflection points; we can determine precisely where the graph is increasing or where it is concave up. Even when using technology, this helps greatly with deciding how to set the graph window.

Many important functions used in the sciences have unspecified parameters in them. For such functions, it can still be very useful to produce "generic" graphs based on the information obtained from calculus. Technology can be used to produce graphs with specific parameter values, but cannot produce such a generic graph.

EXAMPLE 1: Sketch the graph of $f(x) = Axe^{-kx}$, where A and k are parameters, with $A > 0$ and $k > 0$.

SOLUTION: We will look for intercepts and asymptotes and use the derivatives to determine where the function is increasing, decreasing, concave up, and concave down to sketch a generic graph.

Intercepts   Because $f(0) = 0$, the point $(0, 0)$ is both an x- and y-intercept. Setting $f(x) = 0$ we find that there are no other x- intercepts.

First Derivative   We have

$$f'(x) = Ae^{-kx} - Akxe^{-kx}$$
$$= Ae^{-kx}(1 - kx)$$

so that setting $f'(x) = 0$ we get $x = \frac{1}{k}$. We can now use a number-line analysis, with $x = \frac{1}{k}$ as the split point and $x = 0$ and $x = \frac{2}{k}$ as test points. Because $f'(0) = Ae^{-k(0)}(1 - k(0)) = A$ (positive) and $f'(\frac{2}{k}) = Ae^{-k(\frac{2}{k})}(1 - k(\frac{2}{k})) = -Ae^{-2}$ (negative) we conclude that f is increasing on $-\infty < x < \frac{1}{k}$ and decreasing on $\frac{1}{k} < x < \infty$. See Figure 1. From the First Derivative Test we also know that we have a local maximum at $x = \frac{1}{k}$; since there are no other critical numbers, this point is also a global maximum. The maximum value is $f(\frac{1}{k}) = A \cdot (\frac{1}{k})e^{-k(\frac{1}{k})} = \frac{A}{k}e^{-1}$.

Second Derivative   We have

$$f''(x) = Ake^{-kx}(1 - kx) + Ae^{-kx}(-k)$$
$$= Ake^{-kx}(-2 + kx)$$

so that setting $f''(x) = 0$ we get $x = \frac{2}{k}$. Again we can use a number-line analysis, with $x = \frac{2}{k}$ as the split point and $x = 0$ and $x = \frac{3}{k}$ as the test points. We get $f''(0) = Ake^{-k(0)}(-2 + k(0)) = -2Ak$ (negative) and $f''(\frac{3}{k}) = Ake^{-k(\frac{3}{k})}(-2 + k(\frac{3}{k})) = Ake^{-3}$ (positive). Thus, f is concave down on $-\infty < x < \frac{2}{k}$ and concave up on $\frac{2}{k} < x < \infty$. See Figure 2. Finally, we know that there is an inflection point at $x = \frac{2}{k}$; the corresponding y-coordinate is $f(\frac{2}{k}) = \frac{2A}{k}e^{-2}$.

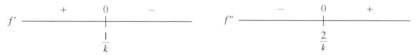

f' $\underset{\frac{1}{k}}{\underline{\quad + \qquad 0 \qquad - \quad}}$ f'' $\underset{\frac{2}{k}}{\underline{\quad - \qquad 0 \qquad + \quad}}$

Figure 1   **Figure 2**

Asymptotes   Because $Axe^{-kx} = A\frac{x}{e^{kx}}$ and because positive exponential functions go to infinity faster than any power of x (see Sections 1.4 and 1.5) we have $\lim_{x \to \infty} f(x) = 0$. Thus, $y = 0$ is a horizontal asymptote as $x \to \infty$. As $x \to -\infty$ both x and e^{-kx} get large in magnitude, so there is no horizontal asymptote on the other side.

Sketch   Using the above information, we can now sketch a graph. Because the parameters A and k are not specified, our graph will not have a scale on the x- or y-axis. What we get is the general shape of the function for any positive values of A and k. See Figure 3. We have included the y-values of the maximum and inflection point this time, but they may not be of interest for every function with parameters.

Finally, we can check our results to a certain extent by generating a computer or calculator graph with *particular* values of A and k. Remember, this is no substitute for sketching a generic graph as in Figure 3, in that there is no guarantee that the shape will be similar for other parameter values without going through the full analysis we have made here. See Figure 4. Notice that dou-

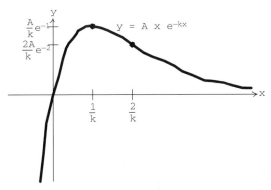

Figure 3

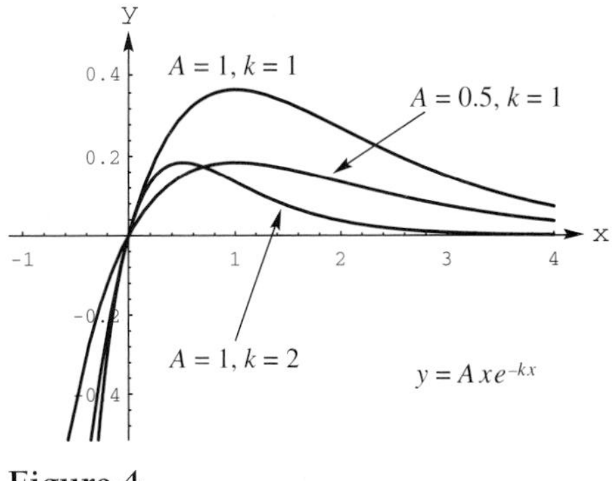

Figure 4

bling k results in the same maximum value as halving A, but at a different x value. The reason for this is easily seen from the generic graph in Figure 3.

EXAMPLE 2: Sketch the graph of $f(x) = x^3 - ax$, where a can be any real number.

SOLUTION: Because $f(-x) = -f(x)$, f is an odd function and therefore its graph is symmetric with respect to the origin. Setting $f(x) = 0$, we find the x-intercepts to be 0 and $\pm\sqrt{a}$. Notice that if $a < 0$ then $\sqrt{a}$ is not a real number; so for this case there is only one x-intercept at $x = 0$. If $a = 0$, then $x = 0$ is again the only intercept. Thus, we must consider the cases $a > 0$, $a = 0$, and $a < 0$ separately. We can go this far without calculus.

First Derivative   When we differentiate f, we obtain

$$f'(x) = 3x^2 - a$$

The critical numbers are $x = \pm\sqrt{\frac{a}{3}}$. Again, we will have to consider the cases $a > 0$, $a = 0$, and $a < 0$.

When $a > 0$ this gives us the intervals $\left(-\infty, -\sqrt{\frac{a}{3}}\right)$, $\left(-\sqrt{\frac{a}{3}}, \sqrt{\frac{a}{3}}\right)$, and $\left(\sqrt{\frac{a}{3}}, \infty\right)$. Using the test points $-2\sqrt{\frac{a}{3}}$, 0, and $2\sqrt{\frac{a}{3}}$, we get the number line in Figure 5. Thus, f is increasing on $\left(-\infty, -\sqrt{\frac{a}{3}}\right)$ and $\left(\sqrt{\frac{a}{3}}, \infty\right)$ and decreasing on $\left(-\sqrt{\frac{a}{3}}, \sqrt{\frac{a}{3}}\right)$. There is a local maximum at $x = -\sqrt{\frac{a}{3}}$ and a local minimum at $x = \sqrt{\frac{a}{3}}$.

Consider now the case of $a < 0$. There are no critical numbers, because $\sqrt{\frac{a}{3}}$ is not a real number. This means that f' is either entirely positive or entirely negative on the whole real line. If we test the single value $x = 0$ (any real number would do) we get $f'(0) = 3(0)^2 - a = -a$, which is a *positive* number because we are assuming for this case that a is negative. Thus, we find that f is increasing for all real x. See Figure 6.

The final case is $a = 0$, which corresponds to the function $f(x) = x^3$. With $f'(x) = 3x^2$ there is one critical number at $x = 0$. A number-line analysis shows that f' is positive for both $x < 0$ and $x > 0$, so $x = 0$ is neither a maximum nor a minimum.

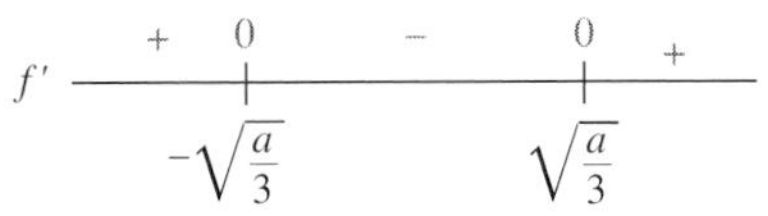

f' ——————————
$\quad\quad +\quad 0\quad\quad -\quad\quad 0\quad +$
$\quad\quad\quad -\sqrt{\frac{a}{3}}\quad\quad\quad \sqrt{\frac{a}{3}}$

Case 1: $a > 0$.

Figure 5

f' ——————————
$\quad\quad\quad\quad\quad\quad +$

Case 2: $a < 0$

Figure 6

Second Derivative   Differentiating again, we get

$$f''(x) = 6x$$

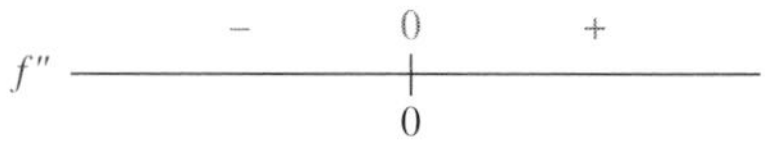

$$f'' \quad \frac{ - 0 + }{0}$$

Figure 7

By studying the sign of $f''(x)$ (Figure 7), we deduce that f is concave upward on $(0, \infty)$ and concave downward on $(-\infty, 0)$. Thus, there is one point of inflection, namely, $(0, 0)$. Notice that the concavity does not depend on the value of a, so we don't need to consider cases as we did for the first derivative.

Sketch   Using the information gained above, we sketch f for each of the cases $a > 0$, $a = 0$, and $a < 0$. Notice that the concavity is not affected by the value of a, but regions of increase and decrease are affected. In Figure 8, we summarize the relevant information and show the corresponding graphs. We leave it to the exercises for you to verify these results by computer or calculator graphs for particular parameter values.   ◄

Value of a	Increasing	Decreasing	Concave Up	Concave Down	x-Intercepts
$a > 0$	$(-\infty, -\sqrt{a/3}), (\sqrt{a/3}, \infty)$	$(\sqrt{a/3}, -\sqrt{a/3})$	$(0, \infty)$	$(-\infty, 0)$	$-\sqrt{a}, 0, \sqrt{a}$
$a < 0$	$(-\infty, \infty)$	nowhere	$(0, \infty)$	$(-\infty, 0)$	0
$a = 0$	$(-\infty, \infty)$	nowhere	$(0, \infty)$	$(-\infty, 0)$	0

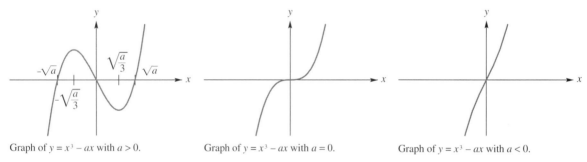

Graph of $y = x^3 - ax$ with $a > 0$.   Graph of $y = x^3 - ax$ with $a = 0$.   Graph of $y = x^3 - ax$ with $a < 0$.

Figure 8

EXAMPLE 3: In Example 3 of Section 4.4 we introduced the problem of finding the dimensions of a handbill that would minimize the amount of paper used for a given amount of printed material and fixed margins. In that problem, we assumed 50 square inches of printed matter, with 4-inch margins at top and bottom and 2-inch margins on each side.

The reason why functions with parameters are so important is because they allow one to solve many problems at once, rather than solving essentially the same problem over and over. We can generalize the problem described above by introducing parameters. Let a represent the combined size of the side margins in inches, let b represent the combined size of the top and bottom margins in inches, and let c represent the amount of printed material in square inches.

From our work on Example 3 of Section 4.4 it should be easy to see that a formula for the area of the handbill would be

$$A = xy = \frac{cx}{x - a} + bx$$

where x represents the length of one side of the handbill, $y = \frac{c}{x - a} + b$ represents the length of the other side, and A represents the area of the handbill. See Figure 9.

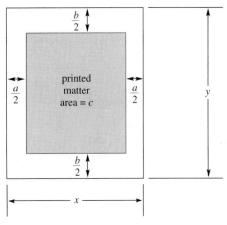

Figure 9

Sketch a graph of A as a function of x using a domain appropriate for the problem. Describe the important features of the function, and explain how the graph could be used to help design handbills or posters.

SOLUTION: From Figure 9 it should be clear that we must have $x > a$, so we use this interval as our domain. Thus we will not have an A-intercept (the A-axis will be the vertical axis in this case). From the problem situation, we can deduce that we will not have any x-intercepts either (why?).

First Derivative   We have

$$\frac{dA}{dx} = \frac{c(x - a) - cx(1)}{(x - a)^2} + b$$

$$= \frac{ac}{(x - a)^2} + b$$

Setting $\dfrac{dA}{dx} = 0$ we get

$$(x - a)^2 = \frac{ac}{b}$$

$$x = a \pm \sqrt{\frac{ac}{b}}$$

Because we are interested only in the domain $x > a$, we use only the solution $x = a + \sqrt{\frac{ac}{b}}$. Rather than a number-line analysis, we will use the Second Derivative Test to test whether this critical number indicates a maximum or minimum, and from that determine the intervals for which the graph is increasing and decreasing.

Second Derivative   We have

$$\frac{d^2A}{dx^2} = \frac{d}{dx}\left[-ac(x-a)^{-2} + b\right]$$

$$= \frac{2ac}{(x-a)^3}$$

For $x > a$, the second derivative is positive, so the graph of A as a function of x is concave up.

Also, we know that the critical number $x = a + \sqrt{\frac{ac}{b}}$ must be a minimum by the Second Derivative Test (because the second derivative is positive for all $x > a$, it must be positive at $x = a + \sqrt{\frac{ac}{b}}$). Because this is the *only* critical number for $x > a$, we know that the graph must be decreasing for $x < a + \sqrt{\frac{ac}{b}}$ and increasing for $x > a + \sqrt{\frac{ac}{b}}$ (why?).

Asymptotes   Because the denominator in the first term of $A = \frac{cx}{x-a} + bx$ is equal to zero at $x = a$, we suspect a vertical asymptote there. Letting $x \to a^+$ we find that the denominator is positive as it approaches zero, and the numerator approaches the finite positive number ca. Thus, the fraction approaches positive infinity, confirming the asymptote at $x = a$.

To see what happens as $x \to \infty$, we can use the trick of dividing the numerator and denominator of the first term of the function by x. We get

$$A = \frac{c}{1 - \dfrac{a}{x}} + bx$$

so that as $x \to \infty$ the first term approaches c. The second term approaches ∞, so the value of A also approaches ∞; thus, we do not have a horizontal asymptote.

We do, however, have what is called an *oblique* or *slant* asymptote. When we replace the first term by its limit c, we get the equation of the oblique asymptote $A = c + bx$. Thus, as $x \to \infty$, the graph of $A = \frac{cx}{x-a} + bx$ approaches the graph of the straight line $A = c + bx$.

Sketch   Summarizing the results from above, we find that the graph of $A = \frac{cx}{x-a} + bx$ on the interval $x > a$ is decreasing on $\left(a, a + \sqrt{\frac{ac}{b}}\right)$ and increasing on $\left(a + \sqrt{\frac{ac}{b}}, \infty\right)$, and that it is concave up for all $x > a$. There is a local and global minimum at $x = a + \sqrt{\frac{ac}{b}}$, and there are no inflection points. There is a vertical asymptote at $x = a$ and an oblique asymptote as $x \to \infty$ with the equation $A = c + bx$. The graph is sketched in Figure 10.

Perhaps the most important feature of the graph is the minimum at $x = a + \sqrt{\frac{ac}{b}}$. For situations where the margins and the amount of printed matter are fixed, this tells us how to make a handbill with the least amount of paper. One dimension would be $x = a + \sqrt{\frac{ac}{b}}$ and the other dimension would be determined by the formula $y = \frac{c}{x-a} + b = \sqrt{\frac{bc}{a}} + b$.

We get some information from the oblique asymptote as well. Because the slope of the asymptote is b, we can see that when b is small, the asymptote will be nearly horizontal; thus, the value of A will not change much as we move off of the minimum at $x = a + \sqrt{\frac{ac}{b}}$. Hence, for a poster or handbill for which the top and bottom margins are relatively small, the choice of the x dimension is not so important. Similarly, when the side margins are relatively small, the choice of the y dimension is not critical (why?). In Figure 11 we illustrate this idea by graphing A as a function of x for $a = 1$, $b = 0.1$, and $c = 25$.

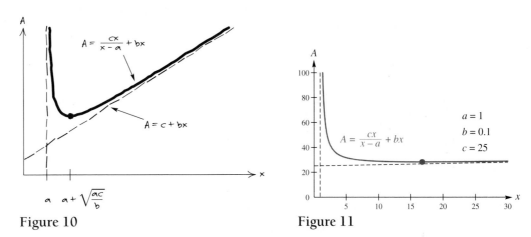

Figure 10   Figure 11

Summary of the Method

In graphing functions with parameters, there is no substitute for common sense. However, the following procedure will be helpful in most cases.

Step 1   Precalculus analysis.
(a) Check the *domain* and *range* of the function to see if any regions of the plane are excluded.
(b) Test for *symmetry* with respect to the y-axis and the origin. (Is the function even or odd?)
(c) Find the *intercepts*.
(d) Try graphing the function with a computer or graphing calculator, using a few selected parameter values to get a feel for what the function might look like.

Step 2   Calculus analysis.
(a) Use the first derivative to find the critical numbers and to find out where the graph is *increasing* and *decreasing*.
(b) Test the critical points for *local maximums* and *minimums*.
(c) Use the second derivative to find out where the graph is *concave upward* and *concave downward* and to locate *inflection points*.
(d) Find the *asymptotes*.

Step 3   Sketch the graph.

Step 4   Check results by graphing the function for key parameter values as indicated by your analysis using a computer or graphing calculator.

Concepts Review

1.   The graph of f is symmetric with respect to the y-axis if $f(-x) = $ _____ ; the graph is symmetric with respect to the origin if $f(-x) = $ _____ .

2.   If $f'(x) < 0$ and $f''(x) > 0$ for all x in an interval I, then the graph of f is both _____ and _____ on I.

3.   The graph of $f(x) = x^3/[(x + a)(x - b)(cx - d)]$ has as vertical asymptotes the lines _____ and as horizontal asymptote the line _____ .

4.   To understand a function with parameters graphically, we can sketch a _____ graph and / or we can sketch several graphs for different _____ .

Problem Set 4.6

In Problems 1–12, make an analysis as suggested in the summary above and then sketch a generic graph. Also sketch graphs for a few selected specific parameter values.

1. $f(x) = x^2 - ax$, a any real number.

2. $f(x) = x^4 - ax$, a any real number.

3. $F(x) = x^3 - ax^2$, a any real number.

4. $F(x) = x^4 - ax^2$, a any real number.

5. $f(x) = Ax^2 e^{-kx}$, with $A > 0$ and $k > 0$.

6. $f(x) = Axe^{-kx^2}$, with $A > 0$ and $k > 0$.

7. $h(x) = \frac{x}{x+a}$, a any real number.

8. $h(x) = \frac{x^2}{x^2+a}$, a any real number.

9. $f(x) = e^{-(x-a)^2}$, a any real number.

10. $f(x) = \frac{x}{x^2-a}$, a any real number.

11. $g(x) = 2x\sqrt{x-a}$, a any real number.

12. $g(x) = \sqrt{a-x} + 1$, a any real number.

13. When an object falls through the air it encounters air resistance. As a result, its velocity eventually levels off and reaches what is called a *terminal* velocity.

One possible model for the velocity of an object subject to air resistance, dropped from rest, is given by $v = A - Ae^{-kt}$. Here, v represents velocity and t represents time. For units we choose feet per second for velocity and seconds for time.

Sketch a generic graph of v as a function of t and explain what this graph means in the context of the problem. Sketch a few other graphs corresponding to particular parameter values, explaining the effect of each parameter on the graph.

14. Equations similar to $y = Ae^{-t} \cos(wt)$, with $A > 0$ and $w > 0$, have been previously used to represent the vertical motion of a car bumper as it oscillates up and down. Here, y represents vertical displacement from the rest position (feet) and t represents time (seconds).

Sketch a generic graph of y as a function of t and explain what this graph means in the context of the problem. Sketch a few other graphs corresponding to particular parameter values, explaining the effect of each parameter on the graph.

15. Verify the results of Example 2 by sketching a few graphs for selected particular parameter values.

16. Sketch a possible graph of a function f that has all the following properties, where $a < b$ are real numbers:
(a) f is everywhere continuous;
(b) $f(a) = -3, f(b) = 1$;
(c) $f'(a) = 0, f'(x) > 0$ for $x \neq a, f'(b) = 3$;
(d) $f''(b) = 0, f''(x) > 0$ for $a < x < b, f''(x) < 0$ for $x > b$.

17. Sketch a possible graph of a function f that has all the following properties, where a is a real number:
(a) f is everywhere continuous;
(b) $f(a) = 1$;
(c) $f'(x) < 0$ for $x < -3, f'(x) > 0$ for $x > a, f''(x) < 0$ for $x \neq a$.

18. Let f be a continuous function with $f(-3) = f(0) = a$, a a real number. If the graph of $y = f'(x)$ is shown in Figure 12, sketch a possible graph for $y = f(x)$.

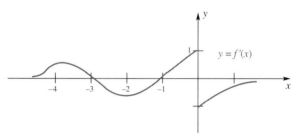

Figure 12

19. Let f be a continuous function with $f(0) = f(2) = b$, b a real number. If the graph of $y = f'(x)$ is shown in Figure 13, sketch a possible graph for $y = f(x)$.

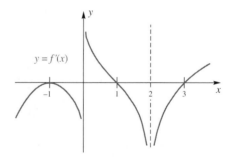

Figure 13

20. Suppose that $f'(x) = (x - a)(x - 1)^2(x + 2)$ and $f(1) = 2$, where a is a real number. Sketch a possible graph of f for each case: $a < -2, -2 < a < 1, a > 1$.

Answers to Concepts Review 1. $f(x), -f(x)$
2. Decreasing; concave up   3. $x = -a, x = b, x = \frac{d}{c}$
4. Generic; parameter values

4.7 The Mean Value Theorem

The Mean Value Theorem is the midwife of calculus—often helping to deliver other theorems that are of major significance. From now on, you will see the phrase "by the Mean Value Theorem" quite regularly, and later in this section we will use it to prove the Monotonicity Theorem, which we left unproved in Section 4.2.

In geometric language, the Mean Value Theorem is easy to state and understand. It says that if the graph of a continuous function has a nonvertical tangent line at every point between A and B, then there is at least one point C on the graph between A and B at which the tangent line is parallel to the secant line AB. In Figure 1, there is just one such point C; in Figure 2, there are several.

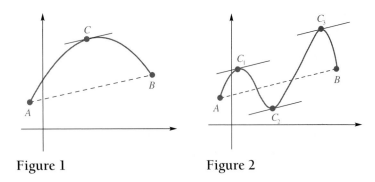

Figure 1   Figure 2

The Theorem Proved

First we state the theorem in the language of functions, then we prove it.

Theorem A

(Mean Value Theorem for Derivatives). If f is continuous on a closed interval $[a, b]$ and differentiable on its interior (a, b), then there is at least one number c in (a, b) where

$$\frac{f(b) - f(a)}{b - a} = f'(c)$$

or, equivalently, where

$$f(b) - f(a) = f'(c)(b - a)$$

Proof Our proof rests on a careful analysis of the function $s(x) = f(x) - g(x)$, introduced in Figure 3. Here $y = g(x)$ is the equation of the line through $(a, f(a))$ and $(b, f(b))$. Because this line has slope $[f(b) - f(a)]/(b - a)$ and goes through $(a, f(a))$, the point-slope form for its equation is

$$g(x) - f(a) = \frac{f(b) - f(a)}{b - a}(x - a)$$

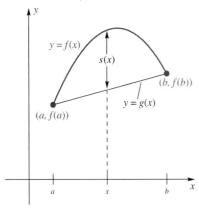

Figure 3

This, in turn, yields a formula for $s(x)$, namely,

$$s(x) = f(x) - g(x) = f(x) - f(a) - \frac{f(b) - f(a)}{b - a}(x - a)$$

It should be clear from the formula and the graph that $s(b) = s(a) = 0$. Taking the derivative of both sides of the formula for $s(x)$ above we see that for x in (a, b)

$$s'(x) = f'(x) - \frac{f(b) - f(a)}{b - a}$$

Now we make a crucial observation. If we knew that there was a number c in (a, b) satisfying $s'(c) = 0$, we would be all finished. For then the last equation would say

$$0 = f'(c) - \frac{f(b) - f(a)}{b - a}$$

which is equivalent to the conclusion of the theorem.

To see that $s'(c) = 0$ for some c in (a, b), reason as follows. Clearly s is continuous on $[a, b]$, being the difference of two continuous functions. Thus, by the Max-Min Existence Theorem (Theorem 4.1A), s must attain both a maximum and a minimum value on $[a, b]$. If both of these happen to be 0, then $s(x)$ is identically 0 on $[a, b]$, and consequently $s'(x) = 0$ for all x in (a, b), much more than we need.

If either the maximum value or the minimum value is different from 0, then that value is attained at an interior point c, because $s(a) = s(b) = 0$. We know that s has a derivative at each point of (a, b), and so by the Critical Point Theorem (Theorem 4.1B), $s'(c) = 0$. That is all we need to know.   ◄

The Theorem Illustrated

EXAMPLE 1: Find the number c guaranteed by the Mean Value Theorem for $f(x) = 2\sqrt{x}$ on $[1, 4]$.

SOLUTION:

$$f'(x) = 2 \cdot \frac{1}{2}x^{-1/2} = \frac{1}{\sqrt{x}}$$

and

$$\frac{f(4) - f(1)}{4 - 1} = \frac{4 - 2}{3} = \frac{2}{3}$$

Then, we must solve

$$\frac{1}{\sqrt{c}} = \frac{2}{3}$$

The single solution is $c = \frac{9}{4}$ (Figure 4).   ◄

EXAMPLE 2: Let $f(x) = x^3 - x^2 - x + 1$ on $[-1, 2]$. Find all numbers c satisfying the conclusion to the Mean Value Theorem.

SOLUTION:

$$f'(x) = 3x^2 - 2x - 1$$

and

$$\frac{f(2) - f(-1)}{2 - (-1)} = \frac{3 - 0}{3} = 1$$

Therefore, we must solve

$$3c^2 - 2c - 1 = 1$$

or, equivalently,

$$3c^2 - 2c - 2 = 0$$

By the Quadratic Formula, there are two solutions, $(2 \pm \sqrt{4 + 24})/6$, which correspond to $c_1 \approx -0.55$ and $c_2 \approx 1.22$. Both of these numbers are in the interval $(-1, 2)$. The appropriate graph is shown in Figure 5. ◄

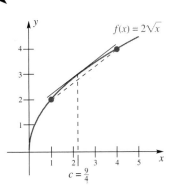

Figure 4

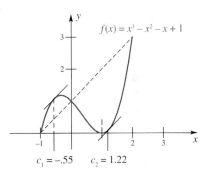

Figure 5

EXAMPLE 3: Let $f(x) = x^{2/3}$ on $[-8, 27]$. Show that the conclusion to the Mean Value Theorem fails and figure out why.

SOLUTION:

$$f'(x) = \frac{2}{3}x^{-1/3}, \qquad x \neq 0$$

and

$$\frac{f(27) - f(-8)}{27 - (-8)} = \frac{9 - 4}{35} = \frac{1}{7}$$

We must solve

$$\frac{2}{3}c^{-1/3} = \frac{1}{7}$$

which gives

$$c = \left(\frac{14}{3}\right)^3 \approx 102$$

But $c = 102$ is not in the interval $(-8, 27)$ as required. The problem is, of course, that $f(x)$ is not differentiable everywhere on $(-8, 27)$; $f'(0)$ fails to exist (see Figure 6). ◄

The Theorem Used　In Section 4.2, we promised a rigorous proof of the Monotonicity Theorem (Theorem 4.2A). This is the theorem that relates the sign of the derivative of a function to whether that function is increasing or decreasing.

Proof of the Monotonicity Theorem　We suppose that f is continuous on I and that $f'(x) > 0$ at each point x in the interior of I. Consider any two points x_1 and x_2 of I with $x_1 < x_2$. By the Mean Value Theorem applied on the interval $[x_1, x_2]$, there is a number c in (x_1, x_2) satisfying

$$f(x_2) - f(x_1) = f'(c)(x_2 - x_1)$$

Because $f'(c) > 0$, we see that $f(x_2) - f(x_1) > 0$—that is, $f(x_2) > f(x_1)$. This is what we mean when we say f is increasing on I.

The case where $f'(x) < 0$ on I is handled similarly.　◄

Our next theorem will be used repeatedly in the next chapter. In words, it says that *two functions with the same derivative differ by a constant*, possibly the zero constant (see Figure 7).

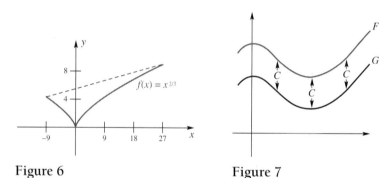

Figure 6　　　　　　　　Figure 7

Theorem B

If $F'(x) = G'(x)$ for all x in (a, b), then there is a constant C such that

$$F(x) = G(x) + C$$

for all x in (a, b).

Proof　Let $H(x) = F(x) - G(x)$. Then because we assumed that $F'(x) = G'(x)$ we have

$$H'(x) = F'(x) - G'(x) = 0$$

for all x in (a, b). Choose x_1 as some (fixed) point in (a, b) and let x be any other point in the interval. The function H satisfies the hypotheses of the Mean Value Theorem on the closed interval with endpoints x_1 and x. Thus, there is a number c between x_1 and x such that

$$H(x) - H(x_1) = H'(c)(x - x_1)$$

But $H'(c) = 0$ by hypothesis. Therefore, $H(x) - H(x_1) = 0$ or, equivalently, $H(x) = H(x_1)$ for all x in (a, b). Because $H(x) = F(x) - G(x)$, we conclude that $F(x) - G(x) = H(x_1)$. Now let $C = H(x_1)$, and we have the conclusion $F(x) = G(x) + C$.

Concepts Review

1. The Mean Value Theorem says that if f is _____ on $[a, b]$ and differentiable on _____ , then there is a point c in (a, b) such that _____ .

2. The function $f(x) = |\sin x|$ would satisfy the hypotheses of the Mean Value Theorem on the interval $[0, 1]$ but would not satisfy them on the interval $[-1, 1]$ because _____ .

3. If two functions F and G have the same derivative on the interval (a, b), then there is a constant C such that _____ .

4. Because $\frac{d}{dx}(x^4) = 4x^3$, it follows that every function F that satisfies $F'(x) = 4x^3$ has the form $F(x) = $ _____ .

Problem Set 4.7

In each of Problems 1–10, a function is defined and a closed interval is given. Decide whether the Mean Value Theorem applies to the given function on the given interval—if so, find all possible values of c; if not, state the reason. In each problem, sketch the graph of the given function on the given interval.

1. $f(x) = x^2 + 2x; [-2, 2]$

2. $f(x) = x^2 + 3x - 1; [-3, 1]$

3. $g(x) = \dfrac{x^3}{3}; [-2, 2]$

4. $g(x) = \dfrac{1}{3}(x^3 + x - 4); [-1, 2]$

5. $F(t) = \dfrac{t + 3}{t - 3}; [-1, 4]$

6. $F(x) = \dfrac{x + 1}{x - 1}; \left[\dfrac{3}{2}, 5\right]$

7. $h(x) = x^{2/3}; [0, 2]$

8. $h(x) = x^{2/3}; [-2, 2]$

9. $\phi(x) = x + \dfrac{1}{x}; \left[-1, \dfrac{1}{2}\right]$

10. $\phi(x) = x + \dfrac{1}{x}; \left[\dfrac{1}{2}, \dfrac{3}{2}\right]$

11. Johnny traveled 112 miles in 2 hours and claimed that he never exceeded 55 miles per hour. Use the Mean Value Theorem to prove that he lied. *Hint*: Let $f(t)$ be the distance traveled in time t.

12. For the function graphed on $[0, 8]$ in Figure 8, find (approximately) all points c that satisfy the conclusion to the Mean Value Theorem.

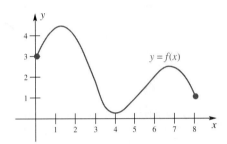

Figure 8

13. **(Rolle's Theorem).** *If f is continuous on $[a, b]$ and differentiable on (a, b) and if $f(a) = f(b)$, then there is at least one number c in (a, b) such that $f'(c) = 0$.* Show that Rolle's Theorem is just a special case of the Mean Value Theorem. (Michel Rolle (1652–1719) was a French mathematician.)

14. Show that if f is the quadratic function defined by $f(x) = \alpha x^2 + \beta x + \gamma, \alpha \neq 0$, then the number c of the Mean Value Theorem is always the midpoint of the given interval $[a, b]$.

15. Prove: If f is continuous on (a, b) and if $f'(x)$ exists and satisfies $f'(x) > 0$ except at one point x_0 in (a, b), then f is increasing on (a, b). *Hint*: Consider f on each of the intervals $(a, x_0]$ and $[x_0, b)$ separately.

16. Use Problem 15 to show that each of the following is increasing on $(-\infty, \infty)$.

(a) $f(x) = x^3$ (b) $f(x) = x^5$

(c) $f(x) = \begin{cases} x^3 & x \leq 0 \\ x & x > 0 \end{cases}$

17. Prove that if $F'(x) = 0$ for all x in (a, b), then there is a constant C such that $F(x) = C$ for all x in (a, b). *Hint*: Let $G(x) = 0$ and apply Theorem B.

18. Suppose that you know that $\cos(0) = 1$, $\sin(0) = 0$, $\frac{d}{dx}\cos x = -\sin x$, and $\frac{d}{dx}\sin x = \cos x$, but nothing else about the sine and cosine functions. Show that $\cos^2 x + \sin^2 x = 1$. *Hint*: Let $F(x) = \cos^2 x + \sin^2 x$ and use Problem 17.

19. Prove that if $F'(x) = D$ for all x in (a, b), then there is a constant C such that $F(x) = Dx + C$ for all x in (a, b). *Hint*: Let $G(x) = Dx$ and apply Theorem B.

20. Suppose $F'(x) = 5$ and $F(0) = 4$. Find a formula for $F(x)$. *Hint*: See Problem 19.

21. Prove: Let f have a derivative on an interval I. Between successive distinct zeros of f', there can be at most one zero of f. *Hint*: Try a proof by contradiction and use Rolle's Theorem (Problem 13).

22. Prove: Let g be continuous on $[a, b]$ and suppose $g''(x)$ exists for all x in (a, b). If there are three values of x in $[a, b]$ for which $g(x) = 0$, then there is at least one value of x in (a, b) such that $g''(x) = 0$.

23. Prove that if $|f'(x)| \leq M$ for all x in (a, b) and if x_1 and x_2 are any two points in (a, b), then

$$|f(x_2) - f(x_1)| \leq M|x_2 - x_1|$$

Note: A function satisfying the above inequality is said to satisfy a *Lipschitz condition* with constant M. (Rudolph Lipschitz (1832–1903) was a German mathematician.)

24. Show that $f(x) = \sin 2x$ satisfies a Lipschitz condition with constant 2 on the interval $(-\infty, \infty)$. See Problem 23.

25. A function f is called **nondecreasing** on an interval I if $x_1 < x_2 \Rightarrow f(x_1) \leq f(x_2)$ for x_1 and x_2 in I. Similarly, f is **nonincreasing** on I if $x_1 < x_2 \Rightarrow f(x_1) \geq f(x_2)$ for x_1 and x_2 in I.

(a) Sketch the graph of a function that is nondecreasing but not increasing.

(b) Sketch the graph of a function that is nonincreasing but not decreasing.

26. Prove that if f is continuous on I and if $f'(x)$ exists and satisfies $f'(x) \geq 0$ on the interior of I, then f is nondecreasing on I. Similarly, if $f'(x) \leq 0$, then f is nonincreasing on I.

27. Prove that if $f(x) \geq 0$ and $f'(x) \geq 0$ on I, then f^2 is nondecreasing on I.

28. Prove that if $g'(x) \leq h'(x)$ for all x in (a, b), then

$$x_1 < x_2 \Rightarrow g(x_2) - g(x_1) \leq h(x_2) - h(x_1)$$

for all x_1 and x_2 in (a, b). *Hint*: Apply Problem 26 with $f(x) = h(x) - g(x)$.

29. Use the Mean Value Theorem to prove that

$$\lim_{x \to \infty}\left(\sqrt{x+2} - \sqrt{x}\right) = 0$$

30. Use the Mean Value Theorem to show that

$$|\sin x - \sin y| \leq |x - y|$$

31. Suppose that in a race, horse A and horse B finished in a dead heat. Prove that their speeds were identical at some instant of the race.

32. In Problem 31, suppose that the two horses crossed the finish line together at the same speed. Show that they had the same acceleration at some instant.

33. Use the Mean Value Theorem to show that the graph of a concave up function f is always above its tangent line, that is, show that

$$f(x) > f(c) + f'(c)(x - c), \qquad x \neq c$$

Answers to Concepts Review: 1. Continuous; (a, b); $f(b) - f(a) = f'(c)(b - a)$ 2. $f'(0)$ does not exist. 3. $F(x) = G(x) + C$ 4. $x^4 + C$

4.8 Chapter Review

Concepts Test

Respond with true or false to each of the following assertions. Be prepared to justify your answer.

1. A continuous function on a closed interval must attain a maximum value on that interval.

2. If a differentiable function f attains a maximum value at an interior point c of its domain, then $f'(c) = 0$.

3. It is possible for a function to have an infinite number of critical numbers.

4. A continuous function that increases for all x must be differentiable everywhere.

5. If $f(x) = 3x^6 + 4x^4 + 2x^2$, then the graph of f is concave up on the whole real line.

6. If f is an increasing differentiable function on an interval I, then $f'(x) > 0$ for all x in I.

7. If $f'(x) > 0$ for all x in I, then f is increasing on I.

8. If $f''(c) = 0$, then f has an inflection point at $(c, f(c))$.

9. A quadratic function has no inflection points.

10. If $f'(x) > 0$ for all x in $[a, b]$, then f attains its maximum value on $[a, b]$ at b.

11. An exponential function of the form $f(x) = e^{bx}$, $b \neq 0$, is increasing for all x.

12. The graph of $y = \dfrac{x^2 - x - 6}{x - 3} = \dfrac{(x + 2)(x - 3)}{x - 3}$ has a vertical asymptote at $x = 3$.

13. The graph of $y = \dfrac{ax^2 + 1}{1 - x^2}$ has a horizontal asymptote of $y = -a$.

14. An exponential function of the form $f(x) = e^{bx}$, $b \neq 0$, is concave up for all x.

15. The function $f(x) = \sqrt{x}$ satisfies the hypotheses of the Mean Value Theorem on $[0, 2]$.

16. On the interval $[-1, 1]$, there will be just one point where the tangent line to $y = x^3$ is parallel to the secant line connecting the endpoints.

17. The graph of $y = \sin x$ has infinitely many points of inflection.

18. If $f'(c) = f''(c) = 0$, then $f(c)$ is neither a maximum nor minimum value.

19. The graph of $y = e^{\sin x}$ has infinitely many points of inflection.

20. Among rectangles of fixed area K, the one with maximum perimeter is a square.

21. If the graph of a differentiable function has three x-intercepts, then it must have at least two points where the tangent line is horizontal.

22. The sum of two increasing functions is an increasing function.

23. The product of two increasing functions is an increasing function.

24. If $f'(0) = 0$ and $f''(x) > 0$ for $x \geq 0$, then f is increasing on $[0, \infty)$.

25. If f is a differentiable function, then f is nondecreasing on (a, b) if and only if $f'(x) \geq 0$ on (a, b).

26. Two differentiable functions have the same derivative on (a, b) if and only if they differ by a constant on (a, b).

27. If $f''(x) > 0$ for all x, then the graph of $y = f(x)$ cannot have a horizontal asymptote.

28. A global maximum value is always a local maximum value.

29. A cubic function $f(x) = ax^3 + bx^2 + cx + d$, $a \neq 0$ can have at most one local maximum value on any open interval.

30. The linear function $f(x) = ax + b$, $a \neq 0$ has no minimum value on any open interval.

Sample Test Problems

In Problems 1–6, a function f and its domain are given. Determine the critical numbers, evaluate f at these values, and find the (global) maximum and minimum values.

1. $f(x) = \dfrac{1}{x^2}$; $\left[-2, -\dfrac{1}{2}\right]$

2. $f(x) = \dfrac{1}{x^2}$; $[-2, 0]$

3. $f(x) = 3x^4 - 4x^3$; $[-2, 3]$

4. $f(x) = xe^{-2x}$; $[-1, 3]$

5. $f(x) = 2x^5 - 5x^4 + 7x$; $[-1, 3]$

6. $f(x) = 2x \sin x$; $[-1, 3]$

In Problems 7–10, a function f is given with domain all real numbers. Indicate where f is increasing and where it is concave down. Sketch the graph of each.

7. $f(x) = x^3 - 3x + 3$

8. $f(x) = e^{1+x-x^2}$

9. $f(x) = x + x^4 - 4x^5$

10. $f(x) = ax^3 - \dfrac{6}{5}x^5$, $a > 0$

11. Find where the function f, defined by $f(x) = x^2(x - a)$, $a > 0$, is increasing and where it is decreasing. Find the local extreme values of f. Find the point of inflection. Sketch the graph.

12. Find the maximum and the minimum values, if they exist, of the function defined by

$$f(x) = \frac{4}{x^2 + 1} + 2$$

In Problems 13–18, sketch the graph of the given function f, labeling all extrema (local and global) and inflection points with their coordinates and showing any asymptotes. Be sure to make use of f' and f''.

13. $f(x) = x^4 - 32x$

14. $f(x) = (x^2 - 1)^2$

15. $f(x) = x \ln(x - 3)$

16. $f(x) = x^2 e^{-x^2}$

17. $f(x) = 3x^4 - 4x^3$

18. $f(x) = \dfrac{x^2 - a^2}{x}$, $a > 0$

19. Sketch a possible graph of a function F that has all the following properties:

(a) is everywhere continuous;

(b) $F(-2) = 3$, $F(2) = -1$;

(c) $F'(x) = 0$ for $x > 2$;

(d) $F''(x) < 0$ for $x < 2$.

20.   Sketch a possible graph of a function F that has all the following properties:

(a) F is everywhere continuous;

(b) $F(-1) = 6$, $F(3) = -2$;

(c) $F'(x) < 0$ for $x < -1$, $F'(-1) = F'(3) = -2$, $F'(7) = 0$

(d) $F''(x) < 0$ for $x < -1$, $F''(x) = 0$ for $-1 < x < 3$, $F''(x) > 0$ for $x > 3$.

21.   A long sheet of metal, 16 inches wide, is to be turned up at both ends to make a horizontal gutter with vertical sides. How many inches should be turned up at each side for maximum carrying capacity?

22.   A fence, 8 feet high, is parallel to the wall of a building and 1 foot from the building. What is the shortest plank that can go over the fence, from the level ground, to prop the wall?

23.   A page of a book is to contain 27 square inches of print. If the margins at the top, bottom, and one side are 2 inches and the margin at the other side is 1 inch, what size page would use the least paper?

24.   A metal water trough with equal semicircular ends and open top is to have a capacity of $128\,\pi$ cubic feet. Determine its radius r and length h if the trough is to require the least material for its construction.

25.   Find the maximum and the minimum of the function defined on the closed interval $[-2, 2]$ by

$$f(x) = \begin{cases} \dfrac{1}{4}(x^2 + 6x + 8) & \text{if } -2 \le x \le 0 \\[2mm] -\dfrac{1}{6}(x^2 + 4x - 12) & \text{if } 0 \le x \le 2 \end{cases}$$

Find where the graph is concave upward and where it is concave downward. Sketch the graph.

26.   For each of the following functions, decide whether the Mean Value Theorem applies on the indicated interval I. If so, find all possible values of c; if not, tell why. Make a sketch.

(a) $f(x) = \dfrac{x^3}{3}$; $I = [-3, 3]$

(b) $F(x) = x^{3/5} + 1$; $I = [-1, 1]$

(c) $g(x) = \dfrac{x + 1}{x - 1}$; $I = [2, 3]$

27.   Find the equations of the tangent lines at the inflection points of the graph of

$$y = x^4 - 6x^3 + 12x^2 - 3x + 1$$

Sketch the graph of the equation, and sketch in the tangent lines you found.

28.   A price function, p, is defined by

$$p(x) = (20 + 4x)e^{-0.1x}$$

where $x \ge 0$ is the number of units.

(a) Find the total revenue function and the marginal revenue function.

(b) On what interval is the total revenue increasing?

(c) For what number x is the marginal revenue a maximum?

CHAPTER 5

The Integral

5.1 | Antiderivatives (Indefinite Integrals)

I may put on my shoes and take them off again. The second operation undoes the first, restoring the shoes to their original position. We say the two operations are *inverse operations*. Mathematics has many pairs of inverse operations: addition and subtraction, multiplication and division, powers and roots, exponentials and logarithms. We have been studying differentiation; its inverse is called *antidifferentiation*.

> **Definition**
>
> We call F an **antiderivative** of f on the interval I if $\frac{d}{dx}F(x) = f(x)$ on I—that is, $F'(x) = f(x)$ for all x in I. (If x is an endpoint of I, $F'(x)$ need only be a one-sided derivative.)

We used *an* antiderivative rather than *the* antiderivative in our definition. You will soon see why.

EXAMPLE 1: Find an antiderivative of the function $f(x) = 4x^3$ on $(-\infty, \infty)$.

SOLUTION: We see a function F satisfying $F'(x) = 4x^3$ for all real x. From our experience with differentiation, we know that $F(x) = x^4$ is one such function. ◄

A moment's thought will suggest other solutions to Example 1. The function $F(x) = x^4 + 6$ also satisfies $F'(x) = 4x^3$; it too is an antiderivative of $f(x) = 4x^3$. In fact, $F(x) = x^4 + C$, where C is any constant, is an antiderivative of $4x^3$ on $(-\infty, \infty)$ (see Figure 1).

281

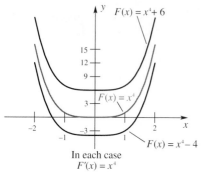

In each case
$F'(x) = x^4$

Figure 1

Now we pose an important question. Is every antiderivative of $f(x) = 4x^3$ of the form $F(x) = x^4 + C$? The answer is yes. This follows from Theorem 4.7B, which says that if two functions have the same derivative, they must differ by a constant.

Our conclusion is this: If a function f has an antiderivative, it will have a whole family of them and each member of this family can be obtained from one of them by the addition of an appropriate constant. We call this family of functions the **general antiderivative** of f. After we get used to this notion, we will often omit the adjective *general*.

EXAMPLE 2: Find the general antiderivative of $f(x) = x^2$ on $(-\infty, \infty)$.

SOLUTION: The function $F(x) = x^3$ will not do, because its derivative is $3x^2$. But that suggests $F(x) = \frac{1}{3}x^3$, which satisfies $f(x) = \frac{1}{3} \cdot 3x^2 = x^2$. However, the general antiderivative is $\frac{1}{3}x^3 + C$. ◀

Notation for Antiderivatives

Leibniz used the symbol $\int f(x)dx$ to represent the antiderivative of a function $f(x)$. He wrote

$$\int x^2 dx = \tfrac{1}{3}x^3 + C$$

and

$$\int 4x^3 dx = x^4 + C$$

We will postpone explaining why Leibniz chose to use the symbols $\int$ and dx until later. For the moment, simply think of $\int(\cdots)dx$ as indicating the antiderivative with respect to x, just as $\frac{d}{dx}(\cdots)$ indicates the derivative with respect to x. Note that $\frac{d}{dx}\int f(x)dx = f(x)$.

Theorem A

(Power Rule). If r is any rational number except -1, then

$$\int x^r dx = \frac{x^{r+1}}{r + 1} + C$$

Proof To establish any result of the form

$$\int f(x)dx = F(x) + C$$

it is sufficient to show that

$$\frac{d}{dx}[F(x) + C] = f(x)$$

In our case,

$$\frac{d}{dx}\left[\frac{x^{r+1}}{r + 1} + C\right] = \frac{1}{r + 1}(r + 1)x^r = x^r$$

◀

We make two comments about Theorem A. First, it is meant to include the case $r = 0$, that is,

$$\int 1\,dx = x + C$$

Second, because no interval I is specified, the conclusion is understood to be valid only on intervals on which x^r is defined. In particular, we must exclude any interval containing $x = 0$ if $r < 0$. A similar understanding holds in what follows.

EXAMPLE 3: Find the general antiderivative of $f(x) = x^{\frac{4}{3}}$.

SOLUTION:

$$\int x^{\frac{4}{3}}\,dx = \frac{x^{\frac{7}{3}}}{\frac{7}{3}} + C = \tfrac{3}{7}x^{\frac{7}{3}} + C$$

Note that *to antidifferentiate a power of x we increase the exponent by 1 and divide by the new exponent.* ◄

Theorem B

$$\int \sin x\,dx = -\cos x + C \qquad \int \cos x\,dx = \sin x + C$$

Proof Simply note that $\frac{d}{dx}(-\cos x) = \sin x$ and $\frac{d}{dx}(\sin x) = \cos x$. ◄

Theorem C

1. $\displaystyle\int e^x\,dx = e^x + C$

2. $\displaystyle\int \frac{1}{x}\,dx = \ln|x| + C$

Proof We have $\frac{d}{dx}e^x = e^x$, which proves 1. For $x > 0$ we have $\ln|x| = \ln x$, and therefore $\frac{d}{dx}\ln|x| = \frac{d}{dx}\ln x = \frac{1}{x}$, which proves 2 for $x > 0$; we leave the case $x < 0$ for the exercises. ◄

Notice that Theorem C completes the missing case from Theorem A. It gives us the antiderivative of x^r when $r = -1$, because $x^{-1} = \frac{1}{x}$.

There is more to be said about notation. Following Leibniz, we shall often use the term **indefinite integral** in place of antiderivative. To antidifferentiate is also to **integrate**. In the symbol $\int f(x)\,dx$, $\int$ is called the **integral sign** and $f(x)$ is called the **integrand**. Thus, we integrate the integrand and thereby obtain the indefinite integral. Perhaps Leibniz used the adjective *indefinite* to suggest that the indefinite integral always involves an arbitrary constant (C).

The Indefinite Integral Is Linear

Recall from Section 3.2 that $\frac{d}{dx}$ is a linear operator. This means two things:

1. $\dfrac{d}{dx}[kf(x)] = k\dfrac{d}{dx}f(x)$

2. $\dfrac{d}{dx}[f(x) + g(x)] = \dfrac{d}{dx}f(x) + \dfrac{d}{dx}g(x)$

From these two properties, a third follows automatically.

3. $\dfrac{d}{dx}[f(x) - g(x)] = \dfrac{d}{dx}f(x) - \dfrac{d}{dx}g(x)$

What is true for derivatives is true for indefinite integrals (antiderivatives).

Theorem D

(Linearity of $\int \cdots dx$). Let f and g have antiderivatives (indefinite integrals) and let k be a constant. Then:

(i) $\int kf(x)dx = k\int f(x)dx$;

(ii) $\int [f(x) + g(x)]dx = \int f(x)dx + \int g(x)dx$; and consequently

(iii) $\int [f(x) - g(x)]dx = \int f(x)dx - \int g(x)dx$.

Proof   To show (i) and (ii), we simply differentiate the right side and observe that we get the integrand on the left side.

$$\frac{d}{dx}\left[k\int f(x)dx\right] = k\frac{d}{dx}\int f(x)dx = kf(x)$$

$$\frac{d}{dx}\left[\int f(x)dx + \int g(x)dx\right] = \frac{d}{dx}\int f(x)dx + \frac{d}{dx}\int g(x)dx$$

$$= f(x) + g(x)$$

Property (iii) follows from (i) and (ii).   ◄

EXAMPLE 4:   Find (a) $\int(3x^2 + 4x)dx$, (b) $\int(u^{\frac{3}{2}} - 3e^u + 14)du$, and (c) $\int\left(\frac{3}{t} + 4\sin t\right)dt$ by using the linearity of $\int$.

SOLUTION:

(a)

$$\int(3x^2 + 4x)dx = \int 3x^2 dx + \int 4x dx$$

$$= 3\int x^2 dx + 4\int x dx$$

$$= 3\left(\frac{x^3}{3} + C_1\right) + 4\left(\frac{x^2}{2} + C_2\right)$$

$$= x^3 + 2x^2 + (3C_1 + 4C_2)$$

$$= x^3 + 2x^2 + C$$

Two arbitrary constants C_1 and C_2 appeared, but they were easily combined in one constant, C, a practice we consistently follow. For our purposes, the value of the constant is not important; what *is* important is the fact that a constant exists.

(b) Note the use of the variable u rather than x. That is fine as long as the corresponding differential symbol is du, because we then have a complete change of notation.

$$\int\left(u^{\frac{3}{2}} - 3e^u + 14\right)du = \int u^{\frac{3}{2}}du - 3\int e^u du + 14\int 1 du$$

$$= \frac{2}{5}u^{\frac{5}{2}} - 3e^u + 14u + C$$

(c) $$\int\left(\frac{3}{t} + 4\sin t\right)dt = \int\left(3\frac{1}{t} + 4\sin t\right)dt$$

$$= \int 3\frac{1}{t}dt + \int 4\sin t dt$$

$$= 3\ln|t| - 4\cos t + C \qquad \blacktriangleleft$$

Guess-and-Check   We can find antiderivatives for slightly more complicated functions by making an educated guess at the antiderivative based on rules that we already know, and then checking the guess to see if any adjustments are necessary. For example, to find an antiderivative for $(2x + 5)^3$ we might guess $\frac{1}{4}(2x + 5)^4$ based on the rule $\int x^3 dx = \frac{1}{4}x^4 + C$. But when we check by taking the derivative we get

$$\frac{d}{dx}\frac{1}{4}(2x + 5)^4 = \frac{1}{4} \cdot 4(2x + 5)^3 \cdot 2$$

$$= 2(2x + 5)^3$$

Because there is an extra factor of 2 resulting from the use of the Chain Rule, we multiply our original guess by $\frac{1}{2}$ to get the new guess

$$\frac{1}{2} \cdot \frac{1}{4}(2x + 5)^4 = \frac{1}{8}(2x + 5)^4$$

This time when we check by taking the derivative we get

$$\frac{d}{dx}\frac{1}{8}(2x + 5)^4 = (2x + 5)^3$$

We conclude that

$$\int(2x + 5)^3 dx = \frac{1}{8}(2x + 5)^4 + C$$

EXAMPLE 5: Find (a) $\int \sin 5x\, dx$ and (b) $\int x\sqrt{x^2 + 1}\, dx$ using guess-and-check.

SOLUTION:

(a) Guess an antiderivative: $-\cos 5x$.

Check: $\dfrac{d}{dx}(-\cos 5x) = 5\sin 5x$.

New guess: $-\dfrac{1}{5}\cos 5x$.

Check: $\dfrac{d}{dx}\left(-\dfrac{1}{5}\cos 5x\right) = \dfrac{1}{5} \cdot 5\sin 5x = \sin 5x$.

Conclusion: $\int \sin 5x\, dx = -\frac{1}{5} \cos 5x + C.$

(b) Even though the integration problem contains two factors, we guess an antiderivative based on the second factor alone. The reason why this approach works in this case is because when we check our guess by taking the derivative, we get a second factor from the Chain Rule. This second factor almost matches the first factor in the original integral (we are off by only the constant factor 2, because the derivative of $x^2 + 1$ is $2x$), and so we have made a good guess.

Rewrite the integral: $\int x(x^2 + 1)^{\frac{1}{2}} dx$

Guess an antiderivative: $\frac{2}{3}(x^2 + 1)^{\frac{3}{2}}$

Check: $\dfrac{d}{dx} \dfrac{2}{3}(x^2 + 1)^{\frac{3}{2}} = \dfrac{2}{3} \cdot \dfrac{3}{2}(x^2 + 1)^{\frac{1}{2}} \cdot (2x)$

$$= 2x(x^2 + 1)^{\frac{1}{2}}$$

New guess: $\frac{1}{2} \cdot \frac{2}{3}(x^2 + 1)^{\frac{3}{2}} = \frac{1}{3}(x^2 + 1)^{\frac{3}{2}}.$

Check: $\dfrac{d}{dx} \dfrac{1}{3}(x^2 + 1)^{\frac{3}{2}} = \dfrac{1}{3} \cdot \dfrac{3}{2}(x^2 + 1)^{\frac{1}{2}} \cdot (2x)$

$$= x(x^2 + 1)^{\frac{1}{2}}$$

Conclusion: $\int x(x^2 + 1)^{\frac{1}{2}} dx = \frac{1}{3}(x^2 + 1)^{\frac{3}{2}} + C.$ ◄

Integration by Computer Algebra

One of the most powerful features that computer algebra systems have is the ability to find antiderivatives for a large class of functions. Indefinite integration is a much less straightforward process than differentiation. Any function that is created by combining elementary functions through addition, subtraction, multiplication, division, or function composition can be differentiated, and the result is again made up of elementary functions. On the other hand, it is quite easy to come up with a function created from elementary functions, for which no antiderivative exists that is itself made up of elementary functions.

For example, the derivative of e^{-x^2} is $-2xe^{-x^2}$, but the antiderivative of e^{-x^2} cannot be written in terms of any of the functions we have encountered so far. Does this mean that $\int e^{-x^2} dx$ does not exist? The answer is no, but we have to define a new function to describe it. If we ask a computer algebra system for the antiderivative of e^{-x^2} we get the answer $\frac{\sqrt{\pi}}{2} \text{erf}(x)$. The special function $\text{erf}(x)$ is called the *error function* and was defined because it occurs frequently in statistics. In fact, we can define $\text{erf}(x)$ as the particular antiderivative of $\frac{2}{\sqrt{\pi}} e^{-x^2}$ that has the value zero at $x = 0$.

There are also functions whose antiderivatives have not been given names yet. If we ask a computer algebra system to integrate the function e^{-x^3} it will come back with the answer $\int e^{-x^3} dx$, that is to say, a restatement of the original problem. This indicates that the computer program was not able to find an antiderivative in terms of any of the functions it knows, either elementary or special. This particular antiderivative is not yet useful enough to have its own name.

Finally, there may be a very small number of times when a fairly simple antiderivative does exist in terms of elementary functions, but your computer algebra system cannot find it. Though these programs are quite sophisticated, and continue to improve with time, they are not perfect. Keep in mind that indefinite integration is generally a much tougher problem for computers (and people!) to attack than differentiation.

EXAMPLE 6: Use your computer algebra system to find $\int xe^{-x^2}\,dx$. Could you have done this problem by guessing an antiderivative as in Example 5? Explain.

SOLUTION: The result from your computer algebra system should be algebraically equivalent to $-\frac{1}{2}e^{-x^2}$, although the exact form of the result may be different (for example, $-\frac{1}{2e^{x^2}}$).

We could also do this problem using guess-and-check. A reasonable guess for the antiderivative of xe^{-x^2} would be e^{-x^2} based on the rule $\int e^x\,dx = e^x + C$ (see Example 5, part b). Because $\frac{d}{dx}e^{-x^2} = -2xe^{-x^2}$, a new guess of $-\frac{1}{2}e^{-x^2}$ would be correct, and consistent with our computer algebra result. ◄

Concepts Review

1. The Power Rule for derivatives says that $d(x^r)/dx =$ _____. The Power Rule for integrals says that $\int x^r dx =$ _____.

2. $\int x^{-1} dx =$ _____.

3. $\int(x^4 + 3x^2 + 1)^8(4x^3 + 6x)dx =$ _____.

4. By linearity, $\int[c_1 f(x) + c_2 g(x)]dx =$ _____.

Problem Set 5.1

For all problems below, attempt to find any antiderivative that you need using the techniques in Examples 1 through 5 of this section. Check your results with a computer algebra system if one is available.

Find the general antiderivative $F(x) + C$ for Problems 1–14.

1. $f(x) = 4$

2. $f(x) = 2x - 4$

3. $f(x) = 3x^2 + \sqrt{2}$

4. $f(x) = \dfrac{5}{x} + \pi$

5. $f(x) = x^{2/3}$

6. $f(x) = x^{-3/4}$

7. $f(x) = 6x^2 - 6x + 1$

8. $f(x) = 3x^2 + 10x - 7$

9. $f(x) = 3e^x + x^3$

10. $f(x) = 2\sin x + e^{-x} + \dfrac{3}{x}$

11. $f(x) = \dfrac{4}{x^5} - \dfrac{3}{x}$

12. $f(x) = \dfrac{1}{x^3} - \dfrac{6}{e^x}$

13. $f(x) = \dfrac{4x^6 + 3x^5 - 8}{x^5}$

14. $f(x) = \dfrac{2x^3 - 3x^2 + 1}{x^2}$

In Problems 15–22, find the indefinite integrals.

15. $\displaystyle\int (x^3 + \sqrt{x})dx$

16. $\displaystyle\int (x^2 + 1)^2 dx$

17. $\displaystyle\int \left(3e^y - \dfrac{2}{y^2} + \dfrac{2}{y}\right)dy$

18. $\displaystyle\int y^2(y^2 - 3)dy$

19. $\displaystyle\int \dfrac{e^x - 2e^{2x} + 1}{e^x}dx$

20. $\displaystyle\int \dfrac{x^3 - 3x^2 + 1}{\sqrt{x}}dx$

21. $\displaystyle\int (3\sin t - 2\cos t)dt$

22. $\displaystyle\int (3t^2 - 2\sin t)dt$

In Problems 23–38, use the method of Example 5 or any other method to find the indefinite integrals.

23. $\displaystyle\int (3x + 1)^4 3\,dx$

24. $\int (x^2 - 4)^3 2x \, dx$

25. $\int (5x^3 - 18)^7 15x^2 \, dx$

26. $\int (x^2 - 3x + 2)^2 (2x - 3) dx$

27. $\int 3x^4 e^{2x^5 + 9} dx$

28. $\int 3xe^{3x^2 + 7} dx$

29. $\int \frac{5x^2 + 2x}{5x^3 + 3x^2 - 8} dx$

30. $\int \tan x \, dx$

31. $\int 3t \sqrt[3]{2t^2 - 11} \, dt$

32. $\int \frac{3y}{\sqrt{2y^2 + 5}} dy$

33. $\int \sin^4 x \cos x \, dx$

34. $\int (\cos^4 2x)(-2 \sin 2x) dx$

35. $\int (\sin^5 x^2)(x \cos x^2) dx$

36. $\int \cos(3x + 1)\sin(3x + 1) dx$

37. $\int (x^2 + 1)^3 x^2 \, dx$

38. $\int (x^4 - 1)x^2 \, dx$

In Problems 39–44, $f''(x)$ is given. Find $f(x)$ by antidifferentiating twice. Note that in this case your answer should involve two arbitrary constants, one from each antidifferentiation. For example, if $f''(x) = x$, then $f'(x) = x^2/2 + C_1$ and $f(x) = x^3/6 + C_1 x + C_2$. The constants C_1 and C_2 cannot be combined.

39. $f''(x) = 3x + 1$

40. $f''(x) = -2x + 3$

41. $f''(x) = \sqrt{x}$

42. $f''(x) = 4e^{2x}$

43. $f''(x) = \frac{1}{x^2}$

44. $f''(x) = 2\sqrt[3]{x} + 1$

45. Prove the formula

$$\int [f(x)g'(x) + g(x)f'(x)] dx = f(x)g(x) + C$$

46. Prove the formula

$$\int \frac{g(x)f'(x) - f(x)g'(x)}{g^2(x)} dx = \frac{f(x)}{g(x)} + C$$

47. Use the formula from Problem 45 to find

$$\int \left[\frac{x^2}{2\sqrt{x - 1}} + 2x\sqrt{x - 1} \right] dx$$

48. Use the formula from Problem 45 to find

$$\int \left[\frac{-x^3}{(2x + 5)^{3/2}} + \frac{3x^2}{\sqrt{2x + 5}} \right] dx$$

49. Prove the formula

$$\int \frac{x^4 + 1}{x^2\sqrt{x^4 - 1}} dx = \frac{\sqrt{x^4 - 1}}{x} + C$$

50. Prove the formula

$$\int \frac{2}{\cos^2 3x} dx = \frac{1}{3} \frac{\sin 3x}{\cos 3x} + C$$

51. Find $\int f''(x) dx$ if $f(x) = x\sqrt{x^3 + 1}$.

52. Prove the formula

$$\int \frac{2g(x)f'(x) - f(x)g'(x)}{2[g(x)]^{\frac{3}{2}}} dx = \frac{f(x)}{\sqrt{g(x)}} + C$$

53. Prove the formula

$$\int f^{m - 1}(x)g^{n - 1}(x)[nf(x)g'(x) + mg(x)f'(x)] dx$$

$$= f^m(x)g^n(x) + C$$

54. Find the indefinite integral

$$\int \sin^3[(x^2 + 1)^4] \cos[(x^2 + 1)^4](x^2 + 1)^3 x \, dx$$

Hint: Guess the antiderivative $\sin^4[(x^2 + 1)^4]$ first, and then check your guess.

55. Find $\int |x| \, dx$.

56. Find $\int \sin^2 x \, dx$.

57. Use a computer algebra system for this one. Let $F_0(x) = x \sin x$ and $F_{n + 1}(x) = \int F_n(x) \, dx$.
(a) Determine $F_1(x)$, $F_2(x)$, $F_3(x)$, and $F_4(x)$.
(b) On the basis of part (a), conjecture the form of $F_{16}(x)$.

58. If $x < 0$, show that $\frac{d}{dx} \ln |x| = \frac{1}{x}$. Recall that for $x < 0$, $|x| = -x$. Now use the Chain Rule.

Answers to Concepts Review: 1. $rx^{r + 1}$; $x^{r + 1}/(r + 1)$, $r \neq -1$ 2. $\ln |x| + C$ 3. $(x^4 + 3x^2 + 1)^9$ 4. $c_1 \int f(x) dx + c_2 \int g(x) dx$

5.2 Introduction to Differential Equations

In the previous section, our task was to integrate (antidifferentiate) a function f to obtain a new function F

$$\int f(x)\,dx = F(x) + C$$

and this was correct, provided that $F'(x) = f(x)$. Because $F'(x) = f(x)$, we can say, in the language of differentials, that $\frac{d}{dx} F(x) = f(x)$ and therefore $dF(x) = f(x)\,dx$ (Section 3.7). Thus, we may look on the boxed formula as saying

$$\int dF(x) = F(x) + C$$

From this perspective, we integrate the differential of a function to obtain the function (plus a constant). This was Leibniz's viewpoint; adopting it will help us solve *differential equations*.

What Is a Differential Equation?   To motivate our answer, we begin with a simple example.

EXAMPLE 1: Find the xy-equation of the curve that passes through $(-1, 2)$ and whose slope at any point on the curve is equal to twice the x-coordinate of that point.

SOLUTION: The condition that must hold at each point (x, y) on the curve is

$$\frac{dy}{dx} = 2x$$

We are looking for a function $y = f(x)$ that satisfies this equation and the additional condition that $y = 2$ when $x = -1$. We suggest two ways of looking at this problem.

Method 1   When an equation has the form $dy/dx = g(x)$, we observe that y must be an antiderivative of $g(x)$, that is,

$$y = \int g(x)\,dx$$

In our case,

$$y = \int 2x\,dx = x^2 + C$$

Method 2   Think of dy/dx as a quotient of two differentials. When we multiply both sides of $dy/dx = 2x$ by dx, we get

$$dy = 2x\,dx$$

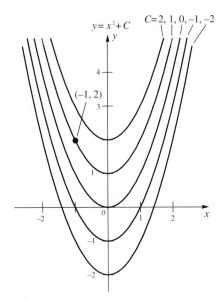

Figure 1

Next we integrate both sides, equate the results, and simplify.

$$\int dy = \int 2x\, dx$$

$$y + C_1 = x^2 + C_2$$

$$y = x^2 + C_2 + C_1$$

$$y = x^2 + C$$

The second method works in a wide variety of problems that are not of the simple form $dy/dx = g(x)$, as we shall see.

The solution $y = x^2 + C$ represents the family of curves illustrated in Figure 1. From this family, we must choose the one for which $y = 2$ when $x = -1$; thus, we want

$$2 = (-1)^2 + C$$

We conclude that $C = 1$ and therefore that $y = x^2 + 1$. ◄

The equation $dy/dx = 2x$ is called a *differential equation*. Other examples are

$$\frac{dy}{dx} = 2xy + \sin x$$

$$y\, dy = (x^3 + 1)\, dx$$

$$\frac{d^2y}{dx^2} + 3\frac{dy}{dx} - 2xy = 0$$

Any equation in which the unknown is a function and which involves derivatives (or differentials) of this unknown function is called a **differential equation**. To solve a differential equation is to find the unknown function. In general, this is a difficult job and one about which many thick books have been written. Here we consider only the simplest case, namely, **first-order separable** differential equations. These are equations involving just the first derivative of the unknown function and are such that the variables can be separated to opposite sides of the equation.

Separation of Variables

Consider the differential equation

$$\frac{dy}{dx} = \frac{x + 3x^2}{y^2}$$

If we multiply both sides by $y^2\, dx$, we obtain

$$y^2\, dy = (x + 3x^2)dx$$

In this form, the differential equation has its variables separated—that is, the y-terms are on one side of the equation and the x-terms are on the other. In separated form, we can solve the differential equation using Method 2 (integrate both sides, equate the results, simplify), as we now illustrate.

EXAMPLE 2: Solve the differential equation

$$\frac{dy}{dx} = \frac{x + 3x^2}{y^2}$$

Then find the solution for which $y = 6$ when $x = 0$.

⟼   ***SOLUTION:*** As noted earlier, the given equation is equivalent to

$$y^2\, dy = (x + 3x^2)\, dx$$

Thus,

$$\int y^2\, dy = \int (x + 3x^2)\, dx$$

$$\frac{y^3}{3} + C_1 = \frac{x^2}{2} + x^3 + C_2$$

$$y^3 = \frac{3x^2}{2} + 3x^3 + (3C_2 - 3C_1)$$

$$= \frac{3x^2}{2} + 3x^3 + C$$

$$y = \sqrt[3]{\frac{3x^2}{2} + 3x^3 + C}$$

To evaluate the constant C, we use the condition $y = 6$ when $x = 0$. This gives

$$6 = \sqrt[3]{C}$$

$$216 = C$$

Thus,

$$y = \sqrt[3]{\frac{3x^2}{2} + 3x^3 + 216} = \left(\frac{3x^2}{2} + 3x^3 + 216\right)^{\frac{1}{3}}$$

The ultimate check on our work is to substitute this result in both sides of the original differential equation to see that this gives an equality. Substituting in the left side, we get

$$\frac{dy}{dx} = \frac{1}{3}\left(\frac{3x^2}{2} + 3x^3 + 216\right)^{-\frac{2}{3}} (3x + 9x^2)$$

$$= \frac{x + 3x^2}{\left(\frac{3}{2}x^2 + 3x^3 + 216\right)^{\frac{2}{3}}}$$

$$= \frac{x + 3x^2}{\left[\left(\frac{3}{2}x^2 + 3x^3 + 216\right)^{\frac{1}{3}}\right]^2}$$

$$= \frac{x + 3x^2}{y^2}$$

which matches the original equation.   ◄

Motion Problems   Recall that if $s(t)$, $v(t)$, and $a(t)$ represent the position, velocity, and acceleration, respectively, at time t of an object moving along a coordinate line, then

$$v(t) = s'(t) = \frac{ds}{dt}$$

$$a(t) = v'(t) = \frac{dv}{dt} = \frac{d^2s}{dt^2}$$

In earlier work (Section 3.5), we assumed that $s(t)$ was known and from this we calculated $v(t)$ and $a(t)$. Now we want to consider the reverse process: given acceleration $a(t)$, find velocity $v(t)$ and position $s(t)$.

EXAMPLE 3: **(Falling-Body Problem).** Near the surface of the earth, the acceleration of a falling body due to gravity is 32 feet per second per second, provided we assume that air resistance can be neglected. If an object is thrown upward from an initial height of 1000 feet (see Figure 2) with a velocity of 50 feet per second, find its velocity and height 4 seconds later.

SOLUTION: Let us assume that the height s is measured positively in the upward direction. Then $v = ds/dt$ is initially positive (s is increasing) but $a = dv/dt$ is negative (the pull of gravity tends to decrease v). Thus, our starting point is the differential equation $dv/dt = -32$, with the additional conditions that $v = 50$ and $s = 1000$ when $t = 0$. Either Method 1 (direct antidifferentiation) or Method 2 (separation of variables) works well.

$$\frac{dv}{dt} = -32$$

$$v = \int -32 \, dt = -32t + C$$

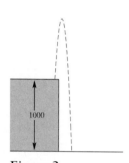

Figure 2

Because $v = 50$ at $t = 0$, we find that $C = 50$ and so

$$v = -32t + 50$$

We know that $v = ds/dt$, and so we have another differential equation,

$$\frac{ds}{dt} = -32t + 50$$

When we integrate, we obtain

$$s = \int (-32t + 50)dt$$

$$s = -16t^2 + 50t + K$$

Because $s = 1000$ at $t = 0$, $K = 1000$ and

$$s = -16t^2 + 50t + 1000$$

Finally at $t = 4$,

$$v = -32(4) + 50 = -78 \text{ feet per second}$$
$$s = -16(4)^2 + 50(4) + 1000 = 944 \text{ feet}$$ ◄

We remark that if $v = v_0$ and $s = s_0$ at $t = 0$, the procedure of Example 3 leads to the well-known falling-body formulas:

$$a = -32$$
$$v = -32t + v_0$$
$$s = -16t^2 + v_0 t + s_0$$

EXAMPLE 4: The acceleration of an object moving along a coordinate line is given as $a(t) = (2t + 3)^{-3}$ in meters per second per second. If the velocity at $t = 0$ is 4 meters per second, find the velocity 2 seconds later.

SOLUTION: To perform the integration in the second line, we use guess-and-check with the rule $\int t^n \, dt = \frac{1}{n+1}t^{n+1} + C$ as the basis for our guess. Do you see why we need the extra factor of $\frac{1}{2}$?

$$\frac{dv}{dt} = (2t + 3)^{-3}$$

$$v = \int (2t + 3)^{-3} dt = \frac{1}{2} \frac{(2t + 3)^{-2}}{-2} + C$$

$$= -\frac{1}{4(2t + 3)^2} + C$$

Because $v = 4$ at $t = 0$,

$$4 = -\frac{1}{4(3)^2} + C$$

which gives $C = \frac{145}{36}$. Thus,

$$v = -\frac{1}{4(2t + 3)^2} + \frac{145}{36}$$

At $t = 2$,

$$v = -\frac{1}{4(49)} + \frac{145}{36} \approx 4.023 \text{ meters per second} \qquad \blacktriangleleft$$

EXAMPLE 5: *(optional)* **(Escape Velocity).**   The gravitational attraction F exerted by Earth on an object of mass m at a distance s from the center of Earth is given by $F = -mgR^2/s^2$, where $-g$ ($g \approx 32$ feet per second per second) is the acceleration of gravity at the surface of Earth and R ($R \approx 3960$ miles) is the radius of Earth (Figure 3). Show that an object projected outward from Earth with an initial velocity $v_0 \geq \sqrt{2gR} \approx 6.93$ miles per second will not fall back to Earth. Neglect air resistance in making this calculation.

SOLUTION: According to Newton's Second Law, $F = ma$; that is,

$$F = m\frac{dv}{dt} = m\frac{dv}{ds}\frac{ds}{dt} = m\frac{dv}{ds}v$$

Thus,

$$mv\frac{dv}{ds} = -mg\frac{R^2}{s^2}$$

Separating variables yields

$$v \, dv = -gR^2 s^{-2} \, ds$$

$$\int v \, dv = -gR^2 \int s^{-2} \, ds$$

$$\frac{v^2}{2} = \frac{gR^2}{s} + C$$

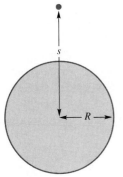

Figure 3

Now $v = v_0$ when $s = R$, and so $C = \frac{1}{2}v_0^2 - gR$. Consequently,

$$v^2 = \frac{2gR^2}{s} + v_0^2 - 2gR$$

Finally, because $2gR^2/s$ gets close to zero with increasing s, we see that v^2 remains nonzero for *all* s if and only if $v_0 \geq \sqrt{2gR}$. If v^2 (and hence v) ever reaches zero, the object will fall back to Earth. ◀

Concepts Review

1. $dy/dx = 3x^2 + 1$ and $dy/dx = x/y^2$ are examples of what is called a _____ .

2. To solve the differential equation $dy/dx = g(x, y)$ is to find the _____ that, when substituted for y, makes the given equation true.

3. To solve the differential equation $dy/dx = x^2y^2$, the first step would be to _____ .

4. To solve a falling body problem near the surface of Earth, we start with the experimental fact that the acceleration a of gravity is -32 feet per second per second—that is, $a = dv/dt = -32$. Solving this differential equation gives $v = ds/dt = $ _____ , and solving the resulting differential equation gives $s = $ _____ .

Problem Set 5.2

In Problems 1–4, show that the indicated function is a solution of the given differential equation; that is, substitute the indicated function for y to see that it produces an equality.

1. $\dfrac{dy}{dx} + \dfrac{x}{y} = 0; y = \sqrt{4 - x^2}$

2. $3y^2 \dfrac{dy}{dx} + x = 0; y = \left(1 - \dfrac{x^2}{2}\right)^{\frac{1}{3}}$

3. $\dfrac{d^2y}{dx^2} + y = 0; y = C_1 \sin x + C_2 \cos x$

4. $\left(\dfrac{dy}{dx}\right)^2 = 1 - y^2; y = \sin(x + C)$

In Problems 5–14, first find the general solution (involving a constant C) for the given differential equation. Then find the particular solution that satisfies the indicated condition. (See Example 2.)

5. $\dfrac{dy}{dx} = 3x^2 + 1; y = 4$ at $x = 1$

6. $\dfrac{dy}{dx} = x^{-2} + 2x; y = 5$ at $x = 1$

7. $\dfrac{dy}{dx} = \dfrac{x}{2y}; y = 3$ at $x = 2$

8. $\dfrac{dy}{dx} = \sqrt[3]{\dfrac{x}{y}}; y = 8$ at $x = 1$

9. $\dfrac{dy}{dt} = 3y; y = 2$ at $t = 0$

10. $\dfrac{dy}{dt} = 3e^{-y}; y = 2$ at $t = 1$

11. $\dfrac{ds}{dt} = 3t^2 + 4t - 1; s = 5$ at $t = 2$

12. $\dfrac{du}{dt} = u(t^2 - 3t); u = 4$ at $t = 0$

13. $\dfrac{dy}{dx} = e^y(2x + 1)^4; y = 6$ at $x = 0$

14. $\dfrac{dy}{dx} = -y^2x(x + 2)^4; y = 1$ at $x = 0$

15. Find the xy-equation of the curve through $(1, 2)$ whose slope at any point is four times its x-coordinate. (See Example 1.)

16. Find the xy-equation of the curve through $(1, 2)$ whose slope at any point is one-half the square of its y-coordinate.

In Problems 17–20, an object is moving along a coordinate line subject to the indicated acceleration a (in centimeters per second per second) with the initial velocity v_0 (in centimeters per second) and directed distance s_0 (in centimeters). Find both the velocity v and directed distance s after 2 seconds. (See Example 4.)

17. $a = t; v_0 = 2, s_0 = 0$

18. $a = (1 + t)^{-3}; v_0 = 4, s_0 = 6$

19. $a = \sin 2t; v_0 = 0, s_0 = 10$

20. $a = 3e^{5t}; v_0 = 0, s_0 = 10$

21. A ball is thrown upward from the surface of the earth with an initial velocity of 96 feet per second. What is the maximum height it reaches? (See Example 3.)

22. A ball is thrown upward from the surface of a planet where the acceleration of gravity is k (a negative constant) feet per second per second. If the initial velocity is v_0, show that the maximum height is $-v_0^2/2k$.

23. On the surface of the moon, the acceleration of gravity is -5.28 feet per second per second. If an object is thrown upward from an initial height of 1000 feet with a velocity of 56 feet per second, find its velocity and height 4.5 seconds later.

24. What is the maximum height that the object of Problem 23 reaches?

25. The rate of change of volume V of a melting snowball is proportional to the surface area S of the ball— that is, $dV/dt = -kS$, where k is a positive constant. If the radius of the ball at $t = 0$ is $r = 2$ and at $t = 10$ is $r = 0.5$, show that $r = -\frac{3}{20}t + 2$.

26. From what height above the earth must a ball be dropped in order to strike the ground with a velocity of -136 feet per second?

27. Determine the escape velocity for each of the following celestial bodies. (See Example 5.) Here $g \approx 32$ feet per second per second.

	Acceleration of Gravity	Radius (miles)
Moon	$-0.165g$	1,080
Venus	$-0.85g$	3,800
Jupiter	$-2.6g$	43,000
Sun	$-28g$	432,000

28. If the brakes of a car, when fully applied, produce a constant deceleration of 11 feet per second per second, what is the shortest distance in which the car can be braked to a halt from a speed of 60 miles per hour?

29. What constant acceleration will cause a car to increase its velocity from 45 to 60 miles per hour in 10 seconds?

30. A block slides down an inclined plane with a constant acceleration of 8 feet per second per second. If the inclined plane is 75 feet long and the block reaches the bottom in 3.75 seconds, what was the initial velocity of the block?

31. A certain rocket shot straight up has an acceleration of $6t$ meters per second per second during the first 10 seconds after blast-off, after which the engine cuts out and the rocket is subject to gravitational acceleration of -10 meters per second per second. How high will the rocket go?

32. Starting at station A, a commuter train accelerates at 3 meters per second per second for 8 seconds, then travels at constant speed v_m for 100 seconds, and finally brakes (decelerates) to a stop at station B at 4 meters per second per second. Find (a) v_m and (b) the distance between A and B.

33. Starting from rest, a bus increases speed at constant acceleration a_1, then travels at a constant speed v_m, and finally brakes to a stop at constant deceleration a_2. It took 4 minutes to travel the 2 miles between stop C and stop D and then 3 minutes to go the 1.4 miles between stop D and stop E.

(a) Sketch the graph of the speed v as a function of time t, $0 \le t \le 7$.

(b) Find the maximum speed v_m.

(c) If $a_1 = a_2 = a$, evaluate a.

34. A hot-air balloon left the ground rising at 4 feet per second. Sixteen seconds later, Helena threw a ball straight up to her friend Janet in the balloon. At what speed did she throw the ball if it just made it to Janet?

35. According to Torricelli's Law, the time rate of change of the volume V of water in a draining tank is proportional to the square root of the water's depth. A cylindrical tank of radius $10/\sqrt{\pi}$ centimeters and height 16 centimeters, which was full initially, took 40 seconds to drain.

(a) Write the differential equation for V at time t and the two corresponding conditions.

(b) Solve the differential equation.

(c) Find the volume of water after 10 seconds.

36. The wolf population P in a certain state has been growing at a rate proportional to the cube root of the population size. The population was estimated at 1000 in 1970 and at 1700 in 1980.

(a) Write the differential equation for P at time t and the two corresponding conditions.

(b) Solve the differential equation.

(c) When will the population reach 4000?

37. At $t = 0$, a ball was dropped from a height of 16 feet. It hit the floor and rebounded to a height of 9 feet.

(a) Find a two-part formula for the velocity $v(t)$ that is valid until the ball hits the floor for a second time.

(b) At what two times was the ball at height 9 feet?

Answers to Concepts Review: 1. Differential equation 2. Function 3. Separate variables 4. $-32t + v_0$; $-16t^2 + v_0 t + s_0$

LAB 8: THE DRAINING CAN

Mathematical Background

The usual model for the fluid flowing out of a can with constant cross section is

$$\frac{dV}{dt} = -k\sqrt{V} \tag{1}$$

where V is the volume of fluid in the can and k is a constant to be determined. This is a special case of Torricelli's Law. We can generalize this model as

$$\frac{dV}{dt} = -kV^n \tag{2}$$

where n is to be determined by the data; if $n = \frac{1}{2}$ we get back the usual model.

Lab Introduction

One assumption that goes into model (1) is that volume is proportional to height; this means that the walls of the can should be vertical (the can should be a cylinder). In this lab we will take data on an actual draining can to determine n and k. The initial volume of water will be 12 ounces, then we will measure the time it takes to reach various other levels.

We need to solve differential equation (2) subject to the initial condition $V(0) = 12$. We can separate variables and solve for the integration constant; the result is

$$V = \left(-kt(-n+1) + 12^{-n+1}\right)^{1/(-n+1)} \tag{3}$$

Equation (3) is our solution to differential equation (2) with the initial condition $V(0) = 12$.

Data

Find a 12-ounce clear plastic cup that is as close to a cylinder as possible. On the side of the cup mark the 12-, 9-, 6-, and 3-ounce levels (fill a measuring cup to each level, pour into the plastic cup, and mark). Punch a hole in the bottom of the cup. Fill the cup to above the 12-ounce mark; start a stopwatch when the level gets to the 12-ounce mark, and stop it at the 9-ounce mark. Repeat for the 6, 3, and empty points. If time permits, take more than one reading for each level and average. From this data construct a table of values of V versus t, where t is time in seconds.

Experiment

(1) Graph V as a function of t [equation (3)] for several values of the parameters k and n. Start with $k = 0.1$ and $n = 0.5$. Then, keeping $n = 0.5$, try one larger and one

smaller value of k to see how k affects the graph; sketch all three graphs on one set of axes. Now go back to $k = 0.1$ and vary n by choosing one larger than 0.5 and one smaller; again, sketch three graphs on one set of axes.

(2) Find the values of n and k that best fit the data. To do this, you could start with the usual model of $n = 0.5$ and use trial and error to estimate k; try $k = 0.1$ to start with. Plot V as a function of t using these parameter values and plot the data on top of the curve to see how good the fit is; then adjust first k and then n to improve the fit if possible.

(3) Use the model you developed to estimate the time when there was 1 ounce of water in the cup.

Discussion

1. Fill in the missing steps in the derivation of equation (3) from equation (2).

2. Describe the effect that each parameter (n and k) has on the shape of the $V(t)$ graph.

3. Explain carefully how you fit the model to the data; be sure to explain how you used the calculator or computer at each step. Were you able to get a good fit to the data? If not, why not?

4. Is the usual model for a draining appropriate for your cup? Explain. Discuss possible sources of error.

5.3 | Area and Riemann Sums

We have used the term **sequence** somewhat informally so far to mean a list of numbers with a predictable pattern. For example, we estimated the limit $\lim_{x \to 0} \frac{\sin x}{x}$ numerically by choosing the sequence of x-values 0.1, 0.01, 0.001, . . . and calculating the corresponding sequence of values of $\frac{\sin x}{x}$.

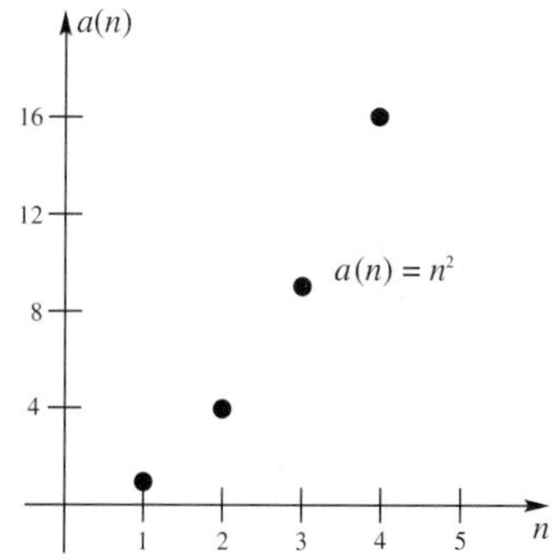

Figure 1

When we want to name a variable that takes only integer values, we typically use letters near the middle of the alphabet, such as i, j, k, m, and (especially) n. For example, we may wish to consider the function determined by $a(n) = n^2$, where n takes positive integer values; its graph is shown in Figure 1. We can now define a sequence to be a function whose domain consists of just the positive integers (or some other subset of the integers). In place of the standard functional notation $a(n)$, it is conventional to use a_n. Thus, we may say: Consider the sequence $\{a_n\}$ determined by $a_n = n^2$ and the sequence $\{b_n\}$ determined by $b_n = 1/n$. Sometimes we indicate a sequence by writing its first few values followed by dots as, for example,

$$a_1, a_2, a_3, a_4, \ldots$$

or even

$$1, 4, 9, 16, \ldots$$

Sequences will be studied in detail in Chapter 9. Here our main interest is in introducing notation for certain of their sums.

Sigma Notation   We first encountered sigma (summation) notation for representing sums in Section 2.6, which we now review. The sum

$$a_1 + a_2 + a_3 + a_4 + \cdots + a_n$$

can be represented in a compact way as

$$\sum_{i=1}^{n} a_i$$

Here Σ (capital Greek sigma) suggests that we are to sum (add) all numbers of the form indicated as the *index i* runs through the positive integers, starting with the integer shown below the Σ and ending with the integer at the top. Thus,

$$\sum_{i=2}^{5} b_i = b_2 + b_3 + b_4 + b_5$$

$$\sum_{k=1}^{4} \frac{k}{k^2 + 1} = \frac{1}{1^2 + 1} + \frac{2}{2^2 + 1} + \frac{3}{3^2 + 1} + \frac{4}{4^2 + 1}$$

$$= \frac{122}{85} \approx 1.435$$

$$\sum_{j=1}^{n} \frac{1}{j} = \frac{1}{1} + \frac{1}{2} + \frac{1}{3} + \cdots + \frac{1}{n}$$

$$\sum_{i=1}^{n} f(x_i) = f(x_1) + f(x_2) + \cdots + f(x_n)$$

Such sums are also called *series*.

EXAMPLE 1:   Let $x_i = \frac{i}{5}$ and let $f(x) = x^2$. Find $\displaystyle\sum_{i=1}^{5} f(x_i)$.

SOLUTION: We have $x_1 = \frac{1}{5}, x_2 = \frac{2}{5}, x_3 = \frac{3}{5}, x_4 = \frac{4}{5}$, and $x_5 = \frac{5}{5} = 1$. Thus,

$$\sum_{i=1}^{5} f(x_i) = f(x_1) + f(x_2) + f(x_3) + f(x_4) + f(x_5)$$

$$= x_1^2 + x_2^2 + x_3^2 + x_4^2 + x_5^2$$

$$= \left(\frac{1}{5}\right)^2 + \left(\frac{2}{5}\right)^2 + \left(\frac{3}{5}\right)^2 + \left(\frac{4}{5}\right)^2 + \left(\frac{5}{5}\right)^2$$

$$= \frac{55}{25} = 2.2$$ ◄

The symbol used for the index does not matter. Thus,

$$\sum_{i=1}^{n} a_i = \sum_{j=1}^{n} a_j = \sum_{k=1}^{n} a_k$$

and all of these are equal to $a_1 + a_2 + \cdots + a_n$. For this reason, the index is sometimes called a **dummy index**.

Area   Two problems, both from geometry, motivate the two biggest ideas in calculus. The problem of the tangent line led us to the *derivative*. The problem of area will lead us to the *definite integral*.

For polygons (closed plane regions bounded by straight line segments), the area is hardly a problem at all. We start by defining the area of a rectangle to be its length times its width, and from this we successively derive the formulas for the area of a parallelogram, a triangle, and any polygon. The sequence of figures in Figure 2 suggests how this is done.

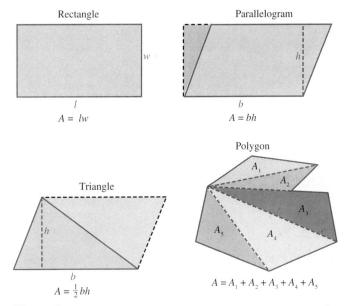

Figure 2

When we consider a region with a curved boundary, the problem of assigning area is significantly more difficult. However, more than 2000 years ago, Archimedes provided the key to a solution. Consider, said he, a sequence of inscribed polygons that approximate the curved region with greater and greater accuracy. For example, for the circle of radius 1, consider regular inscribed polygons $P_1, P_2, P_3, \ldots$ with 4 sides, 8 sides, 16 sides, $\ldots$, as shown in Figure 3. The area of the circle is the limit as $n \to \infty$ of the areas of P_n. Thus, if $A(F)$ denotes the area of region F, then

$$A(\text{circle}) = \lim_{n \to \infty} A(P_n)$$

Archimedes went further, considering also circumscribed polygons $T_1, T_2, T_3, \ldots$ (Figure 4). He showed that you get the same value for the area of the circle of radius 1 (namely, $\pi \approx 3.14159$) whether you use inscribed or circumscribed polygons. It is just a small step from what he did to our modern treatment of area.

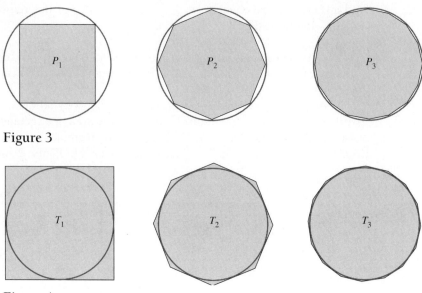

Figure 3

Figure 4

Area by Inscribed Polygons

Consider the region B bounded by the parabola $y = f(x) = x^2$, the x-axis, and the vertical line $x = 2$ (Figure 5). We refer to B as the bounded region under the curve $y = x^2$ between $x = 0$ and $x = 2$. Our aim is to calculate its area $A(B)$.

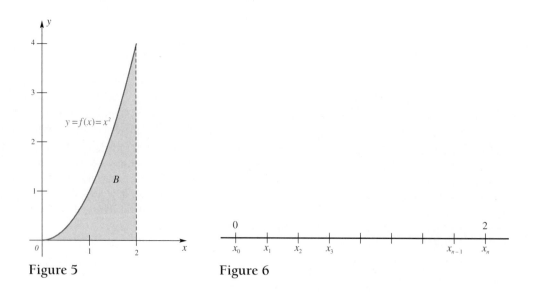

Figure 5

Figure 6

Partition (as in Figure 6) the interval $[0, 2]$ into n subintervals, each of length $\Delta x = 2/n$, by means of points.

$$0 = x_0 < x_1 < x_2 < \cdots < x_{n-1} < x_n = 2$$

Thus,

$$x_0 = 0$$

$$x_1 = \Delta x = \frac{2}{n}$$

$$x_2 = 2 \cdot \Delta x = \frac{4}{n}$$

$$x_3 = 3 \cdot \Delta x = \frac{6}{n}$$

$$\vdots$$

$$x_i = i \cdot \Delta x = \frac{2i}{n}$$

$$x_{n-1} = (n-1) \cdot \Delta x = \frac{(n-1)2}{n}$$

$$x_n = n \cdot \Delta x = n\left(\frac{2}{n}\right) = 2$$

Consider the typical rectangle with base $[x_{i-1}, x_i]$ and height $f(x_{i-1}) = (x_{i-1})^2$, corresponding to the left endpoint of each subinterval. Its area is $f(x_{i-1})\Delta x$ (see upper left part of Figure 7). The union L_n of all such rectangles (L because the *left* endpoint of each interval is used as the height of the rectangle) forms the inscribed polygon shown in the lower right part of Figure 7. Clearly, the area of this inscribed polygon should be less than the area under the curve—that is, $A(L_n) < A(B)$ for each n. Still, we expect that the area of the polygon should get closer to the area under the curve as n gets larger.

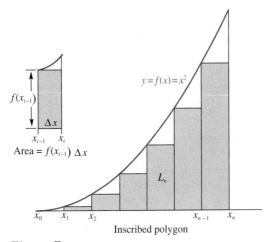

Figure 7

The area $A(L_n)$ can be calculated as follows.

$$A(L_n) = \sum_{i=1}^{n} f(x_{i-1})\,\Delta x$$

$$= f(x_0)\,\Delta x + f(x_1)\,\Delta x + f(x_2)\,\Delta x + \cdots + f(x_{n-1})\,\Delta x$$

When $n = 2$, we are using 2 rectangles. We have $x_0 = 0$ and $x_1 = 1$ and $\Delta x = \frac{2}{2} = 1$. Thus

$$A(L_2) = \sum_{i=1}^{2} f(x_{i-1}) \Delta x$$

$$= f(x_0) \Delta x + f(x_1) \Delta x$$

$$= 0^2 \cdot 1 + 1^2 \cdot 1 = 1$$

Similarly, when $n = 4$ we have $\Delta x = \frac{2}{4} = 0.5$ and $x_0 = 0$, $x_1 = 0.5$, $x_2 = 1$, and $x_3 = 1.5$. Therefore

$$A(L_4) = \sum_{i=1}^{4} f(x_{i-1}) \Delta x$$

$$= f(x_0) \Delta x + f(x_1) \Delta x + f(x_2) \Delta x + f(x_3) \Delta x$$

$$= 0^2 \cdot (0.5) + 0.5^2 \cdot (0.5) + 1^2 \cdot (0.5) + 1.5^2 \cdot (0.5)$$

$$= 1.75$$

n	$A(L_n) = \sum_{i=1}^{n} f(x_{i-1}) \Delta x$
2	1
4	1.75
10	2.28
100	2.6268
1000	2.662668

Figure 8

Calculating such sums by hand becomes tedious as n gets larger. We show the results calculated so far, as well as for $n = 10$, 100, and 1000 in Figure 8. Computers or calculators can perform such calculations quickly, and we will discuss this process shortly.

The area of the polygon does seem to be stabilizing as n gets large. The diagrams in Figure 9 should help you visualize what is happening as n gets larger and larger.

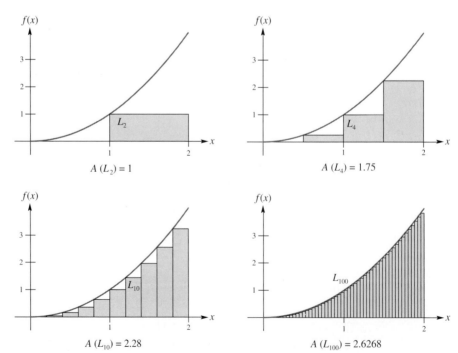

$A(L_2) = 1$ $A(L_4) = 1.75$

$A(L_{10}) = 2.28$ $A(L_{100}) = 2.6268$

Figure 9

Area by Circumscribed Polygons   Consider the rectangle with base $[x_{i-1}, x_i]$ and height $f(x_i) = x_i^2$, corresponding to the right endpoint of each subinterval (shown at the upper left in Figure 10). Its area is $f(x_i)\Delta x$. The union R_n of all such rectangles forms a circumscribed polygon for the region B, as shown at the lower right in Figure 10. In this case, it should be clear that $A(B) < A(R_n)$ for each n.

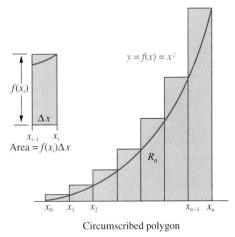

Circumscribed polygon

Figure 10

The area $A(R_n)$ is calculated by analogy with the calculation of $A(L_n)$.

$$A(R_n) = \sum_{i=1}^{n} f(x_i)\,\Delta x$$

$$= f(x_1)\,\Delta x + f(x_2)\,\Delta x + \cdots + f(x_n)\,\Delta x$$

For $n = 2$ we have $\Delta x = 1$ and $x_1 = 1$ and $x_2 = 2$. Thus

$$A(R_2) = \sum_{i=1}^{2} f(x_i)\,\Delta x$$

$$= f(x_1)\,\Delta x + f(x_2)\,\Delta x$$

$$= 1^2 \cdot 1 + 2^2 \cdot 1 = 5$$

For the case $n = 4$, we have $\Delta x = 0.5$ and $x_1 = 0.5$, $x_2 = 1$, $x_3 = 1.5$, and $x_4 = 2$, so

$$A(R_4) = \sum_{i=1}^{4} f(x_i)\,\Delta x$$

$$= f(x_1)\,\Delta x + f(x_2)\,\Delta x + f(x_3)\,\Delta x + f(x_4)\,\Delta x$$

$$= 0.5^2 \cdot (0.5) + 1^2 \cdot (0.5) + 1.5^2 \cdot (0.5) + 2^2 \cdot (0.5)$$

$$= 3.75$$

For larger n-values we again resort to a computer and show the results in Figure 11.

It should be clear from the data we have collected so far that the exact area under the curve $y = x^2$ between $x = 0$ and $x = 2$ lies somewhere be-

n	$A(R_n) = \sum_{i=1}^{n} f(x_i)\,\Delta x$
2	5
4	3.75
10	3.08
100	2.7068
1000	2.670668

Figure 11

tween 2.662668 and 2.670668, giving us almost three-digit accuracy. This is because circumscribed polygons give an overestimate of the area and inscribed polygons give an underestimate of the area. Both estimates seem to be converging to about $2\frac{2}{3}$, but we don't know that for sure yet.

EXAMPLE 2: Estimate the area under the curve $y = 10 - 2x$ between $x = 0$ and $x = 4$ using both circumscribed rectangles and inscribed rectangles. Use $n = 2, 4$, and 8 rectangles and sketch examples of both inscribed and circumscribed rectangles along with the given curve. Also calculate the exact area using simple geometry.

SOLUTION: If we partition the interval $[0, 4]$ into n subintervals, each would have length $\Delta x = \frac{4}{n}$, because the interval has a length of 4 units. The $n + 1$ equally spaced dividing points would be given by

$$x_0 = 0, x_1 = \Delta x, x_2 = 2 \cdot \Delta x, x_3 = 3 \cdot \Delta x, \ldots, x_n = n \cdot \Delta x = 4$$

We calculate $A(L_n) = \sum_{i=1}^{n} f(x_{i-1})\Delta x$ and $A(R_n) = \sum_{i=1}^{n} f(x_i)\Delta x$ for the cases $n = 2, n = 4$, and $n = 8$. As above, L_n represents the union of n rectangles where the height of each rectangle is given by the left endpoint of each subinterval, and R_n represents the corresponding union of rectangles where the heights are given by the right endpoints.

$n = 2$: $\Delta x = \dfrac{4}{2} = 2, x_0 = 0, x_1 = 2, x_2 = 4$

$$A(L_2) = (10 - 2(0)) \cdot 2 + (10 - 2(2)) \cdot 2$$
$$= 20 + 12 = 32$$
$$A(R_2) = (10 - 2(2)) \cdot 2 + (10 - 2(4)) \cdot 2$$
$$= 12 + 4 = 16$$

$n = 4$: $\Delta x = \dfrac{4}{4} = 1, x_0 = 0, x_1 = 0.5, x_2 = 1, x_3 = 1.5, x_4 = 2$

$$A(L_4) = (10 - 2(0)) \cdot 1 + (10 - 2(1)) \cdot 1 + (10 - 2(2)) \cdot 1 + (10 - 2(3)) \cdot 1$$
$$= 10 + 8 + 6 + 4 = 28$$
$$A(R_4) = (10 - 2(1)) \cdot 1 + (10 - 2(2)) \cdot 1 + (10 - 2(3)) \cdot 1 + (10 - 2(4)) \cdot 1$$
$$= 8 + 6 + 4 + 2 = 20$$

$n = 8$: $\Delta x = \dfrac{4}{8} = 0.5, x_0 = 0, x_1 = 0.5, x_2 = 1, \ldots, x_8 = 4$

$$A(L_8) = (10 - 2(0)) \cdot 0.5 + (10 - 2(0.5)) \cdot 0.5 + (10 - 2(1)) \cdot 0.5$$
$$+ (10 - 2(1.5)) \cdot 0.5 + (10 - 2(2)) \cdot 0.5 + (10 - 2(2.5)) \cdot 0.5$$
$$+ (10 - 2(3)) \cdot 0.5 + (10 - 2(3.5)) \cdot 0.5$$
$$= 5 + 4.5 + 4 + 3.5 + 3 + 2.5 + 2 + 1.5 = 26$$
$$A(R_8) = (10 - 2(0.5)) \cdot 0.5 + (10 - 2(1)) \cdot 0.5 + (10 - 2(1.5)) \cdot 0.5$$
$$+ (10 - 2(2)) \cdot 0.5 + (10 - 2(2.5)) \cdot 0.5 + (10 - 2(3)) \cdot 0.5$$
$$+ (10 - 2(3.5)) \cdot 0.5 + (10 - 2(4)) \cdot 0.5$$
$$= 4.5 + 4 + 3.5 + 3 + 2.5 + 2 + 1.5 + 1 = 22$$

In Figure 12 we sketch the graph of $y = 10 - 2x$ on the interval $[0, 4]$ along with $n = 2$ circumscribed rectangles. Clearly, the height of each rectangle corresponds to the value of the function at the *left* endpoint of each subinterval, because the function is decreasing.

Similarly, we will get an inscribed polygon using rectangles where the height is given by the *right* endpoint of each subinterval. In Figure 13 we show the graph of $y = 10 - 2x$ on the interval $[0, 4]$ along with R_8, which consists of $n = 8$ inscribed rectangles.

In Figure 14 we tabulate the results so far. We conclude that the exact area under the curve $y = 10 - 2x$ between $x = 0$ and $x = 4$ must lie between 22 and 26 square units.

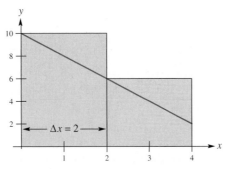

Figure 12 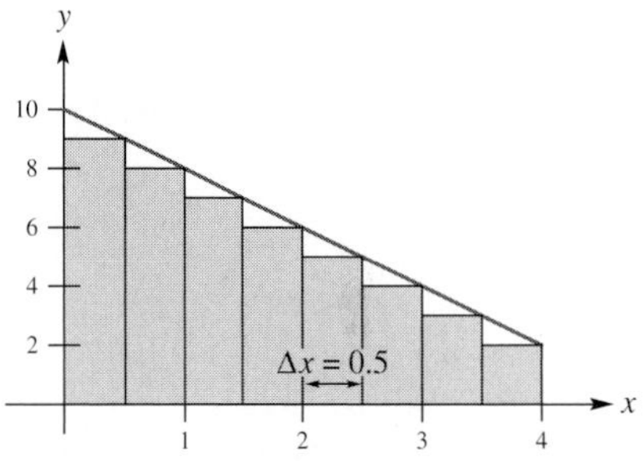   Figure 13

n	$A(L_n) = \sum\limits_{i=1}^{n} f(x_{i-1})\Delta x$	$A(R_n) = \sum\limits_{i=1}^{n} f(x_i)\Delta x$
2	32	16
4	28	20
8	26	22

Figure 14

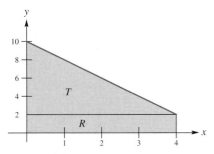

Figure 15

Finally, we can compute the exact area from simple geometry by noticing that the region we are interested in consists of a triangle (call it T) sitting on top of a rectangle (call it R). (See Figure 15.) The total area is clearly

$$A(T) + A(R) = \frac{1}{2} \cdot 4 \cdot 8 + 4 \cdot 2 = 24$$

This result is consistent with our numerical estimates.   ◄

Riemann Sums   Given an interval $[a, b]$, we can partition it into n equally spaced subintervals by means of the points $a = x_0 < x_1 < x_2 < \cdots < x_{n-1} < x_n = b$. If we let $\Delta x = \frac{b-a}{n}$ be the width of each subinterval, then $x_1 = a + \Delta x$, $x_2 = a + 2 \cdot \Delta x$, $x_3 = a + 3 \cdot \Delta x$, and so on. Let $\bar{x}_i$ be *any* point chosen from the interval $[x_{i-1}, x_i]$; we call $\bar{x}_i$ a *sample point*. The sum $\sum_{i=1}^{n} f(\bar{x}_i)\Delta x$ is called a *Riemann sum*. *Note*: Many texts allow subintervals of different lengths in the definition of a Riemann sum, but we don't need that level of generality.

Though the definition of Riemann sum does not demand it, we generally choose the $\bar{x}_i$'s the same way in each subinterval when estimating areas. Two common choices are to let each $\bar{x}_i$ be either the right endpoint or the left endpoint of the interval $[x_{i-1}, x_i]$. In particular, we will refer to the sum $A(L_n) = \sum_{i=1}^{n} f(x_{i-1})\Delta x$ as a left-endpoint Riemann sum (we choose $\bar{x}_i = x_{i-1}$) and $A(R_n) = \sum_{i=1}^{n} f(x_i)\Delta x$ as a right-endpoint Riemann sum (we choose $\bar{x}_i = x_i$). Another common choice is to let $\bar{x}_i = \dfrac{x_i + x_{i-1}}{2}$, the midpoint of the interval $[x_{i-1}, x_i]$; we will investigate this choice in the next section.

In the next example we show how computer algebra systems and graphics calculators can be used to estimate areas by calculating Riemann sums for large values of n.

n	$\sum_{i=1}^{n} f(x_{i-1})\Delta x$
10	0.777817
100	0.749979
1000	0.747140

Left-endpoint Riemann sums.

n	$\sum_{i=1}^{n} f(x_i)\Delta x$
10	0.714605
100	0.743657
1000	0.746508

Right-endpoint Riemann sums.

Figure 16

EXAMPLE 3: Estimate the area under the curve $f(x) = e^{-x^2}$ between $x = 0$ and $x = 1$. Use both left- and right-endpoint Riemann sums with $n = 10, 100,$ and 1000 rectangles.

SOLUTION: We need to compute both left-endpoint Riemann sums and right-endpoint Riemann sums for $n = 10, 100, 1000$. Clearly, this is too tedious to do by hand. For computer algebra systems, we can define a function called RIEMANN that calculates Riemann sums for a given function, interval, number of rectangles, and choice of sample point (left, right, midpoint). For graphing calculators, we can write a program called RIEMANN that performs the same task. (See the technology pages in the solution manual.)

The results for $n = 10$, $n = 100$, and $n = 1000$ are summarized in Figure 16. We see that the exact area under the curve lies between 0.746508 and 0.747140. Thus the area to three decimal places is 0.747.

Notice that because the function is decreasing between $x = 0$ and $x = 1$, the left-endpoint sums provide an overestimate of the area, and the right-endpoint sums provide an underestimate of the area. The case for $n = 10$ rectangles is illustrated in Figure 17. ◄

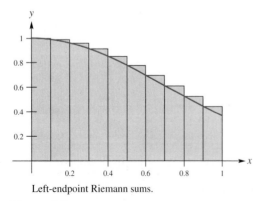

Left-endpoint Riemann sums.

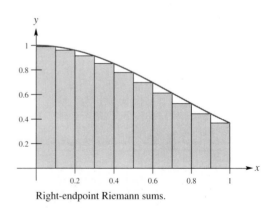

Right-endpoint Riemann sums.

Figure 17

> **Definition of Area**
>
> In the last example we saw that both left- and right-endpoint Riemann sums seemed to be converging to the same value as the number of rectangles got larger and larger. This leads us to define the area A under a continuous, positive function $y = f(x)$ between $x = a$ and $x = b$ as
>
> $$A = \lim_{n \to \infty} A(L_n) = \lim_{n \to \infty} A(R_n)$$
>
> $$= \lim_{n \to \infty} \sum_{i=1}^{n} f(x_{i-1})\Delta x = \lim_{n \to \infty} \sum_{i=1}^{n} f(x_i)\,\Delta x$$

In the next section we will continue this line of reasoning and define the *definite integral* as the limit of Riemann sums for any suitable function and for any choice of sample point $\bar{x}_i$.

Concepts Review

1. $\sum\limits_{i=1}^{4} 2i = $ _____ .

2. The exact area of the region under the curve $y = |x|$ between 0 and 4 is _____ . Similarly, the area under the curve between -4 and -2 is _____ .

3. A sum of the form $\sum\limits_{i=1}^{n} f(\bar{x}_i)\Delta x$ is called a _____ .

4. For a decreasing function, left-endpoint Riemann sums gives an _____ (over, under) estimate of the area under the curve.

Problem Set 5.3

In Problems 1–8, find the value of the indicated sum.

1. $\sum\limits_{k=1}^{5} (3k - 1)$

2. $\sum\limits_{i=1}^{6} 2i^2$

3. $\sum\limits_{i=1}^{5} f(x_i)$ where $x_i = \dfrac{i}{5}$ and $f(x) = \dfrac{2}{x+1}$.

4. $\sum\limits_{i=1}^{5} f(x_i)$ where $x_i = 2i$ and $f(x) = (x+1)^2$.

5. $\sum\limits_{i=1}^{5} (-1)^i 2^{i-1}$

6. $\sum\limits_{i=2}^{4} \dfrac{(-1)^i}{i(2i+1)}$

7. $\sum\limits_{k=1}^{6} f(x_i)$ where $x_i = \dfrac{i}{2}$ and $f(x) = \sin(\pi x)$.

8. $\sum\limits_{k=1}^{7} f(x_i)$ where $x_i = i$, and $f(x) = \cos(\pi x)$.

In Problems 9–16, write the indicated sum in sigma notation.

9. $1 + 2 + 3 + \cdots + 98$

10. $2 + 4 + 6 + \cdots + 100$

11. $1 + \frac{1}{2} + \frac{1}{3} + \cdots + \frac{1}{69}$

12. $1 - \frac{1}{2} + \frac{1}{3} - \frac{1}{4} + \cdots - \frac{1}{50}$

13. $a_1 + a_2 + a_3 + \cdots + a_n$

14. $b_3 + b_4 + b_5 + \cdots + b_{22}$

15. $f(c_1) + f(c_2) + \cdots + f(c_n)$

16. $f(w_1)\Delta x + f(w_2)\Delta x + \cdots + f(w_n)\Delta x$

17. Evaluate $\sum\limits_{i=1}^{10} f(w_i)\Delta x$ if $f(x) = 3x$, $w_i = \frac{i}{5}$, and $\Delta x = \frac{1}{5}$.

For problems 18–21, estimate the area under the curve $y = f(x)$ between $x = a$ and $x = b$ using n circumscribed rectangles and n inscribed rectangles. Sketch both the inscribed and circumscribed rectangles along with the given curve. Also calculate the exact area using simple geometry, if possible. Show your work as in Example 2.

18. $f(x) = 2x + 3; a = -1, b = 2, n = 3$
19. $f(x) = 3x - 2; a = 1, b = 3, n = 4$
20. $f(x) = x^2 + 2; a = 0, b = 2, n = 6$
21. $f(x) = 2x^2 + 1; a = 0, b = 4, n = 8$

In Problems 22–27, estimate the area of the region under the curve $y = f(x)$ over the interval $[a, b]$. Use both left- and right-endpoint Riemann sums with $n = 10, 100$, and 1000 rectangles. Determine the number of digits of accuracy in your estimate. In addition to this, calculate the area using geometry, if possible.

22. $y = x + 1; a = 0, b = 2$

23. $y = \frac{1}{2}x^2 + 1; a = 0, b = 2$

24. $y = e^x; a = 1, b = 4$

25. $y = \ln x; a = 1, b = 4$

26. $y = \sin x; a = 0, b = \pi$

27. $y = \cos x; a = 0, b = \pi$

Sometimes it is desirable to make a change of variable in the index for a sum. For example, the change of variable $k = i - 3$ gives

$$\sum_{i=4}^{13} (i - 3)^3 = \sum_{k=1}^{10} k^3$$

For Problems 28–31, make the indicated change of variable in the index.

28. $\sum_{i=3}^{19} i(i - 2); k = i - 2$

29. $\sum_{k=5}^{14} k2^{k-4}; i = k - 4$

30. $\sum_{k=0}^{10} \frac{k}{k + 1}; i = k + 1$

31. $\sum_{k=4}^{13} (k - 3)\sin\left(\frac{\pi}{k - 3}\right); i = k - 3$

32. Prove the following formula for a **geometric sum**:

$$\sum_{k=0}^{n} ar^k = a + ar + ar^2 + \cdots + ar^n = \frac{a - ar^{n+1}}{1 - r} \quad (r \neq 1)$$

Hint: Let $S = a + ar + ar^2 + \cdots + ar^n$. Simplify $S - rS$ and solve for S.

33. Use Problem 32 to calculate each sum.

(a) $\sum_{k=1}^{10} \left(\frac{1}{2}\right)^k$

(b) $\sum_{k=1}^{10} 2^k$

34. In the song "The Twelve Days of Christmas," my true love gave me 1 gift on the first day, $1 + 2$ gifts on the second day, $1 + 2 + 3$ gifts on the third day, and so on for 12 days.

(a) Find the total number of gifts in 12 days.

(b) Find a simple formula for T_n, the total number of gifts given during a Christmas of n days.

Answers to Concepts Review: 1. 20 2. 8; 6 3. Riemann sum 4. over

---

LAB 9: AREA AND DISTANCE

Mathematical Background

We know what area means and how to find it for a few simple geometric figures, such as a rectangle, a triangle, a trapezoid, or a circle. But what do we mean by area, and how do we go about finding it if the boundary of the figure is a more general curve? In particular, we may want to find the area of a figure whose boundaries are given by functions.

The approach taken is to approximate a general figure with known figures; the simplest approach is to use rectangles. To find the area of the region whose boundaries are a positive function $f(x)$, the x-axis, and two vertical lines at $x = a$ and $x = b$, we divide the x-axis between $x = a$ and $x = b$ into a number of equally long segments, then over each segment we draw in a rectangle whose height approximates the function $f(x)$. By adding up the areas of the rectangles we get an approximation to the area under $f(x)$

between $x = a$ and $x = b$. This area is also referred to as a *definite integral*; the notation for the definite integral of a function between $x = a$ and $x = b$ is

$$\int_a^b f(x)\,dx$$

There are three common ways to get the height of each rectangle: the height of $f(x)$ at the left edge of the segment, the height of $f(x)$ at the right edge of the segment, and the height of $f(x)$ at the midpoint of the segment. We will refer to these as left-endpoint Riemann sums, right-endpoint Riemann sums, and midpoint Riemann sums. This method, using rectangles to find areas, is one of many methods called *numerical integration*. If we now increase the number of rectangles we should get a better approximation to the area under the curve.

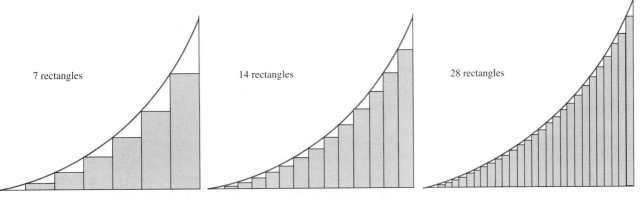

7 rectangles   14 rectangles   28 rectangles

Height of each rectangle is the value of $f(x)$ at the left side of the rectangle.

In the limit as the number of rectangles gets larger and larger, the area of the rectangles should get closer and closer to the area under the curve. As with our previous experience with limiting processes, if the answer repeats in the nth digit we assume n-digit accuracy.

If the function $f(x)$ that represents the upper boundary of the region is a speed-versus-time function (speed is the absolute value of velocity), then the area under the curve is the distance traveled. Here's why: Each rectangle used to approximate the area represents a little part of the trip; the height is the approximate speed, and the length of the base is the time taken for that part of the trip. Thus, the area of the rectangle approximates the distance traveled (speed $\times$ time = distance) during that part of the trip. When these areas are added up, the result should approximate the distance traveled.

Lab Introduction

For parts (1) and (2) of this experiment we are going to approximate the area of a known figure: a quarter-circle. A circle is described by the formula $x^2 + y^2 = r^2$ where r is the radius of the circle. If we solve for y and take the positive square root we get

$$y = \sqrt{r^2 - x^2}$$

which represents the upper half of the circle. Thus, if we let $r = 2$ and find the area under the curve $0 \leq x \leq 2$, we should get one-quarter of the area of a circle with radius 2, which is π, because $\frac{1}{4}\pi 2^2 = \pi$.

In part (3) of the lab we will estimate the distance traveled by the pendulum from Lab 3 (Section 2.3) on one complete swing (back and forth) using midpoint rectangles.

Experiment

(1) Find the area under the function $y = (4 - x^2)^{\frac{1}{2}}$ between $x = 0$ and $x = 2$ using left-endpoint Riemann sums. If possible, have the program display the rectangles at each step; be sure to use a window that allows you to see all of the function and the rectangles (such as $0 \leq x \leq 2$ and $0 \leq y \leq 3$). Start with 5 rectangles and then increase through the sequence, 5, 10, 20, 50, 100, 200, 500, . . . until you get three-digit accuracy (two decimal places). If the program can't display the rectangles, just sketch the rectangles in for the case of 5 rectangles.

(2) Repeat (1) using right-endpoint Riemann sums, and then using midpoint Riemann sums.

(3) Find the distance that the pendulum from Lab 3 traveled from the moment that it was released to the point where it first returned closest to its starting point (read Lab 3 if you didn't do it). On the graph, this is the point where the first maximum occurs; let T represent the t-coordinate of this point. The distance can be approximated by using rectangles as in part (1) to find the area under the *speed* curve.

(a) Graph and find the time of the first maximum (T) of the position function $y = ae^{-kt} \cos(wt)$. You may choose to use numerical or exact methods. Use values for a, k, and w from Lab 3, or use $a = 10$, $k = 0.02$, $w = 3$ if you didn't do that lab.

(b) Graph the speed function corresponding to the distance function. The speed is the absolute value of the derivative of the position function. Use the t-interval $0 \leq t \leq T$ where T is the first maximum from part (a).

(c) Find the distance traveled accurate to one decimal place. (As explained above, this is the area under the speed curve from 0 to T.) Use midpoint Riemann sums. You don't need to display or draw the rectangles.

Discussion

1. What limit do all three estimates from parts (1) and (2) of the experiment section seem to be approaching? How is the limit approached in each case — that is, from above (decreasing sequence) or from below (increasing sequence)? Which technique converges fastest? Each of the three estimates above plays a role. Which is most accurate for a given number of rectangles? Why? Which is an overestimate? Why? Which is an underestimate? Why? Include a sketch if it helps to explain what you mean.

2. Explain the relationship between area and distance traveled when dealing with speed curves.

3. Is there another way to find the distance traveled? If so, find the distance traveled that way too. Do the answers agree to one decimal place?

5.4 The Definite Integral and Numerical Integration

All the preparations have been made; we are ready to define the definite integral. Both Newton and Leibniz introduced early versions of this concept. However, it was Riemann who gave us the modern definition, though ours will be a more restrictive definition than that of Riemann. In formulating this definition, we are guided by the ideas we discussed in the previous section.

We used Riemann sums to estimate the area under a curve in the last section, but we only considered functions that were positive on a given interval. What if we have a function that is negative on an interval, or one that has both positive and negative values on an interval? What is the meaning of a Riemann sum in such cases?

Consider a function f defined on a closed interval $[a, b]$. It may have both positive and negative values on the interval. Its graph might look something like the one in Figure 1.

Now consider a partition of the interval $[a, b]$ into n equally spaced subintervals by means of points $a = x_0 < x_1 < x_2 < \cdots < x_{n-1} < x_n = b$ and let $\Delta x = x_i - x_{i-1} = \dfrac{b-a}{n}$, as before. On each subinterval $[x_{i-1}, x_i]$, pick an arbitrary sample point $\bar{x}_i$ (which may be an endpoint). Then form the Riemann sum $\displaystyle\sum_{i=1}^{n} f(\bar{x}_i)\Delta x$. Its geometric interpretation for one possible choice of partition and sample points (with $n = 6$) is shown in Figure 2. Note that the contribution from a rectangle below the x-axis is the negative of its area, because in this case, $f(\bar{x}_i) < 0$.

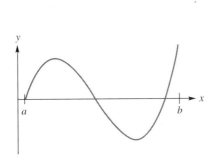

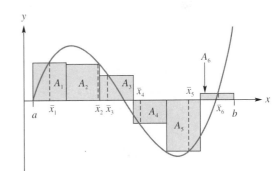

A Riemann sum interpreted as an algebraic sum of areas.

$$\sum_{i=1}^{6} f(\bar{x}_i)\,\Delta x = A_1 + A_2 + A_3 - A_4 - A_5 + A_6$$

Figure 1   Figure 2

EXAMPLE 1: Evaluate the Riemann sum for $f(x) = 10 - x^2$ on the interval $[0, 4]$ using the equally spaced partition points $0 < 1 < 2 < 3 < 4$, with the sample point $\bar{x}_i$ being the midpoint of the ith subinterval. Sketch the function and the corresponding rectangles, and interpret the result as a sum or difference of areas.

SOLUTION:

$$\sum_{i=1}^{4} f(\bar{x}_i)\Delta x$$

$$= f(0.5)(1) + f(1.5)(1) + f(2.5)(1) + f(3.5)(1)$$

$$= [9.75 + 7.75 + 3.75 - 2.25](1)$$

$$= 19$$

The sum of the areas of the rectangles above the x-axis *minus* the area of the rectangle below the x-axis is 19. (See Figure 3.) Note that you don't insert the negative for the last rectangle; it's a result of evaluating the function at the point $x = 3.5$.   ◀

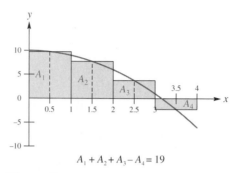

$$A_1 + A_2 + A_3 - A_4 = 19$$

Figure 3

Definition of the Definite Integral

Suppose now that n, Δx, and $\overline{x}_i$ have the meaning discussed above.

Definition

(The Definite Integral). Let f be a function that is defined on the closed interval $[a, b]$. Then if

$$\lim_{n \to \infty} \sum_{i=1}^{n} f(\overline{x}_i)\Delta x$$

exists and has the same value regardless of how the $\overline{x}_i$'s are chosen, we say that f is **integrable** on $[a, b]$. Moreover, $\int_a^b f(x)dx$, called the **definite integral** (or Riemann integral) of f from a to b, is then defined as

$$\int_a^b f(x)dx = \lim_{n \to \infty} \sum_{i=1}^{n} f(\overline{x}_i)\Delta x$$

The heart of the definition is the final line. The concept captured in that equation grows out of our discussion of area in the previous section. However, we have modified the notion presented there. For example, we now allow f to be negative on part or all of $[a, b]$, and we allow $\overline{x}_i$ to be any point on the ith subinterval. We saw from our work in the last section that left- and right-endpoint Riemann sums both converge to the same value as the number of rectangles approaches infinity; as the width of each rectangle gets smaller, the choice of a particular $\overline{x}_i$ within an interval has less and less effect on the sum.

Because we have made these changes, it is important to state precisely how the definite integral relates to area. In general, $\int_a^b f(x)dx$ gives the *signed area* of the region trapped between the curve $y = f(x)$ and the x-axis on the interval $[a, b]$, meaning that a plus sign is attached to areas of parts above the x-axis and a minus sign is attached to areas of parts below the x-axis. In symbols,

$$\int_a^b f(x)\,dx = A_{above} - A_{below}$$

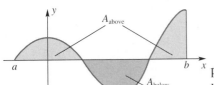

Figure 4

where A_{above} and A_{below} are as shown in Figure 4.

Returning to the symbol $\int_a^b f(x)\,dx$, we might call a the lower endpoint and b the upper endpoint for the integral. However, most authors use the terminology **lower limit** of integration and **upper limit** of integration, which is fine provided that we realize that this usage of the word *limit* has nothing to do with its more technical meaning.

In our definition of $\int_a^b f(x)\,dx$, we implicitly assumed that $a < b$. We remove that restriction with the following definitions.

$$\int_a^a f(x)\,dx = 0$$

$$\int_a^b f(x)\,dx = -\int_b^a f(x)\,dx, \qquad a > b$$

The first definition makes sense in that the area under any curve between $x = a$ and $x = a$ should be equal to zero. The second definition makes sense in that summing a Riemann sum from right to left would correspond to a negative Δx. Thus,

$$\int_2^2 x^3\,dx = 0, \qquad \int_6^2 x^3\,dx = -\int_2^6 x^3\,dx$$

Finally, we point out that x is a **dummy variable** in the symbol $\int_a^b f(x)\,dx$. By this we mean that x can be replaced by any other letter (provided, of course, that it is replaced in each place where it occurs). Thus,

$$\int_a^b f(x)\,dx = \int_a^b f(t)\,dt = \int_a^b f(u)\,du$$

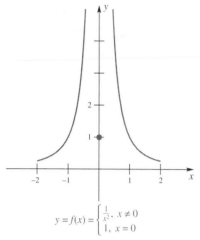

$$y = f(x) = \begin{cases} \frac{1}{x^2}, & x \neq 0 \\ 1, & x = 0 \end{cases}$$

Figure 5

What Functions Are Integrable?

Not every function is integrable. For example, the unbounded function

$$f(x) = \begin{cases} \dfrac{1}{x^2} & \text{if } x \neq 0 \\ 1 & \text{if } x = 0 \end{cases}$$

which is graphed in Figure 5, is not integrable on $[-2, 2]$. This is because the contribution to any Riemann sum from the subinterval containing $x = 0$ can be made arbitrarily large by choosing the corresponding sample point $\bar{x}_i$ sufficiently close to zero. In fact, this reasoning shows that *any function that is integrable on $[a, b]$ must be* **bounded** *there*; that is, there must exist a constant $M > 0$ such that $|f(x)| \leq M$ for all x in $[a, b]$.

Even some bounded functions can fail to be integrable, but they have to be pretty complicated. By all odds, Theorem A (below) is the most important theorem about integrability. Unfortunately, it is too difficult to prove here; we leave that for advanced calculus books.

Theorem A

(Integrability Theorem). If f is bounded on $[a, b]$ and if it is continuous there except at a finite number of points, then f is integrable on $[a, b]$. In particular, if f is continuous on the whole interval $[a, b]$, it is integrable on $[a, b]$.

As a consequence of this theorem, the following functions are integrable on every closed interval $[a, b]$.

1. Polynomial functions.
2. Sine and cosine functions.
3. Rational functions, provided the interval $[a, b]$ contains no points where a denominator is 0.
4. Exponential functions.
5. Logarithmic functions when $a > 0$.

Estimating Definite Integrals

Knowing that a function is integrable allows us to estimate its integral by picking the sample points $\bar{x}_i = x_i$ in any way convenient for us. In the last section we chose either left-endpoint Riemann sums or right-endpoint Riemann sums to estimate areas; we can estimate definite integrals by the same method. These approximations had the advantage that for a function that was either increasing or decreasing, one provided an overestimate of the area and the other provided an underestimate of the region. Neither, however, was very accurate for a small number of rectangles.

It should seem reasonable that a more accurate estimate of a definite integral can be achieved using the midpoint of each interval for each sample point, that is $\bar{x}_i = \dfrac{x_i + x_{i-1}}{2}$.

If you look at a typical rectangle whose height is generated by the midpoint of the interval it sits on, the area above the rectangle and below the curve will tend to balance the area below the rectangle and above the curve. Thus, the area of the rectangle better approximates the area under the curve than if we used the left or right endpoint for the height of the rectangle. (See Figure 6.)

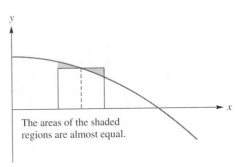

The areas of the shaded
regions are almost equal.

Typical midpoint Riemann sum approximation.

Figure 6

The use of Riemann sums to approximate definite integrals is one example of *numerical integration*. Midpoint Riemann sums provide a simple and reasonably accurate method of numerical integration.

***EXAMPLE* 2:** Estimate the definite integral

$$\int_0^2 \sin(x^2)\,dx$$

numerically using midpoint Riemann sums. Use $n = 10$, 100, and 1000. Write down ten decimal places for each value of n and estimate the number of decimal places of accuracy that you achieve. Sketch the function and give a geometrical interpretation of your result.

SOLUTION: We can use the calculator program RIEMANN or the computer algebra function RIEMANN to calculate the appropriate Riemann sums, as explained in the last section and in the accompanying solutions manual.

In Figure 7 we show a typical calculator screen that results from running the program using $n = 10$ rectangles; in Figure 8 we show a typical computer algebra calculation with $n = 10$ rectangles. The value that is input for R represents the choice of $\overline{x}_i$; use $R = 0$ for left-endpoint sums, $R = 1$ for right-endpoint sums, and $R = 0.5$ for midpoint sums.

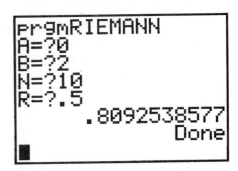

TI-82 screen showing the use of the program RIEMANN.

Figure 7

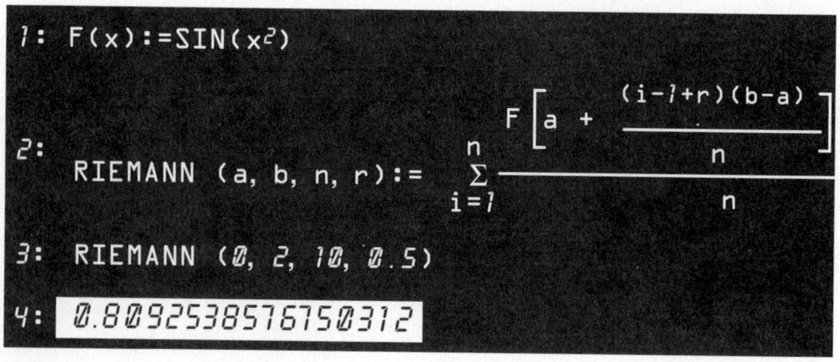

```
1:  F(x):=SIN(x²)

2:
    RIEMANN (a, b, n, r):=   Σ   ─────────────────
                            i=1          n

                                   n   F[a + (i-1+r)(b-a) ]
                                                  n
                                   ─────────────────────────
                                             n

3:  RIEMANN (0, 2, 10, 0.5)

4:  0.80925385767750312
```

Derive screen showing the use of the function RIEMANN.

Figure 8

The results for $n = 10$, as well as for $n = 100$ and $n = 1000$, are shown in Figure 9. It appears that $\int_0^2 \sin(x^2)\,dx \approx 0.805$, accurate to three decimal places (be sure to round where appropriate). In Figure 10 we graph the function $f(x) = \sin(x^2)$ and shade the regions between the curve and the x-axis, with A_1 the area of the region above the x-axis and A_2 the area of the region below the x-axis. Geometrically we have $A_1 - A_2 \approx 0.805$. ◄

n	$\sum_{i=1}^{n} f(\overline{x}_i)\Delta x$
10	0.80925386
100	0.80482008
1000	0.80477693

Midpoint Riemann sums for $f(x) = \sin(x^2)$ on the interval $0 \le x \le 2$.

Figure 9

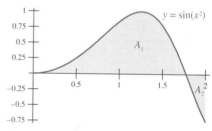

Figure 10

Accuracy and Numerical Integration

In the previous example we used a "powers of ten" convention in choosing the values of n—that is, each value of n was ten times the previous value. We also assumed that if two consecutive values of the Riemann sum agree to k digits, the result is accurate to k digits. You may recall a similar convention we used when estimating limits of functions numerically.

There is a theoretical justification for this convention. We say that a numerical integration method is kth-*order convergent* if for large n-values, multiplication of n by a factor of C results in an approximate reduction of the error by a factor of $\dfrac{1}{C^k}$. This definition of order of convergence is somewhat informal, but sufficient for our purposes; see a numerical analysis text for a more precise definition.

For example, if a method is second-order convergent, then multiplication of n by a factor of 10 results in a reduction of the error by a factor of about $\frac{1}{10^2} = \frac{1}{100}$ for large n-values. When the error is reduced by $\frac{1}{100}$ we gain two additional digits of accuracy. In general, if a method is kth-order convergent, then multiplication of n by a factor of 10 results in approximately k additional digits of accuracy.

Theorem B

Suppose that $f''(x)$ exists on the interval $[a, b]$ and that we use Riemann sums to approximate $\int_a^b f(x)\,dx$. Then

(1)   Midpoint Riemann sums is a second-order convergent method of numerical integration.

(2)   Left- and right-endpoint Riemann sums are both first-order convergent methods of numerical integration.

Proof   See a numerical analysis text.   ◄

Theorem B shows that increasing n by a factor of 10 results in about two more digits of accuracy for midpoint Riemann sums and about one more digit of accuracy for left- or right-endpoint Riemann sums. This is why we used the "powers of ten" convention in the previous example.

Though we will not prove these results, we will demonstrate the first statement numerically. In Figure 11 we have extended the table in Figure 9 of the previous example up to $n = 1,000,000$. The first table in Figure 11 shows midpoint Riemann sums for $\sin(x^2)$ on the interval $0 \le x \le 2$ calculated to twelve digits, and the second table shows the same results rounded to the number of accurate digits. Digits are assumed accurate if they agree with the result for the next higher value of n when rounded. From the second table one sees clearly that about two digits of accuracy are added every time n is increased by a factor of 10.

n	$\displaystyle\sum_{i=1}^{n} f(\overline{x}_i)\Delta x$
10	0.809253857675
100	0.804820077254
1000	0.804776925107
10000	0.804776493701
100000	0.804776489387
1000000	0.804776489344

Midpoint Riemann sums for $\sin(x^2)$ on the interval $0 \le x \le 2$.

n	$\displaystyle\sum_{i=1}^{n} f(\overline{x}_i)\Delta x$
10	0.8
100	0.805
1000	0.80478
10000	0.80477649
100000	0.804776489
1000000	0.804776489344

Midpoint Riemann sums for $\sin(x^2)$ on the interval $0 \le x \le 2$ rounded to the number of accurate digits. About two digits of accuracy are added whenever n is increased by a factor of 10.

Figure 11

Built-In Numerical Integration   Most graphing calculators and all computer algebra systems have built into them the ability to integrate functions numerically. You can expect the method of numerical integration that is used by a particular calculator or computer algebra system to be similar to (but more sophisticated than) simple Riemann sums. As a check, you can always compare the answer you get using the built-in method with the result from using midpoint Riemann sums.

EXAMPLE 3: Use the built-in method of numerical integration from the calculator or computer algebra system you have available to find $\int_0^2 \sin(x^2)dx$, and compare your answer with the results from Example 2. Use the default settings for the number of digits of accuracy.

SOLUTION: The result, accurate to 12 digits, is

$$\int_0^2 \sin(x^2)dx \approx 0.804776489344$$

Your particular calculator or computer algebra system may display fewer digits. With most computer algebra systems you can increase the number of accurate digits reported (see your software documentation for details).   ◄

This result is consistent with our result using midpoint Riemann sums with $n = 1,000,000$ (see Figure 11). The method your calculator or computer algebra system uses will generally achieve similar accuracy much more quickly than with simple Riemann sums. If you would like to learn more about other methods of numerical integration, take a course in numerical analysis when you have finished your calculus courses.

Concepts Review

1. The limit of a Riemann sum $\sum_{i=1}^{n} f(\bar{x}_i)\Delta x$ is called a _____ and is symbolized by _____.

2. Geometrically, the definite integral corresponds to a signed area. In terms of A_{above} and A_{below}, $\int_a^b f(x)dx =$ _____.

3. Thus, the value of $\int_{-1}^4 x\,dx$ is _____.

4. When estimating a definite integral with midpoint Riemann sums, multiplying the number of rectangles by 10 increases the accuracy of the estimate by about _____ digits.

Problem Set 5.4

In Problems 1–4, calculate the Riemann sum $\sum_{i=1}^{n} f(\bar{x}_i)\Delta x$ for the given data. Sketch the function and the corresponding rectangles, and interpret the result as an area, or a sum or difference of areas.

1. $f(x) = x - 5$; partition points $3 < 4 < 5 < 6 < 7$; $\bar{x}_1 = 3, \bar{x}_2 = 4, \bar{x}_3 = 6, \bar{x}_4 = 6.5$.

2. $f(x) = -2x + 3; -3 < -1 < 1 < 3; \bar{x}_1 = -2, \bar{x}_2 = -0.5, \bar{x}_3 = 2.$

3. $f(x) = x^2/2 - 1; [-1, 2]$ is divided into six equal subintervals; $\bar{x}_i$ is midpoint.

4. $f(x) = x^3/3 - 1; [0, 2]$ is divided into eight equal subintervals; $\bar{x}_i$ is right endpoint.

In Problems 5–12, estimate the definite integrals using midpoint Riemann sums. Use $n = 10, 100$, and 1000. Write down ten decimal places for each value of n and estimate the number of decimal places of accuracy that you achieve. Sketch the function and give a geometrical interpretation of your result.

5. $\displaystyle\int_{-2}^{2} (x^3 + 1)dx$

6. $\displaystyle\int_{-1}^{2} \tan\frac{x}{2}\,dx$

7. $\displaystyle\int_{0}^{2} \cos x^2 dx$

8. $\displaystyle\int_{0}^{2} x\cos x^2 dx$

9. $\displaystyle\int_{1}^{3} \ln x\,dx$

10. $\displaystyle\int_{0}^{2} x^2 e^{-x^2}dx$

11. $\displaystyle\int_{-1}^{2} (x^2 - 1)\,dx$

12. $\displaystyle\int_{0}^{4} (x^2 - 2x)\,dx$

13–20. Redo Problems 5–12 using the built-in numerical integration capability of a graphing calculator or computer algebra system. Compare results with what you got using midpoint Riemann sums.

In Problems 21–24, calculate $\displaystyle\int_{0}^{5} f(x)dx$ by using appropriate area formulas from plane geometry. Begin by graphing the given function.

21. $f(x) = \begin{cases} x & \text{if } 0 \leq x < 1 \\ 1 & \text{if } 1 \leq x \leq 3 \\ x - 4 & \text{if } 3 < x < 5 \end{cases}$

22. $f(x) = \begin{cases} x + 2 & \text{if } 0 \leq x < 2 \\ 6 - x & \text{if } 2 \leq x < 5 \end{cases}$

23. $f(x) = \begin{cases} \sqrt{4 - x^2} & \text{if } 0 \leq x \leq 2 \\ 2 & \text{if } 2 < x < 5 \end{cases}$

24. $f(x) = \begin{cases} -\sqrt{9 - x^2} & \text{if } 0 \leq x \leq 3 \\ 2 & \text{if } 3 < x < 5 \end{cases}$

Answers to Concepts Review: 1. Definite integral, $\displaystyle\int_{a}^{b} f(x)dx$ 2. $A_{above} - A_{below}$ 3. $\dfrac{15}{2}$ 4. 2

LAB 10: AREA FUNCTIONS AND THE FUNDAMENTAL THEOREM OF CALCULUS

Mathematical Background

For any functions $f(t)$, we can define an area function for $f(t)$ by

$$A(x) = \int_{a}^{x} f(t)\,dt$$

Thus, for each x, the function $A(x)$ represents the area under $f(t)$ between $t = a$ and $t = x$; these areas can be approximated using midpoint rectangles or the built-in definite integration capability of a calculator or computer.

Lab Introduction

In this lab you are going to look at the function

$$f(x) = xe^{-x} \tag{1}$$

The area function you will investigate will be

$$A(x) = \int_0^x te^{-t}dt \tag{2}$$

It is important to understand that for each x the integral in (2) is a *definite* integral or area.

Experiment

(1) Graph and sketch the function $f(x)$ above for $0 \le x \le 4$ and $0 \le y \le 1$.

(2) Compute numerically the value of the function $A(x)$ above for x starting at 0 and ending at 4 with steps of 0.2; get four-decimal-place accuracy.

(3) Compute the derivative of $A(x)$ numerically by forming a table like the one started below; the $A(x)$ column was computed in part (2).

x	$A(x)$	$A'(x)$
0	0	...
0.2		
0.4		
⋮		
4.0		

Compute $A'(x)$ numerically by using

$$A'(x) \approx \frac{A(x + h) - A(x)}{h}$$

with $h = 0.2$. For example, $A'(1.0) \approx \dfrac{A(1.2) - A(1.0)}{0.2}$. Note that there will be no entry in the table under the $A'(x)$ column for $x = 4$.

(4) Graph $f(x)$, $A(x)$, and $A'(x)$ on the same set of axes by hand—that is, locate the points and connect the dots. Use the table you formed above. Add a new column for $f(x)$; evaluate the function $f(x)$ at the given x-values to complete the table.

Discussion

1. Are there any connections between the graphs of $f(x)$, $A(x)$, and $A'(x)$? What are they? The relationship between $f(x)$ and $A'(x)$ is one form of the Fundamental Theorem of Calculus, the subject of the next section.

2. a. What region does $A(x + h) - A(x)$ represent on the $f(x)$ graph? For example, what region is represented by $A(1.2) - A(1.0)$ (here $x = 1.0$ and $h = 0.2$)? Sketch this region.

    b. Interpret graphically what you get when you divide $A(x + h) - A(x)$ by h.

    c. Discuss why this explains the relationship between $A'(x)$ and $f(x)$ that you found in part (1) above.

3. a. Show that the derivative of $-(1 + x)e^{-x} + 1$ is xe^{-x}. Can you use this fact and the relationship in discussion question 1 to come up with a formula for $A(x)$?

Do so, and graph your guess to see if it looks like the plot of $A(x)$ you made in part (4) of the experiment section.

b. Can you think of a way in which this lab's results can help you in finding the area under a general curve without using numerical integration? Explain. If so, find the area under the curve $f(x) = x^2$ between $x = 0$ and $x = 1$ in two different ways (one using numerical integration and the other using the conclusions of this lab—the Fundamental Theorem of Calculus).

5.5 | The Fundamental Theorem of Calculus

We have been able to estimate definite integrals directly from the definition (using Riemann sums), and we have calculated some exact definite integrals using simple geometry. It would be helpful to have a method of calculating exact definite integrals for a large class of functions; that method is the subject of this section.

We have credited Isaac Newton and Gottfried Leibniz with the simultaneous, but independent, discovery of calculus. Yet the concepts of the slope of the tangent line (derivative) and the area of a curved region (definite integral) were known earlier. Why, then, do Newton and Leibniz figure so prominently in the history of calculus? They do so because they understood and exploited the intimate relationship that exists between antiderivatives and definite integrals, a relationship that enables us to compute easily the exact values of many definite integrals without ever using Riemann sums. This connection is so important that it is called the *Fundamental Theorem of Calculus.*

> ### Is It Fundamental?
>
> The Fundamental Theorem of Calculus is important in providing a powerful tool for evaluating definite integrals. But its deepest significance lies in the link it makes between differentiation (slope) and definite integration (area). This link appears in sparkling clarity when we rewrite the conclusion to the theorem with $f(x)$ replaced by $g'(x)$.
>
> $$\int_a^b g'(x)\,dx = g(b) - g(a)$$

The Fundamental Theorem   You may have met several fundamental theorems before in your mathematical career. The Fundamental Theorem of Arithmetic says that a composite whole number factors uniquely into a product of primes. The Fundamental Theorem of Algebra says that an nth-degree polynomial equation has exactly n solutions, counting multiplicities. Any theorem with the title Fundamental Theorem should be studied carefully and then permanently committed to memory.

Theorem A

(Fundamental Theorem of Calculus). Let f be continuous (hence integrable) on $[a, b]$ and let F by *any* antiderivative of f there, so that $F'(x) = f(x)$. Then,

$$\int_a^b f(x)\,dx = \int_a^b F'(x)\,dx = F(b) - F(a)$$

Proof   Let $a = x_0 < x_1 < x_2 < \cdots < x_{n-1} < x_n = b$ be a partition of $[a, b]$ into equal length intervals. Then

$$F(b) - F(a) = F(x_n) - F(x_{n-1}) + F(x_{n-1}) - F(x_{n-2}) + \cdots + F(x_1) - F(x_0)$$

$$= \sum_{i=1}^{n} [F(x_i) - F(x_{i-1})]$$

because all of the terms of the sum cancel except the first and last. By the Mean Value Theorem applied to F on the interval (x_{i-1}, x_i),

$$F(x_i) - F(x_{i-1}) = F'(\bar{x}_i)(x_i - x_{i-1}) = f(\bar{x}_i)\Delta x$$

for some choice of $\bar{x}_i$ in the open interval (x_{i-1}, x_i). Thus,

$$F(b) - F(a) = \sum_{i=1}^{n} f(\bar{x}_i)\Delta x$$

On the left we have a constant; on the right we have a Riemann sum for f on $[a, b]$. When we take limits of both sides as $n \to \infty$, we obtain

$$F(b) - F(a) = \lim_{n \to \infty} \sum_{i=1}^{n} f(\bar{x}_i)\Delta x = \int_a^b f(x)\, dx \qquad \blacktriangleleft$$

Before going on to examples, ask yourself why we can use the word *any* in the statement of the theorem.

The Meanings of the Word *Integration*

It is important at this point to review and clarify the different types of integration that we have encountered so far. In Section 5.1 we used the term *indefinite integral* as a synonym for the word *antiderivative*. In Section 5.4 we used the term *definite integral* to represent the limit of a Riemann sum as the number of terms in the sum (the number of rectangles) goes to infinity. For a positive function, its definite integral over an interval is the area below the graph of the function and above the x-axis on that interval.

The notations we used for these two different types of integration were very similar. We used the symbol $\int f(x)dx$ to represent the indefinite integral of $f(x)$ and we used the symbol $\int_a^b f(x)dx$ to represent the definite integral of $f(x)$ between $x = a$ and $x = b$. The Fundamental Theorem of Calculus shows why we used these similar notations. Because the notations are so similar, it is easy to miss the significance of the Fundamental Theorem, which links these two very different meanings of the word *integration*.

EXAMPLE 1:   Show that $\int_a^b k\, dx = k(b - a)$, k a constant.

> **SOLUTION:** $F(x) = kx$ is an antiderivative of $f(x) = k$. Thus, by the Fundamental Theorem of Calculus,
>
> $$\int_a^b k\, dx = F(b) - F(a) = kb - ka = k(b - a) \qquad \blacktriangleleft$$

EXAMPLE 2: Show that $\int_a^b x \, dx = \dfrac{b^2}{2} - \dfrac{a^2}{2}$.

SOLUTION: $F(x) = x^2/2$ is an antiderivative of $f(x) = x$. Therefore,

$$\int_a^b x \, dx = F(b) - F(a) = \frac{b^2}{2} - \frac{a^2}{2}$$

◄

EXAMPLE 3: Show that if r is a rational number different from -1, then

$$\int_a^b x^r \, dx = \frac{b^{r+1}}{r+1} - \frac{a^{r+1}}{r+1}$$

SOLUTION: $F(x) = x^{r+1}/(r+1)$ is an antiderivative of $f(x) = x^r$. Thus, by the Fundamental Theorem of Calculus,

$$\int_a^b x^r \, dx = F(b) - F(a) = \frac{b^{r+1}}{r+1} - \frac{a^{r+1}}{r+1}$$

Technical point: If $r < 0$, we require that 0 is not in $[a, b]$. Why?   ◄

It is convenient to introduce a special symbol for $F(b) - F(a)$. We write

$$F(b) - F(a) = \Big[F(x)\Big]_a^b$$

Thus, for example,

$$\int_2^5 x^2 \, dx = \left[\frac{x^3}{3}\right]_2^5 = \frac{125}{3} - \frac{8}{3} = \frac{117}{3} = 39$$

In terms of the symbol for indefinite integrals, we may write the conclusion of the Fundamental Theorem of Calculus as

$$\int_a^b f(x)dx = \left[\int f(x)dx\right]_a^b$$

Note that the C of the indefinite integration cancels out, as it always will, in the definite integration. That is why in the statement of the Fundamental Theorem we could use the phrase *any antiderivative*. In particular, we may always choose $C = 0$ in applying the Fundamental Theorem.

EXAMPLE 4: Evaluate $\int_{-1}^2 (4x - 6x^2)dx$ and interpret the result in terms of area. Include an appropriate sketch.

SOLUTION:

$$\int_{-1}^2 (4x - 6x^2)dx = \Big[2x^2 - 2x^3\Big]_{-1}^2$$
$$= (8 - 16) - (2 + 2) = -12$$

The function $f(x) = 4x - 6x^2$ has roots at 0 and $\frac{2}{3}$. (Why?) In Figure 1 we display three shaded regions with areas A_1, A_2, and A_3. These correspond to the areas between the curve and the x-axis for the intervals $[-1, 0]$, $[0, \frac{2}{3}]$, and $[\frac{2}{3}, 2]$, respectively. Our conclusion is that $-A_1 + A_2 - A_3 = -12$.   ◄

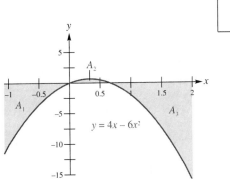

Figure 1

EXAMPLE 5: Evaluate $\int_1^8 \left(x^{\frac{1}{3}} + x^{\frac{4}{3}} \right) dx$.

SOLUTION:

$$\int_1^8 \left(x^{\frac{1}{3}} + x^{\frac{4}{3}} \right) dx = \left[\frac{3}{4} x^{\frac{4}{3}} + \frac{3}{7} x^{\frac{7}{3}} \right]_1^8$$

$$= \left(\frac{3}{4} \cdot 16 + \frac{3}{7} \cdot 128 \right) - \left(\frac{3}{4} \cdot 1 + \frac{3}{7} \cdot 1 \right)$$

$$= \frac{45}{5} + \frac{381}{7} \approx 65.68$$

◄

EXAMPLE 6: Evaluate $\int_0^\pi 3 \sin x \, dx$. Sketch the region whose area corresponds to this definite integral.

SOLUTION:

$$\int_0^\pi 3 \sin x \, dx = [-3 \cos x]_0^\pi = 3 + 3 = 6$$

Because the function $f(x) = 3 \sin x$ is positive on the interval $[0, \pi]$, the result of this definite integral corresponds to the entire area under the curve as pictured in Figure 2.

◄

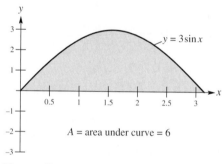

Figure 2

The Definite Integral Is a Linear Operator

Earlier, we learned that $\frac{d}{dx}$ and $\int \cdots dx$ are linear operators. You can add $\int_a^b \cdots dx$ to the list.

Theorem B

(Linearity of the Definite Integral). Suppose that f and g are continuous on $[a, b]$ and that k is a constant. Then:

(i) $\int_a^b k f(x) \, dx = k \int_a^b f(x) \, dx;$

(ii) $\displaystyle\int_a^b [f(x) + g(x)] = \int_a^b f(x)\,dx + \int_a^b g(x)\,dx$; and consequently

(iii) $\displaystyle\int_a^b [f(x) - g(x)] = \int_a^b f(x)\,dx - \int_a^b g(x)\,dx$.

These properties follow from the linearity of antiderivatives (Section 5.1) and the Fundamental Theorem of Calculus.

EXAMPLE 7: Evaluate $\displaystyle\int_1^2 \left(4e^{2x} + \frac{3}{x}\right) dx$.

SOLUTION:

$$\int_1^2 \left(4e^{2x} + \frac{3}{x}\right) dx = 4\int_1^2 e^{2x}\,dx + 3\int_1^2 \frac{1}{x}\,dx$$

$$= 4\left[\frac{1}{2}e^{2x}\right]_1^2 + 3[\ln x]_1^2$$

$$= 4\left(\frac{e^4}{2} - \frac{e^2}{2}\right) + 3(\ln 2 - \ln 1) \approx 96.5$$ ◄

EXAMPLE 8: Evaluate $\displaystyle\int_0^1 \left[x^2 + (x^2 + 1)^4 x\right] dx$.

SOLUTION:

$$\int_0^1 \left[x^2 + (x^2 + 1)^4 x\right] dx = \int_0^1 x^2\,dx + \int_0^1 (x^2 + 1)^4 x\,dx$$

The first integral is easy to do directly. To handle the second, we guess an antiderivative of $\frac{1}{5}(x^2 + 1)^5$ and check by taking the derivative, which is $(x^2 + 1)^4 \cdot 2x$. The modified guess is $\frac{1}{2} \cdot \frac{1}{5}(x^2 + 1)^5$, so that we have

$$\int (x^2 + 1)^4 x\,dx = \frac{1}{2}\frac{(x^2 + 1)^5}{5} + C = \frac{(x^2 + 1)^5}{10} + C$$

Therefore,

$$\int_0^1 x^2\,dx + \int_0^1 (x^2 + 1)^4 x\,dx = \left[\frac{x^3}{3}\right]_0^1 + \left[\frac{(x^2 + 1)^5}{10}\right]_0^1$$

$$= \left(\frac{1}{3} - 0\right) + \left(\frac{32}{10} - \frac{1}{10}\right)$$

$$= \frac{1}{3} + \frac{31}{10} = \frac{103}{30}$$ ◄

Numerical versus Exact Definite Integration   Does the Fundamental Theorem of Calculus mean that we no longer need to estimate definite integrals using Riemann sums or some other method of numerical integration? The answer is a resounding NO. In Section 5.1 we discussed the fact that it is easy to construct a function for which no simple antiderivative exists in terms of elementary functions. This means that for a great many definite integrals of practical interest, we cannot find exact results and must rely on numerical estimates. As in most areas of mathematics, an exact answer is usually preferable to a numerical estimate, but when exact methods fail, numerical estimates can be extremely important.

Definite Integration by Computer Algebra   Computer algebra systems can compute exact definite integrals as well as estimate them numerically as we did in Section 5.4. The computer, however, often runs up against the same problem that we just described. Because it is relatively easy to construct a function for which no simple antiderivative exists in terms of elementary functions, there are a great many functions for which computer algebra systems cannot find exact definite integrals.

It is important to understand roughly what is going on inside the computer when you ask for either an exact definite integral or a numerical approximation to one. To find an exact definite integral, the computer algebra system generally uses the Fundamental Theorem of Calculus, so it must first find an antiderivative. To compute a numerical approximation to a definite integral, the computer algebra system uses a method similar to Riemann sums, hence it only needs to compute a finite sum of numbers. Consequently, when confronted with a difficult definite integral, it is often a good strategy to try for an exact result first, and if that fails, get a numerical approximation. The next example illustrates this idea.

***EXAMPLE* 9:** Attempt to find an exact value for each definite integral below using a computer algebra system. Also find a three-digit approximation to each integral, whether you found an exact value or not.

(1) $\int_0^1 \sin \sqrt{x}\, dx$

(2) $\int_0^1 \sqrt{\sin x}\, dx$

SOLUTION: Computer algebra systems use the word *integrate* for both indefinite and definite integrals. The user then supplies the limits of integration to indicate a definite integral. If the computer algebra system cannot find an exact result, it will return the original problem.

For the integral in (1) we get the exact result

$$\int_0^1 \sin \sqrt{x}\, dx = 2(\cos(1) + \sin(1))$$

For the integral in (2) we get back the original problem—that is, the output of the command is the same as the input. This means that the computer algebra system could not find an exact result.

To find a numerical approximation to the integral in (1), we can simply turn our exact result shown above into a number. Directly after the exact result is generated, issue the command that takes the previous output and turns it into a numerical approximation. We get

$$\int_0^1 \sin \sqrt{x} \, dx \approx 0.602$$

accurate to three digits.

To find a numerical approximation to the integral in (2), we need to use numerical integration. We can use midpoint Riemann sums or the built-in numerical integration capability of the computer algebra system or graphing calculator, as explained in Section 5.4 (see Examples 2 and 3). Either way, we get

$$\int_0^1 \sqrt{\sin x} \, dx \approx 0.643$$

accurate to three digits.   ◄

With both integrals we ended up with a three-digit approximation, but by very different routes. The first integral had the exact symbolic value of $2(\cos(1) \, 1 \, \sin(1))$, which was found using the Fundamental Theorem of Calculus. Turning this symbolic result into a number is a relatively easy task for a computer or calculator. With the second integral we needed to use numerical integration, which involves repeatedly summing a series with a possibly large number of terms. Some numerical integrals may converge slowly, requiring significant computer time.

Concepts Review

1. If f is continuous on $[a, b]$ and if F is any _____ of f there, then $\int_a^b f(x)dx = $ _____ .

2. The symbol $[F(x)]_a^b$ stands for the expression _____ .

3. $\int_1^3 x^2 \, dx = [x^3/3]_1^3 = $ _____ .

4. $\int_{-2}^{-1} x^{-2} \, dx = [$_____$]_{-2}^{-1} = $ _____ .

Problem Set 5.5

In Problems 1–14, use the Fundamental Theorem of Calculus to evaluate each definite integral and interpret the result in terms of area. Include an appropriate sketch. Then check each result with a numerical method (midpoint Riemann sums, or a built-in method of a computer or calculator).

1. $\int_{-1}^2 (3x^2 - 2x + 3) \, dx$

2. $\int_1^2 (4x^3 + 7) \, dx$

3. $\int_{-1}^2 (3e^x + 3) \, dx$

4. $\int_1^2 \left(\frac{3}{x} + 7\right) dx$

5. $\int_{-4}^{-2} \left(y^2 + \frac{1}{y^3}\right) dy$

6. $\int_1^4 \frac{s^4 - 8}{s^2} \, ds$

7. $\displaystyle\int_0^4 \sqrt{e^t}\,dt$

8. $\displaystyle\int_1^8 \sqrt[3]{e^w}\,dw$

9. $\displaystyle\int_2^4 \left(\sqrt{y} + \frac{1}{y}\right)dy$

10. $\displaystyle\int_1^4 \frac{s^4 - 8}{s}\,ds$

11. $\displaystyle\int_0^{\pi/2} \cos x\,dx$

12. $\displaystyle\int_{\pi/6}^{\pi/2} 2\sin t\,dt$

13. $\displaystyle\int_0^1 (2x^4 - 3x^2 + 5)\,dx$

14. $\displaystyle\int_0^1 (x^{4/3} - 2x^{1/3})\,dx$

In Problems 15–30, use the Fundamental Theorem of Calculus combined with guess-and-check, if necessary, to evaluate the given definite integral. Check each result with a numerical method, if possible. Sketch the function and use the sketch to give an interpretation of your result in terms of area.

15. $\displaystyle\int_0^1 (x^2 + 1)^{10}(2x)\,dx$

16. $\displaystyle\int_{-1}^0 \sqrt{x^3 + 1}\,(3x^2)\,dx$

17. $\displaystyle\int_{-1}^3 \frac{1}{(t+2)^2}\,dt$

18. $\displaystyle\int_0^2 y^2 e^{-y^3}\,dy$

19. $\displaystyle\int_0^5 \frac{x}{x^2 + 1}\,dx$

20. $\displaystyle\int_0^3 e^x \cos(e^x)\,dx$

21. $\displaystyle\int_1^{10} \frac{\cos(\ln x)}{x}\,dx$

22. $\displaystyle\int_1^3 \frac{x^2 + 1}{\sqrt{x^3 + 3x}}\,dx$

23. $\displaystyle\int_0^{\pi/2} \cos^2 x \sin x\,dx$

24. $\displaystyle\int_0^{\pi/2} \sin^2 3x \cos 3x\,dx$

25. $\displaystyle\int_0^{\pi/2} (2x + \sin x)\,dx$

26. $\displaystyle\int_0^{\pi/2} [4x + 3 + \cos x]\,dx$

27. $\displaystyle\int_0^4 [\sqrt{x} + \sqrt{2x+1}]\,dx$

28. $\displaystyle\int_{-4}^{-1} \frac{1 - s^4}{2s^2}\,ds$

29. $\displaystyle\int_0^a \cos\left(\frac{2\pi}{a}x\right)dx$

Hint: You will need a generic graph as in Section 4.6.

30. $\displaystyle\int_0^{1/a} e^{-ax}\,dx$

In Problems 31–36, use a computer algebra system to evaluate the given definite integral exactly. If possible, show how the computer came up with the result using the Fundamental Theorem of Calculus.

31. $\displaystyle\int_0^1 x^2 \cos x\,dx$

32. $\displaystyle\int_0^1 x^3 \cos x\,dx$

33. $\displaystyle\int_0^1 x \cos(x^2)\,dx$

34. $\displaystyle\int_0^1 x \cos(x^3)\,dx$

35. $\displaystyle\int_0^1 x^2 \cos(x^3)\,dx$

36. $\displaystyle\int_0^1 x^2 \cos(x^2)\,dx$

37. Continue the process begun in Problems 31–36 as follows. Experiment with a computer algebra system to see what integrals of the form $\int_0^1 x^n \cos(x^m)\, dx$ (n and m positive integers) your system can find an exact result for in terms of functions you know (trig, exponential, and so on). Make a conjecture about which integrals of this type can be found exactly in terms of functions you know, and which cannot. Which cases can you explain?

38. Explain why $(1/n^3)\sum_{i=1}^{n} i^2$ should be a good approximation to $\int_0^1 x^2\, dx$ for large n. Now calculate the summation expression for $n = 10$ and the integral by the Fundamental Theorem of Calculus and compare their values.

39. Show that $\frac{1}{2}x\,|x|$ is an antiderivative of $|x|$ and use this fact to get a simple formula for $\int_a^b |x|\, dx$.

Answers to Concepts Review: 1. Antiderivative; $F(b) - F(a)$ 2. $F(b) - F(a)$ 3. $\frac{26}{3}$ 4. $-1/x; \frac{1}{2}$

5.6 | More Properties of the Definite Integral

Our definition of the definite integral was motivated by the problem of area for curved regions. Consider the two curved regions R_1 and R_2 in Figure 1 and let $R = R_1 \cup R_2$. It is clear that

$$A(R) = A(R_1 \cup R_2) = A(R_1) + A(R_2)$$

which suggests that

$$\int_a^c f(x)\,dx = \int_a^b f(x)\,dx + \int_b^c f(x)\,dx$$

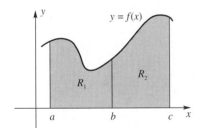

Figure 1

We quickly point out that this does not constitute a proof of the fact about integrals, because—first of all—our discussion of area in Section 5.4 was rather informal, and—second—our diagram supposes that f is positive, which it need not be. Nevertheless, definite integrals do satisfy this interval additive property and they do it no matter how the three points a, b, and c are arranged. We leave the rigorous proof to more advanced works.

Theorem A

(Interval Additive Property). If f is integrable on an interval containing the three points a, b, and c, then

$$\int_a^c f(x)\,dx = \int_a^b f(x)\,dx + \int_b^c f(x)\,dx$$

no matter what the order of a, b, and c.

For example,

$$\int_0^2 x^2 \, dx = \int_0^1 x^2 \, dx + \int_1^2 x^2 \, dx$$

which most people readily believe. But it is also true that

$$\int_0^2 x^2 \, dx = \int_0^3 x^2 \, dx + \int_3^2 x^2 \, dx$$

which may seem surprising. If you mistrust the theorem, you might actually evaluate each of the above integrals to see that the equality holds.

EXAMPLE 1: Find each definite integral using the Fundamental Theorem of Calculus along with the Interval Additive Property.

(a) $\displaystyle\int_0^5 |x - 3| \, dx$

(b) $\displaystyle\int_0^4 f(x) \, dx$ where $f(x) = \begin{cases} x^2 & 0 \le x \le 1 \\ x & 1 \le x \le 4 \end{cases}$

SOLUTION:

(a) We must use the definition of absolute value. When the quantity inside the absolute value symbols is positive, remove the symbols; when the quantity inside the absolute value symbols is negative, remove the symbols and insert a negative sign. Thus

$$|x - 3| = \begin{cases} x - 3 & x - 3 \ge 0 \\ -(x - 3) & x - 3 < 0 \end{cases}$$
$$= \begin{cases} x - 3 & x \ge 3 \\ -x + 3 & x < 3 \end{cases}$$

and so by the Interval Additive Property we get

$$\int_0^5 |x - 3| \, dx = \int_0^3 (-x - 3) \, dx + \int_3^5 (x - 3) \, dx$$

$$= \left[-\frac{x^2}{2} + 3x \right]_0^3 + \left[\frac{x^2}{2} - 3x \right]_3^5$$

$$= \left[-\frac{3^2}{2} + 3(3) \right] - \left[-\frac{0^2}{2} + 3(0) \right]$$

$$+ \left[\frac{5^2}{2} - 3(5) \right] - \left[\frac{3^2}{2} - 3(3) \right]$$

$$= \frac{13}{2} = 6\frac{1}{2}$$

(b) We have

$$\int_0^4 f(x)\,dx = \int_0^1 x^2\,dx + \int_1^4 x\,dx$$

$$= \left[\frac{x^3}{3}\right]_0^1 + \left[\frac{x^2}{2}\right]_1^4$$

$$= \frac{1^3}{3} - \frac{0^3}{3} + \frac{4^2}{2} - \frac{1^2}{2} = 7\frac{5}{6}$$

For part (b), a sketch of the graph and the region under the curve are shown in Figure 2. ◀

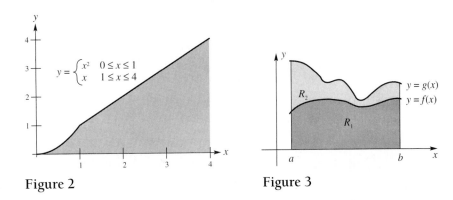

Figure 2   Figure 3

Comparison Properties   Consideration of the areas of the regions R_1 and R_2 in Figure 3 suggests another property of definite integrals.

Theorem B
(Comparison Property). If f and g are integrable on $[a, b]$ and if $f(x) \le g(x)$ for all x in $[a, b]$, then

$$\int_a^b f(x)\,dx \le \int_a^b g(x)\,dx$$

In informal but descriptive language, we say that the definite integral preserves inequalities.

Proof Let $a = x_0 < x_1 < x_2 < \cdots < x_n = b$ be an arbitrary, equally spaced partition of $[a, b]$, and for each i let $\overline{x}_i$ be any sample point on the ith subinterval $[x_{i-1}, x_i]$. We may conclude successively that

$$f(\overline{x}_i) \le g(\overline{x}_i)$$

$$f(\overline{x}_i)\Delta x \le g(\overline{x}_i)\Delta x$$

$$\sum_{i=1}^{n} f(\overline{x}_i)\Delta x \le \sum_{i=1}^{n} g(\overline{x}_i)\Delta x$$

$$\lim_{n\to\infty}\sum_{i=1}^{n} f(\overline{x}_i)\Delta x \le \lim_{n\to\infty}\sum_{i=1}^{n} g(\overline{x}_i)\Delta x$$

$$\int_a^b f(x)\,dx \le \int_a^b g(x)\,dx \qquad \blacktriangleleft$$

Theorem C

(Boundedness Property). If f is integrable on $[a, b]$ and if $m \le f(x) \le M$ for all x in $[a, b]$, then

$$m(b - a) \le \int_a^b f(x)\,dx \le M(b - a)$$

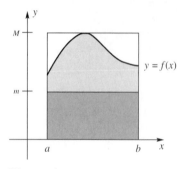

Figure 4

Proof The picture shown in Figure 4 helps us understand the theorem. Note that $m(b - a)$ is the area of the lower, small rectangle, $M(b - a)$ is the area of the large rectangle, and $\displaystyle\int_a^b f(x)\,dx$ is the area under the curve.

To prove the right-hand inequality, let $g(x) = M$ for all x in $[a, b]$. Then, by Theorem B,

$$\int_a^b f(x)\,dx \le \int_a^b g(x)\,dx$$

However,

$$\int_a^b g(x)\,dx = \int_a^b M\,dx = [Mx]_a^b = M(b - a)$$

The left-hand inequality is handled similarly. $\blacktriangleleft$

Differentiating a Definite Integral with Respect to a Variable Upper Limit

The next theorem is important—so important that some authors call it the Second Fundamental Theorem of Calculus.

Let f be integrable on $[a, b]$ and let x be any point in $[a, b]$ (Figure 5). Then for each x, $\displaystyle\int_a^x f(t)\,dt$ is a unique number, and so

$$G(x) = \int_a^x f(t)\,dt$$

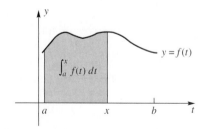

Figure 5

Fundamental Too

You will use Theorem D much less frequently than what we earlier called the Fundamental Theorem of Calculus. Yet it is a fundamental theorem too because it links differentiation and integration in another way. Differentiating a definite integral of a function (with respect to the upper limit) returns the original function. Differentiation and integration are inverse processes; they undo each other.

determines a function G with domain $[a, b]$. We call $\displaystyle\int_a^x f(t)\,dt$ an integral with variable upper limit. Note our use of t rather than x as the dummy variable to avoid confusion with the upper limit.

Theorem D

(Differentiating a Definite Integral). Let f be continuous on the closed interval $[a, b]$ and let x be a (variable) point in (a, b). Then

$$\frac{d}{dx}\left[\int_a^x f(t)\,dt\right] = f(x)$$

This should be learned in words. *The derivative of a definite integral with respect to its upper limit is the integrand at the upper limit.*

Proof If $G(x) = \displaystyle\int_a^x f(t)\,dt$, we must show that

$$G'(x) = \lim_{h \to 0} \frac{G(x + h) - G(x)}{h} = f(x)$$

Now

$$G(x + h) - G(x) = \int_a^{x+h} f(t)\,dt - \int_a^x f(t)\,dt = \int_x^{x+h} f(t)\,dt$$

by Theorem A.

Assume for the moment that $h > 0$ and let m and M be the minimum value and maximum value, respectively, of f on the interval $[x, x + h]$ (Figure 6). By Theorem C,

$$mh \le \int_a^{x+h} f(t)\,dt \le Mh$$

Figure 6

or

$$mh \le G(x + h) - G(x) \le Mh$$

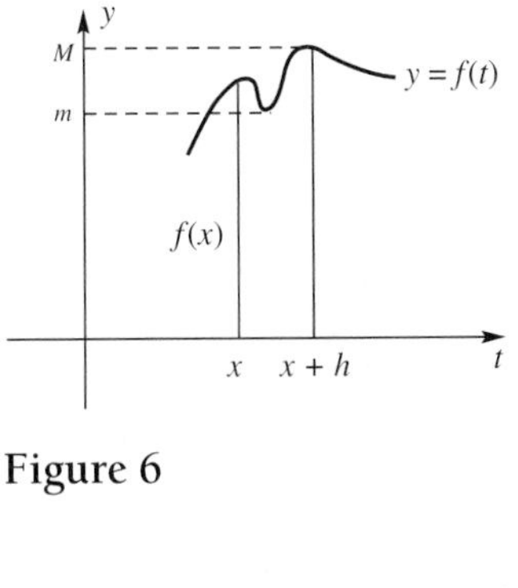

Dividing by h, we obtain

$$m \le \frac{G(x + h) - G(x)}{h} \le M$$

Now m and M really depend on h. Moreover, because f is continuous, both m and M must approach $f(x)$ as $h \to 0$. Because $\dfrac{G(x+h) - G(x)}{h}$ lies between m and M, it must also approach $f(x)$; thus,

$$\lim_{h \to 0} \frac{G(x+h) - G(x)}{h} = f(x)$$

The case where $h < 0$ is handled similarly. ◄

One theoretical consequence of this theorem is that every continuous function f has an antiderivative F given by

$$F(x) = \int_a^x f(t)\,dt$$

However, this fact is not helpful in getting a simple formula for any particular antiderivative.

EXAMPLE 2: Find $\dfrac{d}{dx}\left[\displaystyle\int_1^x t^2\,dt\right]$ two ways.

SOLUTION: First, we will do it the hard way—that is, by evaluating the integral and then taking the derivative.

$$\int_1^x t^2\,dt = \left[\frac{t^3}{3}\right]_1^x = \frac{x^3}{3} - \frac{1}{3}$$

Therefore,

$$\frac{d}{dx}\left[\int_1^x t^2\,dt\right] = \frac{d}{dx}\left[\frac{x^3}{3} - \frac{1}{3}\right] = x^2$$

Now we do it the easy way. By Theorem D,

$$\frac{d}{dx}\left[\int_1^x t^2\,dt\right] = x^2 \qquad ◄$$

EXAMPLE 3: Find $\dfrac{d}{dx}\left[\displaystyle\int_2^x \frac{t^{\frac{3}{2}}}{\sqrt{t^2 + 17}}\,dt\right]$.

SOLUTION: We challenge anyone to do this example by first evaluating the integral. However, by Theorem D, it is a trivial problem:

$$\frac{d}{dx}\left[\int_2^x \frac{t^{\frac{3}{2}}}{\sqrt{t^2 + 17}}\,dt\right] = \frac{x^{\frac{3}{2}}}{\sqrt{x^2 + 17}} \qquad ◄$$

EXAMPLE 4: Find $\dfrac{d}{dx}\left[\displaystyle\int_x^4 \tan^2 u \cos u \; du\right], \dfrac{\pi}{2} < x < \dfrac{3\pi}{2}.$

SOLUTION: Use of the dummy variable u rather than t should not bother anyone. However, the fact that x is the lower limit, rather than the upper limit, is troublesome. Here is how we handle this difficulty.

$$\frac{d}{dx}\left[\int_x^4 \tan^2 u \cos u \; du\right] = \frac{d}{dx}\left[-\int_4^x \tan^2 u \cos u \; du\right]$$

$$= -\frac{d}{dx}\left[\int_4^x \tan^2 u \cos u \; du\right] = -\tan^2 x \cos x$$

The interchange of the upper and lower limits is allowed if we prefix a minus sign (see the definition on p. 313).   ◀

EXAMPLE 5: Find $\dfrac{d}{dx}\left[\displaystyle\int_0^{x^2} (3t - 1)\right] dt.$

SOLUTION: Now we have a new complication; the upper limit is x^2 rather than x (we need x there in order to apply Theorem D). This problem is handled by the Chain Rule. We may think of the expression in brackets as

$$\int_0^u (3t - 1)\,dt \qquad \text{where } u = x^2$$

By the Chain Rule, the derivative with respect to x of this composite function is

$$\frac{d}{dx}\left[\int_0^u (3t - 1)\,dt\right] \cdot \frac{du}{dx} = (3u - 1)(2x) = (3x^2 - 1)(2x) = 6x^3 - 2x$$

If you are skeptical of what we have just done, try doing the problem by first evaluating the integral.   ◀

EXAMPLE 6: Find $\dfrac{d}{dx}\left[\displaystyle\int_{2x}^5 \sqrt{t^2 + 2}\; dt\right].$

SOLUTION: Here we first interchange limits and then use Theorem D in conjunction with the Chain Rule.

$$\frac{d}{dx}\left[\int_{2x}^5 \sqrt{t^2 + 2}\; dt\right] = \frac{d}{dx}\left[-\int_5^{2x} \sqrt{t^2 + 2}\; dt\right]$$

$$= \left(-\sqrt{(2x)^2 + 2}\right)(2) = -2\sqrt{4x^2 + 2}   ◀$$

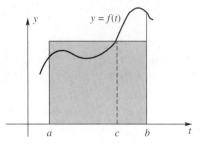

Figure 7

The Mean Value Theorem for Integrals

By this time you realize that the Mean Value Theorem for Derivatives plays an important role in calculus. There is a theorem by the same name for integrals, which is of less importance but still worth knowing. Geometrically, it says that in Figure 7, the area under the curve is equal to the area of the shaded rectangle.

Estimating Integrals

Theorem E with its accompanying Figure 7 suggests a good way to estimate the value of a definite integral. The area of the region under a curve is equal to the area of a rectangle. One can make a good guess at this rectangle by simply "eyeballing" the region. In Figure 7, the area of the shaded part above the curve should match the area of the white part below the curve.

Theorem E

(Mean Value Theorem for Integrals).   If f is continuous on $[a, b]$, there is a number c between a and b such that

$$\int_a^b f(t)\, dt = f(c)(b - a)$$

Proof   Let

$$G(x) = \int_a^x f(t)\, dt \qquad a \le x \le b$$

By the Mean Value Theorem for Derivatives applied to G, there is a point c in (a, b) such that

$$G(b) - G(a) = G'(c)(b - a)$$

that is,

$$\int_a^b f(t)\, dt = G'(c)(b - a)$$

But by Theorem D, $G'(c) = f(c)$. The conclusion follows.   ◄

Note that if we solve for $f(c)$ in the conclusion of Theorem E, we get

$$f(c) = \frac{\displaystyle\int_a^b f(t)\, dt}{b - a}$$

The number $\displaystyle\int_a^b f(x)\, dx/(b - a)$ is called the mean value, or **average value**, of f on $[a, b]$.

To see why it has this name, consider a regular partition $a = x_0 < x_1 < x_2 < \cdots < x_n = b$ with $\Delta x = (b - a)/n$. The average of the n values $f(x_1), f(x_2), \ldots, f(x_n)$ is

$$\frac{f(x_1) + f(x_2) + \cdots + f(x_n)}{n} = \sum_{i=1}^{n} f(x_i)\frac{1}{n}$$

$$= \frac{1}{b - a}\sum_{i=1}^{n} f(x_i)\frac{b - a}{n}$$

$$= \frac{1}{b - a}\sum_{i=1}^{n} f(x_i)\Delta x$$

The sum in the last expression is a Riemann sum for f on $[a, b]$ and therefore approaches $\int_a^b f(x)\,dx$ as $n \to \infty$. Thus $\dfrac{1}{b-a}\left(\int_a^b f(x)\,dx\right)$ appears as the natural extension of the familiar notion of average value.

Concepts Review

1. $\displaystyle\int_1^6 f(x)\,dx = \int_1^8 f(x)\,dx + \int_8^6 f(x)\,dx$ by the

_____ Property, where $a =$ _____ .

2. Because $4 \le x^2 \le 16$ for all x in $[2, 4]$, the Boundedness Property of the definite integral allows us to

say _____ $\le \displaystyle\int_2^4 x^2\,dx \le$ _____ .

3. $\dfrac{d}{dx}\left[\displaystyle\int_1^x \sin^3 t\,dt\right] =$ _____ , but

$\dfrac{d}{dx}\left[\displaystyle\int_1^{x^2} \sin^3 t\,dt\right] =$ _____ .

4. According to the Mean Value Theorem for Integrals, there is a number c between -1 and 3 such that

$\displaystyle\int_{-1}^3 x^2\,dx =$ _____ .

Problem Set 5.6

In Problems 1–6, use the interval additive property to evaluate $\displaystyle\int_0^4 f(x)\,dx$. Begin by drawing a graph of f.

1. $f(x) = \begin{cases} \sqrt{x} & x < 1 \\ x^2 & x \ge 1 \end{cases}$

2. $f(x) = \begin{cases} 1 & 0 \le x < 1 \\ x & 1 \le x < 2 \\ 4 - x & 2 \le x \le 4 \end{cases}$

3. $f(x) = \begin{cases} e^x & 0 \le x < 2 \\ e^2 & 2 \le x \le 4 \end{cases}$

4. $f(x) = |x - 2|$

5. $f(x) = |\cos x|$

6. $f(x) = |e^x - 2|$

In Problems 7–16, find $G'(x)$.

7. $G(x) = \displaystyle\int_{-6}^x (2t + 1)\,dt$

8. $G(x) = \displaystyle\int_0^x (t^2 + t)\,dt$

9. $G(x) = \displaystyle\int_1^x e^{-t^2}\,dt$

10. $G(x) = \displaystyle\int_0^x \sin^4 u \tan u\,du,\ -\pi/2 < x < \pi/2$

11. $G(x) = \displaystyle\int_x^{\pi/4} u \tan u\,du,\ -\pi/2 < x < \pi/2$

12. $G(x) = \displaystyle\int_x^1 x^2\sqrt{u^2 + 1}\,du.$ (Be careful.)

13. $G(x) = \displaystyle\int_1^{2+\sin x} \left(\dfrac{2}{u} + \cos u\right) du$

14. $G(x) = \displaystyle\int_0^{x^2+1} \sqrt{2 + \sin v}\,dv$

15. $G(x) = \displaystyle\int_x^{x^3} \sqrt{1 + s^4}\,ds.$ Hint: $\displaystyle\int_x^{x^3} = \int_x^0 + \int_0^{x^3}.$

16. $G(x) = \displaystyle\int_{\sin x}^{\cos x} u^2\,du$

17. Show that the graph of $y = f(x)$ is concave up everywhere if

$$f(x) = \int_0^x \dfrac{s}{\sqrt{a^2 + s^2}}\,ds, \qquad a \ne 0$$

Hint: Show $f''(x) > 0$ for all real x.

18. Find the interval on which the graph of $y = f(x)$ is concave up if

$$f(x) = \int_0^x \frac{1+t}{1+t^2}\,dt$$

19. Show that $1 \le \int_0^1 \sqrt{1+x^4}\,dx \le \frac{6}{5}$. *Hint:* $1 \le \sqrt{1+x^4} \le 1 + x^4$; use the comparison property (Theorem B).

20. Let f be continuous on $[a, b]$ and thus integrable there. Show that

$$\left| \int_a^b f(x)\,dx \right| \le \int_a^b f(x)\,dx \text{ is false.}$$

Hint: $-|f(x)| \le f(x) \le |f(x)|$; use Theorem B.

In Problems 21–26, find the average value of the given function on the given interval. Use exact methods if possible; otherwise, find a numerical approximation.

21. $f(x) = 4x^3$; $[1, 3]$

22. $f(x) = \dfrac{x}{\sqrt{x^2+16}}$; $[0, 3]$

23. $f(x) = 2 + |x|$; $[-1, 3]$

24. $f(x) = \sin^2 x \cos x$; $[0, \pi/2]$

25. $f(x) = \sin(\cos(x))$; $[0, 10]$

26. $f(x) = \tan(x^4)$; $[-1, 1]$

Sketch the graph of the integrand in Problems 27–30. Then estimate the integral as suggested in the marginal box accompanying Theorem E.

27. $\displaystyle \int_0^2 2^x\,dx$

28. $\displaystyle \int_0^2 \left[1 + \sin(x^2)\right]dx$

29. $\displaystyle \int_{-1}^1 \frac{2}{1+x^2}\,dx$

30. $\displaystyle \int_{10}^{20} \left(1 + \frac{1}{x}\right)^5 dx$

In Problems 31–36, decide whether the given statement is true or false. Then justify your answer.

31. If f is continuous and $f(x) \ge 0$ for all x in $[a, b]$, then $\displaystyle \int_a^b f(x)\,dx \ge 0$.

32. If $\displaystyle \int_a^b f(x)\,dx \ge 0$, then $f(x) \ge 0$ for all x in $[a, b]$.

33. If $\displaystyle \int_a^b f(x)\,dx = 0$, then $f(x) = 0$ for all x in $[a, b]$.

34. If $f(x) \ge 0$ and $\displaystyle \int_a^b f(x)\,dx = 0$, then $f(x) = 0$ for all x in $[a, b]$.

35. If $\displaystyle \int_a^b f(x)\,dx > \int_a^b g(x)\,dx$, then $\displaystyle \int_a^b [f(x) - g(x)]\,dx > 0$.

36. If f and g are continuous and $f(x) > g(x)$ for all x in $[a, b]$, then $\displaystyle \left| \int_a^b f(x)\,dx \right| > \left| \int_a^b g(x)\,dx \right|$.

37. Let $s(t)$ and $v(t)$ denote the position and velocity at time t of an object moving along a coordinate line. In Section 1.3, we defined the average velocity on the interval $[a, b]$ as $[s(b) - s(a)]/(b - a)$; in this section, we defined it as $\displaystyle \int_a^b v(t)\,dt / (b - a)$. Show that the two definitions are equivalent.

38. Give an alternative proof of the Fundamental Theorem of Calculus (Theorem 5.5A) as follows: Let $G(x) = \displaystyle \int_a^x f(t)\,dt$ and let F be any antiderivative of f. Then $F'(x) = G'(x)$. What can we conclude about $G(x) - F(x)$? Now evaluate $G(a) - F(a)$ and then $G(b) - F(b)$.

Answers to Concepts Review: 1. Interval Additive; 8
2. 8; 32   3. $\sin^3 x$; $2x \sin^3(x^2)$ 4. $4c^2$

5.7 | **Substitution**

Up to this point we have attacked integration problems that did not exactly fit any of our integration rules using the method of guess-and-check. The integration rules that we have were used as guidelines to making a reasonable guess at an antiderivative, then we checked the guess by taking its derivative. If necessary, adjustments were made to the original guess, and the process was repeated until a correct antiderivative was found.

The method of integration that we call substitution is just a slightly more formal approach to the same group of problems. It also illustrates a principle in mathematics that is useful in many other contexts: reducing a problem to a simpler problem by making a change of variable.

The Method of Substitution   Consider the problem of finding

$$\int (4x + 6)\cos(x^2 + 3x)\,dx$$

If we make the substitution

$$u = x^2 + 3x$$

we can replace the expression $\cos(x^2 + 3x)$ with the simpler expression $\cos u$. Our integral now becomes

$$\int (4x + 6)\cos u \, dx$$

This integral doesn't really make sense however; x and u are related by the equation $u = x^2 + 3x$, so neither can be considered a constant. We are changing the variable of the problem from x to u, but we can't stop halfway. We must replace *all* the expressions in the integral that involve x with equivalent expressions that involve u.

The key to making the remaining replacements or substitutions is to take the derivative of the equation $u = x^2 + 3x$. The derivative is $\frac{du}{dx} = 2x + 3$, which we can rewrite using the notation of differentials as $du = (2x + 3)dx$. The right-hand side *almost* matches the remaining x-terms $(4x + 6)dx$ in the integral above. To get an exact match we can multiply both sides of the equation $du = (2x + 3)dx$ by 2 to get

$$2du = (4x + 6)dx$$

Switching the order of the first two expressions in the original integral makes it easy to see that the two substitutions

$$x^2 + 3x \rightarrow u$$

$$(4x + 6)dx \rightarrow 2du$$

result in the transformation

$$\int \cos(x^2 + 3x)(4x + 6)\,dx \rightarrow \int \cos u \cdot 2du = \int 2\cos u \, du$$

Clearly the new integration problem involving u is easier to handle than the original problem involving x. We have

$$\int 2\cos u \, du = 2\sin u + C$$

Our final step is to replace u with $x^2 + 3x$ from the original substitution equation, so that the final answer is in terms of the original variable x. We can summarize all of the above steps as follows:

Let $u = x^2 + 3x$, so that $du = (2x + 3)dx$ and therefore $2du = (4x + 6)dx$. Then

$$\int (4x + 6)\cos(x^2 + 3x)dx = \int 2 \cos u \, du$$

$$= 2 \sin u + C$$

$$= 2 \sin(x^2 + 3x) + C$$

This integral could also have been computed using guess-and-check (try it!). The decision whether to use substitution or guess-and-check is somewhat a matter of taste. For integrals where a good guess is fairly obvious, guess-and-check is usually quicker and easier. Substitution is effective when you have no clear guess at an antiderivative.

Summary of the Method of Substitution   We now summarize the steps involved in using substitution to evaluate an indefinite integral. We assume that the variable of integration is x.

(1)   Choose a substitution of the form $u = g(x)$. Take the derivative of the substitution equation and solve for the differential $du = g'(x)dx$.

(2)   Use the two equations in (1) to transform the variable of the integral from x to u. One expression can be substituted for another *only* if they are equal.

(3)   Evaluate the integral in terms of u, if possible. If the integral is still difficult to evaluate, go back to (1) and try another substitution.

(4)   Use the substitution equation $u = g(x)$ to transform your result from (3) from an expression involving u back into an expression involving x.

The most difficult part of a substitution problem is often deciding on the substitution equation $u = g(x)$. One class of integrals on which the method of substitution works well are integrals of the form $\int f(g(x))g'(x)dx$. The next theorem shows why substitution works in this case.

Theorem A

(Substitution Rule for Indefinite Integrals). Let g be a differentiable function and suppose that F is an antiderivative of f. Then if $u = g(x)$,

$$\int f(g(x))g'(x)dx = \int f(u)du = F(u) + C = F(g(x)) + C$$

Proof   It is enough to show that the derivative of the right member is the integrand of the left member. But that is a simple application of the Chain Rule combined with the fact that $F' = f$.

$$\frac{d}{dx}[F(g(x)) + C] = F'(g(x))g'(x) = f(g(x))g'(x) \qquad \blacktriangleleft$$

Take another look at the integration problem $\int (4x + 6)\cos(x^2 + 3x)dx$ that we used to introduce the method of substitution. The expression $4x + 6$ is exactly two times the derivative of $x^2 + 3x$, the argument of the cos function. Thus, we can rewrite the integral as

$2 \int (2x + 3)\cos(x^2 + 3x)\,dx$ so that Theorem A can be used; we have $f(u) = \cos u$ and $g(x) = x^2 + 3x$.

Theorem A and the example that we used above lead us to the following guideline for choosing a substitution equation.

A Guideline for Choosing Substitutions   Check the integrand for an elementary function whose argument is some expression $g(x)$, such as $\cos g(x)$ or $(g(x))^n$. Let $u = g(x)$. Then check to see if the derivative $g'(x)$ matches the rest of the integrand, or at worst, differs from it by a constant multiple.

EXAMPLE 1: Find $\int xe^{-x^2}\,dx$.

> **SOLUTION:** The argument of the exponential function is $-x^2$, so we let $u = -x^2$. Thus, $du = -2x\,dx$ and so
>
> $$\int xe^{-x^2}\,dx = -\frac{1}{2}\int e^{-x^2}(-2x)\,dx$$
>
> $$= -\frac{1}{2}\int e^u\,du$$
>
> $$= -\frac{1}{2}e^u + C$$
>
> $$= -\frac{1}{2}e^{-x^2} + C \qquad\qquad \blacktriangleleft$$

EXAMPLE 2: Evaluate $\displaystyle\int_0^{\sqrt{\pi/2}} x\,\sin^3(x^2)\cos(x^2)\,dx$.

> **SOLUTION:** Recall that $\sin^3(x^2)$ means $(\sin(x^2))^3$. Thus, we let $u = \sin(x^2)$, so that $du = 2x\,\cos(x^2)\,dx$. Then
>
> $$\int x\,\sin^3(x^2)\cos(x^2)\,dx = \frac{1}{2}\int \sin^3(x^2)\cdot 2x\,\cos(x^2)\,dx$$
>
> $$= \frac{1}{2}\int u^3\,du$$
>
> $$= \frac{1}{2}\frac{u^4}{4} + C$$
>
> $$= \frac{1}{8}\sin^4(x^2) + C$$

Then, by the Fundamental Theorem of Calculus,

$$\int_0^{\sqrt{\pi/2}} x\,\sin^3(x^2)\cos(x^2)\,dx = \left[\frac{1}{8}\sin^4(x^2)\right]_0^{\sqrt{\pi/2}}$$

$$= \frac{1}{8}\sin^4\left(\frac{\pi}{4}\right) - \frac{1}{8}\cdot 0 = \frac{1}{32} \qquad\qquad \blacktriangleleft$$

Note that in the two-step procedure illustrated in Example 2, we must be sure to express the indefinite integral in terms of x before we apply the Fundamental Theorem. This is because the limits 0 and $\sqrt{\pi}/2$ apply to x, not u. But what if, in making the substitution $u = \sin(x^2)$, we also made the corresponding changes in the limits of integration to

$u = \sin(0^2) = 0$ and $u = \sin\left[(\sqrt{\pi}/2)^2\right] = \sqrt{2}/2$? Could we then finish the integration with u as the variable? The answer is yes.

$$\int_0^{\sqrt{\pi/2}} x\sin^3(x^2)\cos(x^2)\,dx = \frac{1}{2}\int_0^{\sqrt{2}/2} u^3\,du = \left[\frac{1}{2}\frac{u^4}{4}\right]_0^{\sqrt{2}/2} = \frac{1}{32}$$

Here is the general result, which lets us substitute in the limits of integration as well as in the integrand, thereby producing a one-step procedure.

Theorem B

(Substitution Rule for Definite Integrals). Let g have a continuous derivative on $[a, b]$ and let f be continuous on the range of g. Then

$$\int_a^b f(g(x))g'(x)\,dx = \int_{g(a)}^{g(b)} f(u)\,du$$

Proof   Let F be an antiderivative of f (the existence of F is guaranteed by Theorem 5.6D). Then by the Fundamental Theorem of Calculus,

$$\int_{g(a)}^{g(b)} f(u)\,du = [F(u)]_{g(a)}^{g(b)} = F(g(b)) - F(g(a))$$

On the other hand, by using the Substitution Theorem for Indefinite Integrals (Theorem A),

$$\int f(g(x))g'(x)\,dx = F(g(x)) + C$$

and so, again by the Fundamental Theorem of Calculus,

$$\int_a^b f(g(x))g'(x)\,dx = [F(g(x))]_a^b = F(g(b)) - F(g(a)) \qquad \blacktriangleleft$$

EXAMPLE 3: Evaluate $\displaystyle\int_0^1 \frac{x+1}{(x^2+2x+6)^2}\,dx$.

SOLUTION: Let $u = x^2 + 2x + 6$, so $du = (2x + 2)\,dx = 2(x + 1)\,dx$, and note that $u = 6$ when $x = 0$ and $u = 9$ when $x = 1$. Thus,

$$\int_0^1 \frac{x+1}{(x^2+2x+6)^2}\,dx = \frac{1}{2}\int_0^1 \frac{2x+2}{(x^2+2x+6)^2}\,dx$$

$$= \frac{1}{2}\int_6^9 u^{-2}\,du = \left[-\frac{1}{2}\frac{1}{u}\right]_6^9$$

$$= -\frac{1}{18} - \left(-\frac{1}{12}\right) = \frac{1}{36} \qquad \blacktriangleleft$$

SOLUTION: Let $u = \sqrt{x}$, so $du =$

EXAMPLE 4:   Evaluate $\displaystyle\int_{\pi^2/9}^{\pi^2/4} \frac{\cos\sqrt{x}}{\sqrt{x}}\,dx$.

SOLUTION: Let $u = \sqrt{x}$, so $du = dx/(2\sqrt{x})$. Thus,

$$\int_{\pi^2/9}^{\pi^2/4} \frac{\cos\sqrt{x}}{\sqrt{x}}\,dx = 2\int_{\pi^2/9}^{\pi^2/4} \cos\sqrt{x}\cdot\frac{1}{2\sqrt{x}}\,dx$$

$$= 2\int_{\pi/3}^{\pi/2} \cos u\,du$$

$$= \left[2\sin u\right]_{\pi/3}^{\pi/2} = 2 - \sqrt{3}$$

Note the change in the limits of integration at the second equality. When $x = \pi^2/9$, $u = \sqrt{\pi^2/9} = \pi/3$; when $x = \pi^2/4$, $u = \pi/2$.   ◄

EXAMPLE 5: Find $\displaystyle\int \frac{x}{x^2+1}\,dx$.

SOLUTION: Let $u = x^2 + 1$. Then $du = 2x\,dx$ and so

$$\int \frac{x}{x^2+1}\,dx = \frac{1}{2}\int \frac{x}{x^2+1}\,2x\,dx$$

$$= \frac{1}{2}\int \frac{1}{u}\,du$$

$$= \frac{1}{2}\ln|u| + C$$

$$= \frac{1}{2}\ln(x^2+1) + C$$ ◄

EXAMPLE 6: Find $\displaystyle\int_0^1 x^2 e^{-x^2}\,dx$.

SOLUTION: Based on our experiences in choosing substitution equations so far, a reasonable guess at a substitution would be $u = -x^2$. Then $du = -2x\,dx$. The problem here is that the factor $-2x$ in the formula for du cannot be made to match the factor x^2 in the integration problem simply by multiplying one or the other by a constant.

It is important to be fairly quick to recognize problems where substitution is likely to work, and problems where it is not likely to work. Also, keep in mind that it is easy to come up with a function that has no antiderivative in terms of elementary functions, and hence for which substitution is unlikely to be of much help. The previous examples in this section all followed the pattern described in Theorem A; this example does not.

We must resort at this point either to a numerical estimate of the integral, or to an exact solution by computer algebra. The computer algebra system Maple gives the exact result

$$-\frac{1}{2}\cos(1) + \frac{1}{4}\sqrt{2}\,\sqrt{\pi}\,\mathrm{FresnelC}\!\left(\frac{\sqrt{2}}{\sqrt{\pi}}\right)$$

which is not really of much use if we are not familiar with the special function FresnelC. A numerical estimate of the integral can be found as described in Section 5.4; the result is

$$\int_0^1 x^2 e^{-x^2}\, dx \approx 0.182111$$

accurate to six digits.

Concepts Review

1. Under the substitution $u = x^3 + 1$, the definite integral $\int_0^1 x^2(x^3 + 1)^4\, dx$ transforms to the new definite integral _____ .

2. Substitution would probably be _____ (helpful, unhelpful) for finding the integral $\int x^2 e^{x^3}\, dx$.

3. Substitution would probably be _____ (helpful, unhelpful) for finding the integral $\int x^3 e^{x^3}\, dx$.

4. A good substitution choice for the integral $\int x \cos(x^2)\, dx$ would be _____ .

Problem Set 5.7

Use the method of substitution to find each of the following indefinite integrals. Check using a computer algebra system if one is available.

1. $\displaystyle \int \sqrt{3x + 2}\, dx$

2. $\displaystyle \int \sqrt[3]{2x - 4}\, dx$

3. $\displaystyle \int \cos(3x + 2)\, dx$

4. $\displaystyle \int \sin(2x - 4)\, dx$

5. $\displaystyle \int x e^{x^2 + 4}\, dx$

6. $\displaystyle \int x^2 e^{x^3 + 5}\, dx$

7. $\displaystyle \int x \sin(x^2 + 4)\, dx$

8. $\displaystyle \int x^2 \cos(x^3 + 5)\, dx$

9. $\displaystyle \int \frac{x}{x^2 + 4}\, dx$

10. $\displaystyle \int \frac{x^2}{x^3 + 5}\, dx$

11. $\displaystyle \int \frac{\ln x}{x}\, dx$

12. $\displaystyle \int x^2 \sin(x^3 + 5) e^{\cos(x^3 + 5)}\, dx$

13. $\displaystyle \int \frac{(\sqrt{t} + 4)^3}{\sqrt{t}}\, dt$

14. $\displaystyle \int \left(1 + \frac{1}{t}\right)^{-2} \left(\frac{1}{t^2}\right) dt$

Try the method of substitution to evaluate each of the following definite integrals; if you cannot get the method to work, explain why. Also find each result using a numerical technique (midpoint Riemann sums, or a built-in method).

15. $\displaystyle \int_0^1 (3x + 1)^3\, dx$

16. $\displaystyle \int_0^4 \sqrt{2t + 1}\, dt$

17. $\displaystyle \int_0^2 \frac{t}{(t^2 + 9)}\, dt$

18. $\displaystyle \int_0^{\sqrt{5}} x e^{9 - x^2}\, dx$

19. $\displaystyle \int_0^1 \frac{x}{(x^3 + 4x + 1)^2}\, dx$

20. $\displaystyle\int_0^2 \frac{x^2}{(9-x^3)^{3/2}}\,dx$

21. $\displaystyle\int_0^{\pi/6} \sin^3\theta\cos\theta\,d\theta$

22. $\displaystyle\int_0^{\pi/6} \frac{\theta}{\cos^3\theta}\,d\theta$

23. $\displaystyle\int_0^1 e^x(e^x+1)^{10}\,dx$

24. $\displaystyle\int_1^{10} \frac{\sqrt{\ln x}}{x}\,dx$

25. $\displaystyle\int_0^1 x\sin(\pi x^2)\,dx$

26. $\displaystyle\int_0^{\pi/4} (\cos 2x+\sin 2x)\,dx$

27. $\displaystyle\int_0^1 x^2\sin(\pi x^2)\,dx$

28. $\displaystyle\int_0^1 x\cos^3(x^2)\sin(x^2)\,dx$

29. $\displaystyle\int_1^4 \frac{1}{\sqrt{t}(\sqrt{t}+1)^3}\,dt$

30. $\displaystyle\int_1^2 \left(1+\frac{1}{t}\right)^2\left(\frac{1}{t^2}\right)\,dt$

31. Prove (by a substitution) that

$$\int_a^b f(-x)\,dx = \int_{-b}^{-a} f(x)\,dx$$

32. Calculate $\displaystyle\int_0^{4\pi} |\sin 2x|\,dx$.

33. The temperature T on a certain day satisfied

$$T(t) = 70 + 8\sin\left[\frac{\pi}{12}(t-9)\right]$$

where t was the number of hours after midnight. Find the average temperature from 6 A.M. to 6 P.M.

34. Use the substitution $u = a + b - x$ to show that

$$\int_a^b f(x)\,dx = \int_a^b f(a+b-x)\,dx$$

Deduce that

$$\int_0^{\pi/2} \sin^n x\,dx = \int_0^{\pi/2} \cos^n x\,dx$$

Answers to Concepts Review: 1. $\frac{1}{3}\int_1^2 u^4\,du$ 2. Helpful

3. Unhelpful 4. $u = x^2$

5.8 | Chapter Review

Concepts Test

Respond with true or false to each of the following assertions. Be prepared to justify your answer.

1. The indefinite integral is a linear operator.

2. $\displaystyle\int [f(x)g'(x) + g(x)f'(x)]\,dx = f(x)g(x) + C.$

3. $y = \cos x$ is a solution to the differential equation $(dy/dx)^2 = 1 - y$.

4. $\displaystyle\int f'(x)\,dx = f(x)$ for every differentiable function f.

5. If $s = -16t^2 + v_0 t$ gives the height at time t of a ball thrown straight from the surface of the earth, then the ball will hit the ground with velocity $-v_0$.

6. $\displaystyle\sum_{i=1}^{n}(a_i + a_{i-1}) = a_0 + a_n + 2\sum_{i=1}^{n-1}a_i.$

7. If $\displaystyle\sum_{i=1}^{10} a_i^2 = 100$ and $\displaystyle\sum_{i=1}^{10} a_i = 20$, then $\displaystyle\sum_{i=1}^{10}(a_i+1)^2 = 150$.

8. Any definite integral that can be found exactly can also be numerically approximated.

9. Any definite integral that can be numerically approximated can also be found exactly.

10. If f is integrable on $[a, b]$, then f must be bounded on $[a, b]$.

11. If $\displaystyle\int_a^b f(x)\,dx = 0$, then $f(x) = 0$ for all x in $[a, b]$.

12. If $\displaystyle\int_a^b [f(x)]^2\,dx = 0$, then $f(x) = 0$ for all x in $[a, b]$.

13. When using Riemann sums to approximate definite integrals, left-endpoint Riemann sums generally converge more quickly than midpoint Riemann sums.

14. Because midpoint Riemann sums are a second-order convergent method of numerical integration, doubling the number of rectangles should reduce the error of the approximation by about $\frac{1}{4}$.

15. $\displaystyle\int_1^5 \sin^2 x \, dx = \int_1^7 \sin^2 x \, dx + \int_7^5 \sin^2 x \, dx.$

16. $\displaystyle\frac{d}{dx}\left[\int_0^{x^2} \frac{1}{1+t^2} \, dt\right] = \frac{1}{1+t^4}.$

17. $\displaystyle\int_0^3 |x-2| \, dx = \int_2^3 (x-2) \, dx - \int_0^2 (x-2) \, dx.$

18. $\displaystyle\int_0^{2\pi} |\sin x| \, dx = \int_0^{2\pi} |\cos x| \, dx.$

19. $\displaystyle\int_0^{\pi/2} |\sin x| \, dx = 4\int_0^{\pi/2} \sin x \, dx.$

20. If $f(x) = 4$ on $[0, 3]$, then every Riemann sum for f on the given interval has the value 12.

21. If $F'(x) = G'(x)$, for all x in $[a, b]$, then $F(b) - F(a) = G(b) - G(a)$.

22. If $F'(x) = f(x)$, for all x in $[0, b]$, then $\displaystyle\int_0^b f(x) \, dx = F(b)$.

23. $\displaystyle\int_{-99}^{99} (ax^3 + bx^2 + cx) \, dx = 2\int_0^{99} bx^2 \, dx.$

24. If $f(x) \leq g(x)$, on $[a, b]$, then $\displaystyle\int_a^b |f(x)| \, dx \leq \int_a^b |g(x)| \, dx.$

25. $\displaystyle\left|\sum_{i=1}^n a_i\right| \leq \sum_{i=1}^n |a_i|.$

26. If f is continuous on $[a, b]$, then $\displaystyle\left|\int_a^b f(x) \, dx\right| \leq \int_a^b |f(x)| \, dx.$

27. $\displaystyle\lim_{n\to\infty} \sum_{i=1}^n \sin\left(\frac{2i}{n}\right) \cdot \frac{2}{n} = \int_0^2 \sin x \, dx.$

28. $\displaystyle\lim_{n\to\infty} \sum_{i=1}^n \sin\left(\frac{2i}{n}\right) \cdot \frac{2}{n} = \lim_{n\to\infty} \sum_{i=1}^n \sin\left(\frac{2i-1}{n}\right) \cdot \frac{2}{n}.$

Sample Test Problems

In Problems 1–6, find the indicated integrals. Choose between exact and numerical methods, and explain why you made the choice you did.

1. $\displaystyle\int_0^1 (x^3 - 3x^2 + 3\sqrt{x}) \, dx$

2. $\displaystyle\int \frac{2x^4 - 3x^2 + 1}{x^2} \, dx$

3. $\displaystyle\int_0^2 \frac{t^3}{t^4 + 9} \, dt$

4. $\displaystyle\int_0^{\pi/2} \cos^4 x \sin x \, dx$

5. $\displaystyle\int_0^2 e^{x^2} \, dx$

6. $\displaystyle\int_0^2 x e^{x^2} \, dx$

In Problems 7–10, solve the differential equation subject to the indicated condition.

7. $\displaystyle\frac{dy}{dx} = \frac{1}{\sqrt{x+1}}; y = 18$ at $x = 3$

8. $\displaystyle\frac{dy}{dt} = \sqrt{2t-1}; y = -1$ at $t = \frac{1}{2}$

9. $\displaystyle\frac{dy}{dt} = e^t y^4; y = 1$ at $t = 1$

10. $\displaystyle\frac{dy}{dx} = \frac{6x - x^3}{2y}; y = 3$ at $x = 0$

11. Find the equation of the curve through $(-2, -\frac{1}{3})$ if its slope at each x is the negative reciprocal of the slope of the curve with equation $xy = 2$.

12. If a particle moving on the x-axis has acceleration $a = 15\sqrt{t} + 8$ at time t and if $v_0 = -6$, $x_0 = -44$, find its position x at $t = 4$. Assume x is measured in feet and t in seconds.

13. A ball is thrown directly upward from a tower 448 feet high with an initial velocity of 48 feet per second. In how many seconds will it strike the ground and with what velocity? Assume that $g = 32$ feet per second per second and neglect air resistance.

14. What constant acceleration will cause a car to increase its velocity from 45 to 60 miles per hour in 10 seconds?

15. Let P be a partition of the interval $[0, 2]$ into four equal subintervals, and let $f(x) = x^2 - 1$. Write the Riemann sum for f on P, in which $\bar{x}_i$ is the right endpoint of each subinterval of P, $i = 1, 2, 3, 4$. Find the value of this Riemann sum and make a sketch.

16. If $f(x) = \displaystyle\int_{-2}^x \frac{1}{t+3} \, dt$, $-2 \leq x$, find $f'(7)$.

17. Find $\displaystyle\int_0^3 (2 - \sqrt{x+1})^2 \, dx$.

18. If $f(x) = 3x^2\sqrt{x^3 - 4}$, find the average value of f on $[2, 5]$.

19. Find $\displaystyle\int_2^4 \frac{5x^2 - 1}{x^2} \, dx$.

20.   Evaluate each sum.

(a) $\displaystyle\sum_{m=2}^{4}\left(\frac{1}{m}\right)$ (b) $\displaystyle\sum_{i=1}^{6}(2-i)$

(c) $\displaystyle\sum_{k=0}^{4}\cos\left(\frac{k\pi}{4}\right)$

21.   Write in sigma notation.

(a) $\frac{1}{2}+\frac{1}{3}+\frac{1}{4}+\cdots+\frac{1}{78}$

(b) $x^2+2x^4+3x^6+4x^8+\cdots+50x^{100}$

22.   Sketch the region under the curve $y=16-x^2$ between $x=0$ and $x=3$, showing the inscribed polygon corresponding to a partition of $[0,3]$ into n equal subintervals. Find a formula for the area of this polygon and then estimate the area under the curve by taking a limit numerically.

23.   Evaluate each integral.

(a) $\displaystyle\int_{0}^{4}|x-1|\,dx$ (b) $\displaystyle\int_{0}^{4}|x^2-3x+2|\,dx$

24.   Use midpoint Riemann sums with $n=4$ and $n=12$ to estimate the integral $\displaystyle\int_{0}^{1}\tan(x^2)\,dx$. How many digits of accuracy do you believe you have obtained in your estimate of the integral? Explain your reasoning.

25.   Find c of the Mean Value Theorem for Integrals for $f(x)=3x^2$ and $[-4,-1]$.

26.   Find $G'(x)$ for each function G.

(a) $\displaystyle G(x)=\int_{1}^{x}\frac{1}{t^2+1}\,dt$ (b) $\displaystyle G(x)=\int_{1}^{x^2}\frac{1}{t^2+1}\,dt$

(c) $\displaystyle G(x)=\int_{x}^{x^2}\frac{1}{t^2+1}\,dt$

27.   Evaluate each of the following limits by recognizing it as a definite integral.

(a) $\displaystyle\lim_{n\to\infty}\sum_{i=1}^{n}\sqrt{\frac{4i}{n}}\cdot\frac{4}{n}$ (b) $\displaystyle\lim_{n\to\infty}\sum_{i=1}^{n}\left(1+\frac{2i}{n}\right)^2\frac{2}{n}$

28.   Show that if $f(x)=\displaystyle\int_{2x}^{5x}\frac{1}{t}\,dt$, then f is a constant

CHAPTER 6

Applications of the Integral

6.1 The Area of a Plane Region

The discussion of area in Section 5.3 served to motivate the definition of the definite integral. With that definition now firmly established, we reverse directions and use the definite integral to calculate areas of regions of more and more complicated shapes. As is our practice, we begin with simple cases.

A Region above the x-Axis Let $y = f(x)$ determine a curve in the xy-plane and suppose f is continuous and nonnegative on the interval $a \leq x \leq b$ (as in Figure 1). Consider the region R bounded by the graphs of $y = f(x)$, $x = a$, $x = b$, and $y = 0$. We refer to R as the region under $y = f(x)$ between $x = a$ and $x = b$. Its area $A(R)$ is given by

$$A(R) = \int_a^b f(x)\,dx$$

EXAMPLE 1: Find the area of the region R under $y = x^4 - 2x^3 + 2$ between $x = -1$ and $x = 2$.

SOLUTION: The graph of R is shown in Figure 2. A reasonable estimate for the area of R is its base times an average height, say, $(3)(2) = 6$. The exact value is

$$A(R) = \int_{-1}^{2} (x^4 - 2x^3 + 2)\,dx = \left[\frac{x^5}{5} - \frac{x^4}{2} + 2x \right]_{-1}^{2}$$

$$= \left(\frac{32}{5} - \frac{16}{2} + 4 \right) - \left(-\frac{1}{5} - \frac{1}{2} - 2 \right) = \frac{51}{10} = 5.1$$

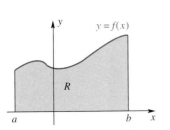

Figure 1

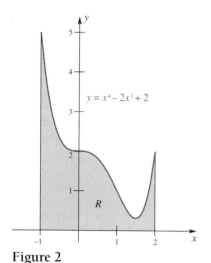

$y = x^4 - 2x^3 + 2$

R

Figure 2

The calculated value 5.1 is close enough to our estimate of 6 to give us confidence in its correctness. ◄

A Region below the *x*-axis

Area is a nonnegative number. If the graph of $y = f(x)$ is below the *x*-axis, then $\int_a^b f(x)\,dx$ is a negative number and therefore cannot be an area. However, it is just the negative of the area of the region bounded by $y = f(x)$, $x = a$, $x = b$, and $y = 0$.

EXAMPLE 2: Find the area of the region R bounded by $y = \dfrac{x^2}{3} - 4$, the *x*-axis, $x = -2$, and $x = 3$.

SOLUTION: The region R is shown in Figure 3. Our estimate for its area is $(5)(3) = 15$. We have

$$\int_{-2}^{3} \left(\frac{x^2}{3} - 4 \right) dx = \left[\frac{x^3}{9} - 4x \right]_{-2}^{3}$$

$$= \left(\frac{27}{9} - 12 \right) - \left(-\frac{8}{9} + 8 \right) = -\frac{145}{9} \approx -16.11$$

and therefore $A(R) = \dfrac{145}{9} \approx 16.11$. We are reassured by the nearness of 16.11 to our estimate. ◄

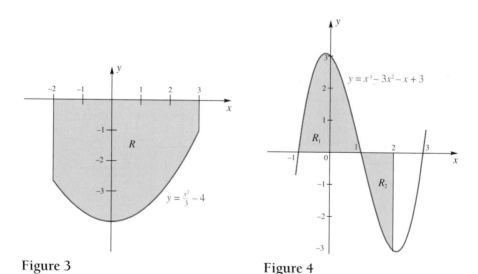

Figure 3

R

$y = \dfrac{x^2}{3} - 4$

Figure 4

$y = x^3 - 3x^2 - x + 3$

R_1

R_2

EXAMPLE 3: Find the area of the region R bounded by $y = x^3 - 3x^2 - x + 3$, the segment of the *x*-axis between $x = -1$ and $x = 2$, and the line $x = 2$.

SOLUTION: The region R is shaded in Figure 4. Note that part of it is above the *x*-axis and part is below. The areas of these two parts, R_1 and R_2, must be calculated separately. Using a calculator- or computer-generated graph along with a numerical root finder (as outlined in Section 2.5), you will find that the curve crosses the *x*-axis at -1, 1, and 3. Thus,

$$A(R) = A(R_1) + A(R_2)$$

$$= \int_{-1}^{1} (x^3 - 3x^2 - x + 3)\,dx - \int_{1}^{2} (x^3 - 3x^2 - x + 3)\,dx$$

$$= \left[\frac{x^4}{4} - x^3 - \frac{x^2}{2} + 3x\right]_{-1}^{1} - \left[\frac{x^4}{4} - x^3 - \frac{x^2}{2} + 3x\right]_{1}^{2}$$

$$= 4 - \left(-\frac{7}{4}\right) = \frac{23}{4}$$

Notice that we could have written this area as one integral using the absolute value symbol.

$$A(R) = \int_{-1}^{2} |x^3 - 3x^2 - x + 3|\,dx$$

But this is no real simplification, because in order to evaluate this integral, we have to split it into two parts just as we did above.   ◄

A Helpful Way of Thinking

So far so good. For simple regions of the type considered above, it is quite easy to write down the correct integral. When we consider more complicated regions (for example, regions between two curves), the task of selecting the right integral is more difficult. However, there is a way of thinking that can be very helpful. It goes back to the definition of area and of the definite integral. Here it is in five steps.

Step 1   Sketch the region.

Step 2   Slice it into thin pieces (strips); label a typical piece.

Step 3   Approximate the area of this typical piece, pretending it is a rectangle.

Step 4   Add up the approximations to the areas of the pieces.

Step 5   Take the limit as the width of the pieces approaches zero, thus getting a definite integral.

To illustrate, we consider yet another simple example.

EXAMPLE 4:   Set up the integral for the area of the region under $y = 1 + \sqrt{x}$ between $x = 0$ and $x = 4$ (Figure 5).

SOLUTION:

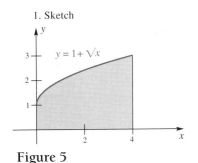

1. Sketch

$y = 1 + \sqrt{x}$

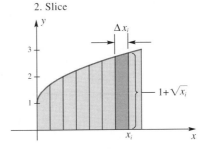

2. Slice

$1 + \sqrt{x_i}$

3. Approximate area of typical piece:

$$\Delta A_i \approx (1 + \sqrt{x_i})\,\Delta x_i$$

4. Add up: $A \approx \sum (1 + \sqrt{x_i})\,\Delta x_i$

5. Take limit: $A = \int_{0}^{4} (1 + \sqrt{x})\,dx$

Figure 5

Once we understand this five-step procedure, we can abbreviate it to three: *slice, approximate, integrate*. Think of the word *integrate* as meaning add up and take the limit as the piece width tends to zero. In this process $\sum \cdots \Delta x$ transforms into $\int \cdots dx$. Figure 6 gives the abbreviated form for the same problem.   ◀

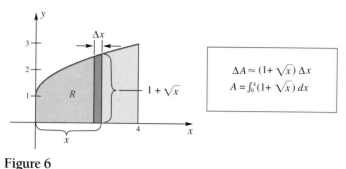

$$\Delta A \approx (1 + \sqrt{x})\,\Delta x$$
$$A = \int_0^4 (1 + \sqrt{x})\,dx$$

Figure 6

A Region between Two Curves

Consider curves $y = f(x)$ and $y = g(x)$ with $g(x) \leq f(x)$ on $a \leq x \leq b$. They determine the region shown in Figure 7. We use the *slice, approximate, integrate* method to find its area. Be sure to note that $f(x) - g(x)$ gives the correct height for the thin slice, even when the graph of g goes below the x-axis. In this case, $g(x)$ is negative; so subtracting $g(x)$ is the same as adding a positive number. You can check that $f(x) - g(x)$ also gives the correct height, even when both $f(x)$ and $g(x)$ are negative.

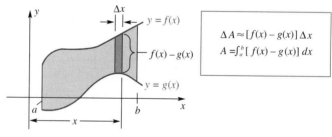

$$\Delta A \approx [f(x) - g(x)]\,\Delta x$$
$$A = \int_a^b [f(x) - g(x)]\,dx$$

Figure 7

EXAMPLE 5: Find the area of the region between the curves $y = x + 2$ and $y = e^{x^2}$.

SOLUTION: We start by finding where the two curves intersect. To do this we need to solve $e^{x^2} = x + 2$, or equivalently $e^{x^2} - x - 2 = 0$. This equation cannot be solved exactly (why?), so we must find approximate solutions by graphing and using a numerical root finder (as in Example 2 of Section 2.5). To four decimals, the approximate roots are $x \approx -0.5876$ and $x \approx 1.0571$.

One job remains—to evaluate the integral. Once again we must resort to a numerical method, because we do not have a simple antiderivative for the term e^{x^2}. We could use midpoint Riemann sums, or the built-in capability of a calculator or computer algebra system (as in Section 5.4). Either way we find that

$$\int_{-0.5876}^{1.0571} x + 2 - e^{x^2}\,dx \approx 1.385$$

Our sketch of the region, together with the corresponding integral, is shown in Figure 8.

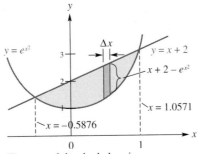

The area of the shaded region is approximated by the integral $\int_{-0.5876}^{1.0571} x + 2 - e^{x^2}\, dx \approx 1.385$.

Figure 8

How much accuracy do we have in our result? Since the limits of integration themselves have been approximated, we can't be sure that we have four-digit accuracy in our area approximation. We could recompute the integral using more accurate limits of integration to see if the answer changes; we leave that investigation to the exercises. ◀

EXAMPLE 6: **(Horizontal Slicing).** Find the area of the region between the parabola $y^2 = 4x$ and the line $4x - 3y = 4$.

SOLUTION: We will need the points of intersection of these two curves. The y-coordinates of these points can be found by writing the second equation as $4x = 3y + 4$ and then equating the two expressions for $4x$.

$$y^2 = 3y + 4$$
$$y^2 - 3y - 4 = 0$$
$$(y - 4)(y + 1) = 0$$
$$y = 4, -1$$

From this, we conclude that the points of intersection are $(4, 4)$ and $(\frac{1}{4}, -1)$. The required region is sketched in Figure 9.

Now imagine slicing this region vertically. We face a problem, because the lower boundary consists of two different curves. Slices at the extreme left extend from the lower branch of the parabola to its upper branch. For the rest of the region, slices extend from the line to the parabola. To do the problem with vertical slices requires that we first split our region into two parts, set up an integral for each part, and then evaluate both integrals.

A far better approach is to slice the region horizontally, as shown in Figure 10, thus using y rather than x as the integration variable. Note that horizontal slices always go from the parabola (at the left to the line (at the right). The width of such a slice is the largest x-value ($x = \frac{1}{4}(3y + 4)$) minus the smallest x-value ($x = \frac{1}{4}y^2$).

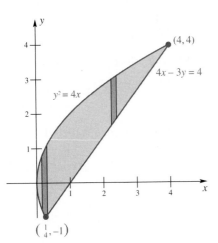

Figure 9

$$A = \int_{-1}^{4}\left[\frac{3y + 4 - y^2}{4}\right] dy = \frac{1}{4}\int_{-1}^{4}(3y + 4 - y^2)\, dy$$

$$= \frac{1}{4}\left[\frac{3y^2}{2} + 4y - \frac{y^3}{3}\right]_{-1}^{4}$$

$$= \frac{1}{4}\left[\left(24 + 16 - \frac{64}{3}\right) - \left(\frac{3}{2} - 4 + \frac{1}{3}\right)\right]$$

$$= \frac{125}{24} \approx 5.21$$

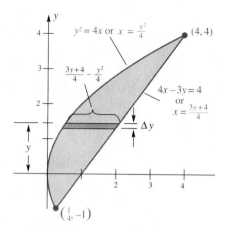

Figure 10

There are two key concepts to keep in mind here: (1) The integrand resulting from a horizontal slicing involves y, not x; and (2) to get the integrand, solve both equations for x and subtract the smaller x-value from the larger.   ◄

Distance and Displacement   Consider an object moving along a straight line with velocity $v(t)$ at time t. If $v(t) \geq 0$, then $\displaystyle\int_a^b v(t)\,dt$ gives the distance traveled during the time interval $a \leq t \leq b$. To see why, look at Figure 11. The area of the shaded rectangle is $v(\bar{t}_i)\Delta t$. If the velocity were constant, this area would represent the distance traveled by the object during the time period Δt. Because the velocity is not constant, the shaded area only approximates the distance traveled for this time period.

Summing the areas of all of the rectangles we get $\displaystyle\sum_{i=1}^{n} v(\bar{t}_i)\Delta t$, an approximation to the total distance traveled between time $t = a$ and time $t = b$. The approximation gets better as the number of rectangles, n, gets larger (and hence the base of each rectangle gets smaller).

Thus, as $n \to \infty$, the Riemann sum $\displaystyle\sum_{i=1}^{n} v(\bar{t}_i)\Delta t$ approaches both the distance traveled by the object and the area under the curve, $\displaystyle\int_a^b v(t)\,dt$.

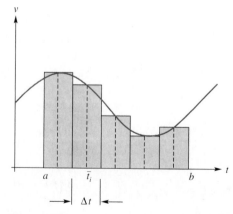

Figure 11

EXAMPLE 7: A car's speedometer measures the velocity of the car (assuming that the car is moving forward). During a car trip, the speedometer reading (v, in miles per hour) as a function of time (t, in hours) is given by the equation

$$v = 25 \cos(\pi (t - 1)) + 25$$

for $0 \leq t \leq 2$. Sketch a graph of the velocity function on the given interval and describe the trip in words. Then determine how far the car travels during the trip.

SOLUTION: The graph is shown in Figure 12. The car starts and ends the trip at zero velocity and reaches a maximum velocity of 50 miles per hour at the midpoint of the trip. At the beginning of the trip the car picks up speed very slowly (smooth start) and at the end of the trip it loses speed very slowly (smooth stop). The trip takes 2 hours to complete.

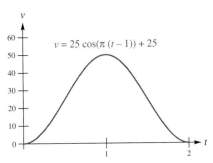

Figure 12

The total distance the car travels during the trip is given by the area under the curve between $t = 0$ and $t = 2$:

$$\int_0^2 v(t)\,dt = \int_0^2 25 \cos(\pi (t - 1)) + 25\,dt$$

$$= \left[\frac{25}{\pi} \sin(\pi (t - 1)) + 25t \right]_0^2$$

$$= \left[\frac{25}{\pi} \sin(\pi) + 50 \right] - \left[\frac{25}{\pi} \sin(-\pi) \right]$$

$$= 50$$

Because the units of velocity are $\frac{miles}{hours}$ and the units of time are *hours,* the units of distance would be $\frac{miles}{hours} \cdot hours = miles$. Thus, the car travels 50 miles.   ◄

If $v(t)$ is sometimes negative (which corresponds to the object moving in reverse), then

$$\int_a^b v(t)\,dt = s(b) - s(a)$$

measures the **displacement** of the object—that is, the directed distance from its starting position $s(a)$ to its ending position $s(b)$. To get the **total distance** that the object traveled during $a \leq t \leq b$, we must calculate $\int_a^b |v(t)|\,dt$, the area between the velocity curve and the t-axis.

Problems 31–33 illustrate these ideas.

Concepts Review

1. Let R be the region between the curve $y = f(x)$ and the x-axis on the interval $[a, b]$. If $f(x) \geq 0$ for all x in $[a, b]$, then $A(R) = $ _____, but if $f(x) \leq 0$ for all x in $[a, b]$, then $A(R) = $ _____.

2. To find the area of the region between two curves, it is wise to think of the following three-word motto: _____.

3. Suppose the curves $y = f(x)$ and $y = g(x)$ bound a region R on which $f(x) \leq g(x)$. Then the area of R is given by $A(R) = \int_a^b$ _____ dx, where a and b are determined by solving the equation _____.

4. If $v(t) \geq 0$ represents the velocity of an object, then $\int_a^b v(t)\, dt$ gives the _____ during the time interval $a \leq t \leq b$.

Problem Set 6.1

In Problems 1–10, use the three-step procedure (*slice, approximate, integrate*) to set up an integral (or integrals) for the area of the indicated region. You must decide between numerical and exact methods, both for integration and for finding intersection points when necessary.

1.

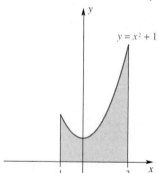

$y = x^2 + 1$

2.

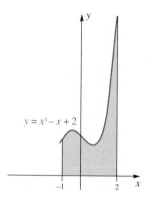

$y = x^3 - x + 2$

3.

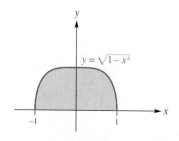

$y = \sqrt{1 - x^4}$

4.

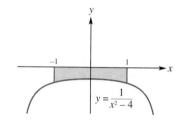

$y = \dfrac{1}{x^2 - 4}$

5.

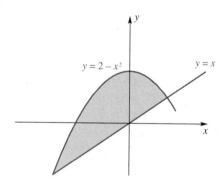

$y = 2 - x^2$ $y = x$

6.

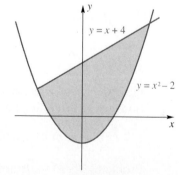

$y = x + 4$ $y = x^2 - 2$

7.

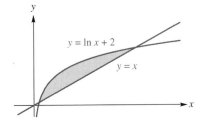

8.

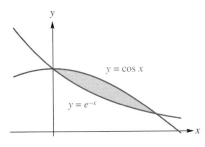

9.

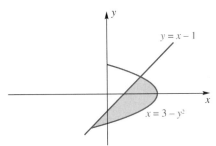

10.

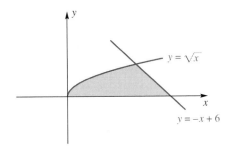

In Problems 11–28, sketch the region bounded by the graphs of the given equations, show a typical rectangle, approximate its area, set up an integral, and calculate the area of the region. Find intersection points and evaluate the integral exactly if possible, otherwise use numerical methods. Make an estimate of the area to confirm your answer.

11. $y = 4 - \frac{1}{3}x^2$, $y = 0$, between $x = 0$ and $x = 3$

12. $y = 4x - x^2$, $y = 0$, between $x = 1$ and $x = 3$

13. $y = e^{-x}$, $y = 0$, between $x = 0$ and $x = 2$

14. $y = e^{-x^2}$, $y = 0$, between $x = 0$ and $x = 2$

15. $y = \sin x$, $y = 0$, $x = -1$, $x = 2$

16. $y = \sin x^2$, $y = 0$, $x = -1$, $x = 2$

17. $y = \sqrt{x - 4}$, $y = 0$, $x = 8$

18. $y = x^2 - 4x + 3$, $x - y - 1 = 0$

19. $y = x^2$, $y = x + 2$

20. $y = 2\sqrt{x}$, $y = 2x - 4$, $x = 0$

21. $y = \sin x$, $y = x^2$

22. $y = e^{-x^2}$, $y = x^2$

23. $x = 6y - y^2$, $x = 0$

24. $x = -y^2 + y + 2$, $x = 0$

25. $x = 4 - y^2$, $x + y - 2 = 0$

26. $x = y^2 - 3y$, $x - y + 3 = 0$

27. $y^2 - 2x = 0$, $y^2 + 4x - 12 = 0$

28. $x = y^4$, $x = 2 - y^4$

29. Sketch the region R bounded by $y = x + 6$, $y = x^3$, and $2y + x = 0$. Then find its area. *Hint:* You will have to divide R into two pieces.

30. Find the area of the triangle with vertices at $(-1, 4)$, $(2, -2)$, and $(5, 1)$ by integration.

31. An object moves along a line so that its velocity at time t is $v(t) = 3t^2 + 36$ feet per second. Find the displacement and total distance traveled by the object for $-1 \le t \le 9$.

32. A car speedometer reads $v(t) = 60 \sin(\pi\sqrt{t})$ miles per hour for the time interval $0 \le t \le 1$ in hours. How far did the car travel? Sketch a graph of velocity versus time and make up a story to correspond to this graph.

33. Starting at $s = 0$ when $t = 0$, an object moves along a line so that its velocity at time t is $v(t) = 2t - 4$ centimeters per second. How long will it take to get to $s = 12$? To travel a total distance of 12 centimeters?

34. Consider the curve $y = 1/x^2$ for $1 \le x \le 6$.

(a) Calculate the area under this curve.

(b) Determine c so that the line $x = c$ bisects the area of part (a).

(c) Determine d so that the line $y = d$ bisects the area of part (a).

35. Calculate areas A, B, C, and D in Figure 13. Check by calculating $A + B + C + D$ in one integration.

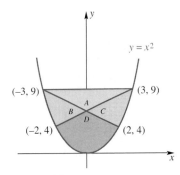

Figure 13

36. Prove Cavalieri's Principle. If two regions have the same width at every x in $[a, b]$, then they have the same area (see Figure 14).

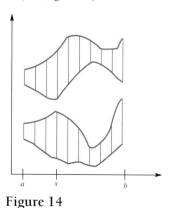

Figure 14

37. Find the area of the region trapped between $y = \sin x$ and $y = \frac{1}{2}, 0 \leq x \leq 17\pi/6$.

38. Redo Example 5 using limits of integration that are accurate to five decimal places and again using limits of integration that are accurate to six decimal places. Do you think the answer we got in Example 5 was actually accurate to four decimal places? Explain.

Answers to Concepts Review: 1. $\displaystyle\int_a^b f(x)\,dx; \ -\int_a^b f(x)\,dx$

2. Slice, approximate, integrate 3. $g(x) - f(x)$; $f(x) = g(x)$ 4. Distance traveled

6.2 | Volumes of Solids: Slabs, Disks, Washers

That the definite integral can be used to calculate areas is not surprising; it was invented for that purpose. But uses of the integral go far beyond that application. If a quantity can be thought of as a result of chopping something into small pieces, approximating each piece, adding up, and taking the limit as the pieces shrink in size, then it probably can be interpreted as a definite integral. In particular, this is true for the volumes of solids that can be cut into thin slices, where the volume of each slice is easy to approximate.

What is volume? We start with simple solids called *right cylinders,* four of which are shown in Figure 1. In each case the solid is generated by moving a plane region (the base) through a distance h in a direction perpendicular to that region. And in each case, the volume of the solid is defined to be the area A of the base times the height—that is,

$$V = A \cdot h$$

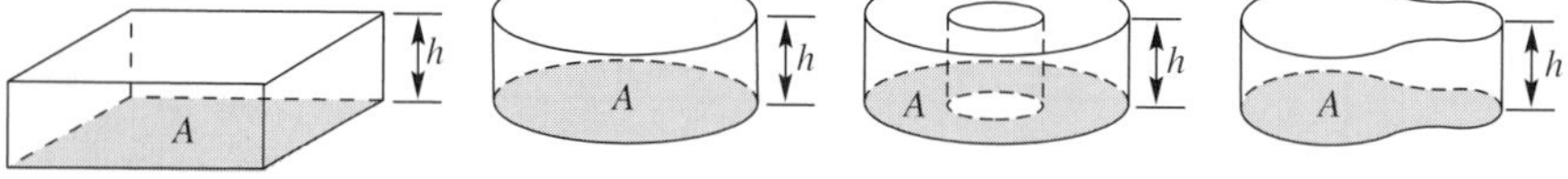

Figure 1

Next, consider a solid that has the property that cross sections perpendicular to a given line have known area. In particular, suppose that line is the x-axis and that the area of the cross section at x is $A(x), a \leq x \leq b$ (Figure 2). Partition the interval $[a, b]$ by inserting points $a = x_0 < x_1 < x_2 < \cdots < x_n = b$ and pass planes through these points perpendicular to the x-axis, thus slicing the solid into thin **slabs** (Figure 3). The "volume" ΔV_i of a slab should be approximately that of a cylinder, namely,

$$\Delta V_i \approx A(\overline{x}_i)\,\Delta x_i, \qquad x_{i-1} \leq \overline{x}_i \leq x_i$$

and the "volume" V of the solid should be given approximately by the Riemann sum

$$V \approx \sum_{i=1}^{n} A(\overline{x}_i)\,\Delta x_i$$

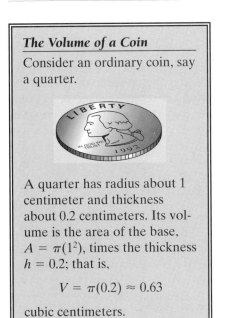

The Volume of a Coin

Consider an ordinary coin, say a quarter.

A quarter has radius about 1 centimeter and thickness about 0.2 centimeters. Its volume is the area of the base, $A = \pi(1^2)$, times the thickness $h = 0.2$; that is,

$$V = \pi(0.2) \approx 0.63$$

cubic centimeters.

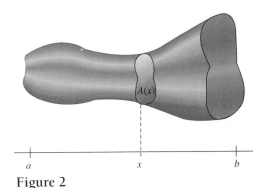

Figure 2

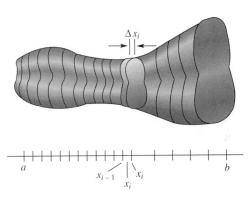

Figure 3

When we let the norm of the partition approach zero, we obtain a definite integral; this integral is defined to be the **volume** of the solid.

$$V = \int_a^b A(x)\, dx$$

Rather than routinely applying the boxed formula to obtain volumes, we suggest that in each problem you go through the process that led to it, at least in summary form. Just as for areas, we call this process *slice, approximate, integrate.* It is illustrated in the examples that follow.

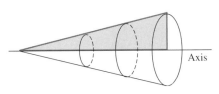

Figure 4

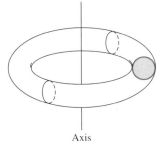

Figure 5

Solids of Revolution: Method of Disks   When a plane region, lying entirely on one side of a fixed line in its plane, is revolved about that line, it generates a **solid of revolution.** The fixed line is called the **axis** of the solid of revolution.

As an illustration, if the region bounded by a semicircle and its diameter is revolved about that diameter, it sweeps out a spherical solid. If the region inside a right triangle is revolved about one of its legs, it generates a conical solid (Figure 4). When a circular region is revolved about a line in its plane that does not intersect the circle (Figure 5), it sweeps out a torus (doughnut). In each case, it is possible to represent the volume as a definite integral.

EXAMPLE 1: Find the volume of the solid of revolution obtained by revolving the plane region R bounded by $y = \sqrt{x}$, the x-axis, and the line $x = 4$ about the x-axis.

SOLUTION: The region R, with a typical slice, is displayed below as the left part of Figure 6. When revolved about the x-axis, this region generates a solid of revolution and the slice generates a disk, a thin coin-shaped object. Recalling that the volume of a circular cylinder is $\pi r^2 h$, we approximate the volume ΔV of this disk, $\Delta V \approx \pi(\sqrt{x})^2 \Delta x$, and then integrate.

$$V = \pi \int_0^4 x\, dx = \pi \left[\frac{x^2}{2} \right]_0^4 = \pi \frac{16}{2} = 8\pi \approx 25.13 \quad \blacktriangleleft$$

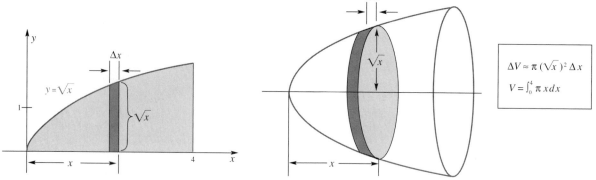

Figure 6

EXAMPLE 2: Find the volume of the solid generated by revolving the region bounded by the curve $y = x^3$, the y-axis, and the line $y = 3$ about the y-axis (Figure 7).

SOLUTION: Here we slice horizontally, which makes y the appropriate choice for the integration variable. Note that $y = x^3$ is equivalent to $x = \sqrt[3]{y}$ and $\Delta V \approx \pi(\sqrt[3]{y})^2 \Delta y$.

$$V = \pi \int_0^3 y^{\frac{2}{3}}\, dy = \pi \left[\frac{3}{5} y^{\frac{5}{3}} \right]_0^3 = \pi \frac{9\sqrt[3]{9}}{5} \approx 11.76 \quad \blacktriangleleft$$

Method of Washers   Sometimes, slicing a solid of revolution results in disks with holes in the middle. We call them **washers**. See the diagram and accompanying volume formula shown in Figure 8.

EXAMPLE 3: Find the volume of the solid generated by revolving the region bounded by the parabolas $y = x^2$ and $y^2 = 8x$ about the x-axis.

SOLUTION: The key words are still *slice, approximate, integrate* (see Figure 9).

$$V = \pi \int_0^2 (8x - x^4)\, dx = \pi \left[\frac{8x^2}{2} - \frac{x^5}{5} \right]_0^2 = \frac{48\pi}{5} \approx 30.16 \quad \blacktriangleleft$$

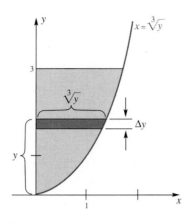

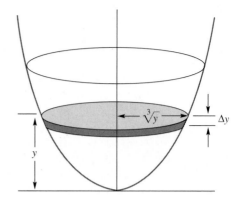

$$\Delta V \approx \pi (\sqrt[3]{y})^2 \Delta y$$
$$V = \int_0^3 \pi y^{2/3}\, dy$$

Figure 7

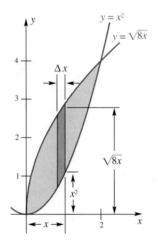

Figure 7

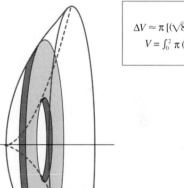

$$\Delta V \approx \pi [(\sqrt{8x})^2 - (x^2)^2]\Delta x$$
$$V = \int_0^2 \pi (8x - x^4)\, dx$$

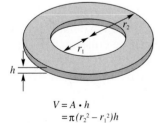

$$V = A \cdot h$$
$$= \pi (r_2{}^2 - r_1{}^2)h$$

Figure 8

EXAMPLE 4: The semicircular region bounded by $x = \sqrt{4 - y^2}$ and the y-axis is revolved about the line $x = -1$. Set up the integral that represents its volume.

SOLUTION: Here the outer radius of the washer is $\sqrt{4 - y^2} + 1$ and the inner radius is 1. Figure 10 exhibits the solution. The integral can be simplified. The part above the x-axis has the same volume as the part below it (which manifests itself in an even integrand). Thus, we may integrate from 0 to 2 and double the result. Also, the integrand itself can be simplified.

$$V = \pi \int_{-2}^{2} \left[\left(1 + \sqrt{4 - y^2} \right)^2 - 1^2 \right] dy$$

$$= 2\pi \int_{0}^{2} \left[2\sqrt{4 - y^2} + 4 - y^2 \right] dy$$

$$\approx 72.99$$

Because we do not currently have a way of finding an antiderivative for the $2\sqrt{4 - y^2}$ term in the integrand (unless we use a computer algebra system), we resorted to a numerical approximation. ◄

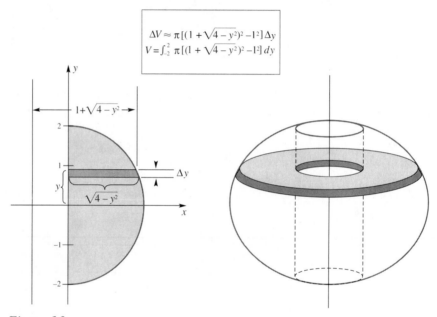

$$\Delta V \approx \pi[(1 + \sqrt{4 - y^2})^2 - 1^2]\Delta y$$
$$V = \int_{-2}^{2} \pi[(1 + \sqrt{4 - y^2})^2 - 1^2]\, dy$$

Figure 10

Other Solids with Known Cross Sections

So far, our solids have had circular cross sections. However, our method works just as well for solids whose cross sections are squares or triangles. In fact, all that is really needed is that the areas of the cross sections can be calculated, because, in this case, we can also calculate the volume of the slice—a slab—with this cross section.

EXAMPLE 5: Let the base of a solid be the first quadrant plane region bounded by $y = 1 - x^2/4$, the x-axis, and the y-axis. Suppose that cross sections perpendicular to the x-axis are squares. Find the volume of the solid.

SOLUTION: When we slice this solid perpendicularly to the x-axis, we get thin square boxes (Figure 11), like slices of cheese.

$$V = \int_{0}^{2} \left(1 - \frac{x^2}{2} + \frac{x^4}{16} \right) dx = \left[x - \frac{x^3}{6} + \frac{x^5}{80} \right]_{0}^{2}$$

$$= 2 - \frac{8}{6} + \frac{32}{80} = \frac{16}{15} \approx 1.07 \qquad \blacktriangleleft$$

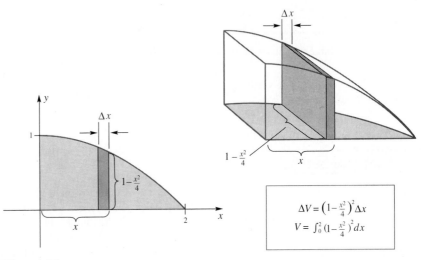

Figure 11

EXAMPLE 6: The base of a solid is the region between one arch of $y = \sin x$ and the x-axis. Each cross section perpendicular to the x-axis is an equilateral triangle sitting on this base. Find the volume of the solid.

SOLUTION: We need the fact that the area of an equilateral triangle of side u is $\sqrt{3}u^2/4$ (see Figure 12). We proceed as shown in Figure 13.

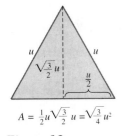

$A = \frac{1}{2}u\sqrt{\frac{3}{2}}u = \sqrt{\frac{3}{4}}u^2$

Figure 12

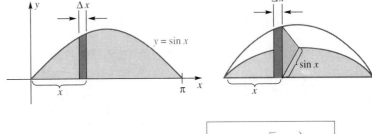

$$\Delta V \approx \left(\sqrt{\tfrac{3}{4}}\sin^2 x\right)\Delta x$$
$$A = \int_0^\pi \left(\sqrt{\tfrac{3}{4}}\sin^2 x\right) dx$$

Figure 13

To perform the indicated integration, we use the half-angle formula $\sin^2 x = (1 - \cos 2x)/2$. We could also, of course, have approximated the integral numerically.

$$V = \frac{\sqrt{3}}{4}\int_0^\pi \frac{1 - \cos 2x}{2}\, dx = \frac{\sqrt{3}}{8}\int_0^\pi (1 - \cos 2x)\, dx$$
$$= \frac{\sqrt{3}}{8}\left[\int_0^\pi 1\, dx - \frac{1}{2}\int_0^\pi \cos 2x \cdot 2\, dx\right]$$
$$= \frac{\sqrt{3}}{8}\left[x - \frac{1}{2}\sin 2x\right]_0^\pi = \frac{\sqrt{3}}{8}\pi \approx 0.68$$

Notice that the substitution $u = 2x$ was used to evaluate the second integral above.   ◄

Concepts Review

1. The volume of a disk of radius r and thickness h is _____ .

2. The volume of a washer of inner radius r, outer radius R, and thickness h is _____ .

3. If the region R bounded by $y = x^2$, $y = 0$, and

$x = 3$ is revolved about the x-axis, the disk at x will have volume $\Delta V \approx$ _____ .

4. If the region R of Question 3 is revolved about the line $y = -2$, the washer at x will have the volume $\Delta V \approx$ _____ .

Problem Set 6.2

In Problems 1–4, find the volume of the solid generated when the indicated region is revolved about the specified axis: *slice, approximate, integrate.*

1. x-axis

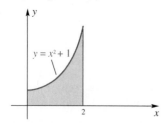

2. x-axis

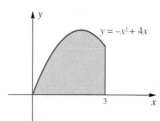

3. (a) x-axis
    (b) y-axis

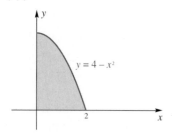

4. (a) x-axis
    (b) y-axis

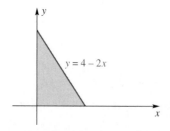

In Problems 5–10, sketch the region R bounded by the graphs of the given equations, showing a typical vertical rectangle. Then find the volume of the solid generated by revolving R about the x-axis. Evaluate the integral exactly if possible; if not, use a numerical approximation.

5. $y = \dfrac{x^2}{4}, x = 4, y = 0$

6. $y = e^{2x} + 1, x = 0, y = 2$

7. $y = \dfrac{1}{x}, x = 1, x = 4, y = 0$

8. $y = \tan x, y = 0, x = 1$

9. $y = \sqrt[3]{4 - x^2}, y = 0$ between $x = -1$ and $x = 2$.

10. $y = x^{2/3}, y = 0$ between $x = 1$ and $x = 8$.

In Problems 11–16, sketch the region R bounded by the graphs of the given equations, showing a typical horizontal rectangle. Then find the volume of the solid generated by revolving R about the y-axis. Evaluate the integral exactly if possible; if not, use a numerical approximation.

11. $x = y^2, x = 0, y = 2$

12. $x = \dfrac{2}{y}, y = 1, y = 6, x = 0$

13. $x = \sqrt{y}, y = 4, x = 0$

14. $x = \ln y, y = 1, y = 2, x = 0$

15. $x = e^y, y = 0, y = 4, x = 0$

16. $x = \sqrt{9 - y^2}, x = 0$

17. Find the volume of the solid generated by revolving about the x-axis the region bounded by the upper half of the ellipse

$$\frac{x^2}{a^2} + \frac{y^2}{b^2} = 1$$

and the x-axis and thus find the volume of a *prolate spheroid*. Here a and b are positive constants, with $a > b$.

18. Find the volume of the solid generated by revolving about the x-axis the region bounded by the line $y = 4x$ and the parabola $y = 4x^2$.

19. Find the volume of the solid generated by revolving about the x-axis the region bounded by the line $2x - y = -4$ and the curve $y = e^x$.

20. Find the volume of the solid generated by revolving about the x-axis the region in the first quadrant bounded by the circle $x^2 + y^2 = r^2$, the x-axis, and the line $x = r - h$, $0 < h < r$, and thus find the volume of a *spherical segment* of height h, radius of sphere r.

21. Find the volume of the solid generated by revolving about the y-axis the region bounded by the line $y = 4x$ and the parabola $y = 4x^2$.

22. Find the volume of the solid generated by revolving about the line $y = 2$ the region in the first quadrant bounded by the parabolas $3x^2 - 16y + 48 = 0$ and $x^2 - 16y + 80 = 0$, and the y-axis.

23. The base of a solid is the region inside the circle $x^2 + y^2 = 4$. Find the volume of the solid if every cross section by a plane perpendicular to the x-axis is a square. *Hint:* See Examples 5 and 6.

24. Do Problem 23 assuming that every cross section by a plane perpendicular to the x-axis is an isosceles triangle with base on the xy-plane and altitude 4.

25. The base of a solid is bounded by one arch of $y = \sqrt{\cos x}$, $-\pi/2 \leq x \leq \pi/2$, and the x-axis. Each cross section perpendicular to the x-axis is a square sitting on this base. Find the volume of the solid.

26. The base of a solid is the region bounded by $y = 1 - x^2$ and $y = 1 - x^4$. Cross sections of the solid that are perpendicular to the x-axis are squares. Find the volume of the solid.

27. Find the volume of one octant (one-eighth) of the solid region common to two right circular cylinders of radius 1 whose axes intersect at right angles. *Hint:* Horizontal cross sections are squares. See Figure 14.

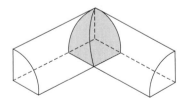

Figure 14

28. The base of a solid is the region R bounded by $y = \sqrt{x}$ and $y = x^2$. Each cross section perpendicular to the x-axis is a semicircle with diameter extending across R. Find the volume of the solid.

29. Find the volume of the solid generated by revolving the region in the first quadrant bounded by the curve $y^2 = x^3$, the line $x = 4$, and the x-axis

(a) about the line $x = 4$;   (b) about the line $y = 8$.

30. Find the volume of the solid generated by revolving the region bounded by the curve $y^2 = x^3$, the line $x = 4$, and the x-axis

(a) about the line $x = 4$;   (b) about the line $y = 8$.

31. An open barrel of radius r and height h is initially full of water. It is tilted until the water level coincides with a diameter of the base and just touches the rim of the top. Find the volume of water left in the barrel. See Figure 15.

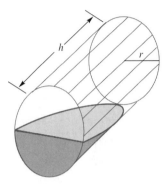

Figure 15

32. A wedge is cut from a right circular cylinder of radius r. (Figure 16). The upper surface of the wedge is in a plane through a diameter of the circular base and makes an angle θ with the base. Find the volume of the wedge.

Figure 16

33. (The Water Clock) A water tank is obtained by revolving the curve $y = kx^4$, $k > 0$, about the y-axis.

(a) Find $V(y)$, the volume of water in the tank as a function of its depth y.

(b) Water drains through a small hole according to Torricelli's Law $(dV/dt = -m\sqrt{y})$. Show that the water level falls at a constant rate.

34. Show that the volume of a general cone (Figure 17) is $\frac{1}{3}Ah$, where A is the area of the base and h

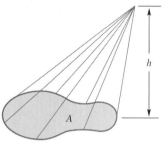

Figure 17

is the height. Use this result to give the formula for the volume of:

(a) a right circular cone of radius r and height h;

(b) a regular tetrahedron with edge length r.

35. State the version of Cavalieri's Principle for volume (see Problem 36 of Section 6.1).

Answers to Concepts Review: 1. $\pi r^2 h$ 2. $\pi(R^2 - r^2)h$
3. $\pi x^4 \Delta x$ 4. $\pi[(x^2 + 2)^2 - 4]\Delta x$

6.3 | Length of a Plane Curve

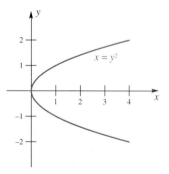

Figure 1

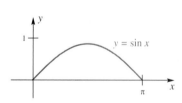

Figure 2

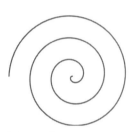

Figure 3

How long is the spiral curve shown in Figure 1? If it were a piece of string, most of us would stretch it taut and measure it with a ruler. But if it is the graph of an equation, that is a little hard to do.

A little reflection suggests a prior question. What is a plane curve? We have used the term *curve* informally until now; it is time to be more precise. We begin with several examples.

The graph of $y = \sin x$, $0 \leq x \leq \pi$, is a plane curve (Figure 2). So is the graph of $x = y^2$, $-2 \leq y \leq 2$ (Figure 3). In both cases, the curve is the graph of a function, the first of the form $y = f(x)$, the second of the form $x = g(y)$. However, the spiral curve does not fit this pattern. Neither does the circle $x^2 + y^2 = a^2$, although in this case we could think of it as the combined graph of the two functions $y = f(x) = \sqrt{a^2 - x^2}$ and $y = g(x) = -\sqrt{a^2 - x^2}$.

The circle suggests another way of thinking about curves. Recall from trigonometry (see Section 0.4) that $x = a \cos t, y = a \sin t, 0 \leq t \leq 2\pi$, describe the circle $x^2 + y^2 = a^2$ (Figure 4). Here t is an auxiliary variable, from now on called a **parameter**. Both x and y are expressed in terms of this parameter. We say that $x = a \cos t, y = a \sin t, 0 \leq t \leq 2\pi$, are **parametric equations** describing the circle.

Note: We have previously used the word *parameter* to represent a constant, as opposed to a variable. In mathematics, as in any language, a word can have different meanings depending on the context.

If we were to graph the parametric equations $x = t \cos t, y = t \sin t$, $0 \leq t \leq 5\pi$, we could get a curve something like the spiral with which we started. And we can even think of the sine curve (Figure 2) and the parabola (Figure 3) in parametric form. We write

$$y = \sin t \qquad x = t \qquad 0 \leq t \leq \pi$$
$$y = t \qquad x = t^2 \qquad -2 \leq t \leq 2$$

In both cases, t is the parameter.

Thus, for us, a **plane curve** is determined by a pair of parametric equations $x = f(t), y = g(t), a \leq t \leq b$, where we assume that f and g are continuous on the given interval. Think of t as measuring time. As t increases from a to b, the point (x, y) traces out a curve in the plane. Here is another example.

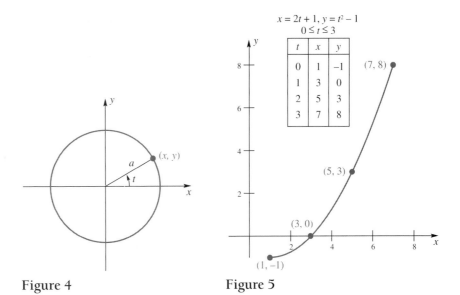

Figure 4   Figure 5

EXAMPLE 1: Sketch the curve determined by the parametric equations $x = 2t + 1$, $y = t^2 - 1, 0 \le t \le 3$.

    SOLUTION: We make a three-column table of values, then plot the ordered pairs (x, y), and—finally—connect these points in the order of increasing t, as shown in Figure 5. ◄

Actually, the definition we have given is too broad for the purposes we have in mind, so we immediately restrict it to what is called a *smooth curve.*

Definition

A plane curve is **smooth** if it is determined by a pair of parametric equations $x = f(t)$, $y = g(t)$, $a \le t \le b$, where f' and g' exist and are continuous on $[a, b]$ and $f'(t)$ and $g'(t)$ are not simultaneously zero on (a, b).

The adjective *smooth* is chosen to indicate that an object moving along the curve so that its position at time t is (x, y) would suffer no sudden changes of direction (continuity of f' and g' ensures this) and would not stop or double back ($f'(t)$ and $g'(t)$ not simultaneously zero ensures this).

Using Technology to Graph Parametric Equations

Most graphing calculators and computer algebra systems have built into them the ability to graph parametric equations. The calculator or computer goes through essentially the same process that you go through by hand; a three-column table of values is calculated, then the (x, y) pairs are connected in order of increasing t.

GRAPHING CALCULATORS On a graphing calculator, the t-values that are used to calculate the three-column table of values are determined by a minimum t-value, a maximum t-value and a t-value step size. The step size determines the change in t from one t-value to the next; thus, if the minimum and maximum t-values are 0 and 10, respectively, and the step size is

0.5, then there will be 20 t-subintervals and hence 21 points plotted. *Note*: The smaller the step size, the more accurate the graph, *and* the more time taken to generate the plot. Thus, you need to strike a compromise between accuracy and the length of plotting time.

We suggest that you use between 20 and 100 points for a typical parametric plot. If n is the number of t-subintervals desired (and therefore $n + 1$ points) then the formula for the step size would be

$$t \text{ step size} = \frac{\text{maximum } t - \text{minimum } t}{n}$$

After generating a parametric plot, you can use the Trace feature; the values of t, x, and y are displayed for each point.

COMPUTER ALGEBRA SYSTEMS With most computer algebra systems, you specify only the minimum and maximum t-values to be used for the plot; the computer generally determines the number of points required to generate the plot.

EXAMPLE 2: Use a graphing calculator or computer algebra system to graph the curve from Example 1 determined by the parametric equations $x = 2t + 1$, $y = t^2 - 1$, and the t-interval $0 \le t \le 3$.

SOLUTION:

Graphing Calculators   For most graphing calculators, you should put the calculator in parametric mode before setting the range values. We generate a plot using 20 t-subintervals (and hence 21 points). Thus,

$$t \text{ step size} = \frac{\text{maximum } t - \text{minimum } t}{n} = \frac{3 - 0}{20} = 0.15$$

Because the graph begins at the point $(1, -1)$ when $t = 0$ and ends at the point $(7, 8)$ when $t = 3$, a reasonable choice for the graph window would be $0 \le x \le 8$ and $-2 \le y \le 9$. Notice that even a very brief table of values calculated by hand can be helpful in determining how to set the window on your calculator.

Computer Algebra Systems   For most computer algebra systems, you enter the two equations that determine the parametric curve as a vector or ordered pair of expressions, and use the appropriate plotting command (Plot or ParametricPlot).

The resulting calculator and computer graphs are shown in Figure 6. ◄

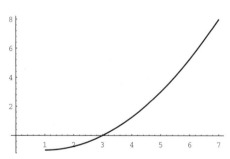

Mathematica graph of $x = 2t + 1$, $y = t^2 - 1$, on the
t-interval $0 \le t \le 3$.

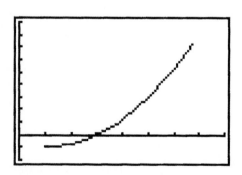

TI-82 graph of $x = 2t + 1$, $y = t^2 - 1$, on the t-interval
$0 \le t \le 3$.

Figure 6

Length   Finally, we are ready for the main question. What is meant by the length of the smooth curve given parametrically by $x = f(t)$, $y = g(t)$, $a \leq t \leq b$?

Partition the interval $[a, b]$ into n subintervals by means of points

$$a = t_0 < t_1 < t_2 < \cdots < t_n = b$$

This cuts the curve into n pieces with corresponding endpoints $Q_0, Q_1, Q_2, \ldots, Q_{n-1}, Q_n$, as shown in Figure 7.

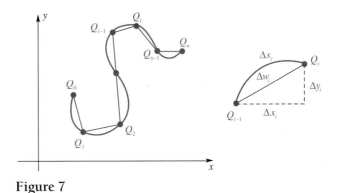

Figure 7

Our idea is to approximate the curve by the indicated polygonal line, calculate its length, and then take the limit as the norm of the partition approaches zero. In particular, we approximate the "length" Δs_i of the typical piece (see Figure 7) by

$$\Delta w_i = \sqrt{(\Delta x_i)^2 + (\Delta y_i)^2}$$
$$= \sqrt{[f(t_i) - f(t_{i-1})]^2 + [g(t_i) - g(t_{i-1})]^2}$$

From the Mean Value Theorem for Derivatives, we know that there are points we can call $\bar{t}_i$ and $\hat{t}_i$ in (t_{i-1}, t_i) such that

$$f(t_i) - f(t_{i-1}) = f'(\bar{t}_i)\Delta t_i$$
$$g(t_i) - g(t_{i-1}) = g'(\hat{t}_i)\Delta t_i$$

where $\Delta t_i = t_i - t_{i-1}$. Thus,

$$\Delta w_i = \sqrt{[f'(\bar{t}_i)\Delta t_i]^2 + [g'(\hat{t}_i)\Delta t_i]^2}$$
$$= \sqrt{[f'(\bar{t}_i)]^2 + [g'(\hat{t}_i)]^2}\,\Delta t_i$$

and the length of the polygonal line is

$$\sum_{i=1}^{n}\Delta w_i = \sum_{i=1}^{n}\sqrt{[f'(\bar{t}_i)]^2 + [g'(\hat{t}_i)]^2}\,\Delta t_i$$

The latter expression is almost a Riemann sum, the only difficulty being that $\bar{t}_i$ and $\hat{t}_i$ are not likely to be the same point. However, it is shown in advanced calculus books that in the limit, this makes no difference. Thus, we may define the length L of the curve to be the limit of the expression above as the norm of the partition approaches zero; that is,

$$L = \int_a^b \sqrt{[f'(t)]^2 + [g'(t)]^2}\, dt = \int_a^b \sqrt{\left(\frac{dx}{dt}\right)^2 + \left(\frac{dy}{dt}\right)^2}\, dt$$

Two special cases are of great interest. If this curve is given by $y = f(x)$, $a \le x \le b$, we treat x as the parameter and the boxed result takes the form

$$L = \int_a^b \sqrt{1 + \left(\frac{dy}{dx}\right)^2}\, dx$$

Similarly, if the curve is given by $x = g(y)$, $c \le y \le d$, we treat y as the parameter, obtaining

$$L = \int_c^d \sqrt{1 + \left(\frac{dx}{dy}\right)^2}\, dy$$

Something would be amiss if these formulas failed to yield the familiar results for circles and line segments. Fortunately, the results are the expected ones, as the following two examples illustrate.

EXAMPLE 3: Find the circumference of the circle $x^2 + y^2 = a^2$.

SOLUTION: We write the equation of the circle in parametric form: $x = a \cos t$, $y = a \sin t$, $0 \le t \le 2\pi$. Then $dx/dt = -a \sin t$, $dy/dt = a \cos t$, and by the first of our formulas,

$$L = \int_0^{2\pi} \sqrt{a^2 \sin^2 t + a^2 \cos^2 t}\, dt = \int_0^{2\pi} a\, dt = [at]_0^{2\pi} = 2\pi a \qquad \blacktriangleleft$$

EXAMPLE 4: Find the length of the line segment from $A(0, 1)$ to $B(5, 13)$.

SOLUTION: The given line segment is shown in Figure 8. Note that the equation of the corresponding line is $y = \frac{12}{5}x + 1$, so $dy/dx = \frac{12}{5}$ and thus by the second of the three length formulas,

$$L = \int_0^5 \sqrt{1 + \left(\frac{12}{5}\right)^2}\, dx = \int_0^5 \sqrt{\frac{5^2 + 12^2}{5^2}}\, dx = \frac{13}{5} \int_0^5 1\, dx$$

$$= \left[\frac{13}{4}x\right]_0^5 = 13$$

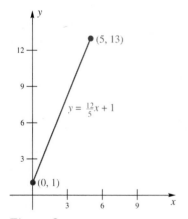

Figure 8

This does agree with the result obtained by use of the distance formula. $\blacktriangleleft$

EXAMPLE 5: Sketch the graph of the curve given parametrically by $x = 2 \cos t$, $y = 4 \sin t$, $0 \le t \le \pi$ and find its length.

SOLUTION: The graph (Figure 9) is drawn, as in Example 1, by first making a three-column table of values, or by using a calculator or computer as in Example 2.

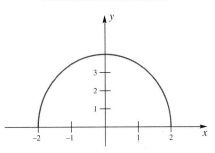

$x = 2 \cos t, \ y = 4 \sin t$
$0 \le t \le \pi$

t	x	y
0	2	0
$\pi/6$	$\sqrt{3}$	2
$\pi/3$	1	$2\sqrt{3}$
$\pi/2$	0	4
$2\pi/3$	−1	$2\sqrt{3}$
$5\pi/6$	$-\sqrt{3}$	2
π	−2	0

Figure 9

The length L of this curve is given by

$$L = \int_0^\pi \sqrt{(-2 \sin t)^2 + (4 \cos t)^2} \, dt$$

$$= 2 \int_0^\pi \sqrt{\sin^2 t + 4 \cos^2 t} \, dt$$

$$= 2 \int_0^\pi \sqrt{1 + 3 \cos^2 t} \, dt \approx 9.688$$

We use a numerical approximation for the integral because we do not know an antiderivative for $\sqrt{1 + 3 \cos^2 t}$. In fact, it has been shown that this function does not have an antiderivative that can be written in terms of the elementary functions of calculus.   ◀

The situation in Example 4 is quite typical. To get an integral that we can compute exactly, we must pick our curves carefully.

Arc-Length Problems

Pick a curve $y = f(x)$ at random and it is almost certain that you will be unable to integrate $\sqrt{1 + [f'(x)]^2}$ by hand. For example, we can find the length of the curve $y = x^r$ by the Fundamental Theorem of Calculus only if $r = 1$ or $r = 1 + 1/n$ for some integer n. Thus, you should expect to require a numerical approximation for most arc-length problems. If the problem "works out," then it was probably invented for a calculus book.

EXAMPLE 6: Find the length of the arc of the curve $y = x^{\frac{3}{2}}$ from the point $(1, 1)$ to the point $(4, 8)$ (see Figure 10).

SOLUTION: We begin by estimating this length. Pretend the curve is a straight line. Then its length would be $\sqrt{(4 - 1)^2 + (8 - 1)^2} = \sqrt{58} \approx 7.5$. We can reason that the actual length must be slightly greater than 7.5.

For the exact calculation, we note that $dy/dx = \frac{3}{2}x^{\frac{1}{2}}$, so

$$L = \int_1^4 \sqrt{1 + \left(\frac{3}{2}x^{\frac{1}{2}}\right)^2}\, dx = \int_1^4 \sqrt{1 + \frac{9}{4}x}\, dx$$

Let $u = 1 + \frac{9}{4}x$; then $du = \frac{9}{4}dx$. Hence,

$$\int \sqrt{1 + \frac{9}{4}x}\, dx = \frac{4}{9}\int \sqrt{u}\, du = \frac{4}{9} \cdot \frac{2}{3} u^{\frac{3}{2}} + C$$

$$= \frac{8}{27}\left(1 + \frac{9}{4}x\right)^{\frac{3}{2}} + C$$

Therefore,

$$\int_1^4 \sqrt{1 + \frac{9}{4}x}\, dx = \left[\frac{8}{27}\left(1 + \frac{9}{4}x\right)^{\frac{3}{2}}\right]_1^4$$

$$= \frac{8}{27}\left(10^{\frac{3}{2}} - \frac{13^{\frac{3}{2}}}{8}\right) \approx 7.63 \quad ◄$$

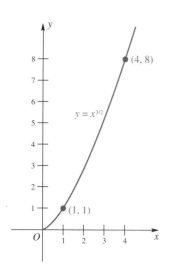

$y = x^{3/2}$

$(4, 8)$

$(1, 1)$

Figure 10

Differential of Arc Length　Let f be continuously differentiable on $[a, b]$. For each x in (a, b), define $s(x)$ by

$$s(x) = \int_a^x \sqrt{1 + [f'(u)]^2}\, du$$

Then $s(x)$ gives the length of the arc of the curve $y = f(u)$ from the point $(a, f(a))$ to $(x, f(x))$ (see Figure 11). By the theorem on differentiating an integral with respect to its upper limit (Theorem 5.6D),

$$s'(x) = \frac{ds}{dx} = \sqrt{1 + [f'(x)]^2} = \sqrt{1 + \left(\frac{dy}{dx}\right)^2}$$

Thus, ds—the differential arc length—can be written as

$$ds = \sqrt{1 + \left(\frac{dy}{dx}\right)^2}\, dx$$

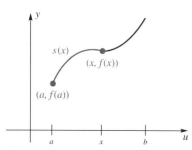

$s(x)$

$(x, f(x))$

$(a, f(a))$

Figure 11

In fact, depending on how a graph is parametrized, we are led to three formulas for ds, namely,

$$ds = \sqrt{1 + \left(\frac{dy}{dx}\right)^2}\, dx$$

$$= \sqrt{1 + \left(\frac{dx}{dy}\right)^2}\, dy$$

$$= \sqrt{\left(\frac{dx}{dt}\right)^2 + \left(\frac{dy}{dt}\right)^2}\, dt$$

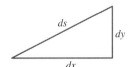

Figure 12

Some people prefer to remember these formulas by writing (see Figure 12)

$$(ds)^2 = (dx)^2 + (dy)^2$$

The three forms arise by dividing and multiplying the right-hand side of this equation by $(dx)^2$, $(dy)^2$, and $(dt)^2$, respectively. For example,

$$(ds)^2 = \left[\frac{(dx)^2}{(dx)^2} + \frac{(dy)^2}{(dx)^2}\right](dx)^2 = \left[1 + \left(\frac{dy}{dx}\right)^2\right](dx)^2$$

which gives the first of the three formulas.

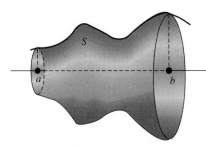

Figure 13

Area of a Surface of Revolution

If a smooth plane curve is revolved about an axis in its plane, it generates a surface of revolution, as illustrated in Figure 13. Our aim is to determine the area of such a surface.

To get started, we introduce the formula for the area of the frustum of a cone. A **frustum** of a cone is the surface between two planes perpendicular to the axis of the cone (shaded in Figure 14). If a frustum has base radii r_1 and r_2 and slant height l, then its area A is given by

$$A = 2\pi\left(\frac{r_1 + r_2}{2}\right)l = 2\pi(\text{average radius}) \cdot (\text{slant height})$$

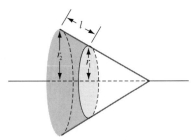

Figure 14

The derivation of this result depends only on the formula for the area of a circle (see Problem 32).

Let $x = f(t)$, $y = g(t)$, and $a \le t \le b$ determine a smooth curve in the upper half of the xy-plane, as shown in Figure 15. Partition the interval $[a, b]$ into n pieces by means of points $a = t_0 < t_1 < \cdots < t_n = b$, thereby also dividing the curve into n pieces. Let Δs_i denote the length of the typical piece and let y_i be the y-coordinate of a point on this piece. When the curve is revolved about the x-axis, it generates a surface, and the typical piece generates a narrow band. The "area" of this band ought to be approximately that of a frustum—that is, approximately $2\pi y_i\, \Delta s_i$. When we add the contributions of all the pieces and

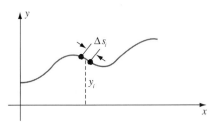

Figure 15

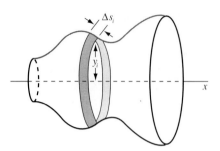

Figure 16

take the limit as the number of pieces approaches infinity, we get what we define to be the area of the surface of revolution. All this is indicated in Figure 16 and the boxed formula below.

$$A = \lim_{n \to \infty} \sum_{i=1}^{n} 2\pi y_i \, \Delta s_i = \int_{*}^{**} 2\pi y \, ds$$

In using the formula for A we must make an appropriate interpretation of y, ds, and the limits * and **. Recall in our discussion of the differential of arc length that ds can have any one of three forms, which in turn correspond to three possible choices for limits of integration. Thus, if the surface is obtained by revolving the curve $y = f(x)$, $a \le x \le b$, about the x-axis, the formula takes the form

$$A = 2\pi \int_{*}^{**} y \, ds = 2\pi \int_{a}^{b} f(x)\sqrt{1 + [f'(x)]^2} \, dx$$

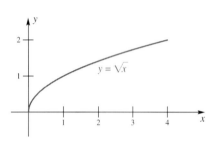

Figure 17

EXAMPLE 7: Find the area of the surface of revolution generated by revolving the curve $y = \sqrt{x}$, $0 \le x \le 4$, about the x-axis (Figure 17).

SOLUTION: Here, $f(x) = \sqrt{x}$ and $f'(x) = 1/(2\sqrt{x})$. Thus,

$$A = 2\pi \int_{0}^{4} \sqrt{x}\sqrt{1 + \frac{1}{4x}} \, dx = 2\pi \int_{0}^{4} \sqrt{x}\sqrt{\frac{4x + 1}{4x}} \, dx$$

$$= \pi \int_{0}^{4} \sqrt{4x + 1} \, dx = \left[\pi \cdot \frac{1}{4} \cdot \frac{2}{3}(4x + 1)^{\frac{3}{2}}\right]_{0}^{4}$$

$$= \frac{\pi}{6}\left(17^{\frac{3}{2}} - 1^{\frac{3}{2}}\right) \approx 36.18 \qquad \blacktriangleleft$$

If the curve is given parametrically by $x = f(t)$, $y = g(t)$, $a \le t \le b$, then the surface area formula becomes

$$A = 2\pi \int_{*}^{**} y \, ds = 2\pi \int_{a}^{b} g(t)\sqrt{[f'(t)]^2 + [g'(t)]^2} \, dt$$

Concepts Review

1. The graph of the parametric equations $x = 4 \cos t$, $y = 4 \sin t$, $0 \le t \le 2\pi$ is a curve called a _____.

2. The curve determined by $y = x^2 + 1$, $0 \le x \le 4$, can be put in a parametric form using t as the parameter by writing $y =$ _____, $x =$ _____.

3. The formula for the length L of the curve $x = f(t)$, $y = g(t)$, $a \le t \le b$ is $L =$ _____.

4. The proof of the formula for the length of a curve depends on an earlier theorem with the name _____.

Problem Set 6.3

1. Use an x-integration to find the length of the segment of the line $y = 3x + 5$ between $x = 1$ and $x = 4$. Check by the distance formula.

2. Use a y-integration to find the length of the segment of the line $2x - 4y + 6 = 0$ between $y = 0$ and $y = 2$. Check by the distance formula.

In Problems 3–8, find the length of the indicated curve. Sketch the curve and indicate on the sketch the part of the curve whose length you find. Try to evaluate the integral exactly; failing that, use a numerical method.

3. $y = 2x^{\frac{3}{2}}$ between $x = \frac{1}{3}$ and $x = 7$.

4. $y = \frac{2}{3}(x^2 + 1)^{\frac{3}{2}}$ between $x = 1$ and $x = 4$.

5. $y = x^2$ between $x = 0$ and $x = 4$.

6. $y = x^3$ between $x = 0$ and $x = 4$.

7. $x = \sin y$ between $y = 0$ and $y = \pi$.

8. $x = e^y$ between $y = 0$ and $y = 2$.

In Problems 9–14, sketch the graph of the given parametric equations and find its length. Try to evaluate the integral exactly; failing that, use a numerical method.

9. $x = t^3, y = t^2; 0 \le t \le 4$

10. $x = 3t^2 + 2, y = 2t^3 - 1; 1 \le t \le 3$

11. $x = 3 \sin t, y = 3 \cos t - 3; 0 \le t \le 2\pi$

12. $x = 4 \cos t + 5, y = 4 \sin t - 1; 0 \le t \le 2\pi$

13. $x = t, y = \ln t; 1 \le t \le e^2$. Compare to Problem 8 and explain any similarities.

14. $x = \cos t, y = t; 0 \le t \le 2\pi$

15. Sketch the graphs of each of the following parametric equations.

(a) $x = 3 \cos t, y = 3 \sin t, 0 \le t \le 2\pi$

(b) $x = 3 \cos t, y = \sin t, 0 \le t \le 2\pi$

(c) $x = t \cos t, y = t \sin t, 0 \le t \le 6\pi$

(d) $x = \cos t, y = \sin 2t, 0 \le t \le 2\pi$

(e) $x = \cos 3t, y = \sin 2t, 0 \le t \le 2\pi$

(f) $x = \cos t, y = \sin \pi t, 0 \le t \le 40$

16. Find the lengths of each of the curves in Problem 15.

17. Using the same axes, draw the graphs of $y = x^n$ on $[0, 1]$ for $n = 1, 2, 4, 10$, and 100. Find the length of each of these curves. Guess at the length when $n = 10,000$.

18. Sketch the graph of the four-cusped *hypocycloid* $x = a \sin^3 t, y = a \cos^3 t; 0 \le t \le 2\pi$, and find its length. *Hint*: By symmetry, you can quadruple the length of the first quadrant portion. Sketch first for a few selected values of the parameter a.

19. A point P on the rim of a wheel of radius a is initially at the origin. As the wheel rolls to the right along the x-axis, P traces out a curve called a *cycloid* (see Figure 18). Derive parametric equations for the cycloid as follows.

(a) Show that $\overline{OT} = a\theta$.

(b) Convince yourself that $\overline{PQ} = a \sin \theta, \overline{QC} = a \cos \theta, 0 \le \theta \le \pi/2$.

(c) Show that $x = a(\theta - \sin \theta), y = a(1 - \cos \theta)$.

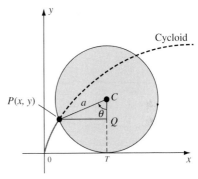

Figure 18

20. Find the length of one arch of the cycloid of Problem 19. *Hint:* First show that

$$\left(\frac{dx}{d\theta}\right)^2 + \left(\frac{dy}{d\theta}\right)^2 = 4\theta^2 \sin^2\left(\frac{\theta}{2}\right)$$

Check using numerical integration and a few selected values of the parameter a.

21. Suppose the wheel of Problem 19 turns at a constant rate $\omega = d\theta/dt$, where t is time. Then $\theta = \omega t$.

(a) Show that the speed ds/dt of P along the cycloid is

$$\frac{ds}{dt} = 2a\omega\left|\sin\frac{\omega t}{2}\right|$$

(b) When is the speed a maximum and when is it a minimum?

(c) Explain why a bug on a wheel of a car going 60 miles per hour sometimes is traveling at 120 miles per hour.

22. Find the length of each curve.

(a) $y = \int_1^x \sqrt{u^3 - 1}\, du, 1 \le x \le 2$

(b) $x = t - \sin t, y = 1 - \cos t; 0 \le t \le 4\pi$

23. Find the length of each curve.

(a) $y = \int_{\pi/6}^x \sqrt{64 \sin^2 u \cos^4 u - 1}\, du, \frac{\pi}{6} \le x \le \frac{\pi}{3}$

(b) $x = a \cos t + at \sin t, y = a \sin t - at \cos t; -1 \le t \le 1$

In Problems 24–31, find the area of the surface generated by revolving the given curve about the x-axis.

24. $y = 6x; 0 \le x \le 1$

25. $y = \sqrt{25 - x^2}; -2 \le x \le 3$

26. $y = x^2; 0 \le x \le 1$

27. $y = x^3; 0 \le x \le 1$. Compare results with Problem 26. Which integral was easier to find exactly? Why?

28. $y = e^{-x^2}; -2 \le x \le 2$

29. $y = \sin x; 0 \le x \le \pi$

30. $x = t, y = t^3; 0 \le t \le 1$

31. $x = 1 - t^2, y = e^t; 0 \le t \le 1$

32.   If the surface of a cone of slant height l and base radius r is cut along a lateral edge and laid flat, it becomes the sector of a circle of radius l and central angle θ (see Figure 19).

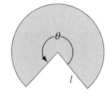

Figure 19

(a) Show that $\theta = 2\pi r/l$ radians.

(b) Use the formula $\frac{1}{2}l^2\theta$ for the area of a sector of radius l and central angle θ to show that the lateral surface area of a cone is πrl.

(c) Use the result of (b) to obtain the formula $A = 2\pi[(r_1 + r_2)/2]l$ for the lateral area of a frustum of a cone with base radii r_1 and r_2 and slant height l (see the discussion on page 373). You may need a computer algebra system for this integral.

33.   Show that the area of the part of the surface of a sphere of radius a between two parallel planes h units apart ($h < 2a$) is $2\pi ah$. Thus, show that if a right circular cylinder is circumscribed about a sphere, then two planes parallel to the base of the cylinder bound regions of the same area on the sphere and the cylinder.

34.   Figure 20 shows one arch of a cycloid. Its parametric equations (see Problem 19) are given by

$$x = a(t - \sin t) \qquad y = a(1 - \cos t)$$
$$0 \le t \le 2\pi$$

(a) Show that the area of the surface generated when this curve is revolved about the x-axis is

$$A = 2\sqrt{2}\,\pi a^2 \int_0^{2\pi} (1 - \cos t)^{\frac{3}{2}}\, dt$$

(b) With the help of the half-angle formula $1 - \cos t = 2\sin^2(t/2)$, evaluate A.

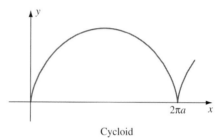

Cycloid

Figure 20

35.   The circle $x = a \cos t, y = a \sin t, 0 \le t \le 2\pi$, is revolved about the line $x = b, 0 < a < b$, thus generating a torus (doughnut). Find its surface area.

Answers to Concepts Review: 1. Circle   2. $t^2 + 1; t$

3. $\int_a^b \{[f'(t)]^2 + [g'(t)]^2\}^{\frac{1}{2}}\, dt$ 4. Mean Value Theorem (for derivatives)

LAB 11: ARC LENGTH

Mathematical Background

There are many functions that do not have antiderivatives given in terms of elementary functions. So there are many integrals for which the fundamental theorem of calculus is not much help. One application where integrals of this form often arise is in finding the length of a curve.

To find the length of a curve given by a function $y = f(x)$ from $x = a$ to $x = b$, we first approximate the curve with line segments.

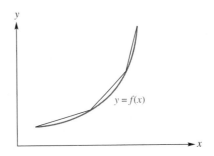

The length of one of these line segments, Δs, can be given in terms of Δx and Δy by using the right triangle pictured below.

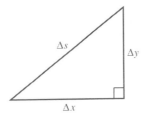

We then have $\Delta s = \sqrt{\Delta x^2 + \Delta y^2} = \left(\sqrt{1 + \left(\dfrac{\Delta y}{\Delta x}\right)^2}\right)\Delta x$. If we then add up all the line segments and let Δx go to zero, $\dfrac{\Delta y}{\Delta x}$ becomes $f'(x)$ and the sum becomes

$$\int_a^b \sqrt{1 + (f'(x))^2}\, dx \tag{1}$$

Lab Introduction

In this lab, we are going to use formula (1) to determine the breaking point for a guitar string.

Here are two models for the shape of a guitar string at its maximum stretch. The first assumes that the string is plucked and the shape is like a tent, and the second assumes that the string is excited by sympathetic vibration and the shape is a sine curve (guitar players will know that a string will vibrate without touching it if another string tuned to the same note is plucked).

For a three-foot-long guitar string, the shape of the string when plucked six inches from the end is given by

$$f(x) = \begin{cases} 2ax & 0 \le x \le 0.5 \\ -\dfrac{2a}{5}x + \dfrac{6a}{5} & 0.5 \le x \le 3 \end{cases}$$

where a is the distance of the pulled string above the rest position, in feet.

For the second model, the shape of a three-foot guitar string vibrating in the fundamental frequency (called the first mode) has the shape $f(x) = a \sin\left(\dfrac{\pi}{3}x\right)$. For the second model, a is called the amplitude, and x and y are measured in feet.

Experiment

(1) For the first model, graph the guitar string for the a values $-0.2, 0, 0.4$ using the window $0 \leq x \leq 3$, $-0.4 \leq y \leq 0.4$. Put all three sketches on one set of axes and label each with the corresponding a value; also indicate on each sketch what distance a represents. Now repeat using the second model.

(2) For the first model find the length of the string for each of the a values in part (1). Repeat for the second model. Use formula (1); try to find the value of the integral exactly, or it that fails, integrate numerically.

(3) Suppose that it has been determined experimentally that the string breaks if it is stretched to a length of 3.1 feet. For each model, determine the amplitude a that will break the string; thus, you want to find the a value for which the length is 3.1 feet. If possible, use equation (1) to find a formula for the length of the string as a function of a, and then set this formula equal to 3.1 and solve for a (whether you do this exactly or numerically is up to you). If this can't be done, guess an a value, compute the length, and guess again until you hit 3.1. Obtain two-decimal-place accuracy.

Discussion

1. a. Which integrals in part 2 could be found exactly (that is, using the Fundamental Theorem of Calculus) and which required a numerical technique? Which integrals in part 3 could be found exactly?

b. Was it helpful to be able to find the exact integral in part 3 in order to find the value of a that breaks the string? Why? Explain carefully how you found a for each model.

c. Discuss the advantages and disadvantages of numerical integration (such as Midpoint Riemann sums) versus exact integration (the Fundamental Theorem of Calculus).

2. Compare the results that you got for the two models. Which had to be stretched farther to break the string? Why?

6.4 Work

We turn now to an application of the definite integral that does not concern length, area, or volume. In physics, we learn that if an object moves a distance d along a line while subjected to a *constant* force F in the direction of the motion, then the work W done by that force is given by

$$\text{work} = (\text{force}) \cdot (\text{distance})$$

that is,

$$W = F \cdot d$$

If force is measured in pounds and distance in feet, then work is in foot-pounds. If force is in dynes (the force required to give a mass of 1 gram an acceleration of 1 centimeter per second per second) and distance is in centimeters, then work is in dyne-centimeters, also called ergs. If force is in newtons and distance is in meters, then work is in newton-meters, also called joules. For example, a worker pushing a cart with a constant

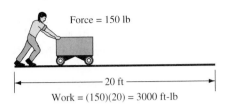

Figure 1

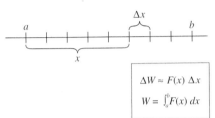

Figure 2

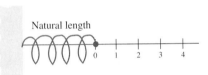

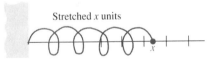

Figure 3

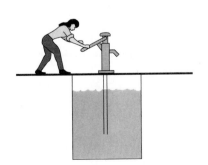

Figure 4

force of 150 pounds through a distance of 20 feet does $(150)(20) = 3000$ foot-pounds of work (Figure 1). A person lifting a weight (force) of 2 newtons a distance of 3 meters does $(2)(3) = 6$ joules of work.

In most practical situations, force is not a constant, but rather varies as the object moves along the line. Suppose, in fact, that the object is being moved along the x-axis from a to b subject to a variable force of magnitude $F(x)$ at the point x, where F is a continuous function. Then how much work is done? Once again, the words *slice, approximate, integrate* lead us to an answer. Here *slice* means to partition the interval $[a, b]$ into small pieces; *approximate* means to suppose that on a typical piece from x to $x + \Delta x$, the force is constant with value $F(x)$, so that the bit of work done is $\Delta W \approx F(x)\Delta x$; *integrate* means to add up all the bits of work corresponding to the pieces Δx and then take the limit as the length of the pieces approaches zero (Figure 2). We conclude that the work done in moving the object from a to b is given by

$$W = \int_a^b F(x)\,dx$$

Application to Springs   According to Hooke's Law in physics, the force $F(x)$ necessary to keep a spring stretched (or compressed) x units beyond (or short of) its natural length (Figure 3) is given by

$$F(x) = kx$$

Here, the constant k, the so-called spring constant, is positive and depends on the particular spring under consideration. The stiffer the spring, the greater the value of k.

EXAMPLE 1: If the natural length of a spring is 10 inches and if it takes a force of 3 pounds to keep it extended 2 inches, find the work done in stretching the spring from its natural length to a length of 15 inches.

SOLUTION: By Hooke's Law, the force $F(x)$ required to keep the spring stretched x inches is given by $F(x) = kx$. To evaluate k for this particular spring, we note that $F(2) = 3$. Thus, $k \cdot 2 = 3$, or $k = \frac{3}{2}$, and so

$$F(x) = \frac{3}{2}x$$

When the spring is at its natural length of 10 inches, $x = 0$; when it is 15 inches long, $x = 5$. Therefore, the work done in stretching the spring is

$$W = \int_0^5 \frac{3}{2}x\,dx = \left[\frac{3}{2} \cdot \frac{x^2}{2}\right]_0^5 = \frac{75}{4} = 18.75 \text{ inch-pounds} \quad \blacktriangleleft$$

Application to Pumping a Liquid   To pump water out of a tank requires work, as anyone who has ever tried a hand pump will know (Figure 4). But how much work? That is the question.

While the problem does not quite fit within the previous discussion, its solution rests on the same basic principles. We illustrate.

EXAMPLE 2: A tank in the shape of a right circular cone (Figure 5) is full of water. If the height of the tank is 10 feet and the radius of its top is 4 feet, find the work done in pumping the water (a) over the top edge of the tank, and (b) to a height 10 feet above the top of the tank.

SOLUTION: (a) Position the tank in a coordinate system, as shown in Figure 5. Both a three-dimensional view and a two-dimensional cross section are shown. Imagine slicing the water into thin horizontal disks, each of which must be lifted over the edge of the tank. A disk of thickness Δy at height y has radius $4y/10$. Thus, its volume is approximately $\pi(4y/10)^2 \, \Delta y$ and its weight is about $\delta\pi(4y/10)^2 \, \Delta y$, where $\delta = 62.4$ is the (weight) density of water in pounds per cubic foot. The force necessary to lift this disk of water is its weight, and it must be lifted a distance $10 - y$. Thus, the work ΔW done on this disk is approximately

$$\Delta W = (\text{force}) \cdot (\text{distance}) \approx \delta\pi\left(\frac{4y}{10}\right)^2 \Delta y \cdot (10 - y)$$

$$W = \int_0^{10} \delta\pi\left(\frac{4y}{10}\right)^2 (10 - y)\,dy = \delta\pi\frac{4}{25}\int_0^{10}(10y^2 - y^3)\,dy$$

$$= \frac{(4\pi)(62.4)}{25}\left[\frac{10y^3}{3} - \frac{y^4}{4}\right]_0^{10} \approx 26{,}100 \text{ foot-pounds}$$

Notice that we rounded our result to three digits because the density of water was given to only three digits.

(b) Part (b) is just like part (a), except that each disk of water must now be lifted a distance $20 - y$, rather than $10 - y$. Thus,

$$W = \delta\pi\int_0^{10}\left(\frac{4y}{10}\right)^2 (20 - y)\,dy = \frac{4\delta\pi}{25}\int_0^{10}(20y^2 - y^3)\,dy$$

$$= \frac{(4\pi)(62.4)}{25}\left[\frac{20y^3}{3} - \frac{y^4}{4}\right]_0^{10} \approx 131{,}000 \text{ foot-pounds}$$

Note that the limits are still 0 and 10 (not 0 and 20). Why? ◄

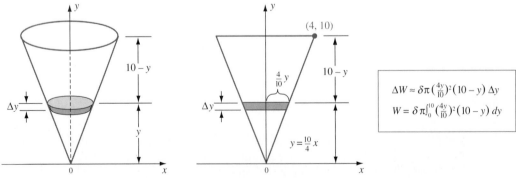

Figure 5

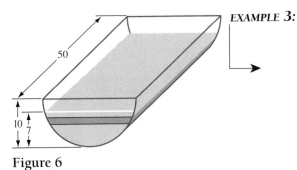

Figure 6

EXAMPLE 3: Find the work done in pumping the water over the rim of a tank that is 50 feet long and has a semicircular end of radius 10 feet, if the tank is filled to a depth of 7 feet (Figure 6).

SOLUTION: We position the end of the tank in a coordinate system, as shown in Figure 7. A typical horizontal slice is shown both on this two-dimensional picture and the three-dimensional one in Figure 6. This slice is approximately a thin box, so we calculate its volume by multiplying length, width, and thickness. Its weight is its density $\delta = 62.4$ times its volume. Finally we note that this slice must be lifted through a distance of $-y$ (the minus sign results from the fact that y is negative in our diagram).

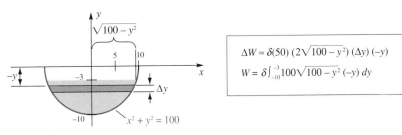

Figure 7

$$\Delta W \approx \delta(50)\,(2\sqrt{100 - y^2})\,(\Delta y)\,(-y)$$
$$W = \delta \int_{-10}^{-3} 100\sqrt{100 - y^2}\,(-y)\,dy$$

$$W = 50\delta \int_{-10}^{-3} (100 - y^2)^{\frac{1}{2}}(-2y)\,dy = \left[(50\delta)\left(\frac{2}{3}\right)(100 - y^2)^{\frac{3}{2}}\right]_{-10}^{-3}$$

$$= \tfrac{100}{3}(91)^{\frac{3}{2}}\,\delta \approx 1{,}810{,}000 \text{ foot-pounds} \qquad \blacktriangleleft$$

Concepts Review

1. The work done by a force F in moving an object along a straight line from a to b is _____ if F is constant but is _____ if $F = F(x)$ is variable.

2. Hooke's Law says that the force F required to keep a spring stretched x units beyond its natural length is _____.

3. The work done in lifting an object weighing 30 pounds from ground level to a height of 10 feet is _____ foot-pounds.

4. The work done in lifting a thin horizontal disk of water of radius 5 feet and thickness Δy feet a distance of $12 - y$ feet is _____ foot-pounds, assuming that water weighs 62.4 pounds per cubic foot. Thus, if a cylindrical tank of radius 5 feet and height 12 feet is full of water, the total work done in pumping all the water over the top rim is given by the integral _____.

Problem Set 6.4

1. A force of 8 pounds is required to keep a spring stretched $\frac{1}{2}$ foot beyond its normal length. Find the value of the spring constant and the work done in stretching the spring $\frac{1}{2}$ foot beyond its natural length. Assume Hooke's Law.

2. For the spring of Problem 1, how much work is done in stretching the spring 1 foot?

3. A force of 200 dynes is required to keep a spring of natural length 10 centimeters compressed to a length of 8 centimeters. Find the work done in compressing the spring from its natural length to a length of 6 centimeters. (Assume Hooke's Law applies to compressing as well as stretching.)

4. It requires 60 ergs (dyne-centimeters) of work

to stretch a spring from a length of 8 centimeters to 9 centimeters and another 120 ergs to stretch it from 9 centimeters to 10 centimeters. Evaluate the spring constant and find the natural length of the spring, assuming Hooke's Law.

5. For any spring obeying Hooke's Law, show that the work done in stretching a spring a distance d is given by $W = \frac{1}{2}kd^2$.

6. Two similar springs S_1 and S_2, each 3 feet long, are such that the force required to keep either of them stretched a distance of s feet is $F = 8s$ pounds. One end of one spring is fastened to an end of the other, and the combination is stretched between the walls of a room 12 feet wide (Figure 8). What work is done in moving the midpoint, P, 1 foot to the right?

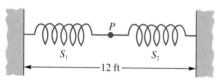

Figure 8

7. Nonlinear springs are springs that do not obey Hooke's Law. Real-world springs generally have an *elastic range* for which they obey Hooke's Law, but beyond which the linear assumption fails to hold. Consider a nonlinear spring with force function $F(x) = 5x - 0.01x^3$ for $0 \le x \le 10$ where the force $F(x)$ is given in pounds and the distance x is given in inches.

(a) Sketch the graph of the force function and from a visual inspection of the graph determine a range of x-values for which Hooke's Law approximately holds (that is, the range for which the function is approximately linear). What is the approximate spring constant k for the linear range?

(b) Calculate the work done to stretch the spring to 2 inches and to 6 inches.

(c) Calculate the work done to stretch the spring up to the point where the force is 30 pounds. Is the spring still in the approximately linear range at this point? Explain.

8. Repeat Problem 7 for the force function $F(x) = 50 \sin (0.1x)$.

In each of Problems 9–12, the vertical end of a tank is shown. Assume that the tank is 10 feet long, that it is full of water, and that the water is to be pumped to a height 5 feet above the top of the tank. Find the work done in emptying the tank.

9.

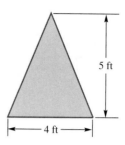

10.

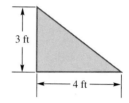

11.

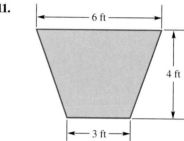

12.

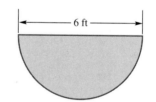

13. Find the work done in pumping all the oil (density $\delta = 50$ pounds per cubic foot) over the edge of a cylindrical tank that stands on end. Assume that the radius of the base is 5 feet, that the height is 10 feet, and that the tank is full of oil.

14. Do Problem 13 assuming that the tank has circular cross sections of radius $5 + x$ feet at height x feet above the base.

15. A volume v of gas is confined in a cylinder, one end of which is closed by a movable piston. If A is the area in square inches of the face of the piston and x is the distance in inches from the cylinder head to the piston, then $v = Ax$. The pressure of the confined gas is a continuous function p of the volume, and $p(v) = p(Ax)$ will be denoted by $f(x)$. Show that the work done by the piston in compressing the gas from a volume $v_1 = Ax_1$ to a volume $v_2 = Ax_2$ is

$$W = A \int_{x_2}^{x_1} f(x)\, dx$$

Hint: The total force on the face of the piston is $p(v) \cdot A = p(Ax) \cdot A = A \cdot f(x)$.

16.   A cylinder and piston, whose cross-sectional area is 1 square inch, contain 16 cubic inches of gas under a pressure of 40 pounds per square inch. If the pressure and the volume of the gas are related adiabatically (that is, without loss of heat) by the law $pv^{1.4} = c$ (a constant), how much work is done by the piston in compressing the gas to 2 cubic inches?

17. If the area of the face of the piston in Problem 16 is 2 square inches, find the work done by the piston.

18. Once cubic foot of air under a pressure of 80 pounds per square inch expands adiabatically to 4 cubic feet according to the law $pv^{1.4} = c$. Find the work done by the gas.

19. A cable weighing 2 pounds per foot is used to haul a 200-pound load to the top of a shaft that is 500 feet deep. How much work is done?

20. A 10-pound monkey hangs at the end of a 20-foot chain that weighs $\frac{1}{2}$ pound per foot. How much work does it do in climbing the chain to the top? Assume that the end of the chain is attached to the monkey.

21. A space capsule weighing 5000 pounds is propelled to an altitude of 200 miles above the surface of the earth. How much work is done against the force of gravity? Assume the earth is a sphere of radius 4000 miles and that the force of gravity is $f(x) = -k/x^2$, where x is the distance from the center of the earth to the capsule (the inverse-square law). Thus the lifting force required is k/x^2, and this equals 5000 when $x = 4000$.

22. According to Coulomb's Law, two like electrical charges repeal each other with a force that is inversely proportional to the square of the distance between them. If the force of repulsion is 10 dynes when they are 2 centimeters apart, find the work done in bringing the charges from 5 centimeters apart to 1 centimeter apart.

23. A bucket weighing 100 pounds is filled with sand weighing 500 pounds. A crane lifts the bucket from the ground to a point 80 feet in the air at a rate of 2 feet per second, but sand simultaneously leaks out through a hole at 3 pounds per second. Neglecting friction and the weight

of the cable, determine how much work is done. *Hint:* Begin by estimating ΔW, the work required to lift the bucket from y to $y + \Delta y$.

24. Center City has just built a new water tower (Figure 9). Its main elements consist of a spherical tank having an inner radius of 10 feet and weighing 10,000 pounds; four 40-foot pillars of linear density $100 \cos(x^2/10{,}000)$ pounds per foot, with x measuring the number of feet from the base; and a 30-foot filler pipe weighing 40 pounds per foot. Initially, all elements were on the ground with the pillars and pipe lying flat. The tank was lifted to position and then the pillars and pipe were pivoted about their bases to the vertical. Find the work done.

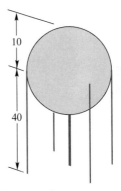

Figure 9

25. The filler pipe for the water tower of Problem 24 has inner diameter 1 foot. Assume that the water was pumped from ground level up through the pipe into the tank. How much work was done in filling the pipe and the tank with water?

26. A conical buoy weighs m pounds and floats with its vertex V down and h feet below the surface of the water (Figure 10). A boat crane lifts the buoy to the deck so that V is 15 feet above the water surface. How much work

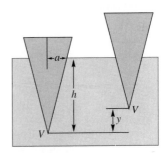

Figure 10

is done? *Hint:* Use Archimedes's Principle, which says that the force required to hold the buoy y feet above its original position ($0 \le y \le h$) is equal to its weight minus the weight of the water displaced by the buoy.

27. Rather than lifting the buoy of Problem 26 and Figure 10 out of the water, suppose we attempt to push it down until its top is even with the water level. Assume that $h = 8$, that the top is originally 2 feet above water level, and that the buoy weighs 300 pounds. How much work is required? *Hint:* You do not need to know a (the radius at water level), but it is helpful to know that $\delta(\frac{1}{3}\pi a^3)(8) = 300$. Archimedes's Principle implies that the force needed to hold the buoy z feet ($0 \le z \le 2$) below floating position is equal to the weight of the additional water displaced.

28. Initially the bottom tank in Figure 11 was full of water and the top tank was empty. Find the work done in pumping all the water into the top tank. The dimensions are in feet.

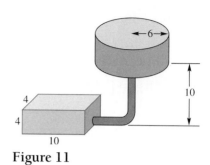

Figure 11

Answers to Concepts Review: 1. $F \cdot (b - a)$, $\displaystyle\int_a^b F(x)\,dx$

2. kx 3. 300

4. $62.4(12 - y)\pi(25)\Delta y$, $62.4\pi(25)\displaystyle\int_0^{12}(12 - y)\,dy$

6.5 | Moments; Center of Mass

Suppose that two masses of sizes m_1 and m_2 are placed on a seesaw at distances d_1 and d_2 from the fulcrum and on opposite sides of it (Figure 1). We will refer to such masses as *point masses*. The seesaw will balance if and only if $d_1 m_1 = d_2 m_2$.

Figure 1

A good mathematical model for this situation is obtained by replacing the seesaw with a horizontal coordinate line having its origin at the fulcrum (Figure 2). Then the coordinate x_1 of m_1 is $x_1 = -d_1$, that of m_2 is $x_2 = d_2$, and the condition for balance is

$$x_1 m_1 + x_2 m_2 = 0$$

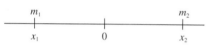

Figure 2

The product of the mass m of a particle and its *directed* distance from a point (the length of its lever arm) is called the **moment** of the particle with respect to that point (Figure 3). It measures the tendency of the mass to produce a rotation about that point. The condition for two masses along a line to balance at a point on that line is that the sum of their moments with respect to the point be zero.

The situation just described can be generalized. The total moment M (with respect to the origin) of a system of n masses of sizes m_1, $m_2, \ldots, m_n$ located at points $x_1, x_2, \ldots, x_n$ along the x-axis is the sum of individual moments—that is,

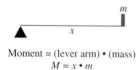

Moment = (lever arm) • (mass)
$$M = x \cdot m$$

Figure 3

$$M = x_1 m_1 + x_2 m_2 + \cdots + x_n m_n = \sum_{i=1}^n x_i m_i$$

The condition for balance at the origin is that $M = 0$. Of course, we should not expect balance at the origin except in special circumstances. But surely any system of masses will balance somewhere. The question is where. What is the x-coordinate of the point where the fulcrum should be placed to make the system in Figure 4 balance?

Figure 4

Call the desired coordinate $\bar{x}$. The total moment with respect to it should be zero—that is,

$$(x_1 - \bar{x})m_1 + (x_2 - \bar{x})m_2 + \cdots + (x_n - \bar{x})m_n = 0$$

or

$$x_1 m_1 + x_2 m_2 + \cdots + x_n m_n = \bar{x}m_1 + \bar{x}m_2 + \cdots + \bar{x}m_n$$

When we solve for $\bar{x}$, we obtain

$$\bar{x} = \frac{M}{m} = \frac{\sum_{i=1}^{n} x_i m_i}{\sum_{i=1}^{n} m_i}$$

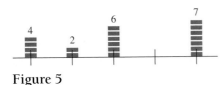

Figure 5

The point $\bar{x}$, called the **center of mass**, is the balance point. Notice that it is just the total moment with respect to the origin divided by the total mass.

EXAMPLE 1: Masses of 4, 2, 6, and 7 pounds are located at points 0, 1, 2, and 4, respectively, along the x-axis (Figure 5). Find the center of mass.

SOLUTION:

$$\bar{x} = \frac{(0)(4) + (1)(2) + (2)(6) + (4)(7)}{4 + 2 + 6 + 7} = \frac{42}{19} \approx 2.21$$

Your intuition should confirm that $x = 2.21$ is about right for the balance point. ◄

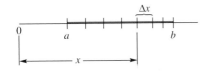

$$\Delta m \approx \delta(x)\, \Delta x \qquad \Delta M \approx x\delta(x)\, \Delta x$$
$$m = \int_a^b \delta(x)\, dx \qquad \Delta M = \int_a^b x\delta(x)\, dx$$

Continuous Mass Distribution Along a Line   Consider now a straight segment of thin wire of varying density (mass per unit length) for which we want to find the balance point. We impose a coordinate line along the wire and suppose that the density at x is $\delta(x)$. By *the density at a point* we mean that the mass of a very small piece of the wire of length Δx centered at x is approximately equal to $\Delta x \cdot \delta(x)$. Following our standard procedure (*slice, approximate, integrate*), we first obtain the total mass m and then the total moment M with respect to the origin (Figure 6). This leads us to the formula

$$\bar{x} = \frac{M}{m} = \frac{\displaystyle\int_a^b x\delta(x)\, dx}{\displaystyle\int_a^b \delta(x)\, dx}$$

Two comments are in order. First, remember this formula by analogy with the point-mass formula from above.

$$\frac{\sum x_i m_i}{\sum m_i} \sim \frac{\sum x \Delta m}{\sum \Delta m} \sim \frac{\int x \delta(x)\, dx}{\int \delta(x)\, dx}$$

Second, note that we have assumed that moments of small pieces of wire add together to give the total moment, just as was the case for point masses. This should seem reasonable to you if you imagine the mass of the typical piece of length Δx to be concentrated at the point x.

0   10

Figure 7

EXAMPLE 2: The density $\delta(x)$ of a wire at the point x centimeters from one end is given by $\delta(x) = 3x^2$ grams per centimeter. Find the center of mass of the piece between $x = 0$ and $x = 10$.

SOLUTION: We expect $\bar{x}$ to be nearer 10 than 0, because the wire is much heavier (denser) toward the right end (Figure 7).

$$\bar{x} = \frac{\displaystyle\int_0^{10} x \cdot 3x^2\, dx}{\displaystyle\int_0^{10} 3x^2\, dx} = \frac{[3x^4/4]_0^{10}}{[x^3]_0^{10}} = \frac{7500}{1000} = 7.5 \text{ cm} \quad \blacktriangleleft$$

Mass Distributions in the Plane

Consider n point masses of sizes $m_1, m_2, \ldots, m_n$ situated at points $(x_1, y_1), (x_2, y_2), \ldots, (x_n, y_n)$ in the coordinate plane (Figure 8). Then the total moments M_y and M_x with respect to the y-axis and x-axis, respectively, are given by

$$M_y = \sum_{i=1}^{n} x_i m_i \qquad M_x = \sum_{i=1}^{n} y_i m_i$$

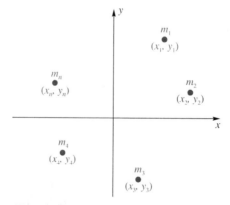

Figure 8

The coordinates $(\bar{x}, \bar{y})$ of the center of mass (balance point) are

$$\bar{x} = \frac{M_y}{m} = \frac{\sum\limits_{i=1}^{n} x_i m_i}{\sum\limits_{i=1}^{n} m_i} \qquad \bar{y} = \frac{M_x}{m} = \frac{\sum\limits_{i=1}^{n} y_i m_i}{\sum\limits_{i=1}^{n} m_i}$$

EXAMPLE 3: Five particles, having masses 1, 4, 2, 3, and 2 units are located at points $(6, -1)$, $(2, 3)$, $(-4, 2)$, $(-7, 4)$, and $(2, -2)$, respectively. Find the center of mass.

SOLUTION:

$$\bar{x} = \frac{(6)(1) + (2)(4) + (-4)(2) + (-7)(3) + (2)(2)}{1 + 4 + 2 + 3 + 2} = \frac{-11}{12}$$

$$\bar{y} = \frac{(-1)(1) + (3)(4) + (2)(2) + (4)(3) + (-2)(2)}{1 + 4 + 2 + 3 + 2} = \frac{23}{12} \qquad \blacktriangleleft$$

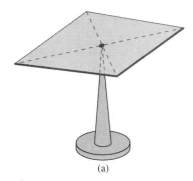

We next consider the problem of finding the center of mass of a lamina (thin planar sheet). For simplicity, we suppose it is homogeneous—that is, that it has constant mass density δ. For a homogeneous rectangular sheet, the center of mass is at the geometric center, as diagrams (a) and (b) in Figure 9 suggest.

Consider the homogeneous lamina bounded by $x = a$, $x = b$, $y = f(x)$, and $y = g(x)$, with $g(x) \le f(x)$. *Slice* this lamina into narrow strips parallel to the y-axis, which are therefore nearly rectangular in shape, and imagine the mass of each strip to be concentrated at its geometric center. Then *approximate* and *integrate* (Figure 10). From this we can calculate the coordinates $(\bar{x}, \bar{y})$ of the center of mass using the formulas

$$\bar{x} = \frac{M_y}{m} \qquad \bar{y} = \frac{M_x}{m}$$

(a)

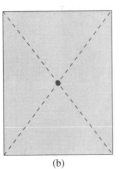

(b)

Figure 9

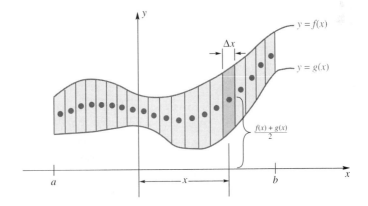

$$\Delta m \approx \delta\,[\,f(x) - g(x)\,]\,\Delta x \qquad\qquad m = \delta \int_a^b [\,f(x) - g(x)\,]\,dx$$

$$\Delta M_y \approx x\,\delta\,[\,f(x) - g(x)\,]\,\Delta x \qquad\qquad M_y = \delta \int_a^b x[\,f(x) - g(x)\,]\,dx$$

$$\Delta M_x \approx \frac{f(x) + g(x)}{2}\,\delta\,[\,f(x) - g(x)\,]\,\Delta x \qquad\qquad M_x = \frac{\delta}{2} \int_a^b [\,f^2(x) - g^2(x)\,]\,dx$$

Figure 10

When we do, the factor δ cancels between numerator and denominator, and we obtain

$$\bar{x} = \frac{\displaystyle\int_a^b x[f(x) - g(x)]\, dx}{\displaystyle\int_a^b [f(x) - g(x)]\, dx} \qquad \bar{y} = \frac{\dfrac{1}{2}\displaystyle\int_a^b [f^2(x) - g^2(x)]\, dx}{\displaystyle\int_a^b [f(x) - g(x)]\, dx}$$

Sometimes, slicing parallel to the x-axis works better than slicing parallel to the y-axis. This leads to formulas for $\bar{x}$ and $\bar{y}$ in which y is the variable of integration. Do not try to memorize all these formulas. It is much better to remember how they were derived.

The center of mass of a homogeneous lamina does not depend on its density or its mass, but only on the shape of the corresponding region in the plane. Thus, our problem becomes a geometric problem rather than a physical one. Accordingly, we often speak of the **centroid** of a planar region rather than the center of mass of a homogeneous lamina.

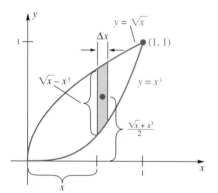

Figure 11

EXAMPLE 4: Find the centroid of the region bounded by the curves $y = x^3$ and $y = \sqrt{x}$.

SOLUTION: Note the diagram in Figure 11.

$$\bar{x} = \frac{\displaystyle\int_0^1 x[\sqrt{x} - x^3]\, dx}{\displaystyle\int_0^1 [\sqrt{x} - x^3]\, dx} = \frac{\left[\dfrac{2}{5}x^{\frac{5}{2}} - \dfrac{x^5}{5}\right]_0^1}{\left[\dfrac{2}{3}x^{\frac{3}{2}} - \dfrac{x^4}{4}\right]_0^1} = \frac{\dfrac{1}{5}}{\dfrac{5}{12}} = \frac{21}{25}$$

$$\bar{y} = \frac{\displaystyle\int_0^1 \frac{1}{2}(\sqrt{x} + x^3)(\sqrt{x} - x^3)\, dx}{\displaystyle\int_0^1 (\sqrt{x} - x^3)\, dx} = \frac{\dfrac{1}{2}\displaystyle\int_0^1 [(\sqrt{x})^2 - (x^3)^2]\, dx}{\displaystyle\int_0^1 (\sqrt{x} - x^3)\, dx}$$

$$= \frac{\dfrac{1}{2}\left[\dfrac{x^2}{2} - \dfrac{x^7}{7}\right]_0^1}{\dfrac{5}{12}} = \frac{\dfrac{5}{28}}{\dfrac{5}{12}} = \frac{3}{7}$$

These answers should seem reasonable to you.   ◄

EXAMPLE 5: Find the centroid of the region under the curve $y = \sin x$, $0 \le x \le \pi$ (Figure 12).

SOLUTION: This region is symmetric about the line $x = \pi/2$, from which we conclude (without integration) that $\bar{x} = \pi/2$. In fact, it is both intuitively obvious and true that if a region has a vertical or horizontal line of symmetry, then the centroid will lie on that line.

Your intuition should also tell you that $\bar{y}$ will be less than $\frac{1}{2}$, because more of the area is near the x-axis. But to find this number exactly, we must calculate

$$\bar{y} = \frac{\displaystyle\int_0^\pi \frac{1}{2}\sin x \cdot \sin x\, dx}{\displaystyle\int_0^\pi \sin x\, dx} = \frac{\dfrac{1}{2}\displaystyle\int_0^\pi \sin^2 x\, dx}{\displaystyle\int_0^\pi \sin x\, dx}$$

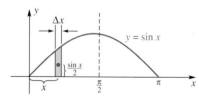

Figure 12

The denominator is easy to calculate; it has value 2. To calculate the numerator exactly, we use the half-angle formula $\sin^2 x = (1 - \cos 2x)/2$. Alternatively, we could have used a numerical approximation with calculator or computer, because our final result must be approximated anyway.

$$\int_0^\pi \sin^2 x \, dx = \frac{1}{2}\left(\int_0^\pi 1 \, dx - \int_0^\pi \cos 2x \, dx\right)$$

$$= \frac{1}{2}\left[x - \frac{1}{2}\sin 2x\right]_0^\pi = \frac{\pi}{2}$$

Thus,

$$\bar{y} = \frac{\dfrac{1}{2}\cdot\dfrac{\pi}{2}}{2} = \frac{\pi}{8} \approx 0.39 \qquad \blacktriangleleft$$

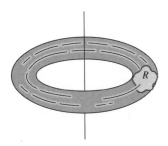

Figure 13

The Theorem of Pappus   About A.D. 300, the Greek geometer Pappus stated a novel result that connects centroids with volumes of solids of revolution (Figure 13).

Theorem A

(Pappus's Theorem). If a region R, lying on one side of a line in its plane, is revolved about that line, then the volume of the resulting solid is equal to the area of R multiplied by the distance traveled by its centroid.

Rather than prove this theorem, which is really quite easy (see Problem 22), we illustrate it.

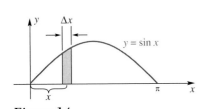

Figure 14

EXAMPLE 6: Illustrate the correctness of Pappus's Theorem for the region under $y = \sin x$, $0 \le x \le \pi$, when it is revolved about the x-axis (Figure 14).

SOLUTION: This is the region of Example 5, for which $\bar{y} = \pi/8$. The area A of this region is

$$A = \int_0^\pi \sin x \, dx = [-\cos x]_0^\pi = 2$$

The volume V of the corresponding solid of revolution is

$$V = \pi\int_0^\pi \sin^2 x \, dx = \frac{\pi}{2}\int_0^\pi [1 - \cos 2x] \, dx$$

$$= \frac{\pi}{2}\left[x - \frac{1}{2}\sin 2x\right]_0^\pi = \frac{\pi^2}{2}$$

We must show that

$$A(2\pi\bar{y}) = V$$

But this amounts to showing

$$2\left(2\pi\,\frac{\pi}{8}\right) = \frac{\pi^2}{2}$$

which is clearly true.   ◀

Concepts Review

1. An object of mass 4 is at $x = 1$ and a second object of mass 6 is at $x = 3$. Simple geometric intuition tells us that the center of mass will be to the _____ of $x = 2$. In fact, it is at $\bar{x} =$ _____.

2. A homogeneous wire lying along the x-axis between $x = 0$ and $x = 5$ will balance at $\bar{x} =$ _____. However, if the wire has density $\delta(x) = 1 + x$, it will balance to the _____ of 2.5. In fact, it will balance at $\bar{x}$, where

$$\bar{x} = \int_0^5 \underline{\hspace{2cm}}\, dx \Big/ \int_0^5 \underline{\hspace{2cm}}\, dx$$

3. The homogeneous rectangular lamina with corner points $(0, 0)$, $(2, 0)$, $(2, 6)$, and $(0, 6)$ will balance at $\bar{x} =$ _____, $\bar{y} =$ _____.

4. A rectangular lamina with corners at $(2, 0)$, $(4, 0)$, $(4, 2)$, and $(2, 2)$ is attached to the lamina of Question 3. Assuming both laminas have the same constant density, the resulting L-shaped lamina will balance at $\bar{x} =$ _____, $\bar{y} =$ _____.

Problem Set 6.5

If integration is required for any problem below, try to get an exact answer first. Failing that, find a numerical approximation.

1. Particles of mass $m_1 = 4$, $m_2 = 6$, and $m_3 = 9$ are located at $x_1 = 2$, $x_2 = -2$, and $x_3 = 1$, respectively, along a line. Where is the center of mass?

2. John and Mary—weighing 150 and 120 pounds, respectively—sit at opposite ends of a 12-foot teeter board with the fulcrum in the middle. Where should their 80-pound son Tom sit in order for the board to balance?

3. A straight wire 9 units long has density $\delta(x) = \sqrt{x}$ at a point x units from one end. Find the distance from this end to the center of mass.

4. Do Problem 3 with $\delta(x) = \ln(1 + x)$.

5. The masses and coordinates of a system of particles in the coordinate plane are given by the following: 3, $(1, 1)$; 2, $(7, 1)$; 4, $(-2, -5)$; 6, $(-1, 0)$; 2, $(4, 6)$. Find the moments of this system with respect to the coordinate axes, and find the coordinates of the center of mass.

6. The masses and coordinates of a system of particles are given by the following: 3, $(-3, 2)$; 6, $(-2, -2)$; 2, $(3, 5)$; 5, $(4, 3)$; 1, $(7, -1)$. Find the moments of this system with respect to the coordinate axes, and find the coordinates of the center of mass.

In Problems 7–10, divide the indicated region into rectangular pieces and assume that the moments M_x and M_y of

the whole region can be found by adding the corresponding moments of the pieces. Use this to find the centroid of each region.

7.

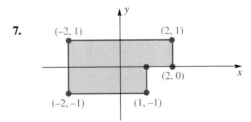

8.

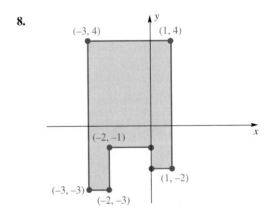

9.

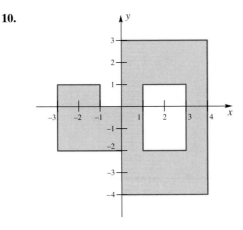

10.

In Problems 11–18, find the centroid of the region bounded by the given curves. Make a sketch and use symmetry where possible.

11. $y = 4 - x^2, y = 0$

12. $y = \frac{1}{2}x^2, y = 0, x = 4$

13. $y = x^3, y = 0, x = 2$

14. $y = \frac{1}{2}(x^2 - 10), y = 0$, and between $x = -2$ and $x = 3$

15. $y = 2x - 4, y = e^{-x}, x = 0$

16. $y = e^x + e^{-x}, y = x + 2$

17. $x = y^2, x = 4$

18. $x = y^2 - 3y - 4, x = -y - 1$

19. Use Pappus's Theorem to find the volume of the solid when the region bounded by $y = x^3$, $y = 0$, and $x = 2$ is revolved about the y-axis (see Problem 13 for the centroid). Do the same problem by using another method to check your answer.

20. Use Pappus's Theorem to find the volume of the torus obtained when the region inside the circle $x^2 + y^2 = a^2$ is revolved about the line $x = 2a$.

21. Use Pappus's Theorem together with the known volume of a sphere to find the centroid of a semicircular region of radius a.

22. Prove Pappus's Theorem by assuming that the region of area A in Figure 15 is to be revolved about the y-axis. *Hint*: $V = 2\pi \int_a^b xh(x)\,dx$ and $\overline{x} = \int_a^b xh(x)\,dx/A$.

23. The region of Figure 15 is revolved about the line $y = e$, generating a solid.

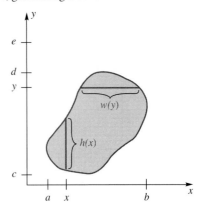

Figure 15

(a) Use the methods of Section 6.2 to write a formula for the volume in terms of $w(y)$.

(b) Show that Pappus's formula, when simplified, gives the same result.

24. Consider the triangle T of Figure 16.
(a) Show that $\overline{y} = h/3$ (and thus that the centroid of a triangle is at the intersection of the medians).
(b) Find the volume of the solid obtained when T is revolved around $y = k$ (Pappus's Theorem).

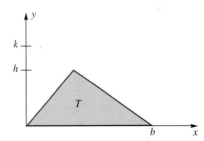

Figure 16

25. A regular polygon P of $2n$ sides is inscribed in a circle of radius r.
(a) Find the volume of the solid obtained when P is revolved about one of its sides.
(b) Check your answer by letting $n \to \infty$.

Answers to Concepts Review: 1. Right; $(4 \cdot 1 + 6 \cdot 3)/10 = 2.2$ 2. 2.5; right; $x(1 + x)$; $1 + x$ 3. 1; 3   4. $\frac{24}{16}; \frac{40}{16}$

6.6 | Chapter Review

Concepts Test

Respond with true or false to each of the following assertions. Be prepared to justify your answer.

1. The area of the region bounded by $y = \cos x$, $y = 0$, $x = 0$, and $x = \pi$ is $\int_0^\pi \cos x \, dx$.

2. The area of a circle of radius a is $4 \int_0^a \sqrt{a^2 - x^2} \, dx$

3. The area of the region bounded by $y = f(x)$, $y = g(x)$, $x = a$, and $x = b$ is either $\int_a^b [f(x) - g(x)] \, dx$ or its negative.

4. All right cylinders whose bases have the same area and whose heights are the same have identical volumes.

5. If two solids with bases in the same plane have cross sections of the same area in all planes parallel to their bases, then they have the same volume.

6. If the radius of the base of a cone is doubled while the height is halved, the volume will remain the same.

7. The area of the region bounded by $y = \sqrt{x}$, $y = 0$, and $x = 1$ is $\dfrac{\pi}{2}$.

8. The solids obtained by revolving the region of Problem 7 about $x = 0$ and $x = 1$ have the same volume.

9. Any smooth curve in the plane that lies entirely within the unit circle will have finite length.

10. The work required to stretch a spring 2 inches beyond its natural length is twice that required to stretch it 1 inch (assume Hooke's Law).

11. It will require the same amount of work to empty a cone-shaped tank and a cylindrical tank of water by pumping it over the rim if both tanks have the same height and volume.

12. Two weights of 100 pounds at distances 10 and 15 feet from the fulcrum will just balance a 200-pound weight on the other side of the fulcrum and 12.5 feet from it.

13. If $\bar{x}$ is the center of mass of a system of masses $m_1, m_2, \ldots, m_n$ distributed along a line at points with coordinates $x_1, x_2, \ldots, x_n$, respectively, then $\sum_{i=1}^n (x_i - \bar{x}) m_i = 0$.

14. The centroid of the region bounded by $y = \cos x$, $y = 0$, $x = 0$, and $x = 2\pi$ is at $(\pi, 0)$.

15. According to the theorem of Pappus, the volume of the solid obtained by revolving the region (of area 2) bounded by $y = \sin x$, $y = 0$, $x = 0$, and $x = \pi$ about the y-axis is $2(2\pi)\left(\dfrac{\pi}{2}\right) = 2\pi^2$.

16. The area of the region bounded by $y = \sqrt{x}$, $y = 0$, and $x = 9$ is $\int_0^3 (9 - y^2) \, dy$.

17. If the density of a wire is proportional to the square of the distance from its midpoint, then its center of mass is at the midpoint.

18. The centroid of a triangle with base on the x-axis has y-coordinate equal to one-third the altitude of the triangle.

Sample Test Problems

Problems 1–7 refer to the plane to the region R bounded by the curve $y = x - x^2$ and the x-axis (Figure 1).

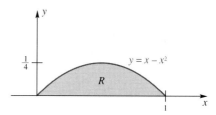

Figure 1

1. Find the area of R.

2. Find the volume of the solid S_1 generated by revolving the region R about the x-axis.

3. Find the volume of the solid S_2 generated by revolving R about the y-axis.

4. Find the volume of the solid S_3 generated by revolving R about the line $y = -1$.

5. Find the volume of the solid S_4 generated by revolving R about the line $x = 2$.

6. Find the coordinates of the centroid of R.

7. Use Pappus's Theorem and Problems 1 and 6 to find the volume of the solids S_1, S_2, S_3 and S_4 above.

8. The natural length of a certain spring is 16 inches, and a force of 8 pounds is required to keep it stretched 4 inches. Find the work done in each case.

(a) Stretching it from a length of 18 inches to a length of 24 inches.

(b) Compressing it from its natural length to a length of 14 inches.

9.   An upright cylindrical tank is 10 feet in diameter and 10 feet high. If water in the tank is 8 feet deep, how much work is done in pumping all the water over the edge of the top of the tank?

10.   An object weighing 300 pounds is suspended from the top of a building by a uniform cable. If the cable is 100 feet long and weighs 120 pounds, how much work is done in pulling the object to the top?

11.   A region R is bounded by the line $y = 3x$ and the parabola $y = x^2$. Find the area of R by (a) taking x as the integration variable, and (b) taking y as the integration variable.

12.   Find the centroid of R in Problem 11.

13.   Find the volume of the solid of revolution generated by revolving the region R of Problem 11 about the x-axis. Check by using Pappus's Theorem.

14.   Find the volume of the solid generated by revolving the region R of Problem 11 about the y-axis. Check by using Pappus's Theorem.

15.   Find or approximate the length of the arc of the curve $y = \sin x$ from $x = 0$ to $x = \pi$.

16.   Sketch the graph of the parametric equations

$$x = t^2, \qquad y = \tfrac{1}{3}(t^3 - 3t)$$

Then find the length of the loop of the resulting curve.

17.   A solid with the semicircular base bounded by $y = \sqrt{4 - x^2}$ and $y = 0$ has cross sections perpendicular to the x-axis that are squares. Find the volume of this solid.

In Problems 18–23, write an expression involving integrals that represents the required concept. Refer to Figure 2.

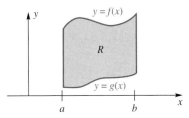

Figure 2

18.   The area of R.

19.   The volume of the solid obtained when R is revolved about the x-axis.

20.   The volume of the solid obtained when R is revolved about $x = a$.

21.   The moments M_x and M_y of a homogeneous lamina with shape R, assuming its density is δ.

22.   The *total* length of the boundary of R.

23.   The *total* surface area of the solid of Problem 19.

CHAPTER 7

Transcendental Functions and Differential Equations

7.1 Inverse Functions and Their Derivatives

We have already used the idea of function inverses to define functions such as $\sin^{-1}x$ (the inverse of $\sin x$), and $\ln x$ (the inverse of e^x). In this section we will study inverse functions in general.

A function f takes a number x from its domain D and assigns it to a single value y from its range R. If we are lucky, as in the case of the two functions graphed in Figures 1 and 2, we can reverse f; that is, for any given y in R, we can unambiguously go back and find the x from which it came. This new function that takes y and assigns x to it is denoted f^{-1}. Note that its domain is R and its range is D. It is called the **inverse** of f, or simply f-inverse. Here we are using the superscript -1 in a new way. The symbol f^{-1} does not denote $1/f$, as you might expect. We, and all mathematicians, use it to name the inverse function.

Sometimes we can give a formula for f^{-1}. If $y = f(x) = 2x$, then $x = f^{-1}(y) = \frac{1}{2}y$ (Figure 1). Similarly, if $y = f(x) = x^3 - 1$, then $x = f^{-1}(y) = \sqrt[3]{y + 1}$ (Figure 2). In each case, we simply solve the equation that determines f for x in terms of y. The result is $x = f^{-1}(y)$.

But life is more complicated than the two examples above indicate. Not every function can be reversed in an unambiguous way. Consider $y = f(x) = x^2$, for example. For each y (except $y = 0$), there are *two* x's that correspond to it (Figure 3). The function $y = g(x) = \sin x$ is even worse. For each y, there are infinitely many x's that correspond to it (Figure 4). Such functions do not have inverses; at least,

395

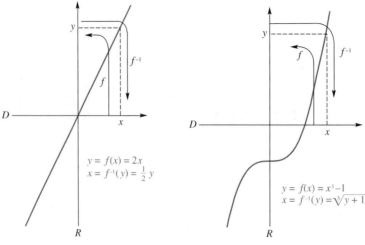

y
f^{-1}
f
D
x
$y = f(x) = 2x$
$x = f^{-1}(y) = \frac{1}{2}y$
R

Figure 1

y
f
f^{-1}
D
x
$y = f(x) = x^3 - 1$
$x = f^{-1}(y) = \sqrt[3]{y+1}$
R

Figure 2

they do not unless we somehow restrict the set of x-values, a subject we will take up later.

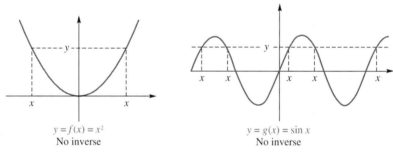

y
x
x
$y = f(x) = x^2$
No inverse

Figure 3

y
x x x x x
$y = g(x) = \sin x$
No inverse

Figure 4

Existence of Inverse Functions

It would be nice to have a simple criterion for deciding whether a function f has an inverse. One such criterion is that the function be **one-to-one**; that is, $x_1 \neq x_2$ implies $f(x_1) \neq f(x_2)$. This is equivalent to the geometric condition that every horizontal line meets the graph of $y = f(x)$ in at most one point. But in a given situation, this criterion may be very hard to apply, because it demands that we have complete knowledge of the graph. A more practical criterion that covers most examples that arise in this book is that a function be **strictly monotonic**. By this we mean that it is either always increasing or always decreasing on its domain.

Theorem A

If f is strictly monotonic on its domain, then f has an inverse.

This is a practical result, because we have an easy way of deciding whether a function f is strictly monotonic. We simply examine the sign of f'.

EXAMPLE 1: Show that $f(x) = x^5 + 2x + 1$ has an inverse.

> **SOLUTION:** $f'(x) = 5x^4 + 2 > 0$ for all x. Thus f is increasing on the whole real line and so has an inverse there. ◄

We do not claim that we can always give a formula for f^{-1}. In the example just considered, this would require that we be able to solve $y = x^5 + 2x + 1$ for x, a task beyond our capabilities (even with computer algebra).

There is a way of salvaging the notion of inverse for functions that do not have inverses on their natural domain. We simply *restrict the domain* to a set on which the graph is either increasing or decreasing. Thus, for $y = f(x) = x^2$, we may restrict the domain to $x \geq 0$ ($x \leq 0$ would also work). For $y = g(x) = \sin x$, we restrict the domain to the interval $[-\pi/2, \pi/2]$. Then both functions have inverses (see Figure 5) and we can give a formula for each one, namely, $f^{-1}(y) = \sqrt{y}$ and $g^{-1}(y) = \sin^{-1}y$.

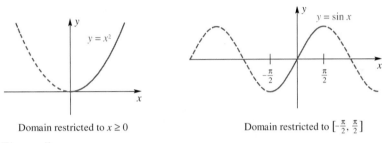

Domain restricted to $x \geq 0$ Domain restricted to $\left[-\frac{\pi}{2}, \frac{\pi}{2}\right]$

Figure 5

If f has an inverse f^{-1}, then f^{-1} also has an inverse, namely, f. Thus, we may call f and f^{-1} a pair of inverse functions. One function undoes (or reverses) what the other did—that is,

$$\boxed{f^{-1}(f(x)) = x \text{ and } f(f^{-1}(y)) = y}$$

EXAMPLE 2: Show that $f(x) = 2x + 6$ has an inverse, find a formula for $f^{-1}(y)$, and verify the results in the box above.

> **SOLUTION:** Because f is an increasing function, it has an inverse. To find $f^{-1}(y)$, we solve $y = 2x + 6$ for x, which gives $x = (y - 6)/2 = f^{-1}(y)$. Finally, note that

$$f^{-1}(f(x)) = f^{-1}(2x + 6) = \frac{(2x + 6) - 6}{2} = x$$

and

$$f(f^{-1}(y)) = f\left(\frac{y - 6}{2}\right) = 2\left(\frac{y - 6}{2}\right) + 6 = y \qquad ◄$$

The Graph of $y = f^{-1}(x)$ Suppose f has an inverse. Then

$$\boxed{x = f^{-1}(y) \text{ if and only if } y = f(x)}$$

Undoing Machines

We may view a function as a machine that accepts an input and produces an output. If the f machine and the f^{-1} machine are hooked together in tandem, they undo each other.

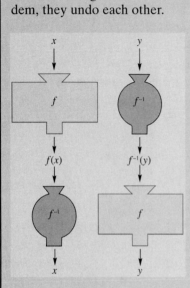

Consequently, $y = f(x)$ and $x = f^{-1}(y)$ determine the same (x, y) pairs, and so have identical graphs. However, it is conventional to use x as the domain variable for functions, so we now inquire about the graph of $y = f^{-1}(x)$. Note that we have interchanged the roles of x and y. A little thought convinces us that to interchange the roles of x and y on a graph is to reflect the graph across the line $y = x$. *Thus, the graph of $y = f^{-1}(x)$ is just the graph of $y = f(x)$ reflected across the line $y = x$* (Figure 6).

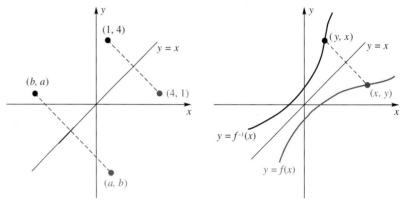

Figure 6

A related matter is that of finding a formula for $f^{-1}(x)$. To do it, we first find $f^{-1}(y)$ and then replace y by x in the resulting formula. Thus, we propose the following three-step process for finding $f^{-1}(x)$.

Step 1 Solve the equation $y = f(x)$ for x in terms of y.

Step 2 Use $f^{-1}(y)$ to name the resulting expression in y.

Step 3 Replace y by x to get the formula for $f^{-1}(x)$.

Before trying the three-step process on a particular function f, you might think we should first verify that f has an inverse. However, if we can actually carry out the first step and get a single x for each y, then f^{-1} does exist. (Note that when we try this for $y = f(x) = x^2$, we get $x = \pm\sqrt{y}$, which immediately shows that f^{-1} does not exist, unless, of course, we have restricted the domain to eliminate one of the two signs, $+$ or $-$.)

EXAMPLE 3: Find a formula for $f^{-1}(x)$ if $y = f(x) = x/(1 - x)$.

SOLUTION: Here are the three steps for this example.

Step 1
$$y = \frac{x}{1 - x}$$
$$(1 - x)y = x$$
$$y - xy = x$$
$$x + xy = y$$
$$x(1 + y) = y$$

$$x = \frac{y}{1 + y}$$

Step 2 $$f^{-1}(y) = \frac{y}{1 + y}$$

Step 3 $f^{-1}(x) = \dfrac{x}{1 + x}$ ◄

Derivatives of Inverse Functions

We conclude this section by investigating the relationship between the derivative of a function and the derivative of its inverse. Consider first what happens to a line l_1 when it is reflected across the line $y = x$. As the left half of Figure 7 makes clear, l_1 is reflected into a line l_2; moreover, their respective slopes m_1 and m_2 are related by $m_2 = 1/m_1$, provided $m_1 \neq 0$. If l_1 happens to be a tangent line to the graph of f at the point (c, d), then l_2 is the tangent line to the graph of f^{-1} at the point (d, c) (see the right half of Figure 7). We are led to the conclusion that

$$(f^{-1})'(d) = m_2 = \frac{1}{m_1} = \frac{1}{f'(c)}$$

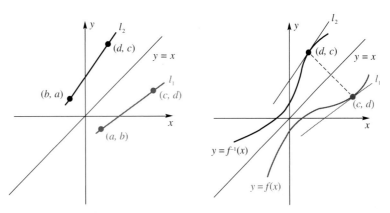

Figure 7

Pictures are sometimes deceptive, so we claim only to have made the following result plausible. For a formal proof, see any advanced calculus book.

Theorem B

(Inverse Function Theorem). Let f be differentiable and strictly monotonic on an interval I. If $f'(x) \neq 0$ at a certain x in I, then f^{-1} is differentiable at the corresponding point $y = f(x)$ in the range of f and

$$(f^{-1})'(y) = \frac{1}{f'(x)}$$

The conclusion to Theorem B is often written symbolically as

$$\frac{dx}{dy} = \frac{1}{dy/dx}$$

EXAMPLE 4: Let $y = f(x) = x^5 + 2x + 1$, as in Example 1. Find $(f^{-1})'(4)$.

SOLUTION: Even though we cannot find a formula for f^{-1} in this case, we note that $y = 4$ corresponds to $x = 1$ and, because $f'(x) = 5x^4 + 2$,

$$(f^{-1})'(4) = \frac{1}{f'(1)} = \frac{1}{5 + 2} = \frac{1}{7}$$ ◄

Concepts Review

1. A function is one-to-one if $x_1 \neq x_2$ implies _____ .

2. A one-to-one function f has an inverse f^{-1} satisfying $f^{-1}(f(x)) =$ _____ and $f(\underline{\hspace{2cm}}) = y$.

3. A useful criterion for f to be one-to-one (and so have an inverse) on a domain is that f be strictly _____ there. This means that f is either _____ or _____ .

4. Let $y = f(x)$, where f has the inverse f^{-1}. The relation connecting the derivatives of f and f^{-1} is _____ .

Problem Set 7.1

In Problems 1–6, the graph of $y = f(x)$ is shown. In each case, decide whether f has an inverse, and if so, estimate $f^{-1}(2)$.

1.

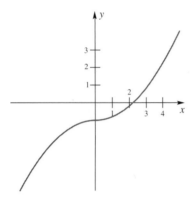

4.

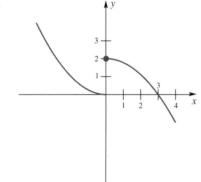

2.

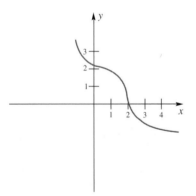

5.

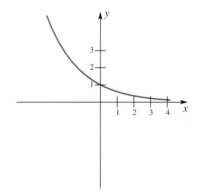

3.

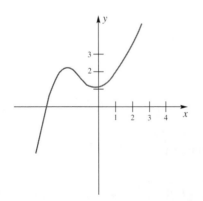

6.

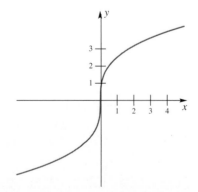

In Problems 7–14, show that f has an inverse by showing that it is strictly monotonic (see Example 1).

7. $f(x) = -3x^5 - x$

8. $f(x) = x^7 + 5x^3$

9. $f(x) = \tan x = \dfrac{\sin x}{\cos x}, -\dfrac{\pi}{2} < x < \dfrac{\pi}{2}$

10. $f(x) = \cos x, 0 \le x \le \pi$

11. $f(x) = 3e^{2x}$

12. $f(x) = \ln(x^2 + 1), x \le 0$

13. $f(x) = \displaystyle\int_0^x \sqrt{t^2 + 2}\, dt$

14. $f(x) = \displaystyle\int_1^x \sin^2 t\, dt$

In Problems 15–28, find a formula for $f^{-1}(x)$, and then verify that $f^{-1}(f(x)) = x$ and $f(f^{-1}(x)) = x$ (see Examples 2 and 3).

15. $f(x) = 3x - 1$

16. $f(x) = -\dfrac{x}{4} + 5$

17. $f(x) = \sqrt{2x + 5}$

18. $f(x) = -\sqrt{2 - x}$

19. $f(x) = \dfrac{1}{x - 5}$

20. $f(x) = \dfrac{1}{\sqrt{x - 3}}$

21. $f(x) = x^2, x \le 0$

22. $f(x) = (x + 1)^2, x \ge -1$

23. $f(x) = (x - 4)^3$

24. $f(x) = x^{3/2}, x \ge 0$

25. $f(x) = 3e^{5x}$

26. $f(x) = 3 \ln(x + 1), x > -1$

27. $f(x) = \dfrac{x^3 + 1}{x^3 + 2}$

28. $f(x) = \left(\dfrac{2x + 1}{3x - 1}\right)^3$

In Problems 29 and 30, restrict the domain of f so that f has an inverse, yet keeping its range as large as possible. Then find $f^{-1}(x)$. *Suggestion*: First graph f.

29. $f(x) = 2x^2 + x - 4$

30. $f(x) = x^2 - 3x + 1$

In each of Problems 31–34, the graph of $y = f(x)$ is shown. Sketch the graph of $y = f^{-1}(x)$ and estimate $(f^{-1})'(3)$.

31.

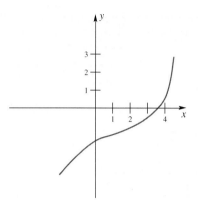

32.

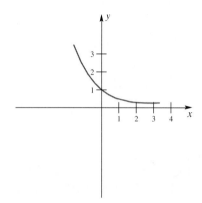

33.

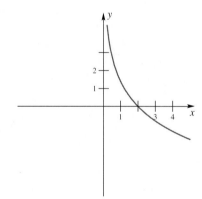

34.

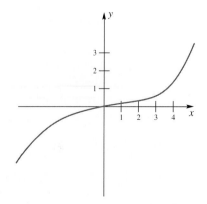

In Problems 35–38, find $(f^{-1})'(2)$ by using Theorem B (see Example 4). Note that you can find the x corresponding to $y = 2$ by inspection.

35. $f(x) = 3x^5 + x - 2$

36. $f(x) = x^5 + 5x - 4$

37. $f(x) = 2 \tan x, \ -\dfrac{\pi}{2} < x < \dfrac{\pi}{2}$

38. $f(x) = \sqrt{x + 1}$

39. Suppose that both f and g have inverses and that $h(x) = f(g(x))$. Show that h has an inverse given by $h^{-1}(y) = g^{-1}(f^{-1}(y))$.

40. Verify the result of Problem 39 for $f(x) = 1/x$, $g(x) = 3x + 2$.

41. If $f(x) = \displaystyle\int_0^x \sqrt{1 + \cos^2 t}\, dt$, then f has an inverse. (Why?) Let $A = f(\pi/2)$ and $B = f(5\pi/6)$. Find (a) $(f^{-1})'(A)$, (b) $(f^{-1})'(B)$, and (c) $(f^{-1})'(0)$.

42. Let $f(x) = \dfrac{ax + b}{cx + d}$ and assume $bc - ad \neq 0$.

(a) Find the formula for $f^{-1}(x)$.

(b) Why is the condition $bc - ad \neq 0$ needed?

(c) What condition on a, b, c, and d will make $f = f^{-1}$?

43. Suppose f is continuous and strictly increasing on $[0, 1]$ with $f(0) = 0$ and $f(1) = 1$. If $\displaystyle\int_0^1 f(x)\,dx = \frac{2}{5}$, calculate $\displaystyle\int_0^1 f^{-1}(y)\,dy$. *Hint:* Draw a picture.

44. Find a function $f(x)$ for which an inverse exists for all x, but for which it is impossible to find a formula for f^{-1} in terms of previously defined functions. *Hint:* Find an $f(x)$ for which $f'(x) > 0$ for all x, but for which $f(x) = y$ cannot be solved algebraically for y.

Answers to Concepts Review: 1. $f(x_1) \neq f(x_2)$
2. $x; f^{-1}(y)$ 3. Monotonic; increasing; decreasing
4. $(f^{-1})'(y) = 1/f'(x)$

7.2 A Different Approach to Logarithmic and Exponential Functions

In Chapter 1 we introduced the natural exponential and natural logarithm functions, because they arise naturally in many different contexts (especially in growth problems). But mathematicians are concerned not only with the usefulness of mathematics in the real world, but also with the development of the subject. This means paying attention to definitions and to the proofs of the theorems.

In Sections 1.4 and 1.5 we left out some details in the proofs of the properties of the log and exponential functions. In this section we show that by using alternate definitions of these functions, we can give more complete proofs of some of these properties. Thus, this section is of more theoretical than practical interest; no new results are presented, but previous results are proven using the new definitions.

We ask you now to put yourself in the position of a student who has not yet encountered the natural logarithm or natural exponential function. The definitions and theorems below present an alternative approach to the development of these functions and their properties. We could not have presented this approach earlier in the book, because it depends on a knowledge of definite integration.

Definition

The **natural logarithm function**, denoted by $f(x) = \ln x$, can be defined by

$$f(x) = \ln x = \int_1^x \frac{1}{t}\, dt, \qquad x > 0$$

Its domain is the set of positive real numbers.

The diagrams in Figure 1 indicate the geometric meaning of ln x. It measures the area under the curve $y = 1/t$ between 1 and x if $x > 1$ and the negative of this area if $0 < x < 1$. Also, ln $1 = 0$. Clearly, ln x is well defined for $x > 0$. And what is the derivative of this function? Just exactly what we expect from our previous work.

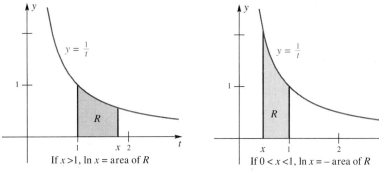

Figure 1

The Derivative of the Natural Logarithm   Recall that the derivative of an integral with respect to its upper limit is the integrand evaluated at the upper limit (Theorem 5.6D). Thus

$$\frac{d}{dx}\ln x = \frac{1}{x} \qquad x > 0$$

This can be combined with the Chain Rule. If $u = f(x) > 0$ and if f is differentiable, then

$$\frac{d}{dx}\ln u = \frac{1}{u}\frac{du}{dx}$$

EXAMPLE 1: Show that

$$\frac{d}{dx}\ln |x| = \frac{1}{x} \qquad x \neq 0$$

SOLUTION: Two cases are to be considered. If $x > 0$, $|x| = x$ and

$$\frac{d}{dx}\ln |x| = \frac{d}{dx}\ln x = \frac{1}{x}$$

If $x < 0$, $|x| = -x$, and so

$$\frac{d}{dx}\ln |x| = \frac{d}{dx}\ln(-x) = \frac{1}{-x}\frac{d}{dx}(-x) = \left(\frac{1}{-x}\right)(-1) = \frac{1}{x} \qquad \blacktriangleleft$$

We know that for every differential formula, there is a corresponding integration formula. Thus, Example 1 implies that

$$\int \frac{1}{x}dx = \ln |x| + C, \qquad x \neq 0$$

or, with u replacing x,

$$\int \frac{1}{u}\,du = \ln|u| + C, \qquad x \neq 0$$

EXAMPLE 2: Evaluate $\displaystyle\int_{-1}^{3} \frac{x}{10 - x^2}\,dx$.

SOLUTION: Let $u = 10 - x^2$, so $du = -2x\,dx$. Then

$$\int \frac{x}{10 - x^2}\,dx = -\frac{1}{2}\int \frac{-2x}{10 - x^2}\,dx = -\frac{1}{2}\int \frac{1}{u}\,du$$

$$= -\frac{1}{2}\ln|u| + C = -\frac{1}{2}\ln|10 - x^2| + C$$

Thus, by the Fundamental Theorem of Calculus,

$$\int_{-1}^{3} \frac{x}{10 - x^2}\,dx = \left[-\frac{1}{2}\ln|10 - x^2| \right]_{-1}^{3}$$

$$= -\frac{1}{2}\ln 1 + \frac{1}{2}\ln 9 = \frac{1}{2}\ln 9$$

For the above calculation to be valid, $10 - x^2$ must never be 0 on the interval $[-1, 3]$. It is easy to see that this is true. ◄

Properties of the Natural Logarithm

Common Logarithms

Properties (ii) and (iii) for common logarithms (base 10 logarithms) were what motivated the invention of logarithms. John Napier (1550–1617) wanted to simplify the complicated calculations that arise in astronomy and navigation. To replace multiplication by addition and division by subtraction was his goal—exactly what (ii) and (iii) accomplish. For over 350 years, common logarithms were an essential aid in computation, but today we use calculators and computers for this purpose. However, natural logarithms retain their importance for other reasons, as you will see.

Theorem A

If a and b are positive numbers and r is any rational number, then

(i) $\ln 1 = 0$;

(ii) $\ln ab = \ln a + \ln b$;

(iii) $\ln \dfrac{a}{b} = \ln a - \ln b$;

(iv) $\ln a^r = r \ln a$

Proof

(i)   See the problems at the end of the section.

(ii) Because, for $x > 0$,

$$\frac{d}{dx}\ln ax = \frac{1}{ax} \cdot a = \frac{1}{x}$$

and

$$\frac{d}{dx}\ln x = \frac{1}{x}$$

it follows from the theorem about two functions with the same derivative (Theorem 4.7B) that

$$\ln ax = \ln x + C$$

To evaluate C, let $x = 1$, obtaining $\ln a = C$. Thus,

$$\ln ax = \ln x + \ln a$$

Finally, let $x = b$.

(iii) See the problems at the end of the section.
(iv) See the problems at the end of the section.

Notice that Theorem A proves that our new definition of the natural logarithm function is consistent with our original definition from Section 1.5. We know this by the following line of reasoning: We have already shown that both definitions lead to the derivative formula $\frac{d}{dx} \ln x = \frac{1}{x}$, and both lead to the conclusion that $\ln 1 = 0$. Two functions with the same derivative differ at most by a constant (Theorem 4.7B); if in addition both functions have the same value at a given point, the constant must be zero, and so the functions must be the same.

The Graph of the Natural Logarithm   The domain of $\ln x$ consists of the positive real numbers, so the graph of $y = \ln x$ is in the right half-plane. Also, for $x > 0$,

$$\frac{d}{dx} \ln x = \frac{1}{x} > 0$$

and

$$\frac{d^2}{dx^2} \ln x = -\frac{1}{x^2} < 0$$

The first formula tells us that the graph is continuous (why?) and rises as x increases; the second tells us that the graph is everywhere concave downward. In Problems 53 and 54, it will be shown that

$$\lim_{x \to \infty} \ln x = \infty$$

and

$$\lim_{x \to 0^+} \ln x = -\infty$$

Finally, $\ln 1 = 0$. These facts imply that the graph of $y = \ln x$ is similar in shape to that shown in Figure 2.

Values of $\ln x$ can be tabulated using numerical integration. For example,

$$\ln 2 = \int_1^2 \frac{1}{x}\, dx \approx 0.6931$$

$$\ln 3 = \int_1^3 \frac{1}{x}\, dx \approx 1.0986$$

$$\ln 10 = \int_1^{10} \frac{1}{x}\, dx \approx 2.3026$$

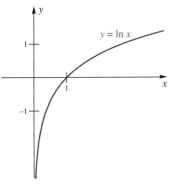

Figure 2

With a scientific or graphing calculator, values for the natural logarithm can be computed using the ln button.

The Exponential Function

The natural logarithm function is differentiable (hence continuous) and increasing on its domain $D = (0, \infty)$; its range is $R = (-\infty, \infty)$. It is, in fact precisely the kind of function studied in Section 7.1 and therefore has an inverse $\ln^{-1}$ with domain $(-\infty, \infty)$ and range $(0, \infty)$. This function is so important that it is given a special name and a special symbol.

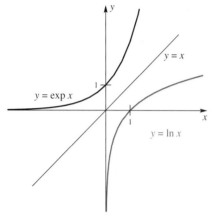

Figure 3

> **Definition**
>
> The inverse of ln is called the **natural exponential function** and is denoted by exp. Thus
>
> $$x = \exp y \text{ if and only if } y = \ln x$$

It follows immediately from this definition that:

(i) $\exp(\ln x) = x$, $x > 0$
(ii) $\ln(\exp y) = y$,   all y

Because exp and ln are inverse functions, the graph of $y = \exp x$ is just the graph of $y = \ln x$ reflected across the line $y = x$ (Figure 3).

But why the name *exponential function*? You will see.

Properties of the Exponential Function

We begin by introducing a new way of defining the constant e, originally defined using limits at the end of Section 1.5. The letter e is appropriate because Leonhard Euler first recognized the significance of this number.

> **Definition**
>
> The letter e denotes the unique positive real number such that $\ln e = 1$.

Because $\ln e = 1$, it is also true that $\exp 1 = e$. The number e, like π, is irrational. Its decimal expansion is known to thousands of places; the first few digits are

$$e \approx 2.718281828459045$$

Now we make a crucial observation, one that depends only on facts already demonstrated—(i) above and Theorem 7.1A. If r is any rational number,

$$e^r = \exp(\ln e^r) = \exp(r \ln e) = \exp r$$

Let us emphasize the result. For rational r, exp r *is identical with* e^r. The natural exponential function, introduced in the most abstract way as the inverse of the natural logarithm, which itself was defined by an integral, has turned out to be just the constant e raised to a variable power.

Possible Definitions of e

Authors choose different ways to define e.

1. $e = \ln^{-1} 1$
   (our new definition)

2. $e = \lim\limits_{h \to 0}(1 + h)^{1/h}$
   (our definition from Section 1.5)

3. $e = \lim\limits_{n \to \infty}$
   $\left(1 + \dfrac{1}{1!} + \dfrac{1}{2!} + \cdots + \dfrac{1}{n!}\right)$

Any of the above statements can be used to define e; the others can then be proved as theorems.

But what if r is irrational? Here we remind you of a gap in our original presentation of the exponential function. Irrational powers were not defined in a totally rigorous manner. What is meant by $e^{\sqrt{2}}$? Previously we defined numbers of this type using limits (Section 1.4), but we did not attempt to prove that these limits exist. Guided by what we learned above, we can now simply define e^x for all x (rational or irrational) by

$$e^x = \exp x$$

Note that (i) and (ii), from the definition of the natural exponential function above, now take the form:

> **(i)′** $e^{\ln x} = x,$ $x > 0$
>
> **(ii)′** $\ln(e^y) = y,$ all y

Also, we can easily prove two of the familiar laws of exponents in the case where the base is the number e.

Theorem B

Let a and b be any real numbers. Then $e^a e^b = e^{a+b}$ and $e^a/e^b = e^{a-b}$.

Proof To prove the first, we write

$$
\begin{aligned}
e^a e^b &= \exp(\ln e^a e^b) && \text{(by (i))}\\
&= \exp(\ln e^a + \ln e^b) && (\text{Theorem 7.1A})\\
&= \exp(a + b) && \text{(by (ii)′)}\\
&= e^{a+b} && (\text{because } \exp x = e^x)
\end{aligned}
$$

For the proof of the second part of Theorem B, see the problems at the end of the section. ◄

The Derivative of e^x

Because exp and ln are inverses, we know from Theorem 7.1B that exp $x = e^x$ is differentiable. To find a formula for $\frac{d}{dx}e^x$, we could use that theorem. Alternatively, let $y = e^x$, so that

$$x = \ln y$$

Now differentiate both sides with respect to x, obtaining

$$1 = \frac{1}{y}\frac{dy}{dx} \qquad \text{(Chain Rule)}$$

Thus,

$$\frac{dy}{dx} = y = e^x$$

We have proved the remarkable fact that e^x is its own derivative; that is,

$$\boxed{\dfrac{d}{dx}e^x = e^x}$$

If $u = f(x)$ is differentiable, then the Chain Rule yields

$$\frac{d}{dx} e^u = e^u \frac{du}{dx}$$

EXAMPLE 3: Find $\dfrac{d}{dx} e^{\sqrt{x}}$.

> **SOLUTION:** Using $u = \sqrt{x}$, we obtain

$$\frac{d}{dx} e^{\sqrt{x}} = e^{\sqrt{x}} \frac{d}{dx} \sqrt{x} = e^{\sqrt{x}} \cdot \frac{1}{2} x^{-\frac{1}{2}} = \frac{e^{\sqrt{x}}}{2\sqrt{x}} \quad \blacktriangleleft$$

The derivative formula $\dfrac{d}{dx} e^x = e^x$ automatically yields the integral formula $\int e^x \, dx = e^x + C$, or with u replacing x,

$$\int e^u \, du = e^u + C$$

EXAMPLE 4: Find $\int x^2 e^{-x^3} \, dx$.

> **SOLUTION:** Let $u = -x^3$, so $du = -3x^2 \, dx$. Then

$$\int x^2 e^{-x^3} \, dx = -\frac{1}{3} \int e^{-x^3} (-3x^2 \, dx)$$

$$= -\frac{1}{3} \int e^u \, du = -\frac{1}{3} e^u + C$$

$$= -\frac{1}{3} e^{-x^3} + C \quad \blacktriangleleft$$

Although the symbol e^y will largely supplant exp y throughout the rest of this book, exp occurs frequently in scientific writing, especially when the exponent y is complicated. For example, in the study of statistics, one often encounters the normal curve, which is the graph of

$$f(x) = \frac{1}{\sigma\sqrt{2\pi}} \exp\left[-\frac{(x - \mu)^2}{2\sigma^2} \right]$$

Concepts Review

1. The function ln can be defined in terms of definite integration by $\ln x =$ _____ . The domain of this function is _____ and its range is _____ .

2. From the preceding definition, it follows that $\dfrac{d}{dx} \ln x =$ _____ for $x > 0$.

3. In this section the number e is defined in terms of ln by _____ ; its value to two decimal places is _____ .

4. Because $e^x = \exp x = \ln^{-1} x$, it follows that $e^{\ln x} =$ _____ and $\ln(e^x) =$ _____ .

Problem Set 7.2

1.   Use the definition of natural logarithm as a definite integral

$$\ln x = \int_1^x \frac{1}{t}\, dt$$

to estimate each of the following. Use a numerical technique for each integral, such as midpoint Riemann sums, or the built-in technique of a computer or calculator.

(a) $\ln 6$ (b) $\ln 1.5$ (c) $\ln 81$

(d) $\ln \sqrt{2}$ (e) $\ln\left(\frac{1}{36}\right)$ (f) $\ln 48$

2.   Use your calculator to make the computations in Problem 1 directly using the ln key.

3.   Calculate

(a) $e^{3 \ln 2}$ (b) $e^{(\ln 64)/2}$

Explain why your answers are not surprising.

4.   Calculate

(a) $\ln \sqrt{e}$ (b) $\ln(e^2 \cdot e^3)$

Explain why your answers are not surprising.

In Problems 5–8, simplify the given expression.

5. $\ln e^{-x+2}$

6. $\ln(x^2 e^{-2x})$

7. $e^{2 \ln x}$

8. $e^{\ln x - 2 \ln y}$

In Problems 9–12, use Theorem A to write the expressions as a logarithm of a single quantity.

9. $2 \ln(x + 1) - \ln x$

10. $\frac{1}{2}\ln(x - 9) + \frac{1}{2}\ln x$

11. $\ln(x - 2) - \ln(x + 2) + 2 \ln x$

12. $\ln(x^2 - 9) - 2 \ln(x - 3) - \ln(x + 3)$

In Problems 13–20, find the indicated derivative. Assume in each case that x is restricted so that ln is defined.

13. $\dfrac{d}{dx} \ln(x - 5)^4$

14. $\dfrac{d}{dx} \ln\sqrt{3x - 25}$

15. $\dfrac{dy}{dx}$ if $y = \ln x^3 + (\ln x)^3$

16. $\dfrac{dy}{dx}$ if $y = \dfrac{1}{\ln x} + \ln\left(\dfrac{1}{x}\right)$

17. $f'(x)$ if $f(x) = \ln\left(x + \sqrt{x^2 - 1}\right)$

18. $f'(x)$ if $f(x) = \ln\left(x + \sqrt{x^2 + 1}\right)$

19. $f'(100)$ if $f(x) = \ln\sqrt[3]{x}$

20. $f'\left(\dfrac{\pi}{2}\right)$ if $f(x) = \ln(\sin x)$

In Problems 21–26, find $\dfrac{dy}{dx}$.

21. $y = e^{\sqrt{x+1}}$

22. $y = e^{\ln x}$

23. $y = x^2 e^x$

24. $y = e^{x^2 \ln x}$

25. $e^{xy} + y = 2$. *Hint:* Use implicit differentiation.

26. $xe^y + 2x - \ln y = 4$

In Problems 27–38, find the integrals. Use a numerical technique where appropriate, if exact methods fail.

27. $\displaystyle \int \frac{\ln x}{x}\, dx$

28. $\displaystyle \int \frac{2}{x(\ln x)^2}\, dx$

29. $\displaystyle \int_0^3 \frac{x^3}{x^4 + 1}\, dx$

30. $\displaystyle \int_0^1 \frac{x + 1}{x^2 + 2x + 2}\, dx$

31. $\displaystyle \int_{0.1}^{20} \cos(\ln x)\, dx$

32. $\displaystyle \int_0^{3\pi} \ln(1.5 + \sin x)\, dx$

33. $\displaystyle \int e^{3x+1}\, dx$

34. $\displaystyle \int xe^{x^2 - 3}\, dx$

35. $\displaystyle \int_1^2 \frac{e^x}{e^x - 1}\, dx$

36. $\displaystyle \int_1^2 \frac{e^{-1/x}}{x^2}\, dx$

37. $\displaystyle \int_{-3}^3 \exp(-1/x^2)\, dx$

38. $\displaystyle \int_0^{8\pi} e^{-0.1^x} \sin x\, dx$

39.   Evaluate $\displaystyle \int_0^{\pi/3} \tan x\, dx$. This integral can be done exactly; use a substitution.

40.   The region bounded by $y = (x^2 + 4)^{-1}$, $y = 0$, $x = 1$, and $x = 4$ is revolved about the y-axis, generating a solid. Find its volume.

41. The region bounded by $y = e^{-x^2}$, $y = 0$, $x = 0$, and $x = 1$ is revolved about the y-axis. Find the volume of the resulting solid.

42. Find the length of the curve given parametrically by $x = e^t \sin t$, $y = e^t \cos t$, $0 \le t \le \pi$.

43. Find the length of the curve $y = x^2/4 - \ln\sqrt{x}$, $1 \le x \le 2$.

44. Let $f(x) = 2x^2 \ln x - x^2$. Find all local extreme values on its domain.

45. Let $f(x) = \ln(1.5 + \sin x)$.

(a) Find the absolute extreme points on $[0, 3\pi]$.

(b) Find any inflection points on $[0, 3\pi]$.

46. Let $f(x) = \cos(\ln x)$.

(a) Find the absolute extreme points on $[0.1, 20]$.

(b) Find the absolute extreme points on $[0.01, 20]$.

47. Draw the graphs of $f(x) = x \ln(1/x)$ and $g(x) = x^2\ln(1/x)$ on $(0, 1]$.

(a) Find the area of the region between these curves on $(0, 1]$.

(b) Find the absolute maximum value of $|f(x) - g(x)|$ on $(0, 1]$.

48. Follow the directions of Problem 47 for $f(x) = x \ln x$ and $g(x) = \sqrt{x} \ln x$.

49. Find the area of the region between the graphs of $y = f(x) = \exp(-x^2)$ and $y = f''(x)$ on $[-3, 3]$.

50. Sketch the graph of $y = \ln \cos x + \ln \sec x$ on $(-\pi/2, \pi/2)$. Explain why you get this graph.

51. Explain why $\lim_{x \to 0} \ln \dfrac{\sin x}{x} = 0$.

52. The rate of transmission in a telegraph cable is observed to be proportional to $x^2\ln(1/x)$, where x is the ratio of the radius of the core to the thickness of the insulation $(0 < x < 1)$. What value of x gives the maximum rate of transmission?

53. Use the fact that $\ln 4 > 1$ to show that $\ln 4^m > m$ for $m > 0$. Conclude that $\ln x$ can be made as large as desired by choosing x sufficiently large. What does this imply about $\lim_{x \to \infty} \ln x$?

54. Use the fact that $\ln x = -\ln(1/x)$ and Problem 53 to show that $\lim_{x \to 0^+} \ln x = -\infty$.

55. Calculate

$$\lim_{n \to \infty} \left[\frac{1}{n+1} + \frac{1}{n+2} + \cdots + \frac{1}{2n} \right]$$

by writing the expression in brackets as

$$\left[\frac{1}{1 + 1/n} + \frac{1}{1 + 2/n} + \cdots + \frac{1}{1 + n/n} \right]\frac{1}{n}$$

and recognizing the latter as a Riemann sum.

56. Evaluate $\lim_{n \to \infty} \dfrac{e^{1/n} + e^{2/n} + \cdots + e^{n/n}}{n}$.

57. A famous theorem (the Prime Number Theorem) says that the number of primes less than n for large n is approximately $n/(\ln n)$. About how many primes are there less than 1,000,000?

58. By appealing to the graph of $y = 1/x$, show that

$$\frac{1}{2} + \frac{1}{3} + \cdots + \frac{1}{n} < \ln n < 1 + \frac{1}{2} + \frac{1}{3} + \cdots + \frac{1}{n}$$

59. Explain why if $a < b$ then $e^{-a} > e^{-b}$.

60. Stirling's Formula says that for large n, we can approximate $n! = 1 \cdot 2 \cdot 3 \cdots \cdot n$ by

$$n! \approx \sqrt{2\pi n}\left(\frac{n}{e}\right)^n$$

(a) Calculate 10! exactly and then approximately using the above formula.

(b) Approximate 60!.

61. If customers arrive at a check-out counter at the average rate of k per minute, then (see books on probability theory) the probability that exactly n customers will arrive in a period of x minutes is given by the formula

$$P_n(x) = \frac{(kx)^n e^{-kx}}{n!}$$

Find the probability that exactly 8 customers will arrive during a 30-minute period if the average rate for this check-out counter is 1 customer every 4 minutes.

62. The **normal curve** with mean μ and standard deviation σ is defined by

$$y = f(x) = \frac{1}{\sigma\sqrt{2\pi}} \exp\left[-\frac{1}{2}\left(\frac{x - \mu}{\sigma}\right)^2\right]$$

Show that it has maximum at $x = \mu$ and inflection points at $x = \mu \pm \sigma$.

63. Draw the graphs of $y = x^p e^{-x}$ for various positive values of p using the same axes. Make conjectures about:

(a) $\lim_{x \to \infty} x^p e^{-x}$;

(b) the x-coordinate of the maximum point for $f(x) = x^p e^{-x}$.

64. Prove parts (i), (iii), and (iv) of Theorem A.

65. Prove the second part of Theorem B.

Answers to Concepts Review: 1. $\int_1^x (1/t)\,dt$; $(0, \infty)$; $(-\infty, \infty)$ 2. $1/x$ 3. $\ln e = 1$; 2.72 4. x; x

LAB 12: FALLING OBJECTS

Mathematical and Physical Background

When an object is dropped, there are two forces acting on it: gravity and air resistance. If there were no air resistance, the object would continue to accelerate—that is, go faster and faster. With air resistance, the object stops accelerating at some point and reaches what is called "terminal velocity." One assumption that seems to work fairly well is that the force due to air resistance is proportional to the velocity raised to some power. The most common model used is to assume a constant times velocity to the first power for the air resistance term (sometimes velocity squared is used). From the assumption on the air resistance force, one can write down a differential equation for the velocity (v) of the object. Then one can solve this differential equation to get the distance of the object above the ground (y) as a function of time (t).

Using Newton's Law, force = mass $\times$ acceleration, we get the differential equation

$$m\frac{dv}{dt} = -kv - mg$$

The term $-kv$ represents the force of air resistance and the term $-mg$ represents the force of gravity. (g is the acceleration due to gravity: $g = 32$ ft/sec^2.) If we divide through by m and let $p = \dfrac{k}{m}$ we get the equation

$$\frac{dv}{dt} = -pv - g \tag{1}$$

Thus, the parameter p is related to the mass m of the object and the air-resistance constant k; p can be determined by experiment for a particular object.

Differential equation (1) can be solved for v using separation of variables; then by integrating again one can obtain the height y. We use the initial conditions $v(0) = 0$ and $y(0) = y_0$.

$$\int \left(\frac{1}{pv + g}\right) dv = -\int dt$$

$$\frac{1}{p}\ln(pv + g) = -t + C_1$$

$$pv + g = C_2 e^{-pt} \qquad (C_2 = e^{pC_1})$$

$$v = C_3 e^{-pt} - \frac{g}{p} \qquad \left(C_3 = \frac{C_2}{p}\right)$$

$$v = \frac{g}{p}(e^{-pt} - 1) \qquad (\text{use } v(0) = 0)$$

The above equation represents the velocity of the object; we integrate again to get the position.

$$\int v\, dt = \int \frac{g}{p}(e^{-pt} - 1)\, dt$$

$$y = -\frac{g}{p}t - \frac{g}{p^2}e^{-pt} + C_3$$

$$y(t) = -\frac{g}{p}t + \frac{g}{p^2}(1 - e^{-pt}) + y_0 \qquad (\text{use } y(0) + y_0)$$

Data

Drop a balloon from six feet and time how long it takes to fall to the ground; call this time T_1. Try to tell from observation whether or not the balloon reached terminal velocity before it hit the ground. Drop the balloon three times and average the results. Now drop the balloon from three feet and time how long it takes to fall (again, do this three times and average). This should be the same as the time it took the balloon to fall halfway to the ground on the first part of the experiment; call this time T_2. Data for this experiment are in the appendix.

Lab Introduction

We can use the data from the above experiment to determine the parameter p in differential equation (1). Once we have obtained p we can then attempt to answer the question "Did the balloon reach terminal velocity before it hit the ground?"

Experiment

(1) Graph the function $y(t)$ from the bottom of page 411 for several values of the parameter p; start with $p = 1$ and try some larger and some smaller values. Sketch the results on one set of axes. Remember that $y_0 = 6$.

(2) Find the parameter p that corresponds to the balloon experiment. Let T_1 be the time at which the balloon hits the ground. From the experiment we know that $y(T_1) = 0$; find the value of p that makes this equation true. Adjust the parameter p so that the graph of $y(t)$ crosses the t-axis at the time T_1; look at your sketches from (1) above to get an idea where to start.

(3) Did the balloon reach terminal velocity before it hit the ground? Using the value of p determined from (2), graph the velocity (see the derivation in the Mathematical Background section for the equation of the velocity). Use this graph to estimate the time T_3 at which the balloon reached its terminal velocity—that is, stopped accelerating. You will have to make some sort of assumption about what it means to "reach" terminal velocity. Finally, graph the velocity for a few other values of p and sketch the results.

(4) At the time T_2 the balloon was three feet above the ground; plot this data point on the graph of $y(t)$ as a check of whether or not the model is a good one [use the p from part (2)].

Discussion

1. How does p affect the shape of $y(t)$ graph? What happens as p gets close to 0? What happens as p gets large? Explain in your own words. Why should it have this effect? Remember, $p = \frac{k}{m}$, where k is the air resistance constant and m is mass.

2. How did you find the p that fits the data? Carefully explain your approach. How accurate is your value? How do you know?

3. Did you encounter a problem in determining when the balloon reached terminal velocity? Explain how you determined the answer to this question.

4. If p is increased, does it take more or less time for the object to reach terminal velocity? Why?

5. Based on the results of plotting the midway data point, does the model seem to be a good one? If not, could p be adjusted, or do we need a new model?

7.3 General Exponential and Logarithmic Functions

In this section we continue the process, started in the previous section, of carefully proving properties of logarithms and exponents based on the definitions of that section. This time, however, we also state and prove some new derivative and integral formulas.

We defined $e^{\sqrt{2}}$, e^{π}, and all other irrational powers of e in the previous section by the formula $e^x = \exp(x)$, where $\exp(x)$ was in turn defined as the inverse of $\ln(x)$. But what about $2^{\sqrt{2}}$, π^{π}, π^{e}, $\sqrt{2}^{\pi}$, and similar irrational powers of other numbers? In fact, we want to give meaning to a^x for $a > 0$ and x any real number. Now if $r = p/q$ is a rational number, $a^r = \left(\sqrt[q]{a}\right)^p$. But we also know that

$$a^r = \exp(\ln a^r) = \exp(r \ln a) = e^{r \ln a}$$

This suggests a definition.

What is 2^{π}?

When asked this equation, one student wrote

$$2^{\pi} = 2 \cdot 2 \cdot 2 \cdot \angle$$

which is nonsense. In Section 1.4 we defined 2^{π} as the limit of the sequence

$$2^3, 2^{3.1}, 2^{3.14}, 2^{3.141}, \ldots$$

though it is tedious to calculate anything this way. With our new definition

$$2^{\pi} = e^{\pi \ln 2}$$

Although calculators vary, it is likely that your calculator uses this formula in finding 2^{π}. This way, the calculator needs to know only one function (e^x) rather than an infinite number of functions (2^x, 3^x, etc.) to make this kind of calculation.

Definition

For $a > 0$ and any real number x,

$$a^x = e^{x \ln a}$$

Of course, this definition will be appropriate only if the usual properties of exponents are valid for it, a matter we take up shortly. To shore up our confidence in the definition, we use it to calculate 3^2 (with a little help from our calculator).

$$3^2 = e^{2 \ln 3} \approx e^{2(1.0986123)} \approx 9.0000002$$

The slight discrepancy is due to the round-off characteristics of our calculator.

Now we can fill a small gap in the properties of the natural logarithm left over from Section 7.2.

$$\boxed{\ln(a^x) = \ln(e^{x \ln a}) = x \ln a}$$

Thus, Property (iv) of Theorem 7.2A holds for all real x, not just rational x as claimed there. We will need this fact in the proof of Theorem A below.

Properties of a^x Theorem A summarizes the familiar properties of exponents, which can all be proved now in a completely rigorous manner. Theorem B shows us how to differentiate and integrate a^x.

Theorem A

If $a > 0$, $b > 0$, and x and y are real numbers,

(i) $a^x a^y = a^{x+y}$;

(ii) $\dfrac{a^x}{a^y} = a^{x-y}$;

(iii) $(a^x)^y = a^{xy}$;

(iv) $(ab)^x = a^x b^x$;

(v) $\left(\dfrac{a}{b}\right)^x = \dfrac{a^x}{b^x}$.

Proof   We content ourselves with proving (ii) and (iii), leaving the others for the exercises.

(ii) $\dfrac{a^x}{a^y} = e^{\ln(a^x/a^y)} = e^{\ln a^x - \ln a^y}$

$\qquad\quad = e^{x\ln a - y\ln a} = e^{(x-y)\ln a} = a^{x-y}$

(iii) $(a^x)^y = e^{y\ln a^x} = e^{yx\ln a} = a^{yx} = a^{xy}$

◄

Theorem B

$$\frac{d}{dx} a^x = a^x \ln a$$

$$\int a^x \, dx = \left(\frac{1}{\ln a}\right) a^x + C, \qquad a \neq 1$$

Proof

$$\frac{d}{dx} a^x = \frac{d}{dx}\left(e^{x\ln a}\right) = e^{x\ln a}\frac{d}{dx}(x\ln a)$$

$$= a^x \ln a$$

◄

The integral formula follows immediately from the derivative formula.

EXAMPLE 1: Find $\dfrac{d}{dx}\left(3^{\sqrt{x}}\right)$.

SOLUTION: We use the Chain Rule with $u = \sqrt{x}$.

$$\frac{d}{dx}\left(3^{\sqrt{x}}\right) = 3^{\sqrt{x}}\ln 3\frac{d}{dx}\sqrt{x} = \frac{3^{\sqrt{x}}\ln 3}{2\sqrt{x}}$$

◄

EXAMPLE 2: Find $\dfrac{dy}{dx}$ if $y = (x^4 + 2)^5 + 5^{x^4 + 2}$.

SOLUTION:

$$\frac{dy}{dx} = 5(x^4 + 2)^4 \cdot 4x^3 + 5^{x^4 + 2}\ln 5 \cdot 4x^3$$

$$= 4x^3[5(x^4 + 2)^4 + 5^{x^4 + 2}\ln 5]$$

$$= 20x^3[(x^4 + 2)^4 + 5^{x^4 + 1}\ln 5]$$

◄

EXAMPLE **3:** Find $\int 2^{x^3} x^2 \, dx$.

> **SOLUTION:** Let $u = x^3$, so $du = 3x^2 \, dx$. Then
>
> $$\int 2^{x^3} x^2 \, dx = \frac{1}{3} \int 2^{x^3} (3x^2 \, dx) = \frac{1}{3} \int 2^u \, du$$
>
> $$= \frac{1}{3} \frac{2^u}{\ln 2} + C = \frac{2^{x^3}}{3 \ln 2} + C \qquad \blacktriangleleft$$

The Function $\log_a$

Finally, we are ready to make a connection with the logarithms you studied in algebra. We note that if $0 < a < 1$, $f(x) = a^x$ is a decreasing function; if $a > 1$, it is an increasing function, as you may check by considering the derivative. In either case, f has an inverse. We call this inverse the **logarithmic function to the base a**. This is equivalent to the following definition.

Definition _____

Let a be a positive number different from 1. Then

$$y = \log_a x \text{ if and only if } x = a^y$$

Historically, the most commonly used base was base 10 and the resulting logarithms were called **common logarithms**. But in calculus and all of advanced mathematics, the significant base is e. Notice that $\log_e$, being the inverse of $f(x) = e^x$, is just another symbol for $\ln$; that is,

$$\log_e x = \ln x$$

We have come full circle (see Figure 1). The function $\ln$, which we introduced in Section 7.2 as an integral, has turned out to be an ordinary logarithm but to a rather surprising base, e. Note that this is how we *defined* the function $\ln$ back in Section 1.5.

Now observe that if $y = \log_a x$ so $x = a^y$, then

$$\ln x = \ln a^y, \text{ so}$$

$$\ln x = y \ln a$$

from which we conclude that

$$\boxed{\log_a x = \frac{\ln x}{\ln a}}$$

From this, it follows that $\log_a$ satisfies the usual properties associated with logarithms (see Theorem 7.2A). Also,

$$\boxed{\frac{d}{dx} \log_a x = \frac{1}{\ln a} \frac{d}{dx} \ln x = \frac{1}{x \ln a}}$$

Why Other Bases? _____

Are bases other than e really needed? No. The formulas

$$a^x = e^{x \ln a}$$

and

$$\log_a x = \frac{\ln x}{\ln a}$$

allow us to turn any problem involving exponential functions or logarithmic functions with base a to corresponding functions with base e. This supports our terminology— natural exponential and natural logarithmic functions. It also explains the universal use of the latter functions in advanced work.

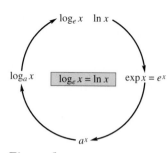

Figure 1

EXAMPLE 4: If $y = \log_{10}(x^4 + 13)$, find $\dfrac{dy}{dx}$.

SOLUTION: Let $u = x^4 + 13$ and apply the Chain Rule.

$$\frac{dy}{dx} = \frac{1}{(x^4 + 13)\ln 10} \cdot 4x^3 = \frac{4x^3}{(x^4 + 13)\ln 10}$$ ◀

The Functions a^x, x^a, and x^x

Begin by comparing the three graphs in Figure 2. More generally, let a be a constant. Do not confuse $f(x) = a^x$, an **exponential function**, with $g(x) = x^a$, a **power function**. And do not confuse their derivatives. We have just learned that

$$\frac{d}{dx}(a^x) = a^x \ln a$$

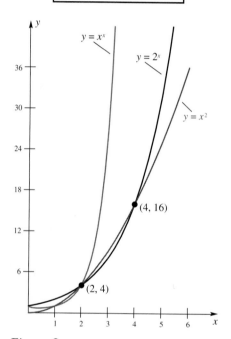

Figure 2

What about $\frac{d}{dx}(x^a)$? For a rational, you proved in the exercises in Section 3.6 that

$$\frac{d}{dx}(x^a) = ax^{a-1}$$

Now we can prove that this is true even if a is irrational. To see this, write

$$\frac{d}{dx}(x^a) = \frac{d}{dx}\left(e^{a \ln x}\right) = e^{a \ln x} \cdot \frac{a}{x}$$

$$= x^a \cdot \frac{a}{x} = ax^{a-1}$$

Finally, we consider $f(x) = x^x$, a variable to a variable power. There is a formula for $\frac{d}{dx}(x^x)$, but we do not recommend that you memorize it. Rather, we suggest that you learn a method for finding it, as illustrated below.

EXAMPLE 5: If $y = x^x, x > 0$, find $\dfrac{dy}{dx}$.

    SOLUTION: We may write

$$y = x^x = e^{x \ln x}$$

Thus, by the Chain Rule,

$$\frac{d}{dx}y = e^{x \ln x}\frac{d}{dx}(x \ln x) = x^x\left(x \cdot \frac{1}{x} + \ln x\right) = x^x(1 + \ln x)$$

◄

EXAMPLE 6: If $y = (x^2 + 1)^\pi + \pi^{\sin x}$, find $\dfrac{dy}{dx}$.

    SOLUTION:

$$\frac{dy}{dx} = \pi(x^2 + 1)^{\pi - 1}(2x) + \pi^{\sin x}\ln \pi \cdot \cos x$$

◄

EXAMPLE 7: If $y = (x^2 + 1)^{\sin x}$, find $\dfrac{dy}{dx}$.

    SOLUTION: We use our new definition of exponentiation for real exponents $a^x = e^{x \ln a}$ to write

$$y = (x^2 + 1)^{\sin x} = e^{\sin x \cdot \ln(x^2 + 1)}$$

and therefore

$$\frac{dy}{dx} = e^{\sin x \cdot \ln(x^2 + 1)}\frac{d}{dx}\left(\sin x \cdot \ln(x^2 + 1)\right)$$

$$= e^{\sin x \cdot \ln(x^2 + 1)}\left(\cos x \cdot \ln(x^2 + 1) + \sin x \cdot \frac{2x}{x^2 + 1}\right)$$

$$= (x^2 + 1)^{\sin x}\left(\cos x \cdot \ln(x^2 + 1) + \sin x \cdot \frac{2x}{x^2 + 1}\right)$$

◄

EXAMPLE 8: Evaluate $\displaystyle\int_{1/2}^{1}\frac{5^{\frac{1}{x}}}{x^2}dx$.

    SOLUTION: Let $u = 1/x$, so $du = (-1/x^2)dx$. Then

$$\int\frac{5^{\frac{1}{x}}}{x^2}dx = -\int 5^{\frac{1}{x}}\left(-\frac{1}{x^2}dx\right) = -\int 5^u\,du$$

$$= -\frac{5^u}{\ln 5} + C = -\frac{5^{\frac{1}{x}}}{\ln 5} + C$$

Thus, by the Fundamental Theorem of Calculus,

$$\int_{1/2}^{1}\frac{5^{\frac{1}{x}}}{x^2}dx = \left[-\frac{5^{\frac{1}{x}}}{\ln 5}\right]_{1/2}^{1} = \frac{1}{\ln 5}(5^2 - 5)$$

$$= \frac{20}{\ln 5} \approx 12.426699$$

◄

Concepts Review

1. In terms of e and $\ln$, $\pi^{\sqrt{3}} = $ _____ .
More generally $a^x = $ _____ .

2. $\ln x = \log_a x$, where $a = $ _____ .

3. $\log_a x$ can be expressed in terms of $\ln$ by $\log_a x = $ _____ .

4. The derivative of the power function $f(x) = x^a$ is $f'(x) = $ _____ ; the derivative of the exponential function $g(x) = a^x$ is $g'(x) = $ _____ .

Problem Set 7.3

In Problems 1–8, solve for x. *Hint*: $\log_a b = c$ if an only if $a^c = b$.

1. $\log_3 9 = x$

2. $\log_4 x = 2$

3. $\log_9 x = \frac{3}{2}$

4. $\log_x 81 = 4$

5. $2\log_{10}\left(\frac{x}{3}\right) = 1$

6. $\log_4\left(\frac{1}{x}\right) = 3$

7. $\log_2(x + 1) - \log_2 x = 2$

8. $\log_6(x + 1) + \log_6 x = 1$

Use your calculator and $\log_a x = (\ln x)/(\ln a)$ to calculate each of the logarithms in Problems 9–12.

9. $\log_5 13$

10. $\log_6(0.12)$

11. $\log_{11}(8.16)^{1/5}$

12. $\log_{10}(91.2)^3$

In Problems 13–16, use natural logarithms to solve each of the exponential equations. *Hint*: To solve $3^x = 11$, take $\ln$ of both sides, obtaining $x \ln 3 = \ln 11$; then $x = (\ln 11)/(\ln 3) \approx 2.1827$.

13. $2^x = 19$

14. $5^x = 12$

15. $4^{3x-1} = 5$

16. $12^{1/(x-1)} = 3$

Find the indicated derivative or integral (see Examples 1–4).

17. $\dfrac{d}{dx}\left(5^{x^2}\right)$

18. $\dfrac{d}{dx}\left(3^{2x^4-4x}\right)$

19. $\dfrac{d}{dx}\log_2 e^x$

20. $\dfrac{d}{dx}\log_{10}(x^2 + 9)$

21. $\dfrac{d}{dx}\left[2^x\ln(x+5)\right]$

22. $\dfrac{d}{dx}\sqrt{\log_{10} x}$

23. $\displaystyle\int x2^{x^2}\,dx$

24. $\displaystyle\int 10^{5x-1}\,dx$

25. $\displaystyle\int_1^4 \frac{5^{\sqrt{x}}}{\sqrt{x}}\,dx$

26. $\displaystyle\int_0^1 (10^{3x} + 10^{-3x})\,dx$

In Problems 27–32, find dy/dx. Be sure to distinguish between problems of the type a^x, x^a, and x^x as in Examples 5–7.

27. $y = 10^{(x^2)} + (x^2)^{10}$

28. $y = \sin^2 x + 2^{\sin x}$

29. $y = x^{\pi+1} + (\pi + 1)^x$

30. $y = 2^{(e^x)} + (2^e)^x$

31. $y = (x^2 + 1)^{\ln x}$

32. $y = (\ln x^2)^{2x+3}$

33. If $f(x) = x^{\sin x}$, find $f'(1)$.

34. Let $f(x) = \pi^x$ and $g(x) = x^\pi$. Which is larger, $f(e)$ or $g(e)$? $f'(e)$ or $g'(e)$?

35. How are $\log_{1/2} x$ and $\log_2 x$ related?

36. Sketch the graphs of $\log_{1/3} x$ and $\log_3 x$ using the same coordinate axes.

37. The magnitude M of an earthquake on the Richter scale is

$$M = 0.67\log_{10}(0.37E) + 1.46$$

where E is the energy of the earthquake in kilowatt-hours. Find the energy of an earthquake of magnitude 7. Of magnitude 8.

38. The loudness of sound is measured in decibels in honor of Alexander Graham Bell (1847–1922), inventor of the telephone. If the variation in pressure P is pounds per square inch, then the loudness L in decibels is

$$L = 20\log_{10}(121.3P)$$

Find the variation in pressure caused by a rock band at 115 decibels.

39. In the equally tempered scale to which keyed instruments have been tuned since the days of J.S. Bach (1685–1750), the frequencies of successive notes C, C#, D, D#, E, F, F#, G, G#, A, A#, B, $\overline{C}$ form a geometric sequence (progression), with $\overline{C}$ having twice the frequency of C. What is the ratio r between the frequencies of successive notes? If the frequency of A is 440, find the frequency of $\overline{C}$.

40. You are suspicious that the xy-data you have collected lie on either an exponential curve, $y = A \cdot b^x$, or a power curve, $y = A \cdot x^b$. To check, you plot $\ln y$ against x and also $\ln y$ against $\ln x$. Explain how this will help you come to a conclusion. See Section 2.7.

41. (An Amusement) Given the problem of finding y' if $y = x^x$, student A did the following:

$$y = x^x$$

WRONG 1 $y' = x \cdot x^{x-1} \cdot 1$ $\begin{pmatrix}\text{misapplying the} \\ \text{Power Rule}\end{pmatrix}$

$$= x^x$$

Student B did this:

$$y = x^x$$

WRONG 2 $y' = x^x \cdot \ln x \cdot 1$ $\begin{pmatrix}\text{misapplying the} \\ \text{Exponential} \\ \text{Function Rule}\end{pmatrix}$

$$= x^x \ln x$$

The sum $x^x + x^x \ln x$ is correct (Example 5), so

WRONG 1 + WRONG 2 = RIGHT

Show that the same procedure yields a correct answer for $y = f(x)^{g(x)}$.

42. Convince yourself that $f(x) = (x^x)^x$ and $g(x) = x^{(x^x)}$ are not the same function. Then find $f'(x)$ and $g'(x)$. *Note*: When mathematicians write x^{x^x}, they mean $x^{(x^x)}$.

43. Consider $f(x) = \dfrac{a^x - 1}{a^x + 1}$ for fixed a, $a > 0$, $a \neq 1$. Show that f has an inverse and find a formula for $f^{-1}(x)$.

44. For fixed $a > 1$, let $f(x) = x^a/a^x$ on $[0, \infty)$. Show:

(a) $\lim\limits_{x \to \infty} f(x) = 0$ (study $\ln f(x)$);

(b) $f(x)$ is maximized at $x_0 = a/\ln a$;

(c) $x^a = a^x$ has two positive solutions if $a \neq e$, only one such solution if $a = e$;

(d) $\pi^e < e^\pi$.

45. Let $f_u(x) = x^u e^{-x}$ for $x \geq 0$. Show that for any fixed $u > 0$:

(a) $f_u(x)$ attains its maximum at $x_0 = u$;

(b) $f_u(u) > f_u(u + 1)$ and $f_{u+1}(u + 1) > f_{u+1}(u)$ imply

$$\left(\frac{u + 1}{u}\right)^u < e < \left(\frac{u + 1}{u}\right)^{u+1}$$

(c) $\dfrac{u}{u + 1} e < \left(\dfrac{u + 1}{u}\right)^u < e$

Conclude from (c) that $\lim\limits_{u \to \infty} \left(1 + \dfrac{1}{u}\right)^u = e$.

46. Find $\lim\limits_{x \to 0^+} x^x$. Also find the coordinates of the minimum point for $f(x) = x^x$ on $[0, 4]$.

47. Draw the graphs of $y = x^3$ and $y = 3^x$ using the same axes and find all their intersection points.

48. Evaluate $\displaystyle\int_0^{4\pi} x^{\sin x}\, dx$.

49. Prove parts (i) and (iv) of Theorem A.

Answers to Concepts Review: 1. $e^{\sqrt{3}\ln \pi}$; $e^{x \ln a}$ 2. e
3. $(\ln x)/(\ln a)$ 4. ax^{a-1}; $a^x \ln a$

7.4 | Exponential Growth and Decay

At the beginning of 1987, the world's population was about 5 billion. It was said at the time that by the year 2010, it would reach 7.7 billion. How are such predictions made?

To treat the problem mathematically, let $y = f(t)$ denote the size of the population at time t, where t is the number of years after 1987. Actually $f(t)$ is an integer, and its graph "jumps" when someone is born or dies. However, for a large population, these jumps are so small relative to the total population that we will not go far wrong if we pretend that f is a nice differentiable function.

It seems reasonable to suppose that the increase (Δy) in population (births minus deaths) during a short time period (Δt) is proportional to the size of the population at the beginning of the period and to the length of that period. Thus, $\Delta y = ky\,\Delta t$, or

$$\frac{\Delta y}{\Delta t} = ky$$

In its limiting form, this yields the differential equation

$$\boxed{\frac{dy}{dt} = ky}$$

The constant k represents the instantaneous change in population per person per year at a particular time t; multiplication by 100 gives the percentage growth rate per year. If $k > 0$, the population is growing; if $k < 0$, it is shrinking. For world population, history indicates that k is about 0.019 people per person per year, or an increase of 1.9% per year (though some statisticians report a considerably lower figure).

Solving the Differential Equation   We began our study of differential equations in Section 5.2, and you might refer to that section now. We want to solve $dy/dt = ky$ subject to the condition that $y = y_0$ when $t = 0$. Separating variables and integrating, we obtain

$$\frac{dy}{y} = k\,dt$$

$$\int \frac{dy}{y} = \int k\,dt$$

$$\ln y = kt + C$$

The condition $y = y_0$ at $t = 0$ yields $C = \ln y_0$. Thus,

$$\ln y - \ln y_0 = kt$$

or

$$\ln \frac{y}{y_0} = kt$$

Changing to exponential form yields

$$\frac{y}{y_0} = e^{kt}$$

or, finally,

$$\boxed{y = y_0 e^{kt}}$$

Notice that the solution to our differential equation is a natural exponential function, which we previously used to model population growth in Section 1.5 and in Chapter 2. Because the differential equation came directly from an assumption about how populations grow, we have even more justification for using exponential functions to model population growth. We also have yet another way to define the natural exponential function—as the solution to a certain differential equation!

Returning to the problem of world population, we choose to measure time t in years after January 1, 1987, and y in billions of people. Thus, $y_0 = 5$ and, because $k = 0.019$,

$$y = 5e^{0.019t}$$

By the year 2010, when $t = 23$, we can predict that y will be about

$$y = 5e^{0.019(23)} \approx 7.7 \text{ billion}$$

EXAMPLE 1: How long will it take world population to double under the assumptions above?

SOLUTION: The question is equivalent to asking how many years after 1987 it will take for the population to reach 10 billion. We need to solve

$$10 = 5e^{0.019t}$$

for t. After dividing both side by 5, we take the logarithms.

$$\ln 2 = 0.019t$$

$$t = \frac{\ln 2}{0.019} \approx 36 \text{ years} \qquad \blacktriangleleft$$

If world population will double in the first 36 years after 1987, it will double in any 36-year period; so for example, it will quadruple in 72 years. More generally, if an exponentially growing quantity doubles in an initial interval of length T, it will double in *any* interval of length T, because

$$\frac{y(t + T)}{y(t)} = \frac{y_0 e^{k(t+T)}}{y_0 e^{kt}} = \frac{y_0 e^{kT}}{y_0} = 2$$

We call the number T the **doubling time**.

EXAMPLE 2: The number of bacteria in a rapidly growing culture was estimated to be 10,000 at noon and 40,000 after two hours. How many bacteria will there be at 5 P.M.?

SOLUTION: We assume that the differential equation $dy/dt = ky$ is applicable, so $y = y_0 e^{kt}$. Now we have two conditions ($y_0 = 10{,}000$ and $y = 40{,}000$ at $t = 2$), from which we conclude that

$$40{,}000 = 10{,}000e^{k(2)}$$

or

$$4 = e^{2k}$$

Taking logarithms yields

$$\ln 4 = 2k$$

or

$$k = \tfrac{1}{2} \ln 4 = \ln\sqrt{4} = \ln 2 \approx 0.693$$

Thus,

$$y = 10{,}000e^{0.693t}$$

and, at $t = 5$, this gives

$$y = 10{,}000e^{0.693(5)} \approx 320{,}000 \qquad \blacktriangleleft$$

A Logistic Model for Population Growth

The exponential model for population growth is flawed because it projects faster and faster growth indefinitely far into the future (Figure 1). In most cases (including that of world population), the limited amount of space and resources will eventually force a slowing of the growth rate. This suggests another model for population growth, called the **logistic model**, in which we assume the rate of growth is proportional both to the population size y and to the quantity $\frac{L - y}{L}$, where L is the maximum population that can be supported by the environment. This leads to the differential equation

$$\frac{dy}{dt} = ky\frac{(L - y)}{L}$$

For small y, $\frac{dy}{dt} \approx ky$, which gives exponential growth. But as $t \to \infty$, $y \to L$, so that $\lim_{t\to\infty} y(t) = L$. And as y nears L, $\frac{dy}{dt}$ gets smaller and smaller, curtailing the growth and producing a curve like the one in Figure 2. This model is explored in Problems 24, 25, and 32 of this section and again in Section 7.5 and 8.4.

Note: Some authors write the logistic differential equation as $\frac{dy}{dt} = ky(L - y)$. The advantage of the form we use above is that the units of the parameter k are the same as for the simple growth equation $\frac{dy}{dt} = ky$; thus, comparisons between the two models are easier to make.

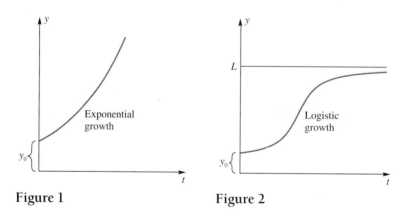

Figure 1   Figure 2

Radioactive Decay

Not everything grows; some things decrease over time. In particular, the radioactive elements decay, and they do it at a rate proportional to the amount present. Thus, their change rates also satisfy the differential equation

$$\frac{dy}{dt} = ky$$

but now with k negative. It is still true that $y = y_0 e^{kt}$ is the solution to this equation. A typical graph appears in Figure 3.

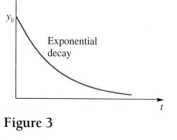

Figure 3

EXAMPLE 3: Carbon-14, one of the three isotopes of carbon, is radioactive and decays at a rate proportional to the amount present. Its **half-life** is 5730 years; that is, it takes 5730 years for a given amount of carbon-14 to decay to one-half its original size. If 10 grams were present originally, how much will be left after 2000 years?

SOLUTION: The half-life of 5730 allows us to determine k, because it implies that

$$\frac{1}{2} = e^{k(5730)}$$

or after taking logarithms,

$$-\ln 2 = 5730k$$

$$k = \frac{-\ln 2}{5730} \approx -0.000121$$

Thus,

$$y = 10e^{-0.000121t}$$

At $t = 2000$, this gives

$$y = 10e^{-0.000121(2000)} \approx 7.85 \text{ grams} \qquad \blacktriangleleft$$

In Problem 13, we show how Example 3 may be used to determine the age of fossils and other once-living things.

Compound Interest   If we put $100 in the bank at 12% annual interest compounded monthly, it will be worth $100(1.01)$ at the end of 1 month, $100(1.01)^2$ at the end of 2 months, and $100(1.01)^{12}$ at the end of 12 months, or 1 year. More generally, if we put A_0 dollars in the bank at $100r$ percent compounded n times per year, it will be worth $A(t)$ dollars at the end of t years, where

$$A(t) = A_0\left(1 + \frac{r}{n}\right)^{nt}$$

EXAMPLE 4: Suppose that John put $500 in the bank at 13% interest, compounded daily. How much will it be worth at the end of two years?

SOLUTION: Here $r = 0.13$ and $n = 365$, so

$$A = 500\left(1 + \frac{0.13}{365}\right)^{365(2)} \approx \$648.43 \qquad \blacktriangleleft$$

Now let us consider what happens when interest is **compounded continuously**; that is, when n, the number of compounding periods in a year, tends to infinity. Then we claim

$$A(t) = \lim_{n\to\infty} A_0\left(1 + \frac{r}{n}\right)^{nt} = A_0 \lim_{n\to\infty}\left[\left(1 + \frac{r}{n}\right)^{n/r}\right]^{rt}$$

$$= A_0\left[\lim_{n\to\infty}(1 + h)^{1/h}\right]^{rt} = A_0 e^{rt}$$

Here we replaced r/n by h and noted that $n \to \infty$ corresponds to $h \to 0$. But the big step is knowing that the expression in brackets is the number e. In Section 1.1, Example 7, we estimated the limit $\lim_{h\to 0}(1 + h)^{1/h}$ numerically, and then used this limit as our definition of the constant e. In this chapter, we defined e as the solution of the equation $\ln x = 1$, so that the statement $\lim_{h\to 0}(1 + h)^{1/h} = e$ now becomes a theorem.

Theorem A

$$\lim_{h\to 0}(1 + h)^{1/h} = e$$

Proof   First recall that if $f(x) = \ln x$, then $f'(x) = 1/x$ and, in particular, $f'(1) = 1$. Then from the definition of derivative and properties of ln, we get

$$1 = f'(1) = \lim_{h \to 0} \frac{f(1 + h) - f(1)}{h} = \lim_{h \to 0} \frac{\ln(1 + h) - \ln 1}{h}$$

$$= \lim_{h \to 0} \frac{1}{h} \ln(1 + h) = \lim_{h \to 0} \ln(1 + h)^{\frac{1}{h}}$$

Thus, $\lim_{h \to 0} \ln(1 + h)^{\frac{1}{h}} = 1$, a result we will use in a moment. Now, $g(x) = e^x = \exp x$ is a continuous function, and it therefore follows that we can pass the limit inside the exponential function in the following argument.

$$\lim_{h \to 0} (1 + h)^{\frac{1}{h}} = \lim_{h \to 0} \exp[\ln(1 + h)^{\frac{1}{h}}] = \exp[\lim_{h \to 0} \ln(1 + h)^{\frac{1}{h}}]$$

$$= \exp 1 = e$$

For another proof of Theorem A, See Problem 45 of Section 7.3.   ◄

EXAMPLE 5: Suppose the bank of Example 4 compounded interest continuously. How much would John then have at the end of two years?

SOLUTION:

$$A(t) = A_0 e^{rt} = 500 e^{(0.13)(2)} \approx \$648.47$$

Note that the difference between the two methods of compounding amounts to just four cents in two years. Though some banks try to get advertising mileage out of offering continuous compounding of interest, the difference in yields between continuous and daily compounding (which many banks offer) is minuscule.   ◄

Here is another approach to the problem of continuous compounding of interest. Let A be the value at time t of A_0 dollars invested at interest rate r. To say that interest is compounded continuously is to say that the instantaneous rate of change of A with respect to time is rA—that is,

$$\frac{dA}{dt} = rA$$

This differential equation was solved at the beginning of the section; its solution is $A = A_0 e^{rt}$.

Concepts Review

1. The rate of change dy/dt of a quantity y growing exponentially satisfies the differential equation $dy/dt =$ _____. In contrast, if y is growing logistically toward an upper bound L, $dy/dt =$ _____.

2. If a quantity growing exponentially doubles after T years, it will be _____ times as large after $3T$ years.

3. The time for an exponentially decaying quantity y to go from size y_0 to size $y_0/2$ is called _____.

4. The number e can be expressed as a limit by $e = \lim_{h \to 0}$ _____.

Problem Set 7.4

In Problems 1–4, solve the given differential equation subject to the given condition. Note that $y(a)$ denotes the value of y at $t = a$.

1. $\dfrac{dy}{dt} = -5y$, $y(0) = 4$

2. $\dfrac{dy}{dt} = 6y$, $y(0) = 5$

3. $\dfrac{dy}{dt} = 0.006y$, $y(10) = 2$

4. $\dfrac{dy}{dx} = -0.003y$, $y(-2) = 4$

5. A bacterial population grows at a rate proportional to its size. Initially it is 10,000 and after 10 days it is 24,000. What is the population after 25 days? See Example 2.

6. How long will it take the population of Problem 5 to double? See Example 1.

7. How long will it take the population of Problem 5 to triple? See Example 1.

8. The population of the United States was 4 million in 1790 and 180 million in 1960. If the rate of growth is assumed proportional to the number present, what estimate would you give for the population in 2020?

9. The population of a certain country is growing at 3.2% per year; that is, if it is A at the beginning of a year, it is $1.032A$ at the end of that year. Assuming that it is 4.5 million now, what will it be at the end of 1 year? 2 years? 10 years? 100 years?

10. Determine the proportionality constant k in $dy/dt = ky$ for Problem 9. Then use $y = 4.5e^{kt}$ to find the population after 100 years.

11. A radioactive substance has a half-life of 810 years. If there were ten grams initially, how much would be left after 300 years?

12. If a radioactive substance loses 15% of its radioactivity in 3 days, what is its half-life?

13. (Carbon Dating) All living things contain carbon-12, which is stable, and carbon-14, which is radioactive. While a plant or animal is alive, the ratio of these two isotopes of carbon remains unchanged since the carbon-14 is contantly renewed; after death, no more carbon-14 is absorbed. The half-life of carbon-14 is 5730 years. If charred logs of an old fort showed only 70% of the carbon-14 expected in living matter, when did the fort burn down? Assume that the fort burned soon after it was built of freshly sawed logs.

14. Human hair from a grave in Africa proved to have only 51% of the carbon-14 of living tissue. When was the body buried? See Problem 13.

15. Newton's law of cooling states that the rate at which an object cools is proportional to the difference in temperature between the object and the surrounding medium. Thus, if an object is taken from an oven at $300°F$ and left to cool in a room at $75°F$, its temperature T after t hours will satisfy the differential equation

$$\frac{dT}{dt} = k(T - 75)$$

If the temperature fell to $200°F$ in $\frac{1}{2}$ hour, what will it be after 3 hours? Compare the form of the solution you obtained in this problem with the model we used in Example 2 of Section 2.3 to describe the cooling of a human body after death.

16. A thermometer registered $-20°C$ outside and then was brought in the house where the temperature was $24°C$. After 5 minutes, it registered $0°C$. When will it register $20°C$? See Problem 15.

17. If $375 is put in the bank today, what will it be worth at the end of 2 years if interest is 9.5% and is compounded as specified?

(a) Annually   (b) Monthly

(c) Daily   (d) Continuously

Hint: See Examples 4 and 5.

18. Do Problem 17 assuming that the interest rate is 14.4%.

19. How long does it take money to double in value for the specified interest?

(a) 12% compounded monthly

(b) 12% compounded continuously

20. Inflation between 1977 and 1981 ran at about 11.5% per year. On this basis, what would you expect a car that cost $4000 in 1977 to cost in 1981?

21. Manhattan Island is said to have been bought from the Indians by Peter Minuit in 1626 for $24. Suppose Minuit had instead put the $24 in the bank at 6% interest compounded continuously. What is that $24 worth in 1996? It would be interesting to compare the result with the actual value of Manhattan Island in 1996.

22. If Methuselah's parents had put $100 in the bank for him at birth and left it there, what would Methuselah have had at his death (969 years later) if interest was 8% compounded annually?

23. It can be shown that for small x, $\ln(1 + x) \approx x$. Use this fact to show that the doubling time for money invested at p percent compounded annually is about $70/p$ years.

24. The equation for logistic growth is

$$\frac{dy}{dt} = ky\,\frac{L - y}{L}$$

Show that this differential equation has the solution

$$y = \frac{Ly_0}{y_0 + (L - y_0)e^{-kt}}$$

Hint: $\frac{1}{y(L - y)} = \frac{1}{Ly} + \frac{1}{L(L - y)}$.

25. Sketch the graph of the solution in Problem 24 when $y_0 = 5$, $L = 16$, $k = 0.0186$ (a *logistic model* for world population—see the discussion in this section). Note that $\lim_{t \to \infty} y = 16$.

26. Find or estimate each of the following limits.

(a) $\lim_{x \to 0} (1 + x)^{1000}$ (b) $\lim_{x \to 0} (1)^{1/x}$

(c) $\lim_{x \to 0^+} (1 + \epsilon)^{1/x}, \epsilon > 0$ (d) $\lim_{x \to 0^-} (1 + \epsilon)^{1/x}, \epsilon > 0$

(e) $\lim_{x \to 0} (1 + x)^{1/x}$

27. Use the fact that $e = \lim_{h \to 0} (1 + h)^{1/h}$ to find each limit.

(a) $\lim_{x \to 0} (1 - x)^{1/x}$ *Hint:* $(1 - x)^{1/x} = [(1 - x)^{1/-x}]^{-1}$

(b) $\lim_{x \to 0} (1 + 3x)^{1/x}$ (c) $\lim_{n \to \infty} \left(\frac{n + 2}{n}\right)^n$

(d) $\lim_{n \to \infty} \left(\frac{n - 1}{n}\right)^{2n}$

28. Show that the differential equation

$$\frac{dy}{dt} = ay + b$$

has solution

$$y = \left(y_0 + \frac{b}{a}\right)e^{at} - \frac{b}{a}$$

Assume $a \neq 0$.

29. Consider a state that had a population of 10 million in 1995, natural exponential growth rate of 1.2% per year, and immigration from other states and countries of 60,000 per year. Use the differential equation of Problem 28 to model this situation and predict the population in 2020. Take $a = 0.012$.

30. Important news is said to diffuse through an adult population of fixed size 100,000 at a time rate proportional to the number of people who have not heard the news. Five days after a scandal in city hall was reported, a poll showed that half the people had heard it. How long will it take for 99% of the people to hear it? *Hint:* Initially one person knew the rumor—the person who started it.

31. Assume that (1) the world population will continue to grow exponentially with growth constant $k = 0.019$ indefinitely into the future, (2) it takes $\frac{1}{2}$ acre of land to supply food for a person, and (3) there are 13,500,000 square miles of arable land in the world. How long will it be before the world reaches the maximum population? *Note:* There were 5 billion people in 1987 and 1 square mile is 640 acres.

32. Using the same axes, draw the graphs for $0 \leq t \leq 100$ of the following two models for the growth of world population (both described in this section).

(a) Exponential growth: $y = 5e^{0.019t}$

(b) Logistic growth: $y = 80/(5 + 11e^{-0.030t})$

Compare what the two models predict for world population in 2000, 2040, and 2090. *Note:* Both models assume world population was 5 billion in 1987 ($t = 0$).

Answers to Concepts Review: 1. ky; $ky\dfrac{L - y}{L}$ 2. 8
3. Half-life 4. $(1 + h)^{1/h}$

7.5 Numerical and Graphical Approaches to Differential Equations

Mathematical models in the physical and social sciences often come in the form of differential equations. These equations often do not have simple, exact solutions as did the differential equations of Sections 6.2 and 7.4. Fortunately, there are numerical techniques that we can use to estimate solutions to differential equations, and there are graphical techniques that we can use to understand differential equations qualitatively.

In what follows we will use t rather than x as our independent variable, because so many differential equations describe processes that depend on time.

Euler's Method   Suppose that we want to solve a first-order differential equation subject to an initial condition

$$\frac{dy}{dt} = f(y, t), \qquad y(t_0) = y_0 \tag{1}$$

The notation $f(y, t)$ indicates that the function f depends on two variables, y and t. Using the definition of the derivative, we can make the approximation

$$\frac{dy}{dt} \approx \frac{y(t + h) - y(t)}{h} \tag{2}$$

where h is a small but nonzero number. Replacing y with $y(t)$ in (1) to emphasize the dependence on t, we can combine (1) and (2) to get

$$\frac{y(t + h) - y(t)}{h} \approx f(y(t), t)$$

Solving for $y(t + h)$ yields

$$y(t + h) \approx hf(y(t), t) + y(t)$$

If we now look at the solution to (1) only at the equally spaced points $t_0, t_1, t_2, \ldots$, where $t_{n+1} = t_n + h$, let $y_n = y(t_n)$, and replace the $\approx$ with $=$, we get

$$\boxed{\begin{aligned} t_{n+1} &= t_n + h \\ y_{n+1} &= hf(y_n, t_n) + y_n \end{aligned}}$$

$$\tag{3}$$

which is Euler's method for solving differential equations numerically. The value h represents the step size, and y_0 and t_0 are the initial y- and t-values. Equations (3) are then used to generate $(t_1, y_1), (t_2, y_2), \ldots$. We use the $=$ instead of the $\approx$ because we are now defining the y_n, but keep in mind that we are only approximating the solution to the differential equation.

EXAMPLE 1: Generate an approximate solution to the differential equation with initial condition given by

$$\frac{dy}{dt} = 2y, \qquad y(0) = 1$$

on the interval $0 \le t \le 1$. Use a step size of $h = 0.1$. Compare your results with the exact solution obtained using separation of variables.

SOLUTION: With $f(y) = 2y$ and $h = 0.1$, equations (3) become

$$t_{n+1} = t_n + 0.1$$
$$y_{n+1} = 2(0.1)y_n + y_n = 1.2y_n$$

Using $t_0 = 0$ and $y_0 = 1$ we get

$$t_1 = t_0 + 0.1 = 0.1, \qquad y_1 = 1.2y_0 = 1.2(1) = 1.2$$
$$t_2 = t_1 + 0.1 = 0.2, \qquad y_2 = 1.2y_1 = 1.2(1.2) = 1.44$$
$$t_3 = t_2 + 0.1 = 0.3, \qquad y_3 = 1.2y_2 = 1.2(1.44) \approx 1.73$$

$$\vdots$$

$$t_{10} = t_9 + 0.1 = 1.0, \qquad y_{10} = 1.2y_9 = 1.2(5.16) \approx 6.19$$

for the Euler approximations.

To compute an exact solution, we solve $\frac{dy}{dt} = 2y$ subject to the initial condition $y(0) = 1$ using separation of variables as in Section 7.4 to get

$$y = e^{2t}$$

We can now evaluate this function at the t-values $t_0 = 0$, $t_1 = 0.1$, $t_2 = 0.2, \ldots$, $t_{10} = 1.0$ and compare with the Euler approximations $y_0, y_1, y_2, \ldots, y_{10}$ at these points. We have

$$e^{2t_0} = e^0 = 1$$

$$e^{2t_1} = e^{0.2} \approx 1.22$$

$$e^{2t_2} = e^{0.4} \approx 1.49$$

$$\vdots$$

$$e^{2t_{10}} = e^{2.0} \approx 7.39$$

Figure 1 shows the complete set of values rounded to two decimal places for t_n, y_n, and e^{2t_n}, $0 \leq n \leq 10$.

			Euler Estimates	Exact Solution	
n	t_n	y_n	e^{2t_n}	error
0	0.0	1.00	1.00	0
1	0.1	1.20	1.22	0.02
2	0.2	1.44	1.49	0.05
3	0.3	1.73	1.82	0.09
4	0.4	2.07	2.23	0.16
5	0.5	2.49	2.72	0.23
6	0.6	2.99	3.32	0.33
7	0.7	3.58	4.06	0.48
8	0.8	4.30	4.95	0.65
9	0.9	5.16	6.05	0.89
10	1.0	6.19	7.39	1.20

Figure 1

We can see that the Euler approximations are somewhat accurate for the first few t-values, but at $t = 1.0$ we don't quite even get one digit of accuracy. We will next explore why this is the case.   ◄

Graphical Interpretation of Euler's Method

Euler's method for solving differential equations of the form $\frac{dy}{dt} = f(y, t)$ with an initial condition $y(t_0) = y_0$ is based on the local linearity property of smooth functions. Even if we can't find an exact solution $y(t)$, we can calculate the slope of its tangent at the initial point (t_0, y_0) by using the differential equation:

$$\frac{dy}{dt}(t_0) = f(y_0, t_0)$$

If we move h units in the t-direction along the tangent line, we get to the point $(t_1, y_1) = (t_0 + h, y_0 + hf(t_0, y_0))$. See Figure 2. Using (t_1, y_1) as the new starting point, we repeat this process to generate (t_2, y_2), (t_3, y_3), $(t_4, y_4), \ldots$.

We can see that when the solution $y(t)$ to the differential equation $\frac{dy}{dt} = f(y, t)$ near $t = t_0$ is concave up, the Euler estimate of $y(t_1)$ is less than the actual value of $y(t_1)$. This is why the Euler estimates of $y(t)$ generated in Example 1 are smaller than the actual values of the solution $y(t) = e^{2t}$. What would happen if $y(t)$ were concave down near $t = t_0$?

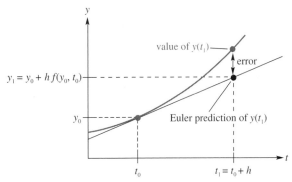

Figure 2

Order of Convergence of Numerical Methods

When using Euler's method, as when using other numerical techniques, it is important to make h sufficiently small. We say that a numerical method for solving differential equations is *kth-order convergent* if for small h values, multiplication of h by a factor of $C < 1$ results in an approximate reduction of the error at any given point by a factor of C^k (compare with the definition of order of convergence for numerical integration methods given in Section 5.4).

Euler's method is a first-order numerical method (we will not attempt to prove this fact). Thus, every time we reduce h through multiplication by $\frac{1}{10}$ (say from 0.2 to 0.02), we should get a reduction factor in the error at any given point of about $\frac{1}{10}$, or one decimal place.

The order of convergence of a numerical method helps us to determine when the step size h is small enough for our desired accuracy. With Euler's method, a good approach is to let h get smaller through powers of 10 such as $1, 0.1, 0.01, \dots$. When two consecutive results agree to n digits, then you have about n-digit accuracy. If you are using Euler's method to generate graphs, then when two consecutive graphs seem to agree, you have enough accuracy for that particular graph window (for a smaller window, you might need to make h smaller).

Euler's Method by Calculator or Computer

We can easily program a graphing calculator or computer algebra system to estimate solutions to differential equations using Euler's method. This becomes critical when h is small, and hence the number of points computed is large. The output can be displayed numerically (in table form) or graphically.

EXAMPLE 2: The logistic differential equation

$$\frac{dy}{dt} = ky\frac{(L - y)}{L}$$

was mentioned in Section 7.4 as an alternative to the pure exponential growth model $\frac{dy}{dt} = ky$. Recall that L represents the maximum sustainable population, and that k represents the growth rate when the population y is close to y_0.

Use a graphing calculator or computer and Euler's method to solve the logistic differential equation for $k = 0.05$, $L = 500$, and with the initial condition $y(0) = 20$. Calculate and graph a solution over the interval $0 \le t \le 200$. Also, numerically estimate the solution at $t = 100$ and at $t = 200$ and indicate how much accuracy you have attained with these estimates. Create a context (a situation) that might plausibly correspond to your solution, and describe your results in that context.

SOLUTION: It's best to start with a value for h that corresponds to a fairly small number of points (say 10 or 20). Because we want to compute a solution on the interval $0 \leq t \leq 200$, a reasonable first choice for the step size h would be $h = 10$ (generating 21 points). We will compare these results with the results for $h = 1$ and if necessary $h = 0.1$.

To implement Euler's method, we need to create a loop that calculates the points (t_1, y_1), (t_2, y_2), (t_3, y_3), ... based on the equations

$$t_{n+1} = t_n + h$$

$$y_{n+1} = hf(y_n, t_n) + y_n$$

and the starting point (t_0, y_0). On a computer algebra system we will save all of the points generated, after which we can plot them or inspect the ones we are interested in. On a graphing calculator, we will plot the points as they are generated, but we will save only the last point. The details for specific calculators and computer algebra systems are given in the accompanying solutions manual.

In Figure 3, we show the graphs generated for $h = 10, 1, 0.1$. We see that the graphs for $h = 1$ and $h = 0.1$ are essentially the same, and hence we are reasonably certain that this graphical solution is fairly accurate. The fact that the curve for $h = 10$ is below the more accurate curves results from the fact that the solution is concave up near $t = 0$, as explained above.

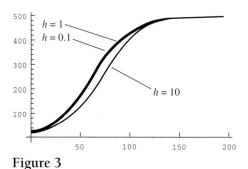

Figure 3

Numerical results at $t = 100$ and at $t = 200$ are given in the table in Figure 4. We can see that at $t = 200$ the estimated value of y converges to 499.5 accurate to four digits. At $t = 100$ the estimated value of y converges to 430 accurate to only two digits. This shows that a given amount of accuracy at one point on the solution curve does not imply the same accuracy for all previous points. This is easily seen from the graph in Figure 3 as well.

h	Euler Estimate of $y(100)$	h	Euler Estimate of $y(200)$
10	407.02	10	499.86
1	428.65	1	499.50
0.1	430.23	0.1	499.46

Figure 4

A possible context for this problem could be as follows. Twenty wild horses break off from their herd and wander into a canyon. This new, smaller herd undergoes rapid growth for the first 50 years. The herd continues to grow, but the rate of growth begins to level off. After 100 years there are about 430 horses in the herd. The finite amount of food in the canyon then severely limits the growth of the herd until the number of horses levels off at about 500, the maximum population size that the canyon can support. ◄

Built-In Numerical Methods for Solving Differential Equations Some graphing calculators and computer algebra systems have built into them sophisticated routines for solving differential equations numerically. The built-in methods, though similar to Euler's method, are generally at least fourth-order convergent; thus, a one-digit reduction in the step size h results in an additional four digits of accuracy (rather than just one digit for Euler's method). Hence, one can attain a given amount of accuracy much more quickly than with Euler's method. If you take a course in numerical analysis you will study these more powerful methods.

EXAMPLE 3: On a small island, the population growth rate resulting from births and deaths is a fairly constant 3% per year. In addition to the internal growth rate, the population is affected by immigration and emigration; people arrive at the island at a rate of 100 per year during the summer months and leave at a rate of 100 per year during the winter months. We can model the rate of change of the population due to immigration/emigration with the term $-100 \cos(2\pi t)$, where t is measured in years and $t = 0$ corresponds to January 1.

Let y represent population on the island, and t represent time in years. The total rate of change of population would be the sum of the internal growth rate and the immigration/emigration term, resulting in the differential equation

$$\frac{dy}{dt} = 0.03y - 100 \cos(2\pi t)$$

If the population of the island starts out at $y = 2000$, estimate the population after 1, 2, 3, and 4 years. Also, graph the population as a function of time for a four-year period. Explain the results in the context of the problem.

SOLUTION: The most important decision that you have with a problem like this is whether or not you can easily obtain an exact solution. Exact solutions are always preferred, because they are more reliable. The only method of finding exact solutions that we currently have at our disposal is the method of separation of variables. Thus, your first question should be "Can I separate variables?"

To be able to separate variables we must be able to write the differential in the form $\frac{dy}{dt} = f(t)g(y)$; then we can divide by $g(y)$, multiply by dt, and integrate both sides. The right-hand side of the equation $\frac{dy}{dt} = 0.03y - 100 \cos(2\pi t)$ is a *sum* of one function of y and another function of t; it cannot be factored into a *product* of two such functions. Thus we can't use separation of variables.

We are left with finding an approximate solution numerically. We could use Euler's method as in Example 2, but if your calculator or computer algebra system has a built-in numerical method it will generally provide a given amount of accuracy much more quickly.

In Figure 5 we show the graph of a numerically generated solution, with points plotted at $t = 1, 2, 3, 4$. In Figure 6 we give the numerically generated y-values for $t = 1, 2, 3, 4$. The accompanying solutions manual shows how to generate the graphs and function values for particular calculators and computer algebra systems. ◄

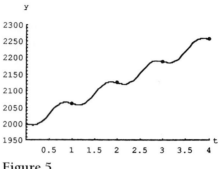

Figure 5

Time	Population
t	y
1	2061
2	2124
3	2188
4	2255

Figure 6

We see that the change in population over time has two distinct features; a general upward trend combined with oscillations within each year. The upward trend results from the $0.03y$ term (internal population growth) in the differential equation, and the oscillations result from the $-100\cos(2\pi t)$ term (immigration/emigration). Also, if we look very carefully at the table in Figure 6, we see that the rate of growth of the population is increasing. In the first year the population grows by 61, in the second year by 63, in the third year by 64, and in the fourth year by 67.   ◀

NOTE: There is, in fact, an exact solution to the differential equation for this problem, though it cannot be obtained by separation of variables. If you take a course in differential equations you will learn how to solve such equations.

Concepts Review

1. If a differential cannot be solved exactly, you may be able to use a _____ technique to approximate the solution. An example of such a technique is _____ method.

2. For the differential equation $\frac{dy}{dt} = 3y$ with initial condition $y(0) = 2$, the first step using Euler's

method with $h = 1$ would give the approximation $y(1) \approx y_1 = $ _____ .

3. The approximation in question 2 would be an _____ (under, over) estimate to the exact solution at $t = 1$.

4. Euler's method is a _____-order-convergent numerical method.

Problem Set 7.5

Problems 1–4 are the same as Problems 1–4 of Section 7.4. Generate an approximate solution to each differential equation with the given initial condition on the given interval using Euler's method. Use a step size of $h = 0.2$. Compare your results with the exact solution obtained using separation of variables from Section 7.4.

1. $\frac{dy}{dt} = -5y,\ y(0) = 4,\ 0 \le t \le 1$.

2. $\frac{dy}{dt} = 6y,\ y(0) = 5,\ 0 \le t \le 1$.

3. $\frac{dy}{dt} = 0.006y,\ y(10) = 2,\ 10 \le t \le 11$.

4. $\frac{dy}{dx} = -0.003y,\ y(-2) = 4,\ -2 \le t \le -1$.

5–8. Use a graphing calculator or computer and Euler's method to solve each of the differential equations in Problems 1–4. Use $h = 0.1$ and $h = 0.01$. Sketch the graphs of your two numerical solutions together on the given interval, and compare with the graph of the exact solution. How much accuracy do you have at the right endpoint of the interval for each case?

9. Redo Example 2 using $k = 0.10$. About how long does it take for the population to reach its limiting value this time? How does this differ from the original example? Why? You may use Euler's method or the built-in numerical method of a graphing calculator or computer algebra system.

10. Redo Example 3 using an internal growth rate of 10% instead of 3%. How is the solution different from that of the original example? Why? You may use either the built-in numerical method of a graphing calculator or computer algebra system, or you may use Euler's method (with sufficiently small h).

In Problems 11–14, use any method to find or estimate $y(1)$. Explain why you chose the method you did.

11. $\dfrac{dy}{dt} = \dfrac{y^2}{5}, y(0) = 1$

12. $\dfrac{dy}{dt} = \dfrac{y^2}{a}, y(0) = 1$, a an undetermined parameter $\neq 1$.

13. $\dfrac{dy}{dt} = \dfrac{y^2}{5} + t, y(0) = 1$

14. $\dfrac{dy}{dt} = \dfrac{y^2 t}{5}, y(0) = 1$

Answers to Concept Review: 1. numerical; Euler's 2. 8 3. under 4. first

---

LAB 13: PLANNING YOUR RETIREMENT NUMERICALLY

Mathematical Background

See the discussion of Euler's method in the text.

Lab Introduction

One way to plan a retirement is to save a sum of money during your working years, invest that sum in high-yield investments (mutual funds, bonds, certificates of deposit, etc.) and live off of the interest during your retirement. If the average rate of return of all of your investments is r (essentially the interest rate), then the differential equation that governs the total worth of your investments (y) is

$$\frac{dy}{dt} = ry \tag{1}$$

where t is time. You can solve this differential equation by separating variables; the result is

$$y = y_0 e^{rt} \tag{2}$$

where y_0 is the initial investment. Thus, your money will grow exponentially.

The problem is that after retirement you must also take money out; let I represent the fixed income you will need after retirement. The differential equation now becomes

$$\frac{dy}{dt} = ry - I \tag{3}$$

This equation can also be solved by separating variables as follows:

$$\int \frac{1}{ry - I} \, dy = \int dt$$

$$\frac{1}{r} \ln(ry - I) = t + C_1$$

$$y = C_2 e^{rt} + \frac{I}{r} \qquad \left(C_2 = \frac{1}{r} e^{rC_1} \right)$$

We can now use the initial condition $y(0) = y_0$ to get

$$y = \left(y_0 - \frac{I}{r}\right)e^{rt} + \frac{I}{r} \tag{4}$$

Finally, to add one more dimension of reality, consider what happens when the rate of return (or interest rate) r fluctuates over time. We could use a trig function to represent fluctuations in the interest rate over time; a reasonable model for the interest rate might be

$$r + a\sin(2\pi wt) \tag{5}$$

where r is now the *average* interest rate over time, and a and w are parameters that can be adjusted to fit the data. The total possible fluctuation in interest rates is $2a$ (a units above r and a units below r) and w is the number of cycles per year for the interest rate (thus, if w is 0.25, there are one-quarter cycles per year, or one cycle every four years). Then the differential equation becomes

$$\frac{dy}{dt} = (r + a\sin(2\pi wt))y - I \tag{6}$$

This model cannot be solved exactly, and so we can use Euler's method or the built-in method of a calculator or computer. Our goal in this lab will be to understand the behavior of the various models above, and to use them to predict the possible effects of fluctuating interest rates on retirement plans.

Note: We have made a simplifying modeling assumption here: y really changes in discrete units rather than continuously; when an investment pays out, that sum goes back into the account, and so y increases by some finite amount. However, with a number of investments paying off at different times, y acts more like a continuously changing variable. Also, the continuous case is usually easier to deal with. Similarly with I; we assume that the withdrawals that constitute I are spread out throughout the year.

Experiment

(1) Hilda-Mae needs a fixed income of $10,000 above her Social Security allowance. She has $105,000 in various investments at age 50; she wants to retire and have enough money to make it to age 90. Use the exact solution (4) to differential equation (3) to graph and sketch Hilda-Mae's principal y as a function of time t on the interval $0 \le t \le 40$. Use units of thousands of dollars; thus, $y_0 = 105$ and $I = 10$. Assume a fixed interest rate of 10% (i.e., $r = 0.1$). Now repeat for the same I and r, but assume that her initial investment is $y_0 = 97$. Estimate the time when Hilda-Mae's money runs out (from the graph) for this case.

(2) Repeat (1) using a numerical method as follows. Solve differential equation (3) numerically with $y_0 = 105$, $I = 10$, and $r = 0.1$; graph the solution on $0 \le t \le 40$. If you use Euler's method, start with $h = 1$ and decrease until the graph stabilizes. Repeat for the same I and r and for $y_0 = 97$. Estimate the time when Hilda-Mae's money runs out. Compare those results to the ones from (1).

(3) Investigate what will happen to Hilda-Mae's investments if interest rates should fluctuate.

   (a) Graph the interest rate function (5) for the values $r = 0.1$, $a = 0.05$, and
        $w = 0.2$ on the interval $0 \le t \le 40$. Choose the y-range $0 \le y \le 0.2$. Find the

t-coordinates of the maximums and minimums on this interval and find the time from one maximum to the next.

(b) Use a numerical method to graph the solution to differential equation (6) on the interval $0 \le t \le 40$ and $0 \le y \le 200$. Use the values $I = 10$ and $y_0 = 105$. Use the interest rate function from (3a). If you use Euler's method, make sure that h is small enough. Compare with the results from (1) and (2). Repeat for $y_0 = 97$ and compare with the results from (1) and (2). Finally, choose $y_0 = 95$ and plot this together with the case $y_0 = 97$ on $0 \le t \le 40$ and $90 \le y \le 110$; estimate the *t*-coordinates of the maximums and minimums.

Discussion

1. Fill in any missing steps in finding the solution (4) to differential equation (3) above.

2. Describe in plain English what happened to Hilda-Mae's money for the two cases in step (1). What initial investment level y_0 will keep the total amount y constant? What happens if y_0 is above or below this value? How do you know? Look at the solution (4) to differential equation (3).

3. Did you get the same graph using a numerical method in step (2) as you did with the exact solution in step (1) for each case? If you used Euler's method, how small did h have to be? Briefly compare exact versus numerical solutions with differential equations; discuss the role of parameters for each. Give examples of differential equations that you can solve exactly but not numerically, and ones that you can solve numerically but not exactly.

4. Describe the curve that represents the interest rate in (3a). How much does the rate vary? How long does one complete cycle last?

5. Describe the numerical solution curves for each case in (3b); discuss the results in terms of the total value of the investments (y). Is the result for the case $y_0 = 97$ surprising? Does this contradict your conclusion from discussion question 2? Explain what happened. Discuss other possible effects of fluctuating interest rates on long-term investments based on what you have learned in this lab.

6. How are the oscillations of the principal y related to the oscillations of the interest rate? Do the two have maximums at the same time? Do they have the same period (length of one whole cycle)? Describe what is happening and then try to explain it in terms of the problem situation.

7.6 The Trigonometric Functions and Their Inverses

The six basic trigonometric functions (sine, cosine, tangent, cotangent, secant, and cosecant) were defined in Section 0.4. We also defined the inverses of sine, cosine, and tangent so that they could be used to solve equations involving trig functions. In Section 3.3 we found derivatives for the sine and cosine functions. In this section we review many of these definitions, add the definition of arcsecant, and find derivatives for all of the trig and inverse trig functions.

Derivatives of the Trigonometric Functions   We learned in Section 3.3 that $\frac{d}{dx}\sin x = \cos x$ and $\frac{d}{dx}\cos x = -\sin x$. The other four trigonometric functions are defined in terms of sine and cosine by

$$\tan x = \frac{\sin x}{\cos x} \qquad \cot x = \frac{\cos x}{\sin x}$$

$$\sec x = \frac{1}{\cos x} \qquad \csc x = \frac{1}{\sin x}$$

These facts, together with the Quotient Rule for derivatives, allow us to find the derivatives of each of these functions. For example,

$$\frac{d}{dx}\cot x = \frac{d}{dx}\left(\frac{\cos x}{\sin x}\right) = \frac{\sin x(-\sin x) - \cos x \cos x}{\sin^2 x}$$

$$= \frac{-1}{\sin^2 x} = -\csc^2 x$$

Here is a summary of the six derivative formulas. They should be memorized.

$$\frac{d}{dx}\sin x = \cos x \qquad\qquad \frac{d}{dx}\cos x = -\sin x$$

$$\frac{d}{dx}\tan x = \sec^2 x \qquad\qquad \frac{d}{dx}\cot x = -\csc^2 x$$

$$\frac{d}{dx}\sec x = \sec x \tan x \qquad\qquad \frac{d}{dx}\csc x = -\csc x \cot x$$

Composite Functions   We can combine the rule above with the Chain Rule and with other differentiation rules in more and more complicated ways. For example, if $u = f(x)$ is differentiable, then

$$\frac{d}{dx}\sin u = \cos u \cdot \frac{du}{dx}$$

Here are two examples, to review the various rules.

EXAMPLE 1: Find $\frac{d}{dx}\sin(3x^2 + 4)$.

    SOLUTION: Let $u = 3x^2 + 4$.

$$\frac{d}{dx}\sin(3x^2 + 4) = \frac{d}{dx}\sin u = \cos u \frac{d}{dx}u$$

$$= [\cos(3x^2 + 4)]6x = 6x\cos(3x^2 + 4) \qquad \blacktriangleleft$$

EXAMPLE 2: Find $\frac{d}{dx}\tan^2(9x)$.

    SOLUTION: This requires a double use of the Chain Rule.

$$\frac{d}{dx}\tan^2(9x) = 2\tan(9x)\frac{d}{dx}\tan(9x)$$

$$= 2\tan(9x)\sec^2(9x)\frac{d}{dx}(9x)$$

$$= 2\tan(9x)\sec^2(9x)9$$

$$= 18\tan(9x)\sec^2(9x) \qquad \blacktriangleleft$$

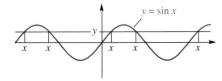

Figure 1

Inverse Sine and Inverse Cosine   Recall that with the sine and cosine functions, each y in the range corresponds to infinitely many x's (Figure 1). Thus, we must restrict the domain of each, keeping the range as large as possible while insisting that the resulting function have an inverse. This can be done in many ways, but the most common procedure is shown in Figures 2 and 3. We show also the graph of the corresponding inverse function, obtained by reflecting across the line $y = x$.

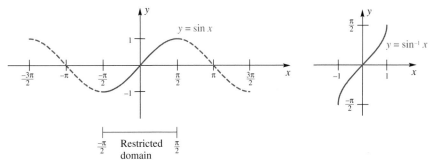

Figure 2

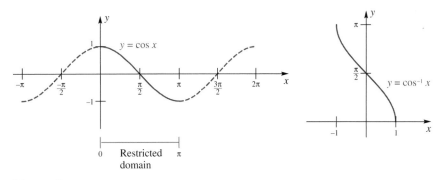

Figure 3

We formalize what we have shown in a definition.

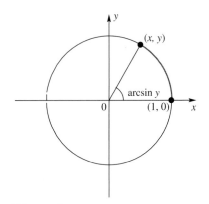

Figure 4

> **Definition**
>
> To obtain inverses for sine and cosine, we restrict their domain to $[-\pi/2, \pi/2]$ and $[0, \pi]$, respectively. Thus,
>
> $$y = \sin^{-1}x \Longleftrightarrow x = \sin y \text{ where } -1 \leq x \leq 1 \text{ and } \frac{-\pi}{2} \leq y \leq \frac{\pi}{2}$$
>
> $$y = \cos^{-1}x \Longleftrightarrow x = \cos y \text{ where } -1 \leq x \leq 1 \text{ and } 0 \leq y \leq \pi$$

The symbol arcsin is often used for $\sin^{-1}$, and arccos is similarly used for $\cos^{-1}$. Think of arcsin as meaning "the arc whose sine is," or "the angle whose sine is" (Figure 4).

EXAMPLE 3: Calculate (a) $\sin^{-1}(\sqrt{2}/2)$, (b) $\sin^{-1}(-\frac{1}{2})$, (c) $\cos^{-1}(\sqrt{3}/2)$, (d) $\cos^{-1}(-\frac{1}{2})$, (e) $\cos(\cos^{-1}0.6)$, and (f) $\sin^{-1}(\sin 3\pi/2)$. Calculate an exact value based on the definitions given above, and then check your results by finding an approximate value using the $\sin^{-1}$ and $\cos^{-1}$ buttons on your calculator. Make sure your calculator is in radian mode.

SOLUTION:

(a) $\sin^{-1}\left(\dfrac{\sqrt{2}}{2}\right) = \dfrac{\pi}{4} \approx 0.785$ (b) $\sin^{-1}\left(-\dfrac{1}{2}\right) = -\dfrac{\pi}{6} \approx -0.524$

(c) $\cos^{-1}\left(\dfrac{\sqrt{3}}{2}\right) = \dfrac{\pi}{6} \approx 0.524$ (d) $\cos^{-1}\left(-\dfrac{1}{2}\right) = \dfrac{2\pi}{3} \approx 2.094$

(e) $\cos(\cos^{-1}0.6) = 0.6$ (f) $\sin^{-1}\left(\sin \dfrac{3\pi}{2}\right) = -\dfrac{\pi}{2} \approx -1.571$

The only one of these that is tricky is (f). Note that it would be wrong to give $3\pi/2$ as the answer, because $\sin^{-1}y$ is always in the interval $[-\pi/2, \pi/2]$. Work the problem in steps, as follows.

$$\sin^{-1}\left(\sin \frac{3\pi}{2}\right) = \sin^{-1}(-1) = -\pi/2 \qquad \blacktriangleleft$$

Another Way to Say It

$$\sin^{-1}x$$

is the number in the interval $[-\pi/2, \pi/2]$ whose sine is x.

$$\cos^{-1}x$$

is the number in the interval $[0, \pi]$ whose cosine is x.

$$\tan^{-1}x$$

is the number in the interval $(-\pi/2, \pi/2)$ whose tangent is x.

Inverse Tangent In Figure 5, we have shown the graph of the tangent function, its restricted domain, and the graph of $y = \tan^{-1}x$.

Definition

To obtain an inverse for tangent, we restrict the domain to $(-\pi/2, \pi/2)$. Thus,

$$y = \tan^{-1}x \Leftrightarrow x = \tan y \text{ where } -\infty < x < \infty \text{ and } -\frac{\pi}{2} < y < \frac{\pi}{2}$$

There is a standard way to restrict the domain of the cotangent function, namely to $(0, \pi)$, so that it has an inverse. However, this function does not play a significant role in calculus, so we will say no more about it.

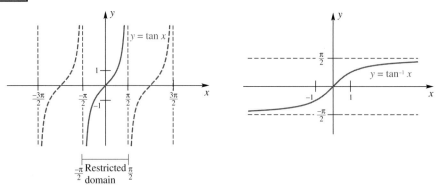

Figure 5

EXAMPLE 4: Calculate   (a) $\tan^{-1}(1)$,   (b) $\tan^{-1}(-\sqrt{3})$,   (c) $\tan^{-1}(-0.145)$,   and   (d) $\tan^{-1}(\tan 5.236)$. Find exact results where reasonable.

SOLUTION:

(a) $\tan^{-1}(1) = \frac{\pi}{4} \approx 0.785$

(b) $\tan^{-1}(-\sqrt{3}) = -\frac{\pi}{3} \approx -1.047$

(c) $\tan^{-1}(-0.145) = -0.1439964$

(d) $\tan^{-1}(\tan 5.236) = -1.0471853$ ◄

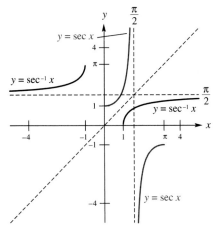

Figure 6

Inverse Secant   To define an inverse for secant, we use the fact that $\sec x = \frac{1}{\cos x}$ to obtain the formula

$$\sec^{-1}x = \cos^{-1}\left(\frac{1}{x}\right)$$

In the exercises you are asked to fill in the missing steps in the derivation of this formula.

Because the domain of the function $\cos^{-1}x$ is $\{x: -1 \le x \le 1\}$, the domain of the function $y = \sec^{-1}x$ is $\{x: x \le -1$ or $x \ge 1\}$ and the range is $\{y: 0 \le y < \frac{\pi}{2}$ or $\frac{\pi}{2} < y \le \pi\}$ (this corresponds to restricting the domain of the secant function to $[0, \frac{\pi}{2})\cup(\frac{\pi}{2}, \pi]$). See Figure 6, where we graph both the secant function on the interval $[0, \pi]$ and its inverse $y = \sec^{-1}x$. We will have no need to define $\csc^{-1}x$, though this can also be done.

EXAMPLE 5: Calculate (a) $\sec^{-1}(-1)$, (b) $\sec^{-1}(2)$, and (c) $\sec^{-1}(-1.32)$.

SOLUTION:

(a) $\sec^{-1}(-1) = \cos^{-1}(-1) = \pi$

(b) $\sec^{-1}(2) = \cos^{-1}\left(\frac{1}{2}\right) = \frac{\pi}{3}$

(c) $\sec^{-1}(-1.32) = \cos^{-1}\left(-\frac{1}{1.32}\right) = \cos^{-1}(-0.7575758) = 2.4303875$ ◄

Four Useful Identities   We will need each of the following identities very shortly. You can recall them by reference to the triangles in Figure 7.

(i) $\sin(\cos^{-1}x) = \sqrt{1 - x^2}$

(ii) $\cos(\sin^{-1}x) = \sqrt{1 - x^2}$

(iii) $\sec(\tan^{-1}x) = \sqrt{1 + x^2}$

(iv) $\tan(\sec^{-1}x) = \begin{cases} \sqrt{x^2 - 1} & x > 1 \\ -\sqrt{x^2 - 1} & x < -1 \end{cases}$

To prove (i), recall that $\sin^2\theta + \cos^2\theta = 1$; in particular, if $0 \le \theta \le \pi$, then

$$\sin \theta = \sqrt{1 - \cos^2\theta}$$

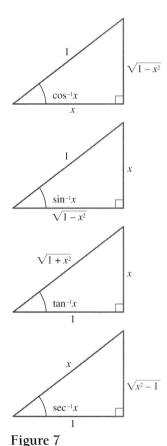

Figure 7

Now apply this with $\theta = \cos^{-1}x$ and use the fact that $\cos(\cos^{-1}x) = x$ to get

$$\sin(\cos^{-1}x) = \sqrt{1 - \cos^2(\cos^{-1}x)} = \sqrt{1 - x^2}$$

We ask you to prove the other identities in the exercises.

Derivatives of Four Inverse Trigonometric Functions

From the Inverse Function Theorem (Theorem 7.1B), we conclude that $\sin^{-1}$, $\cos^{-1}$, $\tan^{-1}$, and $\sec^{-1}$ are differentiable. Our aim is to find formulas for their derivatives. We state the results and then show where they come from.

(i) $\quad \dfrac{d}{dx}\sin^{-1}x = \dfrac{1}{\sqrt{1 - x^2}} \qquad -1 < x < 1$

(ii) $\quad \dfrac{d}{dx}\cos^{-1}x = \dfrac{-1}{\sqrt{1 - x^2}} \qquad -1 < x < 1$

(iii) $\quad \dfrac{d}{dx}\tan^{-1}x = \dfrac{1}{1 + x^2}$

(iv) $\quad \dfrac{d}{dx}\sec^{-1}x = \dfrac{1}{|x|\sqrt{x^2 - 1}} \qquad |x| > 1$

Our derivations follow the same pattern in each case. To see (i), let $y = \sin^{-1}x$, so that

$$x = \sin y$$

Now differentiate both sides with respect to x, using the Chain Rule on the right-hand side. Then

$$1 = \cos y \frac{d}{dx}y = \cos(\sin^{-1}x)\frac{d}{dx}(\sin^{-1}x)$$

$$= \sqrt{1 - x^2}\frac{d}{dx}(\sin^{-1}x)$$

At the last step, we used identity (ii) under Four Useful Identities above. We conclude that $\frac{d}{dx}(\sin^{-1}x) = 1/\sqrt{1 - x^2}$.

Results (ii), (iii), and (iv) are proved similarly (you are asked to prove (ii) and (iii) in the exercises), but (iv) has a little twist—so we will go through its derivation. Let $y = \sec^{-1}x$, so

$$x = \sec y$$

Differentiating both sides with respect to x, we obtain

$$1 = \sec y \tan y \frac{d}{dx}y$$

$$= \sec(\sec^{-1}x)\tan(\sec^{-1}x)\frac{d}{dx}(\sec^{-1}x)$$

$$= x\left(\pm\sqrt{x^2 - 1}\right)\frac{d}{dx}(\sec^{-1}x)$$

At the last step, we used identity (iv) from above. But as indicated, we use the plus sign if $x > 1$ and the minus sign if $x < -1$. Thus,

$$1 = \pm\, x\sqrt{x^2 - 1}\,\frac{d}{dx}\,(\sec^{-1}x)$$

$$= |x|\sqrt{x^2 - 1}\,\frac{d}{dx}\,(\sec^{-1}x)$$

The desired result follows immediately.

EXAMPLE 6: Find $\frac{d}{dx}\sin^{-1}(3x - 1)$.

  SOLUTION: We use (i) and the Chain Rule.

$$\frac{d}{dx}\sin^{-1}(3x - 1) = \frac{1}{\sqrt{1 - (3x - 1)^2}}\frac{d}{dx}(3x - 1)$$

$$= \frac{3}{\sqrt{-9x^2 + 6x}}$$ ◄

EXAMPLE 7: Find $\frac{d}{dx}\tan^{-1}\sqrt{x + 1}$.

  SOLUTION: We use (iii) and the Chain Rule.

$$\frac{d}{dx}\tan^{-1}\sqrt{x + 1} = \frac{1}{1 + \left(\sqrt{x + 1}\right)^2}\frac{d}{dx}\sqrt{x + 1}$$

$$= \frac{1}{x + 2}\cdot\frac{1}{2}(x + 1)^{-\frac{1}{2}}$$

$$= \frac{1}{2(x + 2)\sqrt{x + 1}}$$ ◄

Of course, every differentiation formula leads to an integration formula, a matter we will say much more about in the next chapter. In particular:

(i) $\displaystyle\int\frac{1}{\sqrt{1 - x^2}}\,dx = \sin^{-1}x + C;$

(ii) $\displaystyle\int\frac{1}{1 + x^2}\,dx = \tan^{-1}x + C;$

(iii) $\displaystyle\int\frac{1}{x\sqrt{x^2 - 1}}\,dx = \sec^{-1}|x| + C.$

EXAMPLE 8: Evaluate $\displaystyle\int_0^{1/2}\frac{dx}{\sqrt{1 - x^2}}$.

  SOLUTION:

$$\int_0^{1/2}\frac{1}{\sqrt{1 - x^2}}\,dx = [\sin^{-1}x]_0^{1/2} = \sin^{-1}\frac{1}{2} - \sin^{-1}0$$

$$= \frac{\pi}{6} - 0 = \frac{\pi}{6}$$

Recall that we don't need to add a constant C to the antiderivative, because it subtracts out in a definite integral. ◄

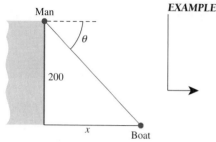

Man

θ

200

x

Boat

Figure 8

EXAMPLE 9: A man standing on top of a vertical cliff is 200 feet above a lake. As he watches, a motorboat moves directly away from the foot of the cliff at a rate of 25 feet per second. How fast is the angle of depression of his line of sight changing when the boat is 150 feet from the foot of the cliff?

SOLUTION: The essential details are shown in Figure 8. Note that θ, the angle of depression, is

$$\theta = \tan^{-1}\left(\frac{200}{x}\right)$$

Thus,

$$\frac{d\theta}{dt} = \frac{1}{1 + (200/x)^2} \cdot \frac{-200}{x^2} \cdot \frac{dx}{dt} = \frac{-200}{x^2 + 40,000} \cdot \frac{dx}{dt}$$

When we substitute $x = 150$ and $dx/dt = 25$, we obtain $d\theta/dt = -0.08$ radians per second. ◄

Concepts Review

1. To obtain an inverse for the cosine function, we restrict the domain to _____. The resulting inverse function is denoted by $\cos^{-1}$ or by _____.

2. To obtain an inverse for the tangent function, we restrict the domain to _____. The resulting inverse function is denoted by $\tan^{-1}$ or by _____.

3. $\frac{d}{dx}\sin(\arcsin x) = $ _____.

4. Because $\frac{d}{dx}\arctan x = 1/(1 + x^2)$, it follows that $4\int_0^1 1/(1 + x^2)\,dx = $ _____. Because $\frac{d}{dx}\arcsin x = 1/\sqrt{1 - x^2}$, it can be shown that $\int_0^{1/2}\left(1/\sqrt{1 - x^2}\right)dx = $ _____.

Problem Set 7.6

In Problems 1–12, find the exact values without use of a calculator. Show your work. Then check with a calculator.

1. $\sin^{-1}\left(\dfrac{\sqrt{3}}{2}\right)$

2. $\cos^{-1}\left(\dfrac{1}{2}\right)$

3. $\arcsin\left(-\dfrac{\sqrt{2}}{2}\right)$

4. $\arccos\left(-\dfrac{\sqrt{2}}{2}\right)$

5. $\tan^{-1}(-\sqrt{3})$

6. $\sec^{-1}\left(-\dfrac{1}{2}\right)$

7. $\arccos\left(-\dfrac{1}{2}\right)$

8. $\arctan\left(-\dfrac{\sqrt{3}}{3}\right)$

9. $\sec^{-1}(-2)$

10. $\sin^{-1}(-1)$

11. $\sin(\sin^{-1}0.541)$

12. $\sin(\cos^{-1}0.6)$

In Problems 13–18, express θ in terms of x using the inverse trigonometric functions $\sin^{-1}$, $\cos^{-1}$, $\tan^{-1}$, and $\sec^{-1}$.

13.

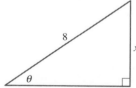

8

x

θ

14.

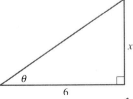

15.

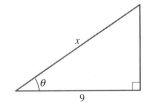

16.

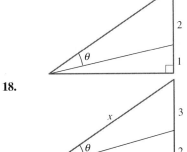

17.

18.

In Problems 19–30, find dy/dx. Check with a computer algebra system if available.

19. $y = \cos^2(x - 2)$

20. $y = \sin\sqrt{2 + 7x}$

21. $y = \cot x \csc x$

22. $y = \dfrac{x}{1 - \tan x}$

23. $y = e^{\cot x}$

24. $y = e^x \cos x$

25. $y = (\tan x)e^{\tan x}$

26. $y = \ln(\sec x + \tan x)$

27. $y = \sin^{-1}(x^2)$

28. $y = \frac{1}{2}\sin^{-1}(e^x)$

29. $y = 7\cos^{-1}\sqrt{2x}$

30. $y = (3x - 1)\cos^{-1}(x^2)$

In Problems 31–42, find each of the integrals. Check with a computer algebra system or numerical technique where appropriate.

31. $\displaystyle\int x \sin(x^2)\,dx$

32. $\displaystyle\int \sin 2x \cos 2x\,dx$

33. $\displaystyle\int \tan x\,dx = \int \frac{\sin x}{\cos x}\,dx$

34. $\displaystyle\int \cot x\,dx$

35. $\displaystyle\int \frac{\sec^2 x}{\tan x}\,dx$

36. $\displaystyle\int \frac{\sin x}{\cos^2 x}\,dx$

37. $\displaystyle\int_0^1 e^{2x}\cos(e^{2x})\,dx$

38. $\displaystyle\int_0^{\pi/2} \sin^2 x \cos x\,dx$

39. $\displaystyle\int_0^{\sqrt{2}/2} \frac{1}{\sqrt{1 - x^2}}\,dx$

40. $\displaystyle\int_{\sqrt{2}}^2 \frac{dx}{x\sqrt{x^2 - 1}}$

41. $\displaystyle\int_{-1}^1 \frac{1}{1 + x^2}\,dx$

42. $\displaystyle\int_0^{\pi/2} \frac{\sin \theta}{1 + \cos^2\theta}\,d\theta$

43. Find the area of the region bounded by $x^2 y + y - 4 = 0$, the coordinate axes, and the line $x = 1$.

44. Find the volume of the solid generated by revolving about the x-axis the region bounded by $y = 5(x^2 + 1)^{-1/2}$, $y = 0$, $x = 0$, and $x = 4$.

45. Where (if at all) on the interval $[0, \infty)$ is $f(x) = 1.25e^x + 0.135 \sin(12x)$ decreasing?

46. A picture 5 feet high is hung on a wall so that its bottom is 8 feet from the floor, as shown in Figure 9. A viewer with eye level at 5.4 feet stands b feet from the wall. Express θ, the vertical angle subtended by the pic-

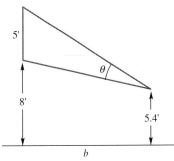

Figure 9

ture at her eye, in terms of b and then find θ if $b = 12.9$ feet.

47. The lower edge of a wall hanging, 10 feet in height, is 2 feet above the observer's eye level. Find the ideal distance b to stand from the wall for viewing the hanging; that is, find b that maximizes the angle subtended for the viewer's eye.

48. Draw the graphs of $y = \arcsin x$ and $y = \arctan(x/\sqrt{1 - x^2})$ using the same axes. Make a conjecture. Prove it.

49. Draw the graph of $y = \pi/2 - \arcsin x$. Make a conjecture. Prove it.

50. Draw the graph of $y = \sin(\arcsin x)$ on $[-1, 1]$. Then draw the graph of $y = \arcsin(\sin x)$ on $[-2\pi, 2\pi]$. Explain the differences you observe.

51. Show

$$\int \frac{dx}{\sqrt{a^2 - x^2}} = \sin^{-1}\frac{x}{a} + C, \qquad a > 0$$

by writing $a^2 - x^2 = a^2[1 - (x/a)^2]$ and making the substitution $u = x/a$.

52. Demonstrate the result in Problem 51 by differentiating the right-hand side to get the integrand.

53. Show

$$\int \frac{dx}{a^2 + x^2} = \frac{1}{a}\tan^{-1}\frac{x}{a} + C, \qquad a \neq 0$$

54. Show

$$\int \frac{dx}{x\sqrt{x^2 - a^2}} = \frac{1}{a}\sec^{-1}\frac{|x|}{a}, \qquad a > 0$$

55. Show, by differentiating the right-hand side, that

$$\int \sqrt{a^2 - x^2}\, dx = \frac{x}{2}\sqrt{a^2 - x^2} + \frac{a^2}{2}\sin^{-1}\frac{x}{a} + C, \qquad a > 0$$

56. Use the result of Problem 55 to show

$$\int_{-a}^{a} \sqrt{a^2 - x^2}\, dx = \frac{\pi a^2}{2}$$

Why is this result expected?

57. Express $d\theta/dt$ in terms of x, dx/dt, and the constants a and b.

(a)

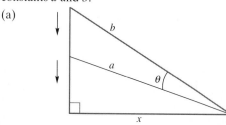

(b)

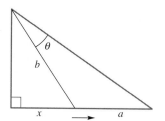

58. The structural steel work of a new office building is finished. Across the street, 60 feet from the ground floor of the freight elevator shaft in the building, a spectator is standing and watching the freight elevator ascend at a constant rate of 15 feet per second. How fast is the angle of elevation of the spectator's line of sight to the elevator increasing 6 seconds after his line of sight passes horizontal?

59. An airplane is flying at a constant altitude of 2 miles and a constant speed of 600 miles per hour on a straight course that will take it directly over an observer on the ground. How fast is the angle of elevation of the observer's line of sight increasing when the distance from her to the plane is 3 miles? Give your result in radians per minute.

60. A revolving beacon light is located on an island and is 2 miles away from the nearest point P on the straight shoreline of the mainland. The beacon throws a spot of light that moves along the shoreline as the beacon revolves. If the speed of the spot of light on the shoreline is 5π miles per minute when the spot is 1 mile from P, how fast is the beacon revolving?

61. A man on a dock is pulling in a rope attached to a rowboat at a rate of 5 feet per second. If the man's hands are 8 feet higher than the point where the rope is attached to the boat, how fast is the angle of depression of the rope changing when there are still 17 feet of rope out?

62. A visitor from outer space is approaching Earth (radius = 6376 kilometers) at 2 kilometers per second. How fast is the angle θ subtended by Earth at her eye increasing when she is 3000 kilometers from the surface?

63. An object starts at the origin and moves along the y-axis so that its velocity in centimeters per second is $dy/dt = (0.2)\cos^2(0.1y)$. How long will it take to get to $y = 2.5\pi$ centimeters? To get $y = 5\pi$ centimeters?

64. Fill in the missing steps in the derivation of formula $\sec^{-1}x = \cos^{-1}\left(\frac{1}{x}\right)$ given in the text.

65. Prove parts (ii), (iii), and (iv) under the heading Four Useful Identities in the text.

66. Prove parts (ii) and (iii) under the heading Derivatives of Four Inverse Trigonometric Functions in the text.

Answers to Concepts Review:　1. $[0, \pi]$; arccos
2. $(-\pi/2, \pi/2)$; arctan　3. 1　4. π; $\pi/6$

7.7 | The Hyperbolic Functions and Their Inverses

In both mathematics and science, certain combinations of e^x and e^{-x} occur so often that they are given special names.

Definition

(The hyperbolic functions). The hyperbolic sine, hyperbolic cosine, and four related functions are defined by

$$\sinh x = \frac{1}{2}\left(e^x - e^{-x}\right) \qquad \cosh x = \frac{1}{2}\left(e^x + e^{-x}\right)$$

$$\tanh x = \frac{\sinh x}{\cosh x} \qquad \coth x = \frac{\cosh x}{\sinh x}$$

$$\operatorname{sech} x = \frac{1}{\cosh x} \qquad \operatorname{csch} x = \frac{1}{\sinh x}$$

The terminology suggests that there must be some connection with the trigonometric functions; there is. First, the fundamental identity for the hyperbolic functions (reminiscent of $\cos^2 x + \sin^2 x = 1$ in trigonometry) is

$$\boxed{\cosh^2 x - \sinh^2 x = 1}$$

To verify, we write

$$\cosh^2 x - \sinh^2 x = \frac{e^{2x} + 2 + e^{-2x}}{4} - \frac{e^{2x} - 2 + e^{-2x}}{4} = 1$$

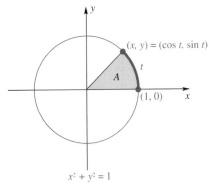

Figure 1

Second, recall that the trigonometric functions are intimately related to the unit circle (Figure 1), so much so that they are sometimes called the circular functions. In fact, the parametric equations $x = \cos t$, $y = \sin t$ describe the unit circle. In parallel fashion, the parametric equations $x = \cosh t$, $y = \sinh t$ describe the right branch of the unit hyperbola $x^2 - y^2 = 1$ (Figure 2). Moreover, in both cases the parameter t is related to the shaded area A by $t = 2A$, though this is not obvious in the second case (see Problem 32).

Because $\sinh(-x) = -\sinh x$, sinh is an odd function; $\cosh(-x) = \cosh x$, so cosh is an even function. Correspondingly, the graph of $y = \sinh x$ is symmetric with respect to the origin and the graph of $y = \cosh x$ is symmetric with respect to the y-axis. The graphs are shown in Figure 3.

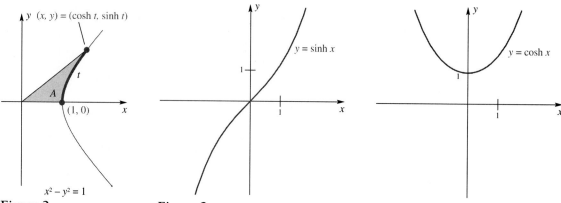

Figure 2   **Figure 3**

Derivatives of Hyperbolic Functions

We can find $\dfrac{d}{dx}\sinh x$ and $\dfrac{d}{dx}\cosh x$ directly from the definitions.

$$\frac{d}{dx}\sinh x = \frac{d}{dx}\left(\frac{e^x - e^{-x}}{2}\right) = \frac{e^x + e^{-x}}{2} = \cosh x$$

and

$$\frac{d}{dx}\cosh x = \frac{d}{dx}\left(\frac{e^x - e^{-x}}{2}\right) = \frac{e^x + e^{-x}}{2} = \sinh x$$

Note that these facts confirm the character of the graphs we drew. For example, because $\frac{d}{dx}(\sinh x) = \cosh x > 0$, the graph of hyperbolic sine is always increasing. Similarly $\frac{d^2}{dx^2}(\cosh x) = \cosh x > 0$, which means that the graph of hyperbolic cosine is concave upward.

The derivatives of the other four hyperbolic functions follow from those for the first two, combined with the Quotient Rule. Here is a list of all six differentiation formulas.

$$\frac{d}{dx}\sinh x = \cosh x \qquad \frac{d}{dx}\cosh x = \sinh x$$

$$\frac{d}{dx}\tanh x = \operatorname{sech}^2 x \qquad \frac{d}{dx}\coth x = -\operatorname{csch}^2 x$$

$$\frac{d}{dx}\operatorname{sech} x = -\operatorname{sech} x \tanh x$$

$$\frac{d}{dx}\operatorname{csch} x = -\operatorname{csch} x \coth x$$

EXAMPLE 1: Find $\dfrac{d}{dx}\tanh(\sin x)$.

SOLUTION: $\dfrac{d}{dx}\tanh(\sin x) = \operatorname{sech}^2(\sin x)\dfrac{d}{dx}(\sin x)$

$$= \cos x \cdot \operatorname{sech}^2(\sin x) \qquad \blacktriangleleft$$

EXAMPLE 2: Find $\dfrac{d}{dx}\cosh^2(3x - 1)$.

SOLUTION: We apply the Chain Rule twice.

$$\frac{d}{dx}\cosh^2(3x - 1) = 2\cosh(3x - 1)\frac{d}{dx}\cosh(3x - 1)$$

$$= 2\cosh(3x - 1)\sinh(3x - 1)\frac{d}{dx}(3x - 1)$$

$$= 6\cosh(3x - 1)\sinh(3x - 1)$$ ◄

EXAMPLE 3: Find $\displaystyle\int \tanh x \, dx$.

SOLUTION: Let $u = \cosh x$, so $du = \sinh x \, dx$.

$$\int \tanh x \, dx = \int \frac{\sinh x}{\cosh x} \, dx = \int \frac{1}{u} \, du$$

$$= \ln|u| + C = \ln|\cosh x| + C = \ln(\cosh x) + C$$

We were able to drop the absolute value signs because $\cosh x > 0$.   ◄

Inverse Hyperbolic Functions

Because hyperbolic sine and hyperbolic tangent have positive derivatives, they are increasing functions and automatically have inverses. To obtain inverses for hyperbolic cosine and hyperbolic secant, we restrict their domains to $x \geq 0$. Thus,

$$x = \sinh^{-1}y \Leftrightarrow y = \sinh x$$

$$x = \cosh^{-1}y \Leftrightarrow y = \cosh x \text{ and } x \geq 0$$

$$x = \tanh^{-1}y \Leftrightarrow y = \tanh x$$

$$x = \text{sech}^{-1}y \Leftrightarrow y = \text{sech} \, x \text{ and } x \geq 0$$

Because the hyperbolic functions are defined in terms of e^x and e^{-x}, it is not surprising that the inverse hyperbolic functions can be expressed in terms of the natural logarithm. For example, consider $y = \cosh x$ for $x \geq 0$; that is, consider

$$y = \frac{e^x + e^{-x}}{2}, \qquad x \geq 0$$

Our goal is to solve this equation for x, which will give $\cosh^{-1}y$. Multiplying both members by $2e^x$, we get $2ye^x = e^{2x} + 1$, or

$$(e^x)^2 - 2ye^x + 1 = 0, \qquad x \geq 0$$

If we solve this quadratic equation in e^x, we obtain

$$e^x = \frac{2y \pm \sqrt{(2y)^2 - 4}}{2} = y \pm \sqrt{y^2 - 1}, \qquad y \geq 1$$

or, after taking natural logarithms,

$$x = \ln\left(y \pm \sqrt{y^2 - 1}\right), \qquad y \geq 1$$

The condition $x \geq 0$ forces us to choose the plus sign, so

$$x = \cosh^{-1}y = \ln\left(y + \sqrt{y^2 - 1}\right), \qquad y \geq 1$$

Similar arguments apply to each of the inverse hyperbolic functions. We obtain the following results (note that the roles of x and y have been interchanged).

$$\sinh^{-1}x = \ln\left(x + \sqrt{x^2 + 1}\right)$$

$$\cosh^{-1}x = \ln\left(x + \sqrt{x^2 - 1}\right), \qquad x \geq 1$$

$$\tanh^{-1}x = \frac{1}{2}\ln\frac{1 + x}{1 - x}, \qquad -1 < x < 1$$

$$\operatorname{sech}^{-1}x = \ln\left(\frac{1 + \sqrt{1 - x^2}}{x}\right), \qquad 0 < x \leq 1$$

Each of these functions is differentiable. In fact,

$$\frac{d}{dx}\sinh^{-1}x = \frac{1}{\sqrt{x^2 + 1}}$$

$$\frac{d}{dx}\cosh^{-1}x = \frac{1}{\sqrt{x^2 = 1}}, \qquad x > 1$$

$$\frac{d}{dx}\tanh^{-1}x = \frac{1}{1 - x^2}, \qquad -1 < x < 1$$

$$\frac{d}{dx}\operatorname{sech}^{-1}x = \frac{-1}{x\sqrt{1 - x^2}}, \qquad 0 < x < 1$$

EXAMPLE 4: Show that $\dfrac{d}{dx}\sinh^{-1}x = 1/\sqrt{x^2 + 1}$ by two different methods.

SOLUTION:

Method 1   Let $y = \sinh^{-1}x$, so

$$x = \sinh y$$

Now differentiate both sides with respect to x.

$$1 = (\cosh y)\frac{d}{dx}y$$

Thus,

$$\frac{d}{dx}y = \frac{d}{dx}(\sinh^{-1}x) = \frac{1}{\cosh y} = \frac{1}{\sqrt{1 + \sinh^2 y}} = \frac{1}{\sqrt{1 + x^2}} \qquad \blacktriangleleft$$

Method 2   Use the logarithmic expression for $\sinh^{-1}x$.

$$\frac{d}{dx}(\sinh^{-1}x) = \frac{d}{dx}\ln\left(x + \sqrt{x^2 + 1}\right)$$

$$= \frac{1}{x + \sqrt{x^2 + 1}}\frac{d}{dx}\left(x + \sqrt{x^2 + 1}\right)$$

$$= \frac{1}{x + \sqrt{x^2 + 1}}\left(1 + \frac{x}{\sqrt{x^2 + 1}}\right)$$

$$= \frac{1}{\sqrt{x^2 + 1}}$$

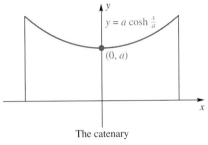

$y = a \cosh \frac{x}{a}$

$(0, a)$

The catenary

Figure 4

Applications: The Catenary

If a homogeneous flexible cable or chain is suspended between two fixed points at the same height, it forms a curve called a **catenary** (Figure 4). Furthermore (see Problem 29), a catenary can be placed in a coordinate system so that its equation takes the form

$$y = a \cosh \frac{x}{a}$$

EXAMPLE 5: Find the length of the catenary $y = a \cosh(x/a)$ between $x = -a$ and $x = a$.

SOLUTION: The desired length (see Section 6.3) is given by

$$\int_{-a}^{a} \sqrt{1 + \left(\frac{dy}{dx}\right)^2}\, dx = \int_{-a}^{a} \sqrt{1 + \sinh^2\left(\frac{x}{a}\right)}\, dx$$

$$= \int_{-a}^{a} \sqrt{\cosh^2\left(\frac{x}{a}\right)}\, dx$$

$$= 2\int_{0}^{a} \cosh\left(\frac{x}{a}\right) dx$$

$$= 2a\int_{0}^{a} \cosh\left(\frac{x}{a}\right)\left(\frac{1}{a}\right) dx$$

$$= \left[2a \sinh \frac{x}{a} \right]_0^a$$

$$= 2a \sinh 1 \approx 2.35a \qquad \blacktriangleleft$$

Concepts Review

1. sinh and cosh are defined by sinh $x = $ _____ and cosh $x = $ _____ .

2. In *hyperbolic* trigonometry, the identity corresponding to $\sin^2 x + \cos^2 x = 1$ is _____ .

3. Because of the identity in Question 2, the graph of the parametric equations $x = \cosh t$, $y = \sinh t$ is _____ .

4. The graph of $y = a \cosh(x/a)$ is a curve called a _____ ; this curve is important as a model for _____ .

Problem Set 7.7

In Problems 1–6, verify that the given equations are identities. For Problems 1, 2, and 6, also verify graphically by graphing both sides.

1. $e^x = \cosh x + \sinh x$

2. $e^{-x} = \cosh x - \sinh x$

3. $\sinh(x + y) = \sinh x \cosh y + \cosh x \sinh y$

4. $\cosh(x + y) = \cosh x \cosh y + \sinh x \sinh y$

5. $\tanh(x + y) = \dfrac{\tanh x + \tanh y}{1 + \tanh x \tanh y}$

6. $\sinh 2x = 2 \sinh x \cosh x$

In Problems 7–20, find $\frac{dy}{dx}$.

7. $y = \sinh^2 x$

8. $y = 5 \sinh^3 x$

9. $y = \cosh(x^2 - 1)$

10. $y = \ln(\coth x)$

11. $y = x^2 \sinh x$

12. $y = e^x \cosh x$

13. $y = \sinh 4x \cosh 2x$

14. $y = \sinh 3x \cosh 5x$

15. $y = \cosh^{-1}(x^3)$

16. $y = \tanh^{-1}(2x^5 - 1)$

17. $y = x \sinh^{-1}(-2x)$

18. $y = \ln(\sinh^{-1} x)$

19. $y = \sinh(\cos x)$

20. $y = \sinh^{-1}(\sin x)$

In Problems 21–24, find the integrals.

21. $\displaystyle\int x \cosh(x^2 + 3)\,dx$

22. $\displaystyle\int \frac{\sinh \sqrt{x}}{\sqrt{x}}\,dx$

23. $\displaystyle\int e^x \sinh e^x\,dx$

24. $\displaystyle\int \tanh x \ln(\cosh x)\,dx$

25. Find the area of the region bounded by $y = \cosh 2x$, $y = 0$, $x = 0$, and $x = \ln 3$.

26. Find the area of the region bounded by $y = \sinh x$, $y = 0$, and $x = \ln 2$.

27. The region bounded by $y = \cosh x$, $y = 0$, $x = 0$, and $x = 1$ is revolved about the x-axis. Find the volume of the resulting solid. *Hint*: $\cosh^2 x = (1 + \cosh 2x)/2$.

28. The curve $y = \cosh x$, $0 \le x \le 1$, is revolved about the x-axis. Find the area of the resulting surface.

29. To derive the equation of a hanging cable (catenary), we consider the section AP from the lowest point A to a general point $P(x, y)$ (see Figure 5) and imagine the rest of the cable to have been removed.

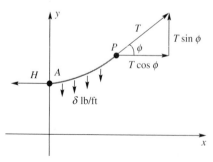

Figure 5

The forces acting on the cable are:

1. H = horizontal tension pulling at A;

2. T = tangential tension pulling at P;

3. $W = \delta s$ = weight of s feet of cable of density δ pounds per foot.

To be in equilibrium, the horizontal and vertical components of T must just balance H and W, respectively. Thus, $T \cos \phi = H$ and $T \sin \phi = W = \delta s$, and so

$$\frac{T \sin \phi}{T \cos \phi} = \tan \phi = \frac{\delta s}{H}$$

But because $\tan \phi = dy/dx$, we get

$$\frac{dy}{dx} = \frac{\delta s}{H}$$

and therefore,

$$\frac{d^2 y}{dx^2} = \frac{\delta}{H}\frac{ds}{dx} = \frac{\delta}{H}\sqrt{1 + \left(\frac{dy}{dx}\right)^2}$$

Show that $y = a \cosh(x/a) + C$ satisfies this differential equation with $a = H/\delta$.

30. Call the graph of $y = b - a \cosh(x/a)$ an inverted catenary and imagine it to be an arch sitting on the x-axis. Show that if the width of this arch along the x-axis is $2a$, then each of the following is true.

(a) $b = a \cosh 1 \approx 1.54308a$

(b) The height of the arch is approximately $0.54308a$.

(c) The height of an arch of width 48 is approximately 13.

31. A farmer built a large hayshed of length 100 feet and width 48 feet. A cross section has the shape of an inverted catenary (see Problem 30) with equation $y = 37 - 24 \cosh(x/24)$.

(a) Draw a picture of this shed.

(b) Find the volume of the shed.

(c) Find the surface area of the roof of the shed.

32. Show that $A = t/2$, where A denotes the area in Figure 2 of this section. *Hint*: At some point you will need to use Formula 44 from the back of the book.

33. Demonstrate for r any real number:

(a) $(\sinh x + \cosh x)^r = \sinh rx + \cosh rx$

(b) $(\cosh x - \sinh x)^r = \cosh rx - \sinh rx$

34. Show that the area under the curve $y = \cosh t$, $0 \le t \le x$ is equal to its arc length.

35. Find the equation of the Gateway Arch in St. Louis, Missouri, given that is an inverted catenary. Assume that it stands on the x-axis, that it is symmetric with respect to the y-axis, and that it is 630 feet wide at the base and 630 feet high at the center.

36. Draw the graphs of $y = \sinh x$, $y = \ln(x + \sqrt{x^2 + 1})$, and $y = x$ using the same axes and scaled so that $-3 \le x \le 3$ and $-3 \le y \le 3$. What does this demonstrate?

Answers to Concepts Review: 1. $(e^x - e^{-x})/2$, $(e^x + e^{-x})/2$ 2. $\cosh^2 x - \sinh^2 x = 1$ 3. One branch of a hyperbola 4. Catenary; a hanging chain

7.8 Chapter Review

Concepts Test

Respond with true or false to each of the following assertions. Be prepared to justify your answer.

1. $\ln|x|$ is defined for all real x.

2. The graph of $y = \ln x$ has no inflection points.

3. $\displaystyle\int_1^{e^3} \frac{1}{t}\, dt = 3$

4. The graph of an invertible function $y = f(x)$ is intersected exactly once by every horizontal line.

5. The domain of $\ln^{-1}$ is the set of all real numbers.

6. $\ln x / \ln y = \ln x - \ln y$

7. $(\ln x)^4 = 4 \ln x$

8. $\ln(2e^{x+1}) - \ln(e^x) = 1$ for all x

9. The functions $f(x) = 4 + e^x$ and $g(x) = \ln(x - 4)$ are a pair of inverse functions.

10. $\exp x + \exp y = \exp(x + y)$

11. $\displaystyle\lim_{x \to 0^+} (\ln \sin x - \ln x) = 0$

12. $\pi^{\sqrt 2} = e^{\sqrt 2 \ln x}$

13. $\dfrac{d}{dx}(\ln \pi) = \dfrac{1}{\pi}$

14. $\displaystyle\int \frac{1}{x}\, dx = \ln|x| + C$

15. $\dfrac{d}{dx}(x^e) = ex^{e-1}$

16. If $f(x) \cdot \exp[g(x)] = 0$ for $x = x_0$, then $f(x_0) = 0$.

17. $\dfrac{d}{dx}(x^x) = x^x \ln x$.

18. The differential equation with initial condition $\frac{dy}{dt} = ty$, $y(0) = 1$ can be solved exactly for $y(1)$ using separation of variables or approximated to a desired number of digits using Euler's method.

19. The differential equation with initial condition $\frac{dy}{dt} = t + y$, $y(0) = 1$ can be solved exactly for $y(1)$ using separation of variables or approximated to a desired number of digits using Euler's method.

20. The differential equation with initial condition $\frac{dy}{dt} = ay$, $y(0) = 1$, a an unknown parameter, can be solved exactly for $y(1)$ using separation of variables or approximated to a desired number of digits using Euler's method.

21. $\arcsin(\sin x) = x$ for all real numbers x

22. $\tan^{-1} x = \dfrac{\sin^{-1} x}{\cos^{-1} x}$

23. $\cosh(\ln 3) = \frac{5}{6}$

24. $\displaystyle\lim_{x \to -\infty} \tan^{-1} x = -\frac{\pi}{2}$

25. $\sin^{-1}(\cosh x)$ is defined for all real x

26. Both $y = \sinh x$ and $y = \cosh x$ satisfy the differential equation $y'' + y = 0$.

27. $\ln(2x^2 - 18) - \ln(x - 3) - \ln(x + 3) = \ln 2$ for all real x

28. If y is growing exponentially and if y triples between $t = 0$ and $t = t_1$, then y will also triple between $t = 2t_1$ and $t = 3t_1$.

29. It is to a saver's advantage to have money invested at 11% compounded continuously rather than 12% compounded monthly.

30. If $\dfrac{d}{dx}(a^x) = a^x$ with $a > 0$, then $a = e$.

Sample Test Problems

In Problems 1–24, differentiate the indicated functions.

1. $\ln \dfrac{x^4}{2}$

2. $\sin^2(x^3)$

3. $e^{x^2 - 4x}$

4. $\log_{10}(x^5 - 1)$

5. $\tan(\ln e^x)$

6. $e^{\ln \cot x}$

7. $2 \tanh \sqrt x$

8. $\tanh^{-1}(\sin x)$

9. $\sinh^{-1}(\tan x)$

10. $2 \sin^{-1}\sqrt{3x}$

11. $\sec^{-1} e^x$

12. $\ln \sin^2\left(\dfrac{x}{2}\right)$

13. $3 \ln(e^{5x} + 1)$

14. $\ln(2x^3 - 4x + 5)$

15. $\cos e^{\sqrt{x}}$

16. $\ln(\tanh x)$

17. $2 \cos^{-1}\sqrt{x}$

18. $4^{3x} + (3x)^4$

19. $2 \csc e^{\ln\sqrt{x}}$

20. $(\log_{10} 2x)^{2/3}$

21. $4 \tan 5x \sec 5x$

22. $x \tan^{-1}\dfrac{x^2}{2}$

23. x^{1+x}

24. $(1 + x^2)^e$

In Problems 25–34, find the antiderivatives of the indicated functions and verify your results by differentiation.

25. e^{3x-1}

26. $6 \cot 3x$

27. $e^x \sin e^x$

28. $\dfrac{6x + 3}{x^2 + x - 5}$

29. $\dfrac{e^{x+2}}{e^{x+3} + 1}$

30. $4x \cos x^2$

31. $\dfrac{4}{\sqrt{1 - 4x^2}}$

32. $\dfrac{\cos x}{1 + \sin^2 x}$

33. $\dfrac{-1}{x + x(\ln x)^2}$

34. $\operatorname{sech}^2(x - 3)$

In Problems 35–37, find the intervals on which f is increasing and the intervals on which f is decreasing. Find where the graph of f is concave upward and where it is concave downward. Find any extreme values and points of inflection. Sketch the graph of f and label the maximums, minimums, and inflection points.

35. $f(x) = x^2 - 4 \sin x, \, -\pi \le x \le \pi$

36. $f(x) = \dfrac{x^2}{e^x}, \, x$ real

37. Let $f(x) = \dfrac{\ln(ax)}{x}, \, 0 < x < \infty, \, a$ real

38. A certain radioactive substance has a half-life of 10 years. How long will it take for 100 grams to decay to 1 gram?

39. If $100 is put in the bank today at 12% interest, how much will it be worth at the end of 1 year if interest is compounded as indicated?

(a) Annually   (b) Monthly

(c) Daily   (d) Continuously

40. Suppose that $100 is put in the bank today where the interest rate $r(t)$ varies as a function of time (t in years). If $r(t) = 0.12 + 0.02 \sin(t^2)$, how much will the investment be worth at the end of 1 year if the interest is compounded continuously? Use Euler's method or the built-in differential equation solving capability of a calulator or computer.

41. Find the equation of the tangent line to $y = (\cos x)^{\sin x}$ at $(0, 1)$.

42. A town grew exponentially from 10,000 in 1970 to 14,000 in 1980. Assuming the same type of growth continues, what will the population be in 2000?

CHAPTER 8

Techniques of Integration

8.1 Substitution and Tables of Integrals

Our repertoire of functions now includes all the so-called elementary functions. These are the constant functions, the power functions, the logarithmic and exponential functions, the trigonometric and inverse trigonometric functions, and all functions obtained from them by addition, subtraction, multiplication, division, and function composition. Thus,

$$f(x) = \frac{e^x + e^{-x}}{2} = \cosh x$$

$$g(x) = (1 + \cos^4 x)^{\frac{1}{2}}$$

$$h(x) = \frac{3^{x^2 - 2x}}{\ln(x^2 + 1)} - \sin[\cos(\cosh x)]$$

are elementary functions.

Differentiation of an elementary function is straightforward, requiring only a systematic use of the rules we have learned. And the result is always an elementary function. Integration (antidifferentiation) is a far different matter. It involves a few techniques and a large bag of tricks; what is worse, it does not always yield an elementary function. For example, it is known that the antiderivatives of e^{-x^2} and $(\sin x)/x$ are not elementary functions. Thus, an example of a well-defined function that is not elementary would be

$$f(x) = \int_0^x e^{-t^2} \, dt$$

453

The two principal techniques for integration are *substitution* and *integration by parts.* The method of substitution was introduced in Section 5.7; we have used it occasionally in the intervening chapters. Here, we review the method and apply it in a wide variety of situations.

Standard Forms   Effective use of the method of substitution depends on the ready availability of a table of known integrals. One such list (but too long to memorize) appears inside the back of this book. The short list shown below is so useful that we think every calculus student should memorize it.

Standard Integral Forms

Constants, Powers

1. $\displaystyle\int k\,du = ku + C$

2. $\displaystyle\int u^r du = \begin{cases} \dfrac{u^{r+1}}{r+1} + C & r \neq -1 \\ \ln|u| + C & r = -1 \end{cases}$

Exponentials

3. $\displaystyle\int e^u\,du = e^u + C$

4. $\displaystyle\int a^u\,du = \dfrac{a^u}{\ln a} + C,\, a \neq 1, a > 0$

Trigonometric Functions

5. $\displaystyle\int \sin u\,du = -\cos u + C$

6. $\displaystyle\int \cos u\,du = \sin u + C$

7. $\displaystyle\int \sec^2 u\,du = \tan u + C$

8. $\displaystyle\int \csc^2 u\,du = -\cot u + C$

9. $\displaystyle\int \sec u \tan u\,du = \sec u + C$

10. $\displaystyle\int \csc u \cot u\,du = -\csc u + C$

Algebraic Functions

11. $\displaystyle\int \frac{du}{\sqrt{a^2 - u^2}} = \sin^{-1}\left(\frac{u}{a}\right) + C$

12. $\displaystyle\int \frac{du}{a^2 + u^2} = \frac{1}{a}\tan^{-1}\left(\frac{u}{a}\right) + C$

Substitution in Indefinite Integrals   Suppose you face an indefinite integration. If it is a standard form, simply write the answer. If not, look for a substitution that will change it to a standard form. If the first substitution you try does not work, try another. Skill at this, like most worthwhile activities, depends on practice.

The method of substitution was given in Theorem 5.7A and is restated here for easy reference.

Theorem A

(Substitution Rule for Indefinite Integrals). Let g be a differentiable function and suppose that F is an antiderivative of f. Then, if $u = g(x)$,

$$\int f(g(x))g'(x)dx = \int f(u)du = F(u) + C = F(g(x)) + C$$

EXAMPLE 1: Find $\int \dfrac{x}{\cos^2(x^2)}\, dx$.

SOLUTION: Stare at this integral for a few moments. Because $1/\cos^2 x = \sec^2 x$, you may be reminded of the standard form $\int \sec^2 u\, du$. Let $u = x^2$, $du = 2x\, dx$. Then

$$\int \frac{x}{\cos^2(x^2)}\, dx = \frac{1}{2}\int \frac{1}{\cos^2(x^2)}\cdot 2x\, dx = \frac{1}{2}\int \sec^2 u\, du$$

$$= \frac{1}{2}\tan u + C = \frac{1}{2}\tan(x^2) + C \qquad \blacktriangleleft$$

EXAMPLE 2: Find $\int \dfrac{3}{\sqrt{5-9x^2}}\, dx$.

SOLUTION: Think of $\int \dfrac{du}{\sqrt{a^2-u^2}}$. Let $u = 3x$, so $du = 3\, dx$. Then,

$$\int \frac{3}{\sqrt{5-9x^2}}\, dx = \int \frac{1}{\sqrt{5-u^2}}\, du = \sin^{-1}\left(\frac{u}{\sqrt{5}}\right) + C$$

$$= \sin^{-1}\left(\frac{3x}{\sqrt{5}}\right) + C \qquad \blacktriangleleft$$

EXAMPLE 3: Find $\int \dfrac{6e^{\frac{1}{x}}}{x^2}\, dx$.

SOLUTION: Think of $\int e^u\, du$. Let $u = 1/x$, so $du = (-1/x^2)\, dx$. Then,

$$\int \frac{6e^{\frac{1}{x}}}{x^2}\, dx = -6\int e^{\frac{1}{x}}\left(\frac{-1}{x^2}\, dx\right) = -6\int e^u\, du$$

$$= -6e^u + C = -6e^{\frac{1}{x}} + C \qquad \blacktriangleleft$$

EXAMPLE 4: We can use substitution to find the integrals of some trigonometric functions that are not so obvious. Find the following integrals.

(a) $\displaystyle\int \tan x\, dx$

(b) $\displaystyle\int \sec x\, dx$

SOLUTION:
(a) We use the definition of $\tan x$ to write the integral as

$$\int \frac{\sin x}{\cos x}\, dx$$

Think of $\int \dfrac{1}{u}\, du$ and let $u = \cos x$, so that $du = -\sin x\, dx$. Then

$$\int \frac{\sin x}{\cos x}\, dx = -\int \frac{1}{u}\, du = -\ln|u| + C = -\ln|\cos x| + C$$

(b) This one requires a trick that is not at all obvious. We rewrite the integral as

$$\int \sec x\, dx = \int \sec x\, \frac{\tan x + \sec x}{\tan x + \sec x}\, dx$$

$$= \int \frac{\sec x \tan x + \sec^2 x}{\sec x + \tan x}\, dx$$

We can now proceed as in part (a), letting $u = \sec x + \tan x$ and hence $du = (\sec x \tan x + \sec^2 x)\, dx$. Therefore we can finish the problem, getting

$$= \int \frac{\sec x \tan x + \sec^2 x}{\sec x + \tan x}\, dx$$

$$= \int \frac{1}{u}\, du = \ln|u| + C = \ln|\sec x + \tan x| + C \qquad \blacktriangleleft$$

In the exercises you are asked to derive formulas for $\int \cot x\, dx$ and $\int \csc x\, dx$.

Substitution in Definite Integrals

This topic was covered in Section 5.7. It is just like substitution in indefinite integrals, but we must remember to make the appropriate change in the limits of integration.

EXAMPLE 5: Evaluate $\displaystyle\int_4^5 t\sqrt{t - 4}\, dt$.

SOLUTION: Let $u = t - 4$, so $du = dt$. The integrand now has an "extra" t in it, so we must solve our substitution equation for t to get $t = u + 4$. Finally, note that $u = 0$ when $t = 4$ and $u = 1$ when $t = 5$. Thus,

$$\int_4^5 t\sqrt{t - 4}\, dt = \int_0^1 (u + 4)u^{\frac{1}{2}}\, du$$

$$= \int_0^1 u^{\frac{3}{2}} + 4u^{\frac{1}{2}}\, du$$

$$= \left[\frac{2}{5}u^{\frac{5}{2}} + 4 \cdot \frac{2}{3}u^{\frac{3}{2}} \right]_0^1$$

$$= \left(\frac{2}{5} + \frac{8}{3} \right) - 0 \approx 3.067 \qquad \blacktriangleleft$$

Note that because we changed the limits of integration from t values to u values, we don't need to substitute again after the integration.

Use of Tables of Integrals   Our list of standard forms is very short (12 formulas); the list inside the back cover of this book is longer (113 formulas) and potentially more useful. Notice that the integrals listed there are grouped according to type. Much more extensive tables of integrals may be found in most libraries. One of the better known is *Standard Mathematical Tables*, published by the Chemical Rubber Company.

One might ask why tables of integrals are needed, given that computer algebra systems can handle integration. As we have pointed out before, when you do a problem more than one way, each becomes a check on the other. Also, as of this writing, tables of integrals are more extensive and a bit more reliable than computer algebra system results (occasional incorrect results for the major systems are still being found). And, of course, it may simply be easier and quicker to consult a table than to go to a computer.

For some problems we will be able to use the table of integrals directly; in other cases we must make an algebraic manipulation or substitution first. We illustrate each of these cases below.

EXAMPLE 6: Find $\int \sqrt{6x - x^2} \, dx$.

SOLUTION: We use Formula 102 with $a = 3$.

$$\int \sqrt{6x - x^2} \, dx = \frac{x - 3}{2} \sqrt{6x - x^2} + \frac{9}{2} \sin^{-1}\left(\frac{x - 3}{3}\right) + C \qquad \blacktriangleleft$$

EXAMPLE 7: Find $\int \dfrac{dx}{\sqrt{2x^2 + 1}}$.

SOLUTION: This integral is close to fitting the form of Formula 45, but we must treat the factor of 2 in front of the x^2 term. There are two approaches we can use; perform the substitution $u^2 = 2x^2$, or factor out the 2. We use the factoring approach below, and leave the substitution method for the exercises.

$$\int \frac{dx}{\sqrt{2x^2 + 1}} = \int \frac{dx}{\sqrt{2\left(x^2 + \frac{1}{2}\right)}}$$

$$= \int \frac{dx}{\sqrt{2}\sqrt{x^2 + \frac{1}{2}}}$$

$$= \frac{1}{\sqrt{2}} \int \frac{dx}{\sqrt{x^2 + \frac{1}{2}}}$$

We can now complete the problem using Formula 45 with $u = x$ and $a^2 = \frac{1}{2}$. We get

$$\frac{1}{\sqrt{2}} \int \frac{dx}{\sqrt{x^2 + \frac{1}{2}}} = \frac{1}{\sqrt{2}} \ln\left|x + \sqrt{x^2 + \frac{1}{2}}\right| + C \qquad \blacktriangleleft$$

EXAMPLE 8: Find $\int_0^{\pi/2} (\cos x)\sqrt{6\sin x - \sin^2 x} \, dx$.

SOLUTION: This time we must perform a substitution first. Let $u = \sin x$, so $du = \cos x \, dx$. Then apply Formula 102 as in Example 6.

$$\int_0^{\pi/2} \cos x \sqrt{6 \sin x - \sin^2 x}\, dx = \int_0^1 \sqrt{6u - u^2}\, du$$

$$= \left[\frac{u-3}{2}\sqrt{6u - u^2} + \frac{9}{2}\sin^{-1}\left(\frac{u-3}{3}\right) \right]_0^1$$

$$= -\sqrt{5} + \frac{9}{2}\sin^{-1}\left(\frac{-2}{3}\right) - \frac{9}{2}\sin^{-1}(-1)$$

$$\approx 1.55 \qquad \blacktriangleleft$$

Checking Results with Technology   You can use a graphing calculator or a computer algebra system to check the results of an integration. For a definite integral, use numerical integration with a calculator or computer.

An indefinite integral is harder to check. If a computer algebra system is available, you can check the integral directly (though there can be problems with this approach, as we will see). Another possibility is to choose an interval of integration (or two) arbitrarily and check the corresponding definite integral. Of course, you can always check an indefinite integral by differentiating the result (by hand or on computer) to see if you get back to the original expression.

EXAMPLE 9: Check the integration problem from Example 8.

SOLUTION: We could use the built-in integration command of a graphing calculator or a computer algebra system, as outlined in Section 5.4. In Figure 1 we give the results of using midpoint Riemann sums (Section 5.3) to approximate the integral $\int_0^{\pi/2} (\cos x)\sqrt{6 \sin x - \sin^2 x}\, dx$ for $n = 10, 100, 1000$.

n	$\displaystyle\sum_{i=1}^{n} f(\overline{x}_i)\Delta x$
10	1.56031
100	1.54906
1000	1.54875

Integration by midpoint Riemann sums with $f(x) = (\cos x)\sqrt{6 \sin x - \sin^2 x}$ and limits of integration 0 and $\frac{\pi}{2}$.

Figure 1

We see that, accurate to four digits, we get

$$\int_0^{\pi/2} (\cos x)\sqrt{6 \sin x - \sin^2 x}\, dx \approx 1.549$$

consistent with our result from Example 8. $\blacktriangleleft$

EXAMPLE 10: Check the integration problem from Example 7.

SOLUTION: We use the Integrate command from three computer algebra systems and show the results in Figure 2.

None of the results from any of the three computer algebra systems appears to be identical to the result

$$\int \frac{dx}{\sqrt{2x^2 + 1}} = \frac{1}{\sqrt{2}}\ln\left| x + \sqrt{x^2 + \tfrac{1}{2}} \right| + C$$

we got in Example 7. The results from Maple and Derive seem to be closer to the form of the result we computed, however. This points out one of the weaknesses of computer algebra; many algebraic forms that are actually the same appear to be quite different.

```
> int (1/sqrt(2*x^2+1), x);
```

$$\frac{1}{2}\sqrt{2}\ln(x\sqrt{2}+\sqrt{2x^2+1})$$ Maple screen

In[21]:=

Integrate [1/Sqrt [2 x^2 + 1], x]

Out[21]=

$$\frac{\text{ArcSinh [Sqrt [2] x]}}{\text{Sqrt [2]}}$$ Mathematica screen

Derive screen

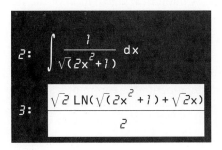

The integral $\int \dfrac{1}{\sqrt{2x^2+1}}\,dx$ with various computer algebra systems.

Figure 2

An approach that sometimes works in a case like this is to subtract the two forms that you would like to show are equivalent, and then use the Simplify command on the result. We show the result of this approach, using the computer algebra system Derive in Figure 3.

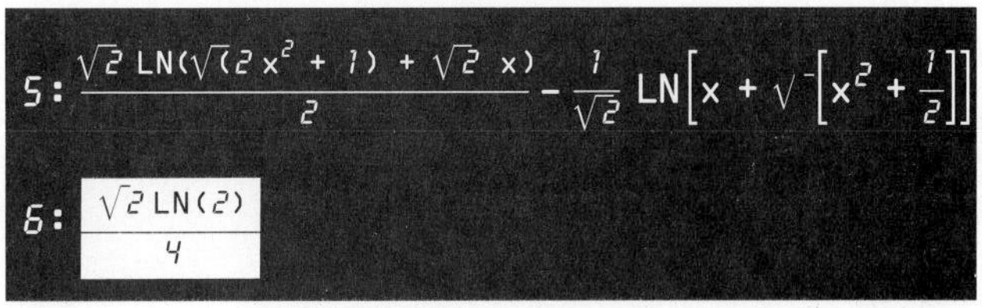

5: $\dfrac{\sqrt{2}\,LN(\sqrt{(2x^2+1)}+\sqrt{2}\,x)}{2} - \dfrac{1}{\sqrt{2}}\,LN\left[x+\sqrt{\left[x^2+\dfrac{1}{2}\right]}\right]$

6: $\dfrac{\sqrt{2}\,LN(2)}{4}$

Derive screen. Expression 5 is simplified to yield expression 6.

Figure 3

Notice that the result of simplifying the difference between our result and that of Derive is not zero, but the constant expression $\dfrac{\sqrt{2}\ln(2)}{4}$. This is to be expected; recall that *two indefinite integrals may differ by a constant.* In the exercises you are asked to perform this simplification yourself.

A final comment is needed for this approach. You should have noticed that we dropped the absolute value function from the result we obtained from integral tables. The purpose of the absolute value symbols in many of the integral table entries is to extend the domain of definition of the resulting integral; the function without absolute value symbols is still an indefinite integral, but one whose domain is not as large. Computer algebra systems, however, do not understand this point, and they will not simplify the difference of the two integrals if we do not first remove the absolute value symbols.

A second approach we can use is to choose an interval of integration and calculate the corresponding definite integral numerically. Then we can use the Fundamental Theorem of Calculus to check our indefinite integral, as follows. This approach does not require a computer algebra system.

Because our integrand $\dfrac{1}{\sqrt{2x^2+1}}$ is defined for all real numbers, we can choose any interval of integration. For simplicity, we choose the interval $0 \le x \le 1$. A numerical calculation of the definite integral yields

$$\int_0^1 \frac{dx}{\sqrt{2x^2+1}} \approx 0.810497$$

(check this on your calculator or computer). If our indefinite integral is correct, the Fundamental Theorem of Calculus would give us

$$\int_0^1 \frac{dx}{\sqrt{2x^2+1}} = \frac{1}{\sqrt{2}} \ln\left|x+\sqrt{x^2+\tfrac{1}{2}}\right|\Big|_0^1$$

$$= \frac{1}{\sqrt{2}}\ln\left|1+\sqrt{1^2+\tfrac{1}{2}}\right| - \frac{1}{\sqrt{2}}\ln\left|0+\sqrt{0^2+\tfrac{1}{2}}\right|$$

$$\approx 0.810497$$

which provides another check on our result.

We leave it for the student to perform yet another check by differentiating the expression $\dfrac{1}{\sqrt{2}}\ln\left|x+\sqrt{x^2+\tfrac{1}{2}}\right|$ and showing that the result is $\dfrac{1}{\sqrt{2x^2+1}}$. ◀

Concepts Review

1. The substitution $u = 1 + x^3$ transforms $\int 3x^2(1+x^3)^5\,dx$ to _____.

2. The substitution $u =$ _____ transforms $\int e^x/(4+e^{2x})\,dx$ to $\int 1/(4+u^2)\,du$.

3. The substitution $u = 1 + \sin x$ transforms $\int_0^{\pi/2}(1+\sin x)^3\cos x\,dx$ to _____.

4. One way to check an indefinite integral is to choose an _____ and check the corresponding _____ using a _____ technique.

Problem Set 8.1

In Problems 1–62, perform the indicated integrations. Use a substitution and/or the integral tables in the back of the book. Check each definite integral using a numerical technique (see Example 9). Check each indefinite integral using either a computer algebra system or by choosing an interval of integration and using a numerical technique (Example 10).

1. $\int (x-1)^4\,dx$

2. $\int \sqrt{2x}\,dx$

3. $\displaystyle\int x(x^2+1)^4\,dx$

4. $\displaystyle\int x\sqrt{x^2+2}\,dx$

5. $\displaystyle\int\frac{dx}{x^2+1}$

6. $\displaystyle\int\frac{e^x}{1+2e^x}\,dx$

7. $\displaystyle\int\frac{x}{x^2+1}\,dx$

8. $\displaystyle\int\frac{x^2}{x^2+1}\,dx$

9. $\displaystyle\int 3t\sqrt{2+t^2}\,dt$

10. $\displaystyle\int\frac{dt}{\sqrt{1+t}}$

11. $\displaystyle\int\frac{\tan z}{\cos z}\,dz$

12. $\displaystyle\int\frac{e^{\sin z}}{\sec z}\,dz$

13. $\displaystyle\int\frac{\cos\sqrt{x}}{\sqrt{x}}\,dx$

14. $\displaystyle\int\frac{x}{\sqrt{1-x^4}}\,dx$

15. $\displaystyle\int_0^{\pi/2}\frac{\cos x}{1+\sin^2 x}\,dx$

16. $\displaystyle\int_0^{3/4}\frac{\cos\sqrt{1-x}}{\sqrt{1-x}}\,dx$

17. $\displaystyle\int\frac{2x^2+x}{x+1}\,dx$

18. $\displaystyle\int\frac{x^3+2x^2}{x-2}\,dx$

19. $\displaystyle\int\frac{\sqrt{\tan x}}{1-\sin^2 x}\,dx$

20. $\displaystyle\int e^{\sin\theta\cos\theta}\cos 2\theta\,d\theta$

21. $\displaystyle\int\frac{\cos(\ln 4x^2)dx}{x}$

22. $\displaystyle\int\frac{\csc^2(\ln x)}{x}\,dx$

23. $\displaystyle\int\frac{5e^x}{\sqrt{1-e^{2x}}}\,dx$

24. $\displaystyle\int\frac{x}{x^4+1}\,dx$

25. $\displaystyle\int\frac{5e^{2x}}{\sqrt{1-e^{2x}}}\,dx$

26. $\displaystyle\int\frac{x^3}{x^4+1}\,dx$

27. $\displaystyle\int_0^1 x10^{x^2}\,dx$

28. $\displaystyle\int_0^{\pi/6}2^{\sin x}\cos x\,dx$

29. $\displaystyle\int\frac{\sin x-\cos x}{\sin x}\,dx$

30. $\displaystyle\int\frac{\sin(4t-1)}{1-\sin^2(4t-1)}\,dt$

31. $\displaystyle\int\frac{z+2}{\cot(z^2+4z-3)}\,dz$

32. $\displaystyle\int\csc 2t\,dt$

33. $\displaystyle\int e^x\sec e^x\,dx$

34. $\displaystyle\int e^x\sec^2(e^x)\,dx$

35. $\displaystyle\int\frac{\sec^3 x+e^{\sin x}}{\sec x}\,dx$

36. $\displaystyle\int\frac{(6t-1)\sin\sqrt{3t^2-t-1}}{\sqrt{3t^2-t-1}}\,dt$

37. $\displaystyle\int\frac{t^2\cos(t^3-2)}{\sin^2(t^3-2)}\,dt$

38. $\displaystyle\int\frac{1+\cos 2x}{\sin^2 2x}\,dx$

39. $\int \dfrac{t^2 \cos^2(t^3 - 2)}{\sin^2(t^3 - 2)}\, dt$

40. $\int \dfrac{\csc^2 2t}{\sqrt{1 + \cot 2t}}\, dt$

41. $\int \dfrac{e^{\tan^{-1} 2t}}{1 + 4t^2}\, dt$

42. $\int (t + 1)e^{-t^2 - 2t - 5}\, dt$

43. $\int \dfrac{y}{\sqrt{16 - 9y^4}}\, dy$

44. $\int \dfrac{\sec^2 2y}{9 + \tan^2 2y}\, dy$

45. $\int \dfrac{\sec x \tan x}{1 + \sec^2 x}\, dx$

46. $\int \dfrac{5}{\sqrt{9 - 4x^2}}\, dx$

47. $\int \dfrac{e^{3t}}{\sqrt{4 - e^{6t}}}\, dt$

48. $\int \dfrac{dt}{2t\sqrt{4t^2 - 1}}$

49. $\int_0^{\pi/2} \dfrac{\sin x}{16 + \cos^2 x}\, dx$

50. $\int_0^1 \dfrac{e^{2x} - e^{-2x}}{e^{2x} + e^{-2x}}\, dx$

51. $\int \dfrac{t\, dt}{\sqrt{2t^2 - 9}}$

52. $\int \dfrac{\tan x}{\sqrt{\sec^2 x - 4}}\, dx$

53. $\int x\sqrt{3x + 2}\, dx$

54. $\int 2t\sqrt{3 - 4t}\, dt$

55. $\int \dfrac{dx}{9 - 16x^2}$

56. $\int \dfrac{dx}{5x^2 - 11}$

57. $\int x^2\sqrt{9 - 2x^2}\, dx$

58. $\int \dfrac{\sqrt{16 - 3t^2}}{t}\, dt$

59. $\int \dfrac{dx}{\sqrt{5 + 3x^2}}$

60. $\int t^2\sqrt{3 + 5t^2}\, dt$

61. $\int \dfrac{x + 1}{\sqrt{x^2 + 2x - 3}}\, dx$

62. $\int \dfrac{\sin t \cos t}{\sqrt{3 \sin t + 5}}\, dt$

63. Find $\int \dfrac{dx}{\sqrt{2x^2 + 1}}$ by performing a substitution and then using formula 45 in the back of the book (compare results with Example 7).

64. Find the length of the curve $y = \ln(\cos x)$ between $x = 0$ and $x = \pi/4$.

65. Verify the simplification in Figure 3 of this section.

66. Derive formulas for $\int \cot x\, dx$ and $\int \csc x\, dx$.

See Example 4. These are formulas 13 and 15 in the table of integrals in the back of this book.

Answers to Concepts Review: 1. $\int u^5\, du$ 2. e^x

3. $\int_1^2 u^3\, du$ 4. Interval of integration; definite integral; numerical

8.2 Integration by Parts

If integration by substitution fails, it may be possible to use a double substitution, better known as *integration by parts*. This method is based on the integration of the formula for the derivative of a product of two functions.

Let $u = u(x)$ and $v = v(x)$. Then

$$\frac{d}{dx}[u(x)v(x)] = u(x)v'(x) + v(x)u'(x)$$

By integrating both members of this equation, we obtain

$$u(x)v(x) = \int u(x)v'(x)\,dx + \int v(x)u'(x)\,dx$$

or, after rearrangement,

$$\int u(x)v'(x)\,dx = u(x)v(x) - \int v(x)u'(x)\,dx$$

Because $dv = v'(x)\,dx$ and $du = u'(x)\,dx$, the preceding equation is often written symbolically as follows.

Integration by Parts—Indefinite Integrals

$$\int u\,dv = uv - \int v\,du$$

The corresponding formula for the definite integrals is

$$\int_a^b u(x)v'(x)\,dx = [u(x)v(x)]_a^b - \int_a^b v(x)u'(x)\,dx$$

We abbreviate this as shown at the top of page 464 (see also Figure 1).

A Geometric Interpretation
of
Integration by Parts

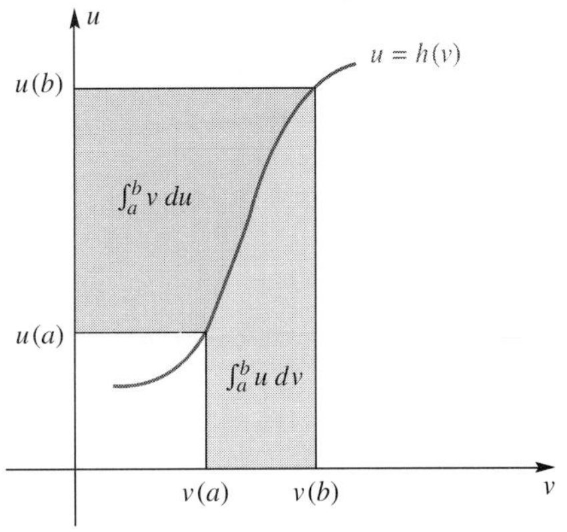

$\int_a^b u\,dv = u(b)v(b) - u(a)v(a) - \int_a^b v\,du$

Figure 1

Integration by Parts—Definite Integrals

$$\int_a^b u \, dv = [uv]_a^b - \int_a^b v \, du$$

These formulas allow us to shift the problem of integrating $u \, dv$ to that of integrating $v \, du$. Success depends on the proper choice of u and dv, which comes with practice.

Simple Examples

EXAMPLE 1: Find $\int x \cos x \, dx$.

SOLUTION: We wish to write $x \cos x \, dx$ as $u \, dv$. One possibility is to let $u = x$ and $dv = \cos x \, dx$. Then $du = dx$ and $v = \int \cos x \, dx = \sin x$ (we can omit the arbitrary constant at this stage). Here is a summary of this double substitution in a convenient format.

$$u = x \qquad dv = \cos x \, dx$$
$$du = dx \qquad v = \sin x$$

The integration-by-parts formula gives

$$\int \underbrace{x}_{u} \underbrace{\cos x \, dx}_{dv} = \underbrace{x}_{u} \underbrace{\sin x}_{v} - \int \underbrace{\sin x}_{v} \underbrace{dx}_{du}$$

$$= x \sin x + \cos x + C$$

We were successful on our first try. Another substitution would be

$$u = \cos x \qquad dv = x \, dx$$
$$du = -\sin x \, dx \qquad v = \frac{x^2}{2}$$

The integration-by-parts formula gives

$$\int \underbrace{\cos x}_{u} \underbrace{x \, dx}_{dv} = \underbrace{(\cos x)}_{u} \underbrace{\frac{x^2}{2}}_{v} - \int \underbrace{\frac{x^2}{2}}_{v} \underbrace{(\sin x \, dx)}_{du}$$

$$= \frac{x^2}{2} \cos x + \frac{1}{2} \int x^2 \sin x \, dx$$

which is correct but not helpful. The new integral on the right-hand side is more complicated than the original one. Thus, we see the importance of a wise choice for u and dv. ◄

EXAMPLE 2: Find $\int_1^2 \ln x \, dx$.

SOLUTION: We make the following substitutions.

$$u = \ln x \qquad dv = dx$$

$$du = \left(\frac{1}{x}\right) dx \qquad v = x$$

Then, according to integration-by-parts formula,

$$\int_1^2 \ln x \, dx = [x \ln x]_1^2 - \int_1^2 x \frac{1}{x} dx$$

$$= 2 \ln 2 - \int_1^2 dx$$

$$= 2 \ln 2 - 1 \qquad \blacktriangleleft$$

EXAMPLE 3: Find $\int \arcsin x \, dx$.

SOLUTION: We make the substitutions

$$u = \arcsin x \qquad dv = dx$$

$$du = \frac{1}{\sqrt{1 - x^2}} dx \qquad v = x$$

Then

$$\int \arcsin x \, dx = x \arcsin x - \int \frac{x}{\sqrt{1 - x^2}} dx$$

We can now complete the problem by using a substitution on the resulting integral. With $u = x^2$, $du = 2x \, dx$ we get

$$\int \frac{x}{\sqrt{1 - x^2}} dx = \frac{1}{2} \int \frac{du}{\sqrt{1 - u}}$$

$$= \frac{1}{2} \int (1 - u)^{-\frac{1}{2}} du$$

$$= -\frac{1}{2} \cdot 2(1 - u)^{\frac{1}{2}} + C$$

$$= -\left(1 - x^2\right)^{\frac{1}{2}} + C$$

so that our final result is

$$\int \arcsin x \, dx = x \arcsin x + \left(1 - x^2\right)^{\frac{1}{2}} + C \qquad \blacktriangleleft$$

Repeated Integration by Parts   Sometimes it is necessary to apply integration by parts several times.

EXAMPLE 4: Find $\int x^2 \sin x \, dx$.

SOLUTION: Let

$$u = x^2 \qquad\qquad dv = \sin x \, dx$$
$$du = 2x \, dx \qquad\quad v = -\cos x$$

Then

$$\int x^2 \sin x \, dx = -x^2 \cos x + 2 \int x \cos x \, dx$$

We have improved our situation (the exponent on x in the integrand has gone from 2 to 1), which suggests reapplying integration by parts to the integral on the right. Actually, we did this integration in Example 1, so we will make use of the result obtained there.

$$\int x^2 \sin x \, dx = -x^2 \cos x + 2(x \sin x + \cos x + C)$$

$$= -x^2 \cos x + 2x \sin x + 2 \cos x + K \qquad\blacktriangleleft$$

EXAMPLE 5: A car bumper is pushed down and released. Suppose that the vertical *velocity v* (in feet per second) of the bumper can be described by the equation $v = e^{-t} \sin t$ where t is time (in seconds). Find the total vertical distance traveled by the bumper over a five-second period.

SOLUTION: In Section 6.1 we saw that if the velocity of an object as a function of time is given by $v(t)$, then between $t = a$ and $t = b$ the displacement is given by $\int_a^b v(t) \, dt$, but the total distance traveled is given by $\int_a^b |v(t)| \, dt$. Thus, we want to find $\int_0^5 |e^{-t} \sin t| \, dt$.

We need to determine the intervals where $e^{-t} \sin t$ is positive and where this function is negative. This is most easily done by graphing $v = e^{-t} \sin t$. In Figure 2 we show the graphs of both velocity and the absolute value of velocity versus time.

From the velocity graph we see that $v = 0$ at around 3 seconds; by setting $v = e^{-t} \sin t = 0$ we see that in fact $v = 0$ at $t = \pi \approx 3.14$. Then, because $v \geq 0$ for $0 \leq t \leq \pi$ and $v \leq 0$ for $\pi \leq t \leq 5$ (again from the velocity graph) we have

$$\int_0^5 |e^{-t} \sin t| \, dt = \int_0^\pi e^{-t} \sin t \, dt + \int_\pi^5 (-e^{-t} \sin t) \, dt$$

$$= \int_0^\pi e^{-t} \sin t \, dt - \int_\pi^5 e^{-t} \sin t \, dt$$

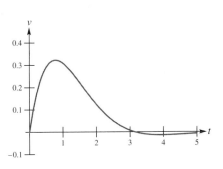

Graph of velocity versus time.

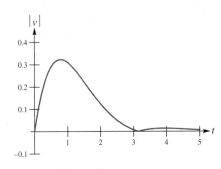

Graph of |velocity| versus time.

Figure 2

To finish the problem we need to find the indefinite integral $\int e^{-t}\sin t\, dt$.

We try integration by parts with $u = e^{-t}$ and $dv = \sin t\, dt$. Then $du = -e^{-t}dt$ and $v = -\cos t$. Thus,

$$\int e^{-t}\sin t\, dt = -e^{-t}\cos t - \int e^{-t}\cos t\, dt$$

which does not seem to have improved things. But do not give up; let us integrate by parts again. In the right integral, let $u = e^{-t}$ and $dv = \cos t\, dt$, so $du = -e^{-t}\, dt$ and $v = \sin t$. Then

$$\int e^{-t}\cos t\, dt = e^{-t}\sin t + \int e^{-t}\sin t\, dt$$

When we substitute this in our first result, we get

$$\int e^{-t}\sin t\, dt = -e^{-t}\cos t - \left(e^{t}\sin t + \int e^{-t}\sin t\, dt \right)$$

$$= -e^{-t}\cos t - e^{-t}\sin t - \int e^{-t}\sin t\, dt$$

By transposing the last term to the left side and combining terms, we obtain

$$2\int e^{-t}\sin t\, dt = -e^{-t}(\cos t + \sin t)$$

from which

$$\int e^{-t}\sin t\, dt = -\frac{1}{2}e^{-t}(\cos t + \sin t) + C$$

What made integration by parts work was the fact that the integral we wanted to find reappeared on the right side.

We can now finish the problem to get

$$\int_0^5 \left| e^{-t}\sin t \right| dt \quad = \int_0^\pi e^{-t}\sin t \, dt - \int_\pi^5 e^{-t}\sin t \, dt$$

$$= \left[-\frac{1}{2}e^{-t}(\cos t + \sin t) \right]_0^\pi - \left[-\frac{1}{2}e^{-t}(\cos t + \sin t) \right]_\pi^5$$

$$= \left[\frac{1}{2}e^{-\pi} - \left(-\frac{1}{2}e^0 \right) \right]$$

$$\qquad - \left[-\frac{1}{2}e^{-5}(\cos 5 + \sin 5) - \left(\frac{1}{2}e^{-\pi} \right) \right]$$

$$\approx 0.540939$$

Thus, the bumper travels about 0.541 feet during the first 5 seconds. This number also corresponds to the area under the second graph in Figure 2. ◄

Reduction Formulas A formula of the form

$$\int f^n(x) \, dx = g(x) + \int f^k(x) \, dx$$

where $k < n$ is called a **reduction formula** (the exponent on f is reduced). Formulas 33 through 38 in the tables of integrals in the back of this book are examples. Such formulas can often be obtained through integration by parts.

EXAMPLE 6: Prove the reduction formula

$$\int \sin^n x \, dx = -\frac{1}{n}\sin^{n-1}x \cos x + \frac{n-1}{n}\int \sin^{n-2}x \, dx$$

which is Formula 33 in the tables of integrals (with u replaced by x).

SOLUTION: Let $u = \sin^{n-1}x$ and $dv = \sin x \, dx$. Then

$$du = (n-1)\sin^{n-2}x \cos x \, dx \quad \text{and} \quad v = -\cos x$$

from which

$$\int \sin^n x \, dx = -\sin^{n-1}x \cos x + (n-1)\int \sin^{n-2}x \cos^2 x \, dx$$

If we replace $\cos^2 x$ by $1 - \sin^2 x$ in the last integral, we obtain

$$\int \sin^n x \, dx = -\sin^{n-1}x \cos x + (n-1)\int \sin^{n-2}x \, dx - (n-1)\int \sin^n x \, dx$$

After combining the first and last integrals above, and solving for $\int \sin^n x \, dx$, we get the reduction formula (valid for $n \geq 2$)

$$\int \sin^n x \, dx = -\frac{1}{n}\sin^{n-1}x \cos x + \frac{n-1}{n}\int \sin^{n-2}x \, dx$$

which completes the proof. In the exercises at the end of this section you are asked to prove similar reduction formulas for $\int \cos^n x \, dx$, $\int \tan^n x \, dx$, and $\int \sec^n x \, dx$. ◄

EXAMPLE 7: Use the reduction formula above to evaluate

$$\int_0^{\pi/2} \sin^8 x \, dx$$

SOLUTION: Note first that

$$\int \sin^n x \, dx = \left[\frac{-\sin^{n-1} x \cos x}{n}\right]_0^{\pi/2} + \frac{n-1}{n}\int_0^{\pi/2} \sin^{n-2} x \, dx$$

$$= 0 + \frac{n-1}{n}\int_0^{\pi/2} \sin^{n-2} x \, dx$$

Thus,

$$\int_0^{\pi/2} \sin^8 x \, dx = \frac{7}{8}\int_0^{\pi/2} \sin^6 x \, dx$$

$$= \frac{7}{8}\cdot\frac{5}{6}\int_0^{\pi/2} \sin^4 x \, dx$$

$$= \frac{7}{8}\cdot\frac{5}{6}\cdot\frac{3}{4}\int_0^{\pi/2} \sin^2 x \, dx$$

$$= \frac{7}{8}\cdot\frac{5}{6}\cdot\frac{3}{4}\cdot\frac{1}{2}\int_0^{\pi/2} 1 \, dx$$

$$= \frac{7}{8}\cdot\frac{5}{6}\cdot\frac{3}{4}\cdot\frac{1}{2}\cdot\frac{\pi}{2} = \frac{35}{256}\pi \qquad \blacktriangleleft$$

The general formula for $\displaystyle\int_0^{\pi/2} \sin^n x \, dx$ can be found in a similar way (Formula 113 at the back of the book).

Choosing a Method of Integration

When presented with the need to perform an indefinite integration, how do we know which of the methods presented so far is the "right" one? Should we try a substitution, try an integration by parts, use a table of integrals, or go to a computer algebra system?

There is no easy answer. Substitution and integration by parts are "first principles" methods; they are useful if the problem is fairly simple or if you want to prove one of the integration formulas in a table of integrals. If the problem is a little harder but fits one of the forms in a table of integrals, by all means use the table. If you simply can't seem to get a good handle on an integration problem, try the integration command of a computer algebra system; sometimes seeing the result will help you see what other methods of integration will work.

Perhaps most important: When possible, use more than one method of integration so that each method is a check on the other. If a computer algebra system is available, you can always use it to check your results, as outlined in Section 8.1.

As for deciding between substitution and integration by parts, see the lab project at the end of this section. This project will also help you to determine when a given function doesn't have an indefinite integral in terms of elementary functions.

Concepts Review

1. The integration-by-parts formula says that

$$\int u \, dv = \underline{\hspace{3cm}}.$$

2. To apply this formula to $\int x \sin x \, dx$, let $u = \underline{\hspace{2cm}}$ and $dv = \underline{\hspace{2cm}}$.

3. Applying the integration-by-parts formula yields the value $\underline{\hspace{2cm}}$ for $\int_0^{\pi/2} x \sin x \, dx$.

4. A formula that expresses $\int f^n(x) \, dx$ in terms of $\int f^k(x) \, dx$, where $k < n$, is called a $\underline{\hspace{2cm}}$ formula.

Problem Set 8.2

In Problems 1–24, use integration by parts to perform the indicated operations. Check each definite integral using a numerical technique. Check each indefinite integral using either a computer algebra system or by choosing an interval of integration and using a numerical technique (see Section 8.1).

1. $\displaystyle\int xe^x \, dx$

2. $\displaystyle\int xe^{3x} \, dx$

3. $\displaystyle\int x \sin 3x \, dx$

4. $\displaystyle\int \ln 3x \, dx$

5. $\displaystyle\int \arctan x \, dx$

6. $\displaystyle\int x\sqrt{x + 1} \, dx$

7. $\displaystyle\int t \sec^2 5t \, dt$

8. $\displaystyle\int \arctan(1/t) \, dt$

9. $\displaystyle\int \sqrt{x} \, \ln x \, dx$

10. $\displaystyle\int z^3 \ln z \, dz$

11. $\displaystyle\int x \arctan x \, dx$

12. $\displaystyle\int t \cos 4t \, dt$

13. $\displaystyle\int w \ln w \, dw$

14. $\displaystyle\int 3x \sin 5x \, dx$

15. $\displaystyle\int_{\pi/6}^{\pi/2} x \csc^2 x \, dx$

16. $\displaystyle\int_{\pi/4}^{\pi/2} \csc^3 x \, dx$

17. $\displaystyle\int xa^x \, dx$

18. $\displaystyle\int 4x \, 2^{3x} \, dx.$ *Hint:* $\sin^3 x = \left(1 - \cos^2 x\right) \sin x.$

19. $\displaystyle\int x^2 e^x \, dx$

20. $\displaystyle\int \ln^2 x \, dx$

21. $\displaystyle\int e^t \cos t \, dt$

22. $\displaystyle\int x^2 \cos x \, dx$

23. $\displaystyle\int \sin(\ln x) \, dx$

24. $\int (\ln x)^3 \, dx$. *Hint*: Use Problem 20.

25. Find the area of the region bounded by the curve $y = \ln x$, the x-axis, and the line $x = e$.

26. Find the volume of the solid generated by revolving the region of Problem 25 about the x-axis.

27. Find the area of the region bounded by the curves $y = 3xe^{-x/3}$, $y = 0$, and $x = 9$. Make a sketch.

28. Find the volume of the solid generated by revolving the region described in Problem 27 about the x-axis.

29. Derive the reduction formula (see Example 6).

$$\int \cos^n x \, dx = \frac{\cos^{n-1} x \sin x}{n} + \frac{n-1}{n} \int \cos^{n-2} x \, dx$$

which is number 34 in the tables in the back of the book.

30. Use Problem 29 to find a nice reduction formula for $\int_0^{\pi/2} \cos^n x \, dx$.

31. Evaluate $\int_0^{\pi/2} \sin^7 x \, dx$ (see Example 7).

32. Evaluate $\int_0^{\pi/2} \cos^6 x \, dx$ (see Problem 30).

33. Prove the reduction formula

$$\int \tan^n x \, dx = \frac{\tan^{n-1} x}{n-1} - \int \tan^{n-2} x \, dx, \, n \neq 1$$

which is number 35 in the tables in the back of the book. *Hint*: Use the identity $\tan^2 x = \sec^2 x - 1$ to write

$$\int \tan^n x \, dx = \int (\sec^2 x - 1)\tan^{n-2} x \, dx$$

$$= \int \sec^2 x \, \tan^{n-2} x \, dx - \int \tan^{n-2} x \, dx$$

Then use integration by parts on the first integral on the right-hand side with $u = \tan^{n-2} x$ and $dv = \sec^2 x \, dx$.

34. Prove the reduction formula

$$\int \sec^n x \, dx = \frac{\sec^{n-2} x \tan x}{n-1} + \frac{n-2}{n-1} \int \sec^{n-2} x \, dx$$

Hint: Write the integral as

$$\int \sec^n x \, dx = \int \sec^2 x \, \sec^{n-2} x \, dx$$

Then use integration by parts with $u = \sec^{n-2} x$ and $dv = \sec^2 x \, dx$.

35. Derive the reduction formula $\int x^n e^x \, dx =$

$$x^n e^x - n \int x^{n-1} e^x \, dx \text{ and use it to find } \int x^3 e^x \, dx.$$

36. If $p(x)$ is a polynomial of degree n and G_1, $G_2, \ldots, G_{n+1}$ are successive antiderivatives of g, then by repeated integration by parts

$$\int p(x)g(x) \, dx = p(x)G_1(x) - p'(x)G_2(x)$$

$$+ p''(x)G_3(x) - \cdots + (-1)^n p^{(n)}(x)G_{n+1}(x) + C$$

Use this result to find each of the following.

(a) $\int (x^3 - 2x)e^x \, dx$

(b) $\int (x^2 - 3x + 1) \sin x \, dx$

37. Graph $y = x \sin x$ for $x \geq 0$.
(a) Find a formula for the area of the nth hump.
(b) The second hump is revolved about the y-axis. Find the volume of the resulting solid.

38. The Fourier coefficient, normally written as $a_n = \dfrac{1}{\pi} \displaystyle\int_{-\pi}^{\pi} f(x) \sin nx \, dx$, plays an important role in applied mathematics. Show that if $f'(x)$ is continuous on $[-\pi, \pi]$, then $\lim\limits_{n \to \infty} a_n = 0$. *Hint*: Integrate by parts.

39. Find the error in the following "proof" that $0 = 1$. In $\int (1/t) \, dt$, set $u = 1/t$ and $dv = dt$. Then $du = -t^{-2} \, dt$ and $uv = 1$. Integration by parts gives

$$\int (1/t) \, dt = 1 - \int (-1/t) \, dt$$

or $0 = 1$.

Answers to Concepts Review: 1. $uv - \displaystyle\int v \, du$ 2. x;

$\sin x \, dx$ 3. 1 4. Reduction

LAB 14: INTEGRATION BY PARTS

Mathematical Background

There is no "product rule" for integration. This makes integration of products of functions difficult. One technique that can sometimes be used is substitution; another common approach is integration by parts. The formula for integration by parts can be written several ways; two common ones are

$$\int f'(x)g(x)\,dx = f(x)g(x) - \int f(x)g'(x)\,dx$$

$$\int u\,dv = uv - \int v\,du$$

If we use $\frac{d}{dx}$ for derivative and $\int (\ )\,dx$ for antiderivative, then the rule can also be written

$$\int f(x)g(x)\,dx = \left(\int f(x)\,dx\right)g(x) - \int\left(\left(\int f(x)\,dx\right)\frac{d}{dx}g(x)\right)dx$$

In words, choose one of the functions and take the derivative, then take the antiderivative of the other function, and integrate the product of the two. If the new product is easier to integrate than the original product, then you have made a good choice; if not, try another choice for f and g. Sometimes you must integrate by parts several times to get an answer.

Lab Introduction

In this lab you will experiment with integration by parts and try to discover some general principles for how to choose the two functions and when to use the technique. The basic idea is to split the function you are integrating into a product of two functions, take the derivative of one and integrate the other. Then put each part into the formula above and see if you can do the new integral. If you cannot, try reversing the roles of the two functions, or try integrating by parts again.

A computer algebra system will come handy in *checking* $\int f(x)\,dx$ and $\frac{d}{dx}g(x)$ for the problem below; do not, however, use an antiderivative rule for this step that you do not already know. This is the only place you will need the computer; do the rest by hand. For some you may not need the computer at all. You can, of course, check your answer with the computer when you are done with a problem.

Experiment

(1) Try integrating by parts on each of the following.

(a) $\int xe^{-x}\,dx$ (b) $\int 4x^2e^{3x}\,dx$

(c) $\displaystyle\int x^{1/2}e^x\,dx$ (d) $\displaystyle\int x^2 e^{x^2}\,dx$

Note which integrals worked out and which did not. Now use the indefinite integration capability of a computer algebra system on each problem to check your answers. Note which problems your computer could not do (that is, for which the result either contained another integration or contained unfamiliar functions).

(2) Same instructions as part (1).

(a) $\displaystyle\int 3x\sin(7x)\,dx$ (b) $\displaystyle\int x^2\sin\!\left(\frac{x}{2}\right)dx$

(c) $\displaystyle\int x^{1/3}\cos(x)\,dx$ (d) $\displaystyle\int x\cos(x^3)\,dx$

(3) Same instructions as part (1).

(a) $\displaystyle\int x\ln x\,dx$ (b) $\displaystyle\int x^2\ln(3x)\,dx$

(c) $\displaystyle\int x^{1/2}\ln(2x)\,dx$ (d) $\displaystyle\int \ln x^2\,dx$

Hint for (d): write $\ln x^2$ as $1\ln x^2$.

Discussion

1. For the first set of integrals, why did the method fail in some cases and not in others? Come up with a rule of thumb to explain when integration by parts works and when it does not when a power of x is multiplied by an exponential function.

2. Same as Discussion question 1, for when a power of x is multiplied by a sin or cos function.

3. Same as Discussion question 1, for when a power of x is multiplied by a logarithmic function.

4. Were there cases where the integration command of your computer algebra system did not give a final result in terms of known functions? Were there cases where integration by parts was not helpful, but computer algebra did give a final answer in terms of known functions? Can you tell just by looking at the answer given by your computer algebra system when integration by parts *would* have been helpful? Explain.

8.3   Some Trigonometric Integrals

Substitution and integration by parts are the primary general integration techniques you will need in order to attack most of the integration problems you are likely to encounter in the future. However, there are some specific substitutions and some algebraic techniques that are required frequently enough to make them worthy of study. These substitutions and techniques are the focus of the next two sections.

When we combine the method of substitution with a clever use of trigonometric identities, we can integrate a wide variety of trigonometric forms.

Useful Identities

Some trigonometric identities needed in this section are the following. Recall that a more complete list of identities is printed at the end of Section 0.4.

Pythagorean Identities

$$\sin^2 x + \cos^2 x = 1$$

$$1 + \tan^2 x = \sec^2 x$$

$$1 + \cot^2 x = \csc^2 x$$

Half-Angle Identities

$$\sin^2 x = \frac{1 - \cos 2x}{2}$$

$$\cos^2 x = \frac{1 + \cos 2x}{2}$$

Product Identities

$$\sin mx \cos nx = \tfrac{1}{2}[\sin(m + n)x + \sin(m-n)x]$$

$$\sin mx \sin nx = -\tfrac{1}{2}[\cos(m + n)x - \cos(m - n)x]$$

$$\cos mx \cos nx = \tfrac{1}{2}[\cos(m + n)x + \cos(m - n)x]$$

EXAMPLE 1: Find $\displaystyle\int \cos^3 x \, dx$.

SOLUTION: One standard technique used when dealing with integrals involving trig functions is to use the basic Pythagorean identity $\sin^2 x + \cos^2 x = 1$ to replace some of the sines with cosines or vice versa. Because

$$\cos^3 x = \cos x \cos^2 x$$

we can replace $\cos^2 x$ with $1 - \sin^2 x$ to get

$$\int \cos^3 x \, dx = \int \cos x \cos^2 x \, dx$$

$$= \int \cos x (1 - \sin^2 x) dx$$

$$= \int (\cos x - \cos x \sin^2 x) dx$$

$$= \int \cos x \, dx - \int \cos x \sin^2 x \, dx$$

$$= \sin x - \frac{1}{3} \sin^3 x + C$$

where the second integral can be evaluated with the help of the substitution $u = \sin x$ (you should verify this).   ◄

EXAMPLE 2: Find $\int \sin^2 x \, dx$.

SOLUTION: Here we make use of a half-angle identity.

$$\int \sin^2 x \, dx = \int \frac{1 - \cos 2x}{2} \, dx$$

$$= \frac{1}{2} \int dx - \frac{1}{4} \int (\cos 2x)(2) \, dx$$

$$= \frac{1}{2} \int dx - \frac{1}{2} \int \cos 2x \, dx$$

$$= \frac{1}{2} x - \frac{1}{2} \cdot \frac{1}{2} \sin 2x + C$$

$$= \frac{1}{2} x - \frac{1}{4} \sin 2x + C \qquad ◄$$

Examples 1 and 2 are typical in that integration of a sine raised to an odd power can be performed by replacing all of the sines with cosines except one (and similarly with a cosine raised to an odd power). Sines or cosines raised to an even power require the use of a half-angle identity.

Are They Different?

Indefinite integrations by alternate methods may lead to answers that seem to be different. By one method, with $u = \cos x$ and $du = -\sin x \, dx$,

$$\int \sin x \cos x \, dx = -\int u \, du = -\frac{u^2}{2} = -\frac{1}{2} \cos^2 x + C$$

By a second method, letting $u = \sin x$ and $du = \cos x \, dx$,

$$\int \sin x \cos x \, dx = \int u \, du = \frac{u^2}{2} = \frac{1}{2} \sin^2 x + C$$

But two such answers should differ by at most an additive constant. Note, however, that

$$\frac{1}{2} \sin^2 x + C = \frac{1}{2}(1 - \cos^2 x) + C = -\frac{1}{2} \cos^2 x + \left(\frac{1}{2} + C\right)$$

Now reconcile these answers with a third answer.

$$\int \sin x \cos x \, dx = \frac{1}{2} \int \sin 2x \, dx = -\frac{1}{4} \cos 2x + C$$

EXAMPLE 3: Find $\int \sin 2x \cos 3x \, dx$.

SOLUTION: We use the first product identity listed above, with $m = 2$ and $n = 3$, to get

$$\int \sin 2x \cos 3x \, dx = \frac{1}{2} \int [\sin 5x + \sin(-x)] \, dx$$

$$= \frac{1}{2} \int \sin 5x \, dx - \frac{1}{2} \int \sin x \, dx$$

$$= -\frac{1}{10} \cos 5x + \frac{1}{2} \cos x + C \qquad \blacktriangleleft$$

Example 3 illustrates a type of integral often encountered in Fourier Analysis (important in the study of vibrations). Example 4 contains a key result used in this area of mathematics.

EXAMPLE 4: If m and n are positive integers, show that

$$\int_{-\pi}^{\pi} \sin mx \sin nx \, dx = \begin{cases} 0 & \text{if } n \neq m \\ \pi & \text{if } n = m \end{cases}$$

SOLUTION: If $m \neq n$,

$$\int_{-\pi}^{\pi} \sin mx \sin nx \, dx = -\frac{1}{2} \int_{-\pi}^{\pi} [\cos(m + n)x - \cos(m - n)x] \, dx$$

$$= -\frac{1}{2} \left[\frac{1}{m + n} \sin(m + n)x - \frac{1}{m - n} \sin(m - n)x \right]_{-\pi}^{\pi}$$

$$= 0$$

because the sine of any integral multiple of π is equal to zero.
If $m = n$,

$$\int_{-\pi}^{\pi} \sin mx \sin nx \, dx = -\frac{1}{2} \int_{-\pi}^{\pi} [\cos 2mx - 1] \, dx$$

$$= -\frac{1}{2} \left[\frac{1}{2m} \sin 2mx - x \right]_{-\pi}^{\pi}$$

$$= -\frac{1}{2} [-2\pi] = \pi \qquad \blacktriangleleft$$

Integrands Involving $\sqrt{a^2 - x^2}$, $\sqrt{a^2 + x^2}$, and $\sqrt{x^2 - a^2}$

Some integrals involving square roots can be simplified with the help of a trigonometric substitution. The table below will help you determine which trig function to substitute for each of the following radical expressions.

	Radical	Substitution	Restriction on t
1.	$\sqrt{a^2 - x^2}$	$x = a \sin t$	$-\pi/2 \leq t \leq \pi/2$
2.	$\sqrt{a^2 + x^2}$	$x = a \tan t$	$-\pi/2 \leq t \leq \pi/2$
3.	$\sqrt{x^2 - a^2}$	$x = a \sec t$	$0 \leq t \leq \pi, t \neq \pi/2$

Now note the simplifications that these substitutions achieve.

1. $\sqrt{a^2 - x^2} = \sqrt{a^2 - a^2 \sin^2 t} = \sqrt{a^2 \cos^2 t} = |a \cos t| = a \cos t$

2. $\sqrt{a^2 + x^2} = \sqrt{a^2 + a^2 \tan^2 t} = \sqrt{a^2 \sec^2 t} = |a \sec t| = a \sec t$

3. $\sqrt{x^2 - a^2} = \sqrt{a^2 \sec^2 t - a^2} = \sqrt{a^2 \tan^2 t} = |a \tan t| = \pm a \tan t$

The restrictions on t allow us to remove the absolute value signs in the first two cases, but they also achieve something else. These restrictions are exactly the ones we introduced in Section 7.6 in order to define the inverses of the sine, tangent, and secant functions. This means that we can solve the substitution equations for t in each case, which will allow us to write our final answers in the following example in terms of x.

In addition to these trigonometric substitutions, we may need to use the formulas for

$\int \tan x \, dx$ and $\int \sec x \, dx$ from Example 4 of Section 8.1, the reduction formulas for

$\int \sin^n x \, dx, \int \cos^n x \, dx, \int \tan^n x \, dx,$ and $\int \sec^n x \, dx$ developed in Example 6 and the

problems of Section 8.2, and the Four Useful Identities from Section 7.6. The integration formulas are all listed in the table of integrals in the back of the book.

EXAMPLE 5: Find each integral. Use either a simple substitution or a trigonometric substitution.

(a) $\int \sqrt{x - 1} \, dx$

(b) $\int \sqrt{x^2 - 1} \, dx$

(c) $\int x\sqrt{x^2 - 1} \, dx$

SOLUTION:

(a) Let $u = x - 1$ and so $du = dx$. Then

$$\int \sqrt{x - 1} \, dx = \int \sqrt{u} \, du = \frac{2}{3} u^{\frac{3}{2}} + C = \frac{2}{3}(x - 1)^{\frac{3}{2}} + C$$

(b) Let $x = \sec t$ so that $dx = \sec t \tan t \, dt$. Then

$$\int \sqrt{x^2 - 1} \, dx = \int \tan t \sec t \tan t \, dt$$

$$= \int \tan^2 t \sec t \, dt$$

$$= \int (\sec^2 t - 1)\sec t \, dt$$

$$= \int \sec^3 t \, dt - \int \sec t \, dt$$

We can now finish by using a reduction formula for $\int \sec^3 t \, dt$ from Problem 34 of Section 8.2 (formula 37 from the tables in the back of the book) and the formula for $\int \sec t \, dt$ from Example 4 of Section 8.1 (formula 14 in the tables in the back of the book). We have

$$\int \sec^3 t \, dt - \int \sec t \, dt = \left(\frac{1}{2} \sec t \tan t + \frac{1}{2} \int \sec t \, dt \right) - \int \sec t \, dt$$

$$= \frac{1}{2} \sec t \tan t - \frac{1}{2} \int \sec t \, dt$$

$$= \frac{1}{2} \sec t \tan t - \frac{1}{2} \ln \left| \sec t + \tan t \right|$$

$$= \frac{1}{2} x \sqrt{x^2 - 1} - \frac{1}{2} \ln \left| x + \sqrt{x^2 - 1} \right|$$

In the last line, we used identity (iv) under Four Useful Identities from Section 7.6.

(c) Let $u = x^2 - 1$ so that $du = 2x \, dx$. Then

$$\int x \sqrt{x^2 - 1} \, dx = \frac{1}{2} \int \sqrt{u} \, du = \frac{1}{3} u^{\frac{3}{2}} + C = \frac{1}{3} (x^2 - 1)^{\frac{3}{2}} + C \qquad \blacktriangleleft$$

Notice that only (b) required a trigonometric substitution. Carefully inspect the three integrals in this example to get a feel for when a trig substitution would be appropriate.

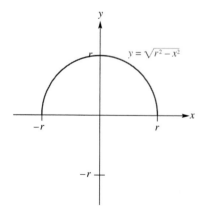

Figure 1

EXAMPLE 6: Use calculus to derive the formula for the area of a circle.

SOLUTION: The equation of a circle of radius r centered at the origin is $x^2 + y^2 = r^2$. The top half of the circle is given by the function $y = \sqrt{r^2 - x^2}$. See Figure 1. The area of the upper-right quarter of the circle would be given by $\int_0^r \sqrt{r^2 - x^2} \, dx$, so that the area of the entire circle would be

$$\text{area} = 4 \int_0^r \sqrt{r^2 - x^2} \, dx$$

We first evaluate the corresponding indefinite integral

$$\int \sqrt{r^2 - x^2} \, dx$$

Start with the substitution

$$x = r \sin t, \qquad -\frac{\pi}{2} \le t \le \frac{\pi}{2}.$$

Then $dx = r \cos t \, dt$ and $\sqrt{r^2 - x^2} = r \cos t$. Thus,

$$\int \sqrt{r^2 - x^2} \, dx = \int r \cos t \cdot r \cos t \, dt = r^2 \int \cos^2 t \, dt$$

$$= \frac{r^2}{2} \int (1 + \cos 2t) \, dt$$

$$= \frac{r^2}{2} \left(t + \frac{1}{2} \sin 2t \right) + C$$

$$= \frac{r^2}{2} (t + \sin t \cos t) + C$$

Now, $x = r \sin t$ is equivalent to $x/r = \sin t$ and, because t was restricted so sine is invertible,

$$t = \sin^{-1}\left(\frac{x}{r}\right)$$

Also, from identity (ii) under Four Useful Identities in Section 7.6,

$$\cos t = \cos \left[\sin^{-1}\left(\frac{x}{r}\right) \right] = \sqrt{1 - \frac{x^2}{r^2}} = \frac{1}{r} \sqrt{r^2 - x^2}$$

a fact that you can see from the triangle in Figure 2. Thus,

$$\int \sqrt{r^2 - x^2} \, dx = \frac{r^2}{2} \left(\sin^{-1}\left(\frac{x}{r}\right) + \frac{x}{r} \cdot \frac{1}{r} \sqrt{r^2 - x^2} \right) + C$$

$$= \frac{r^2}{2} \sin^{-1}\left(\frac{x}{r}\right) + \frac{x}{2} \sqrt{r^2 - x^2} + C$$

We can now finish the problem by substituting in the limits of integration to get

$$\text{area} = 4 \int_0^r \sqrt{r^2 - x^2} \, dx$$

$$= 4 \left[\frac{r^2}{2} \sin^{-1}\left(\frac{x}{r}\right) + \frac{x}{2} \sqrt{r^2 - x^2} \right]_0^r$$

$$= 4 \cdot \frac{r^2}{2} \cdot \frac{\pi}{2} = \pi r^2 \qquad \blacktriangleleft$$

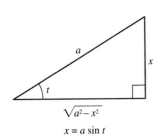

$x = a \sin t$

Figure 2

Concepts Review

1. To handle $\int \cos^2 x \, dx$, we first rewrite it as
_____ .

2. To handle $\int \cos^3 x \, dx$, we first rewrite it as
_____ .

3. To handle an integral involving $\sqrt{4 + x^2}$, make the substitution $x =$ _____ .

4. To handle an integral involving $\sqrt{x^2 - 4}$, make the substitution $x =$ _____ .

Problem Set 8.3

In Problems 1–16, perform the indicated integrations.

1. $\displaystyle\int \cos^2 x\, dx$

2. $\displaystyle\int \sin^4 5x\, dx$

3. $\displaystyle\int \cos^3 x\, dx$

4. $\displaystyle\int \sin^3 x\, dx$

5. $\displaystyle\int_0^{\pi/2} \sin^5 t\, dt$

6. $\displaystyle\int_0^{\pi/2} \cos^6 t\, dt$

7. $\displaystyle\int \sin^7 3x \cos^2 3x\, dx$

8. $\displaystyle\int (\sin^3 t)\sqrt{\cos t}\, dt$

9. $\displaystyle\int \cos^3\theta \sin^{-2}\theta\, d\theta$

10. $\displaystyle\int \sin^{1/2}\theta \cos^3\theta\, d\theta$

11. $\displaystyle\int \sin^4 2t \cos^2 2t\, dt$

12. $\displaystyle\int \sin^6 t \cos^2 t\, dt$

13. $\displaystyle\int \sin 4y \cos 5y\, dy$

14. $\displaystyle\int \cos y \cos 4y\, dy$

15. $\displaystyle\int \sin 3t \sin t\, dt$

16. $\displaystyle\int_{-\pi}^{\pi} \cos mx \cos nx\, dx,\ m \neq n$

In Problems 17–28, perform the indicated integrations. You may need to use a simple substitution or a trigonometric substitution as in Examples 5 and 6. Try to get a re-sult without use of any formula in the tables of integrals that has not been proven in the text or by you in the problems.

17. $\displaystyle\int \sqrt{x + 4}\, dx$

18. $\displaystyle\int x\sqrt{x + 4}\, dx$

19. $\displaystyle\int \sqrt{x^2 + 4}\, dx$

20. $\displaystyle\int x\sqrt{x^2 + 4}\, dx$

21. $\displaystyle\int_1^4 \frac{dx}{\sqrt{x + 2}}$

22. $\displaystyle\int_0^4 \frac{\sqrt{t}}{t + 1}\, dt$

23. $\displaystyle\int \frac{\sqrt{1 - x^2}}{x}\, dx$

24. $\displaystyle\int \frac{x^2\, dx}{\sqrt{9 - x^2}}$

25. $\displaystyle\int \frac{dx}{x\sqrt{x^2 + 9}}$

26. $\displaystyle\int \frac{dx}{(x^2 + 9)^{\frac{3}{2}}}$

27. $\displaystyle\int_5^8 \frac{dx}{\sqrt{x^2 - 16}}$

28. $\displaystyle\int_2^5 \frac{\sqrt{t^2 - 4}}{t^3}\, dt$

29. The region bounded by $y = x + \sin x$, $y = 0$, $x = \pi$, is revolved about the x-axis. Find the volume of the resulting solid.

30. Let $f(x) = \displaystyle\sum_{n=1}^{N} a_n \sin(nx)$. Use Example 4 to show each of the following.

(a) $\displaystyle\frac{1}{\pi}\int_{-\pi}^{\pi} f(x)\sin(mx)\, dx = \begin{cases} a_m & \text{if } m \leq N \\ 0 & \text{if } m > N \end{cases}$

(b) $\displaystyle\frac{1}{\pi}\int_{-\pi}^{\pi} f^2(x)\, dx = \sum_{n=1}^{N} a_n^2$

Note: Integrals of this type occur in a subject called *Fourier series,* which has applications to heat, vibrating strings, and other physical phenomena.

31. The shaded region between one arch of $y = \sin x$, $0 \le x \le \pi$, and the line $y = k$, $0 \le k \le 1$, is revolved about the line $y = k$, generating a solid S. Determine k so that S has (a) minimum volume and (b) maximum volume.

32. Find $\displaystyle\int \frac{x\,dx}{x^2 + 9}$ by (a) an algebraic substitution and (b) a trigonometric substitution. Then reconcile your answers.

33. Find $\displaystyle\int \frac{\sqrt{4 - x^2}}{x}\,dx$ by (a) the substitution $u = \sqrt{4 - x^2}$ and (b) a trigonometric substitution. Then reconcile your answers.

34. Starting at $(a, 0)$, an object is pulled along by a string of length a with the pulling end moving along the positive y-axis. The path of the object is a curve called a **tractrix** and has the property that the string is always tangent to the curve (see Figure 3). Set up a differential equation for the curve and solve it.

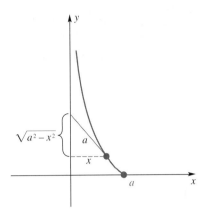

Figure 3

Answers to Concepts Review: 1. $\displaystyle\int [1 + \cos 2x)/2]\,dx$

2. $\displaystyle\int (1 - \sin^2 x)\cos x\,dx$ 3. $2\tan t$ 4. $2\sec t$

8.4 | Integration of Rational Functions

A **rational function** is the quotient of two polynomial functions. Examples are

$$f(x) = \frac{2}{(x + 1)^3}, \quad g(x) = \frac{2x + 2}{x^2 - 4x + 8}, \quad h(x) = \frac{x^5 + 2x^3 - x + 1}{x^3 + 5x}$$

Of these, f and g are **proper rational functions**, meaning that the degree of the numerator is less than that of the denominator. An improper (not proper) rational function can always be written as a sum of a polynomial function and a proper rational function. Thus, for example,

$$h(x) = \frac{x^5 + 2x^3 - x + 1}{x^3 + 5x} = x^2 - 3 + \frac{14x + 1}{x^3 + 5x}$$

$$\begin{array}{r} x^2 - 3 \\ x^3 + 5x\,\overline{\smash{\big)}\,x^5 + 2x^3 - x + 1} \\ \underline{x^5 + 5x^3} \\ -3x^3 - x \\ \underline{-3x^3 - 15x} \\ 14x + 1 \end{array}$$

Figure 1

a result obtained by long division (Figure 1). Because polynomials are easy to integrate, the problem of integrating rational functions is really that of integrating proper rational functions. But can we always integrate proper rational functions? In theory, the answer is yes, though the practical details may overwhelm us. Consider first the cases of f and g above.

EXAMPLE 1: Find $\displaystyle\int \frac{2}{(x + 1)^3}\,dx$.

SOLUTION: Think of the substitution $u = x + 1$.

$$\int \frac{2}{(x + 1)^3}\,dx = 2\int (x + 1)^{-3}\,dx = \frac{2(x + 1)^{-2}}{-2} + C$$

$$= \frac{-1}{(x + 1)^2} + C$$

◄

Partial Fraction Decomposition (Linear Factors)

Adding fractions is a standard algebraic exercise. For example,

$$\frac{2}{x-1} + \frac{3}{x+1} = \frac{5x-1}{(x-1)(x+1)} = \frac{5x-1}{x^2-1}$$

It is the reverse process of *decomposing* a fraction into a sum of simpler fractions that interests us now.

EXAMPLE 2: Decompose $(3x-1)/(x^2-x-6)$ and then find its indefinite integral.

SOLUTION: Because the denominator factors as $(x+2)(x-3)$, it seems reasonable to hope for a decomposition of the the following form.

(1) $$\frac{3x-1}{(x+2)(x-3)} = \frac{A}{x+2} + \frac{B}{x-3}$$

Our job is, of course, to determine A and B so (1) is an identity, a task we find easier after we have multiplied both sides by $(x+2)(x-3)$. We obtain

(2) $$3x - 1 = A(x-3) + B(x+2)$$

or, equivalently,

(3) $$3x - 1 = (A + B)x + (-3A + 2B)$$

However, (3) is an identity if and only if the coefficients of the like powers of x on both sides are equal—that is,

$$A + B = 3$$
$$-3A + 2B = -1$$

By solving this pair of equations for A and B, we obtain $A = \frac{7}{5}$, $B = \frac{8}{5}$. Consequently,

$$\frac{3x-1}{x^2-x-6} = \frac{3x-1}{(x+2)(x-3)} = \frac{\frac{7}{5}}{x+2} + \frac{\frac{8}{5}}{x-3}$$

and

$$\int \frac{3x-1}{x^2-x-6}\, dx = \frac{7}{5}\int \frac{1}{x+2}\, dx + \frac{8}{5}\int \frac{1}{x-3}\, dx$$

$$= \frac{7}{5} \ln|x+2| + \frac{8}{5} \ln|x-3| + C \qquad \blacktriangleleft$$

If there was anything difficult about this process, it was the determination of A and B. We found their values by "brute force"; there is an easier way. In (2), which we wish to be an identity, substitute the convenient values $x = 3$ and $x = -2$, obtaining

$$8 = A \cdot 0 + B \cdot 5$$
$$-7 = A \cdot (-5) + B \cdot 0$$

This immediately gives $B = \frac{8}{5}$ and $A = \frac{7}{5}$.

You have just witnessed an odd, but correct, mathematical maneuver. Equation (1) turns out to be an identity (true for all x except

---

Solve This DE

"… often, there is little resemblance between a differential equation and its solution. Who would suppose that an expression as simple as

$$\frac{dy}{dx} = \frac{1}{a^2 - x^2}$$

could be transformed into

$$y = \frac{1}{2a}\log_e\left(\frac{a+x}{a-x}\right) + C$$

This resembles the transformation of a chrysalis into a butterfly."

Silvanus P. Thompson

The method of partial fractions makes this an easy transformation. Do you see how it is done?

−2 and 3) if and only if the essentially equivalent equation (2) is true precisely at −2 and 3. Ask yourself why this is so. Ultimately it depends on the fact that the two sides of equation (2), both linear polynomials, are identical if they have the same values at any two points.

EXAMPLE 3: Find $\int \dfrac{x}{(x-3)^2}\, dx$.

SOLUTION: This time we try a decomposition of the form

$$\frac{x}{(x-3)^2} = \frac{A}{x-3} + \frac{B}{(x-3)^2}$$

with A and B to be determined. After clearing of fractions, we get

$$x = A(x-3) + B$$

If we now substitute the convenient value $x = 3$ and any other value, such as $x = 0$, we obtain $B = 3$ and $A = 1$. Thus,

$$\int \frac{x}{(x-3)^2}\, dx = \int \frac{1}{x-3}\, dx + 3\int \frac{1}{(x-3)^2}\, dx$$

$$= \ln|x-3| - \frac{3}{x-3} + C \qquad \blacktriangleleft$$

Partial Fractions by Computer Algebra   Partial fraction decompositions can be performed using computer algebra systems. The decomposition from Example 3 is done on three such systems in Figure 2.

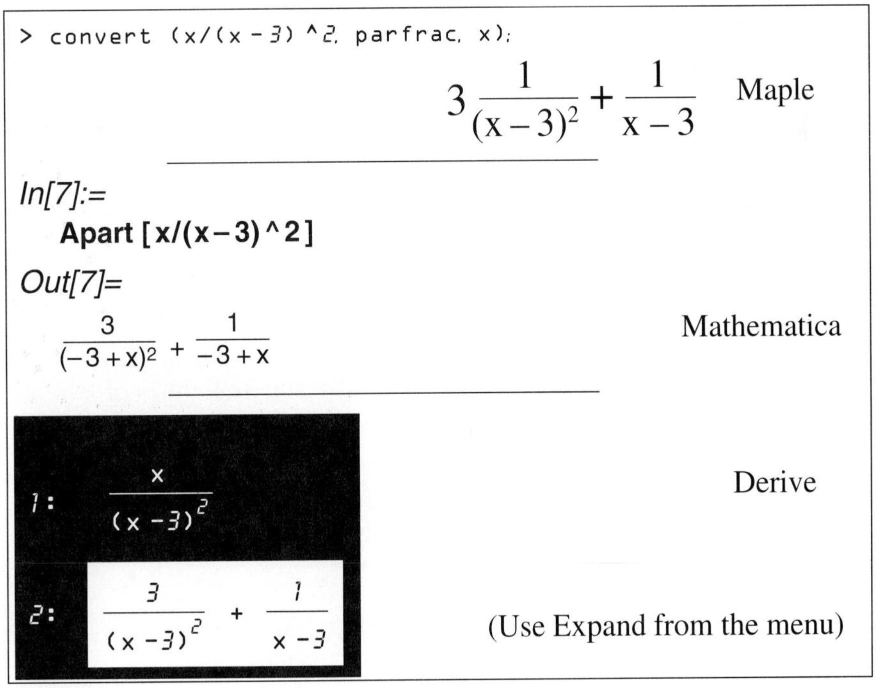

```
> convert (x/(x - 3) ^2, parfrac, x);
```

$$3\,\frac{1}{(x-3)^2} + \frac{1}{x-3} \qquad \text{Maple}$$

$In[7]:=$

   Apart [x/(x−3)^2]

$Out[7]=$

$$\frac{3}{(-3+x)^2} + \frac{1}{-3+x} \qquad \text{Mathematica}$$

Derive

$1:\quad \dfrac{x}{(x-3)^2}$

$2:\quad \dfrac{3}{(x-3)^2} + \dfrac{1}{x-3}$

(Use Expand from the menu)

Partial fractions by computer algebra.

Figure 2

Quadratic Factors   Not all polynomials can be factored into a product of linear factors with real number coefficients. For example, the expression $x^3 + x$ can be factored as $x(x^2 + 1)$, which consists of the linear factor x and the quadratic factor $x^2 + 1$. This quadratic expression cannot itself be factored without the use of complex numbers. Such a factor is called an *irreducible* quadratic factor.

A partial fraction decomposition can still be accomplished when irreducible quadratic factors appear, but the algebra gets a bit more tedious. Computer algebra systems can handle this case as well. We leave examples of such types of problems for the exercises, and we summarize below how to determine the form of a partial fraction decomposition.

Summary   To decompose a rational function $f(x) = p(x)/q(x)$ into partial fractions, proceed as follows.

Step 1   If $f(x)$ is improper—that is, if $p(x)$ is of degree at least that of $q(x)$—divide $p(x)$ by $q(x)$, obtaining

$$f(x) = \text{a polynomial} + \frac{N(x)}{D(x)}$$

where the degree of $N(x)$ is less than the degree of $D(x)$.

Step 2   Factor $D(x)$ into a product of linear and/or irreducible quadratic factors with real coefficients. By a theorem of algebra, this is always (theoretically) possible.

Step 3   For each factor of the form $(ax + b)^k$, expect the decomposition to have the terms

$$\frac{A_1}{(ax + b)} + \frac{A_2}{(ax + b)^2} + \cdots + \frac{A_k}{(ax + b)^k}$$

Step 4   For each factor of the form $(ax^2 + bx + c)^m$, expect the decomposition to have the terms

$$\frac{B_1 x + C_1}{ax^2 + bx + c} + \frac{B_2 x + C_2}{(ax^2 + bx + c)^2} + \cdots + \frac{B_m x + C_m}{(ax^2 + bx + c)^m}$$

Step 5   Set $N(x)/D(x)$ equal to the sum of all the terms found in Steps 3 and 4. The number of constants to be determined should equal the degree of the denominator, $D(x)$.

Step 6   Multiply both sides of the equation found in Step 5 by $D(x)$ and solve for the unknown constants. This can be done by either of two methods: (1) Equate coefficients of like-degree terms; or (2) assign convenient values to the variable x.

The Logistic Differential Equation Revisited   In Section 7.4 we first encountered the logistic differential equation for population growth

$$\frac{dy}{dt} = ky\frac{(L - y)}{L}$$

and in Section 7.5 we solved it numerically using Euler's method. Recall that y represents the population size, t represents time, L represents the maximum population size that can be supported by the environment, and k represents the percentage growth rate per unit of time when the population is small.

We can now solve this equation exactly, using separation of variables and a partial fraction decomposition.

EXAMPLE 4: Use separation of variables to solve the logistic differential equation for y subject to the initial condition $y(0) = y_0$.

SOLUTION: Separating variables on the equation $\frac{dy}{dt} = ky\frac{L-y}{L}$ and integrating both sides, we get

$$\int \frac{L\,dy}{y(L-y)} = \int k\,dt$$

We can easily evaluate the integral on the right-hand side to get

$$\int k\,dt = kt + C$$

but we need a partial fraction expansion on the left-hand side. Because we have two distinct linear factors, as in Example 2, we write

$$\frac{L}{y(L-y)} = \frac{A}{y} + \frac{B}{L-y}$$

$$L = A(L-y) + By$$

Letting $y = 0$ in the second equation we get $A = 1$, and letting $y = L$ in the same equation we get $B = 1$. We can now integrate to get

$$\int \frac{L}{y(L-y)}\,dy = \int \frac{1}{y}\,dy + \int \frac{1}{L-y}\,dy$$

$$= \ln y - \ln(L-y)$$

$$= \ln\left(\frac{y}{L-y}\right)$$

(We used one of the properties of logarithms in the last step.) Equating the two integrals, we get

$$\ln\left(\frac{y}{L-y}\right) = kt + C \tag{1}$$

We have two tasks left: solve for y in terms of t, and use the initial condition $y(0) = y_0$ to eliminate the constant of integration C. It is convenient to solve for C first, so letting $t = 0$ and $y = y_0$ in equation (1) we get

$$\ln\left(\frac{y_0}{L-y_0}\right) = k(0) + C$$

$$C = \ln\left(\frac{y_0}{L-y_0}\right)$$

We substitute this result back into equation (1) to get

$$\ln\left(\frac{y}{L-y}\right) = kt + \ln\left(\frac{y_0}{L-y_0}\right) \tag{2}$$

To solve equation (2) for y, we first apply the natural exponential function to both sides of (2):

$$\ln\left(\frac{y}{L-y}\right) = kt + \ln\left(\frac{y_0}{L-y_0}\right)$$

$$e^{\ln\left(\frac{y}{L-y}\right)} = e^{kt + \ln\left(\frac{y_0}{L-y_0}\right)} = e^{kt}e^{\ln\left(\frac{y_0}{L-y_0}\right)}$$

$$\frac{y}{L-y} = e^{kt}\frac{y_0}{L-y_0}$$

Next we clear fractions and collect the y-terms on the left-hand side:

$$y(L-y_0) = e^{kt}y_0(L-y) = e^{kt}y_0L - e^{kt}y_0y$$

$$y(L-y_0) + e^{kt}y_0y = e^{kt}y_0L$$

Finally we factor out the y from the left-hand side and divide to get

$$y(L-y_0 + e^{kt}y_0) = e^{kt}y_0L$$

$$y = \frac{e^{kt}y_0L}{L-y_0 + e^{kt}y_0}$$

$$y = \frac{y_0L}{(L-y_0)e^{-kt} + y_0}$$

The last step involved multiplying numerator and denominator by e^{-kt} to make the final form a bit simpler. ◄

   Compare our result from the last example with the logistic model of population growth we used in Example 6 of Section 4.4. The two functions are exactly the same (in that example we used m for the maximum supportable population). Thus, we have tied together a *function* that we referred to as a logistic model and a *differential equation* that we also called logistic by showing that the function is the *solution* to the differential equation.

   In the sciences, mathematical modeling usually occurs at the level of differential equations. The differential equation helps us to see the assumptions that go into a model. In the logistic population model, the assumption is that the rate of growth of the population is proportional to both the population size and amount by which the population differs from the maximum supportable population.

Concepts Review

1. If the degree of the polynomial $p(x)$ is less than the degree of $q(x)$, then $f(x) = p(x)/q(x)$ is called a _____ rational function.

2. To integrate the improper rational function $f(x) = (x^2 + 4)/(x + 1)$, we first rewrite it as $f(x) = $ _____.

3. If $(x - 1)(x + 1) + 3x + x^2 = ax^2 + bx + c$,

then $a = $ _____ , $b = $ _____ , and $c = $ _____ .

4. To evaluate $\displaystyle\int \frac{L\,dy}{y(L-y)}$ we can decompose $\dfrac{L}{y(L-y)}$ in the form _____ .

Problem Set 8.4

In Problems 1–12, use the method of partial fraction decomposition to perform the required integration. Check both your partial fraction decomposition and your integration with a computer algebra system if one is available.

1. $\displaystyle\int \frac{2}{x^2 + 2x}\, dx$

2. $\displaystyle\int \frac{2}{x^2 - 1}\, dx$

3. $\displaystyle\int \frac{5x + 3}{x^2 - 9}\, dx$

4. $\displaystyle\int \frac{x - 6}{x^2 - 2x}\, dx$

5. $\displaystyle\int \frac{x - 11}{x^2 + 3x - 4}\, dx$

6. $\displaystyle\int \frac{3x - 13}{x^2 + 3x - 10}\, dx$

7. $\displaystyle\int \frac{2x^2 + x - 4}{x^3 - x^2 - 2x}\, dx$

8. $\displaystyle\int \frac{6x^2 + 22x - 23}{(2x - 1)(x^2 + x - 6)}\, dx$

9. $\displaystyle\int \frac{3x^3}{x^2 + x - 2}\, dx$

10. $\displaystyle\int \frac{x^4 + 8x^2 + 8}{x^3 - 4x}\, dx$

11. $\displaystyle\int \frac{x + 1}{(x - 3)^2}\, dx$

12. $\displaystyle\int \frac{5x + 7}{x^2 + 4x + 4}\, dx$

Problems 13 and 14 contain irreducible quadratic factors. Use the guidelines given in the text or a computer algebra system to find a partial fraction decompositon of each integrand, and then find the integral.

13. $\displaystyle\int \frac{1}{(x^2 + 1)(x - 1)}\, dx$

14. $\displaystyle\int \frac{x}{(x^2 + 1)(x - 1)}\, dx$

15. Let us suppose that the earth will not support a population of more than 16 billion and that there were 2 billion people in 1925 and 4 billion people in 1975. Then if y is the population t years after 1925, an appropriate model is the logistic differential equation

$$\frac{dy}{dt} = ky\frac{16 - y}{16}$$

(a) Solve this differential equation.

(b) Find the population in 2015.

(c) When will the population be 9 billion?

16. Do Problem 15 assuming the upper limit for the population is 10 billion.

17. The Law of Mass Action in chemistry results in the differential equation

$$\frac{dx}{dt} = k(a - x)(b - x) \quad k > 0, a > 0, b > 0$$

where x is the amount of a substance at time t resulting from the reaction of two others. Assume that $x = 0$ when $t = 0$.

(a) Solve this differential equation in the case $b > a$.

(b) Show that $x \to a$ as $t \to \infty$ (if $b > a$).

(c) Suppose that $a = 2$ and $b = 4$ and that 1 gram of the substance is formed in 20 minutes. How much will be present in 1 hour?

(d) Solve the differential equation if $a = b$.

18. The differential equation

$$\frac{dy}{dt} = k(y - m)(M - y)$$

with $k > 0$ and $0 \le m < y_0 < M$ has been used to model some growth problems. Solve the equation and find $\lim\limits_{t \to \infty} y$.

19. As a model for the production of trypsin from trypsinogen in digestion, biochemists have proposed the model

$$\frac{dy}{dt} = k(A - y)(B + y)$$

where $k > 0$, A is the initial amount of trypsinogen, and B is the initial amount of trypsin. Solve this differential equation.

Answers to Concepts Review: 1. Proper   2. $x - 1 + \dfrac{5}{x + 1}$

3. $2; 3; -1$ 4. $\dfrac{A}{y} + \dfrac{B}{L - y}$

LAB 15: POPULATION MODELS

Mathematical Background

To model population growth, one usually starts with the assumption that population growth is proportional to population size. Letting y represent population and t represent time, we get the differential equation

$$\frac{dy}{dt} = ky$$

The solution to this equation (using separation of variables) is $y = Ce^{kt}$, which grows exponentially. This is unrealistic, and so we need to add something to the equation that will limit growth. A common equation used is called the logistic differential equation, which is given by

$$\frac{dy}{dt} = ky\frac{L-y}{L} \tag{1}$$

The idea is that the $\frac{L-y}{L}$ term limits the growth of the population; without this term the population would experience unrestricted growth as in the first case. Subject to the initial condition $y(0) = y_0$ we can solve this equation for y to get

$$y = \frac{y_0 L}{(L-y_0)e^{-kt} + y_0} \tag{2}$$

as is shown in the text.

Lab Introduction

You will graph y as a function of t for various values of k, L, and y_0 in order to discover what they represent. Then you will see how well this equation fits the growth of a sunflower plant, which can be regarded as a population of cells.

We will need the first and second derivatives of y if we want to find maximums, minimums, and inflection points. Equation (1) gives the first derivative as a function of y; if we now take its derivative with respect to t we get

$$\frac{d^2y}{dt^2} = k\frac{dy}{dt}\frac{L-y}{L} - k\frac{y}{L}\left(\frac{dy}{dt}\right)$$

which we can factor and use equation (1) to get

$$\frac{d^2y}{dt^2} = \frac{k^2}{L^2}y(L-y)(L-2y) \tag{3}$$

This is simpler than taking two derivatives of (2). Thus, we can use (1) and (3) to find extreme values and inflection points by solving for y first; then use (2) to find the corresponding t values. This is just the reverse of the usual way of finding inflection points, where one has a formula for the second derivative in terms of the independent variable that one can set equal to zero and solve. The difference here is that we *started* with a differential equation that gives the first derivative in terms of the *dependent* variable, and so it was simpler to work from there.

Data

The following data represent the mean height in centimeters for sunflower plants (Reed and Holland, *Proc. Nat. Acad. Sci.*, vol. 5, 1919, p. 140).

Days	Height
7	17.93
14	36.36
21	67.76
28	98.10
35	131.00
42	169.50
49	205.50
56	228.30
63	247.10
70	250.50
77	253.80
84	254.50

Calculator Experiment

(1) (a) Graph the function $y(t)$ for the parameter values $y_0 = 30$, $L = 140$, $k = 0.001$; use the window $0 \leq t \leq 200$ and $0 \leq y \leq 300$. Find any maximums, minimums, or inflection points (both y and t values). For inflection points use (3) to solve for y and then use (2) to find t.

(b) Now see what happens when the parameters are varied; for each case sketch the graph and find the inflection point(s) (both t and y values). First vary y_0; choose $y_0 = 0.01$ and $y_0 = 150$. Now go back to the original three values and vary L; choose $L = 10$ and $L = 200$. Finally, after returning to the original three values again, vary k; choose $k = 0.0001$ and $k = 0.01$. Label the inflection point(s) on each sketch.

(2) Determine the values of y_0, k, and L that best fit the data given above. In this case y represents the height of the sunflower plant; if we think of a plant as a population of cells, then each centimeter would represent a certain number of cells. As the population grows, the plant grows. Use what you learned in part (1) above: You can approximate y_0 and L, then just fiddle to find k (or you could use a data point to write an equation with k as the unknown and solve). Be sure to plot both the data and the graph of the equation together. Find the inflection point(s) for these values. Make a sketch of the final graph, with the parameter values listed and the data plotted on it.

Discussion

1. Describe the effect that each parameter—y_0, k, or L—has on the shape of the logistic curve, and explain what each represents in terms of population growth. Refer to the sketches you made.

2. Explain what the inflection points mean in terms of population growth. What effect does each parameter have on the inflection point(s)? Discuss both y and t values.

3. Does the growth of a sunflower plant seem to follow logistic growth? According to your model, what was the initial height, and what will be the final height? When does the growth *rate* of a sunflower plant stop increasing and start decreasing?

4. Give some examples of populations that could *not* be modeled with the logistic equation. Sketch an example of a population as a function of time that could not be made to fit the logistic model. What factors might cause a population to follow a nonlogistic path?

8.5 | Indeterminate Forms

Here are three familiar limit problems:

$$\lim_{x \to 0} \frac{\sin x}{x}, \qquad \lim_{x \to \infty} \frac{x}{e^x}, \qquad \lim_{h \to 0} (1 + h)^{\frac{1}{h}}$$

The first was treated at length in Section 3.3, and the third was used as the definition of the constant e in Section 1.1. The three limits have a common feature. An attempt to find the limits by direct substitution leads to the undefined answers of $\frac{0}{0}, \frac{\infty}{\infty}$, or 1^∞, respectively. We are not saying that the limits above do not exist, only that we cannot directly determine them.

In fact, we have already found or estimated values for all three limits. You may recall that an intricate geometric argument led us to the conclusion $\lim_{x \to 0} \frac{\sin x}{x} = 1$. We argued that because e^x grows faster than x as x approaches infinity, we get $\lim_{x \to \infty} \frac{x}{e^x} = 0$. A numerical approach led to the approximation $\lim_{h \to 0} (1 + h)^{\frac{1}{h}} \approx 2.7183$.

How Indeterminate Forms Arise   A limit is said to have an *indeterminate form* when two competing forces are at work, tending to pull the limit in opposite directions. For example, consider the limit $\lim_{x \to 0^+} \frac{x}{x^2}$. Because the numerator is approaching 0, there is a tendency for the fraction to get small. But because the denominator is also approaching 0, there is an opposing tendency for the fraction to get large. Which force will win the battle? In this case it is easy to tell; we just use some simple algebra to get

$$\lim_{x \to 0^+} \frac{x}{x^2} = \lim_{x \to 0^+} \frac{1}{x} = \infty$$

This tells us that the denominator approaches 0 *faster* than the numerator does.

A similar argument applies to $\lim_{x \to \infty} \frac{x}{x^2}$. Both numerator and denominator approach infinity. Thus, the numerator pulls the fraction toward infinity, but the denominator pulls the fraction toward 0. Algebra shows that $\lim_{x \to \infty} \frac{x}{x^2} = \lim_{x \to \infty} \frac{1}{x} = 0$. The denominator approaches infinity faster than the numerator does.

Indeterminate Forms of Type $\frac{0}{0}$ and $\frac{\infty}{\infty}$ If both $\lim_{x \to u} f(x) = 0$ and $\lim_{x \to u} g(x) = 0$, then $\lim_{x \to u} \frac{f(x)}{g(x)}$ is said to be indeterminate of type $\frac{0}{0}$. If both $\lim_{x \to u} f(x) = \infty$ (or $\lim_{x \to u} f(x) = -\infty$) and $\lim_{x \to u} g(x) = \infty$ (or $\lim_{x \to u} g(x) = -\infty$), then $\lim_{x \to u} \frac{f(x)}{g(x)}$ is said to be indeterminate of type $\frac{\infty}{\infty}$.

Would it not be nice to have a standard procedure for handling all problems of this type? That is too much to hope for. However, there is a simple rule that works beautifully on a wide variety of such problems. It is known as *l'Hôpital's Rule* (pronounced lō′pē - täl).

L'Hôpital's Rule   In 1696, Guillaume François Antoine de l'Hôpital published the first textbook on differential calculus; it included the following rule, which he had learned from his teacher Johann Bernoulli.

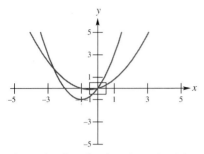

The graphs of two functions $f(x)$ and $g(x)$ for which $\lim\limits_{x \to 0} f(x) = 0$ and $\lim\limits_{x \to 0} g(x) = 0$. The boxed section is shown in the next graph.

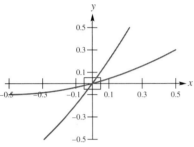

The graphs of $f(x)$ and $g(x)$ zoomed in by a factor of 10. The boxed section is shown in the next graph.

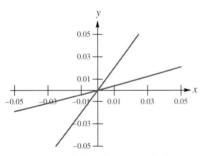

When zoomed in by a factor of 100, the curves $f(x)$ and $g(x)$ are nearly linear, and therefore of the approximate form $f(x) \approx px$ and $g(x) \approx qx$.

Figure 1

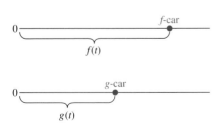

Figure 2

> ### Theorem A
>
> **(L'Hôpital's Rule).** Suppose that both $\lim\limits_{x \to u} f(x) = 0$ and $\lim\limits_{x \to u} g(x) = 0$, or that both $\lim\limits_{x \to u} f(x) = \infty$ (or $\lim\limits_{x \to u} f(x) = -\infty$) and $\lim\limits_{x \to u} g(x) = \infty$ (or $\lim\limits_{x \to u} g(x) = -\infty$). Thus we have an indeterminate form of type $\frac{0}{0}$ or type $\frac{\infty}{\infty}$. If $\lim\limits_{x \to u} \frac{f'(x)}{g'(x)}$ exists in either the finite or infinite sense (that is, if this limit is a finite number or $-\infty$ or $+\infty$), then
>
> $$\lim_{x \to u} \frac{f(x)}{g(x)} = \lim_{x \to u} \frac{f'(x)}{g'(x)}$$
>
> Here u may stand for any of the symbols a a^-, a^+, $-\infty$, or $+\infty$.

Justification of l'Hôpital's Rule

What follows are not formal proofs of l'Hôpital's Rule, but rather arguments that should help you understand why the rule is true for certain cases. In fact, informal arguments such as these form the core ideas of the more formal proofs you will find in analysis books.

THE $\frac{0}{0}$ FORM: Suppose that both functions $f(x)$ and $g(x)$ are linear; we assume that $f(x) = px$ and $g(x) = qx$. Then

$$\lim_{x \to 0} \frac{f(x)}{g(x)} = \lim_{x \to 0ç} \frac{px}{qx} = \frac{p}{q} = \lim_{x \to 0} \frac{f'(x)}{g'(x)}$$

Now suppose that $\lim\limits_{x \to 0} f(x) = 0$ and $\lim\limits_{x \to 0} g(x) = 0$. Recall that if $f(x)$ and $g(x)$ are smooth (differentiable) then they are *locally* linear. Therefore near 0, both functions can be approximated by functions of the form px and qx, and the same argument holds. See Figure 1.

It is not difficult to adapt this argument to the case where $u \neq 0$ is a finite number, by translating the functions to the origin. The case where $u = \pm\infty$ is a bit less obvious (it's harder to "linearize" at infinity).

THE $\frac{\infty}{\infty}$ FORM: We consider the case where $u = \infty$. Imagine that $f(t)$ and $g(t)$ represent the positions of two cars on the t-axis at time t (Figure 2). These two cars, the f-car and the g-car, are on endless journeys with respective velocities $f'(t)$ and $g'(t)$. Now, if

$$\lim_{t \to \infty} \frac{f'(t)}{g'(t)} = L$$

then ultimately the f-car travels about L times as fast as the g-car. It is therefore reasonable to say that in the long run, it will travel about L times as far; that is,

$$\lim_{t \to \infty} \frac{f(t)}{g(t)} = L$$

EXAMPLE 1: Use l'Hôpital's Rule to show that

$$\lim_{x \to 0} \frac{\sin x}{x} = 1 \text{ and } \lim_{x \to 0} \frac{1 - \cos x}{x} = 0$$

SOLUTION: We worked pretty hard to demonstrate these two facts in Section 3.3. After noting that both limits have the $\frac{0}{0}$ form, we can now establish the desired results in two lines. By l'Hôpital's Rule,

$$\lim_{x \to 0} \frac{\sin x}{x} = \lim_{x \to 0} \frac{\dfrac{d}{dx}\sin x}{\dfrac{d}{dx}x} = \lim_{x \to 0} \frac{\cos x}{1} = 1$$

$$\lim_{x \to 0} \frac{1 - \cos x}{x} = \lim_{x \to 0} \frac{\dfrac{d}{dx}(1 - \cos x)}{\dfrac{d}{dx}x} = \lim_{x \to 0} \frac{\sin x}{1} = 0$$

Notice that we are *not* using the quotient rule for derivatives here; we just differentiate both numerator and denominator.   ◄

EXAMPLE 2: Find $\lim\limits_{x \to \infty} \dfrac{x}{e^x}$.

SOLUTION: The limit is of the form $\frac{\infty}{\infty}$, so by l'Hôpital's Rule,

$$\lim_{x \to \infty} \frac{x}{e^x} = \lim_{x \to \infty} \frac{1}{e^x} = 0 \qquad ◄$$

Sometimes $\lim f'(x)/g'(x)$ also has the indeterminate form $\frac{0}{0}$ or $\frac{\infty}{\infty}$. Then we may apply l'Hôpital's Rule again, as we now illustrate. Each application of l'Hôpital's Rule is flagged with the symbol Ⓛ.

EXAMPLE 3: Find $\lim\limits_{x \to 0} \dfrac{\sin x - x}{x^3}$. Check your result numerically as in Section 1.1.

SOLUTION: By l'Hôpital's Rule applied three times in succession,

$$\lim_{x \to 0} \frac{\sin x - x}{x^3} \overset{Ⓛ}{=} \lim_{x \to 0} \frac{\cos x - 1}{3x^2}$$

$$\overset{Ⓛ}{=} \lim_{x \to 0} \frac{-\sin x}{6x}$$

$$\overset{Ⓛ}{=} \lim_{x \to 0} \frac{-\cos x}{6}$$

$$= -\frac{1}{6}$$

x	$\dfrac{\sin x - x}{x^3}$
1	-0.158529
0.1	-0.1665834
0.01	-0.1666658
0.001	-0.1666667

Figure 3

To estimate the limit numerically, we form a table of values letting x approach 0 in powers of $\frac{1}{10}$. See Figure 3. Our numerical result of -0.16667 (to five digits) is consistent with our exact result using l'Hôpital's Rule.   ◄

EXAMPLE 4: Show that if a is any positive real number, $\lim\limits_{x\to\infty}\dfrac{x^a}{e^x}=0$.

SOLUTION: Suppose as a special case that $a = 2.5$. Then three applications of l'Hôpital's Rule give

$$\lim_{x\to\infty}\frac{x^{2.5}}{e^x} \overset{(L)}{=} \lim_{x\to\infty}\frac{2.5x^{1.5}}{e^x} \overset{(L)}{=} \lim_{x\to\infty}\frac{(2.5)(1.5)x^{0.5}}{e^x} \overset{(L)}{=} \lim_{x\to\infty}\frac{(2.5)(1.5)(0.5)}{x^{0.5}e^x}=0$$

A similar argument works for any $a > 0$. ◀

EXAMPLE 5: Show that if a is any positive real number, $\lim\limits_{x\to\infty}\dfrac{\ln x}{x^a}=0$.

SOLUTION: Both $\ln x$ and x^a tend to ∞ as $x \to \infty$. Hence by one application of l'Hôpital's Rule,

$$\lim_{x\to\infty}\frac{\ln x}{x^a} \overset{(L)}{=} \lim_{x\to\infty}\frac{1/x}{ax^{a-1}} = \lim_{x\to\infty}\frac{1}{ax^a}=0 \qquad ◀$$

Examples 4 and 5 say something that is worth remembering, namely, *for large x, e^x grows faster as x increases than any constant power of x, whereas $\ln x$ grows more slowly than any constant power of x.* For example, e^x grows faster than x^{100} and $\ln x$ grows more slowly than $\sqrt[100]{x}$. The chart in the margin box and Figure 4 offer additional illustration.

Just because we have an elegant rule does not mean we should use it indiscriminately. In particular, we must always make sure it applies. Otherwise we will be led into all kinds of errors, as we now illustrate.

EXAMPLE 6: Find $\lim\limits_{x\to 0}\dfrac{1-\cos x}{x^2+3x}$.

SOLUTION:

$$\lim_{x\to 0}\frac{1-\cos x}{x^2+3x} \overset{(L)}{=} \lim_{x\to 0}\frac{\sin x}{2x+3} \overset{(L)}{=} \lim_{x\to 0}\frac{\cos x}{2}=\frac{1}{2} \qquad \text{WRONG}$$

The first application of l'Hôpital's Rule was correct; the second was not, because at that stage the limit did not have the $\frac{0}{0}$ form. Here is what we should have done.

$$\lim_{x\to 0}\frac{1-\cos x}{x^2+3x} \overset{(L)}{=} \lim_{x\to 0}\frac{\sin x}{2x+3}=0 \qquad \text{RIGHT}$$

We stop differentiating as soon as either the numerator or denominator has a nonzero limit. ◀

The Indeterminate Form 1^∞

The limit $\lim\limits_{h\to 0}(1+h)^{\frac{1}{h}}$ from the beginning of this section illustrates another type of indeterminancy. The term $1 + h$ tends toward 1, which drives the limit toward 1 because

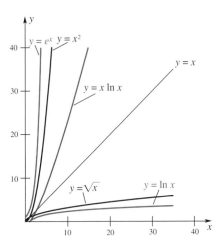

Figure 4

$1^x = 1$ for any finite number x. On the other hand, the exponent tends toward ∞, which drives the limit toward ∞ if $h \to 0^+$ or 0 if $h \to 0^-$. This is because a number greater than 1 raised to a large power is large ($x^\infty = \infty$ if $x > 1$) whereas a number between 0 and 1 raised to a large power is small ($x^\infty = 0$ if $0 \le x < 1$). We call this an indeterminate form of type 1^∞. In this case, the resulting limit of $2.718\ldots$ lies somewhere between the extreme possibilities of 0 and ∞.

Here the trick is to consider not the original expression, but rather its logarithm. Usually l'Hôpital's Rule will apply to the logarithm.

EXAMPLE 7: Find $\lim\limits_{h \to 0}(1 + h)^{\frac{1}{h}}$.

SOLUTION: Because this takes the indeterminate form 1^∞, we let $y = (1 + h)^{\frac{1}{h}}$ and take the natural logarithm of both sides.

$$\ln y = \frac{1}{h}\ln(1 + h) = \frac{\ln(1 + h)}{h}$$

By l'Hôpital's Rule for $\frac{0}{0}$ forms,

$$\lim_{h \to 0} \ln y = \lim_{h \to 0} \frac{\ln(1 + h)}{h} \overset{\text{(L)}}{=} \lim_{h \to 0} \frac{\dfrac{1}{1 + h}}{1} = 1$$

Now $y = e^{\ln y}$, and because the exponential function $f(x) = e^x$ is continuous,

$$\lim_{x \to 0^+} y = \lim_{x \to 0^+} \exp(\ln y) = \exp\left(\lim_{x \to 0^+} \ln y\right) = \exp 1 = e \qquad \blacktriangleleft$$

Note: Recall that in Section 1.1 we actually defined the constant e as $\lim\limits_{h \to 0}(1 + h)^{\frac{1}{h}}$, so this result is not surprising. It does show, however, that our various definitions and theorems are *consistent*, a very important consideration in mathematics.

Summary

We have classified certain limit problems as indeterminate forms, using the symbols $\frac{0}{0}$, $\frac{\infty}{\infty}$, and 1^∞. Each involves a competition of opposing forces, which means that the result is not obvious. However, with the help of l'Hôpital's Rule, which applies directly only to the $\frac{0}{0}$ and $\frac{\infty}{\infty}$ forms, we can often determine the correct limit.

There are other indeterminate forms that we have not covered, such as $0 \cdot \infty$, $\infty - \infty$, and 0^0. To handle these forms we can often convert the expression to one of the forms $\frac{0}{0}$ or $\frac{\infty}{\infty}$. We will not address such forms explicitly in this section.

There are many other possibilities symbolized by, for example, $\frac{0}{\infty}$, $\frac{\infty}{0}$, and ∞^∞. Why don't we call these indeterminate forms? Because, in each of these cases, the forces are in collusion, not competition.

x	$\sin x^{\frac{1}{x}}$
1	0.84147098
0.1	9.8347E-11
0.01	9.983E-201

Figure 5

EXAMPLE 8: Find $\lim\limits_{x \to 0^+}(\sin x)^{\frac{1}{x}}$.

SOLUTION: We might call this a 0^∞ form, but it is not indeterminate. Note that $\sin x$ is approaching zero, and raising it to the exponent $\frac{1}{x}$, an increasingly large number, serves only to make it approach zero faster. Thus,

$$\lim_{x \to 0^+}(\sin x)^{\frac{1}{x}} = 0$$

A numerical check provides support for our conclusion. See Figure 5. $\qquad \blacktriangleleft$

Concepts Review

1. L'Hôpital's Rule is useful in finding $\lim_{x \to a} [f(x)/g(x)]$, where both _____ and _____ are zero.

2. If $\lim_{x \to a} f(x) = \lim_{x \to a} g(x) = \infty$, then $\lim_{x \to a} f(x)/g(x) = \lim_{x \to a}$ _____ .

3. From l'Hôpital's Rule, we can conclude that $\lim_{x \to 0} \frac{\tan x}{x} = \lim_{x \to 0}$ _____ = _____ , but l'Hôpital's Rule gives us no information about $\lim_{x \to 0} (\cos x)/x$ because _____ .

4. e^x grows faster than any power of x but _____ grows more slowly than any positive power of x.

Problem Set 8.5

In Problems 1–30, find the indicated limit. Make sure l'Hôpital's Rule applies before you use it. Check each limit numerically as in Example 3.

1. $\lim_{x \to 0} \dfrac{\sin x - 2x}{x}$

2. $\lim_{x \to \pi/2} \dfrac{\cos x}{x - \frac{1}{2}\pi}$

3. $\lim_{x \to 0} \dfrac{x - 2\sin x}{\tan x}$

4. $\lim_{x \to 0} \dfrac{\sin^{-1}x}{3\tan^{-1}x}$

5. $\lim_{x \to \infty} \dfrac{\ln(x^{100})}{x}$

6. $\lim_{x \to \infty} \dfrac{\ln x}{2^x}$

7. $\lim_{x \to \infty} \dfrac{x^{10}}{e^x}$

8. $\lim_{x \to \infty} \dfrac{2x}{\ln(3x + e^x)}$

9. $\lim_{x \to 1^+} \dfrac{x^2 - 2x + 2}{x^2 - 1}$

10. $\lim_{x \to 1} \dfrac{\ln x}{x^2 - 1}$

11. $\lim_{x \to \pi/2} \dfrac{\ln \sin x}{\frac{1}{2}\pi - x}$

12. $\lim_{x \to 0} \dfrac{e^x - e^{-x}}{4\sin x}$

13. $\lim_{x \to \pi/2} \dfrac{\sec x + 1}{\tan x}$

14. $\lim_{x \to 0^+} \dfrac{\ln \sin x}{\ln \tan x}$

15. $\lim_{x \to \infty} \dfrac{\ln(\ln x)}{\ln x}$

16. $\lim_{x \to (\frac{1}{2})^-} \dfrac{\ln(1 - 2x)}{\tan \pi x}$

17. $\lim_{t \to 1} \dfrac{\sqrt{t} - t}{\ln t}$

18. $\lim_{x \to 0^+} \dfrac{8^{\sqrt{x}} - 1}{3^{\sqrt{x}} - 1}$

19. $\lim_{x \to 0^+} (2x)^{x^2}$

20. $\lim_{x \to 0} (\cos x)^{\cot x}$

21. $\lim_{x \to (\frac{1}{2})\pi^-} (\cos x)^{\tan x}$

22. $\lim_{x \to 0} (x + e^{x/2})^{2/x}$

23. $\lim_{x \to 0} \dfrac{\ln \cos 3x}{2x^2}$

24. $\lim_{x \to 0} x^x$

25. $\lim_{x \to 0} x^{e/x}$

26. $\lim_{x \to 0^+} \dfrac{2\sin x}{\sqrt{x}}$

27. $\lim_{x \to 0} \dfrac{\displaystyle\int_0^x \sqrt{1 + \sin t}\, dt}{x}$

28. $\lim_{x \to 0^+} \dfrac{\displaystyle\int_0^x \sqrt{t}\cos t\, dt}{x^2}$

29. $\lim_{x \to \infty} \dfrac{\displaystyle\int_1^x \sqrt{1 + e^{-t}}\, dt}{x}$

30. $\lim_{x \to 1^+} \dfrac{\displaystyle\int_1^x \sin t\, dt}{x - 1}$

31. In Section 3.4, we worked very hard to prove that $\lim\limits_{x\to 0}(\sin x)/x = 1$; l'Hôpital's Rule allows us to show this fact in one line. However, even if we had l'Hôpital's Rule available at that stage, it would not have helped us. Explain why.

32. Find $\lim\limits_{x\to 0}\dfrac{x^2\sin(1/x)}{\tan x}$

Hint: Begin by deciding why l'Hôpital's Rule is not applicable. Then find or estimate the limit by other means.

33. Determine constants a, b, and c so that

$$\lim_{x\to 1}\frac{ax^4 + bx^3 + 1}{(x-1)\sin\pi x} = c$$

34. Find each limit.

(a) $\lim\limits_{n\to\infty}\sqrt[n]{a}$

(b) $\lim\limits_{n\to\infty}\sqrt[n]{n}$

(c) $\lim\limits_{n\to\infty}n(\sqrt[n]{a}-1)$

(d) $\lim\limits_{n\to\infty}n(\sqrt[n]{n}-1)$

Hint: Transform to problems involving a continuous variable x.

35. Find each limit.

(a) $\lim\limits_{x\to 0^+}x^x$

(b) $\lim\limits_{x\to 0^+}(x^x)^x$

(c) $\lim\limits_{x\to 0^+}x^{(x^x)}$

(d) $\lim\limits_{x\to 0^+}((x^x)^x)^x$

(e) $\lim\limits_{x\to 0^+}x^{(x^{(x^x)})}$

36. Graph $y = x^{1/x}$ for $x > 0$. Show what happens for very small x and very large x. Indicate the maximum value.

37. Consider $f(x) = n^2xe^{-nx}$.

(a) Graph $f(x)$ for $n = 1, 2, 3, 4, 5, 6$ on $[0, 1]$ using the same axes.

(b) For $x > 0$, find $\lim\limits_{n\to\infty}f(x)$.

(c) Evaluate $\displaystyle\int_0^1 f(x)\,dx$ for $n = 1, 2, 3, 4, 5, 6$.

(d) Guess at $\lim\limits_{n\to\infty}\displaystyle\int_0^1 f(x)\,dx$. Then justify your answer rigorously.

38. Find the absolute maximum and minimum points (if they exist) for $f(x) = (x^{25} + x^3 + 2x)e^{-x}$ on $[0, \infty)$.

Answers to Concepts Review: 1. $\lim\limits_{x\to a}f(x)$; $\lim\limits_{x\to a}g(x)$

2. $f'(x)/g'(x)$ 3. $\sec^2 x$; 1; $\lim\limits_{x\to 0}\cos x \neq 0$ 4. $\ln x$

8.6 | Improper Integrals

Integrals with Infinite Limits   In the definition of $\displaystyle\int_a^b f(x)\,dx$, it was assumed that the interval $[a, b]$ was finite. However, there are many applications in physics, economics, and probability in which we wish to allow a or b (or both) to be infinite. We must therefore find a way to give meaning to symbols like

$$\int_0^\infty \frac{1}{1+x^2}\,dx, \qquad \int_{-\infty}^{-1} xe^{-x^2}\,dx, \qquad \int_{-\infty}^\infty x^2e^{-x^2}\,dx$$

These integrals are called **improper integrals** with infinite limits.

One Infinite Limit   The graph of $f(x) = e^{-x}$ on $[0, \infty)$ is shown in Figure 1. The integral $\displaystyle\int_0^b e^{-x}\,dx$ makes perfectly good sense no matter how large we make b; in fact, we can evaluate this integral explicitly.

$$\int_0^b e^{-x}\,dx = \left[-e^{-x}\right]_0^b = 1 - e^{-b}$$

Now $\lim\limits_{b\to\infty}\left(1 - e^{-b}\right) = 1$, so it seems natural to define

$$\int_0^\infty e^{-x}\,dx = 1$$

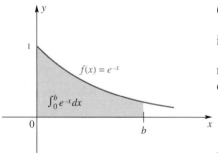

Figure 1

Here is the general definition.

> **Definition**
>
> $$\int_{-\infty}^{b} f(x)\, dx = \lim_{a \to -\infty} \int_{a}^{b} f(x)\, dx$$
>
> $$\int_{a}^{\infty} f(x)\, dx = \lim_{b \to \infty} \int_{a}^{b} f(x)\, dx$$
>
> If the limits on the right exist and have finite values, then we say the corresponding improper integrals **converge** and have those values. Otherwise, the integrals are said to **diverge.**

EXAMPLE 1: Find, if possible, $\int_{0}^{\infty} xe^{-x}\, dx$. Sketch the region whose area is represented by this integral. Also estimate the integral numerically (using a computer or calculator) as a check.

SOLUTION: Using integration by parts we have

$$\int xe^{-x}\, dx = -xe^{-x} + \int e^{-x}\, dx = -xe^{-x} - e^{-x}$$

so that

$$\int_{0}^{b} xe^{-x}\, dx = -xe^{-x} - e^{-x}\Big|_{0}^{b}$$

$$= (-be^{-b} - e^{-b}) - (-0 \cdot e^{-0} - e^{-0})$$

$$= 1 - be^{-b} - e^{-b}$$

and hence

$$\int_{0}^{\infty} xe^{-x}\, dx = \lim_{b \to \infty}(1 - be^{-b} - e^{-b})$$

$$= \lim_{b \to \infty}\left(1 - \frac{b}{e^{b}} - \frac{1}{e^{b}}\right)$$

$$\overset{\text{\textcircled{L}}}{\underset{\downarrow}{}}$$

$$= 1$$

Notice that the middle limit $\lim\limits_{b \to \infty} \frac{b}{e^{b}}$ is indeterminate of form $\frac{\infty}{\infty}$ and so requires l'Hôpital's Rule (see Example 2 of Section 8.5).

To check the integral numerically, we use the definition given above. We have

$$\int_0^\infty xe^{-x}\,dx = \lim_{b\to\infty}\int_0^b xe^{-x}\,dx$$ ◀

We can form a table of values with b getting larger and larger, and evaluate each integral numerically using a computer or calculator.

For this integral we choose the sequence 10, 20, 30, ... for b. This sequence is more helpful in seeing the convergence of the integral to 1 than the sequence 10, 100, 1000, ..., which we recommended in Section 1.2 for infinite limits. Use your own judgment in choosing a sequence of values for b that approaches infinity. In Figure 2 we show the result, which checks with our exact value of 1 for the improper integral. In this table we calculated the integral numerically, accurate to 15 digits, but you would normally not require this many digits.

See Figure 3 for a sketch of the corresponding region. Of course we cannot sketch the entire region since it extends infinitely far out on the x-axis.

b	$\displaystyle\int_0^b xe^{-x}\,dx$
10	0.999500600772613
20	0.999999956715774
30	0.999999999997099

Figure 2

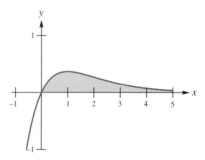

Graph of $y = x\,e^{-x}$. Area of shaded region (which extends infinitely far to the right) is equal to 1.

Figure 3

EXAMPLE 2: Find, if possible, $\displaystyle\int_0^\infty \sin x\,dx$.

SOLUTION:

$$\int_0^\infty \sin x\,dx = \lim_{b\to\infty}\int_0^b \sin x\,dx = \lim_{b\to\infty}\bigl[-\cos x\bigr]_0^b$$

$$= \lim_{b\to\infty}\bigl[1 - \cos b\bigr]$$

The latter limit does not exist; we conclude that the given integral diverges.

Think about the geometric meaning of $\displaystyle\int_0^\infty \sin x\,dx$ to confirm this result (Figure 4).   ◀

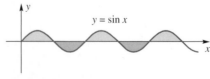

$y = \sin x$

Figure 4

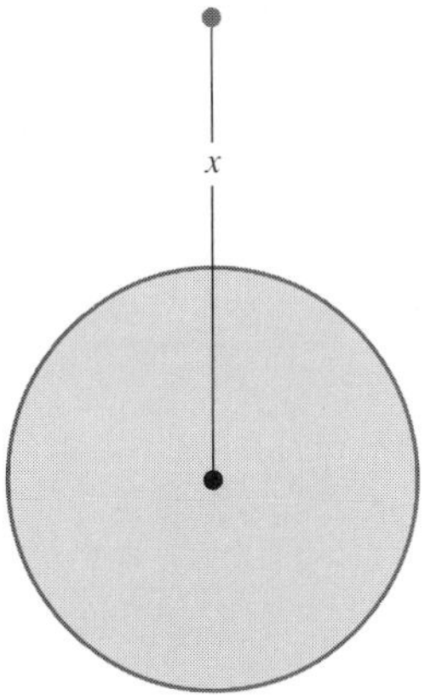

Figure 5

EXAMPLE 3: According to Newton's Inverse-Square Law, the force exerted by the earth on a space capsule is $-k/x^2$, where x is the distance (in miles, for instance) from the capsule to the center of the earth (Figure 5). The force $F(x)$ required to lift the capsule is therefore $F(x) = k/x^2$. How much work is done in propelling a 1000-pound capsule out of the earth's gravitational field?

SOLUTION: We can evaluate k by noting that at $x = 3960$ miles (the radius of the earth), $F = 1000$ pounds. This yields $k = 1000(3960)^2 \approx 1.568 \times 10^{10}$. The work done in mile-pounds is therefore

$$1.568 \times 10^{10} \int_{3960}^{\infty} \frac{1}{x^2}\, dx = \lim_{b \to \infty} 1.568 \times 10^{10} \left[-\frac{1}{x} \right]_{3960}^{b}$$

$$= \lim_{b \to \infty} 1.568 \times 10^{10} \left[-\frac{1}{b} + \frac{1}{3960} \right]$$

$$= \frac{1.568 \times 10^{10}}{3960} \approx 3.96 \times 10^6 \qquad \blacktriangleleft$$

Both Limits Infinite   First we need a definition.

Definition

If both $\displaystyle\int_{-\infty}^{0} f(x)\,dx$ and $\displaystyle\int_{0}^{\infty} f(x)\,dx$ converge, then $\displaystyle\int_{-\infty}^{\infty} f(x)\,dx$ is said to converge and have value

$$\int_{-\infty}^{\infty} f(x)\,dx = \int_{-\infty}^{0} f(x)\,dx + \int_{0}^{\infty} f(x)\,dx$$

Otherwise, $\displaystyle\int_{-\infty}^{\infty} f(x)\,dx$ diverges.

EXAMPLE 4: The function $f(x) = k/(1 + x^2)$ occurs in probability theory, where it is called the Cauchy density function. The constant k is to be chosen so that the total area under the curve on $(-\infty, \infty)$ is 1 (see Figure 6). Determine k.

SOLUTION: We want

$$\int_{-\infty}^{\infty} \frac{k}{1 + x^2}\, dx = 1$$

Now,

$$\int_{0}^{\infty} \frac{k}{1 + x^2}\, dx = \lim_{b \to \infty} \int_{0}^{\infty} \frac{k}{1 + x^2}\, dx$$

$$= \lim_{b \to \infty} \left[k \tan^{-1}x \right]_{0}^{b}$$

$$= \lim_{b \to \infty} k \tan^{-1}b = k\frac{\pi}{2}$$

Because $f(x)$ is an even function (see Figure 6, or graph it yourself for a few values of k) we can use symmetry to get

$$\int_{-\infty}^{0} \frac{k}{1+x^2} \, dx = k\frac{\pi}{2}$$

Thus,

$$\int_{-\infty}^{\infty} \frac{k}{1+x^2} \, dx = \int_{-\infty}^{0} \frac{k}{1+x^2} \, dx + \int_{0}^{\infty} \frac{k}{1+x^2} \, dx = k\pi$$

We conclude that $k = 1/\pi$.   ◀

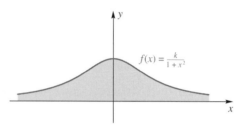

$f(x) = \frac{k}{1+x^2}$

Figure 6

EXAMPLE 5: Show that $\displaystyle\int_{1}^{\infty} 1/x^p \, dx$ diverges for $p \leq 1$ and converges for $p > 1$.

SOLUTION: If $p \neq 1$,

$$\int_{1}^{\infty} \frac{1}{x^p} \, dx = \lim_{b \to \infty} \int_{1}^{b} x^{-p} \, dx = \lim_{b \to \infty} \left[\frac{x^{-p+1}}{-p+1} \right]_{1}^{b}$$

$$= \lim_{b \to \infty} \left[\frac{1}{1-p} \right]\left[\frac{1}{b^{p-1}} - 1 \right] = \begin{cases} \infty & \text{if } p < 1 \\ \frac{1}{1-p} & \text{if } p > 1 \end{cases}$$

If $p = 1$, then

$$\int_{1}^{\infty} \frac{1}{x} \, dx = \lim_{b \to \infty} \int_{1}^{b} \frac{1}{x} \, dx = \lim_{b \to \infty} [\ln x]_{1}^{b}$$

$$= \lim_{b \to \infty} [\ln b - 0] = \infty$$

The conclusion follows.   ◀

Integrals with Infinite Integrands
Considering the many complicated integrations we have done, here is one that looks simple enough.

$$\int_{-2}^{1} \frac{1}{x^2} \, dx = \left[-\frac{1}{x} \right]_{-2}^{1} = -1 - \frac{1}{2} = -\frac{3}{2} \qquad ???$$

One glance at the graph in Figure 7 tells us that something is terribly wrong. The answer (if there is one) has to be a positive number. (Why?)

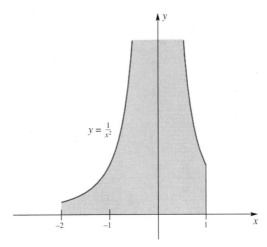

$$y = \frac{1}{x^2}$$

Figure 7

Where is our mistake? To answer, we refer back to Section 5.4. Recall that for a function to be integrable in the standard (or proper) sense it must be bounded. Our function, $f(x) = 1/x^2$, is not bounded, so it is not integrable in the proper sense. We say that $\displaystyle\int_{-2}^{1} x^{-2}\, dx$ is an improper integral with an infinite integrand (*unbounded integrand* is a more accurate but less colorful term).

Until now, we have carefully avoided infinite integrands in all our examples and problems. We could continue to do this, but that would be avoiding a kind of integral that has important applications. Our task for this section is to define and analyze this new kind of integral.

Integrands That Are Infinite at an Endpoint

We give the definition for the case where f tends to infinity at the right endpoint of the interval of integration. There is a completely analogous definition for the case where the left endpoint is the troublesome one.

Definition

Let f be continuous on the half-open interval $[a, b)$ and suppose $\displaystyle\lim_{x \to b^-} |f(x)| = \infty$. Then,

$$\int_a^b f(x)\,dx = \lim_{t \to b^-} \int_a^t f(x)\,dx$$

provided this limit exists and is finite, in which case we say the integral converges. Otherwise, we say the integral diverges.

Note the geometric interpretation shown in Figure 8.

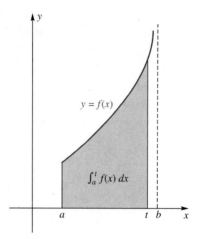

Figure 8

EXAMPLE 6: Evaluate, if possible, the improper integral $\int_0^2 \dfrac{dx}{\sqrt{4-x^2}}$. Sketch the corresponding region. Check your result numerically using the definition of improper integral.

SOLUTION: Note that the integrand tends to infinity at 2.

$$\int_0^2 \frac{dx}{\sqrt{4-x^2}} = \lim_{t\to 2^-} \int_0^t \frac{dx}{\sqrt{4-x^2}} = \lim_{t\to 2^-} \left[\sin^{-1}\left(\frac{x}{2}\right)\right]_0^t$$

$$= \lim_{t\to 2^-} \left[\sin^{-1}\left(\frac{t}{2}\right) - \sin^{-1}\left(\frac{0}{2}\right)\right] = \frac{\pi}{2}$$

To check numerically, we use the definition

$$\int_0^2 \frac{dx}{\sqrt{4-x^2}} = \lim_{t\to 2^-} \int_0^t \frac{dx}{\sqrt{4-x^2}}$$

We form a table of values using a computer or calculator with t approaching 2 from the left in powers of $\frac{1}{10}$. See Figure 9. To two decimal places we get a value for the integral of 1.57, consistent with our exact result of $\frac{\pi}{2}$. Check these results with your own computer or calculator. ◀

The region is sketched in Figure 10.

t	$\int_0^t \dfrac{dx}{\sqrt{4-x^2}}$
1.9	1.2532359
1.99	1.4707546
1.999	1.5391722
1.9999	1.5607963
1.99999	1.5676340
1.999999	1.5697963
1.9999999	1.5704801

Figure 9

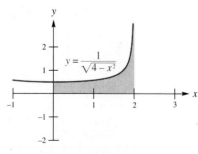

Area of shaded unbounded region is $\frac{\pi}{2}$.

Figure 10

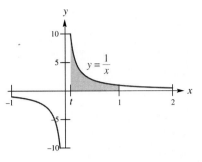

Area of shaded region is –ln t, which approaches
∞ as t approaches 0.

Figure 11

EXAMPLE 7: Evaluate, if possible, $\displaystyle\int_0^1 \frac{1}{x}\, dx$.

SOLUTION:

$$\int_0^1 \frac{1}{x}\, dx = \lim_{t \to 0^+} \int_t^1 \frac{1}{x}\, dx = \lim_{t \to 0^+} \left[\ln x\right]_t^1$$

$$= \lim_{t \to 0^+} \left[-\ln t\right] = \infty$$

See the sketch in Figure 11. We conclude that the integral diverges. We leave a numerical check of this result to the reader. ◄

EXAMPLE 8: Show that $\displaystyle\int_0^1 \frac{1}{x^p}\, dx$ converges if $p < 1$, but diverges if $p \geq 1$.

SOLUTION: Example 7 took care of the case $p = 1$. If $p \neq 1$,

$$\int_0^1 \frac{1}{x^p}\, dx = \lim_{t \to 0^+} \int_t^1 x^{-p}\, dx = \lim_{t \to 0^+} \left[\frac{x^{-p+1}}{-p+1}\right]_t^1$$

$$= \lim_{t \to 0^+} \left[\frac{1}{1-p} - \frac{1}{1-p} \cdot \frac{1}{t^{p-1}}\right] = \begin{cases} \dfrac{1}{1-p} & \text{if } p < 1 \\ \infty & \text{if } p > 1 \end{cases} \quad ◄$$

Three Key Examples

From Example 5, we learned that

$$\int_1^\infty \frac{1}{x^p}\, dx$$

converges if and only if $p > 1$. From Examples 7 and 8, we learn that

$$\int_0^1 \frac{1}{x^p}\, dx$$

converges if and only if $p < 1$. The first has an infinite limit; the second has an infinite integrand. If you feel at home with these two integrals, you should also be at ease with any other improper integrals you may meet.

Integrands That Are Infinite at an Interior Point

The integral $\displaystyle\int_{-2}^1 1/x^2\, dx$ has an integrand that tends to infinity at $x = 0$, an interior point of the interval $[-2, 1]$. Here is the appropriate definition to give meaning to such an integral.

Definition

Let f be continuous on $[a, b]$ except at a number c, where $a < c < b$, and suppose $\lim\limits_{x \to c} |f(x)| = \infty$. Then we define

$$\int_a^b f(x)\,dx = \int_a^c f(x)\,dx + \int_c^b f(x)\,dx$$

provided both integrals on the right converge. Otherwise, we say $\int_a^b f(x)\,dx$ diverges.

EXAMPLE 9: Show that $\int_{-2}^1 \dfrac{1}{x^2}\,dx$ diverges.

SOLUTION:

$$\int_2^1 \frac{1}{x^2}\,dx = \int_{-2}^0 \frac{1}{x^2}\,dx + \int_0^1 \frac{1}{x^2}\,dx$$

The second of the integrals on the right diverges by Example 7. This is enough to give the conclusion.   ◄

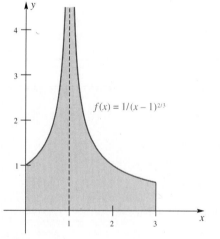

Figure 12

$f(x) = 1/(x-1)^{2/3}$

EXAMPLE 10: Evaluate, if possible, the improper integral $\int_0^3 \dfrac{dx}{(x-1)^{\frac{2}{3}}}$.

SOLUTION: The integrand tends to infinity at $x = 1$ (see Figure 12). Thus,

$$\int_0^3 \frac{dx}{(x-1)^{\frac{2}{3}}} = \int_0^1 \frac{dx}{(x-1)^{\frac{2}{3}}} + \int_1^3 \frac{dx}{(x-1)^{\frac{2}{3}}}$$

$$= \lim_{t \to 1^-} \int_0^t \frac{dx}{(x-1)^{\frac{2}{3}}} + \lim_{s \to 1^+} \int_s^3 \frac{dx}{(x-1)^{\frac{2}{3}}}$$

$$= \lim_{t \to 1^-} \left[3(x-1)^{\frac{1}{3}} \right]_0^t + \lim_{s \to 1^+} \left[3(x-1)^{\frac{1}{3}} \right]_s^3$$

$$= 3 \lim_{t \to 1^-} \left[(t-1)^{\frac{1}{3}} + 1 \right] + 3 \lim_{s \to 1^+} \left[2^{\frac{1}{3}} - (s-1)^{\frac{1}{3}} \right]$$

$$= 3 + 3\left(2^{\frac{1}{3}}\right) \approx 6.78$$ ◄

Concepts Review

1. $\int_a^\infty f(x)\,dx$ is said to _____ if $\lim\limits_{b \to \infty} \int_a^b f(x)\,dx$ exists and is finite.

2. $\int_0^\infty \cos x\,dx$ does not converge because _____ does not exist.

3. $\int_1^\infty (1/x^p)\,dx$ converges if and only if _____ .

4. The improper integral $\int_0^1 (1/x^p)\,dx$ converges if and only if _____ .

Problem Set 8.6

In Problems 1–30, evaluate the given improper integral or show that it is divergent. Sketch the function and shade the region whose area you found. Estimate each integral numerically using the appropriate definition of improper integral, as in Examples 1 and 6.

1. $\int_1^\infty e^x \, dx$

2. $\int_{-\infty}^{-2} \frac{dx}{x^5}$

3. $\int_4^\infty xe^{-x^2} \, dx$

4. $\int_{-\infty}^0 e^{3x} \, dx$

5. $\int_3^\infty \frac{x \, dx}{\sqrt{9 + x^2}}$

6. $\int_1^\infty \frac{dx}{\sqrt{3x}}$

7. $\int_1^2 \frac{dx}{(x-1)^{\frac{1}{3}}}$

8. $\int_3^7 \frac{dx}{\sqrt{x-3}}$

9. $\int_0^1 \frac{dx}{\sqrt{1-x^2}}$

10. $\int_0^9 \frac{dx}{\sqrt{9-x}}$

11. $\int_1^\infty \frac{dx}{x^{1.01}}$

12. $\int_2^\infty \frac{x}{1+x^2} \, dx$

13. $\int_1^\infty \frac{dx}{x^{0.99}}$

14. $\int_{-3}^2 \frac{1}{x^4} \, dx$

15. $\int_{-1}^{27} x^{-\frac{2}{3}} \, dx$

16. $\int_2^4 \frac{dx}{(3-x)^{\frac{2}{3}}}$

17. $\int_0^3 \frac{x}{9-x^2} \, dx$

18. $\int_2^\infty \frac{x}{(1+x^2)^2} \, dx$

19. $\int_2^\infty \frac{dx}{x \ln x}$

20. $\int_1^\infty \frac{\ln x}{x} \, dx$

21. $\int_2^\infty \frac{dx}{x(\ln x)^2}$

22. $\int_0^\infty xe^{-x} \, dx$

23. $\int_{-\infty}^0 \frac{dx}{(2x-1)^3}$

24. $\int_2^\infty \frac{dx}{(1-x)^{\frac{2}{3}}}$

25. $\int_{-\infty}^\infty \frac{x}{\sqrt{x^2+4}} \, dx$

26. $\int_{-\infty}^\infty \frac{x}{(x^2+4)^2} \, dx$

27. $\int_0^\infty e^{-x} \cos x \, dx$. *Hint*: Use table of integrals.

28. $\int_0^\infty e^{-x} \sin x \, dx$

29. $\int_0^1 \frac{\ln x}{x} \, dx$

30. $\int_0^1 \ln x \, dx$

31. Find the area of the region under the curve $y = 2/(4x^2 - 1)$ to the right of $x = 1$. *Hint*: Use partial fractions.

32. Find the area of the region under the curve $y = 1/(x^2 + x)$ to the right of $x = 1$.

33. Suppose that Newton's law for the force of gravity had the form $-k/x$ rather than $-k/x^2$ (see Example 3). Show that it would then be impossible to send anything out of the earth's gravitational field.

34. If a 1000-pound capsule weighs only 165

pounds on the moon (radius 1080 miles), how much work is done in propelling this capsule out of the moon's gravitational field? (See Example 3.)

35. Suppose a company expects its annual profits t years from now to be $f(t)$ dollars and that interest is considered to be compounded continuously at an annual rate r. Then the present value of all future profits can be shown to be

$$FP = \int_0^\infty e^{-rt} f(t)\,dt$$

Find FP if $r = 0.08$ and $f(t) = 100,000$.

36. Do Problem 35 assuming $f(t) = 100,000 + 1000t$.

37. In probability theory, *waiting times* tend to have the exponential *density* (probability per unit of x) given for $\alpha > 0$ by $f(x) = \alpha e^{-\alpha x}$ on $[0, \infty)$. Show each of the following.

(a) $\int_0^\infty f(x)\,dx = 1$

(b) $\int_0^\infty x f(x)\,dx = 1/\alpha$ (the mean)

38. In electromagnetic theory, the magnetic potential u at a point on the axis of a circular coil is given by

$$u = Ar \int_a^\infty \frac{dx}{(r^2 + x^2)^{\frac{3}{2}}}$$

where A, r, and a are constants. Evaluate u.

39. There is a subtlety in the definition of $\int_{-\infty}^\infty f(x)\,dx$ illustrated by the following. Explain why (a) $\int_{-\infty}^\infty \sin x\,dx$ diverges but (b) $\lim_{a\to\infty} \int_{-a}^a \sin x\,dx = 0$.

40. Consider an infinitely long wire coinciding with the positive x-axis and having mass density $\delta(x) = (1 + x^2)^{-1}$ at x, $0 \le x < \infty$.
(a) Calculate the total mass of the wire. (See Example 4).
(b) Show that this wire does not have a center of mass. (In probability language, the Cauchy distribution does not have a mean.)

41. Give an example of a region in the first quadrant that gives a solid of finite volume when revolved about the x-axis but gives a solid of infinite volume when revolved about the y-axis.

42. Calculate $\int_{-a}^a \frac{1}{\pi}(1 + x^2)^{-1}\,dx$ for $a = 10, 50,$ and 100 and thereby add numerical support to the result of Example 4.

43. If $f(x)$ tends to infinity at both a and b, then we define

$$\int_a^b f(x)\,dx = \int_a^c f(x)\,dx + \int_c^b f(x)\,dx$$

where c is any point between a and b, provided of course that both the latter integrals converge. Otherwise, we say the given integral diverges. Use this to evaluate $\int_{-3}^3 \frac{x}{\sqrt{9 - x^2}}\,dx$ or show that it diverges.

44. Evaluate $\int_{-2}^2 \frac{1}{4 - x^2}\,dx$ or show that it diverges. See Problem 43.

45. Show that $\int_0^\infty \frac{1}{x^p}\,dx$ diverges for all p. *Hint:* Write $\int_0^\infty = \int_0^1 + \int_1^\infty$.

46. Suppose that f is continous on $[0, \infty)$ except at $x = 1$, where $\lim_{x\to 1}|f(x)| = \infty$. How would you define $\int_0^\infty f(x)\,dx$?

47. Find the area of the region between the curves $y = (x - 8)^{-\frac{2}{3}}$ and $y = 0$ for $0 \le x < 8$.

48. Let R be the region in the first quadrant below the curve $y = x^{-\frac{2}{3}}$ and to the left of $x = 1$.
(a) Show that the area of R is finite by finding its value.
(b) Show that the volume of the solid generated by revolving R about the x-axis is infinite.

49. Find b so that $\int_0^b \ln x\,dx = 0$.

50. Is $\int_0^1 \frac{\sin x}{x}\,dx$ an improper integral? Explain.

51. (Gamma Function) Let $\Gamma(n) = \int_0^\infty x^{n-1}e^{-x}\,dx$, $n > 0$. This integral can be shown to converge. Show each of the following.
(a) $\Gamma(1) = 1$
(b) $\Gamma(n + 1) = n\Gamma(n)$
(c) $\Gamma(n + 1) = n!$, n a positive integer.

52. Evaluate $\int_0^\infty x^{n-1}e^{-x}\,dx$ for $n = 1, 2, 3, 4,$ and 5, thereby confirming Problem 51(c).

Answers to Concepts Review: 1. Converge
2. $\lim_{b\to\infty} \int_0^b \cos x\,dx$ 3. $p > 1$ 4. $p < 1$

LAB 16: PROBABILITY AND IMPROPER INTEGRALS

Mathematical Background

In probability theory, functions $f(x)$ called Probability Density Functions (or pdf's for short) are used to represent probabilities. The most probable x-values are the ones where the pdf $f(x)$ is greatest. For instance, men's heights in inches might follow the pdf shown below (called the normal curve).

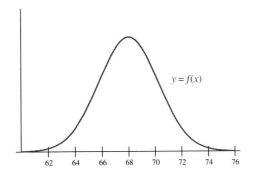

This means that heights near 68 are most likely; heights near 64 or 72 are less likely; heights near 60 or 76 are very unlikely.

If the pdf $f(x)$ is known, then the probability of an x-value between a and b is given by the area under $f(x)$ between a and b, that is, the

$$\text{probability that } a \leq x \leq b = \int_a^b f(x)\,dx$$

Thus, for the example above, if a man is chosen at random, the

$$\text{probability that } 68 \leq \text{ man's height} \leq 72 = \int_{68}^{72} f(x)\,dx$$

This is equivalent to finding the percentage of men whose heights lie between 68 inches and 72 inches.

To estimate properties of a large group (referred to as a population) one often looks at similar properties of a smaller group (called a sample). Two quantities of interest are called the mean and standard deviation. For a set of numbers $\{x_1, x_2, x_3, \ldots, x_n\}$ the sample mean is defined as

$$m = \frac{1}{n}\sum_{i=1}^{n} x_i$$

and the sample standard deviation is defined as

$$s = \sqrt{\frac{1}{n-1}\sum_{i=1}^{n}(x_i - m)^2}$$

The mean measures the "middle" of the data (actually the balance point) and the standard deviation measures the "spread" of the data (as a rule of thumb, the distance from the mean to the largest or smallest piece of data is usually about two standard deviations).

Lab Introduction

We are going to estimate the pdf for heights of either adult females or adult males using the heights of the females or males in class. We must choose either females or males, because the two groups follow different height distributions. A reasonable model for the pdf for heights of adult females or males is given by

$$f(x) = \frac{1}{\sigma\sqrt{2\pi}}e^{-\frac{(x-\mu)^2}{2\sigma^2}} \tag{1}$$

where μ and σ are parameters to be determined; this is called the normal probability density function. The parameters μ and σ represent the mean and standard deviation of the heights of *all* adult females or males (a population); we will estimate them with the mean m and standard deviation s of the heights of adult females or males in class.

For this lab you will need to devise a numerical strategy for finding improper integrals; you will use midpoint rectangles, but because you can't use an upper limit of ∞ on your calculator, you will need to choose upper limits that get larger and larger.

Data

Find the mean m and standard deviation s of the heights of adult females or males in class (in inches) and use these values as an approximation to μ and σ in (1). Find m and s using the statistical part of your calculator; just enter the data into the statistical memories and use the sample mean and sample standard deviation keys to get the results. Sample data for this experiment are contained in Appendix A, but you should really use the data from your class!

Experiment

(1) Graph the pdf $f(x)$; choose appropriate x and y scales. The curve should *appear* to touch the x-axis. Sketch this pdf and include the scale.

(2) Find the probability that an adult female or male chosen at random from the United States is over 6 feet (72 inches) tall; thus you want to find

$$\int_{72}^{\infty} f(x)\,dx = \lim_{t\to\infty}\int_{72}^{t} f(x)\,dx.$$

Devise a strategy using midpoint rectangles to find an estimate of this improper integral that is accurate to three digits. It may be helpful to display the rectangles for the first few N values (eliminating the display of the rectangles speeds up the program). You must integrate from 72 to t where you make t larger and larger, but you must also make sure that N is large enough. Choose $t = 80, 90, 100, \ldots$ and choose $N = 20, 50, 100, 200, \ldots$. Try each of the following strategies:

(a) Fix $t = 80$ and let N get larger until the result stabilizes; then choose the next t value and repeat. Continue until two consecutive t values give the same result; this is the probability that you want.

(b) Fix $N = 20$ and let t get larger and larger (you may have to choose very large values for t to get the result to stabilize). Does the answer seem right this time? Repeat with $N = 50$. Displaying the rectangles helps here to understand what is happening.

(3) If there are about 100 million adult females or males in the United States, about how many are over 6 feet tall according to this model? Over 6′6″? Get three-digit accu-

racy for each result. Use midpoint rectangles and the technique from (2) that worked. Be sure to use sufficiently large N and t as you determined in part (2).

(4) Estimate the height of the tallest adult female or male in the United States. To do this, find the value A for which $\int_A^\infty f(x)\,dx = \dfrac{1}{100,000,000}$. Use the technique you developed in (2) to do the integrals; use trial-and-error for A.

Discussion

1. Explain what is happening in each strategy in part (2) above. Which strategy works and which doesn't? Why?

2. Did you get three-digit accuracy for your answers in part (3) above? How do you know? Do these answers seem reasonable? Explain.

3. Explain how you went about answering part (4) above. Does your answer seem reasonable? Explain.

4. Discuss possible sources of error in this lab. Which source of error do you think is most responsible for any possible inaccuracies?

8.7 | Chapter Review

Concepts Test

Respond with true or false to each of the following assertions. Be prepared to justify your answer.

1. To find $\int x\,\sin(x^2)\,dx$, try the substitution $u = x^2$.

2. To find $\int x^2\,\sin(x)\,dx$, try the substitution $u = x^2$.

3. To find $\int \dfrac{x}{1 + x^4}\,dx$, try the substitution $u = x^2$.

4. To find $\int \dfrac{x^3}{1 + x^4}\,dx$, try the substitution $u = x^2$.

5. To find $\int \dfrac{1}{\sqrt{4 - 5x^2}}\,dx$, try the substitution $u = \sqrt{5}x$.

6. To find $\int \dfrac{2t - 1}{t^2 - t}\,dt$, try a partial fraction decomposition.

7. To find $\int \dfrac{2t - 1}{t^2 - t}\,dt$, try the substitution $u = t^2 - t$.

8. To find $\int x^2 \cos x\,dx$, use integration by parts.

9. To find $\int \cos^2 x\,dx$, use half-angle formulas.

10. To find $\int \dfrac{e^x}{1 + e^x}\,dx$, use integration by parts.

11. To find $\int \dfrac{x + 2}{\sqrt{x^2 + 4}}\,dx$, use a trigonometric substitution.

12. To find $\int x^2\sqrt[3]{3 - 2x}\,dx$, let $u = \sqrt[3]{3 - 2x}$.

13. To find $\int \sin^2 x\,\cos^5 x\,dx$, rewrite the integrand as $\sin^2 x(1 - \sin^2 x)^2 \cos x$.

14. To find $\int \sqrt{x^2 - 9}\,dx$, try a trigonometric substitution.

15. To find $\int x^2 \ln x\,dx$, try integration by parts.

16. To find $\int \sin 2x \cos 4x\,dx$, use half-angle formulas.

17. $\dfrac{x}{x^2 - 1}$ can be expressed in the form $\dfrac{A}{x - 1} + \dfrac{B}{x + 1}$.

18. $\dfrac{x^2 + 2}{x(x^2 - 1)}$ can be expressed in the form $\dfrac{A}{x} + \dfrac{B}{x - 1} + \dfrac{C}{x + 1}$.

19. The integration by parts formula is the integral version of the Product Rule for derivatives.

20. $\displaystyle\lim_{x\to\infty} \dfrac{x^{100}}{e^x} = 0.$

21. $\displaystyle\lim_{x\to\infty} \dfrac{x^{1/10}}{\ln x} = \infty.$

22. $\displaystyle\lim_{x\to\infty} \dfrac{1000x^4 + 1000}{0.001x^4 + 1} = \infty.$

23. $\displaystyle\lim_{x\to\infty} xe^{-1/x} = 0.$

24. If $\displaystyle\lim_{x\to a} f(x) = \lim_{x\to a} g(x) = \infty$, then it can be shown that $\displaystyle\lim_{x\to a}\dfrac{f(x)}{g(x)} = 1.$

25. If $\displaystyle\lim_{x\to a} f(x) = 1$ and $\displaystyle\lim_{x\to a} g(x) = \infty$, then it can be shown that $\displaystyle\lim_{x\to a}[f(x)]^{g(x)} = 1.$

26. If $\displaystyle\lim_{x\to\infty} \ln f(x) = 2$, then $\displaystyle\lim_{x\to\infty} f(x) = e^2.$

27. If $p(x)$ is a polynomial, then $\displaystyle\lim_{x\to\infty}\dfrac{p(x)}{e^x} = 0.$

28. If $p(x)$ is a polynomial, then $\displaystyle\lim_{x\to 0}\dfrac{p(x)}{e^x} = p(0).$

29. $\displaystyle\int_0^1 \dfrac{1}{x^{1.001}}\,dx$ converges.

30. $\displaystyle\int_0^\infty \dfrac{1}{x^p}\,dx$ diverges for all $p > 0$.

31. If f is continuous on $[0, \infty]$ and $\displaystyle\lim_{x\to\infty} f(x) = 0$, then $\displaystyle\int_0^\infty f(x)\,dx$ converges.

32. $\displaystyle\int_0^\pi \dfrac{\sin x}{x}\,dx$ is an improper integral.

33. We can always express the indefinite integral of an elementary function in terms of elementary functions.

Sample Test Problems

In Problems 1–36, perform the indicated integrations by any correct method. For the definite integrals, choose between an exact and a numerical method, and provide a reason for your choice. Identify which integrals are improper integrals.

1. $\displaystyle\int_0^4 \dfrac{t}{\sqrt{9 + t^2}}\,dt$

2. $\displaystyle\int \cot^2(2\theta)\,d\theta$

3. $\displaystyle\int_0^{\pi/2} e^{\cos x}\sin x\,dx$

4. $\displaystyle\int_0^{\pi/4} x\sin 2x\,dx$

5. $\displaystyle\int_0^{\pi/2} \sin^3(2t)\,dt$

6. $\displaystyle\int_0^{\pi/2} \sin(2t^3)\,dt$

7. $\displaystyle\int \dfrac{y^3 + y}{y + 1}\,dy$

8. $\displaystyle\int_0^{3/2} \dfrac{dy}{\sqrt{2y + 1}}$

9. $\displaystyle\int \dfrac{e^{2t}}{e^t - 2}\,dt$

10. $\displaystyle\int x^2 e^x\,dx$

11. $\displaystyle\int_0^1 \dfrac{dy}{\sqrt{2 + 3y^2}}$

12. $\displaystyle\int_0^1 \dfrac{dy}{\sqrt{2 + 3y^3}}$

13. $\displaystyle\int x^3 \sinh x\,dx$

14. $\displaystyle\int \dfrac{(\ln y)^5}{y}\,dy$

15. $\displaystyle\int \dfrac{\sin\sqrt{x}}{\sqrt{x}}\,dx$

16. $\displaystyle\int \dfrac{\ln t^2}{t}\,dt$

17. $\displaystyle\int e^{t/3}\sin 3t\,dt$

18. $\displaystyle\int \sin\dfrac{3x}{2}\cos\dfrac{x}{2}\,dx$

19. $\displaystyle\int_0^\infty \dfrac{dx}{(x + 1)^2}$

20. $\int_0^\infty \dfrac{dx}{1+x^2}$

21. $\int \dfrac{\sqrt{x}}{1+\sqrt{x}}\,dx$

22. $\int \dfrac{e^{2y}\,dy}{\sqrt{9-e^{2y}}}$

23. $\int \cos^3 x\sqrt{\sin x}\,dx$

24. $\int e^{\ln(3\cos x)}dx$

25. $\int_{-2}^0 \dfrac{dx}{2x+3}$

26. $\int_0^\infty \dfrac{dx}{e^{x/2}}$

27. $\int \dfrac{\sqrt{9-y^2}}{y}\,dy$

28. $\int \dfrac{e^{4x}}{1+e^{8x}}\,dx$

29. $\int x^3 \ln x\,dx$

30. $\int a^3 \ln x\,dx$

31. $\int_{-\infty}^1 e^{2x}\,dx$

32. $\int_{-1}^1 \dfrac{dx}{1-x}$

33. $\int_0^\infty \dfrac{dx}{x+1}$

34. $\int_{-\infty}^1 \dfrac{dx}{(2-x)^2}$

35. $\int_{-\infty}^\infty \dfrac{x}{(x^2+1)}\,dx$

36. $\int_{-\infty}^\infty \dfrac{x}{1+x^4}\,dx$

Find the limits in Problems 37–48.

37. $\lim\limits_{x\to0}\dfrac{4x}{\tan x}$

38. $\lim\limits_{x\to0}\dfrac{\tan 2x}{\sin 3x}$

39. $\lim\limits_{x\to0}\dfrac{\sin x-\tan x}{\frac13 x^2}$

40. $\lim\limits_{x\to0}\dfrac{\cos x}{x^2}$

41. $\lim\limits_{x\to1^-}\dfrac{\ln(1-x)}{\cot \pi x}$

42. $\lim\limits_{t\to\infty}\dfrac{\ln t}{t^2}$

43. $\lim\limits_{x\to\infty}\dfrac{2x^3}{\ln x}$

44. $\lim\limits_{x\to0^+}(\sin x)^{1/x}$

45. $\lim\limits_{x\to0^+} x^x$

46. $\lim\limits_{x\to0}(1+\sin x)^{2/x}$

47. $\lim\limits_{t\to\infty} t^{1/t}$

48. $\lim\limits_{x\to\pi/2}\dfrac{\tan 3x}{\tan x}$

49. The velocity of a person walking is given as a function of time by the equation $v=10e^{-t}\sin t$ with v in feet per second and t in seconds. Find the distance walked by this person from the start ($t=0$) to the point at which the person stops and turns around. Sketch the curve and give a geometric interpretation of the distance traveled.

50. Find the area of the region bounded by the curves $s=t/\sqrt{t+1}$, $s=0,t=0$, and $t=1$. Show that you can get the same result using either integration by parts or substitution. Sketch the region.

51. Find the volume of the solid generated by revolving the region bounded by the curves $y=x\cos x$, $y=0,x=0,x=\frac{\pi}{2}$ about the x-axis. Make a sketch.

52. Find the length of the segment of the curve $y=\ln(\sin x)$ from $x=\pi/6$ to $x=\pi/3$.

CHAPTER 9

Infinite Series

9.1 Infinite Sequences and Dynamical Systems

The word *sequence* has been used informally from time to time in this text. In simple language, a sequence

$$a_1, a_2, a_3, a_4, \ldots$$

is an ordered arrangement of real numbers, one for each positive integer. More formally, an **infinite sequence** is a function whose domain is the set of positive integers and whose range is a set of real numbers. We may denote a sequence by $a_1, a_2, a_3, \ldots$, by $\{a_n\}_{n=1}^{\infty}$, or simply by $\{a_n\}$. Occasionally, we will extend the notion slightly by allowing the domain to consist of all integers greater than or equal to a specified integer, as in $b_0, b_1, b_2, \ldots$ and $c_8, c_9, c_{10}, \ldots$, which are also denoted by $\{b_n\}_{n=0}^{\infty}$ and $\{c_n\}_{n=8}^{\infty}$.

A sequence may be specified by giving enough initial terms to establish a pattern, as in

$$1, 4, 7, 10, 13, \ldots$$

or by **explicit formula** for the nth term, as in

$$a_n = 3n - 2, \qquad n \geq 1$$

or by a **recursion formula**, which defines a term by the previous term, as in

$$a_n = a_{n-1} + 3, \qquad n \geq 2, a_1 = 1$$

513

Sequences defined by a recursion formula are also referred to as **discrete dynamical systems**.

Note that each of our three illustrations describes the same sequence. Here are four more explicit formulas and the first few terms of the sequences they generate.

(1) $a_n = 1 - \dfrac{1}{n}$, $n \geq 1$: $0, \dfrac{1}{2}, \dfrac{2}{3}, \dfrac{3}{4}, \dfrac{4}{5}, \cdots$

(2) $b_n = 1 + (-1)^n \dfrac{1}{n}$, $n \geq 1$: $0, \dfrac{3}{2}, \dfrac{2}{3}, \dfrac{5}{4}, \dfrac{4}{5}, \dfrac{7}{6}, \dfrac{6}{7}, \cdots$

(3) $c_n = (-1)^n \dfrac{1}{n}$, $n \geq 1$: $0, \dfrac{3}{2}, \dfrac{-2}{3}, \dfrac{5}{4}, \dfrac{-4}{5}, \dfrac{7}{6}, \dfrac{-6}{7}, \cdots$

(4) $d_n = 0.999$, $n \geq 1$: $0.999, 0.999, 0.999, 0.999, \ldots$

Patterns

Someone is sure to argue that there are many different sequences that begin

1, 4, 7, 10, 13

and we agree. For example, the formula

$$3n - 2 + (n - 1) \cdot$$
$$(n - 2) \cdot \cdots \cdot (n - 5)$$

generates those five numbers. Who but an expert would think of this formula? When we ask you to look for a pattern, we mean a simple and obvious pattern.

Convergence   Consider the four sequences just defined. Each of them has values that pile up near 1 (see the diagrams in Figure 1). But do they all *converge* to 1? The correct response is that sequences $\{a_n\}$ and $\{b_n\}$ converge to 1, but $\{c_n\}$ and $\{d_n\}$ do not.

For a sequence to converge to 1 means first that values of the sequence should get close to 1. But they must do more than get close; they must remain close, which rules out sequence $\{c_n\}$. And close means arbitrarily close—that is, within *any* specified degree of accuracy, which rules out sequence $\{d_n\}$. While sequence $\{d_n\}$ does not converge to 1, it is correct to say that it converges to 0.999. Sequence $\{c_n\}$ does not converge at all; we say it diverges.

Definition

The sequence $\{a_n\}$ is said to **converge** to L and we write

$$\lim_{n \to \infty} a_n = L$$

if for each positive number ε, there is a corresponding positive number N such that

$$|a_n - L| < \varepsilon \text{ whenever } n \geq N$$

In other words, no matter how small the distance to the limit L that you specify, I can give you a term in the sequence beyond which all the rest of the terms in the sequence are even closer to the limit L.

A sequence that fails to converge to any finite number L is said to **diverge**, or to be divergent.

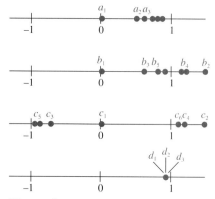

Figure 1

To see a relationship with something studied earlier (Section 4.6), consider graphing $a_n = 1 - 1/n$ and $a(x) = 1 - 1/x$. The only difference is that in the sequence case, the domain is restricted to the positive integers. In the first case, we write $\lim_{n \to \infty} a_n = 1$; in the second, $\lim_{n \to \infty} a(x) = 1$. Note the interpretation of ε and N in the diagrams in Figure 2.

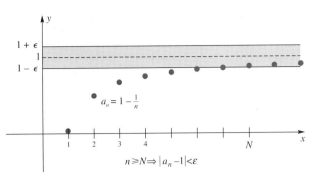
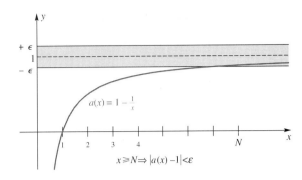

Figure 2

EXAMPLE 1: Show that

$$\lim_{n\to\infty} \frac{1}{n^2} = 0$$

SOLUTION: We need to show that for each positive number ε there exists a positive number N for which $\left|\dfrac{1}{n^2} - 0\right| < \varepsilon$ whenever $n \geq N$. Now, the following inequalities are all equivalent:

$$\left|\frac{1}{n^2} - 0\right| < \varepsilon$$

$$\frac{1}{n^2} < \varepsilon$$

$$n^2 > \frac{1}{\varepsilon}$$

$$n > \frac{1}{\sqrt{\varepsilon}}$$

Thus, $\left|\dfrac{1}{n^2} - 0\right| < \varepsilon$ whenever $n > \dfrac{1}{\sqrt{\varepsilon}}$. This means that whatever value of ε we are given, we can choose $N = \dfrac{1}{\sqrt{\varepsilon}}$.   ◄

All the limit theorems that were stated in the Main Limit Theorem of Section 3.2 hold for convergent sequences as well. We state several of them without proof.

Theorem A

Let $\{a_n\}$ and $\{b_n\}$ be convergent sequences and k a constant. Then

1. $\displaystyle\lim_{n\to\infty} k = k$;

2. $\displaystyle\lim_{n\to\infty} k a_n = k \lim_{n\to\infty} a_n$;

3. $\displaystyle\lim_{n\to\infty} (a_n \pm b_n) = \lim_{n\to\infty} a_n \pm \lim_{n\to\infty} b_n$;

4. $\displaystyle\lim_{n\to\infty} (a_n \cdot b_n) = \lim_{n\to\infty} a_n \cdot \lim_{n\to\infty} b_n$;

5. $\displaystyle\lim_{n\to\infty} \frac{a_n}{b_n} = \frac{\displaystyle\lim_{n\to\infty} a_n}{\displaystyle\lim_{n\to\infty} b_n}$ provided $\displaystyle\lim_{n\to\infty} b_n \neq 0$.

n	$\dfrac{3n^2}{7n^2 + 1}$
10	0.4279601
100	0.4285653
1000	0.4285714
10000	0.4285714

Figure 3

EXAMPLE 2: Find both a numerical estimate and an exact value for $\lim\limits_{n\to\infty} \dfrac{3n^2}{7n^2 + 1}$.

SOLUTION: We can estimate the limit numerically as we did with functions in Sections 1.1 and 1.2. In Figure 3 we show a table of values with n approaching infinity in powers of 10. We conclude that the limit is 0.4285714, accurate to seven decimal places.

For an exact value for the limit, we can divide numerator and denominator by the largest power of n that occurs in the denominator. This justifies our first step below; the others are justified by appealing to statements from Theorem A as indicated by the circled numbers.

$$\lim_{n\to\infty} \frac{3n^2}{7n^2 + 1} = \lim_{n\to\infty} \frac{3}{7 + (1/n^2)}$$

$$\overset{\text{⑤}}{=} \frac{\lim\limits_{n\to\infty} 3}{\lim\limits_{n\to\infty}[7 + (1/n^2)]}$$

$$\overset{\text{③}}{=} \frac{\lim\limits_{n\to\infty} 3}{\lim\limits_{n\to\infty} 7 + \lim\limits_{n\to\infty} 1/n^2}$$

$$\overset{\text{①}}{=} \frac{3}{7 + \lim\limits_{n\to\infty} 1/n^2} = \frac{3}{7 + 0} = \frac{3}{7} \approx 0.4285714$$

This result is consistent with our numerical estimate. ◀

n	$\dfrac{\ln n}{e^n}$
10	0.00010454
100	1.7132E-43
1000	3.506E-434

Figure 4

EXAMPLE 3: Does the sequence $\{(\ln n)/e^n\}$ converge, and if so, to what number?

SOLUTION: For a numerical estimate, see Figure 4. It appears that the limit is zero.

Again, we can also find the limit exactly. Here and in many sequence problems, it is convenient to use the following almost obvious fact (see Figure 2).

$$\boxed{\text{If } \lim_{x\to\infty} f(x) = L, \text{ then } \lim_{n\to\infty} f(n) = L}$$

This is convenient because we can apply l'Hôpital's Rule to the continuous variable problem. In particular, by l'Hôpital's Rule,

$$\lim_{x\to\infty} \frac{\ln x}{e^x} = \lim_{x\to\infty} \frac{1/x}{e^x} = 0$$

Thus,

$$\lim_{n\to\infty} \frac{\ln n}{e^n} = 0$$

that is, $\{(\ln n)/e^n\}$ converges to 0. ◀

What happens to the number in the sequence $\{0.999^n\}$ as $n \to \infty$? We suggest that you calculate 0.999^n for $n = 10, 100, 1000,$ and $10,000$ on your calculator to make a good guess. Then note the following example.

EXAMPLE 4: Show that if $-1 < r < 1$, then $\lim\limits_{n\to\infty} r^n = 0$.

> **SOLUTION:** Rather than give a formal proof based on the definition of limit of a sequence (as we did in Example 1), we give an intuitive justification for the result.
>
> If r is a positive rational number then $r = \frac{p}{q}$ where p and q are positive integers and $p < q$. Thus, $r^n = \frac{p^n}{q^n}$ and as n gets large, the denominator gets large faster than the numerator; therefore the fraction approaches zero. If r is negative, the terms of the sequence alternate between positive and negative but still approach zero. If r is irrational, we know there must be a rational number s for which $|r| < s < 1$ (why?). Because s^n must approach zero as we have just argued, r^n, which is smaller in magnitude, must also approach zero. ◄

What if $r > 1$; for example, $r = 1.5$? Then r^n will march off toward ∞. In this case, we write

$$\lim_{n\to\infty} r^n = \infty, \qquad r > 1$$

However, we say the sequence $\{r^n\}$ diverges. To converge, it would have to approach a *finite* limit. The sequence $\{r_n\}$ also diverges when $r \le -1$, and it converges when $r = 1$ (why?).

Monotonic Sequences

Consider now an arbitrary **nondecreasing sequence** $\{a_n\}$, by which we mean $a_n \le a_{n+1}$, $n \ge 1$. One example is the sequence $a_n = n^2$; another is $a_n = 1 - 1/n$. If you think about it a little, you may convince yourself that such a sequence can do only one of two things. Either it marches off to infinity or, if it cannot do that because it is bounded above, then it must bump against a lid (see Figure 5). Here is the formal statement of this very important result.

Theorem B

(Monotonic Sequence Theorem). If U is an upper bound for a nondecreasing sequence $\{a_n\}$, then the sequence converges to a limit A that is less than or equal to U. Similarly, if L is a lower bound for a nonincreasing sequence $\{b_n\}$, then the sequence $\{b_n\}$ converges to a limit B that is greater than or equal to L.

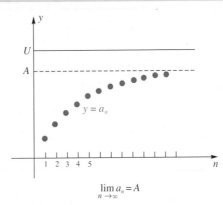

$$\lim_{n\to\infty} a_n = A$$

Figure 5

The phrase **monotonic sequence** is used to describe either a nondecreasing or nonincreasing sequence; hence the name for this theorem.

Theorem B describes a very deep property of the real number system. It is equivalent to the *completeness property* of the real numbers, which in simple language says that the real line has no "holes" in it. It is this property that distinguishes the real number line from the rational number line (which is full of holes). A great deal more could be said about this topic; we hope Theorem B appeals to your intuition and that you will accept it on faith until you take a more advanced course.

We make one more comment about Theorem B. It is not necessary that the sequences $\{a_n\}$ and $\{b_n\}$ be monotonic initially, only that they be monotonic from some point on—that is, for $n \geq K$. In fact, *the convergence or divergence of a sequence does not depend on the character of the initial terms but rather on what is true for large n.*

Discrete Dynamical Systems

As mentioned above, sequences that are generated by a recursion formula are often referred to as *discrete dynamical systems*. They can be used to model processes that increase or decrease in individual (discrete) steps, rather than continuously. For example, a population increases by one when a birth occurs and decreases by one when a death occurs, so discrete dynamical systems are often used to model population growth. Of course, such sequences are of interest in their own right as well.

We can describe any discrete dynamical system in the form

$$a_{n+1} = g(a_n)$$

where the initial point a_1 (or a_0 if one prefers) must also be given. For the sequence $a_n = a_{n-1} + 3, n \geq 2, a_1 = 1$ given near the beginning of this section, we would have $g(x) = x + 3$.

To estimate the limit of such a sequence, we can simply use the definition of the sequence and the process of *iteration*. Given a_1, we let $a_2 = g(a_1)$, $a_3 = g(a_2)$, and so on.

EXAMPLE 5: Estimate the limit of the sequence given by $a_{n+1} = e^{-a_n}$ with $a_1 = 0.5$ by using iteration.

SOLUTION: For the first three values in the sequence, we have

$$a_1 = 0.5$$

$$a_2 = e^{-a_1} = e^{-0.5} \approx 0.6065307$$

$$a_3 = e^{-a_2} = e^{-0.6065307} \approx 0.5452392$$

The first 27 values in the sequence (obtained on a calculator) are shown in the accompanying table.

n	a_n	n	a_n	n	a_n
1	0.5	10	0.5675596	19	0.5671408
2	0.6065307	11	0.5669072	20	0.5671447
3	0.5452392	12	0.5672772	21	0.5671425
4	0.5797031	13	0.5670674	22	0.5671438
5	0.5600646	14	0.5671864	23	0.5671430
6	0.5711721	15	0.5671189	24	0.5671434
7	0.5648629	16	0.5671571	25	0.5671432
8	0.5684380	17	0.5671354	26	0.5671433
9	0.5664095	18	0.5671477	27	0.5671433

Although it took 27 steps to get a repetition of the first seven digits, the process did produce a sequence that converges to the approximate value 0.5671433. ◄

Intuitively, a sequence a_n that converges to a particular value L becomes "fixed" at L as n gets larger and larger. We define a **fixed point** of the function g to be a number p for which $p = g(p)$. It should seem reasonable, therefore, that a fixed point of the function g would be a candidate for the limiting value of the sequence defined by $a_{n+1} = g(a_n)$.

EXAMPLE 6: Find any fixed points for the function $g(x) = e^{-x}$ and relate your results to those of Example 5.

SOLUTION: We need to solve the equation

$$p = e^{-p}$$

This equation cannot be solved using basic algebra, so we resort to a numerical approach as outlined in Section 2.5. Any solutions to the above equation would correspond to intersections of the graphs of the equations $y = x$ and $y = e^{-x}$ (why?). See Figure 6. We see that there can only be one intersection, and it should be between $x = 0$ and $x = 1$. Resorting to the built-in numerical root-finder of a graphics calculator or computer algebra system (or the bisection program given in the technology pages in the solutions manual) we get $x \approx 0.5671433$, so that the only solution to the equation $p = e^{-p}$ is $p \approx 0.5671433$.

We see that the fixed point of the function $g(x) = e^{-x}$ is the same as the limiting value of the dynamical system given by $a_{n+1} = e^{-a_n}$, $a_1 = 0.5$ from Example 5.   ◄

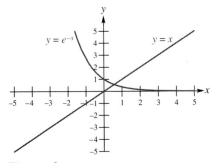

Figure 6

The question arises, do recursively defined sequences always converge to fixed points? Let's look at another example.

EXAMPLE 7: Determine the limiting behavior of the sequence defined by $a_{n+1} = 2a_n(1 - a_n)$ for the starting values $a_1 = -0.1, 0.1, 0.7, 1.1$. Also determine all fixed points for the function $f(x) = 2x(1 - x)$ and discuss how the fixed points are related to the limiting values of the sequence.

SOLUTION: The results of iterating the sequence for the various starting values are shown in Figure 7. We see that for the starting values 0.1 and 0.7 the se-

n	a_n	n	a_n
1	0.700	1	0.100
2	0.420	2	0.180
3	0.487	3	0.295
4	0.500	4	0.416
5	0.500	5	0.486
6	0.500	6	0.500
7	0.500	7	0.500
8	0.500	8	0.500
9	0.500	9	0.500
10	0.500	10	0.500

n	a_n	n	a_n
1	−0.100	1	1.100
2	−0.220	2	−0.220
3	−0.537	3	−0.537
4	−1.650	4	−1.650
5	−8.744	5	−8.744
6	−170.411	6	−170.411
7	−58420.603	7	−58420.603
8	−6826050523	8	−6.826E+09
9	−9.31899E+19	9	−9.319E+19
10	−1.73687E+40	10	−1.737E+40

Figure 7   The sequence $a_{n+1} = 2a_n(1 - a_n)$ for various starting values.

quence appears to converge to the limit $L = 0.5$. For the starting values -0.1 and 1.1 the sequence appears to diverge.

The fixed points are easily determined using basic algebra. We have

$$p = 2p(1 - P)$$
$$p = 2p - 2p^2$$
$$2p^2 - p = 0$$
$$p(2p - 1) = 0$$

so that $p = 0$ or $p = \frac{1}{2} = 0.5$.

We see that one of the fixed points, $p = 0.5$, is the limit of the corresponding sequence for the starting values 0.1 and 0.7. The other fixed point, $p = 0$, is not the limit of the sequence for any starting value, even though the starting values of -0.1 and 0.1 are close to this fixed point. ◄

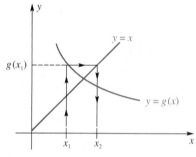

Figure 8

Convergence of the Sequence Sometimes the sequence converges to a fixed point; sometimes it does not. We can get a pretty good idea about which is which by executing an appropriate number of steps of the algorithm. But wouldn't it be nice to have a way of telling in advance whether there will be convergence? And to be sure of our conclusion?

To get a feeling for the problem, let's look at it geometrically. Note that we can get x_2 from x_1 by locating $g(x_1)$, sending it horizontally to the line $y = x$, and projecting down to the x-axis (see Figure 8). When we use this process repeatedly, we are faced with one of the situations in Figure 9.

What determines convergence or divergence? It appears to depend on the slope of the curve $y = g(x)$, namely, $g'(x)$, near the fixed point p. If $|g'(x)|$ is too large, the sequence diverges; if $|g'(x)|$ is small enough, the sequence converges. Here is a general result.

Theorem C

Let g be a function with a single fixed point p on the interval (a, b); thus $g(p) = p$ and $a < p < b$. If, in addition, g is differentiable and g' is continuous and also satisfies $|g'(x)| \leq M < 1$ for all x in $[a, b]$, M a constant, then the sequence defined by

$$a_{n+1} = g(a_n), \qquad a_1 \text{ in } [a, b]$$

converges to p as $n \to \infty$.

Proof Suppose $|g'(x)| \leq M < 1$ for all x in $[a, b]$ and let p be a single fixed point of g on (a, b). By the Mean Value Theorem for Derivatives, we may write

$$g(x) - g(p) = g'(c)(x - p)$$

with c some point between x and p. Thus,

$$|g(x) - g(p)| = |g'(c)||x - p| \leq M|x - p|$$

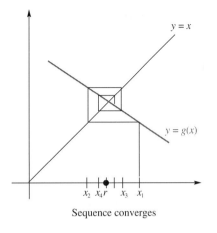

Sequence converges

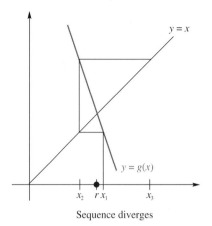

Sequence diverges

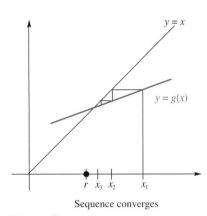

Sequence converges

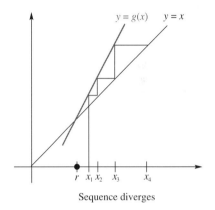

Sequence diverges

Figure 9

Applying this inequality successively to $a_1, a_2, \ldots$ yields

$$|a_2 - p| = |g(a_1) - g(p)| \leq M|a_1 - p|$$

$$|a_3 - p| = |g(a_2) - g(p)| \leq M|a_2 - p| \leq M^2|a_1 - p|$$

$$|a_4 - p| = |g(a_3) - g(p)| \leq M|a_3 - p| \leq M^3|a_1 - p|$$

$$\vdots$$

$$|a_n - p| = |g(a_{n-1}) - g(p)| \leq M|a_{n-1} - p| \leq M^{n-1}|a_1 - p|$$

Because $M^{n-1} \to 0$ as $n \to \infty$, we conclude that $a_n \to p$ as $n \to \infty$.   ◀

Now we can understand the behavior in Example 7. The sequence defined by $a_{n+1} = 2a_n(1 - a_n)$ can be written as $a_{n+1} = g(a_n)$ where

$$g(x) = 2x(1 - x) = 2x - 2x^2$$

Thus

$$g'(x) = 2 - 4x$$

For the fixed point $p = 0$ we have $g'(p) = g'(0) = 2 - 4(0) = 2$. Because $|g'(0)| > 1$, Theorem C does not apply, so we are not surprised that the sequence fails to converge to this fixed point. On the other hand, for the fixed point $p = 0.5$, we have $g'(p) = g'(0.5) = 2 - 4(0.5) = 0$. Because $|g'(0.5)| < 1$, and in fact $|g'(x)| < 1$ for $0.25 < x < 0.75$ (why?), Theorem C applies. Thus, we expect that if the starting point is close enough to $p = 0.5$, the sequence should converge to this fixed point.

Notice that one of the convergent starting points ($a_1 = 0.7$) we used in Example 7 lies in the interval $0.25 < x < 0.75$ for which $|g'(x)| < 1$, and the other convergent starting point ($a_1 = 0.1$) does not. The theorem guarantees only that if we start the sequence close enough to $p = 0.5$, then the sequence will converge to this fixed point, but it does not exclude the possibility that a sequence with a distant starting point could also converge to this fixed point.

Note: In Theorem C we specified that we must have $|g'(x)| < 1$ for all x in some interval containing the fixed point p. In fact, if $g'(x)$ is continuous, and $|g'(p)| < 1$, then such an interval *must* exist. Therefore, we can conclude that if the starting point of a sequence is close enough to a fixed point p for which $|g'(p)| < 1$, the sequence will converge to p.

Definition

A fixed point p of a function $g(x)$ is called an *attracting fixed point* if $|g'(p)| < 1$ (we assume that $g'(x)$ is continuous).

Thus, if p is an attracting fixed point of the function $g(x)$, then the sequence $a_{n+1} = g(a_n)$ will converge to p if the starting value of the sequence is sufficiently close to p.

EXAMPLE 8: Find all fixed points for the function $g(x) = \cos x$ and determine which ones are attracting. Determine at least one starting value for the sequence $a_{n+1} = \cos a_n$ that results in convergence to an attracting fixed point.

SOLUTION: To find any fixed points of the function $g(x) = \cos x$ we need to solve the equation $p = g(p)$ or, in this case, $p = \cos(p)$. If we graph $y = p$ and $y = \cos(p)$ and look for points of intersection, it becomes clear that there can be only one fixed point, somewhere between 0 and 1 (see Figure 10). Using a numerical method (with a calculator or computer) to solve the equation $p = \cos(p)$ or $p - \cos(p) = 0$, we obtain the fixed point $p = 0.739$.

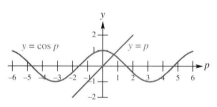

Figure 10

To test whether $p = 0.739$ is an attracting fixed point, we evaluate $g'(p)$.

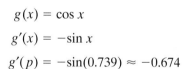

$$g(x) = \cos x$$

$$g'(x) = -\sin x$$

$$g'(p) = -\sin(0.739) \approx -0.674$$

Because $|-0.674| < 1$, we know that $p = 0.739$ is an attracting fixed point. We try a starting point close to 0.739. Using $a_0 = 0.5$ and $a_{n+1} = \cos a_n$, we get the results shown in Figure 11.

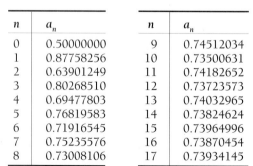

n	a_n	n	a_n
0	0.50000000	9	0.74512034
1	0.87758256	10	0.73500631
2	0.63901249	11	0.74182652
3	0.80268510	12	0.73723573
4	0.69477803	13	0.74032965
5	0.76819583	14	0.73824624
6	0.71916545	15	0.73964996
7	0.75235576	16	0.73870454
8	0.73008106	17	0.73934145

Figure 11

Two successive iterations are the same, accurate to three digits. As expected, the point of convergence is our attracting point, $p = 0.739$. ◀

Concepts Review

1. An arrangement of numbers $a_1, a_2, a_3, \ldots$ is called _____ .

2. We say the sequence $\{a_n\}$ converges if _____ .

3. A fixed point of a function $f(x)$ is a point p for which _____ .

4. The dynamical system $a_{n+1} = f(a_n)$ will converge to the fixed point p if the initial value a_0 is close enough to p and if _____ .

Problem Set 9.1

In Problems 1–16, an explicit formula for a_n is given. Write out the first five terms of each sequence and attempt to find $\lim_{n \to \infty} a_n$ using a numerical approach, an exact approach, or both. See Examples 2 and 3.

1. $a_n = \dfrac{n}{2n - 1}$

2. $a_n = \dfrac{3n + 1}{n + 2}$

3. $a_n = \dfrac{4n^2 + 1}{n^2 - 2n + 3}$

4. $a_n = \dfrac{3n^2 + 2}{n + 4}$

5. $a_n = (-1)^n \dfrac{n}{n + 1}$

6. $a_n = \dfrac{n \sin(n\pi/2)}{2n + 1}$

7. $a_n = e^{-n}\cos n$

8. $a_n = \dfrac{e^n}{2^n}$

9. $a_n = \dfrac{(-\pi)^n}{4^n}$

10. $a_n = \left(\frac{1}{2}\right)^n + 2^n$

11. $a_n = 1 + (0.9)^n$

12. $a_n = \dfrac{n^3}{e^n}$

13. $a_n = \dfrac{\ln n}{n}$

14. $a_n = \dfrac{\ln(1/n)}{\sqrt{n}}$

15. $a_n = \left(1 + \dfrac{1}{n}\right)^n$

16. $a_n = n^{1/n}$

In Problems 17–22, find an explicit formula $a_n =$ _____ for each sequence, determine if the sequence converges or diverges, and if it converges, find $\lim_{n \to \infty} a_n$. Again, you may use a combination of numerical and exact approaches.

17. $\frac{1}{2}, \frac{2}{3}, \frac{3}{4}, \frac{4}{5}, \ldots$

18. $\dfrac{1}{2^2}, \dfrac{2}{2^3}, \dfrac{3}{2^4}, \dfrac{4}{2^5}, \ldots$

19. $-1, \frac{2}{3}, -\frac{3}{5}, \frac{4}{7}, -\frac{5}{9}, \ldots$

20. $1, \dfrac{1}{1 - \frac{1}{2}}, \dfrac{1}{1 - \frac{2}{3}}, \dfrac{1}{1 - \frac{3}{4}}, \ldots$

21. $1, \dfrac{2}{2^2 - 1^2}, \dfrac{3}{3^2 - 2^2}, \dfrac{4}{4^2 - 3^2}, \ldots$

22. $\dfrac{1}{2 - \frac{1}{2}}, \dfrac{2}{3 - \frac{1}{3}}, \dfrac{3}{4 - \frac{1}{4}}, \dfrac{4}{5 - \frac{1}{5}}, \ldots$

For each dynamical system in Problems 23–30, estimate the limit numerically (or determine that the limit does not exist). If the limit exists, try to find an exact value for the limit by finding an appropriate fixed point.

23. $a_{n+1} = 1 + \frac{1}{2}a_n, a_1 = 1$

24. $a_{n+1} = \dfrac{1}{2}\left(a_n + \dfrac{2}{a_n}\right), a_1 = 2$

25. $u_{n+1} = \sqrt{3 + u_n}, u_1 = \sqrt{3}$

26. $x_{n+1} = \sqrt{3 + x_n}; x_1 = 507$

27. $u_{n+1} = 1.1^{u_n}, u_1 = 0$

28. $x_{n+1} = \dfrac{1}{10}e^{-2x_n}; x_1 = 1$

29. $x_{n+1} = 3 \tan^{-1}x_n; x_1 = 2$

30. $x_{n+1} = \sqrt{2.5 + x_n}; x_1 = 1$

31–36. Find *all* of the fixed points for the dynamical systems in Problems 23–28, and determine which ones are attracting and which ones are not. Relate your findings to the results of Problems 23–28.

In Problems 37–40, write the first four terms of the sequence $\{a_n\}$. Then use Theorem B to show that the sequence converges.

37. $a_n = \dfrac{4n - 3}{2^n}$

38. $a_n = \dfrac{n}{n + 1}\left(2 - \dfrac{1}{n^2}\right)$

39. $a_n = \left(1 - \dfrac{1}{4}\right)\left(1 - \dfrac{1}{9}\right)\cdots\left(1 - \dfrac{1}{n^2}\right), n \geq 2$

40. $a_n = 1 + \dfrac{1}{2!} + \dfrac{1}{3!} + \cdots + \dfrac{1}{n!}$

41. Using the definition of limit, prove that $\lim_{n \to \infty} n/(n + 1) = 1$; that is, for a given $\varepsilon > 0$ find N such that $n \geq N \Rightarrow |n/(n + 1) - 1| < \varepsilon$.

42. As in Problem 41, prove that $\lim_{n \to \infty} n/(n^2 + 1) = 0$.

43. Prove that if $\{a_n\}$ converges and $\{b_n\}$ diverges, then $\{a_n + b_n\}$ diverges.

44. If $\{a_n\}$ and $\{b_n\}$ both diverge, does it follow that $\{a_n + b_n\}$ diverges?

45. A famous sequence $\{f_n\}$, called the *Fibonacci Sequence* after Leonardo Fibonacci, who introduced it around A.D. 1200, is defined by the recursion formula

$$f_{n+2} = f_{n+1} + f_n, \quad f_1 = f_2 = 1$$

(a) Find f_3 through f_{10}.
(b) Let $\phi = \frac{1}{2}(1 + \sqrt{5}) \approx 1.618034$. The Greeks called this number the *golden ratio*, claiming that a rectangle whose dimensions were in this ratio was perfect. It can be shown that

$$f_n = \frac{1}{\sqrt{5}}\left[\left(\frac{1 + \sqrt{5}}{2}\right)^n - \left(\frac{1 - \sqrt{5}}{2}\right)^n\right]$$

$$= \frac{1}{\sqrt{5}}[\phi^n - (-1)^n\phi^{-n}]$$

You should check that this gives the right result for $n = 1$ and $n = 2$. The general result can be proved by induction (it is a nice challenge). More in line with this section, though, use this explicit formula to prove that $\lim_{n \to \infty} f_{n+1}/f_n = \phi$.
(c) Show, using the limit just proved, that ϕ satisfies the equation $x^2 - x - 1 = 0$. Then, in another interesting twist, use the Quadratic Formula to show that the two roots of this equation are ϕ and $-1/\phi$, the two numbers that occur in the explicit formula for f_n.

46. Consider an equilateral triangle containing $1 + 2 + 3 + \cdots + n = n(n + 1)/2$ circles each with diameter 1 and stacked as indicated in Figure 12 for the case $n = 4$. Find $\lim_{n \to \infty} A_n/B_n$, where A_n is the total area of the circles and B_n is the area of the triangle.

Dynamical systems is an area of contemporary research and serves as a possible model for turbulence, one of the least understood phenomena in science. The remaining problems will introduce you to this exciting area, but you will need a programmable calculator or a computer to carry out the calculations. Each problem deals with the function

$$f(x) = \lambda(x - x^2)$$

Consider the dynamical system

$$x_{n+1} = \lambda(x_n - x_n^2)$$

as we gradually increase λ from 2 to 4.

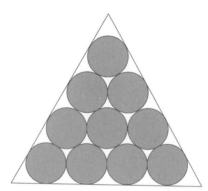

Figure 12

47. ($\lambda = 2.5$) Sketch $y = x$ and $y = 2.5(x - x^2)$ using the same axes and find the limit of the dynamical system $x_{n+1} = \lambda(x_n - x_n^2)$ by iteration for a few starting values x_0.

48. Find a fixed point of $f(x)$ for $\lambda = 2.5$ by simple algebra, thus confirming your results from Problem 47.

49. ($\lambda = 3.1$) Sketch $y = x$ and $y = f(x) = 3.1(x - x^2)$ using the same axes and attempt to find the limit of the dynamical system $x_{n+1} = \lambda(x_n - x_n^2)$ by iteration (note that $|f'(x)| > 1$ at the root). You will find that x_n bounces back and forth but gets closer and closer to two values r_1 and r_2, called *attractors*. Find r_1 and r_2 to five decimal places. Superimpose the graph of $y = g(x) = f(f(x)) = 9.61x - 39.40x^2 + 59.58x^3 - 29.79x^4$ on your earlier graph and observe that r_1 and r_2 appear to be the two roots of $x = g(x)$, where $|g'(x)| < 1$. Thus, r_1 and r_2 are fixed points of $g(x) = f(f(x))$.

50. In Problem 49, note that $f(r_1) = r_2$ and $f(r_2) = r_1$. Use this to show that $g'(r_1) = g'(r_2)$.

51. ($\lambda = 3.5$) In this case, use iteration to obtain four attractors, s_1, s_2, s_3, and s_4. Guess for what function they are the fixed points.

52. ($\lambda = 3.56$) Use iteration to get eight attractors.

53. ($\lambda = 3.57$) As you keep increasing λ by smaller and smaller amounts, you will double the number of attractors at each stage until at approximately 3.57 you should get chaos. Beyond $\lambda = 3.57$, other strange things happen, which you may want to investigate.

54. Try similar experiments on $x = \lambda \sin \pi x$.

For very readable accounts of this strange phenomenon, see: Robert M. May, "Simple mathematical models with very complicated dynamics," *Nature*, vol. 261

(1976), 459–467; and Douglas R. Hofstadter, "Strange attractors: Mathematical patterns delicately poised between order and chaos," *Scientific American*, vol. 245 (November 1981), 22–43.

Answers to Concepts Review: 1. A sequence 2. $\lim\limits_{n \to \infty} a_n$ exists (finite sense) 3. $p = f(p)$ 4. $|f'(p)| < 1$

LAB 17: DISCRETE POPULATION MODELS

Mathematical Background

One way of generating sequences of numbers is through the process of *iteration*. If we have a function $f(x)$, what we mean by iteration is to take an x-value and use it as input for $f(x)$; the number we get out is then used as input again, and the process is repeated. This can be represented as $x_{n+1} = f(x_n)$. Thus, for example, if we iterate the function $f(x) = x^2$ with the starting value $x_0 = 2$ we get the sequence $2, 4, 16, 256, \ldots$. This sequence is referred to as the orbit of $x_0 = 2$ under the function $f(x) = x^2$.

When iterating a function $f(x)$, a *fixed point* p is defined as a number for which $p = f(p)$. For the function $f(x) = x^2$ the fixed points would be solutions to $p = p^2$; the only solutions to this equation are $p = 0$ and $p = 1$.

A fixed point is called *attracting* if points nearby approach the fixed point in the limit as the number of iterations goes to infinity. For the function $f(x) = x^2$ the point 0 is attracting and the point 1 is not. To see this, start with a point near $p = 0$ such as 0.1; under iteration of $f(x) = x^2$ we get $0.1, 0.01, 0.0001, 0.00000001, \ldots$, which is approaching the fixed point $p = 0$. If we choose a point near $p = 1$ (such as 1.1) and iterate, we get (to two decimals) $1.10, 1.21, 1.46, 2.14, 4.59, 21.11, \ldots$, which is clearly not approaching $p = 1$ in the limit. When points close to a fixed point move away from the fixed point we call the fixed point *repelling*.

One can show (using the Mean Value Theorem) that if p is a fixed point and if $|f'(p)| < 1$ then the point p is an attracting fixed point, and if $|f'(p)| > 1$ then p is a repelling fixed point. For the function $f(x) = x^2$ we have $f'(x) = x^2$; thus $|f'(0)| = 0 < 1$ and $|f'(1)| = 2 > 1$. This confirms what we found before, that $p = 0$ is attracting and $p = 1$ is repelling.

Lab Introduction

In Lab 15 we looked at a model for population growth called the logistic equation. That model was given by the following differential equation:

$$\frac{dy}{dt} = ky\frac{(L - y)}{L} \tag{1}$$

where y is the population at time t, L is the maximum supportable population, and k is a growth constant. This equation is separable; in that lab you found y explicitly and discovered the effects of the constants k and L and the intial population. Basically all populations tended to L as time tended to ∞.

One of the assumptions we made in that model was that the population was growing continuously. That assumption is not a good one for populations such as locust populations or gypsy moths, where mating and/or death occur periodically. For these populations, we need to make the continuous model discrete; that is, we want to look at populations at discrete times.

If we fix a time period Δt and approximate the derivative in (1) we get

$$\frac{y(t + \Delta t) - y(t)}{\Delta t} = ky(t)\frac{L - P(t)}{L}$$

which can be rearranged to give

$$y(t + \Delta t) = y(t) + \Delta tky(t)\frac{L - y(t)}{L}$$

If we now think of t as the nth time period and $t + \Delta t$ as the $(n + 1)$th time period, then letting $y_n = y(t)$ and $y_{n+1} = y(t + \Delta t)$ we get

$$y_{n+1} = y_n + ky_n\frac{L - y_n}{L}\Delta t$$

Finally, if we replace y by x and let $\Delta t = 1$ (to represent reproduction, say, once a year), we get

$$x_{n+1} = x_n + \frac{k}{L}x_n(L - x_n)$$

Thus, we want to iterate the function

$$f(x) = x + \frac{k}{L}x(L - x) \tag{2}$$

with x_0 as the starting value. Our goal is to determine the behavior of the orbits as x_0, L, and k vary, as we did in Lab 15. We can then compare the discrete and continuous models.

Experiment

(1) Start with the same values for x_0, L, and k that you used in Lab 15 (in that lab x_0 was called y_0); those values were $k = 0.1$, $L = 140$, and $x_0 = 30$. Iterate the function $f(x)$ given in (2) for 100 iterations, plot the values of this sequence against the iteration number using a calculator or computer program, and then sketch the result. Now try the following variations: $k = 0.01$, $L = 10$, and $x_0 = 0.01$ (each time go back to the original values for the other parameters). Compare results with those of Lab 15.

(2) Follow the orbits for $M = 140$ and $k = 1.8, 1.9, 2.0, \ldots, 2.7, 2.8$. Choose the starting value $x_0 = 30$. Look both at the numbers and at the plots of the results. Note what happens to each orbit and look for convergence (fixed points) or other predictable behavior. Sketch the plots.

(3) When there is predictable behavior, write down the final pattern (the final value(s) of x) and plot them against k; plot the final x (or x's) on the vertical axis and k on the horizontal axis.

(4) Determine the fixed points p of $f(x)$ by setting $p = f(p)$ and solving this equation. Find conditions on L and k that tell you when the fixed points are attracting, using the condition given in the Mathematical Background section. For the case $L = 140$, determine the values of k that make each fixed point an attractor.

Discussion

1. Describe what happens in part (1) above for each choice of x_0, k, and L. Discuss similarities between the discrete and continuous case (Lab 14).

2. Describe in words what happened to the orbits for the k given in (2) above. Relate the results to what is happening in terms of populations. What kinds of behaviors are possible with the discrete logistic equation that were not possible with the continuous one? Could you see these behaviors in the example of sunflower growth from the previous lab? What kind of population might exhibit these behaviors?

3. Write down your theory on how the patterns of orbits change as k changes; use your graph from (3) above to determine your theory. Can you test your theory? If so, do it. If not, go back and develop one that you can test.

4. Does your calculation in (4) above agree with your results from (3)? Can you determine *exactly* where the first important change in behavior occurs as k increases from 1.8 to 2.8?

5. Can you find a k that corresponds to a population with a three-year cycle? Is there more than one k that will work? *Hint*: Look between 2.8 and 2.9.

9.2 | Infinite Series

In a famous paradox announced some 2400 years ago, Zeno of Elea said that a runner cannot finish a race because he must first cover half the distance, then half the remaining distance, then half the still remaining distance, and so on, forever. Because the runner's time is finite, he cannot traverse the infinite number of segments of the course. Yet we all know that runners do finish races.

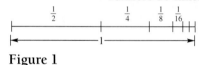

Figure 1

Imagine a racecourse to be 1 mile long. The segments of Zeno's argument would then have length $\frac{1}{2}$ mile, $\frac{1}{4}$ mile, $\frac{1}{8}$ mile, and so on (Figure 1). In mathematical language, finishing the race would amount to evaluating the sum

$$\tfrac{1}{2} + \tfrac{1}{4} + \tfrac{1}{8} + \tfrac{1}{16} + \tfrac{1}{32} + \cdots$$

which might seem impossible. But wait. Up to now, the word *sum* has been defined only for the addition of a finite number of terms. The indicated "infinite sum" has, as yet, no meaning for us.

Consider the partial sums

$$S_1 = \frac{1}{2}$$

$$S_2 = \frac{1}{2} + \frac{1}{4} = \frac{3}{4}$$

$$S_3 = \frac{1}{2} + \frac{1}{4} + \frac{1}{8} = \frac{7}{8}$$

$$S_4 = \frac{1}{2} + \frac{1}{4} + \frac{1}{8} + \cdots + \frac{1}{2^n} = 1 - \frac{1}{2^n}$$

Clearly, these partial sums get increasingly close to 1. In fact

$$\lim_{n \to \infty} S_n = \lim_{n \to \infty} \left(1 + \frac{1}{2^n}\right) = 1$$

This we define to be the value of the infinite sum.

More generally, consider

$$a_1 + a_2 + a_3 + a_4 + \cdots$$

which is also denoted by $\sum_{k=1}^{\infty} a_k$ or $\sum a_k$ and is called an **infinite series** (or series for short).

Then S_n, the **nth partial sum**, is given by

$$S_n = a_1 + a_2 + a_3 + \cdots + a_n = \sum_{k=1}^{n} a_k$$

Note: It is sometimes convenient to start a series with the term labeled a_0 instead of the term labeled a_1. We would then write the series as

$$\sum_{k=0}^{\infty} a_k = a_0 + a_1 + a_2 + a_3 + \cdots$$

We make the following formula definition.

Definition

The infinite series $\sum_{k=1}^{\infty} a_k$ **converges** and has sum S if the sequence of partial sums $\{S_n\}$ converges to S. If $\{S_n\}$ diverges, then the series **diverges**. A divergent series has no sum.

Geometric Series   A series of the form

$$\sum_{k=1}^{\infty} ar^{k-1} = a + ar + ar^2 + ar^3 + \cdots$$

where $a \neq 0$ is called a **geometric series**. If we start with the 0th term, we would write $\sum_{k=0}^{\infty} ar^k = a + ar + ar^2 + ar^3 + \cdots$.

EXAMPLE 1: Show that a geometric series converges with sum $S = a/(1 - r)$ if $|r| < 1$, but diverges if $|r| \geq 1$.

SOLUTION: Let $S_n = a + ar + ar^2 + \cdots + ar^{n-1}$. If $r = 1$, $S_n = na$, which grows without bound, and so $\{S_n\}$ diverges. If $r \neq 1$, we may write

$$S_n - rS_n = (a + ar + \cdots + ar^{n-1}) - (ar + ar^2 + \cdots ar^n) = a - ar^n$$

and so

$$\boxed{\; S_n = \frac{a - ar^n}{1 - r} = \frac{a}{1 - r} - \frac{a}{1 - r} r^n \;}$$

If $|r| < 1$, then $\lim_{n \to \infty} r^n = 0$ (Section 9.1, Example 4) and thus

$$S = \lim_{n \to \infty} S_n = \frac{a}{1 - r}$$

If $|r| > 1$ or $r = -1$, the sequence $\{r^n\}$ diverges, and consequently so does $\{S_n\}$.   ◄

EXAMPLE 2: Use the result of Example 1 to sum the following two geometric series.

(a) $\dfrac{4}{3} + \dfrac{4}{9} + \dfrac{4}{27} + \dfrac{4}{81} + \cdots$.

(b) $0.515151 \ldots = \dfrac{51}{100} + \dfrac{51}{10,000} + \dfrac{51}{1,000,000} + \cdots$.

SOLUTION:

(a) Because

$$\frac{4}{3} + \frac{4}{9} + \frac{4}{27} + \frac{4}{81} + \cdots = \frac{4}{3}\left(1 + \frac{1}{3} + \frac{1}{9} + \frac{1}{27} + \cdots\right)$$

$$= \frac{4}{3}\left(1 + \frac{1}{3} + \left(\frac{1}{3}\right)^2 + \left(\frac{1}{3}\right)^3 + \cdots\right)$$

we have $a = \frac{4}{3}$ and $r = \frac{1}{3}$. Therefore

$$S = \frac{a}{1 - r} = \frac{\frac{4}{3}}{1 - \frac{1}{3}} = \frac{\frac{4}{3}}{\frac{2}{3}} = 2$$

(b) Because

$$\frac{51}{100} + \frac{51}{10,000} + \frac{51}{1,000,000} + \cdots = \frac{51}{100}\left(1 + \frac{1}{100} + \frac{1}{10,000} + \cdots\right)$$

$$= \frac{51}{100}\left(1 + \frac{1}{100} + \left(\frac{1}{100}\right)^2 + \cdots\right)$$

we have $a = \frac{51}{100}$ and $r = \frac{1}{100}$. Therefore

$$S = \frac{\frac{51}{100}}{1 - \frac{1}{100}} = \frac{\frac{51}{100}}{\frac{99}{100}} = \frac{51}{99} = \frac{17}{33}$$

Incidentally, the procedure in (b) shows that any repeating decimal represents a rational number.   ◄

EXAMPLE 3: The diagram in Figure 2 represents an equilateral triangle containing infinitely many circles, tangent to the triangle and each other and reaching into the corners. What fraction of the area of the triangle is occupied by the circles?

SOLUTION: Suppose for convenience that the triangle has sides of length $2\sqrt{3}$, which gives it an altitude of 3. Concentrate attention on the vertical stack of circles. With a bit of geometric reasoning (the center of the large circle is two-thirds of the way from the upper vertex to the base), we

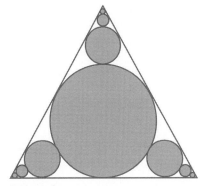

Figure 2

see that the radii of these circles are $1, \frac{1}{3}, \frac{1}{9}, \ldots$ and conclude that the vertical stack has area

$$\pi\left[1^2 + \left(\frac{1}{3}\right)^2 + \left(\frac{1}{9}\right)^2 + \left(\frac{1}{27}\right)^2 + \cdots\right]$$

$$= \pi\left[1 + \frac{1}{9} + \frac{1}{81} + \frac{1}{729} + \cdots\right] = \pi\left[\frac{1}{1 - \frac{1}{9}}\right] = \frac{9\pi}{8}$$

The total area of all the circles is three times this number minus twice the area of the big circle—that is, $27\pi/8 - 2\pi$, or $11\pi/8$. Because the triangle has area $3\sqrt{3}$, the fraction of this area occupied by the circles is

$$\frac{11\pi}{24\sqrt{3}} \approx 0.83 \qquad \blacktriangleleft$$

A General Test for Divergence

Consider the geometric series $a + ar + ar^2 + \cdots ar^{n-1} + \cdots$ once more. Its nth term a_n is given by $a_n = ar^{n-1}$. Example 1 shows that a geometric series converges *if and only if* $\lim_{n \to \infty} a_n = 0$. Could this possibly be true of all series? The answer is no, although half of the statement (the "only-if" half) is correct. This leads to an important divergence test for series.

> **Theorem A**
>
> **(nth-Term Test for Divergence).** If the series $\displaystyle\sum_{n=1}^{\infty} a_n$ converges, then $\lim_{n \to \infty} a_n = 0$.
>
> Equivalently, if $\lim_{n \to \infty} a_n \neq 0$ (or $\lim_{n \to \infty} a_n$ does not exist), the series diverges.

Proof   Let S_n be the nth partial sum and $S = \lim_{n \to \infty} S_n$. Because $a_n = S_n - S_{n-1}$, it follows that

$$\lim_{n \to \infty} a_n = \lim_{n \to \infty} S_n - \lim_{n \to \infty} S_{n-1} = S - S = 0 \qquad \blacktriangleleft$$

EXAMPLE 4: Show that $\displaystyle\sum_{n=1}^{\infty} \frac{n^3}{3n^3 + 2n^2}$ diverges.

SOLUTION:

$$\lim_{n \to \infty} a_n = \lim_{n \to \infty} \frac{n^3}{3n^3 + 2n^2} = \lim_{n \to \infty} \frac{1}{3 + 2/n} = \frac{1}{3}$$

Thus by the nth-term Test, the series diverges. $\qquad \blacktriangleleft$

The Harmonic Series

Students invariably want to turn Theorem A around and make it say that $a_n \to 0$ implies convergence of $\sum a_n \to 0$. The **harmonic series**

$$\sum_{n=1}^{\infty} \frac{1}{n} = 1 + \frac{1}{2} + \frac{1}{3} + \cdots + \frac{1}{n} + \cdots$$

shows that this is false. Clearly $\lim\limits_{n \to \infty} a_n = \lim\limits_{n \to \infty} (1/n) = 0$. However, the series diverges, as we now show.

EXAMPLE 5: Show that the harmonic series diverges.

> **SOLUTION:** We show that S_n grows without bound. Imagine n to be large and write

$$S_n = 1 + \frac{1}{2} + \frac{1}{3} + \frac{1}{4} + \frac{1}{5} + \cdots + \frac{1}{n}$$

$$= 1 + \frac{1}{2} + \left(\frac{1}{3} + \frac{1}{4}\right) + \left(\frac{1}{5} + \frac{1}{6} + \frac{1}{7} + \frac{1}{8}\right) + \left(\frac{1}{9} + \cdots + \frac{1}{16}\right) + \cdots + \frac{1}{n}$$

$$> 1 + \frac{1}{2} + \frac{2}{4} + \frac{4}{8} + \frac{8}{16} + \cdots + \frac{1}{n}$$

$$= 1 + \frac{1}{2} + \frac{1}{2} + \frac{1}{2} + \frac{1}{2} + \cdots + \frac{1}{n}$$

> It is clear that by taking n sufficiently large, we can introduce as many $\frac{1}{2}$'s into the last expression as we wish. Thus, $\{S_n\}$ diverges; hence, so does the harmonic series. ◄

Properties of Convergent Series

Convergent series behave much like finite sums; what you expect to be true usually is true.

Theorem B

(Linearity of Convergent Series). If $\displaystyle\sum_{k=1}^{\infty} a_k$ and $\displaystyle\sum_{k=1}^{\infty} b_k$ both converge and c is a constant, then $\displaystyle\sum_{k=1}^{\infty} ca_k$ and $\displaystyle\sum_{k=1}^{\infty} (a_k + b_k)$ also converge and, moreover,

(i) $\displaystyle\sum_{k=1}^{n} ca_k = c\sum_{k=1}^{n} a_k;$

(ii) $\displaystyle\sum_{k=1}^{\infty} (a_k + b_k) = \sum_{k=1}^{\infty} a_k + \sum_{k=1}^{\infty} b_k.$

Proof   This theorem introduces a subtle shift in language. The symbol $\displaystyle\sum_{k=1}^{\infty} a_k$ is now being used both for the infinite series $a_1 + a_2 + \cdots$ and for the sum of this series, which is a number.

By hypothesis, $\lim\limits_{n \to \infty} \displaystyle\sum_{k=1}^{n} a_k$ and $\lim\limits_{n \to \infty} \displaystyle\sum_{k=1}^{n} b_k$ both exist. Thus, use the properties of sums with finitely many terms and the properties of limits.

(i) $\displaystyle\sum_{k=1}^{\infty} ca_k = \lim_{n\to\infty} \sum_{k=1}^{n} ca_k = \lim_{n\to\infty} c \sum_{k=1}^{n} a_k$

$\displaystyle = c \lim_{n\to\infty} \sum_{k=1}^{n} a_k = c \sum_{k=1}^{n} a_k$

(ii) $\displaystyle\sum_{k=1}^{\infty} (a_k + b_k) = \lim_{n\to\infty} \sum_{k=1}^{n} (a_k + b_k) = \lim_{n\to\infty} \left[\sum_{k=1}^{n} a_k + \sum_{k=1}^{n} b_k \right]$

$\displaystyle = \lim_{n\to\infty} \sum_{k=1}^{n} a_k + \lim_{n\to\infty} \sum_{k=1}^{n} b_k = \sum_{k=1}^{\infty} a_k + \sum_{k=1}^{\infty} b_k$ ◀

EXAMPLE 6: Calculate $\displaystyle\sum_{k=1}^{\infty} \left[3\left(\tfrac{1}{8}\right)^k - 5\left(\tfrac{1}{3}\right)^k \right]$.

SOLUTION: By Theorem B and Example 1,

$$\sum_{k=1}^{\infty} \left[3\left(\frac{1}{8}\right)^k - 5\left(\frac{1}{3}\right)^k \right] = 3\sum_{k=1}^{\infty} \left(\frac{1}{8}\right)^k - 5\sum_{k=1}^{\infty} \left(\frac{1}{3}\right)^k$$

$$= 3\frac{\frac{1}{8}}{1 - \frac{1}{8}} - 5\frac{\frac{1}{3}}{1 - \frac{1}{3}} = \frac{3}{7} - \frac{5}{2} = -\frac{29}{14}$$ ◀

Numerical Estimation of Series   As is the case with equation solving and definite integration, there are many series that cannot be summed exactly, but for which a numerical estimate of the sum can be made.

The procedure for estimating the sum of an infinite series follows directly from the definition; simply keep adding in terms until the sum stabilizes. We recommend the following procedure: find the sum of the first 10 terms, then the first 20 terms, then the first 30 terms, and so on. When you get the same result accurate to *n* decimal places two times in a row, then you have an estimate accurate to about *n* decimal places.

This procedure is not foolproof. Some series that diverge do so so slowly that they may appear to converge by this method. Numerical estimation of any sort is subject to this type of problem, but when exact methods fail, you may have no other choice.

Technology and Numerical Estimation of Series   Computer algebra systems have built into them the ability to find numerical estimates of infinite series. This is also true of some graphing calculators; with others a short program (explained in the technology pages of the solutions manual) can be used to sum series numerically.

EXAMPLE 7: For each series below, find a three-digit numerical estimation of the sum. If the series diverges, say so. Also, find the exact sum, if possible.

(1) $\displaystyle\sum_{k=0}^{\infty} \frac{3}{2^k}$

(2) $\displaystyle\sum_{k=1}^{\infty} \frac{3}{k}$

(3) $\displaystyle\sum_{k=1}^{\infty} \frac{3}{k^2}$

→   *SOLUTION:*

(1)   The sum of the finite series $\sum_{k=0}^{n} \frac{3}{2^k}$ for $n = 10, 20, 30$ is shown in Figure 3. The sum appears to be converging to 6 (accurate to five decimal places by the table). Because this is a geometric series, we can find an exact sum. We have $\sum_{k=0}^{\infty} \frac{3}{2^k} = \sum_{k=0}^{\infty} 3\left(\frac{1}{2}\right)^k$ so that $a = 3$ and $r = \frac{1}{2}$. Thus $\sum_{k=0}^{\infty} \frac{3}{2^k} = \frac{3}{1 - \frac{1}{2}} = 6$, which confirms our numerical conjecture.

(2)   The sum of the infinite series $\sum_{k=1}^{n} \frac{3}{k}$ for various n-values up to 100,000 is shown in Figure 4. The sum is either converging very slowly or is not converging; it is hard to tell which is the case. However, because

$$\sum_{k=1}^{\infty} \frac{3}{k} = \frac{3}{1} + \frac{3}{2} + \frac{3}{3} + \cdots$$

$$= 3\left(1 + \frac{1}{2} + \frac{1}{3} + \cdots\right)$$

and because $1 + \frac{1}{2} + \frac{1}{3} + \cdots$ is a harmonic series and therefore divergent, the series diverges.

(3)   The sum of the finite series $\sum_{k=1}^{n} \frac{3}{k^2}$ for various n-values up to 100,000 is shown in Figure 5. The series converges slowly, but does appear to converge to 4.93 accurate to three digits. This series does not fit the form of a geometric series or a harmonic series, and so we do not have a method available for summing this series exactly.

n	$\sum_{k=0}^{n} \frac{3}{2^k}$
10	5.99707031250
20	5.99999713897
30	5.99999999720

Figure 3

n	$\sum_{k=1}^{n} \frac{3}{k}$
10	8.7869047619
20	10.7932189714
30	11.9849613927
100	15.5621325529
1,000	22.4564125816
10,000	29.3628181081
100,000	36.2704383899

Figure 4

n	$\sum_{k=1}^{n} \frac{3}{k^2}$
10	4.64930319349
20	4.78848973173
30	4.83645035280
100	4.90495170055
1,000	4.93180370004
10,000	4.93450221554
100,000	4.93477220069

Figure 5

Concepts Review

1.   An expression of the form $a_1 + a_2 + a_3 \cdots$ is called _____ .

2.   A series $a_1 + a_2 + \cdots$ is said to converge if the sequence $\{S_n\}$ converges, where $S_n = $ _____ .

3.   The geometric series $a + ar + ar^2 + \cdots$ converges if _____ ; in this case the sum of the series is _____ .

4.   If $\lim_{n \to \infty} a_n \neq 0$, we can be sure that the series $\sum_{n=1}^{\infty} a_n$ _____ .

Problem Set 9.2

In Problems 1–16, indicate whether the given series converges or diverges. If it converges, estimate its sum numerically as in Example 7. Also find an exact sum if possible.

1. $\displaystyle\sum_{k=0}^{\infty} \left(\tfrac{1}{5}\right)^k$

2. $\displaystyle\sum_{k=1}^{\infty} \left(-\tfrac{1}{3}\right)^{k-1}$

3. $\displaystyle\sum_{k=0}^{\infty} \left[2\left(\tfrac{1}{3}\right)^k + 3\left(\tfrac{1}{6}\right)^k\right]$

4. $\displaystyle\sum_{k=0}^{\infty} \left[3\left(\tfrac{1}{4}\right)^k - 2\left(\tfrac{1}{5}\right)^k\right]$

5. $\displaystyle\sum_{k=1}^{\infty} \frac{k-3}{k}$

6. $\displaystyle\sum_{k=1}^{\infty} \left(\tfrac{4}{3}\right)^k$

7. $\displaystyle\sum_{k=1}^{\infty} \left(\frac{1}{k} - \frac{1}{k+1}\right)^k$

8. $\displaystyle\sum_{k=1}^{\infty} \frac{2}{k}$

9. $\displaystyle\sum_{k=1}^{\infty} \frac{2}{k^3}$

10. $\displaystyle\sum_{k=0}^{\infty} \frac{k}{10^k}$

11. $\displaystyle\sum_{k=0}^{\infty} \frac{k^2}{10^k}$

12. $\displaystyle\sum_{k=0}^{\infty} \frac{2^k}{10^k}$

13. $\displaystyle\sum_{k=3}^{\infty} \frac{2}{(k-1)(k)}$

14. $\displaystyle\sum_{k=1}^{\infty} \frac{3^{k+1}}{5^{k-1}}$

15. $\displaystyle\sum_{k=1}^{\infty} \left(\frac{3}{(k+1)^2} - \frac{3}{k^2}\right)$

16. $\displaystyle\sum_{k=4}^{\infty} \frac{4}{k-3}$

In Problems 17–20, write the given decimal as an infinite series, then find the sum of the series, and finally use the result to write the decimal as a ratio of two integers.

17. 0.013013013 …

18. 0.125125125 …

19. 0.49999 …

20. 0.36717171 …

21. Evaluate $\displaystyle\sum_{k=0}^{\infty} r(1-r)^k, 0 < r < 2$. *Hint:* Make the substitution $s = 1 - r$.

22. Evaluate $\displaystyle\sum_{k=0}^{\infty} (-1)^k x^k, -1 < x < 1$.

23. A ball is dropped from a height of 100 feet. Each time it hits the floor, it rebounds to $\tfrac{2}{3}$ its previous height. Find the total distance it travels.

24. Three people, A, B, and C, divide an apple as follows. First they divide it into fourths, each taking a quarter. Then they divide the leftover quarter in fourths, each taking a quarter, and so on. Show that each gets a third of the apple.

25. Suppose the government pumps an extra $1 billion into the economy. Assume that each business and individual saves 25% of its income and spends the rest, so that of the initial $1 billion, 75% is spent by individuals and businesses. Of that amount, 75% is spent, and so forth. What is the total increase in spending due to the government action? (This is called the *multiplier effect* in economics.)

26. Do Problem 25 assuming that only 10% of the income is saved at each stage.

27. Assume square *ABCD* (Figure 6) has sides of length 1 and that *E, F, G,* and *H* are midpoints of their respective sides. If the indicated pattern is continued indefinitely, what will be the area of the painted region?

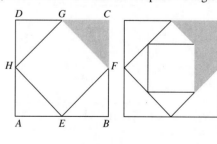

Figure 6

28. If the pattern shown in Figure 7 is continued indefinitely, what fraction of the original square will be painted?

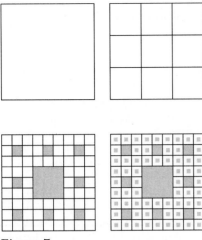

Figure 7

29. Each triangle in the descending chain (Figure 8) has its vertices at the midpoints of the sides of the next larger one. If the indicated pattern of painting is continued indefinitely, what fraction of the area of the original triangle will be painted? Does the original triangle need to be equilateral for this to be true?

Figure 8

30. Circles are inscribed in the triangles of Problem 29 as indicated in Figure 9. If the original triangle is equilateral, what fraction of the area is painted?

31. In another version of Zeno's paradox, Achilles can run 10 times as fast as the tortoise, but the tortoise has a 100-yard head start. Achilles cannot catch the tortoise, says Zeno, because when he runs 100 yards, the tortoise will have moved 10 yards ahead, when Achilles runs another 10 yards, the tortoise will have moved 1 yard ahead, and so on. Convince Zeno that Achilles will catch the tortoise and tell him exactly how many yards Achilles will have to run to do it.

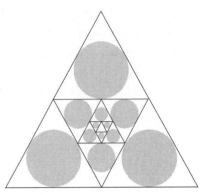

Figure 9

32. Tom and Joel are good runners, both able to run at a constant speed of 10 miles per hour. Their amazing dog Trot can do even better; he runs at 20 miles per hour. Starting from towns 60 miles apart, Tom and Joel run toward each other while Trot runs back and forth between them. How far does Trot run by the time the boys meet? Assume Trot starts with Tom running toward Joel and that he is able to make instant turnarounds. Solve the problem two ways.

(a) Use a geometric series.

(b) Find a simple way to do the problem.

33. Justify: If $\sum_{k=1}^{\infty} a_k$ diverges, so does $\sum_{k=1}^{\infty} ca_k$ for $c \neq 0$.

34. Use Problem 33 to conclude that $\frac{1}{2} + \frac{1}{4} + \frac{1}{6} + \frac{1}{8} + \cdots$ diverges.

35. Suppose one has an unlimited supply of identical blocks each 1 unit long.

(a) Convince yourself that they may be stacked as in Figure 10 without toppling. *Hint*: Consider centers of mass.

(b) How far can one make the top block protrude to the right of the bottom block using this method of stacking?

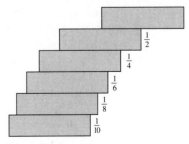

Figure 10

36. How large must N be for $S_N = \sum_{k=1}^{\infty} (1/k)$ just to exceed 4? *Note*: Computer calculations show that for S_N to exceed 20, $N = 272,400,600$ and for S_N to exceed 100, $N \approx 1.5 \times 10^{43}$.

37. Show that if $\sum a_n$ diverges and $\sum b_n$ converges, then $\sum (a_n + b_n)$ diverges.

38. Show that it is possible for $\sum a_n$ and $\sum b_n$ both to diverge, and yet for $\sum (a_n + b_n)$ to converge.

39. By looking at the region in Figure 11 first vertically and then horizontally, conclude that

$$1 + \tfrac{1}{2} + \tfrac{1}{4} + \tfrac{1}{8} + \cdots = \tfrac{1}{2} + \tfrac{2}{4} + \tfrac{3}{8} + \tfrac{4}{16} + \cdots$$

and use this fact to calculate:

(a) $\displaystyle\sum_{k=1}^{\infty} \frac{k}{2^k}$

(b) $\bar{x}$, the horizontal coordinate of the centroid of the region

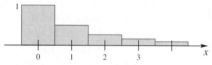

Figure 11

40. Many drugs are eliminated from the body in an exponential manner. Thus, if a drug is given in dosages of size C at time intervals of length t, the amount of A_n of the drug in the body just after the $(n + 1)$st dose is

$$A_n = C + Ce^{-kt} + Ce^{-2kt} + \cdots + Ce^{-nkt}$$

where k is a positive constant that depends on the type of drug.

(a) Derive a formula for A, the amount of drug in the body just after a dose if a person has been on the drug for a very long time (assume an infinitely long time).

(b) Evaluate A if it is known that one-half of a dose is eliminated from the body in 6 hours and doses of size 2 milligrams are given every 12 hours.

Answers to Concepts Review: 1. An infinite series 2. $a_1 + a_2 + a_3 + \cdots + a_n$ 3. $|r| < 1$; $a/(1 - r)$ 4. Diverges

---

LAB 18: BOUNCING BALLS AND INFINITE SERIES

Mathematical and Physical Background

Assume that a ball dropped from a height H will rebound to a height rH, where r is a constant that depends on the ball. For a silicon ball, r is about 0.8; for a squash ball, r is about 0.05. The constant r is called the rebound ratio and can be determined by experiment; because rebound height = $r \times$ original height, we get

$$r = \frac{\text{rebound height}}{\text{original height}} \tag{1}$$

On the next bounce the ball should rebound to a height $r(rH) = r^2H$, and on the bounce after that the ball should rebound to a height $r(r^2H) = r^3H$. Because the ball must travel both up and down after the initial drop, the total up and down distance traveled by the ball is

$$H + 2rH + 2r^2H + 2r^3H + \cdots \tag{2}$$

If air resistance is neglected, the time it takes an object to fall from a height H is given by $\sqrt{\dfrac{2H}{g}}$, where g is the acceleration of gravity. The value of g is about 9.8 meters per sec^2 or 32 ft per sec^2. The time it takes to rebound to a given height is the same as the time it takes to fall from the height; thus, reasoning as in (2), we get the following series for the total time the ball takes before it stops:

$$\sqrt{\frac{2H}{g}} + 2\sqrt{\frac{2rH}{g}} + 2\sqrt{\frac{2r^2H}{g}} + 2\sqrt{\frac{2r^3H}{g}} + \cdots \tag{3}$$

Lab Introduction

In this lab we want to understand and test the mathematical model for the bouncing ball described above; it is the model that appears in most calculus textbooks. We will drop the ball from several heights and measure how high the ball rebounds to estimate the rebound ratio r. It is important to understand that the model assumes that the rebound ratio r remains constant as the ball bounces—that is, if the ball returns to 60% of its original height on the first bounce then it will return to 60% of its previous height on every bounce after that. For each drop, we will also measure the time it takes for the ball to stop bouncing for that drop height. Because the infinite series (3) above predicts the time it takes to stop, we will have two ways of checking the modeling assumption that the rebound ratio is independent of drop height (direct measurement for different drop heights, and time it takes to stop). We will also try to approximate how many bounces the ball makes before it stops (the model assumes infinitely many bounces).

If the data do not support the modeling assumption of constant r, we will attempt to come up with a modified model that reconciles all of the data. In particular, we will assume that the rebound ratio depends on the drop height (thus we get a function $r(H)$), and try to get an idea of what this function looks like.

Data

Drop a ball from 60 inches, from 30 inches, and from 15 inches, and measure the rebound height in each case (we used a lacrosse ball, but any ball that bounces should do). This is most easily done by marking various heights with parallel lines in chalk on a wall or door; height zero should correspond to the *top* of the ball, and the other heights measured from that point. One can then step back and get a pretty good idea of the height the top of the ball reaches on the rebound. Also measure the time it takes for the ball to stop bouncing for each case, using a stopwatch. As usual, do more than one trial and average the results for more accurate data. For each drop height, determine the rebound ratio using equation (1) above. One could also use a distance probe with a computer or calculator, such as the CBL unit from Texas Instruments, to take the data.

Experiment

(1) Find a formula for the general term for series (2), and then write the series in $\sum$ notation (be sure you notice that the first term does not follow the pattern of the other terms). For each of the three drop heights, find the sum of this series accurate to eight digits; this represents the total distance traveled by the ball for each case. The r and H values may be different for each case. You have two possible strategies for finding the sum; you can sum numerically using a calculator or computer algebra system, or you can find the exact sum of the infinite series using what you know about the geometric series (or both as a check).

(2) For each of the three cases in (1), estimate the number of bounces that the ball bounced before it stopped. Each term in series (2) represents one bounce; thus, we want to estimate the number of terms in the series it takes to approximate the sum of the infinite series. The point here is that the mathematical model assumes that the ball bounces an infinite number of times, whereas the real ball must bounce a finite (though possibly large) number of times. You need a criterion for when the finite sum is "sufficiently" close to the infinite sum in a real world "practical" sense; use one-decimal-

place accuracy as your criterion (this corresponds to an error of 0.05 inch). Use a calculator program or a computer algebra system.

(3) Find a formula for the general term of series (3), and then write the series in $\sum$ notation. Find the sum of the series for each drop height to eight digits numerically, using a calculator or computer or using the geometric series, as in (1). The values of r and H may be different for each case. These numbers represent the time it takes the ball to stop bouncing; compare them with the measured times. Now reverse the process; use the measured stopping time values to determine the theoretical value of r that would give you this stopping time for each of the three cases. If you are summing the series numerically, you will have to vary r to fit the data.

(4) Based on all the information gathered above, develop a qualitative model for how the rebound ratio (r) varies with height (H). By this we mean sketch what you think a graph of r as a function of H would look like.

Discussion

1. Were you able to sum series (2) by hand? If so, explain how, and show that you get the same result as you do with the calculator. Are the three estimates of distance traveled consistent with each other? Explain. How can the distance traveled by the ball be finite if you assume an infinite number of bounces?

2. Explain how you determined the number of bounces the ball took before it stopped for each case in (2) above. Discuss the relationship between the finite series and the infinite series, both in terms of the mathematics and in terms of the ball.

3. Were you able to sum series (3) without a calculator or computer? Explain.

4. Discuss how you use the information gathered to determine the approximate relationship between r and H. Does your model now reconcile the various data collected?

9.3 Alternating Series and Absolute Convergence

In the last section, we considered series of nonnegative terms. Now we remove that restriction, allowing some terms to be negative. In particular, we study **alternating series**—that is, series of the form

$$a_1 - a_2 + a_3 - a_4 + \cdots$$

where $a_n > 0$ for all n. An important example is the **alternating harmonic series**

$$1 - \tfrac{1}{2} + \tfrac{1}{3} - \tfrac{1}{4} + \cdots$$

We have seen that the harmonic series diverges; we shall soon see that the alternating harmonic series converges.

A Convergence Test   Let us suppose that the sequence $\{a_n\}$ is decreasing; that is, $a_{n+1} < a_n$ for all n. Also let S_n have its usual meaning,

$$S_1 = a_1$$
$$S_2 = a_1 - a_2 = S_1 - a_2$$
$$S_3 = a_1 - a_2 + a_3 = S_2 + a_3$$
$$S_4 = a_1 - a_2 + a_3 - a_4 = S_3 - a_4$$

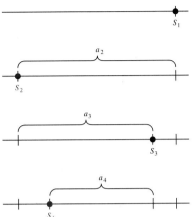

Figure 1

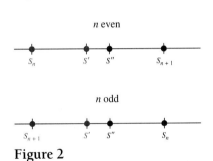

Figure 2

and so on. A geometric interpretation of these partial sums is shown in Figure 1. Note that the even-numbered terms $S_2, S_4, S_6, \ldots$ are increasing and bounded above and hence must converge to a limit; call it S'. Similarly, the odd-numbered terms $S_1, S_3, S_5, \ldots$ are decreasing and bounded below. They also converge, say to S''.

Both S' and S'' are between S_n and S_{n+1} for all n (see Figure 2) and so

$$\left| S'' - S' \right| \le \left| S_{n+1} - S_n \right| = a_{n+1}$$

Thus, the condition $a_{n+1} \to 0$ as $n \to \infty$ will guarantee that $S' = S''$ and, consequently, will assure the convergence of the series to the common value of S' and S'', which we call S. Finally, we note that because S is between S_n and S_{n+1},

$$\left| S - S_n \right| \le \left| S_{n+1} - S_n \right| = a_{n+1}$$

That is, the error made by using S_n as an approximation to the sum S of the whole series is not more than the magnitude of the first neglected term. We have proved the following theorem.

Theorem A

(Alternating-Series Test). Let

$$a_1 - a_2 + a_3 - a_4 + \cdots$$

be an alternating series with $a_n > a_{n+1} > 0$. If $\lim_{n\to\infty} a_n = 0$, then the series converges. Moreover, the error made by using the sum S_n of the first n terms to approximate the sum S of the series is not more than a_{n+1}.

EXAMPLE 1: Show that the alternating harmonic series

$$1 - \tfrac{1}{2} + \tfrac{1}{3} - \tfrac{1}{4} + \cdots = \sum_{k=1}^{\infty} \frac{1}{k}(-1)^{k+1}$$

converges. How many terms of this series would we need to take in order to get a partial sum S_n within 0.005 (two decimal places) of the sum S of the whole series? Estimate the sum for this amount of error.

SOLUTION: The alternating harmonic series satisfies the hypotheses of Theorem A and so converges. We want $\left| S - S_n \right| \le 0.005$, and this will hold if $a_{n+1} \le 0.005$. Because $a_{n+1} = 1/(n+1)$, we require $1/(n+1) \le 0.005$, which is satisfied if $n \ge 199$. Thus we need to take 199 terms to get two decimals of accuracy. This gives you an idea of how slowly the alternating harmonic series converges. The sum of the first 199 terms of the series done on a computer or calculator as outlined in Section 9.2 would be

$$\sum_{k=1}^{99} \frac{1}{k}(-1)^{k+1} = 0.695653$$

Because our result has a possible error of 0.005, we still cannot decide between 0.69 and 0.70 as our two-decimal approximation to the infinite series. (See Problem 41 for a clever way to find the exact sum of this series.)   ◀

EXAMPLE 2: Show that

$$\frac{1}{1!} - \frac{1}{2!} + \frac{1}{3!} - \frac{1}{4!} + \cdots$$

converges. Here $n!$ (read n **factorial**) is the standard symbol for the product of the first n positive integers—that is,

$$n! = n(n-1)(n-2) \cdots 3 \cdot 2 \cdot 1$$

Calculate S_5 and estimate the error made by using this as a value for the sum of the whole series.

SOLUTION: The Alternating-Series Test (Theorem A) applies and guarantees convergence.

$$S_5 = 1 - \frac{1}{2} + \frac{1}{6} - \frac{1}{24} + \frac{1}{120} \approx 0.6333$$

$$\text{error} = |S - S_5| \le a_6 = \frac{1}{6!} \approx 0.0014 \qquad \blacktriangleleft$$

Comparing a Series with Itself
We briefly return to series of positive terms to introduce a useful test, which we can then extend to general series as well.

> **Theorem B**
>
> **(Ratio Test).** Let $\sum a_n$ be a series of positive terms and suppose
>
> $$\lim_{n \to \infty} \frac{a_{n+1}}{a_n} = p$$
>
> **(i)** If $p < 1$, the series converges.
> **(ii)** If $p > 1$, the series diverges.
> **(iii)** If $p = 1$, the test is inconclusive.

Sketch of Proof Because $\lim_{n \to \infty} a_{n+1}/a_n = p$, $a_{n+1} \approx pa_n$; that is, the series behaves like a geometric series with ratio p. A geometric series converges when its ratio is less than 1 and diverges when its ratio is greater than 1. $\blacktriangleleft$

The Ratio Test will always fail for a series whose nth term is a rational expression in n, because in this case $p = 1$. However for a series whose nth term involves $n!$ or r^n, the Ratio Test usually works beautifully.

EXAMPLE 3: Test for convergence or divergence: $\sum_{n=1}^{\infty} \frac{2^n}{n!}$. If the series converges, estimate the sum.

SOLUTION:

$$p = \lim_{n \to \infty} \frac{a_{n+1}}{a_n} = \lim_{n \to \infty} \frac{2^{n+1}}{(n+1)!} \frac{n!}{2^n} = \lim_{n \to \infty} \frac{2}{n+1} = 0$$

We conclude by the Ratio Test that the series converges. We estimate the series numerically by calculating $\sum_{n=1}^{N} \frac{2^n}{n!}$ for $N = 10, 20, 30$ in Figure 3. We get a sum of 6.38906 accurate to six digits. $\blacktriangleleft$

N	$\sum_{n=1}^{N} \frac{2^n}{n!}$
10	6.38899
20	6.38906
30	6.38906

Figure 3

EXAMPLE 4: Test for convergence or divergence: $\displaystyle\sum_{n=1}^{\infty} \frac{2^n}{n^{20}}$. If the series converges, estimate the sum.

SOLUTION:

$$p = \lim_{n\to\infty} \frac{a_{n+1}}{a_n} = \lim_{n\to\infty} \frac{2^{n+1}}{(n+1)^{20}} \frac{n^{20}}{2^n} = \lim_{n\to\infty} \left(\frac{n}{n+1}\right)^{20} \cdot 2 = 2$$

We conclude that the given series diverges.   ◄

EXAMPLE 5: Test for convergence or divergence: $\displaystyle\sum_{n=1}^{\infty} \frac{n!}{n^n}$. If the series converges, estimate the sum.

SOLUTION: We will need the definition of the constant e that we used in Section 1.4.

$$\lim_{n\to\infty} \left(1 + \frac{1}{n}\right)^n = e$$

Thus, we may write

$$p = \lim_{n\to\infty} \frac{a_{n+1}}{a_n} = \lim_{n\to\infty} \frac{(n+1)!}{(n+1)^{n+1}} \frac{n^n}{n!} = \lim_{n\to\infty} \left(\frac{n}{n+1}\right)^n$$

$$= \lim_{n\to\infty} \frac{1}{((n+1)/n)^n} = \lim_{n\to\infty} \frac{1}{(1+1/n)^n} = \frac{1}{e} < 1$$

N	$\displaystyle\sum_{n=1}^{N} \frac{n!}{n^n}$
10	1.87963
20	1.87985
30	1.87985

Figure 4

Therefore, the given series converges. We estimate the series numerically be calculating $\displaystyle\sum_{n=1}^{N} \frac{n!}{n^n}$ for $N = 10, 20, 30$ in Figure 4. Our result is 1.87985.   ◄

Absolute and Conditional Convergence

If the series $\displaystyle\sum_{k=0}^{\infty} |a_k|$ converges, then we say that the series $\displaystyle\sum_{k=0}^{\infty} a_k$ converges absolutely. If $\displaystyle\sum_{k=0}^{\infty} |a_k|$ diverges but $\displaystyle\sum_{k=0}^{\infty} a_k$ converges, we say that $\displaystyle\sum_{k=0}^{\infty} a_k$ is conditionally convergent.

It should be clear that if $\displaystyle\sum_{k=0}^{\infty} |a_k|$ converges, then $\displaystyle\sum_{k=0}^{\infty} a_k$ converges as well (why?). The reverse is not true, as we have seen: the alternating harmonic series $1 - \frac{1}{2} + \frac{1}{3} - \frac{1}{4} + \cdots$ converges, but the harmonic series $1 + \frac{1}{2} + \frac{1}{3} + \frac{1}{4} + \cdots$ does not. Thus the alternating harmonic series is conditionally convergent.

Theorem C

(Absolute Ratio Test). Let $\displaystyle\sum u_n$ be a series of nonzero terms and suppose that

$$\lim_{n\to\infty} \frac{|u_{n+1}|}{|u_n|} = p$$

(i)   If $p < 1$, the series converges absolutely (hence converges).
(ii) If $p > 1$, the series diverges.
(iii) If $p = 1$, the test is inconclusive.

Proof   Only (ii) requires proof. From the original Ratio Test, we could conclude that $\sum |u_n|$ diverges, but here we are claiming more—namely, that $\sum u_n$ diverges. Because

$$\lim_{n \to \infty} \frac{|u_{n+1}|}{|u_n|} > 1$$

it follows that for n sufficiently large, say $n \geq N$, $|u_{n+1}| > |u_n|$. This, in turn, implies that $|u_n| > |u_N| > 0$ for all $n \geq N$ and so $\lim_{n \to \infty} u_n$ cannot be 0. We conclude by the nth-term Test that $\sum u_n$ diverges.   ◀

EXAMPLE 6: Show that $\displaystyle\sum_{n=1}^{\infty} (-1)^{n+1}\frac{3^n}{n!}$ converges absolutely. Estimate the sum of the series.

SOLUTION:

$$\rho = \lim_{n \to \infty} \frac{|u_{n+1}|}{|u_n|} = \lim_{n \to \infty} \frac{3^{n+1}}{(n+1)!} \times \frac{n!}{3^n}$$

$$= \lim_{n \to \infty} \frac{3}{|n+1|} = 0$$

N	$\displaystyle\sum_{n=1}^{N} (1)^{n+1}\frac{3^n}{n!}$
10	0.94667
20	0.95021
30	0.95021

Figure 5

We conclude from the Absolute Ratio Test that the series converges absolutely (and therefore converges). We estimate the series numerically by calculating $\displaystyle\sum_{n=1}^{N} (-1)^{n+1}\frac{3^n}{n!}$ for $N = 10, 20, 30$ in Figure 5. We get a sum of 0.95021 accurate to five digits.   ◀

Concepts Review

1. The alternating series $a_1 - a_2 + a_3 - \cdots$ will converge, provided the terms are decreasing in size and _____ .

2. If $\sum |u_k|$ converges, we say the series $\sum u_k$ converges _____; if $\sum u_k$ converges but $\sum |u_k|$ diverges, we say $\sum u_k$ converges _____ .

3. The premier example of a conditionally convergent series is _____ .

4. The Absolute Ratio Test says that if $\lim_{n \to \infty} \frac{|u_{n+1}|}{|u_n|} = p$, the series $\sum u_n$ converges absolutely if _____ and diverges if _____ .

Problem Set 9.3

In Problems 1–6, show that each of the alternating series converges and then estimate the error made by using the partial sum S_9 as an approximation to the sum S of the series (see Examples 1 and 2). Then find S_9 and a numerical approximation to the infinite sum accurate to two decimal places, and the difference between them. Discuss any differences between these two estimates of the error.

1. $\displaystyle\sum_{n=1}^{\infty} (-1)^{n+1}\frac{2}{3n+1}$

2. $\displaystyle\sum_{n=1}^{\infty} (-1)^{n+1}\frac{1}{\sqrt{n}}$

3. $\displaystyle\sum_{n=0}^{\infty} (-1)^n \frac{1}{e^n}$

4. $\displaystyle\sum_{n=1}^{\infty} (-1)^{n+1}\frac{n}{n^2+1}$

5. $\displaystyle\sum_{n=1}^{\infty} (-1)^{n+1}\frac{\ln n}{n}$

6. $\displaystyle\sum_{n=1}^{\infty} (-1)^{n+1}\frac{\ln n}{\sqrt{n}}$

In Problems 7–12, show that each of the series converges absolutely, and estimate each sum numerically.

7. $\displaystyle\sum_{n=1}^{\infty}\left(-\tfrac{3}{4}\right)^{n}$

8. $\displaystyle\sum_{n=1}^{\infty}(-1)^{n}\,-\frac{1}{n\sqrt{n}}$

9. $\displaystyle\sum_{n=1}^{\infty}(-1)^{n+1}\frac{n}{2^{n}}$

10. $\displaystyle\sum_{n=1}^{\infty}(-1)^{n+1}\frac{n^{2}}{e^{n}}$

11. $\displaystyle\sum_{n=1}^{\infty}(-1)^{n+1}\frac{1}{n(n+1)}$

12. $\displaystyle\sum_{n=1}^{\infty}(-1)^{n+1}\frac{2^{n}}{n!}$

In Problems 13–30, determine whether the series is convergent or divergent. Estimate the sum of each convergent series numerically. If the series converges *very* slowly, make a less accurate estimate.

13. $\displaystyle\sum_{n=1}^{\infty}(-1)^{n+1}\frac{1}{5n}$

14. $\displaystyle\sum_{n=1}^{\infty}\frac{1}{5n}$

15. $\displaystyle\sum_{n=1}^{\infty}(-1)^{n+1}\frac{n}{10n+1}$

16. $\displaystyle\sum_{n=1}^{\infty}(-1)^{n+1}\frac{n}{10n^{1.1}+1}$

17. $\displaystyle\sum_{n=1}^{\infty}(-1)^{n+1}\frac{1}{n\ln n}$

18. $\displaystyle\sum_{n=1}^{\infty}(-1)^{n+1}\frac{\sqrt{n}}{3^{n}}$

19. $\displaystyle\sum_{n=1}^{\infty}\frac{n^{4}}{2^{n}}$

20. $\displaystyle\sum_{n=1}^{\infty}\frac{1}{n!}$

21. $\displaystyle\sum_{n=1}^{\infty}(-1)^{n+1}\frac{n}{n!}$

22. $\displaystyle\sum_{n=1}^{\infty}(-1)^{n+1}\frac{n-1}{n}$

23. $\displaystyle\sum_{n=1}^{\infty}\frac{\cos n\pi}{n}$

24. $\displaystyle\sum_{n=1}^{\infty}\frac{\sin(n\pi/2)}{n^{2}}$

25. $\displaystyle\sum_{n=1}^{\infty}\frac{\sin n}{n\sqrt{n}}$

26. $\displaystyle\sum_{n=1}^{\infty}n\sin\!\left(\frac{1}{n}\right)$

27. $\displaystyle\sum_{n=1}^{\infty}(-1)^{n+1}\frac{1}{\sqrt{n(n+1)}}$

28. $\displaystyle\sum_{n=1}^{\infty}\frac{(-1)^{n+1}}{\sqrt{n+1}+\sqrt{n}}$

29. $\displaystyle\sum_{n=1}^{\infty}\frac{(-3)^{n+1}}{n^{2}}$

30. $\displaystyle\sum_{n=1}^{\infty}(-1)^{n+1}\sin\frac{\pi}{n}$

31. Explain why if $\sum a_{n}$ diverges, so does $\sum|a_{n}|$.

32. Give an example of two series $\sum a_{n}$ and $\sum b_{n}$, both convergent, such that $\sum a_{n}b_{n}$ diverges.

33. Explain why the positive terms of the alternating harmonic series form a divergent series. Do the same for the negative terms.

34. Explain why the results in Problem 33 hold for any conditionally convergent series.

35. Show that the alternating harmonic series

$$1-\tfrac{1}{2}+\tfrac{1}{3}-\tfrac{1}{4}+\tfrac{1}{5}-\tfrac{1}{6}+\cdots$$

(whose sum is actually $\ln 2\approx 0.69$) can be rearranged to converge to 1.3 by using the following steps.

(a) Take enough of the positive terms $1+\tfrac{1}{3}+\tfrac{1}{5}+\cdots$ to just exceed 1.3.

(b) Now add enough of the negative terms $-\tfrac{1}{2}-\tfrac{1}{4}-\tfrac{1}{6}-\cdots$ so that the partial sum S_{n} falls just below 1.3.

(c) Add just enough more positive terms to again exceed 1.3, and so on.

36. Use your calculator to help you find the first 20 terms of the series described in Problem 35. Calculate S_{20}.

37. Convince yourself that a conditionally convergent series can be rearranged to converge to any given number.

38. Show that a conditionally convergent series can be rearranged so as to diverge.

39. Show that $\lim_{n \to \infty} a_n = 0$ is not sufficient to guarantee the convergence of the alternating series $\sum (-1)^{n+1} a_n$. *Hint:* Alternate the terms of $\sum 1/n$ and $\sum -1/n^2$.

40. Discuss the convergence or divergence of

$$\frac{1}{\sqrt{2} - 1} - \frac{1}{\sqrt{2} + 1} + \frac{1}{\sqrt{3} - 1} - \frac{1}{\sqrt{3} + 1}$$

$$+ \frac{1}{\sqrt{4} - 1} - \frac{1}{\sqrt{4} + 1} + \cdots$$

41. Note that

$$1 - \frac{1}{2} + \frac{1}{3} - \frac{1}{4} + \cdots - \frac{1}{2n} = 1 + \frac{1}{2} + \frac{1}{3} + \cdots + \frac{1}{2n}$$

$$- \left(1 + \frac{1}{2} + \frac{1}{3} + \cdots + \frac{1}{n} \right)$$

$$= \frac{1}{n + 1} + \frac{1}{n + 2} + \cdots + \frac{1}{2n}$$

Recognize the latter expression as a Riemann sum and use it to find the sum of the alternating harmonic series.

Answers to Concepts Review: 1. $\lim_{n \to \infty} a_n = 0$ 2. Absolutely; conditionally 3. The alternating harmonic series 4. $p < 1$; $p > 1$

9.4 Taylor's Approximation to Functions

Throughout this book we have used a variety of methods for finding solutions to problems: exact methods, graphical methods, and the methods of numerical approximation. For example, we know that $\int_0^\pi \sin x \, dx$ is easy to calculate exactly using the Fundamental Theorem of Calculus, but that to find $\int_0^\pi \sin (x^2) \, dx$ we must resort to a numerical approximation.

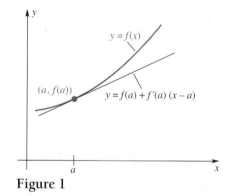

Figure 1

In this section we look at methods for approximating complicated functions by simpler ones—in particular, polynomials. This in turn leads to a deeper understanding of numerical methods used for definite integration and equation solving.

Linear Approximation The idea behind the differential approximation used in Section 3.7 was to approximate a curve near a point by its tangent line at that point. Our idea here is exactly the same, only we do not use differential notation. Examine the diagram in Figure 1. The equation of the tangent line to the curve $y = f(x)$ at $(a, f(a))$ is

$$y = f(a) + f'(a)(x - a)$$

This leads directly to the linear approximation for values of $f(x)$ near $x = a$:

$$f(x) \approx f(a) + f'(a)(x - a)$$

The linear polynomial $P_1(x) = f(a) + f'(a)(x - a)$ is called the **Taylor polynomial of order 1 based at a** for $f(x)$, after the English mathematician Brook Taylor (1685–1731). Clearly, we can expect $P_1(x)$ to be a good approximation to $f(x)$ only near $x = a$.

EXAMPLE 1: Find $P_1(x)$ based at $a = 1$ for $f(x) = \ln x$ and use it to approximate $\ln(0.9)$ and $\ln(1.5)$.

SOLUTION: Because $f(x) = \ln x$, $f'(x) = 1/x$; thus, $f(1) = 0$ and $f'(1) = 1$. Therefore,

$$P_1(x) = 0 + 1(x - 1) = x - 1$$

Consequently (see Figure 2),

$$\ln x \approx x - 1$$

and

$$\ln(0.9) \approx 0.9 - 1 = -0.1$$
$$\ln(1.5) \approx 1.5 - 1 = 0.5$$

These approximations should be compared with the correct four-place values of -0.1054 and 0.4055. As expected, the approximation is much better in the first case, because 0.9 is closer to 1 than is 1.5.   ◄

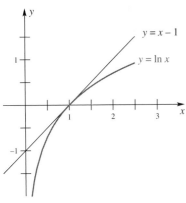

Figure 2

Taylor Polynomials of Order n   Polynomials are the easiest of all functions to evaluate, because they involve only three arithmetic operations: addition, subtraction, and multiplication. This is why we use polynomials to approximate other functions. If a linear polynomial gives a certain approximation to $f(x)$, we should expect a quadratic polynomial (with its curved graph) to do better, a cubic polynomial to do still better, and so on. Our aim is to find the polynomial of degree n that does the best job of approximating $f(x)$. We consider the quadratic case first.

A significant observation to make about the linear case is that f and its approximation P_1, as well as their derivatives f' and P_1', agree at $x = a$. In generalizing to the quadratic polynomial P_2, we impose three conditions, namely,

$$f(a) = P_2(a), \qquad f'(a) = P_2'(a), \qquad f''(a) = P_2''(a)$$

The unique quadratic polynomial satisfying these conditions (the **Taylor polynomial of order 2 based at a**) is

$$P_2(x) = f(a) + f'(a)(x - a) + \frac{f''(a)}{2}(x - a)^2$$

as you may check. The corresponding quadratic approximation is

$$\boxed{f(x) \approx f(a) + f'(a)(x - a) + \frac{f''(a)}{2}(x - a)^2}$$

EXAMPLE 2: Find $P_2(x)$ based at $a = 1$ for $f(x) = \ln x$ and use it to approximate $\ln(0.9)$ and $\ln(1.5)$.

SOLUTION: Here $f(x) = \ln x, f'(x) = 1/x$, and $f''(x) = -1/x^2$, and so $f(1) = 0$, $f'(1) = 1$, and $f''(1) = -1$. Thus,

$$P_2(x) = 0 + 1(x - 1) - \tfrac{1}{2}(x - 1)^2$$

Consequently,

$$\ln x \approx (x - 1) - \tfrac{1}{2}(x - 1)^2$$

and

$$\ln(0.9) \approx (0.9 - 1) - \tfrac{1}{2}(0.9 - 1)^2 = -0.1050$$
$$\ln(1.5) \approx (1.5 - 1) - \tfrac{1}{2}(1.5 - 1)^2 = 0.3750$$

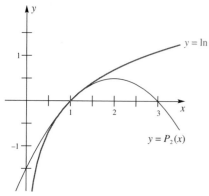

Figure 3

As expected, these are better approximations than we got in the linear case (Example 1). Note the diagram in Figure 3. ◀

With a little effort, we could now derive $P_3(x)$. And from this, it is a simple matter to guess the form of the **Taylor polynomial of order n based at a**; that is, the nth order polynomial P_n, which—together with its first n derivatives—agrees with f and its derivatives at $x = a$. It is

$$P_n(x) = f(a) + f'(a)(x - a) + \frac{f''(a)}{2!}(x - a)^2 + \cdots + \frac{f^{(n)}(a)}{n!}(x - a)^n$$

The corresponding approximation is

$$f(x) \approx f(a) + f'(a)(x - a) + \frac{f''(a)}{2!}(x - a)^2 + \cdots + \frac{f^{(n)}(a)}{n!}(x - a)^n$$

Maclaurin Polynomials   In the case $a = 0$, the Taylor polynomial of order n simplifies to the **Maclaurin polynomial of order n**, named in honor of the Scottish mathematician Colin Maclaurin (1698–1746). It gives a particularly useful approximation valid near $x = 0$, namely,

$$f(x) \approx f(0) + f'(0)x + \frac{f''(0)}{2!}x^2 + \cdots + \frac{f^{(n)}(0)}{n!}x^n$$

EXAMPLE 3: Find the Maclaurin polynomials of order n for e^x and $\cos x$. Approximate $e^{0.2}$ and $\cos(0.2)$ using $n = 4$, and determine the error involved in the approximation. Then sketch the graphs of each function along with its Maclaurin approximation of order 4 on the interval $-3 \leq x \leq 3$.

SOLUTION: The calculation of the required derivatives is shown in the table.

		At $x = 0$		At $x = 0$
$f(x)$	e^x	1	$\cos x$	1
$f'(x)$	e^x	1	$-\sin x$	0
$f''(x)$	e^x	1	$-\cos x$	-1
$f^{(3)}(x)$	e^x	1	$\sin x$	0
$f^{(4)}(x)$	e^x	1	$\cos x$	1
$f^{(5)}(x)$	e^x	1	$-\sin x$	0
$\vdots$	$\vdots$	$\vdots$	$\vdots$	$\vdots$

It follows that

$$e^x \approx 1 + x + \frac{1}{2!}x^2 + \frac{1}{3!}x^3 + \frac{1}{4!}x^4 + \cdots + \frac{1}{n!}x^n$$

$$\cos x \approx 1 - \frac{1}{2!}x^2 + \frac{1}{4!}x^4 - \cdots + (-1)^{n/2}\frac{1}{n!}x^n \qquad (n \text{ even})$$

Thus, using $n = 4$ and $x = 0.2$, we obtain

$$e^{0.2} \approx 1 + 0.2 + \frac{(0.2)^2}{2} + \frac{(0.2)^3}{6} + \frac{(0.2)^4}{24} = 1.2214000000$$

$$\cos(0.2) \approx 1 - \frac{(0.2)^2}{2} + \frac{(0.2)^4}{24} \approx 0.9800666667$$

The correct 10-figure values are 1.221402758 and 0.9800665778. The corresponding absolute errors are 0.000003 and 0.00000009 to one figure each. In Figure 4 we show the graph of each function together with its fourth-order Maclaurin polynomial.   ◀

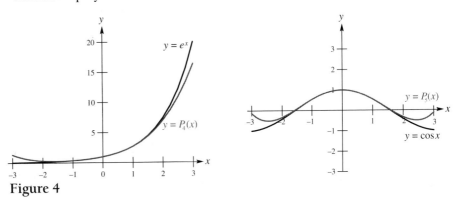

Figure 4

Order versus Degree

We have chosen the terminology Taylor (and Maclaurin) polynomial of *order n* because the highest derivative involved in its construction is of order n. Note that this polynomial can have degree less than n if $f^{(n)}(a) = 0$. If n is odd in Example 3, then the Maclaurin polynomial of order n for $\cos x$ will be of degree $n - 1$. For example, the Maclaurin polynomial of order 5 for $\cos x$ is

$$1 - \tfrac{1}{2}x^2 + \tfrac{1}{24}x^4$$

a polynomial of degree 4. We have sketched the graphs of $P_1(x)$ through $P_5(x)$ and $P_8(x)$ in Figure 5.

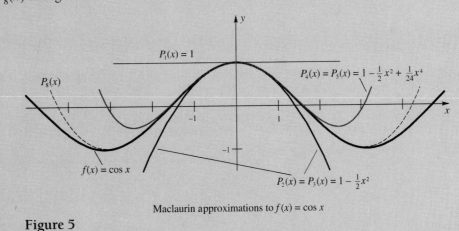

Maclaurin approximations to $f(x) = \cos x$

Figure 5

Error in Numerical Approximation

In Example 3, we used the Maclaurin polynomial of order 4 to approximate $\cos(0.2)$ as follows:

$$\cos(0.2) \approx 1 - \frac{1}{2!}(0.2)^2 + \frac{1}{4!}(0.2)^4 \approx 0.9800667$$

with "first error" marked over the $\frac{1}{2!}(0.2)^2$ term and "second error" marked over the $\frac{1}{4!}(0.2)^4$ term.

This example illustrates the two kinds of errors that occur in approximation processes. First, there is the **error of the method**. In this case, we approximated cos x by a fourth-degree polynomial. Second, there is the **error of calculation**. This includes errors due to round-off, as when we replaced the unending decimal 0.0000666... by 0.0000667 in the last term above. It may also include errors due to the characteristics of the calculating device we are using. For example, the exponentiation key tends to be inaccurate on some hand-held calculators.

Now notice a sad fact of the numerical analyst's life. We can reduce the error of the method by taking more terms in the Maclaurin expansion. But taking more terms means more calculations, which potentially increases the error of calculation. To be a good numerical analyst is to know how to compromise between these two types of error. That unfortunately is more of an art than a science. However, we can say something definite about the first type of error, the subject to which we now turn.

The Error in Taylor's Approximation   For the problem of approximating a function by its Taylor polynomial, we can actually give a formula for the error. This formula is due to the French-Italian mathematician Joseph-Louis Lagrange (1736–1813).

Theorem A

(Taylor's Formula with Remainder). Let f be a function whose $(n + 1)$th derivative, $f^{(n+1)}(x)$, exists for each x in an open interval I containing a. Then for each x in I,

$$f(x) = f(a) + f'(a)(x - a) + \frac{f''(a)}{2!}(x - a)^2 + \cdots + \frac{f^{(n)}(a)}{n!}(x - a)^n + R_n(x)$$

where the remainder (or error) $R_n(x)$ is given by the formula

$$R_n(x) = \frac{f^{(n+1)}(c)}{(n + 1)!}(x - a)^{n+1}$$

where c is some point between x and a.

We postpone the proof until later in the section, choosing to begin with an illustration. The case $a = 0$ occurs most often in practice, and Taylor's Formula is then called Maclaurin's Formula.

EXAMPLE 4: Approximate $e^{0.8}$ with an error of less than 0.001 using a Maclaurin polynomial and Taylor's formula with remainder.

SOLUTION: For $f(x) = e^x$, Maclaurin's Formula gives the remainder

$$R_n(x) = \frac{f^{(n+1)}(c)}{(n + 1)!}x^{n+1} = \frac{e^c}{(n + 1)!}x^{n+1}$$

and so

$$R_n(0.8) = \frac{e^c}{(n + 1)!}(0.8)^{n+1}$$

where $0 < c < 0.8$. Our goal is to choose n large enough so $|R_n(0.8)| < 0.001$. Now, $e^c < e^{0.8} < 3$ (because e^x is an increasing function) and $(0.8)^{n+1} < (1)^{n+1}$, and so

$$|R_n(0.8)| < \frac{3(1)^{n+1}}{(n + 1)!} = \frac{3}{(n + 1)!}$$

It is easy to check that $3/(n + 1)! < 0.001$ when $n \geq 6$, and so we can obtain the desired accuracy by using the Maclaurin polynomial of order 6.

$$e^{0.8} \approx 1 + (0.8) + \frac{(0.8)^2}{2!} + \frac{(0.8)^3}{3!} + \frac{(0.8)^4}{4!} + \frac{(0.8)^5}{5!} + \frac{(0.8)^6}{6!}$$

Our calculator gives 2.2254948.

Can we be sure that this answer is within 0.001 of the true result? Certainly the error of the method is less than 0.001. But could the error of calculation have distorted our answer? Possibly so; however, so few calculations are involved that we feel confident in reporting an answer of 2.2255 accurate within 0.001. ◀

Note: When using Taylor's formula with remainder to make approximations, we need to choose a particular value for c in $f^{(n+1)}(c)$. In the above example we choose the right end-point of the interval for c because for an increasing function, that is where the function is largest. For the trigonometric functions sine and cosine we can use the fact that $|\sin x| \leq 1$ and $|\cos x| \leq 1$ for all x to replace $\sin c$ or $\cos c$ with 1.

Proof of Taylor's Formula

Recall that $R_n(x)$ is defined on I by

$$R_n(x) = f(x) - f(a) - f'(a)(x - a) - \frac{f''(a)(x - a)^2}{2!} - \cdots$$

$$- \frac{f^{(n)}(a)}{n!}(x - a)^n$$

Now think of x as a constant and define a new function g on I by

$$g(t) = f(x) - f(t) - f'(t)(x - t) - \frac{f''(t)(x - t)^2}{2!} - \cdots$$

$$- \frac{f^{(n)}(t)}{n!}(x - t)^n - R_n(x)\frac{(x - t)^{n+1}}{(x - a)^{n+1}}$$

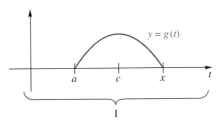

Clearly $g(x) = 0$, because the $(x - t)$ terms all become $(x - x)$. Using the definition of $R_n(x)$, it is almost as easy to see that $g(a) = 0$ (Figure 6). By the Mean Value Theorem for Derivatives applied to g (on the interval with endpoints a and x), there is a point c between a and x such that $g'(c) = 0$. Now we differentiate the expression for $g(t)$ with respect to t (keeping x fixed), and we find that most of the terms cancel. The result is

Figure 6

$$g'(t) = -\frac{f^{(n+1)}(t)}{n!}(x - t)^n + R_n(x)(n + 1)\frac{(x - t)^n}{(x - a)^{n+1}}$$

When we set $g'(c) = 0$, we obtain Taylor's Formula for $R_n(x)$.

Concepts Review

1. If $P_2(x)$ is the Taylor polynomial of order 2 based at 1 for $f(x)$, then $P_2(1) = $ _____, $P_2'(1) = $ _____, and $P_2''(1) = $ _____.

2. The Maclaurin polynomial of order n is just a special name for the Taylor polynomial of order n based at _____ .

3. The Maclaurin polynomial of order 4 for e^x is _____ .

4. The remainder $R_n(x)$ in Taylor's Formula has the form _____ .

Problem Set 9.4

In Problems 1–8, find the Maclaurin polynomial of order 4 for $f(x)$ and use it to approximate $f(0.23)$. Also, sketch each function along with its Maclaurin approximation of order 4 on an interval that shows both where the approximation is good and where it is not (see Example 3).

1. $f(x) = e^{2x}$
2. $f(x) = e^{-3x}$
3. $f(x) = \sin 2x$
4. $f(x) = \tan x$
5. $f(x) = \ln(1 + x)$
6. $f(x) = \sqrt{1 + x}$
7. $f(x) = \tan^{-1} x$
8. $f(x) = \sinh x$

In Problems 9–14, find the Taylor polynomial of order 3 based at a for the given function. Also, sketch each function along with its Taylor approximation on an interval that shows both where the approximation is good and where it is not.

9. e^x; $a = 2$
10. $\sin x$; $a = \dfrac{\pi}{6}$
11. $\tan x$; $a = \dfrac{\pi}{4}$
12. $\sec x$; $a = \dfrac{\pi}{6}$
13. $\tan^{-1} x$; $a = 1$
14. $\sqrt{x}$; $a = 1$

In Problems 15–20, approximate the given quantity with an error of less than 0.001 using a Maclaurin polynomial and Taylor's formula with remainder (see Example 4).

15. e^1
16. e^3
17. $\cos 0.5$
18. $\cos 1$
19. $\sin 1$
20. $\sin 3$

21. Find the Taylor polynomial of order 3 based at 1 for $f(x) = x^3 - 2x^2 + 3x + 5$ and show that it is an exact representation of $f(x)$.

22. Find the Taylor polynomial of order 4 based at 2 for $f(x) = x^4$ and show that it represents $f(x)$ exactly.

23. Find the Maclaurin polynomial of order n (n odd) for $\sin x$. Then use it with $n = 5$ to approximate each of the following.

(a) $\sin(0.1)$ (b) $\sin(0.5)$
(c) $\sin(1)$ (d) $\sin(10)$

Find the error involved in each case. Use graphs to show why the approximation works well in some cases and not in others.

24. Use a Maclaurin polynomial to obtain the approximation $A \approx r^2 t^3 / 12$ for the area of the shaded region in Figure 7. First express A exactly; then approximate.

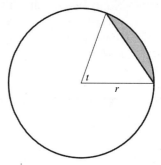

Figure 7

25. If an object of rest mass m_0 has velocity v, then (according to the theory of relativity) its mass m is given by $m = m_0 / \sqrt{1 - v^2/c^2}$, where c is the velocity of light. Show how physicists get the approximation

$$m \approx m_0 + \frac{m_0}{2}\left(\frac{v}{c}\right)^2$$

26. If money is invested at interest rate r compounded monthly, it will double in n years, where n satisfies

$$\left(1 + \frac{r}{12}\right)^{12n} = 2$$

(a) Show that

$$n = \ln 2\left[\frac{1}{12 \ln (1 + r/12)}\right]$$

(b) Use the Maclaurin polynomial of order 2 for $\ln(1 + x)$ and a partial fraction decomposition to obtain the approximation

$$n \approx \frac{0.693}{r} + 0.029$$

(c) Some people use the *rule of 70*, $n \approx 70/100r$, to approximate n. Fill in the table to compare the values obtained from these three formulas:

r	n (exact)	n (approx.)	n (rule 70)
0.05			
0.10			
0.15			
0.20			

27. The author of a biology text claimed that the smallest positive solution to $x = 1 - e^{-(1+k)x}$ is approximately $x = 2k$, provided k is very small. Show how she reached this conclusion and check on it for $k = 0.01$.

28. Show that if x is in $[0, \pi/2]$, the error in using

$$\sin x \approx x - \frac{x^3}{3!} + \frac{x^5}{5!} - \frac{x^7}{7!} + \frac{x^9}{9!}$$

is less than 5×10^{-6} and, therefore, that this formula is good enough to build a five-place sine table.

29. Use Maclaurin's Formula rather than l'Hôpital's Rule to find

(a) $\displaystyle\lim_{x \to 0} \frac{\sin x - x + x^3/6}{x^5}$

(b) $\displaystyle\lim_{x \to 0} \frac{\cos x - 1 + x^2/2 - x^4/24}{x^6}$

30. Let $g(x) = p(x) + x^{n+1}f(x)$, where $p(x)$ is a polynomial of degree at most n and f has derivatives through order n. Show that $p(x)$ is the Maclaurin polynomial of order n for g.

31. Note that the fourth-order Maclaurin polynomial for $\sin x$ is really of third degree since the coefficient of x^4 is 0. Thus,

$$\sin x = x - \frac{x^3}{6} + R_4(x)$$

Show that if $0 \le x \le 0.5$, $|R_4(x)| \le 0.0002605$. Use this result to approximate $\displaystyle\int_0^{0.5} \sin x \, dx$ and give a bound for the error.

Answers to Concepts Review: 1. $f(1); f'(1); f''(1)$ 2. 0
3. $1 + x + \frac{1}{2}x^2 + \frac{1}{6}x^3 + \frac{1}{24}x^4$
4. $f^{(n+1)}(c)(x-a)^{n+1}/(n+1)!$

LAB 19: TAYLOR SERIES AND FOURIER SERIES

Mathematical Background

From ancient times people have speculated whether nature (time, the universe) is fundamentally cyclical or progressive. Temperature seems to follow a cyclical course, repeating the cycle every year. Population seems to follow a progressive course, growing larger every year. The trigonometric functions, $\sin x$ and $\cos x$, have been used to describe cyclical phenomena such as the temperature cycle. Polynomials, $a + bx + cx^2 + dx^3 + \cdots$, are often used to describe quantities that grow progressively larger or smaller, such as the population or the amount of food produced in the world. Thus, the trig functions seem to have a fundamentally different nature than the polynomials. Quite amazingly, on a closed interval each can be made to look like the other!

If $f(x)$ is a function, its Taylor series (centered at $x = 0$) is given by

$$\sum_{k=0}^{\infty} a_n x^n = a_0 + a_1 x + a_2 x^2 + \cdots$$

where

$$\boxed{a_n = \frac{f^{(n)}(0)}{n!}} \tag{1}$$

This is essentially an infinite polynomial. Taylor series are useful in analyzing the behavior of a function near a particular point (in the above case near $x = 0$).

The Fourier sine series of $f(x)$ (on the interval $(0, 1)$) is given by

$$\sum_{n=0}^{\infty} b_n \sin n\pi x = b_1\sin(\pi x) + b_2\sin(2\pi x) + b_3\sin(3\pi x) + \cdots$$

where

$$b_n = 2 \int_0^1 f(x)\sin(n\pi x)\, dx \tag{2}$$

This is an infinite series of sine functions. There are also Fourier series consisting only of cosines, and of both sines and cosines. Fourier series are helpful in analyzing the behavior of a function over an interval.

One of the uses of Fourier series is in the analysis of waveforms. For instance, when a guitar string vibrates, its shape can be analyzed in terms of its Fourier series. The first term corresponds to the "fundamental mode" of vibration, the second term corresponds to the first overtone or harmonic, and so on.

Lab Introduction

We will take a simple combination of trig functions, $\sin \pi x + \cos \pi x$, and find its Taylor series (polynomial) approximations for the first five terms at the point $x = 0$. We want to see where the approximation is good and where it is not, and in what sense the approximation gets better as more terms are added.

We will then reverse the process and start with a polynomial $f(x) = 10x^3 - 14x^2 + 4x$, and find its Fourier sine series approximations for the first five terms. In this case we will assume that the polynomial describes the shape of a guitar string, and we will find which vibrational modes contribute most to its shape. Again, we will see where the approximation is good and where it is not, and in what sense the approximation gets better as more terms are added.

Experiment

(1) Find the first five terms (up through a_4) of the Taylor series (centered at $x = 0$) for $f(x) = \sin \pi x + \cos \pi x$. You could use the built-in Taylor series command of a computer algebra system, or you could calculate the coefficients a_n by hand using (1); find a four-digit approximation to these coefficients.

(2) Graph the function $f(x) = \sin \pi x + \cos \pi x$ on the interval $[-2, 2]$. Now graph its Taylor series, first for one term, then for two terms, on up to five terms. For each case, sketch the Taylor approximation and the function together and estimate from the graph the x-interval on which the function and its approximation agree.

(3) Find the first terms of the Fourier sine series (on the interval $[0, 1]$) for $f(x) = 10x^3 - 14x^2 + 4x$. Use formula (2) to calculate the coefficients b_n; you could do the integrations numerically, or exactly using integration by parts from Section 8.2. Get a three-decimal-place approximation to these coefficients.

(4) Graph each term of the Fourier sine series separately on the interval $[0, 1]$ and sketch the results.

(5) Graph the polynomial in (3) on the interval $[0, 1]$. Now graph its Fourier sine series, first for one term, then for two terms, on up to five terms. For each case, sketch the Fourier approximation and the function together.

(6) Graph and sketch the polynomial in (3) and its five-term Fourier sine approximation on the interval $[-2, 2]$.

Discussion

1. Describe in what sense the approximation in (2) from the experiment section gets better when each term is added. For each case tell the x-values for which the approximation seems to work.

2. Describe in what sense the approximation in (5) from the experiment section gets better when each term is added. For each case tell the x-values for which the approximation seems to work.

3. Which term of the Fourier approximation looks most like the original polynomial? How does this show up in the five-term Fourier approximation?

4. Compare and contrast Taylor series approximations with Fourier series approximations. Discuss how, and in what sense, each approaches the function it is approximating.

9.5 | Applications of Taylor's Formula

Taylor's Formula with Remainder, Theorem A of the last section, forms the basis of much of the subject called numerical analysis. We can use it to gain a deeper understanding of the error involved in some of the numerical techniques we have used in this book; we can also use it to develop and analyze new techniques.

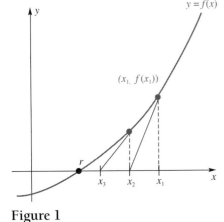

Figure 1

Newton's Method   Consider the problem of solving the equation $f(x) = 0$ for a root r. Suppose that f is differentiable, so that the graph of $y = f(x)$ has a tangent line at each point. If we can find a first approximation x_1 to r by graphing or any other means, then a better approximation x_2 ought to lie at the intersection of the tangent at $(x_1, f(x_1))$ with the x-axis (see Figure 1). Using x_2 as an approximation, we can then find a still better approximation x_3, and so on. This method was explained geometrically in Section 2.5, but no algorithm was given at that time for actually finding roots.

The process can be mechanized so that it is easy to do on a calculator. The equation of the tangent at $(x_1, f(x_1))$ is

$$y - f(x_1) = f'(x_1)(x - x_1)$$

and its x-intercept x_2 is found by setting $y = 0$ and solving for x. The result is

$$x_2 = x_1 - \frac{f(x_1)}{f'(x_1)}$$

More generally, we have the following algorithm, given as a recursion formula or discrete dynamical system (see Section 9.1).

$$x_{n+1} = x_n - \frac{f(x_n)}{f'(x_n)}$$

EXAMPLE 1: Use Newton's Method to find the real root r of $f(x) = x^3 - 3x - 5 = 0$ to seven decimal places.

SOLUTION: We graph the function $f(x)$ first and look for intersections with the x-axis; see Figure 2. Based on the graph, let's use $x_1 = 2.5$ as our first approximation to r. Because $f(x) = x^3 - 3x - 5$ and $f'(x) = 3x^2 - 3$, the algorithm is

$$x_{n+1} = x_n - \frac{x_n^3 - 3x_n - 5}{3x_n^2 - 3} = \frac{2x_n^3 + 5}{3x_n^2 - 3}$$

We obtain the data in the following table.

n	x_n
1	2.5
2	2.3015873
3	2.2792907
4	2.2790188
5	2.2790188

After just four steps, we get a repetition of the first eight digits. We feel confident in reporting that $r \approx 2.2790188$, with perhaps some question about the last digit. ◀

Convergence of Newton's Method It is not always obvious that Newton's Method yields approximations that converge to the root r, though our example gives evidence for that assertion. Could it be that we were just lucky in our choice of example? As a matter of fact, the method does not always lead to convergence, as the diagram in Figure 3 shows (see also Problem 18). In this case, the difficulty is that x_1 is not close enough to r to get a convergent process started. Another obvious difficulty arises if $f'(x)$ is zero at or near r, because $f'(x_n)$ occurs in the denominator of the algorithm. However, we have the following theorem.

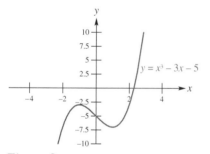

$y = x^3 - 3x - 5$

Figure 2

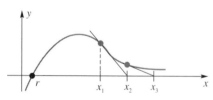

Figure 3

Theorem A

Let f be twice differentiable on an interval I, having as its midpoint a root r of $f(x) = 0$. Suppose there are positive numbers m and M such that $|f'(x)| \geq m$ and $|f''(x)| \leq M$ on I. If x_1 is in I and sufficiently close to r (the distance $|x_1 - r| < 2m/M$ will do), then

(i) $|x_{n+1} - r| \leq \dfrac{M}{2m}(x_n - r)^2$;

(ii) x_n converges to r as $n \to \infty$.

Proof From Taylor's Formula with Remainder, there is a number c between x_n and r such that

$$f(r) = f(x_n) + f'(x_n)(r - x_n) + \frac{f''(c)}{2}(r - x_n)^2$$

After dividing both sides by $f'(x_n)$ and using the fact that $f(r) = 0$, we obtain

$$0 = \frac{f(x_n)}{f'(x_n)} + r - x_n + \frac{f''(c)}{2f'(x_n)}(r - x_n)^2$$

$$x_n - \frac{f(x_n)}{f'(x_n)} - r = \frac{f''(c)}{2f'(x_n)}(r - x_n)^2$$

and, because $x_{n+1} = x_n - \dfrac{f(x_n)}{f'(x_n)}$,

$$\left| x_{n+1} - r \right| = \left| \frac{f''(c)}{2f'(x_n)} \right| (x_n - r)^2$$

$$\left| x_{n+1} - r \right| \leq \frac{M}{2m}(x_n - r)^2$$

which is (i).

Notice that we have

$$\left| x_n - r \right| \leq \frac{M}{2m}\left| x_{n-1} - r \right|^2 \leq \left(\frac{M}{2m} \right)^2 \left| x_{n-1} - r \right|^3 \leq \left(\frac{M}{2m} \right)^3 \left| x_{n-1} - r \right|^4 \leq \cdots$$

so that

$$\left| x_n - r \right| \leq \frac{2m}{M}\left(\frac{M}{2m}\left| x_1 - r \right| \right)^{2^{n-1}}$$

Because $(M/2m)\left| x_1 - r \right| < 1$, the right-hand side of the last inequality approaches 0 as $n \to \infty$. This implies that $\left| x_n - r \right|$ also tends to 0 as $n \to \infty$, which is equivalent to (ii). ◄

The speed of convergence of Newton's Method is truly remarkable, tending in fact to double the number of decimal places of accuracy at each step. To see why this is so, suppose $M/2m \leq 2$. Then if the error $\left| x_n - r \right|$ at the nth step is less than 0.005, the error $\left| x_{n+1} - r \right|$ at the next step satisfies (by (i))

$$\left| x_{n+1} - r \right| \leq \frac{M}{2m}\left| x_n - r \right|^2 \leq 2(0.005)^2 = 0.00005$$

Thus, the accuracy of x_n to two decimal places is doubled to an accuracy of x_{n+1} to four decimal places. Of course, we should not expect quite such spectacular results if $M/2m$ is substantially greater than 2.

Numerical Integration and Error

We know that if f is continuous on a closed interval $[a, b]$, then the definite integral $\displaystyle\int_a^b f(x)\,dx$ must exist. Existence is one thing; evaluation is a very different matter. As we have seen, there are many definite integrals that cannot be evaluated by exact methods—that is, by use of the Fundamental Theorem of Calculus. Even when elementary indefinite integrals can be found, it is often advantageous to use approximation methods, because they are well suited to use on a calculator or computer.

The Error in Riemann Sum Approximations

Back in Section 5.4 we made the claim that left and right Riemann sum approximations were first-order-convergent methods of numerical integration, and that midpoint Riemann sum approximation was a second-order-convergent method of numerical integration. Using Taylor's Theorem with Remainder we can now show why. We consider the case of midpoint approximations below.

Theorem B

If f'' is continuous on $[a, b]$, then there exists a number C for which the error in using a midpoint Riemann sum approximation $\sum_1^n f(\overline{x}_i)\Delta x$ to $\int_a^b f(x)\,dx$ with n subintervals is less than $\dfrac{C(b-a)^3}{n^2}$.

The key to this theorem is the n^2 in the denominator of the error bound. This term is what says that the method is a second-order-convergent method.

Proof   Because f'' is continuous on $[a, b]$ we know that it is bounded there, and so there exists a constant M for which $|f''(x)| \le M$ on $[a, b]$. Therefore

$$\left| \sum_1^n f(\overline{x}_i)\Delta x - \int_a^b f(x)\,dx \right| = \left| \sum_1^n f(\overline{x}_i)\Delta x - \sum_1^n \int_{x_{i-1}}^{x_i} f(x)\,dx \right|$$

(by the additive property of integration)

$$= \left| \sum_1^n \left(f(\overline{x}_i)\Delta x - \int_{x_{i-1}}^{x_i} f(x)\,dx \right) \right|$$

$$\le \sum_1^n \left| f(\overline{x}_i)\Delta x - \int_{x_{i-1}}^{x_i} f(x)\,dx \right|$$

(by the triangle inequality)

$$= \sum_1^n \left| \int_{x_{i-1}}^{x_i} f(\overline{x}_i)\,dx - \int_{x_{i-1}}^{x_i} f(x)\,dx \right|$$

(because $f(\overline{x}_i)$ is a constant)

$$= \sum_1^n \left| \int_{x_{i-1}}^{x_i} f(\overline{x}_i) - f(x)\,dx \right|$$

(linear property of integration)

$$= \sum_1^n \left| \int_{x_{i-1}}^{x_i} f'(\overline{x}_i)(\overline{x}_i - x) + \frac{1}{2}f''(c_i)(\overline{x}_i - x)^2\,dx \right|$$

(using Taylor's formula with remainder)

$$= \sum_1^n \left| \int_{x_{i-1}}^{x_i} \frac{1}{2}f''(c_i)(\overline{x}_i - x)^2\,dx \right|$$

(because $\displaystyle\int_{x_{i-1}}^{x_i} f'(\overline{x}_i)(\overline{x}_i - x)\,dx = 0$; see the problem set)

$$\leq \sum_{1}^{n} \int_{x_{i-1}}^{x_i} \left| \frac{1}{2} f''(c_i)(\overline{x}_i - x)^2 \right| dx$$

$$\leq \sum_{1}^{n} \int_{x_{i-1}}^{x_i} M \left| \frac{1}{2}(\overline{x}_i - x)^2 \right| dx$$

(because $\left| f''(c_i) \right| \leq M$)

$$= M \sum_{1}^{n} \left| \int_{x_{i-1}}^{x_i} (\overline{x}_i - x)^2\,dx + \int_{\overline{x}_i}^{x_i} (\overline{x}_i - x)^2\,dx \right|$$

(by the additive property of integration)

$$= M \sum_{1}^{n} \left| \frac{1}{3}(\overline{x}_i - x)^3 \Big|_{x_{i-1}}^{x_i} + \frac{1}{3}(\overline{x}_i - x)^3 \Big|_{\overline{x}_i}^{x_i} \right|$$

$$= M \sum_{1}^{n} \left| -\frac{1}{3}\left(\frac{x_i + x_{i-1}}{2} - x_{i-1}\right)^3 + \frac{1}{3}\left(\frac{x_i + x_{i-1}}{2} - x_i\right)^3 \right|$$

(using $\overline{x}_i = \dfrac{x_i + x_{i-1}}{2}$)

$$= M \sum_{1}^{n} \left| \frac{1}{3}\left(\frac{x_{i-1} - x_i}{2}\right)^3 + \frac{1}{3}\left(\frac{x_{i-1} - x_i}{2}\right)^3 \right|$$

$$= \sum_{1}^{n} \left| M \frac{1}{12}(x_{i-1} - x_i)^3 \right| = \frac{M}{12} n \Delta x^3 = \frac{M}{12} n \frac{(b-a)^3}{n^3}$$

(because $\Delta x = \dfrac{b-a}{n}$)

$$= \frac{M}{12} \frac{(b-a)^3}{n^2}$$

Letting $C = \dfrac{M}{12}$ we are done.   ◄

The Parabolic Rule (Simpson's Rule)

Using midpoint Riemann sums, we approximated the curve $y = f(x)$ by horizontal line segments. If we approximate the curve with straight-line segments that are not horizontal, we get a numerical integration method called the *Trapezoidal Rule*, but it turns out that this method does not converge any faster than the midpoint method. Perhaps we could do better using parabolic segments—that is, pieces of parabolas designed to pass through points on the curve. Just as before, partition the interval $[a, b]$ into n subintervals of length $h = (b - a)/n$, but this time with n an *even* number. Then fit parabolic segments to neighboring triples of points, as shown in Figure 4.

Using the area formula in Figure 5 (see Problem 20 for the derivation) leads to an approximation caled the **Parabolic Rule**. It is also called **Simpson's Rule** after the English mathematician Thomas Simpson (1710–1761).

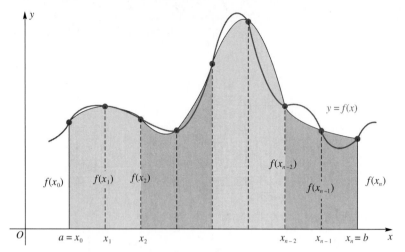

Figure 4

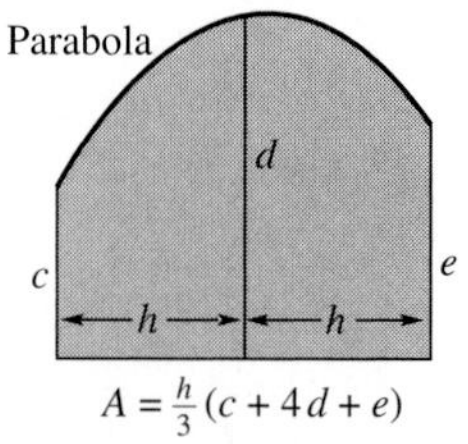

Figure 5

Parabolic Rule (n even)

$$\int_a^b f(x)\,dx \approx \frac{h}{3}\big[f(x_0) + 4f(x_1) + 2f(x_2) + \cdots + 4f(x_{n-1}) + f(x_n)\big]$$

The pattern of coefficients is 1, 4, 2, 4, 2, 4, 2, ..., 2, 4, 1.

EXAMPLE 2: Use the Parabolic Rule with $n = 8$ and $n = 16$ to approximate

$$\int_1^3 x^4 \, dx$$

SOLUTION: We have

$$x_0 = 1.000 \qquad f(x_0) = (1.000)^4 = 1.0000$$
$$x_1 = 1.250 \qquad f(x_1) = (1.250)^4 \approx 2.4414$$
$$x_2 = 1.500 \qquad f(x_2) = (1.500)^4 = 5.0625$$
$$x_3 = 1.750 \qquad f(x_3) = (1.750)^4 \approx 9.3789$$
$$x_4 = 2.000 \qquad f(x_4) = (2.000)^4 = 16.0000$$
$$x_5 = 2.250 \qquad f(x_5) = (2.250)^4 \approx 25.6289$$
$$x_6 = 2.500 \qquad f(x_6) = (2.500)^4 = 39.0625$$
$$x_7 = 2.750 \qquad f(x_7) = (2.750)^4 \approx 57.1914$$
$$x_8 = 3.000 \qquad f(x_8) = (3.000)^4 = 81.0000$$

so that

$$\int_1^3 x^4 \, dx \approx \frac{0.25}{3} [1.0000 + 4(2.4414) + 2(5.0625) + 4(9.3789)$$

$$+ 2(16.0000) + 4(25.6289) + 2(39.0625) + 4(57.1914)$$

$$+ 81.0000]$$

$$\approx 48.4010$$

Similarly, with $n = 16$ we get

$$x_0 = 1.000 \quad f(x_0) = (1.000)^4 = 1.0000$$
$$x_1 = 1.125 \quad f(x_1) = (1.125)^4 \approx 1.6018$$
$$x_2 = 1.250 \quad f(x_2) = (1.250)^4 \approx 2.4414$$
$$x_3 = 1.375 \quad f(x_3) = (1.375)^4 \approx 3.5745$$
$$x_4 = 1.500 \quad f(x_4) = (1.500)^4 = 5.0625$$
$$x_5 = 1.625 \quad f(x_5) = (1.625)^4 \approx 6.9729$$
$$x_6 = 1.750 \quad f(x_6) = (1.750)^4 \approx 9.3789$$
$$x_7 = 1.875 \quad f(x_7) = (1.875)^4 \approx 12.3596$$
$$x_8 = 2.000 \quad f(x_8) = (2.000)^4 = 16.0000$$
$$x_9 = 2.125 \quad f(x_9) = (2.125)^4 \approx 20.3909$$
$$x_{10} = 2.250 \quad f(x_{10}) = (2.250)^4 \approx 25.6289$$
$$x_{11} = 2.375 \quad f(x_{11}) = (2.375)^4 \approx 31.8167$$
$$x_{12} = 2.500 \quad f(x_{12}) = (2.500)^4 = 39.0625$$
$$x_{13} = 2.625 \quad f(x_{13}) = (2.625)^4 \approx 47.4807$$
$$x_{14} = 2.750 \quad f(x_{14}) = (2.750)^4 \approx 57.1914$$
$$x_{15} = 2.875 \quad f(x_{15}) = (2.875)^4 \approx 68.3206$$
$$x_{16} = 3.000 \quad f(x_{16}) = (3.000)^4 = 81.0000$$

Thus,

$$\int_1^3 x^4 \, dx \approx \frac{0.125}{3} [1.0000 + 4(1.6018) + 2(2.4414)$$

$$+ 4(3.5745) + 2(5.0625) + 4(6.9729)$$

$$+ 2(9.3789) + 4(12.3596) + 2(16.0000)$$

$$+ 4(20.3909) + 2(25.6289) + 4(31.8167)$$

$$+ 2(39.0625) + 4(47.4807) + 2(57.1914)$$

$$+ 4(68.3206) + 81.0000]$$

$$\approx 48.40008$$

◄

Because it requires so little extra work, it is usually more efficient to use the Parabolic Rule rather than the Midpoint Rule when writing computer or calculator programs. That its error term is generally smaller is borne out by the following theorem (note the factor of n^4 in the denominator). For a proof, see J. M. H. Olmsted, *Advanced Calculus* (New York: Prentice Hall, 1961), pp. 118–19.

Theorem C

Suppose that the fourth derivative $f^{(4)}(x)$ exists on $[a, b]$. Then the error E_n in the Parabolic Rule is given by

$$E_n = -\frac{(b-a)^5}{180n^4}f^{(4)}(c)$$

for some c between a and b.

Again, the important part of this theorem is the n^4 in the denominator of the error term. This says that Simpson's Method is a fourth-order-convergent method. Thus, if the number of intervals used in the approximation is doubled, the error is reduced by about $\frac{1}{16}$, or more than one decimal place. If the number of intervals is increased by a factor of 10, the error is reduced by a factor of $\frac{1}{10,000}$, or four additional decimal places.

Computers and Integration

Computer packages that claim to evaluate definite integrals do so by some form of numerical integration. They may use one of the rules discussed in this section or they may use something more sophisticated. In any case, they use approximation methods, so you should not expect the answers they give to be exact.

EXAMPLE 3: Find the error in using the Parabolic Rule with $n = 8$ and $n = 16$ to approximate

$$\int_1^3 x^4 \, dx$$

Discuss the rate of convergence for the Parabolic Rule used with this integral.

SOLUTION: The exact value of the integral is easily calculated to be 48.4. In Example 2 we calculated the parabolic approximations with $n = 8$ and $n = 16$ and got 48.4010 and 48.40008, respectively. Thus, the errors are 0.001 and 0.00008. We have increased our accuracy by better than one decimal place, as we expect with the Parabolic Rule. With a fourth-order-convergent method, we expect that by doubling the number of intervals we should reduce the error by about $\frac{1}{16}$, which is more than one decimal place. The actual reduction in error is slightly less than $\frac{1}{16}$ in this case, but would get closer to $\frac{1}{16}$ if we continued to double the value of n. ◄

Concepts Review

1. Newton's Method can fail to yield a root of $f(x) = 0$. This can happen if _____ is too far from the root r or if _____ .

2. We can use Taylor's Formula with remainder to show that the number of decimals of accuracy in Newton's Method tends to _____ with each step.

3. The pattern of coefficients in the Parabolic Rule is _____ .

4. The error formula in using midpoint Riemann sums has n^2 in the denominator, whereas the error in the Parabolic Rule has _____ in the denominator, so we expect the latter to give a better approximation to a definite integral.

Problem Set 9.5

In Problems 1–10, use Newton's Method to approximate the indicated root of the given equation accurate to ten decimal places. Begin by sketching a graph. Discuss the rate of convergence of Newton's Method by looking at the number of accurate digits at each step.

1. The largest root of $x^3 + 6x^2 + 9x + 2 = 0$.

2. The real root of $7x^3 + x - 6 = 0$.

3. The root of $x - 2 + \ln x = 0$.

4. The smallest positive root of $\cos x - e^{-x} = 0$.

5. The root of $\cos x = x$.

6. The root of $x \ln x = 1$.

7. All real roots of $x^4 - 8x^3 + 22x^2 - 24x + 6 = 0$.

8. All real roots of $x^4 + x^3 - 3x^2 + 4x - 28 = 0$.

9. The positive root of $2x^2 - \sin^{-1}x = 0$.

10. The positive root of $2 \tan^{-1}x = x$.

In Problems 11–14, use the Parabolic Rule with $n = 8$ and $n = 16$ to approximate each integral. Then calculate the integral using the Fundamental Theorem of Calculus and find the error in using the Parabolic Rule. Discuss the rate of convergence for the Parabolic Rule for each integral.

11. $\displaystyle\int_1^2 \frac{1}{x^2}\, dx$

12. $\displaystyle\int_1^2 \frac{1}{x}\, dx$

13. $\displaystyle\int_0^4 \sqrt{x}\, dx$

14. $\displaystyle\int_0^2 x\sqrt{x^2 + 4}\, dx$

15. Show that $\displaystyle\int_{x_{i-1}}^{x_i} f'(\overline{x}_i)(\overline{x}_i - x)\,dx = 0$ where $\overline{x}_i = \dfrac{x_i + x_{i-1}}{2}$ is the midpoint of the interval $[x_{i-1}, x_i]$ (see the proof to Theorem B).

16. Approximate π by calculating

$$\int_0^1 \frac{4}{1 + x^2}\, dx$$

using the Parabolic Rule with $n = 10$.

17. Consider finding the real root of $(1 + \ln x)/x = 0$ by Newton's Method. Show that this leads to the algorithm

$$x_{n+1} = 2x_n + \frac{x_n}{\ln x_n}$$

Apply this algorithm with $x_1 = 1.2$. Next try it with $x_1 = 0.5$. Finally, graph $y = (1 + \ln x)/x$ to understand your results.

18. Sketch the graph of $y = x^{1/3}$. Obviously its only x-intercept is zero. Convince yourself that Newton's Method fails to converge. Explain this failure.

19. In installment buying, one would like to figure out the real interest rate (effective rate), but unfortunately this involves solving a complicated equation. If one buys an item worth $\$P$ today and agrees to pay for it with payments of $\$R$ at the end of each month for k months, then

$$P = \frac{R}{i}\left[1 - \frac{1}{(1 + i)^k}\right]$$

where i is the interest rate per month. Tom bought a used car for $\$2000$ and agreed to pay for it with $\$100$ payments at the end of each of the next 24 months.

(a) Show that i satisfies the equation

$$20i(1 + i)^{24} - (1 + i)^{24} + 1 = 0$$

(b) Derive Newton's Algorithm for this equation, namely,

$$i_{n+1} = i_n - \left[\frac{20i_n^2 + 19i_n - 1 + (1 + i_n)^{-23}}{500i_n - 4}\right]$$

(c) Find i accurate to five decimal places starting with $i_1 = 0.012$ and then give the annual rate r as a percent ($r = 1200i$).

20. Let $f(x) = ax^2 + bx + c$. Show that

$$\int_{m-h}^{m+h} f(x)dx \text{ and } (h/3)[f(m-h) + 4f(m) + f(m+h)]$$

both have the value $(h/3)[a(6m^2 + 2h^2) + b(6m) + 6c]$. This establishes the area formula on which the Parabolic Rule is based.

21. Show that the Parabolic Rule is exact for any cubic polynomial in two different ways.

(a) By direct calculation.

(b) By showing that $E_n = 0$.

Answers to Concepts Review: 1. $x_1; f'(r) = 0$ 2. Double 3. 1, 4, 2, 4, 2, …, 4, 1 4. n^4

LAB 20: NEWTON'S METHOD

Mathematical Background

Newton's Method is a way of approximating the roots (zeros) of a function. Thus, it can be used to solve any equation with one unknown. The tangent line to a curve $f(x)$ at the point $x = x_0$ can be written

$$y = f(x_0) + f'(x_0)(x - x_0)$$

If we let $y = 0$ and solve for x, we get

$$x = x_0 - \frac{f(x_0)}{f'(x_0)}$$

This represents the x-coordinate of the point where the tangent hits the x-axis. Thus, it is an approximation of the root of $f(x)$. Call this point x_1. If we now use the tangent at $x = x_1$ to approximate the root of $f(x)$, we get another estimate; call it x_2. Using the reasoning above, the formula for x_2 would be

$$x_2 = x_1 - \frac{f(x_1)}{f'(x_1)}$$

If we keep repeating this process, we are *iterating* the function

$$g(x) = x - \frac{f(x)}{f'(x)} \tag{1}$$

which just means that $x_{n+1} = g(x_n)$. This gives us a sequence $x_0, x_1, x_2, \ldots$, which should get closer and closer to the root of $f(x)$.

Lab Introduction

In this lab, you are going to explore how fast the sequence above approaches the root and how the function could cause the observed behavior. There are also times when Newton's Method fails to converge. That is explored in the fourth and fifth parts of the experiment.

Experiment

(1) Graph the cubic polynomial $f(x) = x^3 + 3x^2 - 4$ with $-3 \le x \le 3$, and determine the roots of $f(x)$ by inspecting the graph and checking your answer.

(2) Use Newton's Method (that is, iterate formula (1)) and a calculator or computer algebra system to find the roots of $f(x)$. For starting values, use $x = -3, -1, 0, 1, 2, 3$ and record the results to five digits. For each starting value get ten iterations if

possible, write down the resulting sequence, and note which starting values go to which roots.

(3) For the starting values $x = -3$ and $x = 3$ record the results to ten digits and determine the number of digits of accuracy of *each iteration* by comparing with the exact root obtained in (1).

(4) Use Newton's Method to find the roots of the cubic polynomial $f(x) = x^3 - 0.64x - 0.36$ with the starting value of $x = 0$. Write out the results to three digits and iterate 20 times.

(5) Use Newton's Method to find the roots of the quadratic $f(x) = x^2 + x + 3$ using the starting values $x = 2, -1.457$. Write out the results to three digits, and iterate 30 times.

Discussion

1. For the first cubic:
   - (a) Explain why each starting value went to the root that it did; include a sketch of the function $f(x)$ in your direction.
   - (b) Explain what happened graphically for the case that failed.
   - (c) Can you predict where any starting value will go? Explain.

2. (a) Compare the speed with which the sequence approaches the root for the cases $x = -3$ and $x = 3$ for the first cubic. Look at both the graph of the iterates and the number of digits of accuracy. Now look at the graph of the function $f(x)$ and explain how the picture is different at each root.
   - (b) Predict the speed of convergence for Newton's Method for the functions $\cos(x) - 1$ and $e^x - 1$ and check your predictions. *Hint*: Graph them first.

3. For the second cubic:
   - (a) Explain the behavior you found.
   - (b) Can you show graphically how Newton's Method gives the sequence you found? Sketch the function and the associated tangent lines.
   - (c) Could you get convergence for a different starting value? Explain.

4. First describe, and then explain what happens for the quadratic. Could you get convergence for a different starting value? Explain.

9.6 | General Power Series

So far, we have been studying what might be called *series of constants*—that is, series of the form $\sum u_n$, where each u_n is a number. Now we consider *series of functions*, series of the form $\sum u_n(x)$. A typical example of such a series is

$$\sum_{n=1}^{\infty} \frac{\sin nx}{n^2} = \frac{\sin x}{1} + \frac{\sin 2x}{4} + \frac{\sin 3x}{9} + \cdots$$

Of course, as soon we substitute a value for x (such as $x = 2.1$), we are back to familiar territory; we have a series of constants.

There are two important questions to ask about a series of functions.

1.   For what x's does the series converge?

2.   To what function does it converge; that is, what is the sum $S(x)$ of the series?

The general situation is a proper subject for an advanced calculus course. However, even in elementary calculus, we can learn a good deal about the special case of a power series. **A power series in x** has the form

$$\sum_{n=0}^{\infty} a_n x^n = a_0 + a_1 x + a_2 x^2 + \cdots$$

(Here we interpret $a_0 x^0$ to be a_0 even if $x = 0$.) We can immediately answer our two questions for one such power series.

EXAMPLE 1: For what x's does the power series

$$\sum_{n=0}^{\infty} ax^n = a + ax + ax^2 + ax^3 + \cdots$$

converge, and what is its sum? Assume $a \neq 0$.

SOLUTION: We actually studied this series in Example 1 of Section 9.2 (with r in place of x) and called it a geometric series. It converges for $-1 < x < 1$ and has sum $S(x)$ given by

$$S(x) = \frac{a}{1 - x}, \qquad -1 < x < 1 \qquad \blacktriangleleft$$

Fourier Series

The series of sine functions mentioned in the introduction is an example of a *Fourier series*, named after Jean Baptiste Joseph Fourier (1768–1830). Fourier series are of immense importance in the study of wave phenomena, because they allow us to represent a complicated wave as a sum of its fundamental components (called the pure tones in the case of sound waves.) It is a large field, which we investigated briefly in Lab 19.

The Convergence Set   We call the set on which a power series converges its **convergence set**. What kind of set can be a convergence set? Example 1 shows that it can be an open interval (see Figure 1). Are there other possibilities?

Convergence set

Figure 1

EXAMPLE 2: Find the convergence set for

$$\sum_{n=0}^{\infty} \frac{x^n}{(n + 1)2^n} = 1 + \frac{1}{2} \cdot \frac{x}{2} + \frac{1}{3} \cdot \frac{x^2}{2^2} + \frac{1}{4} \cdot \frac{x^3}{2^3} + \cdots$$

Graph the partial sum polynomials $P_N(x) = \sum_{n=0}^{N} \frac{x^n}{(n + 1)2^n}$ for $N = 1, 2, 3, 4, 5$

on one set of axes and relate these graphs to the convergence set.

SOLUTION: Note that some of the terms may be negative (if x is negative). Let's test for absolute convergence using the Absolute Ratio Test (Theorem 9.3C).

$$p = \lim_{n \to \infty} \left| \frac{x^{n+1}}{(n + 2)2^{n+1}} \div \frac{x^n}{(n + 1)2^n} \right| = \lim_{n \to \infty} \frac{|x|}{2} \cdot \frac{n + 1}{n + 2} = \frac{|x|}{2}$$

The series converges absolutely (hence converges) when $p = |x|/2 > 1$ and diverges when $|x|/2 > 1$. Consequently, it converges when $|x| < 2$ and diverges when $|x| > 2$.

If $x = 2$ or $x = -2$, the Ratio Test fails. However, when $x = 2$, the series is the harmonic series, which diverges; and when $x = -2$, it is the alternating harmonic series, which converges. We conclude that the convergence set for the given series is the interval $-2 \le x < 2$ (Figure 2).

Convergence set

Figure 2

In Figure 3 we graph the polynomials.

$$P_1(x) = 1 + \frac{x}{4}$$

$$P_2(x) = 1 + \frac{x}{4} + \frac{x^2}{12}$$

$$P_3(x) = 1 + \frac{x}{4} + \frac{x^2}{12} + \frac{x^3}{32}$$

$$P_4(x) = 1 + \frac{x}{4} + \frac{x^2}{12} + \frac{x^3}{32} + \frac{x^4}{80}$$

$$P_5(x) = 1 + \frac{x}{4} + \frac{x^2}{12} + \frac{x^3}{32} + \frac{x^4}{80} + \frac{x^5}{192}$$

On the interval $-2 < x < 2$ the graphs are stabilizing (converging), but outside that interval the graphs are rapidly going off to infinity or negative infinity. Thus, one can estimate the convergence set by viewing the graphs of the first few partial sums. *Note*: It is difficult to determine convergence at the endpoints $x = 2$ and $x = -2$ from graphs.   ◄

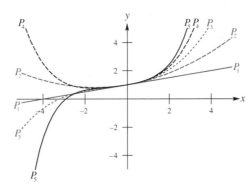

Figure 3

EXAMPLE 3: Find the convergence set for $\displaystyle\sum_{n=0}^{\infty} \frac{x^n}{n!}$.

SOLUTION:

$$p = \lim_{n \to \infty} \left| \frac{x^{n+1}}{(n+1)!} \div \frac{x^n}{n!} \right| = \lim_{n \to \infty} \frac{|x|}{n+1} = 0$$

We conclude from the Absolute Ratio Test that the series converges for all x (Figure 4). ◄

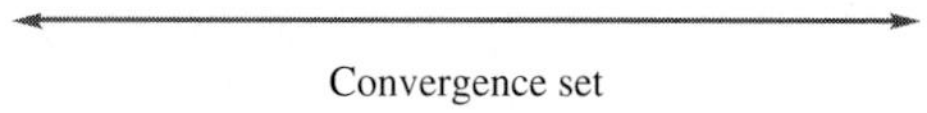

Convergence set

Figure 4

EXAMPLE 4: Find the convergence set for $\displaystyle\sum_{n=0}^{\infty} n!x^n$.

SOLUTION:

$$p = \lim_{n\to\infty} \left| \frac{(n+1)!x^{n+1}}{n!x^n} \right| = \lim_{n\to\infty} (n+1)|x| = \begin{cases} 0 & \text{if } x = 0 \\ \infty & \text{if } x \neq 0 \end{cases}$$

We conclude that the series converges only at $x = 0$ (Figure 5). ◄

0

Convergence set

Figure 5

In each of our examples, the convergence set was an interval (a single point, called a *degenerate* interval, in the last example). This will always be the case. For example, it is impossible for a power series to have a convergence set consisting of two disconnected parts (like $[0, 1] \cup [2, 3]$). Our next theorem tells the whole story.

Theorem A

The convergence set for a power series $\sum a_n x^n$ is always an interval of one of the following three types.

(i) The single point $x = 0$.
(ii) An interval $(-R, R)$, plus (possibly) one or both endpoints.
(iii) The whole real line.

In (i), (ii), and (iii), the series is said to have **radius of convergence** 0, R, and ∞, respectively.

Proof Suppose that the series converges at $x = x_1 \neq 0$. Then $\lim_{n\to\infty} a_n x_1^n = 0$, and so there is certainly a number N such that $|a_n x_1^n| < 1$ for $n \geq N$. Then, for any x for which $|x| < |x_1|$,

$$|a_n x^n| = |a_n x_1^n| \left| \frac{x}{x_1} \right|^n < \left| \frac{x}{x_1} \right|^n$$

where $n \geq N$. Now $\sum |x/x_1|^n$ converges, because it is a geometric series with ratio less than 1. Thus, $\sum |a_n x^n|$ converges, because all of the corresponding terms are smaller. We have shown that if a power series converges at x_1, it converges (absolutely) for all x such that $|x| < |x_1|$.

On the other hand, suppose a power series diverges at x_2. Then it must diverge for all x for which $|x| > |x_2|$. Because if it converged at x such that $|x_1| > |x_2|$, then (by what we have already shown) it would converge at x_2, contrary to hypothesis.

These two paragraphs together eliminate all convergence sets except the three types mentioned in the theorem.   ◄

Power Series in $x - a$ A series of the form

$$\sum a_n(x - a)^n = a_0 + a_1(x - a) + a_2(x - a)^2 + \cdots$$

is called a **power series in $x - a$**. All that we have said about power series in x applies equally well for power series in $x - a$. In particular, its convergence set is always one of the following kinds of intervals.

1.   The single point $x = a$.
2.   An interval $(a - R, a + R)$, plus possibly one or both endpoints (Figure 6).
3.   The whole real line.

Figure 6

In many practical applications we are interested primarily in the open interval $(a - R, a + R)$ of convergence rather than the endpoints of the interval. We call $(a - R, a + R)$ the *largest open interval of convergence.*

EXAMPLE 5: Find the convergence set for $\displaystyle\sum_{n=0}^{\infty} \frac{(x - 1)^n}{(n + 1)^2}$.

SOLUTION: We apply the Absolute Ratio Test.

$$p = \lim_{n\to\infty} \left| \frac{(x + 1)^{n+1}}{(n + 2)^2} \div \frac{(x - 1)^n}{(n + 1)^2} \right| = \lim_{n\to\infty} |x - 1| \frac{(n + 1)^2}{(n + 2)^2}$$

$$= |x - 1|$$

Thus, the series converges if $|x - 1| < 1$—that is, if $0 < x < 2$; it diverges if $|x - 1| > 1$. It also converges (even absolutely) at both of the endpoints 0 and 2, as we see by substitution of these values. The largest open interval of convergence is $(0, 2)$, and the convergence set is the closed interval $[0, 2]$ (Figure 7).   ◄

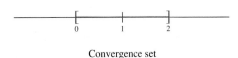

Convergence set

Figure 7

Taylor and Maclaurin Series   We saw in Section 9.4 that a function can be approximated on an interval by a finite Taylor or Maclaurin polynomial. If we include an infinite number of terms, we get a Taylor or Maclaurin *series*. Thus, the *Taylor series* of a function $f(x)$ based at a would be defined as

$$f(a) + f'(a)(x - a) + \frac{f''(a)}{2!}(x - a)^2 + \frac{f'''(a)}{3!}(x - a)^3 + \cdots$$

and the *Maclaurin series* of a function $f(x)$ would be the Taylor series based at 0, or

$$f(0) + f'(0)x + \frac{f''(0)}{2!}x^2 + \frac{f'''(0)}{3!}x^3 + \cdots$$

These series are examples of power series and so can be studied with the methods of this section.

EXAMPLE 6: Find the Maclaurin series for the function $f(x) = \ln(1 + x)$ and determine the largest open interval of convergence. Graph the function $f(x)$ together with the Maclaurin polynomials $P_N(x) = \sum_{n=0}^{N} \frac{f^{(n)}(0)}{n!} x^n$ for $N = 1, 2, 3, 4, 5$ and discuss the interval of convergence in terms of these graphs.

SOLUTION: We have

$$f(x) = \ln(1 + x) \qquad\qquad f(0) = 0$$

$$f'(x) = \frac{1}{1 + x} \qquad\qquad f'(0) = 1$$

$$f''(x) = -\frac{1}{(1 + x)^2} \qquad\qquad f''(0) = -1$$

$$f'''(x) = \frac{2}{(1 + x)^3} \qquad\qquad f'''(0) = 2$$

$$f^{(4)}(x) = -\frac{2 \cdot 3}{(1 + x)^4} \qquad\qquad f^{(4)}(0) = -2 \cdot 3$$

$$\vdots \qquad\qquad\qquad \vdots$$

$$f^{(n)}(x) = (-1)^{n+1} \frac{(n - 1)!}{(1 + x)^n} \qquad f^{(n)}(0) = (-1)^{n+1}(n - 1)!$$

The last formula holds only for $n \geq 1$. Because $f(0) = 0$, the Maclaurin series for $f(x) = \ln(1 + x)$ would be

$$\sum_{n=1}^{\infty} \frac{f^{(n)}(0)}{n!} x^n = \sum_{n=1}^{\infty} \frac{(-1)^{n+1}(n - 1)!}{n!} x^n$$

$$= \sum_{n=1}^{\infty} \frac{(-1)^{n+1}}{n} x^n$$

$$= x - \frac{x^2}{2} + \frac{x^3}{3} - \frac{x^4}{4} + \cdots$$

Using the Absolute Ratio Test to determine the interval of convergence we have

$$p = \lim_{n \to \infty} \left| \frac{(-1)^{n+2}}{n + 1} x^{n+1} \div \frac{(-1)^{n+1}}{n} x^n \right|$$

$$= \lim_{n \to \infty} \left| \frac{n}{n + 1} x \right| = |x|$$

Thus the largest open interval of convergence is $|x| < 1$ or $-1 < x < 1$. This is not unreasonable, because $f(x) = \ln(1 + x)$ is not defined for $x \le -1$.

The graphs of the first five Maclaurin polynomials together with the function $f(x) = \ln(1 + x)$ is shown in Figure 8. We see that the graphs of the polynomials are approaching the graph of $f(x)$ on the interval $-1 < x < 1$, but outside that interval the graphs of the polynomials are rapidly approaching infinity. In particular, notice that even though $f(x)$ is well defined and finite for $x \ge 1$, the Maclaurin polynomials are *not* approximating $f(x)$ there.　◄

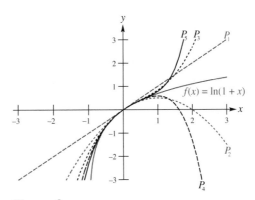

Figure 8

It certainly appears from the previous example that on the interval of convergence of the Taylor series of a function $f(x)$, the series and the function are equal. If this is the case, we say that the series *represents* the function on the interval of convergence.

It turns out that for most functions one encounters in practice, the Taylor series of a function does represent the function on the interval of convergence, though one can concoct functions for which this is not true. The next theorem gives a criterion for when a Taylor series represents the function used to generate it.

Theorem B

(Taylor's Theorem). Let f be a function with derivatives of all orders in some interval $(a - r, a + r)$. A necessary and sufficient condition that the Taylor series

$$f(a) + f'(a)(x - a) + \frac{f''(a)}{2!}(x - a)^2 + \frac{f'''(a)}{3!}(x - a)^3 + \cdots$$

represents the function f on that interval is that

$$\lim_{n \to \infty} R_n(x) = 0$$

where $R_n(x)$ is the remainder in Taylor's Formula; that is,

$$R_n(x) = \frac{f^{(n+1)}(c)}{(n + 1)!}(x - a)^{n+1}$$

c being some point in $(a - r, a + r)$.

Proof We need only recall Taylor's Formula (Theorem 9.4A),

$$f(x) = f(a) + f'(a)(x - a) + \cdots + \frac{f^{(n)}(a)}{n!}(x - a)^n + R_n(x)$$

and the result is obvious.   ◄

Power Series and Computer Algebra

Computer algebra systems can be used to generate and work with Taylor polynomials to any order (subject to hard disk and memory limitations, of course). Thus it becomes easy to approximate a function very closely on the interval of convergence of the corresponding Taylor series.

EXAMPLE 7: Find the 20th-order Maclaurin polynomial for the function $f(x) = \sqrt{x + 1}$, and graph the polynomial together with the function $f(x)$. Estimate the largest open interval of convergence of the Maclaurin series for $f(x)$ based on the graphs.

SOLUTION: In Figure 9 we show screen shots of three computer algebra systems for generating the 20th-order Maclaurin polynomial (Taylor polynomial at $a = 0$), and then plotting the polynomial and function together. The corresponding graph of the polynomial together with the graph of $f(x) = \sqrt{x + 1}$ is shown in Figure 10.

f = Normal [Series[Sqrt[x + 1], {x, 0, 20}]]
Plot[{Sqrt[x + 1], f}, {x, −2, 2}, Plot Range → {−2, 2}]

(Mathematica)

> f:=convert(taylor(sqrt(x+1), x=0, 21),polynom);
> plot({f,sqrt(x+1)},x=2..2,y=−2..2);

(Maple)

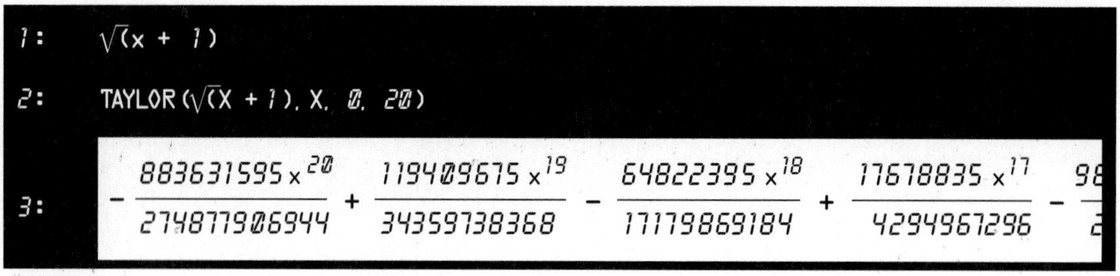

(Derive: Use Calculus—Taylor, Simplify, and then Plot expressions 1 and 3)

Figure 9

We see from the graph that the 20th-order Maclaurin polynomial $P_{20}(x)$ approximates the function $f(x) = \sqrt{x + 1}$ quite well on the interval $(-1, 1)$, but departs from it quite rapidly outside that interval. ($f(x)$ is not defined for $x < -1$, though the polynomial is.) Thus, we estimate the largest open interval of convergence to be $(-1, 1)$. In Problem 24 you are asked to verify this.   ◄

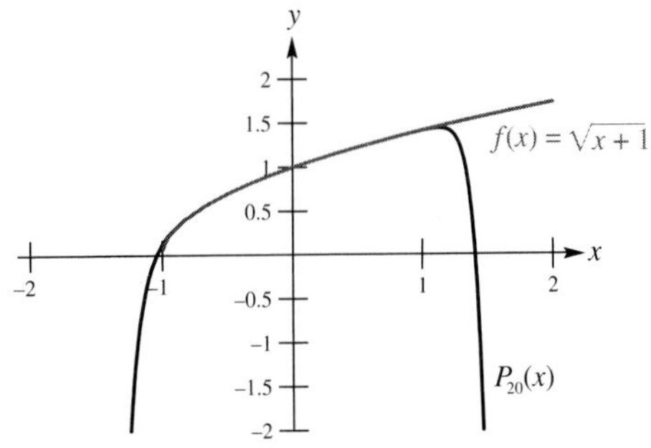

Figure 10

Concepts Review

1. A series of the form $a_0 + a_1 x + a_2 x^2 + \cdots$ is called a _____ .

2. A power series always converges on an _____ , which may or may not include its _____ .

3. The series $1 + x + x^2 + x^3 + \cdots$ converges on the interval _____ .

4. The Taylor Series for a function will represent the function for those x for which the remainder $R_n(x)$ in Taylor's Formula satisfies _____ .

Problem Set 9.6

In Problems 1–20, find the largest open interval of convergence of the given power series. Graph the first five partial sum polynomials on one set of axes and relate these graphs to the interval of convergence. *Hint*: First find a formula for the nth term; then use the Absolute Ratio Test to find the interval of convergence.

1. $\dfrac{x}{1 \cdot 2} - \dfrac{x^2}{2 \cdot 3} + \dfrac{x^3}{3 \cdot 4} - \dfrac{x^4}{4 \cdot 5} + \dfrac{x^5}{5 \cdot 6} - \cdots$

2. $1 + x + \dfrac{x^2}{2!} + \dfrac{x^3}{3!} + \dfrac{x^4}{4!} + \cdots$

3. $x - \dfrac{x^3}{3!} + \dfrac{x^5}{5!} - \dfrac{x^7}{7!} + \dfrac{x^9}{9!} \cdots$

4. $1 - \dfrac{x^2}{2!} + \dfrac{x^4}{4!} - \dfrac{x^6}{6!} + \dfrac{x^8}{8!} - \dfrac{x^{10}}{10!} + \cdots$

5. $x + 2x^2 + 3x^3 + 4x^4 + \cdots$

6. $x + 2^2 x^2 + 3^2 x^3 + 4^2 x^4 + \cdots$

7. $1 - x + \dfrac{x^2}{2} - \dfrac{x^3}{3} + \dfrac{x^4}{4} - \cdots$

8. $1 + x + \dfrac{x^2}{\sqrt{2}} + \dfrac{x^3}{\sqrt{3}} + \dfrac{x^4}{\sqrt{4}} + \dfrac{x^5}{\sqrt{5}} + \cdots$

9. $1 - \dfrac{x}{1 \cdot 3} + \dfrac{x^2}{2 \cdot 4} - \dfrac{x^3}{3 \cdot 5} + \dfrac{x^4}{4 \cdot 6} - \cdots$

10. $\dfrac{x}{2^2 - 1} + \dfrac{x^2}{3^2 - 1} + \dfrac{x^3}{4^2 - 1} + \dfrac{x^4}{5^2 - 1} + \cdots$

11. $1 - \dfrac{x}{2} + \dfrac{x^2}{2^2} - \dfrac{x^3}{2^3} + \dfrac{x^4}{2^4} - \cdots$

12. $1 + 2x + 2^2 x^2 + 2^3 x^3 + 2^4 x^4 + \cdots$

13. $1 + 2x + \dfrac{2^2 x^2}{2!} + \dfrac{2^3 x^3}{3!} + \dfrac{2^4 x^4}{4!} + \cdots$

14. $\dfrac{x}{2} + \dfrac{2x^2}{3} + \dfrac{3x^3}{4} + \dfrac{4x^4}{5} + \dfrac{5x^5}{6} + \cdots$

15. $\dfrac{(x-1)}{1} + \dfrac{(x-1)^2}{2} + \dfrac{(x-1)^3}{3} +$

$\dfrac{(x-1)^4}{4} + \cdots$

16. $1 + (x+2) + \dfrac{(x+2)^2}{2!} + \dfrac{(x+2)^3}{3!} + \cdots$

17. $1 + \dfrac{(x+1)}{2} + \dfrac{(x+1)^2}{2^2} + \dfrac{(x+1)^3}{2^3} + \cdots$

18. $\dfrac{(x-2)}{1^2} + \dfrac{(x-2)^2}{2^2} + \dfrac{(x-2)^3}{3^2} +$

$\dfrac{(x-2)^4}{4^2} + \cdots$

19. $\dfrac{(x+5)}{1 \cdot 2} + \dfrac{(x+5)^2}{2 \cdot 3} + \dfrac{(x+5)^3}{3 \cdot 4} +$

$\dfrac{(x+5)^4}{4 \cdot 5} + \cdots$

20. $(x+3) - 2(x+3)^2 + 3(x+3)^3 - 4(x+3)^4 + \cdots$

In Problems 21–30, find the Taylor series in $x - a$ and determine the largest open interval of convergence. Sketch the function together with the Taylor polynomials $P_N(x) = \sum\limits_{n=0}^{N} \dfrac{f^{(n)}(a)}{n!}(x-a)^n$ for $N = 1, 2, 3, 4, 5$ and relate the interval of convergence to these graphs.

21. $\sinh x, a = 0$

22. $\cosh x, a = 0$

23. $\ln x, a = 5$

24. $\sqrt{x+1}, a = 0$

25. $\dfrac{1}{1+x}, a = 0$

26. $e^x, a = 1$

27. $\sin x, a = \dfrac{\pi}{6}$

28. $\cos x, a = \dfrac{\pi}{3}$

29. $1 + x^2 + x^3, a = 1$

30. $2 - x + 3x^2 - x^3, a = -1$

In Problems 31–36, use a computer algebra system to find the 20th-order Maclaurin polynomial for the given function $f(x)$ and graph the polynomial together with the function $f(x)$. Estimate the largest open interval of convergence of the Maclaurin series for $f(x)$ based on the graphs.

31. $f(x) = \tan x$

32. $f(x) = \tanh x$

33. $f(x) = e^x \sin x$

34. $f(x) = e^{-x} \cos x$

35. $f(x) = \cos x \ln(1+x)$

36. $f(x) = (\sin x)\sqrt{1+x}$

37. From Example 3, we know that $\sum x^n/n!$ converges for all x. Why can we conclude that $\lim\limits_{n\to\infty} x^n/n! = 0$ for all x?

38. Find the radius of convergence of
$$\sum_{n=1}^{\infty} \frac{1 \cdot 2 \cdot 3 \cdots n}{1 \cdot 3 \cdot 5 \cdots (2n-1)} x^{2n+1}$$

39. Find the radius of convergence of
$$\sum_{n=1}^{\infty} \frac{(pn)!}{(n!)^p} x^n$$
where p is a positive integer.

40. Find the sum $S(x)$ of $\sum\limits_{n=0}^{\infty} (x-3)^n$. Where is it valid?

41. Suppose $\sum\limits_{n=0}^{\infty} a_n(x-3)^n$ converges at $x = -1$. Why can you conclude that it converges at $x = 6$? Can you be sure it converges at $x = 7$? Explain.

42. Suppose that $a_{n+3} = a_n$ and let $S(x) = \sum\limits_{n=0}^{\infty} a_n x^n$. Show that the series converges for $|x| < 1$ and give a formula for $S(x)$.

43. Follow the directions of Problem 42 for the case where $a_{n+p} = a_n$ for some fixed positive integer p.

44. Let
$$f(x) = \begin{cases} e^{-1/x^2} & x \neq 0 \\ 0 & x = 0 \end{cases}$$

(a) Show that $f'(0) = 0$ by using the definition of the derivative.

(b) Show that $f''(0) = 0$.

(c) Assuming the known fact that $f^{(n)}(0) = 0$ for all n, find the Maclaurin series for $f(x)$.

(d) Does the Maclaurin series represent $f(x)$?

(e) What is the remainder in Maclaurin's Formula in this case?

This shows that a Maclaurin series may exist and yet not represent the given function. (The remainder does not tend to 0 as $n \to \infty$.)

Answers to Concepts Review: 1. Power series 2. Interval; endpoints 3. $(-1, 1)$ 4. $\lim\limits_{n\to\infty} R_n(x) = 0$

LAB 21: THE GAMMA FUNCTION AND TAYLOR SERIES APPROXIMATIONS

Mathematical Background

Some functions that prove to be useful in scientific applications can't be expressed in a simple form. One such function, which is useful both in engineering problems and in probability theory, is called the Gamma function. It is defined by

$$\Gamma(t) = \int_0^\infty x^{t-1} e^{-x} dx$$

The problem is that the antiderivative of $x^{t-1}e^{-x}$ cannot be found for most values of t (try it) and so a numerical approximation must be used. The Gamma function is defined for all $t > 0$. For $0 < t < 1$, $\Gamma(t)$ is particularly difficult to find, because it is improper at both endpoints.

One of the interesting properties of this function (which can be shown using integration by parts) is that $\Gamma(t) = (t-1)\,\Gamma(t-1)$ and $\Gamma(1) = 1$; hence if t is an integer, then $\Gamma(t) = (t-1)(t-2)(t-3)\cdots(3)(2)(1) = (t-1)!$. So the Gamma function is essentially the same as a factorial, but can be applied to fractions as well as whole numbers.

Lab Introduction

If we interpret the Gamma function as a factorial function—that is, $\Gamma(t) = (t-1)!$—then $\Gamma(0.5)$ would be $(-0.5)!$. The exclamation point is particularly appropriate since we don't know how to take the factorial of a negative number. The point here is that the Gamma function is defined for values for which the factorial function is not; thus it extends the idea of factorial to new numbers. In this lab we will investigate "factorials" for negative values between -1 and 0 by investigating the Gamma function between 0 and 1.

We will use a Taylor series to help to evaluate the integral that defines $\Gamma(t)$. Midpoint rectangles (or Simpson's rule) do not work well in this case because the function being integrated is not defined at $x = 0$ when $t < 1$. The idea is to approximate $x^{t-1}e^{-x}$ with the first few terms of a power series and then integrate the series from 0 to 2 (the value 2 for the cut-off point is somewhat arbitrary). The rest of the integral (from 2 to ∞) can be approximated using Midpoint Riemann sums (or another numerical technique such as Simpson's Rule).

Experiment

(1) Find $\Gamma(0.5) = \displaystyle\int_0^\infty x^{-0.5} e^{-x} dx$ accurate to three decimal places as follows.

    (a) Split the integral into two pieces, the integral from 0 to 2 plus the integral from 2 to ∞

$$\int_0^\infty x^{-0.5} e^{-x}\, dx = \int_0^2 x^{-0.5} e^{-x}\, dx + \int_2^\infty x^{-0.5} e^{-x}\, dx \tag{1}$$

Approximate the second integral numerically; integrate from 2 to t where $t = 6, 8, 10, ...$; keep increasing the upper limit until you get three-decimal-place accuracy. Of course, for each t you must choose n large enough when using midpoint rectangles or Simpson's Rule.

(b) For the first integral in (1) we can use a Taylor series to approximate e^{-x} and then multiply by $x^{-0.5}$. The Taylor series at $x = 0$ for e^{-x} is

$$\sum_{n=0}^{\infty} \frac{(-x)^n}{n!} = \sum_{n=0}^{\infty} \frac{x^n}{n!} (-1)^n$$

and so a power series for $x^{-0.5}e^{-x}$ is

$$x^{-0.5} \sum_{n=0}^{\infty} \frac{(-x)^n}{n!} (-1)^n = \sum_{n=0}^{\infty} \frac{x^{n-0.5}}{n!} (-1)^n \tag{2}$$

We can now integrate (2) from 0 to 2 to get the following series approximation to the first integral in (1):

$$\sum_{n=0}^{\infty} (-1)^n \frac{2^{n+0.5}}{n!(n + 0.5)} \tag{3}$$

Use this series to get three-decimal-place accuracy for the corresponding integral. Approximate this infinite series numerically; you could just keep increasing the upper limit until you get convergence or you could take advantage of the fact that it is an alternating series to determine ahead of time the number of terms it will take to get three decimals. (For an alternating series the error is no more than absolute value of the first neglected term.)

(c) For the three-term and six-term Taylor approximations, graph the function $x^{-0.5}e^{-x}$ and the approximation to it (that is, series (2) for three and six terms) together on the interval $0 \leq x \leq 5$; a good choice for the y-interval is $-2 \leq y \leq 2$.

(d) Why not use midpoint rectangles (or any other standard numerical integration method such as Simpson's Rule for the entire integral from 0 to ∞? Try it and note what happens. Why not use a power series approach for the entire integral? Try it and note what happens.

(e) Put your answers from (a) and (b) together to estimate $\Gamma(0.5)$.

(2) Find $\Gamma(0.1)$, $\Gamma(0.3)$, $\Gamma(0.7)$, and $\Gamma(0.9)$ as in part (1). Use these values and the relationship $\Gamma(t) = (t - 1)\Gamma(t - 1)$ to sketch the graph of $\Gamma(t)$ as a function of t on the interval $0 < t < 2$.

Discussion

1. Discuss the technique of using Taylor series to approximate hard-to-find integrals. What are the advantages and disadvantages as compared to midpoint rectangles?

2. Discuss what goes wrong in each part of (1d). When you explain what happens when you try to do the integral from 0 to ∞ with a power series, refer to the graphs of $x^{-0.5}e^{-x}$ and the approximations to it from part (1c) in your discussion.

3. Describe the Gamma function on the interval $0 < t < 2$. Is this surprising if you think of $\Gamma(t)$ as a "factorial" function? Is there a minimum in this interval?

4. Would it have been easier to find $\Gamma(1.5)$ than $\Gamma(0.5)$? Why? Can you use $\Gamma(1.5)$ to find $\Gamma(0.5)$? How?

9.7   Representing Functions with Power Series

We know from the previous section that the convergence set of power series $\sum a_n x^n$ is an interval I. This interval is the domain for a new function $S(x)$, the sum of the series. An obvious question to ask about $S(x)$ is whether we can give a simple formula for it. We have done this for the geometric series

$$\sum_{n=0}^{\infty} ax^n = \frac{a}{1-x}, \qquad -1 < x < 1$$

which gives us a power series representation for the function $\dfrac{a}{1-x}$.

In the case where the power series is generated from a function $f(x)$ as a Taylor series about some point $x = a$, we have shown that on the interval of convergence, the series $S(x)$ represents the function $f(x)$ as long as the remainder $R_n(x)$ in Taylor's Formula approaches zero as $n \to \infty$.

This section is devoted to finding power series representations of functions, using a variety of methods.

EXAMPLE 1: Find power series representations for the functions $\sin x$, $\cos x$, $\sinh x$, and $\cosh x$, and determine where the representations are valid.

SOLUTION: We use a Maclaurin series approach. Letting $f(x) = \sin x$ we get

$$
\begin{aligned}
f(x) &= \sin x & f(0) &= 0 \\
f'(x) &= \cos x & f'(0) &= 1 \\
f''(x) &= -\sin x & f''(0) &= 0 \\
f'''(x) &= -\cos x & f'''(0) &= -1 \\
f^{(4)}(x) &= \sin x & f^{(4)}(0) &= 0 \\
&\ \ \vdots & &\ \ \vdots
\end{aligned}
$$

Thus,

$$\sin x = x - \frac{x^3}{3!} + \frac{x^5}{5!} - \frac{x^7}{7!} + \cdots$$

and this is valid for all x, provided we can show

$$\lim_{n\to\infty} R_n(x) = \lim_{n\to\infty} \frac{f^{(n+1)}(c)}{(n+1)!} x^{n+1} = 0$$

Now, $|f^{(n+1)}(x)| = |\cos x|$ or $|f^{(n+1)}(x)| = |\sin x|$ and so

$$|R_n(x)| \le \frac{|x|^{n+1}}{(n+1)!}$$

But $\lim\limits_{n \to \infty} x^n/n! = 0$, because $x^n/n!$ is the nth term of a convergence series (see Example 3 and Problem 37 of Section 9.6). As a consequence, we see that $\lim\limits_{n \to \infty} R_n(x) = 0$.

Similarly, we find the following representations, valid for all x:

$$\cos x = 1 - \frac{x^2}{2!} + \frac{x^4}{4!} - \frac{x^6}{6!} + \frac{x^8}{8!} - \cdots$$

$$\sinh x = x + \frac{x^3}{3!} + \frac{x^5}{5!} + \frac{x^7}{7!} + \frac{x^9}{9!} + \cdots$$

$$\cosh x = 1 + \frac{x^2}{2!} + \frac{x^4}{4!} + \frac{x^6}{6!} + \frac{x^8}{8!} + \cdots$$

◀

Term-by-Term Differentiation and Integration
Think of a power series as a polynomial with infinitely many terms. It behaves just like a polynomial under both integration and differentiation; these operations can be performed term by term, as follows.

Theorem A

Suppose that $S(x)$ is the sum of a power series on an interval I; that is,

$$S(x) = \sum_{n=0}^{\infty} a_n x^n = a_0 + a_1 x + a_2 x^2 + a_3 x^3 + \cdots$$

Then, if x is interior to I,

(i) $S'(x) = \sum\limits_{n=0}^{\infty} \frac{d}{dx}(a_n x^n) = \sum\limits_{n=1}^{\infty} n a_n x^{n-1}$

$$= a_1 + 2a_2 x + 3a_3 x^2 + \cdots$$

(ii) $\int_0^x S(t)dt = \sum\limits_{n=0}^{\infty} \int_0^x a_n t^n \, dt = \sum\limits_{n=0}^{\infty} \frac{a_n}{n+1} x^{n+1}$

$$= a_0 x + \frac{1}{2} a_1 x^2 + \frac{1}{3} a_2 x^3 + \frac{1}{4} a_3 x^4 + \cdots$$

The theorem entails several things. It asserts that S is both differentiable and integrable; it shows how the derivative and integral may be calculated; and it implies that the radius of convergence of both the differentiated and integrated series is the same as for the original series (though it says nothing about the endpoints of the interval of convergence). The theorem is hard to prove. We leave the proof to more advanced books.

A nice consequence of Theorem A is that we can apply it to a power series with a known sum formula to obtain sum formulas for other series.

EXAMPLE 2: Apply Theorem A to the geometric series

$$\frac{1}{1-x} = 1 + x + x^2 + x^3 + \cdots, \qquad -1 < x < 1$$

to obtain formulas for two new series.

SOLUTION: Differentiating term by term yields

$$\frac{1}{1-x^2} = 1 + 2x + 3x^2 + 4x^3 + \cdots, \qquad -1 < x < 1$$

Integrating term by term yields

$$\int_0^x \frac{1}{1-t}\,dt = \int_0^x 1\,dt + \int_0^x t\,dt + \int_0^x t^2\,dt + \cdots$$

That is,

$$-\ln(1-x) = x + \frac{x^2}{2} + \frac{x^3}{3} + \cdots, \qquad -1 < x < 1$$

If we replace x by $-x$ in the latter and multiply both sides by -1, we obtain

$$\ln(1+x) = x - \frac{x^2}{2} + \frac{x^3}{3} - \frac{x^4}{4} + \cdots, \qquad -1 < x < 1$$

We obtained this result in Example 6 of Section 9.6 using a Maclaurin series approach. In fact, it can be shown that this series also converges at $x = 1$, though we will not attempt to do so (also see the note in the box).

An Endpoint Result

The question of what is true at an endpoint of the interval of convergence of a power series is tricky. One result is due to Norway's greatest mathematician, Niels Henrik Abel (1802–1829). Suppose

$$f(x) = \sum_{n=0}^{\infty} a_n x^n$$

for $|x| < R$. If f is continuous at an endpoint (R or $-R$) and if the series converges there, then the formula also holds at that endpoint.

EXAMPLE 3: Find a power series representation for $\tan^{-1}x$.

SOLUTION: Recall that

$$\int \frac{1}{1+t^2}\,dt = \tan^{-1}t + C$$

and hence

$$\int_0^x \frac{1}{1+t^2}\,dt = \tan^{-1}t\Big|_0^x = \tan^{-1}x - \tan^{-1}0$$

$$= \tan^{-1}x$$

From the geometric series for $1/(1-x)$ with x replaced by $-t^2$, we get

$$\frac{1}{1+t^2} = 1 - t^2 + t^4 - t^6 + \cdots, \qquad -1 < t < 1$$

Thus,

$$\tan^{-1}x = \int_0^x (1 - t^2 + t^4 - t^6 + \cdots)\,dt$$

That is,

$$\tan^{-1}x = x - \frac{x^3}{3} + \frac{x^5}{5} - \frac{x^7}{7} + \cdots, \qquad -1 < x < 1$$

(By the note in the box about Niels Abel, this also holds at $x = \pm 1$.)   ◄

Note: This series can also be obtained as the Maclaurin series for $\tan^{-1}x$, but it requires a great deal more work because the derivatives of $\tan^{-1}x$ become quite complicated.

EXAMPLE 4: Find a formula for the sum of the series

$$S(x) = 1 + x + \frac{x^2}{2!} + \frac{x^3}{3!} + \cdots$$

SOLUTION: We saw earlier (Section 9.6, Example 3) that this series converges for all x. Differentiating term by term, we obtain

$$S'(x) = 1 + x + \frac{x^2}{2!} + \frac{x^3}{3!} + \cdots$$

That is, $S'(x) = S(x)$ for all x. Furthermore, $S(0) = 1$. This differential equation has the unique solution $S(x) = e^x$ (see Section 7.4). Thus,

$$e^x = 1 + x + \frac{x^2}{2!} + \frac{x^3}{3!} + \cdots$$

Of course, we could have obtained this same formula using the Maclaurin series for e^x.   ◄

EXAMPLE 5: Obtain the power series representation for e^{-x^2}.

SOLUTION: Simply substitute $-x^2$ for x in the series for e^x.

$$e^{-x^2} = 1 - x^2 + \frac{x^4}{2!} - \frac{x^6}{3!} + \cdots$$

Again, a straightforward use of Taylor series would give the same result, but with a great deal more work.   ◄

Algebraic Operations   Power series behave like polynomials under the operations of addition and subtraction, as we know from Theorem 9.2B. The same is true for multiplication and division, as we now illustrate.

EXAMPLE 6: Multiply and divide the power series for $\ln(1 + x)$ by that for e^x.

SOLUTION: We refer to Examples 1 and 3 for the required series. The key to multiplication is to find first the constant term, then the x-term, then the x^2-term, and so on. We arrange our work as follows.

$$\left(0 + x - \frac{x^2}{2} + \frac{x^3}{3} - \frac{x^4}{4} + \cdots\right)$$

$$\times \left(1 + x + \frac{x^2}{2!} + \frac{x^3}{3!} + \frac{x^4}{4!} + \cdots\right)$$

The product is

$$0 + (0 + 1)x + \left(0 + 1 - \frac{1}{2}\right)x^2 + \left(0 + \frac{1}{2!} - \frac{1}{2} + \frac{1}{3}\right)x^3$$

$$+ \left(0 + \frac{1}{3!} - \frac{1}{2!2} + \frac{1}{3} - \frac{1}{4}\right)x^4 + \cdots$$

$$= 0 + x + \frac{1}{2}x^2 + \frac{1}{3}x^3 + 0 \cdot x^4 + \cdots$$

Here is how division is done.

$$
\begin{array}{r}
x - \frac{3}{2}x^2 + \frac{4}{5}x^3 - x^4 \quad + \cdots \\
1 + x + \frac{1}{2}x^2 + \frac{1}{6}x^3 + \cdots \overline{\smash{\big)}\, x - \frac{1}{2}x^2 + \frac{1}{3}x^3 - \frac{1}{4}x^4 + \cdots} \\
\underline{x + x^2 \quad + \frac{1}{2}x^3 + \frac{1}{6}x^4 + \cdots} \\
-\frac{3}{2}x^2 - \frac{1}{6}x^3 - \frac{5}{12}x^4 + \cdots \\
\underline{-\frac{3}{2}x^2 - \frac{3}{2}x^3 - \frac{3}{4}x^4 + \cdots} \\
\frac{4}{3}x^3 + \frac{1}{3}x^4 + \cdots \\
\underline{\frac{1}{3}x^3 + \frac{1}{3}x^4 + \cdots} \\
-x^4 + \cdots
\end{array}
$$

◄

The real question relative to Example 5 is whether the two series we have obtained converge to $[\ln(1 + x)]e^x$ and $[\ln(1 + x)]/e^x$, respectively. Our next theorem, stated without proof, answers this question.

Theorem B

Let $f(x) = \sum a_n x^n$ and $g(x) = \sum b_n x^n$, with both of these series converging at least for $|x| < r$. If the operations of addition, subtraction, and multiplication are performed on these series as if they were polynomials, the resulting series will converge for $|x| < r$ and represent $f(x) + g(x)$, $f(x) - g(x)$, and $f(x) \cdot g(x)$, respectively. If $b_0 \neq 0$, the corresponding result holds for division, but we can guarantee its validity only for $|x|$ sufficiently small.

We mention that the operation of substituting one power series in another is also legitimate for $|x|$ sufficiently small, provided the constant term of the substituted series is zero. Here is an illustration.

EXAMPLE 7: Find the power series for $e^{\tan^{-1}x}$ through terms of degree 4.

SOLUTION: Because

$$e^u = 1 + u + \frac{u^2}{2!} + \frac{u^3}{3!} + \frac{u^4}{4!} + \cdots$$

$$e^{\tan^{-1}x} = 1 + \tan^{-1}x + \frac{(\tan^{-1}x)^2}{2!} + \frac{(\tan^{-1}x)^3}{3!} + \frac{(\tan^{-1}x)^4}{4!} + \cdots$$

Now substitute the series for $\tan^{-1}x$ from Example 2 and combine like terms.

$$e^{\tan^{-1}x} = 1 + \left(x - \frac{x^3}{3} + \cdots\right) + \frac{\left(x - \frac{x^3}{3} + \cdots\right)^2}{2!} + \frac{\left(x - \frac{x^3}{3} + \cdots\right)^3}{3!}$$

$$+ \frac{\left(x - \frac{x^3}{3} + \cdots\right)^4}{4!} + \cdots$$

$$= 1 + \left(x - \frac{x^3}{3} + \cdots\right) + \frac{(x^2 - \frac{2}{3}x^4 + \cdots)}{2} + \frac{(x^3 + \cdots)}{6}$$

$$+ \frac{(x^4 + \cdots)}{24} + \cdots$$

$$= 1 + x + \frac{x^2}{2} - \frac{x^3}{6} - \frac{7x^4}{24} + \cdots \qquad \blacktriangleleft$$

S. Ramanujan (1887–1920)

One of the most remarkable people of the early 20th century was the Indian mathematician Srinivasa Ramanujan. Largely self-educated, Ramanujan left at his death a number of notebooks in which he had recorded his discoveries. These notebooks are only now being thoroughly studied. In them are many strange and wonderful formulas, some for the sums of infinite series. Here is one:

$$\frac{1}{\pi} = \frac{\sqrt{8}}{9801} \sum_{n=0}^{\infty} \frac{(4n)![1103 + 23{,}369n]}{(n!)^4(396)^{411}}$$

Formulas like this were used in 1995 to calculate the decimal expansion of π to over four billion places. (See Problem 41.)

Power Series in $x - a$ We have stated the theorems of this section for power series in x, but with obvious modifications they are equally valid for power series in $x - a$.

The next example shows an application of power series to numerical integration; the method illustrated there can be effective when other numerical techniques are not.

EXAMPLE 8: Compute $\displaystyle\int_0^{0.4} \sqrt{1 + x^4}\, dx$ to five decimal places using a Maclaurin series.

SOLUTION: We can obtain a Maclaurin series for $\sqrt{1 + x}$ on the interval $-1 < x < 1$ (why this interval?) and then replace x with x^4. We have

$$f(x) = \sqrt{1 + x} \qquad\qquad f(0) = 1$$

$$f'(x) = \frac{1}{2}(1 + x)^{-\frac{1}{2}} \qquad\qquad f'(0) = \frac{1}{2}$$

$$f''(x) = -\frac{1}{2} \cdot \frac{1}{2}(1 + x)^{-\frac{3}{2}} \qquad\qquad f''(0) = -\frac{1}{2} \cdot \frac{1}{2}$$

$$f'''(x) = \frac{3}{2} \cdot \frac{1}{2} \cdot \frac{1}{2}(1 + x)^{-\frac{5}{2}} \qquad\qquad f'''(0) = \frac{3}{2} \cdot \frac{1}{2} \cdot \frac{1}{2}$$

$$f^{(4)}(x) = -\frac{5}{2} \cdot \frac{3}{2} \cdot \frac{1}{2} \cdot \frac{1}{2}(1 + x)^{-\frac{7}{2}} \qquad f^{(4)}(0) = -\frac{5}{2} \cdot \frac{3}{2} \cdot \frac{1}{2} \cdot \frac{1}{2}$$

so that

$$\sqrt{1+x} = 1 + \frac{1}{2}x - \frac{1}{2} \cdot \frac{1}{2} \cdot \frac{x^2}{2!} + \frac{3}{2} \cdot \frac{1}{2} \cdot \frac{1}{2} \cdot \frac{x^3}{3!} - \frac{5}{2} \cdot \frac{3}{2} \cdot \frac{1}{2} \cdot \frac{1}{2} \cdot \frac{x^4}{4!} + \cdots$$

$$= 1 + \frac{1}{2}x - \frac{1}{8}x^2 + \frac{1}{16}x^3 - \frac{5}{128}x^4 + \cdots$$

and therefore

$$\sqrt{1+x^4} = 1 + \frac{1}{2}x^4 - \frac{1}{8}x^8 + \frac{1}{16}x^{12} - \frac{5}{128}x^{16} + \cdots$$

Thus,

$$\int_0^{0.4} \sqrt{1+x^4}\, dx = \left[x + \frac{x^5}{10} - \frac{x^9}{72} + \frac{x^{13}}{208} + \cdots \right]_0^{0.4} \approx 0.40102$$

How can we be sure of having five-decimal-place accuracy? We can repeat this process with fewer (or more) terms in the series to see if we get the same result. We leave it to the reader to show that even if we use only the first two terms of the series we get the same five-decimal-place value.   ◄

Summary   We conclude our discussion of series with a list of the important Maclaurin series we have found. These series will be useful in doing the problem set, but what is more significant, they find application throughout mathematics and science.

Important Maclaurin Series

1. $\dfrac{1}{1-x} = 1 + x + x^2 + x^3 + x^4 + \cdots$ $-1 < x < 1$

2. $\ln(1+x) = x - \dfrac{x^2}{2} + \dfrac{x^3}{3} - \dfrac{x^4}{4} + \dfrac{x^5}{5} - \cdots$ $-1 < x \le 1$

3. $\tan^{-1} x = x - \dfrac{x^3}{3} + \dfrac{x^5}{5} - \dfrac{x^7}{7} + \dfrac{x^9}{9} + \cdots$ $-1 \le x \le 1$

4. $e^x = 1 + x + \dfrac{x^2}{2!} + \dfrac{x^3}{3!} + \dfrac{x^4}{4!} + \cdots$

5. $\sin x = x - \dfrac{x^3}{3!} + \dfrac{x^5}{5!} - \dfrac{x^7}{7!} + \dfrac{x^9}{9!} - \cdots$

6. $\cos x = 1 - \dfrac{x^2}{2!} + \dfrac{x^4}{4!} - \dfrac{x^6}{6!} + \dfrac{x^8}{8!} - \cdots$

7. $\sinh x = x + \dfrac{x^3}{3!} + \dfrac{x^5}{5!} + \dfrac{x^7}{7!} + \dfrac{x^9}{9!} + \cdots$

8. $\cosh x = 1 + \dfrac{x^2}{2!} + \dfrac{x^4}{4!} + \dfrac{x^6}{6!} + \dfrac{x^8}{8!} + \cdots$

Concepts Review

1. A power series may be differentiated or _____ term by term on the _____ of its interval of convergence.

2. The first five terms in the power series expansion for $\ln(1-x)$ are _____ .

3. The first four terms in the power series expansion for $\exp(x^2)$ are _____ .

4. The first five terms in the power series expansion for $\exp(x^2) - \ln(1-x)$ are _____ .

Problem Set 9.7

In Problems 1–10, find the power series representation for $f(x)$ and specify the radius of convergence. Each is somehow related to a geometric series (see Examples 2 and 3).

1. $f(x) = \dfrac{1}{1 + x}$

2. $f(x) = \dfrac{1}{(1 + x)^2}$ *Hint*: Differentiate Problem 1.

3. $f(x) = \dfrac{1}{(1 - x)^3}$

4. $f(x) = \dfrac{x}{(1 + x)^2}$

5. $f(x) = \dfrac{1}{2 - 3x} = \dfrac{\frac{1}{2}}{1 - \frac{3}{2}x}$

6. $f(x) = \dfrac{1}{3 + 2x}$

7. $f(x) = \dfrac{x^2}{1 - x^4}$

8. $f(x) = \dfrac{x^3}{2 - x^3}$

9. $f(x) = \displaystyle\int_0^x \ln(1 + t)\, dt$

10. $f(x) = \displaystyle\int_0^x \tan^{-1} t\, dt$

In Problems 11–16, find the terms through x^5 in the Maclaurin series for $f(x)$. *Hint*: It may be easiest to use known Maclaurin series and then perform multiplications, divisions, and so on. For example, $\tan x = (\sin x)/(\cos x)$.

11. $f(x) = \tan x$

12. $f(x) = \tanh x$

13. $f(x) = e^x \sin x$

14. $f(x) = e^{-x} \cos x$

15. $f(x) = \cos x \ln(1 + x)$

16. $f(x) = (\sin x)\sqrt{1 + x}$

In Problem 17–20, use the result of Example 4 to find power series in x for the given functions.

17. $f(x) = e^{-x}$

18. $f(x) = xe^{x^2}$

19. $f(x) = e^x + e^{-x}$

20. $f(x) = e^{2x} - 1 - 2x$

In Problems 21–28, use the methods of Example 6 to find power series in x for each function f.

21. $f(x) = e^{-x} \cdot \dfrac{1}{1 - x}$

22. $f(x) = e^x \tan^{-1} x$

23. $f(x) = \dfrac{\tan^{-1} x}{e^x}$

24. $f(x) = \dfrac{e^x}{1 + \ln(1 + x)}$

25. $f(x) = (\tan^{-1} x)(1 + x^2 + x^4)$

26. $f(x) = \dfrac{\tan^{-1} x}{1 + x^2 + x^4}$

27. $f(x) = \displaystyle\int_0^x \dfrac{e^t}{1 + t}\, dt$

28. $f(x) = \displaystyle\int_0^x \dfrac{\tan^{-1} t}{t}\, dt$

29. Use the method of substitution (Example 7) to find power series through terms of degree 3 for $\tan^{-1}(e^x - 1)$.

30. Use the method of substitution to find power series through terms of degree 3 for $e^{e^x} - 1$.

31. Recall that

$$\sin^{-1} x = \int_0^x \dfrac{1}{\sqrt{1 + t^2}}\, dt$$

Find the first four nonzero terms in the Maclaurin series for $\sin^{-1} x$.

32. Given that

$$\sinh^{-1} x = \int_0^x \dfrac{1}{\sqrt{1 + t^2}}\, dt$$

Find the first four nonzero terms in the Maclaurin series for $\sinh^{-1} x$.

33. Use a Taylor series to calculate, accurate to four decimal places,

$$\int_0^1 \cos(x^2)\, dx$$

34. Use a Taylor series to calculate, accurate to five decimal places,

$$\int_0^{0.5} \sin\sqrt{x}\, dx$$

35. Let

$$f(t) = \begin{cases} 0 & t < 0 \\ t^4 & t \geq 0 \end{cases}$$

Explain why $f(t)$ cannot be represented by a Maclaurin series. Also show that if $g(t)$ gives the distance traveled by a car that is stationary for $t < 0$ and moving ahead for $t \geq 0$, then $g(t)$ cannot be represented by a Maclaurin series.

36. Find the sum of each of the following series by recognizing how it is related to something familiar.

(a) $x - x^2 + x^3 - x^4 + x^5 - \cdots$

(b) $\dfrac{1}{2!} + \dfrac{x}{3!} + \dfrac{x^2}{4!} + \dfrac{x^3}{5!} + \cdots$

(c) $2x + \dfrac{4x^2}{2} + \dfrac{8x^3}{3} + \dfrac{16x^4}{4} + \cdots$

37. Follow the directions of Problem 36.

(a) $1 + x^2 + x^4 + x^6 + x^8 + \cdots$

(b) $\cos x + \cos^2 x + \cos^3 x + \cos^4 x + \cdots$

(c) $\dfrac{x^2}{2} + \dfrac{x^4}{4} + \dfrac{x^6}{6} + \dfrac{x^8}{8} + \cdots$

38. Find the sum of $\displaystyle\sum_{n=1}^{\infty} nx^n$.

39. Find the power series representation of $x/(x^2 - 3x + 2)$. *Hint*: Use partial fractions.

40. Let $y = y(x) = x - \dfrac{x^3}{3!} + \dfrac{x^5}{5!} - \dfrac{x^7}{7!} + \cdots$. Show that y satisfies the differential equation $y'' + y = 0$ with the conditions $y(0) = 0$ and $y'(0) = 0$. From this, guess at a simple formula for y.

41. Did you ever wonder how people find the decimal expansion of π to a large number of places? One method depends on the following identity.

$$\pi = 16 \tan^{-1}\left(\tfrac{1}{5}\right) - 4 \tan^{-1}\left(\tfrac{1}{239}\right)$$

Find the first six digits of π using this identity and the series for $\tan^{-1}x$. (You will need terms through $x^9/9$ for $\tan^{-1}\left(\tfrac{1}{5}\right)$ but only the first term for $\tan^{-1}\left(\tfrac{1}{239}\right)$.) In 1706, John Machin used this method to calculate the first 100 digits of π, while in 1973, Jean Guilloud and Martine Bouyer found the first million digits using the related identity

$$\pi = 48 \tan^{-1}\left(\tfrac{1}{18}\right) + 32 \tan^{-1}\left(\tfrac{1}{57}\right) - 20 \tan^{-1}\left(\tfrac{1}{239}\right)$$

In 1995, mathematicians from Simon Fraser University discovered a formula for computing isolated *binary* digits of π, leading to speculation that this may be possible for decimal digits as well.

Answers to Concepts Review: 1. Integrated; interior
2. $-x - \tfrac{1}{2}x^2 - \tfrac{1}{3}x^3 - \tfrac{1}{4}x^4 - \tfrac{1}{5}x^5$ 3. $1 + x^2 + \tfrac{1}{2}x^4 + \tfrac{1}{6}x^6$
4. $1 + x + \tfrac{3}{2}x^2 + \tfrac{1}{3}x^3 + \tfrac{3}{4}x^4$

9.8 | Chapter Review

Concepts Test

Respond with true or false to each of the following assertions. Be prepared to justify your answer.

1. The sequence $a_n = (-1)^n$ converges to both -1 and 1.

2. The sequence $a_n = \sin^n\left(\dfrac{\pi}{100}n\right)$ converges to zero.

3. The sequence $a_n = \sin^n\left(\dfrac{\pi}{4} + n\pi\right)$ converges to zero.

4. The sequence given by $a_{n+1} = ra_n$, $a_0 = 1$ is the same as the sequence $a_n = r^n$ for any value of r.

5. The sequence $a_{n+1} = a_n - a_n\dfrac{|\sin a_n|}{2}$, $a_0 = 1$ converges.

6. If $\{a_n\}$ and $\{b_n\}$ both diverge, then $\{a_n + b_n\}$ diverges.

7. If $\{a_n\}$ converges, then $\{a_n/n\}$ converges to 0.

8. If $p = f(p)$ and $f'(p) < 1$, $f'(x)$ continuous, then the sequence defined by $a_{n+1} = f(a_n)$ will converge to the point p for any starting value a_0.

9. If $p = f(p)$ and $f'(p) < 1$, $f'(x)$ continuous, then the sequence defined by $a_{n+1} = f(a_n)$ will converge to the point p for starting values a_0 that are close enough to p.

10. $\displaystyle\sum_{n=0}^{\infty}\left(\dfrac{\pi}{e}\right)^n$ converges.

11. $\displaystyle\sum_{n=0}^{\infty}\left(\dfrac{e}{\pi}\right)^n$ converges.

12. $\displaystyle\sum_{n=1}^{\infty}\dfrac{e}{n}$ converges.

13. If we know that a series converges, we can usually find the exact sum of the series.

14. If we know that a series converges, we can usually estimate the sum of the series numerically.

15. The Ratio Test will not help in determining the convergence or divergence of $\displaystyle\sum_{n=1}^{\infty}\dfrac{2n+3}{3n^4 + 2n^3 + 3n + 1}$.

16. $\displaystyle\sum_{n=1}^{\infty}\left(1 - \dfrac{1}{n}\right)^n$ converges.

17. $\displaystyle\sum_{n=1}^{\infty}\dfrac{\sin^2(n\pi/2)}{n}$ converges.

18. $\dfrac{1}{3} + \left(\dfrac{1}{3}\right)^2 + \left(\dfrac{1}{3}\right)^3 + \cdots + \left(\dfrac{1}{3}\right)^{1000} < \dfrac{1}{2}$.

19. If $\sum_{n=1}^{\infty} a_n$ converges, then $\sum_{n=1}^{\infty} (-1)^n a_n$ converges.

20. If $0 \le a_n$ for all n and $\sum_{n=1}^{\infty} a_n$ converges, then $\sum_{n=1}^{\infty} (-1)^n a_n$ converges.

21. $\left| \sum_{n=1}^{\infty} (-1)^{n+1} \frac{1}{n} - \sum_{n=1}^{99} (-1)^{n+1} \frac{1}{n} \right| < 0.01.$

22. If $P(x)$ is the Maclaurin polynomial of order 2 for $f(x)$, then $P(0) = f(0)$, $P'(0) = f'(0)$, and $P''(0) = f''(0)$.

23. $f(x) = x^{5/2}$ has a second-order Maclaurin polynomial.

24. The Maclaurin polynomial of order 3 for $f(x) = 2x^3 - x^2 + 7x - 11$ is an exact representation of $f(x)$.

25. The Maclaurin polynomial of order 16 for $\cos x$ involves only even powers of x.

26. The error involved in approximating $\int_0^2 \cos(x^3) dx$ using the Parabolic Rule with $n = 20$ will be about $\frac{1}{16}$ of the error with $n = 10$.

27. Newton's Method will produce a convergent sequence for the function $f(x) = x^{1/3}$.

28. If the power series $\sum_{n=0}^{\infty} a_n(x - 3)^n$ converges at $x = -1.1$, it also converges at $x = 7$.

29. If $\sum_{n=1}^{\infty} a_n x^n$ converges at $x = -2$, it also converges at $x = 2$.

30. If $f(x) = \sum_{n=0}^{\infty} a_n x^n$ and the series converges at $x = 1.5$, then $\int_0^1 f(x) dx = \sum_{n=0}^{\infty} a_n/(n + 1)$.

31. Every power series converges for at least two values of the variable.

32. If $f(0)$, $f'(0)$, $f''(0)$, ... all exist, then the Maclaurin series for $f(x)$ converges to $f(x)$ in an open interval containing $x = 0$.

33. The function $f(x) = 1 + x + x^2 + x^3 + \cdots$ satisfies the differential equation $y' = y^2$ on the interval $(-1, 1)$.

34. The function $f(x) = \sum_{n=0}^{\infty} (-1)^n x^n / n!$ satisfies the differential equation $y' + y = 0$ on the whole real line.

Sample Test Problems

In Problems 1–8, determine whether the given sequence converges or diverges and, if it converges, find $\lim_{n \to \infty} a_n$.

Choose between exact and numerical methods, and explain the reasons for your choice.

1. $a_n = \dfrac{9n}{\sqrt{9n^2 + 1}}$

2. $a_n = \dfrac{\ln n}{\sqrt{n}}$

3. $a_n = \left(1 + \dfrac{4}{n} \right)^n$

4. $a_n = \dfrac{n!}{3^n}$

5. $a_{n+1} = 3.5 a_n^2, \; a_0 = 0.1$

6. $a_{n+1} = 3.5(a_n - a_n^2), \; a_0 = 0.1$

7. $a_n = \dfrac{\sin^2 n}{\sqrt{n}}$

8. $a_n = \cos\left(\dfrac{n\pi}{6} \right)$

In Problems 9–18, determine whether the given series converges or diverges and, if it converges, find its sum. Use exact or numerical methods, and explain your choice.

9. $\sum_{k=1}^{\infty} \cos k\pi$

10. $\sum_{k=0}^{\infty} e^{-2k}$

11. $\sum_{k=0}^{\infty} \left(\dfrac{3}{2^k} + \dfrac{4}{3^k} \right)$

12. $0.91919191\ldots = \sum_{k=1}^{\infty} 91 \left(\dfrac{1}{100} \right)^k$

13. $\sum_{k=1}^{\infty} \left(\dfrac{1}{\ln 2} \right)^k$

14. $1 - \dfrac{2^2}{2!} + \dfrac{2^4}{4!} - \dfrac{2^6}{6!} + \cdots$

15. $1 - \dfrac{1}{1!} + \dfrac{1}{2!} - \dfrac{1}{3!} + \dfrac{1}{4!} - \cdots$

16. $\sum_{k=1}^{\infty} 3 \left(\dfrac{2}{k} \right)$

17. $\sum_{k=1}^{\infty} 3 \left(\dfrac{2}{k} \right)^k$

18. $\sum_{k=1}^{\infty} 3 \left(\dfrac{2}{k} \right)^0$

In Problems 19–28, indicate whether the given series converges or diverges and give a reason for your conclusion. If the series converges, estimate or find the sum.

19. $\displaystyle\sum_{n=1}^{\infty}(-1)^{n+1}\frac{1}{\sqrt[3]{n}}$

20. $\displaystyle\sum_{n=1}^{\infty}(-1)^{n+1}\frac{1}{\sqrt[n]{3}}$

21. $\displaystyle\sum_{n=1}^{\infty}\frac{2^n+3^n}{4^n}$

22. $\displaystyle\sum_{n=1}^{\infty}\frac{n}{e^{n^2}}$

23. $\displaystyle\sum_{n=1}^{\infty}(-1)^{n+1}\frac{n+1}{10n+12}$

24. $\displaystyle\sum_{n=1}^{\infty}\frac{\sqrt{n}}{n^2+7}$

25. $\displaystyle\sum_{n=1}^{\infty}\frac{n^2}{n!}$

26. $\displaystyle\sum_{n=2}^{\infty}\left(1-\frac{1}{n}\right)^n$

27. $\displaystyle\sum_{n=1}^{\infty}n^2\left(\tfrac{2}{3}\right)^n$

28. $\displaystyle\sum_{n=1}^{\infty}\frac{(-1)^n}{1+\ln n}$

29. Find the Maclaurin polynomial of order 4 for $f(x)$ and use it to approximate $f(0.1)$.
(a) $f(x)=xe^x$ (b) $f(x)=\cosh x$

30. Find the Taylor polynomial of order 3 based at 2 for $g(x)=x^3-2x^2+5x-7$ and show that it is an exact representation of $g(x)$.

31. Use the result of Problem 30 to calculate $g(2.1)$.

32. Find the Taylor polynomial of order 4 based at 1 for $f(x)=1/(x+1)$.

33. Obtain an expression for the error term $R_4(x)$ in Problem 32 and find a bound for it if $x=1.2$.

34. If $f(x)=\ln x$, then $f^{(n)}(x)=(-1)^{n-1}(n-1)!/x^n$. Thus, the Taylor polynomial of order n based at 1 for $\ln x$ is

$$\ln x=(x-1)-\frac{1}{2}(x-1)^2+\frac{1}{3}(x-1)^3+\cdots$$

$$+\frac{(-1)^{n-1}}{n}(x-1)^n+R_n(x)$$

How large would n have to be for us to know $\left|R_n(x)\right|\le 0.00005$ if $0.8\le x\le 1.2$?

35. Refer to Problem 34. Use the Taylor polynomial of order $n=4$ based at 1 to find

$$\int_{0.8}^{1.2}\ln x\,dx$$

and give a good bound for the error that is made.

36. Use Newton's Method to solve $3x-\cos 2x=0$ accurate to six decimal places. Use $x_1=0.5$.

37. Use Newton's Method to find the solution of $x-\tan x=0$ in the interval $(\pi,2\pi)$ accurate to four decimal places.

In Problems 38–43, find the largest open interval of convergence for the power series.

38. $\displaystyle\sum_{n=0}^{\infty}\frac{x^n}{n^3+1}$

39. $\displaystyle\sum_{n=0}^{\infty}\frac{(-2)^{n+1}x^n}{2n+3}$

40. $\displaystyle\sum_{n=0}^{\infty}\frac{(-1)^n(x-4)^n}{n+1}$

41. $\displaystyle\sum_{n=0}^{\infty}\frac{3^n x^{3n}}{(3n)!}$

42. $\displaystyle\sum_{n=0}^{\infty}\frac{(x-3)^n}{2^n+1}$

43. $\displaystyle\sum_{n=0}^{\infty}\frac{n!(x+1)^n}{3^n}$

44. By differentiating the geometric series

$$\frac{1}{1+x}=1-x+x^2-x^3+x^4-\cdots,\qquad |x|<1$$

find a power series that represents $1/(1+x)^2$. What is its interval of convergence?

45. Find a power series that represents $1/(1+x)^3$ on the interval $(-1,1)$.

46. Find the Maclaurin series for $\sin^2x$. For what values of x does the series represent the function?

47. Find the first five terms of the Taylor series for e^x based at the point $x=2$.

48. Write the Maclaurin series for $f(x)=\sin x+\cos x$. For what values of x does it represent f?

49. Write the Maclaurin series for $f(x)=\cos x^2$ and use it to approximate

$$\int_{0}^{1}\cos x^2\,dx$$

How many terms of the series are needed to compute the value of this integral correct to four decimal places?

50. Calculate the following integral correct to five decimal places using a power series. Check using midpoint Riemann sums.

$$\int_0^{0.2} \frac{e^x - 1}{x} \, dx$$

51. How many terms do we have to take in the convergent series

$$1 - \frac{1}{\sqrt{2}} + \frac{1}{\sqrt{3}} - \frac{1}{\sqrt{4}} + \frac{1}{\sqrt{5}} - \frac{1}{\sqrt{6}} + \cdots$$

to be sure that we have approximated its sum to within 0.001?

52. Give a good bound for the maximum error made in approximating $\cos x$ by $1 - x^2/2$ for $-0.1 \le x \le 0.1$.

53. Use the simplest method you can think of to find the first three nonzero terms of the Maclaurin series for each of the following.

(a) $\dfrac{1}{1 - x^3}$

(b) $\sqrt{1 + x^2}$

(c) $e^{-x} - 1 + x$

(d) $x \sec x$

(e) $e^{-x} \sin x$

(f) $\dfrac{1}{1 + \sin x}$

54. Use Simpson's Rule to estimate $\displaystyle\int_0^1 e^{x^3} dx$ using $n = 4$.

55. Do Problem 54 again using $n = 8$. About how many digits of accuracy do you think you have this time?

Conics, Polar Coordinates, and Parametric Curves

10.1 Conic Sections

Take a right circular cone with two nappes and pass planes through it at various angles, as shown in Figure 1. The cross sections will be curves called, respectively, an ellipse, a parabola, and a hyperbola. (You may also obtain various limiting forms of those cross sections: a circle, a

Ellipse

Parabola

Hyperbola

Figure 1

587

point, intersecting lines, and one line.) These curves are called *conic sections*, or simply *conics*. This definition, which comes from the Greek mathematicians, is cumbersome, and we shall immediately adopt a different one. It can be shown that the two notions are consistent.

In the plane, let *l* be a fixed line (the **directrix**) and *F* be a fixed point (the **focus**) not on the line, as shown in Figure 2. A set of points *P* is called a **conic** if the ratio of the distance $|PF|$ from the focus to the distance $|PL|$ from the line is a positive constant *e* (the **eccentricity**)—that is,

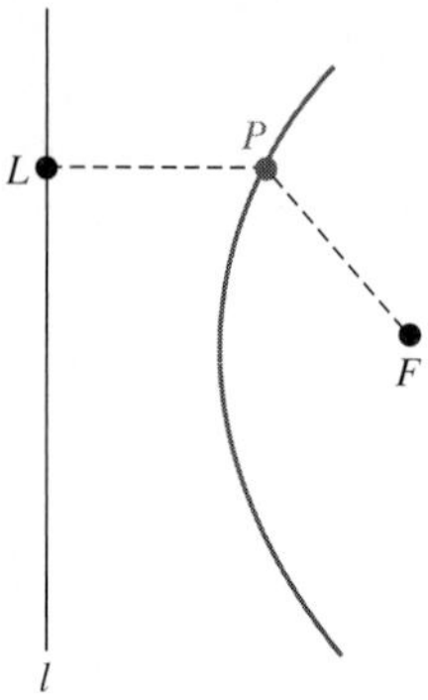

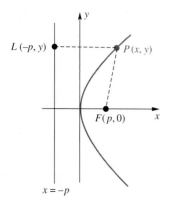

Figure 2

$$\frac{|PF|}{|PL|} = e \text{ or } |PF| = e|PL|$$

If $0 < e < 1$, the conic is an **ellipse**; if $e = 1$, it is a **parabola**; if $e > 1$, it is a **hyperbola**.

When we draw the curves corresponding to $e = \frac{1}{2}$, $e = 1$, and $e = 2$, we get the three curves shown in Figure 3.

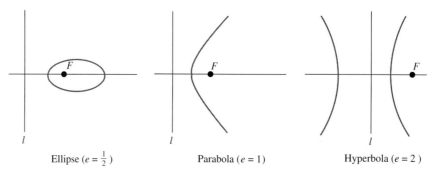

Ellipse $(e = \frac{1}{2})$ Parabola $(e = 1)$ Hyperbola $(e = 2)$

Figure 3

In each case, the curves are symmetric with respect to the line through the focus perpendicular to the directrix. We call this line the **major axis** (or simply the *axis*) of the conic. A point where the conic crosses the axis is called a **vertex**. The parabola has one vertex, while the ellipse and hyperbola have two vertices each.

The Parabola ($e = 1$)

A **parabola** is the set of points *P* that are equidistant from the directrix *l* and the focus *F*—that is,

$$|PF| = |PL|$$

From this definition, we wish to derive the *xy*-equation, and we want it to be as simple as possible. The position of the coordinate axes has no effect on the curve, but it does affect the simplicity of the curve's equation. Because a parabola is symmetric with respect to its axis, it is natural to place one of the coordinate axes—for instance, the *x*-axis—along the axis of the parabola. Let the focus *F* be to the right of the origin, say at $(p, 0)$, and the directrix to the left with equation $x = -p$. Then the vertex is at the origin. All this is shown in Figure 4.

From the condition $|PF| = |PL|$ and the distance formula, we get

Figure 4

$$\sqrt{(x - p)^2 + (y - 0)^2} = \sqrt{(x + p)^2 + (y - y)^2}$$

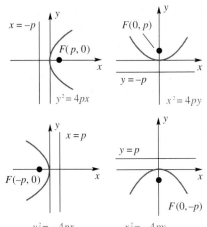

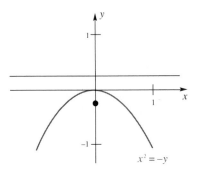

Figure 5

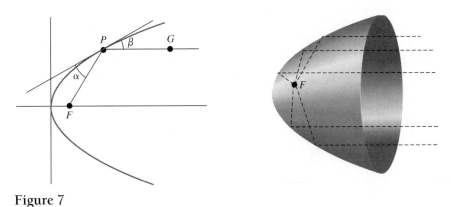

Figure 6

After squaring both sides and simplifying, we obtain

$$y^2 = 4px$$

This is called the **standard equation** of a horizontal parabola (horizontal axis) opening to the right. Note that $p > 0$ and that p is the distance from the focus to the vertex.

EXAMPLE 1: Find the focus and directrix of the parabola with equation $y^2 = 12x$.

SOLUTION: Because $y^2 = 4(3)x$, we see that $p = 3$. The focus is at $(3, 0)$; the directrix is the line $x = -3$.   ◄

There are three variants of the standard equation. If we interchange the roles of x and y, we obtain the equation $x^2 = 4py$. It is the equation of a vertical parabola with focus at $(0, p)$ and directrix $y = -p$. Finally, introducing a minus sign on one side of the equation causes the parabola to open in the opposite direction. All four cases are shown in Figure 5.

EXAMPLE 2: Determine the focus and directrix of the parabola $x^2 = -y$ and sketch the graph.

SOLUTION: We write $x^2 = -4\left(\frac{1}{4}\right)y$, from which we conclude that $p = \frac{1}{4}$. The form of the equation tells us that the parabola is vertical and opens down. The focus is at $(0, -\frac{1}{4})$; the directrix is line $y = \frac{1}{4}$. The graph is shown in Figure 6.   ◄

The Optical Property   A simple geometric property of a parabola is the basis of many important applications: If F is the focus and P is any point on the parabola, the tangent line at P makes equal angles with FP and the line GP, which is parallel to the axis of the parabola (see Figure 7).

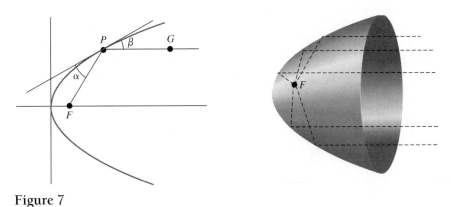

Figure 7

A principle from physics says that when a light ray strikes a reflecting surface, the angle of incidence is equal to the angle of reflection. It follows that if a parabola is revolved about its axis to form a hollow reflecting shell (a *paraboloid*), all light rays from the focus that hit the shell are reflected outward parallel to the axis. This property of the parabola is used in designing searchlights, with the light source placed at the focus. The same idea works the other way around: Certain telescopes have reflectors shaped like paraboloids, so that incoming parallel rays from a distant star will be focused at a single point.

Sound obeys the same laws of reflection as light, and paraboloidal microphones are used to pick up and concentrate sounds from, for example, a distant part of a football stadium. Radar and radio telescopes and satellite dishes are also based on these same principles.

There are many other applications of parabolas. For example, the path of a projectile is a parabola if air resistance and other minor factors are neglected. The cable of an evenly loaded suspension bridge takes the form of a parabola. Arches are often parabolic. The paths of a few comets are parabolic.

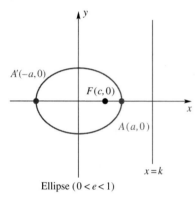

Ellipse $(0 < e < 1)$

Figure 8

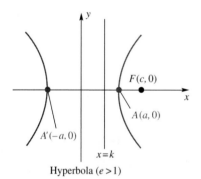

Hyperbola $(e > 1)$

Figure 9

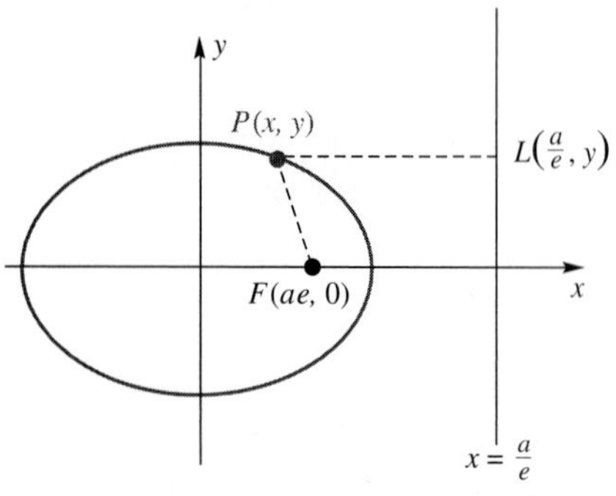

Ellipses and Hyperbolas

The conic determined by the condition $|PF| = e|PL|$ is an **ellipse** if $0 < e < 1$ and a **hyperbola** if $e > 1$. In either case, the conic has two vertices, which we label A' and A. Call the point on the major axis midway between A' and A the **center** of the conic. Ellipses and hyperbolas are symmetric with respect to their centers (as we shall soon demonstrate) and are, therefore, called *central conics*.

To derive the equation of a central conic, place the x-axis along the major axis with the origin at the center. We may suppose the focus to be $F(c, 0)$, the directrix $x = k$, and the vertices $A'(-a, 0)$ and $A(a, 0)$, with c, k, and a all positive. The two possible arrangements are shown in Figures 8 and 9.

The defining condition $|PF| = e|PL|$ applied first with $P = A$ and then $P = A'$ yields

$$a - c = e(k - a) = ek - ea$$
$$a + c = e(k + a) = ek + ea$$

When these two equations are solved for c *and* k, we get

$$c = ea \qquad \text{and} \qquad k = \frac{a}{e}$$

Now let $P(x, y)$ be any point on the ellipse (or hyperbola). Then $L(a/e, y)$ is its projection on the directrix (see Figure 10 for the case of the ellipse). The condition $|PF| = e|PL|$ becomes

$$\sqrt{(x - ea)^2 + y^2} = e\sqrt{\left(x - \frac{a}{e}\right)^2}$$

Squaring both members and collecting terms, we obtain the equivalent equation (why is it equivalent?)

$$x^2 - 2eax + e^2a^2 + y^2 = e^2\left(x^2 - \frac{2a}{e}x + \frac{a^2}{e^2}\right)$$

or

$$(1 - e^2)x^2 + y^2 = a^2(1 - e^2)$$

Figure 10

or

$$\frac{x^2}{a^2} + \frac{y^2}{a^2(1 - e^2)} = 1$$

Because this last equation contains x and y only to even powers, it corresponds to a curve that is symmetric with respect to both the x- and y-axis and to the origin. Also, because of this symmetry, there must be a second focus at $(-ea, 0)$ and a second directrix at $x = -a/e$. The axis containing the two vertices (and the two foci) is the **major axis** and the axis perpendicular to it (through the center) is the **minor axis**.

The Standard Equation of the Ellipse

For the ellipse, $0 < e < 1$, and so $(1 - e^2)$ is positive. To simplify notation, let $b = a\sqrt{1 - e^2}$. Then the equation derived above takes the form

$$\frac{x^2}{a^2} + \frac{y^2}{b^2} = 1$$

which is called the **standard equation of the ellipse**. The number $2a$ is the **major diameter**, whereas $2b$ is the **minor diameter**. Moreover, because $c = ea$, the numbers a, b, and c satisfy the Pythagorean relationship $a^2 = b^2 + c^2$, as is easily verified. All this is summarized in Figure 11. Note the role of the shaded triangle, which captures the condition $a^2 = b^2 + c^2$.

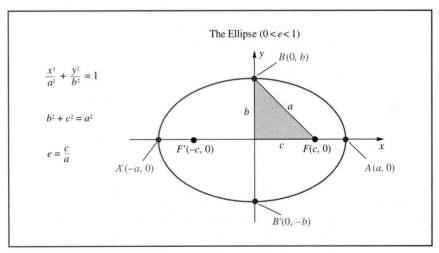

Figure 11

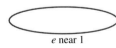
e near 1

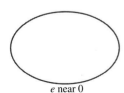
e near 0

Figure 12

Consider the effect of changing the value of e. If e is near 1, then $b = a\sqrt{1 - e^2}$ is small relative to a; the ellipse is thin and very eccentric. On the other hand, if e is near 0 (near zero eccentricity), b is almost as large as a; the ellipse is fat and well rounded (Figure 12). In the limiting case where $b = a$, the equation takes the form

$$\frac{x^2}{a^2} + \frac{y^2}{a^2} = 1$$

which is equivalent to $x^2 + y^2 = a^2$. This is the equation of a circle of a radius a centered at the origin. In other words, *a circle is a special case of an ellipse.*

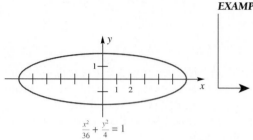

Figure 13

EXAMPLE 3: Sketch the graph of

$$\frac{x^2}{36} + \frac{y^2}{4} = 1$$

and determine its foci and eccentricity.

SOLUTION: Because $a = 6$ and $b = 2$, we calculate

$$c = \sqrt{a^2 - b^2} = \sqrt{36 - 4} = 4\sqrt{2} \approx 5.66$$

The foci are at $(\pm c, 0) = (\pm 4\sqrt{2}, 0)$ and $e = c/a \approx 0.94$. The graph is sketched in Figure 13.   ◄

We call the ellipses sketched so far *horizontal ellipses* because the major axis is the x-axis. If we interchange the roles of x and y, we obtain a vertical ellipse with equation

$$\frac{x^2}{b^2} + \frac{y^2}{a^2} = 1$$

The Standard Equation of the Hyperbola

For the hyperbola, $e > 1$ and so $e^2 - 1$ is positive. If we let $b = a\sqrt{e^2 - 1}$, then the equation $x^2/a^2 + y^2/(1 - e^2)a^2 = 1$, which was derived earlier, takes the form

$$\boxed{\frac{x^2}{a^2} - \frac{y^2}{b^2} = 1}$$

This is called the **standard equation of the hyperbola**. Because $c = ea$, we now obtain $c^2 = a^2 + b^2$. (Note how this differs from the corresponding relationship for an ellipse, in which $a^2 = b^2 + c^2$.)

To interpret b, observe that if we solve for y in terms of x, we get

$$y = \pm \frac{b}{a} \sqrt{x^2 - a^2}$$

For large x, $\sqrt{x^2 - a^2}$ behaves like x—that is, $(\sqrt{x^2 - a^2} - x) \to 0$ as $x \to \infty$—and hence y behaves like

$$y = \pm \frac{b}{a} x$$

More precisely, the graph of the given hyperbola has these two lines as asymptotes.

The important facts for the hyperbola are summarized in Figure 14. Once again, there is an important triangle (shaded in our diagram); it determines the asymptotes mentioned above.

EXAMPLE 4: Sketch the graph of

$$\frac{x^2}{9} - \frac{y^2}{16} = 1$$

showing the asymptotes. What are the equations of the asymptotes? What are the foci?

SOLUTION: We begin by determining the fundamental triangle; it has horizontal leg 3 and vertical leg 4. After drawing it, we can indicate the asymptotes and sketch the graph (Figure 15). The asymptotes are $y = \pm\frac{4}{3}x$. Because $c = \sqrt{a^2 + b^2} = \sqrt{9 + 16} = 5$, the foci are at $(\pm 5, 0)$.   ◄

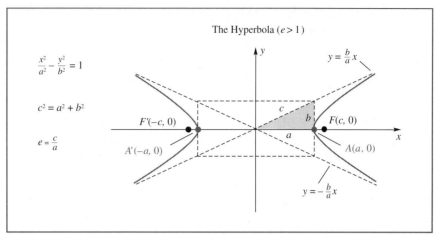

Figure 14

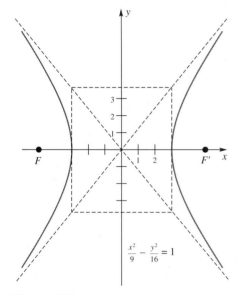

Figure 15

Again, we should consider the effect of interchanging the roles of x and y. The equation takes the form

$$\frac{y^2}{a^2} - \frac{x^2}{b^2} = 1$$

This is the equation of a vertical hyperbola (vertical major axis). Its vertices are at $(0, \pm a)$; its foci are at $(0, \pm c)$.

For both the ellipse and the hyperbola, a is always the distance from the center to a vertex. For the ellipse, $a > b$; for the hyperbola, there is no such requirement.

The String Properties of Ellipses and Hyperbolas
So far we have given the so-called eccentricity definitions of the ellipse and hyperbola. The condition $|PF| = e|PL|$ determines an ellipse if $0 < e < 1$ and a hyperbola if $e > 1$. This definition allowed us to treat these two curves in a unified way. Many authors prefer to introduce these curves by the following alternative definitions.

An **ellipse** is the set of all points P in the plane, the *sum* of whose distances from two fixed points (the foci) is a given positive constant $2a$.

A **hyperbola** is the set of all points P in the plane, the *difference* of whose distances from two fixed points (the foci) is a given positive constant $2a$. Here, the word *difference* is taken to mean the larger distance minus the smaller distance.

To interpret these definitions geometrically, study Figures 16 and 17. For the ellipse,

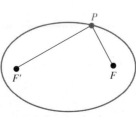

Ellipse: $|PF'| + |PF| = 2a$

Figure 16

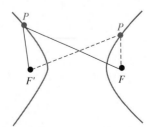

Hyperbola: $||PF'| - |PF'|| = 2a$

Figure 17

imagine a string of length $2a$ tacked down at points F and F'. If a pencil is held tight against the string with its tip at P, it can be used to trace the ellipse. We refer to the properties described in the new definitions as the *string properties* of the ellipse and hyperbola. These properties can be shown to be consequences of the eccentricity definition.

EXAMPLE 5: Find the equation of the set of points, the sum of whose distances from $(\pm 3, 0)$ is equal to 10.

    SOLUTION: This is a horizontal ellipse with $c = 3$. Because $2a = 10$, $a = 5$. Thus $b = \sqrt{a^2 - c^2} = 4$, and the equation is

$$\frac{x^2}{25} + \frac{y^2}{16} = 1 \qquad \blacktriangleleft$$

EXAMPLE 6: Find the equation of the set of points, the difference of whose distances from $(0, \pm 6)$ is equal to 4.

    SOLUTION: This is a vertical hyperbola with $a = 2$ and $c = 6$. Thus, $b = \sqrt{c^2 - a^2} = \sqrt{32} = 4\sqrt{2}$, and the equation is

$$-\frac{x^2}{32} + \frac{y^2}{4} = 1 \qquad \blacktriangleleft$$

Lenses

The optical properties of the conics have been used in the grinding of lenses for hundreds of years. A recent innovation is the introduction of variable lenses to replace bifocal lenses in eyeglasses. Starting from the top, these lenses are ground so that the eccentricity varies continuously from small to large, thus producing cross sections from ellipses to parabolas to hyperbolas and presumably allowing perfect viewing of objects at any distance by tilting the head.

Optical Properties   Consider mirrors with the shapes of an ellipse and a hyperbola, respectively. If a light ray emanating from one focus strikes the mirror, it will be reflected back to the other focus in the case of the ellipse and directly away from the other focus in the case of the hyperbola. These facts are shown in Figure 18.

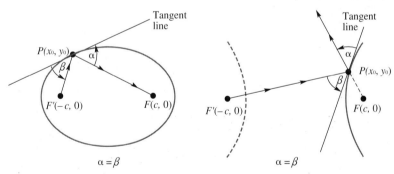

Figure 18

Applications
The reflecting property of the ellipse is the basis of the "whispering gallery" effect that can be observed, for example, in the U.S. Capitol and the Mormon Tabernacle. A speaker standing at one focus can be heard whispering by a listener at the other focus, even though his or her voice is inaudible in other parts of the room. This works because sound waves from one focus, striking the elliptical walls at many different points, are all reflected directly to the other focus. The same property is used in the nonsurgical treatment of kidney stones, where ultrasound waves are used to shatter the kidney stones, which can then easily pass through the patient's system.

    The optical properties of the parabola and hyperbola are combined in one design for a reflecting telescope (Figure 19). The parallel rays from a star are finally focused at the eyepieces at F'.

The string property of the hyperbola is used in navigation. A ship at sea can determine the difference $2a$ in its distance from the fixed transmitters by measuring the difference in reception times of synchronized radio signals. This puts its path on a hyperbola, with the two transmitters F and F' as foci. If another pair of transmitters G and G' are used, the ship must lie at the intersection of the two corresponding hyperbolas (see Figure 20). LORAN, a system of long-range navigation, is based on this principle.

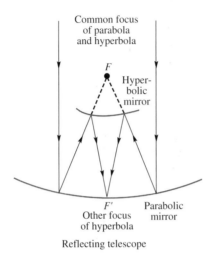

Figure 19   **Figure 20**

Concepts Review

1. The standard equation of the horizontal ellipse centered at $(0, 0)$ is _____ .

2. The hyperbola $x^2/9 - y^2/4 = 1$ has asymptotes _____ .

3. A ray from a light source at one focus of an elliptical mirror will be reflected _____ .

4. A ray from a light source at the focus of a parabolic mirror will be reflected _____ .

Problem Set 10.1

In Problems 1–12, sketch the graph of the given equation indicating vertices, foci, directrices, and asymptotes (if it is a hyperbola).

1. $\dfrac{x^2}{9} + \dfrac{y^2}{16} = 1$

2. $\dfrac{x^2}{10} - \dfrac{y^2}{4} = 1$

3. $\dfrac{-x^2}{16} + \dfrac{y^2}{9} = 1$

4. $\dfrac{x^2}{10} + \dfrac{y^2}{4} = 1$

5. $y^2 = 16x$

6. $y^2 = -28x$

7. $x^2 + 4y^2 = 36$

8. $25x^2 + 4y^2 = 100$

9. $x^2 - 6y = 0$

10. $2x^2 + 7y = 0$

11. $4x^2 - 25y^2 = 100$

12. $x^2 - 4y^2 = 4$

In Problems 13–22, find the equation of the given conic. Assume the center is at the origin.

13. The parabola with focus at $(3, 0)$.

14. The parabola with directrix $x = 2$.

15. The parabola with directrix $y + 4 = 0$.

16. The parabola with focus at $(0, -\frac{1}{3})$.

17. The ellipse with a focus at $(-3, 0)$ and a vertex at $(6, 0)$.

18. The ellipse with a focus at $(6, 0)$ and eccentricity $\frac{2}{3}$.

19.   The hyperbola with a vertex at $(0, -4)$ and a focus at $(0, -5)$.

20.   The hyperbola with a vertex at $(0, -3)$ and eccentricity $\frac{3}{2}$.

21.   The ellipse with foci $(\pm 2, 0)$ and directrices $x = \pm 8$.

22.   The hyperbola with foci $(\pm 4, 0)$ and directrices $x = \pm 1$.

In Problems 23–26, find the equation of the set of points P satisfying the given conditions.

23.   The sum of the distances of P from $(0, \pm 4)$ is 10.

24.   The sum of the distances of P from $(\pm 12, 0)$ is 26.

25.   The difference of the distances of P from $(\pm 5, 0)$ is 8.

26.   The difference of the distances of P from $(0, \pm 13)$ is 10.

27.   The slope of the tangent to the parabola $y^2 = 5x$ at a certain point on the parabola is $\sqrt{5}/4$. Find the coordinates of that point. Make a sketch.

28.   The slope of the tangent to the parabola $x^2 = -14y$ at a certain point on the parabola is $-2\sqrt{7}/7$. Find the coordinates of that point.

29.   Find the equation of the tangent to the parabola $y^2 = -18x$ that is parallel to the line $3x - 2y + 4 = 0$.

30.   Prove that the vertex is the point on a parabola closest to the focus.

In Problems 31–36, find the equation of the tangent line to the given curve at the given point.

31. $\dfrac{x^2}{9} + \dfrac{y^2}{27} = 1$ at $(\sqrt{6}, 3)$

32. $\dfrac{x^2}{16} + \dfrac{y^2}{24} = 1$ at $(-2, 3\sqrt{2})$

33. $\dfrac{x^2}{8} - \dfrac{y^2}{4} = 1$ at $(-4, 2)$

34. $-\dfrac{x^2}{36} + \dfrac{y^2}{8} = 1$ at $(6, 4)$

35. $x^2 + y^2 = 25$ at $(3, 4)$

36. $x^2 - y^2 = 9$ at $(5, 4)$

37.   A spaceship from outer space is sighted from Earth moving on a parabolic path with Earth at the focus. When the line from Earth to the spaceship first makes an angle of $90°$ with the axis of the parabolas, it is measured to be 40 million miles away. How close will the spaceship come to Earth? Treat Earth as a point.

38.   Work Problem 37 assuming the angle is $75°$ rather than $90°$.

39.   The cables for the central span of a suspension bridge take the shape of a parabola. If the towers are 800 meters apart and the cables are attached to them at points 400 meters above the floor of the bridge, how long must the vertical strut be that is 100 meters from the tower? Assume that the cable touches the floor at the midpoint of the bridge (Figure 21).

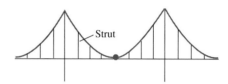

Figure 21

40.   The slope of the tangent to the hyperbola $2x^2 - 7y^2 - 35 = 0$ at a certain point on the hyperbola is $-\frac{2}{3}$. What are the coordinates of the point of tangency (two solutions)?

41.   Find the equations of the tangents to the ellipse $x^2 + 2y^2 - 2 = 0$ that are parallel to the line $3x - 3\sqrt{2}y - 7 = 0$.

42.   Find the area of the ellipse $b^2x^2 + a^2y^2 = a^2b^2$.

43.   Find the volume of the solid obtained by revolving the ellipse $b^2x^2 + a^2y^2 = a^2b^2$ about the y-axis.

44.   The region bounded by the hyperbola $b^2x^2 - a^2y^2 = a^2b^2$ and a vertical line through a focus is revolved about the x-axis. Find the volume of the resulting solid.

45.   Find the dimensions of the rectangle having the greatest possible area that can be inscribed in the ellipse $b^2x^2 + a^2y^2 = a^2b^2$. Assume that the sides of the rectangle are parallel to the axes of the ellipse.

46.   Consider a bridge deck weighing δ pounds per lineal foot and supported by a cable, which is assumed to be of negligible weight compared to the bridge deck. The section OP from the lowest point (the origin) to a general point $P(x, y)$ is shown in Figure 22.

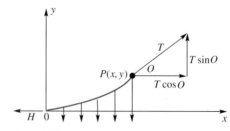

Figure 22

The forces acting on this section of cable are:

H = horizontal tension pulling at 0

T = tangential tension pulling at P

$W = \delta x$ = weight of x feet of bridge deck

For equilibrium, the horizontal and vertical components of T must balance H and W, respectively. Thus,

$$\frac{T \sin \phi}{T \cos \phi} = \tan \phi = \frac{\delta x}{H}$$

that is,

$$\frac{dy}{dx} = \frac{\delta x}{H}, \qquad y(0) = 0$$

Solve this differential equation to show that the cable hangs in the shape of a parabola. (Compare this result with that for the unloaded hanging cable of Problem 29 of Section 7.8.)

47. A door has the shape of an elliptical arch (a half-ellipse) that is 10 feet wide and 4 feet high at the center. A box 2 feet high is to be pushed through the door. How wide can the box be?

48. How high is the arch of Problem 47 at a distance 2 feet from the center?

49. Halley's comet has an elliptical orbit with major and minor diameters of 36.18 AU and 9.12 AU, respectively (1 AU is 1 astronomical unit, Earth's mean distance from the sun). What is its closest approach to the sun (assuming the sun is at a focus)?

50. The orbit of the comet Kahoutek is an ellipse with eccentricity $e = 0.999925$ with the sun at a focus. If its minimum distance to the sun is 0.13 AU, what is its maximum distance from the sun?

51. In October 1957, Russia launched Sputnik I. Its elliptical orbit around Earth reached maximum and minimum distances from Earth of 583 miles and 132 miles respectively. Assuming Earth is a sphere of radius 4000 miles, find the eccentricity of the orbit.

52. The wheel in Figure 23 is turning at t radians per second so that Q has coordinates $(a \cos t, a \sin t)$. Find the coordinates (x, y) of R at time t and show that it is traveling an elliptical path. *Note*: PQR is a right triangle.

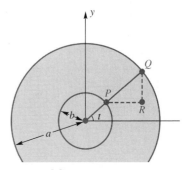

Figure 23

53. Let P be a point on a ladder of length $a + b$, P being a units from the top end. As the ladder slides with its top end on the y-axis and its bottom end on the x-axis, P traces out a curve. Find the equation of this curve.

54. Using the same axes, draw the conics $y = \pm(ax^2 + 1)^{1/2}$ for $-2 \le x \le 2$ and $-2 \le y \le 2$ using $a = -2, -1, -0.5, -0.1, 0, 0.1, 0.6, 1$. Make a conjecture.

55. A ball placed at a focus of an elliptical billiard table is shot with tremendous force so that it continues to bounce off the cushions indefinitely. What is its ultimate path? *Hint*: Draw a picture.

56. If the ball of Problem 55 is initially on the major axis between a focus and the neighboring vertex, what can you say about its path?

57. Describe a string apparatus for constructing a hyperbola. (There are several possibilities).

Answers to Concepts Review: 1. $x^2/a^2 + y^2/b^2 = 1$
2. $y = \pm\frac{2}{3}x$ 3. To the other focus.   4. In a direction parallel to the axis.

10.2 Translation of Axes

So far we have placed the conics in the coordinate system in very special ways—always with the major axis along one of the coordinate axes and either the vertex (in the case of a parabola) or the center (in the case of an ellipse or hyperbola) at the origin. Now we place our conics in a more general position, though we still require that the major axis be parallel to one of the coordinates axes.

The case of a circle is instructive. The circle of radius 5 centered at (2, 3) has equation

$$(x - 2)^2 + (y - 3)^2 = 25$$

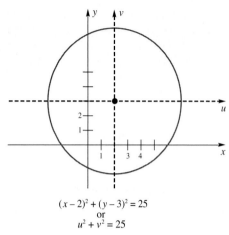

$$(x - 2)^2 + (y - 3)^2 = 25$$
or
$$u^2 + v^2 = 25$$

Figure 1

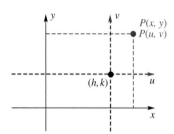

Figure 2

or, in equivalent expanded form,

$$x^2 + y^2 - 4x - 6y = 12$$

The same circle with its center at the origin of the uv-coordinate system (Figure 1) has the simple equation

$$u^2 + v^2 = 25$$

The introduction of new axes does not change the shape or size of a curve, but it may greatly simplify its equation. It is this so-called translation of axes and the corresponding change of variables in an equation that we wish to investigate.

Translations If new axes are chosen in the plane, every point will have two sets of coordinates: the old ones, (x, y), relative to the old axes and the new ones, (u, v), relative to the new axes. The original coordinates are said to undergo a **transformation**. If the new axes are parallel, respectively, to the original axes and have the same directions, the transformation is called a **translation of axes**.

From Figure 2, it is easy to see how the new coordinates (u, v), relate to the old ones (x, y). Let (h, k) be the old coordinates of the new origin. Then,

$$\boxed{u = x - h, \quad v = y - k}$$

or equivalently,

$$\boxed{x = u + h, \quad y = v + k}$$

EXAMPLE 1: Find the new coordinates of $P(-6, 5)$ after a translation of axes to a new origin at $(2, -4)$.

SOLUTION: Because $h = 2$ and $k = -4$, it follows that

$$u = x - h = -6 - 2 = -8 \qquad v = y - k = 5 - (-4) = 9$$

The new coordinates are $(-8, 9)$. ◀

EXAMPLE 2: Given the equation $4x^2 + y^2 + 40x - 2y + 97 = 0$, find the equation of its graph after a translation with new origin $(-5, 1)$.

SOLUTION: In the equation, we replace x by $u + h = u - 5$ and y by $v + k = v + 1$. We obtain

$$4(u - 5)^2 + (v + 1)^2 + 40(u - 5) - 2(v + 1) + 97 = 0$$

or

$$4u^2 - 40u + 100 + v^2 + 2v + 1 + 40u - 200 - 2v - 2 + 97 = 0$$

This simplifies to

$$4u^2 + v^2 = 4$$

or

$$u^2 + \frac{v^2}{4} = 1$$

which we recognize as the equation of an ellipse. ◀

Completing the Square   Given a complicated second-degree equation, how do we know what translation will simplify the equation and bring it to a recognizable form? Here a familiar algebraic process called **completing the square** provides the answer. In particular, we can use this process to eliminate the first-degree terms (that is, the Dx and Ey terms) of any expression of the form

$$Ax^2 + Cy^2 + Dx + Ey + F = 0, \qquad A \neq 0, C \neq 0$$

EXAMPLE 3: Make a translation that will eliminate the first-degree terms of

$$4x^2 + 9y^2 + 8x - 90y + 193 = 0$$

and use this information to sketch the graph of the given equation.

SOLUTION: Recall that to complete the square of $x^2 + ax$, we must add $a^2/4$ (the square of half the coefficient of x). Using this, we rewrite the given equation by adding the same numbers to both sides.

$$4(x^2 + 2x \quad) + 9(y^2 - 10y \quad) = -193$$
$$4(x^2 + 2x + \mathbf{1}) + 9(y^2 - 10y + \mathbf{25}) = -193 + \mathbf{4} + \mathbf{225}$$
$$4(x + 1)^2 + 9(y - 5)^2 = 36$$
$$\frac{(x + 1)^2}{9} + \frac{(y - 5)^2}{4} = 1$$

The translation $u = x + 1$ and $v = y - 5$ transforms this to

$$\frac{u^2}{9} + \frac{v^2}{4} = 1$$

which is the standard form of a horizontal ellipse. The graph is shown in Figure 3.   ◄

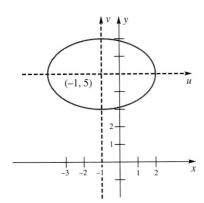

Figure 3

EXAMPLE 4: Use a translation to simplify

$$y^2 - 4x - 12y + 28 = 0$$

Then determine which conic it represents, list the important characteristics of this conic, and sketch its graph.

SOLUTION: We complete the square.

$$y^2 - 12y \qquad = 4x - 28$$
$$y^2 - 12y + \mathbf{36} = 4x - 28 + \mathbf{36}$$
$$(y - 6)^2 = 4(x + 2)$$

The translation $u = x + 2$, $v = y - 6$ transforms this to $v^2 = 4u$, which we recognize as a horizontal parabola opening right with $p = 1$ (Figure 4).   ◄

General Second-Degree Equations   Now we ask an important question. Is the graph of an equation of the form

$$Ax^2 + Cy^2 + Dx + Ey + F = 0$$

always a conic? The answer is no, unless we admit certain limiting forms. The table below indicates the possibilities with a sample equation for each.

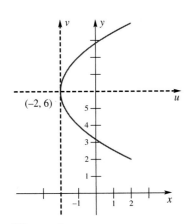

Figure 4

Conics	Limiting Forms
1. ($AC = 0$) Parabola: $y^2 = 4x$	Parallel lines: $y^2 = 4$ Single line: $y^2 = 0$ Empty set: $y^2 = -1$
2. ($AC > 0$) Ellipse: $\frac{x^2}{9} + \frac{y^2}{4} = 1$	Circle: $x^2 + y^2 = 4$ Point: $2x^2 + y^2 = 0$ Empty set: $2x^2 + y^2 = -1$
3. ($AC < 0$) Hyperbola: $\frac{x^2}{9} - \frac{y^2}{4} = 1$	Intersecting lines: $x^2 - y^2 = 0$

Thus, the graphs of the general quadratic equation above fall into three general categories but yield nine different possibilities, including limiting forms.

EXAMPLE 5: Use a translation to simplify

$$4x^2 - y^2 - 8x - 6y - 5 = 0$$

and sketch its graph.

SOLUTION: We rewrite the equation as follows.

$$4(x^2 - 2x \quad) - (y^2 + 6y \quad) = 5$$

$$4(x^2 - 2x + \mathbf{1}) - (y^2 + 6y + \mathbf{9}) = 5 + \mathbf{4} - \mathbf{9}$$

$$4(x - 1)^2 - (y + 3)^2 = 0$$

Let $u = x - 1$ and $v = y + 3$, which results in

$$4u^2 - v^2 = 0$$

or

$$(2u - v)(2u + v) = 0$$

This is the equation of two intersecting lines (Figure 5).   ◀

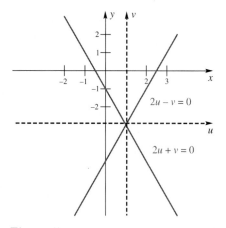

Figure 5

EXAMPLE 6 Write the equation of a hyperbola with foci at $(1, 1)$ and $(1, 11)$ and vertices at $(1, 3)$ and $(1, 9)$.

SOLUTION: The hyperbola is vertical and the center is $(1, 6)$, midway between the vertices. Thus, $a = 3$ and $c = 5$, and so $b = \sqrt{c^2 - a^2} = 4$. The equation is

$$\frac{(y - 6)^2}{9} - \frac{(x - 1)^2}{16} = 1$$ ◀

Summary   Consider the general equation

$$Ax^2 + Cy^2 + Dx + Ey + F = 0$$

If both A and C are zero, we have the equation of a line (provided, of course, that D and E are not both zero). If at least one of A and C is different from zero, we may apply the process of completing the square. We obtain one of several forms, the most typical being:

(1)
$$(y - k)^2 = \pm 4\,p(x - h)$$

(2)
$$\frac{(x - h)^2}{a^2} + \frac{(y - k)^2}{b^2} = 1$$

(3)
$$\frac{(x - h)^2}{a^2} - \frac{(y - k)^2}{b^2} = 1$$

These can be recognized even in this form as the equations of a horizontal parabola with vertex at (h, k), a horizontal ellipse (if $a^2 > b^2$) with center at (h, k), and a horizontal hyperbola with center at (h, k). But to remove any doubt, we may translate the axes by the substitutions $u = x - h$, $v = y - k$, thereby obtaining:

(1)
$$v^2 = \pm 4pu$$

(2)
$$\frac{u^2}{a^2} + \frac{v^2}{b^2} = 1$$

(3)
$$\frac{u^2}{a^2} - \frac{v^2}{b^2} = 1$$

Our work may yield these equations with u and v interchanged, or we may get one of the six limiting forms illustrated in the table on page 602. There are no other possibilities.

Concepts Review

1. The quadratic from $x^2 + ax$ is made a square by adding _____ .

2. $x^2 + 6x + 2(y^2 - 2y) = 3$ is (after completing the squares) equivalent to $(x + 3)^2 + 2(y - 1)^2 =$ _____ , which is the equation of a(n) _____ .

3. Besides … circles, parabolas, and hyperbolas, … other possible graphs for a second-degree equation in x and y are _____ .

4. The graph of $4x^2 - 9y^2 = 0$ is _____ .

Problem Set 10.2

In Problems 1–16, name the conic or limiting form represented by the given equation. Usually you will need to use the process of completing the square (see Examples 3–5).

1. $x^2 + y^2 - 2x + 4y + 4 = 0$

2. $x^2 + y^2 - 6x + 2y + 6 = 0$

3. $4x^2 + 9y^2 - 16x + 72y + 124 = 0$

4. $9x^2 - 16y^2 + 90x + 192y - 495 = 0$

5. $4x^2 + 9y^2 - 16x + 72y + 160 = 0$

6. $9x^2 + 16y^2 + 90x + 192y + 1000 = 0$

7. $y^2 - 10x - 8y - 14 = 0$

8. $4x^2 + 4y^2 + 16x - 20y - 10 = 0$

9. $x^2 + y^2 - 2x + 4y + 20 = 0$

10. $4x^2 - 4y^2 + 16x - 20y - 10 = 0$

11. $4x^2 - 4y^2 + 16x - 20y - 9 = 0$

12. $4x^2 - 16x + 16 = 0$

13. $4x^2 - 16x + 15 = 0$

14. $25x^2 + 4y^2 + 150x - 8y + 129 = 0$

15. $25x^2 - 4y^2 + 150x - 8y + 129 = 0$

16. $3x + 4y - 16 = 0$

In Problems 17–30, sketch the graph of the given equation.

17. $\dfrac{(x + 3)^2}{4} + \dfrac{(y + 2)^2}{16} = 1$

18. $(x + 3)^2 + (y - 4)^2 = 25$

19. $\dfrac{(x + 3)^2}{4} - \dfrac{(y + 2)^2}{16} = 1$

20. $4(x + 3) = (y + 2)^2$

21. $(x + 2)^2 = 8(y - 1)$

22. $(x + 2)^2 = 4$

23. $(y - 1)^2 = 16$

24. $\dfrac{(x + 3)^2}{4} + \dfrac{(y - 2)^2}{8} = 0$

25. $x^2 + 4y^2 - 2x + 16y + 1 = 0$

26. $25x^2 + 9y^2 + 150x - 18y + 9 = 0$

27. $9x^2 - 16y^2 + 54x + 64y - 127 = 0$

28. $x^2 - 4y^2 - 14x - 32y - 11 = 0$

29. $4x^2 + 16x - 16y + 32 = 0$

30. $x^2 - 4x + 8y = 0$

31. Find the focus and directrix of the parabola

$$2y^2 - 4y - 10x = 0$$

32. Determine the distance between the vertices of

$$-9x^2 + 18x + 4y^2 + 24y = 9$$

33. Find the foci of the ellipse

$$16(x - 1)^2 + 25(y + 2)^2 = 400$$

34. Find the focus and directrix of the parabola

$$x^2 - 6x + 4y + 3 = 0$$

In Problems 35–44, find the equation of the given conic.

35. The horizontal ellipse with center $(5, 1)$, major diameter 10, minor diameter 8.

36. The hyperbola with center $(2, -1)$, vertex at $(4, -1)$, and focus at $(5, -1)$.

37. The parabola with vertex $(2, 3)$ and focus $(2, 5)$.

38. The ellipse with center $(2, 3)$ passing through $(6, 3)$ and $(2, 5)$.

39. The hyperbola with vertices at $(0, 0)$ and $(0, 6)$ and a focus at $(0, 8)$.

40. The ellipse with foci at $(2, 0)$ and $(2, 12)$ and a vertex at $(2, 14)$.

41. The parabola with focus $(2, 5)$ and directrix $x = 10$.

42. The parabola with focus $(2, 5)$ and vertex $(2, 6)$.

43. The ellipse with foci $(\pm 2, 2)$ that passes through the origin.

44. The hyperbola with foci $(0, 0)$ and $(0, 4)$ that passes through $(12, 9)$.

45. A curve C goes through the three points, $(-1, 2)$, $(0, 0)$, and $(3, 6)$. Find the equation for C if C is:

(a) a vertical parabola;   (b) a horizontal parabola;

(c) a circle.

46. The ends of an elastic string with a knot at $K(x, y)$ are attached to a fixed point $A(a, b)$ and a point P on the rim of a wheel of radius r centered at $(0, 0)$. As the wheel turns, K traces a curve C. Find the equation of C. Assume the string stays taut and stretches uniformly (that is, $\alpha = |KP| / |AP|$ is constant).

47. Show that the equation of the parabola and hyperbola with vertex $(a, 0)$ and focus $(c, 0)$, $c > a > 0$, can be written as $y^2 = 4(c - a)(x - a)$ and $y^2 = (b^2/a^2)(x^2 - a^2)$, respectively. Then use these expressions for y^2 to show that the parabola is always "inside" the right branch of the hyperbola.

Answers to Concepts Review: 1. $a^2/4$ 2. 14; ellipse 3. A line, parallel lines, intersecting lines, a point, the empty set   4. Intersecting lines

---

$\boxed{10.3}$ The Polar Coordinate System and Graphs

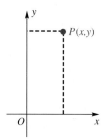

Cartesian Coordinates

Figure 1

Two Frenchmen, Pierre Fermat and René Descartes, introduced what we now call the *Cartesian*, or *rectangular*, coordinate system. Their idea was to specify each point P in the plane by giving two numbers (x, y), the directed distances from a pair of perpendicular axes (Figure 1). This notion is by now so familiar that we use it almost without thinking. Yet it is the fundamental idea in analytic geometry and makes possible the development of calculus as we have given it so far.

Giving the directed distances from a pair of perpendicular axes is not the only way to specify a point. Another way to do this is by giving the polar coordinates.

Polar Coordinates   We start with a fixed half-line, called the **polar axis**, emanating from a fixed point 0, called the **pole** or **origin**. By custom, the polar axis is chosen to

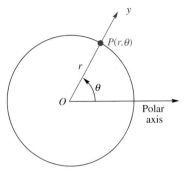

Figure 2

be horizontal and pointing to the right and may therefore be identified with the positive x-axis in the rectangular coordinate system. Any point P (other than the pole) is the intersection of a unique circle with center at 0 and a unique ray emanating from 0. If r is the radius of the circle and θ is the counterclockwise angle the ray makes with the polar axis, then (r, θ) is a pair of **polar coordinates** for P (Figure 2).

Points specified by polar coordinates are easiest to plot if we use polar graph paper. The grid on such paper consists of concentric circles and rays emanating from their common center. A simple version of a polar grid is shown in Figure 3, where we have also plotted several points.

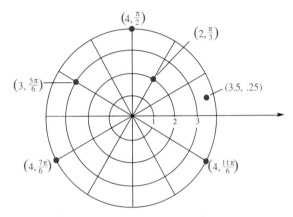

Figure 3

Notice a phenomenon that did not occur with Cartesian coordinates. Each point has many sets of polar coordinates, due to the fact that the angles $\theta + 2\pi n$, $n = 0, \pm 1, \pm 2, \dots$, have the same terminal sides. For example, the point with polar coordinates $(4, \pi/2)$ also has coordinates $(4, 5\pi/2)$, $(4, 9\pi/2)$, $(4, -3\pi/2)$, and so on. There are even more descriptions for a single point if we also allow r to be negative. In this case, (r, θ) is on the ray oppositely directed from the terminal side of θ and $|r|$ units from the origin. Thus, the point with polar coordinates $(-3, \pi/6)$ is as shown in Figure 4, and $(-4, 3\pi/2)$ is another set of coordinates for $(4, \pi/2)$. The origin has coordinates $(0, \theta)$, where θ is any angle.

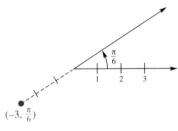

Figure 4

Polar Equations   Examples of polar equations are

$$r = 8 \sin \theta \quad \text{and} \quad r = \frac{2}{1 - \cos \theta}$$

Polar equations, like rectangular ones, are best visualized from their graphs. The **graph of a polar equation** is the set of points, each of which has at least one pair of polar coordinates that satisfy the equation. The most basic way to sketch a graph is to construct a table of values, plot the corresponding points, and then connect these points with a smooth curve.

EXAMPLE 1: Sketch the graph of the polar equation $r = 8 \sin \theta$.

SOLUTION: We substitute multiples of $\pi/6$ for θ and calculate the corresponding r-values. Note that as θ increases from 0 to 2π, the graph is traced twice (Figure 5). ◄

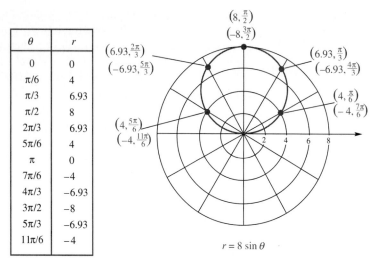

θ	r
0	0
$\pi/6$	4
$\pi/3$	6.93
$\pi/2$	8
$2\pi/3$	6.93
$5\pi/6$	4
π	0
$7\pi/6$	−4
$4\pi/3$	−6.93
$3\pi/2$	−8
$5\pi/3$	−6.93
$11\pi/6$	−4

$r = 8 \sin \theta$

Figure 5

For this equation, if we used values of θ larger than those in the table, we would just trace out the same figure (why?).

EXAMPLE 2: Sketch the graph of $r = \dfrac{2}{1 - \cos \theta}$.

   SOLUTION: See Figure 6.

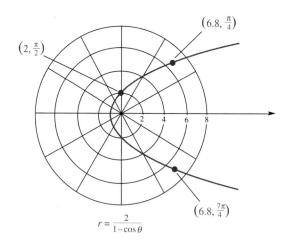

θ	r
0	–
$\pi/4$	6.8
$\pi/2$	2
$3\pi/4$	1.2
π	1
$5\pi/4$	1.2
$3\pi/2$	2
$7\pi/4$	6.8
$2/\pi$	–

$r = \dfrac{2}{1-\cos \theta}$

Figure 6

  Note a phenomenon that does not occur with rectangular coordinates. The coordinates $(-2, 3\pi/2)$ do not satisfy the equation. Yet the point $P(-2, 3\pi/2)$ is on the graph, due to the fact that $(2, \pi/2)$ specifies the same point and does satisfy the equation. we conclude that *in polar coordinates, failure of a particular set of coordinates to satisfy a given equation is no guarantee that the corresponding point is not on the graph of that equation.* This fact causes many difficulties; we must learn to live with them.

Relation to Cartesian Coordinates   We suppose that the polar axis coincides with the positive *x*-axis of the Cartesian system. Then the polar coordinates (r, θ) of a point P and the Cartesian coordinates (x, y) of the same point are related by the equations

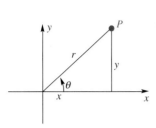

Figure 7

$$x = r \cos \theta \qquad r^2 = x^2 + y^2$$

$$y = r \sin \theta \qquad \tan \theta = \frac{y}{x}$$

That this is true for a point P in the first quadrant is clear from Figure 7 and is easy to show for points in the other quadrants.

Note: Many calculators have a key for converting between rectangular and polar coordinates directly.

EXAMPLE 3: Find the Cartesian coordinates corresponding to $(4, \pi/6)$ and polar coordinates corresponding to $\left(3, \sqrt{3}\right)$.

SOLUTION: If $(r, \theta) = (4, \pi/6)$, then

$$x = 4 \cos \frac{\pi}{6} = 4 \cdot \frac{\sqrt{3}}{2} = 2\sqrt{3}$$

$$y = 4 \sin \frac{\pi}{6} = 4 \cdot \frac{1}{2} = 2$$

If $(x, y) = \left(3, \sqrt{3}\right)$, then (see Figure 8)

$$r^2 = (-3)^2 + \left(\sqrt{3}\right)^2 = 12$$

$$\tan \theta = \frac{\sqrt{3}}{3}$$

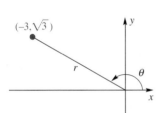

Figure 8

One value of (r, θ) is $\left(2\sqrt{3}, 5\pi/6\right)$. Another is $\left(-2\sqrt{3}, -\pi/6\right)$. Notice that $\left(-2\sqrt{3}, 5\pi/6\right)$ is *not* a possible value, because $\left(-3, \sqrt{3}\right)$ is in the second quadrant, but $\left(-2\sqrt{3}, 5\pi/6\right)$ is in the fourth quadrant. Thus when converting to polar coordinates, you must use some geometric common sense in addition to the formulas.

Sometimes we can identify the graph of a polar equation by finding its equivalent Cartesian form. Here is an illustration.   ◀

EXAMPLE 4: Show that the graph of $r = 8 \sin \theta$ (Example 1) is a circle and that the graph of $r = 2/(1 - \cos \theta)$ (Example 2) is a parabola by changing to Cartesian coordinates.

SOLUTION: If we multiply $r = 8 \sin \theta$ by r, we get

$$r^2 = 8r \sin \theta$$

Translating to Cartesian coordinates, we replace r^2 with $x^2 + y^2$ and $r \sin \theta$ with y to get

$$x^2 + y^2 = 8y$$

Caution

Because r can be 0, there is a potential danger in multiplying both sides of a polar equation by r or in dividing both sides by r. In the first case, we might add the pole to the graph; in the second, we might delete the pole from the graph. In Example 4, we multiplied both sides of $r = 8 \sin \theta$ by r but no harm was done because the pole was already on the graph as the point with θ-coordinate 0.

which may be written successively as

$$x^2 + y^2 - 8y = 0$$

$$x^2 + y^2 - 8y + 16 = 16$$

$$x^2 + (y - 4)^2 = 16$$

The latter is the equation of a circle of radius 4 centered at $(0, 4)$. The second equation is handled by the following steps.

$$r = \frac{2}{1 - \cos \theta}$$

$$r - r \cos \theta = 2$$

$$r - x = 2$$

$$r = x + 2$$

$$r^2 = x^2 + 4x + 4$$

$$x^2 + y^2 = x^2 + 4x + 4$$

$$y^2 = 4(x + 1)$$

We recognize the last equation as that of a parabola with vertex at $(-1, 0)$ and focus at the origin. ◄

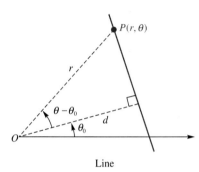

Figure 9

Polar Equations for Lines, Circles, and Conics

If a line passes through the pole, it has the simple equation $\theta = \theta_0$. If the line does not go through the pole, it is some distance $d > 0$ from it. Let θ_0 be the angle from the polar axis to the perpendicular from the pole to the given line (Figure 9). Then, if $P(r, \theta)$ is any point on the line, $\cos(\theta - \theta_0) = d/r$, or

$$\boxed{\text{Line:} \quad r = \frac{d}{\cos(\theta - \theta_0)}}$$

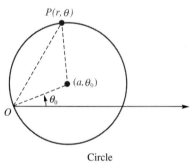

Figure 10

If a circle of radius a is centered at the pole, its equation is simply $r = a$. If it is centered at (r_0, θ_0), its equation is quite complicated unless we choose $r_0 = a$, as in Figure 10. Then, by the Law of Cosines, $a^2 = r^2 + a^2 - 2ra \cos(\theta - \theta_0)$, which simplifies to

$$\boxed{\text{Circle:} \quad r = 2a \cos(\theta - \theta_0)}$$

The cases $\theta_0 = 0$ and $\theta_0 = \pi/2$ are particularly nice. The first gives $r = 2a \cos \theta$; the second gives $r = 2a \cos(\theta - \pi/2)$; that is, $r = 2a \sin \theta$. The latter should be compared with Example 1.

Finally, if a conic (ellipse, parabola, or hyperbola) is placed so that its focus is at the pole, and its directrix is d units away, as in Figure 11, then the familiar defining equation $|PF| = e|PL|$ takes the form

$$r = e[d - r \cos(\theta - \theta_0)]$$

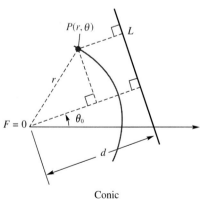

Figure 11

or, equivalently,

$$\text{Conic:} \quad r = \frac{ed}{1 + e\cos(\theta - \theta_0)}$$

Again, there is special interest in the cases $\theta_0 = 0$ and $\theta_0 = \pi/2$. Note in particular that if $e = 1$, $d = 2$, and $\theta_0 = \pi$, we have the equation of Example 2, because $\cos(\theta - \pi) = -\cos\theta$.

Our results are summarized in the following chart.

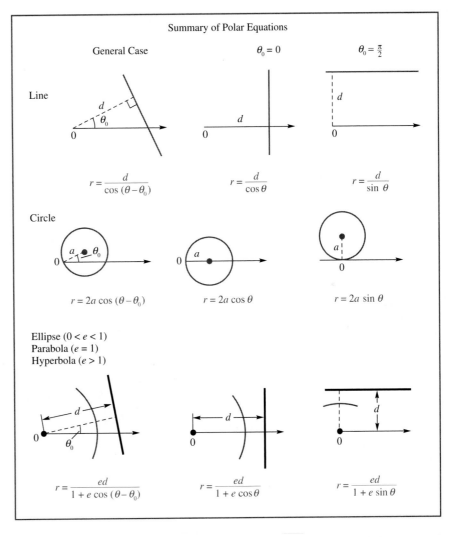

EXAMPLE 5: Find the equation of the horizontal ellipse with eccentricity $\frac{1}{2}$, focus at the pole, and vertical directrix 10 units to the right of the pole.

SOLUTION: Because $\theta_0 = 0$, we have

$$r = \frac{\frac{1}{2} \cdot 10}{1 + \frac{1}{2}\cos\theta} = \frac{10}{2 + \cos\theta}$$

◀

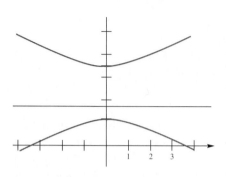

EXAMPLE 6: Identify and sketch the graph of $r = \dfrac{7}{2 + 4\sin\theta}$.

SOLUTION: We put this equation in standard form.

$$r = \frac{7}{2 + 4\sin\theta} = \frac{\left(\frac{7}{2}\right)}{1 + 2\sin\theta} = \frac{2\left(\frac{7}{4}\right)}{1 + 2\sin\theta}$$

which we recognize as the polar equation of hyperbola with $e = 2$, focus at the pole, and horizontal directrix $\frac{7}{4}$ units above the polar axis (Figure 12).   ◄

Concepts Review

1. Every point in the plane has a unique pair (x, y) of Cartesian coordinates but _____ pairs (r, θ) of polar coordinates.

2. The relations $x =$ _____ and $y =$ _____ connect Cartesian and polar coordinates; also _____ = $x^2 + y^2$.

3. The graph of the polar equation $r = 5$ is a _____ ; the graph of $\theta = 5$ is a _____ .

4. The graph of the polar equation $r = ed/(1 + e\cos\theta)$ is a _____ .

Problem Set 10.3

1. Plot the points whose polar coordinates are $\left(4, \frac{1}{3}\pi\right)$, $\left(2, \frac{1}{2}\pi\right)$, $\left(5, \frac{1}{6}\pi\right)$, $\left(0, \frac{11}{7}\pi\right)$, $\left(3, \frac{3}{2}\pi\right)$, $\left(\frac{7}{2}, \frac{2}{3}\pi\right)$, and $(4, 0)$.

2. Plot the points whose polar coordinates are $(3, \pi)$, $(7, 2\pi)$, $\left(2, \frac{11}{6}\pi\right)$, $\left(4, -\frac{1}{3}\pi\right)$, $(0, 0)$, $\left(0, \frac{2}{3}\pi\right)$, and $\left(3, -\frac{3}{2}\pi\right)$.

3. Plot the points whose polar coordinates are $\left(-5, \frac{1}{4}\pi\right)$, $\left(5, -\frac{3}{4}\pi\right)$, $\left(2, -\frac{1}{3}\pi\right)$, $\left(-2, \frac{2}{3}\pi\right)$, $(-6, 0)$, $(3, -\pi)$, $\left(-4, -\frac{2}{3}\pi\right)$, and $(-3, \pi)$.

4. Plot the points whose polar coordinates are $\left(2, \frac{5}{8}\pi\right)$, $\left(-4, \frac{17}{6}\pi\right)$, $\left(5, -\frac{7}{3}\pi\right)$, $(7, 1)$, $(-3, -2)$, $(3, 0)$, $(-3, -\pi)$, and $\left(0, -\frac{5}{4}\pi\right)$.

5. Plot the points whose polar coordinates follow. For each, give four other pairs of polar coordinates, two with positive r and two with negative r.
(a) $\left(4, \frac{1}{3}\pi\right)$ (b) $\left(-3, \frac{1}{4}\pi\right)$
(c) $\left(-5, \frac{1}{6}\pi\right)$ (d) $\left(7, -\frac{2}{3}\pi\right)$

6. Plot the points whose polar coordinates follow. For each, give four other pairs of polar coordinates, two with positive r and two with negative r.
(a) $(-2, \pi)$ (b) $\left(5, -\frac{1}{12}\pi\right)$
(c) $\left(4, \frac{19}{4}\pi\right)$ (d) $\left(-7, -\frac{3}{2}\pi\right)$

7. Find the Cartesian coordinates of the points in Problem 5.

8. Find the Cartesian coordinates of the points in Problem 6.

9. Find polar coordinates of the points whose Cartesian coordinates are as given.
(a) $\left(-2\sqrt{3}, -2\right)$ (b) $\left(1, \sqrt{3}\right)$
(c) $\left(\sqrt{2}, -\sqrt{2}\right)$ (d) $(0, 0)$

10. Find polar coordinates of the points whose Cartesian coordinates are as given.
(a) $\left(-\sqrt{2}/2, \sqrt{2}/2\right)$ (b) $\left(-7\sqrt{3}/2, 7/2\right)$
(c) $(0, -4)$ (d) $(5, -12)$

In each of Problems 11–16, sketch the graph of the given Cartesian equation and then find a polar equation for it.

11. $x - 4y + 2 = 0$

12. $x = 0$

13. $y = -5$

14. $x + y = 0$

15. $x^2 + y^2 = 16$

16. $y^2 = 4px$

In Problems 17–22, find the Cartesian equations of the graphs of the given polar equations.

17. $\theta = \frac{1}{3}\pi$

18. $r = 2$

19. $r\cos\theta + 6 = 0$

20. $r - 6\cos\theta = 0$

21. $r \sin \theta - 4 = 0$

22. $r^2 - 8r \cos \theta - 4r \sin \theta + 11 = 0$

In Problems 23–36, name the curve with the given polar equation. If it is a conic, give its eccentricity. Sketch the graph.

23. $r = 6$

24. $\theta = \dfrac{2\pi}{3}$

25. $r = \dfrac{3}{\sin \theta}$

26. $r = \dfrac{4}{\cos \theta}$

27. $r = 4 \sin \theta$

28. $r = -4 \cos \theta$

29. $r = \dfrac{4}{1 + \cos \theta}$

30. $r = \dfrac{4}{1 + 2 \sin \theta}$

31. $r = \dfrac{6}{2 + \sin \theta}$

32. $r = \dfrac{6}{4 - \cos \theta}$

33. $r = \dfrac{4}{2 + 2 \cos \theta}$

34. $r = \dfrac{4}{2 + 2 \cos(\theta - \pi/3)}$

35. $r = \dfrac{4}{\frac{1}{2} + \cos(\theta - \pi)}$

36. $r = \dfrac{4}{3 \cos(\theta - \pi/3)}$

37. Show that the polar equation of the circle with center (c, α) and radius a can be expressed as $r^2 + c^2 - 2rc \cos(\theta - \alpha) = a^2$.

38. Prove that $r = a \sin \theta + b \cos \theta$ represents a circle and find its center and radius.

39. Let r_1 and r_2 be the minimum and maximum distances (**perihelion** and **aphelion**) of the ellipse $r = ed/[1 + e \cos(\theta - \theta_0)]$ from a focus. Show that:

(a) $r_1 = ed/(1 + e), r_2 = ed/(1 - e)$,

(b) major diameter $= 2ed/(1 - e^2)$ and minor diameter $= 2ed/\sqrt{1 - e^2}$.

40. The perihelion and aphelion for the asteroid Icarus are 17 and 183 million miles, respectively. What is the eccentricity of its elliptical orbit?

41. Earth's orbit around the sun is an ellipse of eccentricity 0.0167 and major diameter 185.8 million miles. Find its perihelion.

42. The orbit of a certain comet is a parabola with the sun at a focus. The angle between the axis of the parabola (assumed pointing into the concave side) and a ray from the sun to the comet is 120° when the comet is 100 million miles from the sun. How close does the comet get to the sun?

43. The position of a comet with a highly eccentric elliptical orbit (e very near 1) is measured with respect to a fixed polar axis (sun is at focus but polar axis is not axis of ellipse) at two times, giving the two points $(4, \pi/2)$ and $(3, \pi/4)$ of the orbit. Here distances are measured in astronomical units ($1\,AU \approx 93$ million miles). For the part of the orbit near the sun, assume $e = 1$, so the orbit is given by $r = d/[1 + \cos(\theta - \theta_0)]$.

(a) The two points give two conditions for d and θ_0. Use them to show that $4.24 \cos \theta_0 - 3.67 \sin \theta_0 - 2 = 0$.

(b) Solve for θ_0 using Newton's Method.

(c) How close does the comet get to the sun?

Answers to Concepts Review: 1. Infinitely many 2. $r \cos \theta$; $r \sin \theta$; r^2 3. Circle; line 4. Conic

---

Lab 22: GRAPHS IN POLAR COORDINATES

Mathematical Background

If we let r represent the distance to a point and θ the angle between the x-axis and the line that connects the point to the origin, then any point in the xy-plane can be described by the pair (r, θ) called polar coordinates. We can also let r assume negative values; in this case the point is located r units in the direction opposite the one represented by θ. Thus, a point can be represented in more than one way; for instance, the point with rectangular (xy) coordinates $(0, 1)$ can be represented by $r = 1$ and $\theta = \pi/2$ or by $r = -1$ and $\theta = 3\pi/2$. (See Figure 1.)

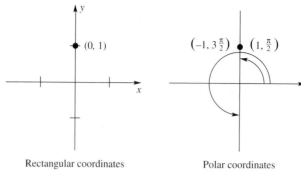

Rectangular coordinates   Polar coordinates

Figure 1

Many interesting curves can be represented more easily in polar coordinates than in rectangular (*xy*) coordinates. Also, many problems in physics are solved more easily in polar coordinates than in rectangular ones. For instance, the potential energy of a planet is most easily described by the function $-\dfrac{gMm}{r}$ where *g* is the universal gravitational constant, *M* and *m* are the masses of the sun and the planet, and *r* is the distance of the planet from the sun. This function can be used to derive the equations of the motion of a planet.

Lab Introduction

We will look at curves of the form $r = f(\theta)$. To graph such a curve, we form a table of *r*- and θ-values by picking the θ's, calculating the *r*'s, and plotting the results. This is what the computer does for us. We need to specify the function $f(\theta)$ and the first and last θ-values, and the computer or calculator does the rest. Generally, one specifies θ to run from 0 to some multiple of π.

In order to understand the graphs that result from the computer or calculator, you will be asked to label some of the graphs with the θ-values that correspond to each part of the curve. For this, it may be helpful to form a table of *r*- versus θ-values; in particular, it is important to determine the θ-values for which $r = 0$. (This is where the graph goes through the origin.)

Experiment

(1) Graph $r = a$ for a few values of *a* between 0 to 3. When using a graphing calculator, set the *x*- and *y*-ranges so that one unit is the same length on each axis, such as $-6 \le x \le 6$ and $-4 \le y \le 4$ for a calculator with a screen of 95 pixels by 63 pixels. Also use $0 \le \theta \le 2\pi$. Sketch and describe the results. For the $a = 3$ curve, show which θ-values correspond to which parts of the curve.

(2) Graph $r = a \cos \theta$ for the same values of *a* and the same θ-, *x*-, and *y*-ranges as in (1). Sketch and describe the results. For the $a = 3$ curve, show which θ-values correspond to which parts of the curve.

(3) Graph $r = \dfrac{a}{\cos \theta}$ for the same values of *a* and the same *x*-, and *y*-ranges as in (1) and for θ-range $-0.25\pi \le \theta \le 0.25\pi$. Sketch and describe the results. For the $a = 3$ curve, show which θ-values correspond to which parts of the curve.

(4) Graph $r = a + b \cos \theta$ for selected a and b between 0 and 3 and the same θ-, x-, and y-ranges as in (1). Choose some $a < b$ and some $b < a$. Describe how the shape changes, depending on whether a or b is larger. For one case where $a < b$, show which θ-values correspond to which parts of the curve; do the same for one case where $b < a$.

(5) Graph $r = a \cos (b\theta)$ for selected a and b and the same θ-, x-, and y- ranges as in (1). Choose a as in (1), and choose various $b \geq 2$ (some even and some odd). Describe the results. For one case where b is even and for one case where b is odd, show which θ-values correspond to which parts of the curve.

(6) Graph $r = a\theta$ for selected a between -1 and 1 and for $-30 \leq x \leq 30$, $-20 \leq y \leq 20$, $0 \leq \theta \leq 6\pi$. Describe the results. For $a = 1$, show which θ-values correspond to which parts of the curve.

(7) Go wild; invent an interesting curve. Try combining some of the above, or invent some new functions. For example, a spiral [as in (6)] multiplied by any of the other curves is usually interesting. Be careful though; for a spiral, a large number of points must be plotted to get a true picture.

Discussion

1. For parts (1)–(3) and part (6) in Experiment section, describe how the parameter a affects the shape of the graph; if a has a particular geometrical meaning, describe it. Explain *why* this is the case. Could you get the same graph with a smaller θ-range for any of these? Explain.

2. For part (4) above, discuss how the parameters a and b affect the shape of the graph. Can you predict when you will get an inner loop? *Why* is this the case?

3. For part (5) above, discuss how the parameters a and b affect the shape of the graph. Be sure to discuss what happens for even and odd values of b. Explain *why* this should be the case. Could you get the whole graph with a smaller θ-range?

4. Describe your figure for part (7) and explain its shape in terms of its equation.

10.4 Technology and Graphs of Polar Equations

The polar equations considered in the previous section led to familiar graphs—mainly lines, circles, and conics. Now we turn our attention to more exotic graphs—cardioids, limaçons, roses, and spirals. The polar equations for these curves are still rather simple; the corresponding Cartesian equations are quite complicated. Thus, we see one of the advantages of having more than one coordinate system available. Some curves have simple equations in one system; other curves have simple equations in a second system. You will exploit this later, in multivariable calculus, when the solution of a problem often begins with the choice of a convenient coordinate system.

Using Technology to Graph in Polar Coordinates
As the graphs of equations given in polar coordinates become more complicated, it becomes less practical to graph by forming a table of values. As with graphs in rectangular coordinates, graphing calculators and computer algebra systems make graphs of equations given in polar coordinates by forming an internal table of values, and then plotting these values. To form this table of values, we must specify a starting and ending value for θ. The points are then connected with

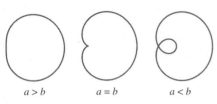

Figure 1

short straight line segments, so that the graph looks like a smooth curve if enough points are plotted.

Cardioids and Limaçons   We consider equations of the form

$$r = a \pm b \cos \theta \qquad r = a \pm b \sin \theta$$

with a and b positive. Their graphs are called **limaçons**, with the special cases in which $a = b$ referred to as **cardioids**. Typical graphs are shown in Figure 1.

EXAMPLE 1: Use a graphing calculator or computer algebra system to sketch the graph of the equation $r = 2 + 4 \cos \theta$. Identify the polar coordinates of several key points on the graph, and describe what happens to r as θ goes from its minimum value to its maximum value.

SOLUTION: With either a graphing calculator or a computer algebra system, one must specify the range of θ-values used to plot the points that generate the graph. For functions that are periodic on the interval $0 \leq \theta \leq 2\pi$ (such as $\sin n\theta$ or $\cos n\theta$ where n is an integer) then the interval $0 \leq \theta \leq 2\pi$ is sufficient to generate the complete graph. (Why?)

With graphing calculators (and some computer algebra systems), one must also specify the θ step size (increment) for the plotted points. If the step size is too large, the graph will appear to consist of line segments instead of being a smooth curve. In Figure 2 we show calculator graphs with θ step sizes of $2\pi/10 = 0.2\pi \approx 0.628$ (too large—10 plotted points) and $2\pi/40 = 0.05\pi \approx 0.157$ (acceptable—40 plotted points).

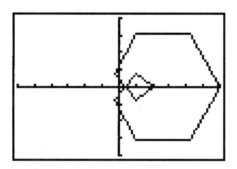

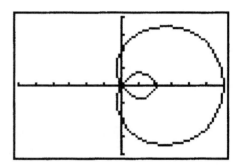

Graph of $r = 2 + 4 \cos \theta$ with θ- range $0 \leq \theta \leq 2\pi, \theta$ step size 0.628 and graph window $-6 \leq x \leq 6, -4 \leq x \leq 4$

Graph of $r = 2 + 4 \cos \theta$ with θ- range $0 \leq \theta \leq 2\pi, \theta$ step size 0.157 and graph window $-6 \leq x \leq 6, -4 \leq x \leq 4$.

Figure 2

In order to understand which parts of the graph correspond to which θ-values, it is useful to generate a table of function values. With a graphing calculator, the Trace feature can be used to get the polar coordinates of points on the graph; with a computer algebra system you can use the Table command (or a similar command) to generate the table of function values. In Figure 3 we show a table with a θ-increment of $0.1\pi \approx 0.314$, given in two formats.

The second table in Figure 3 is much easier to read because the θ-values are given as multiples of π, rather than in pure radians. To convert from the first format to the second, divide the θ values by π (you must do this if you use the Trace of a calculator). Also note that when working with technology it is

θ	r	θ	r
0.000	6	0 Pi	6
0.314	5.80423	0.1 Pi	5.80423
0.628	5.23607	0.2 Pi	5.23607
0.942	4.35114	0.3 Pi	4.35114
1.257	3.23607	0.4 Pi	3.23607
1.571	2	0.5 Pi	2
1.885	0.763932	0.6 Pi	0.763932
2.199	−0.351141	0.7 Pi	−0.351141
2.513	−1.23607	0.8 Pi	−1.23607
2.827	−1.80423	0.9 Pi	−1.80423
3.142	−2	1 Pi	−2
3.456	−1.80423	1.1 Pi	−1.80423
3.770	−1.23607	1.2 Pi	−1.23607
4.084	−0.351141	1.3 Pi	−0.351141
4.398	0.763932	1.4 Pi	0.763932
4.712	2	1.5 Pi	2
5.027	3.23607	1.6 Pi	3.23607
5.341	4.35114	1.7 Pi	4.35114
5.655	5.23607	1.8 Pi	5.23607
5.969	5.80423	1.9 Pi	5.80423
6.283	6	2 Pi	6

Figure 3

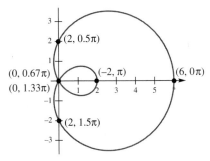

Figure 4

easier to think in decimal multiples of π, rather than fractional multiples as we did in Section 10.3.

We see from the table that as θ goes from 0 to 2π, the r-values decrease from 6 to 0, then from 0 to −2, then increase from −2 to 0 and back finally to 6. The range where r is negative corresponds to the inner loop of the figure. We can determine exactly where $r = 0$ by solving $2 + 4 \cos \theta = 0$; we get $\cos \theta = -\frac{1}{2}$ so that $\theta = \frac{2\pi}{3} \approx 0.67\pi$ or $\theta = \frac{4\pi}{3} \approx 1.33\pi$. Combining some of the values in the table with the points where $r = 0$, we get the picture in Figure 4.

Notice that without the table we could have easily misidentified the point on the inner loop as $(2, 0\pi)$ rather than as $(-2, \pi)$. Although both descriptions fit this point $(2, 0\pi)$ does *not* satisfy the equation $r = 2 + 4 \cos \theta$. ◄

Lemniscates   The graphs of

$$r^2 = \pm a \cos 2\theta \qquad r^2 = \pm a \sin 2\theta$$

are figure-eight shaped curves called **lemniscates**.

***EXAMPLE* 2:** Sketch the graph of the equation $r^2 = 8 \cos 2\theta$, identify the polar coordinates of several key points on the graph, and describe what happens to r as θ goes from its minimum value to its maximum value.

SOLUTION: We must consider both of the equations $r = \sqrt{8 \cos 2\theta}$ and $r = -\sqrt{8 \cos 2\theta}$. As in Example 2, we need only consider the interval $0 \leq \theta \leq 2\pi$. Of course, due to the square root in both functions, not all θ-values result in well-defined real r-values. In Figure 5 we show a table of values for both $r = \sqrt{8 \cos 2\theta}$ and $r = -\sqrt{8 \cos 2\theta}$.

θ	$r = \sqrt{8 \cos 2\theta}$	θ	$r = -\sqrt{8 \cos 2\theta}$
0	2.8284271	0	−2.8284271
0.1 Pi	2.5440393	0.1 Pi	−2.5440393
0.2 Pi	1.5723028	0.2 Pi	−1.5723028
0.3 Pi	undefined	0.3 Pi	undefined
0.4 Pi	undefined	0.4 Pi	undefined
0.5 Pi	undefined	0.5 Pi	undefined
0.6 Pi	undefined	0.6 Pi	undefined
0.7 Pi	undefined	0.7 Pi	undefined
0.8 Pi	1.5723028	0.8 Pi	−1.5723028
0.9 Pi	2.5440393	0.9 Pi	−2.5440393
1 Pi	2.8284271	1 Pi	−2.8284271
1.1 Pi	2.5440393	1.1 Pi	−2.5440393
1.2 Pi	1.5723028	1.2 Pi	−1.5723028
1.3 Pi	undefined	1.3 Pi	undefined
1.4 Pi	undefined	1.4 Pi	undefined
1.5 Pi	undefined	1.5 Pi	undefined
1.6 Pi	undefined	1.6 Pi	undefined
1.7 Pi	undefined	1.7 Pi	undefined
1.8 Pi	1.5723028	1.8 Pi	−1.5723028
1.9 Pi	2.5440393	1.9 Pi	−2.5440393
2 Pi	2.8284271	2 Pi	−2.8284271

Figure 5

If you look closely at the table in Figure 5, you will find that the points in the $r = \sqrt{8 \cos 2\theta}$ table are precisely the same points as those in the $r = -\sqrt{8 \cos 2\theta}$ table. For example, the point $(2.54, 0.1\pi)$ in the $r = \sqrt{8 \cos 2\theta}$ table describes the same point as $(-2.54, 1.1\pi)$ in the $r = -\sqrt{8 \cos 2\theta}$ table. Thus, we need graph only the equation $r = \sqrt{8 \cos 2\theta}$.

In addition to the points in the table, we should determine points where $r = 0$. We need to solve $8 \cos 2\theta = 0$ or

$$\cos 2\theta = 0$$

Thus,

$$2\theta = 0.5\pi, 1.5\pi, 2.5\pi, \text{ or } 3.5\pi$$

and so

$$\theta = 0.25\pi, 0.75\pi, 1.25\pi, \text{ or } 1.75\pi$$

These values could also be determined by looking at a rectangular plot of $y = \cos 2x$ (try this). Putting these values together with the values in the $r = \sqrt{8 \cos 2\theta}$ table, we see that as θ goes from 0 to 0.25π, r decreases from about 2.83 to 0. For θ between 0.25π and 0.75π the r value is undefined (no graph). Then as θ goes from 0.75π to π, r increases from 0 to about 2.83. This pattern is repeated as θ goes from π to 2π.

In Figure 6 we show a computer algebra graph of $r = \sqrt{8 \cos 2\theta}$. Notice that there is a missing piece in the graph near the origin (calculator graphs will also have a missing piece). This is a result of numerical round off, and our analysis from above indicates that we should close the loops in our sketch. In Figure 7 we show our final sketch, with some points labeled. ◄

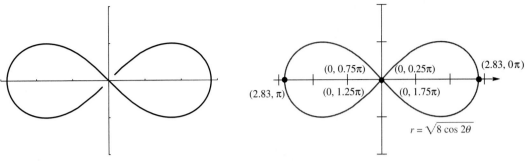

Figure 6   Figure 7

Roses   Polar equations of the form

$$r = a \cos n\theta \qquad r = a \sin n\theta$$

represent flower-shaped curves called **roses**.

EXAMPLE 3: Sketch the graph of $r = 4 \sin 2\theta$ and identify the polar coordinates of several key points on the graph. Describe what happens to r as θ goes from its minimum value to its maximum value.

SOLUTION: We proceed as in Examples 1 and 2. A table of values is shown in Figure 8. Because $\sin \theta$ has extreme values (max's or min's) at $\theta = 0.5\pi, 1.5\pi, 2.5\pi, 3.5\pi, \ldots$ the function $\sin 2\theta$ has its extreme values at $\theta = 0.25\pi, 0.75\pi, 1.25\pi, 1.75\pi, \ldots$. This gives us the points $(4, 0.25\pi), (-4, 0.75\pi), (4, 1.25\pi), (-4, 1.75\pi)$ in addition to those in the table.

As θ goes from 0 to 0.5π, r increases from 0 to 4 and then back to 0. As θ goes from 0.5π to π, r goes from 0 to -4 and then back to 0. This pattern repeats as θ goes from π to 2π. See the sketch in Figure 9.

θ	r
0	0
0.1 Pi	2.351141
0.2 Pi	3.804226
0.3 Pi	3.804226
0.4 Pi	2.351141
0.5 Pi	4.9E-16
0.6 Pi	−2.35114
0.7 Pi	−3.80423
0.8 Pi	−3.80423
0.9 Pi	−2.35114
1 Pi	−9.8E-16
1.1 Pi	2.351141
1.2 Pi	3.804226
1.3 Pi	3.804226
1.4 Pi	2.351141
1.5 Pi	1.47E-15
1.6 Pi	−2.35114
1.7 Pi	−3.80423
1.8 Pi	−3.80423
1.9 Pi	−2.35114
2 Pi	−2E-15

Figure 8

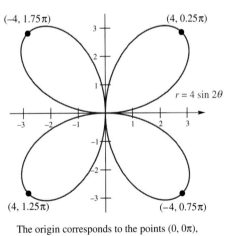

The origin corresponds to the points $(0, 0\pi)$, $(0, 0.5\pi)$, $(0, \pi)$, and $(0, 1.5\pi)$.

Figure 9

When we put the graph together with the information from the table, we can see the order in which the graph is traced out as θ increases from 0 to 2π. The loop in the first quadrant is traced first, then the loop in the fourth quadrant (bottom right), then the loop in the third quadrant (bottom left), and finally the loop in the second quadrant (top left). Of course, this is easily seen as well by using the Trace feature of a graphing calculator.   ◄

Spirals The graph of $r = a\theta$ is called a **spiral of Archimedes**; the graph of $r = ae^{b\theta}$ is called a **logarithmic spiral**.

EXAMPLE 4: Sketch the graph of $r = \theta$ for $\theta \geq 0$.

SOLUTION: We don't really need a table this time, because the equation $r = \theta$ is so simple; the r and θ coordinates are always equal. We would have points such as $(0, 0)$, $(0.5\pi, 0.5\pi)$, (π, π), ... in the table. Note that because this function is not periodic, we must use more than just the interval $0 \leq \theta \leq 2\pi$. In Figure 10 we show the graph with $0 \leq \theta \leq 10\pi$.

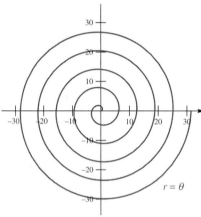

Figure 10

Intersection of Curves in Polar Coordinates In Cartesian coordinates, all points of intersection of two curves can be found by solving the equations of the curves simultaneously. But in polar coordinates, this is not always the case. This is because a point P has many pairs of polar coordinates, and one pair may satisfy the polar equation of one curve and a different pair may satisfy the polar equation of the other curve. For instance (see Figure 11), the circle $r = 4 \cos \theta$ intersects the line $\theta = \pi/3$ at two points, the pole and $(2, \pi/3)$, and yet only the latter is a common solution of the two equations. This happens because the coordinates of the pole that satisfy the equation of the line are $(0, \pi/3)$ and those that satisfy the equation of the circle are $(0, \pi/2)$.

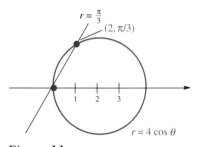

Figure 11

Our conclusion is this: In order to find all intersections of two curves whose polar equations are given, solve the equations simultaneously; then graph the two equations carefully to discover other possible points of intersection.

EXAMPLE 5: Find the points of intersection of the two cardioids $r = 1 + \cos \theta$ and $r = 1 - \sin \theta$.

SOLUTION: If we eliminate r between the two equations, we get $1 + \cos \theta = 1 - \sin \theta$. Thus, $\cos \theta = -\sin \theta$, or $\tan \theta = -1$. We conclude that $\theta = \frac{3}{4}\pi$ and $\theta = \frac{7}{4}\pi$, which yields the two intersection points $(1 - \frac{1}{2}\sqrt{2}, \frac{3}{4}\pi)$ and $(1 + \frac{1}{2}\sqrt{2}, \frac{7}{4}\pi)$. The graphs in Figure 12 show, however, that we have missed a third intersection point, namely, the pole. We missed it because $r = 0$ in $r = 1 + \cos \theta$ when $\theta = \pi$, but $r = 0$ in $r = 1 - \sin \theta$ when $\theta = \pi/2$.

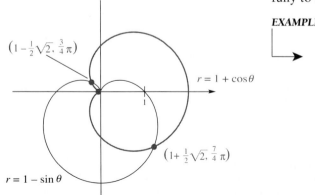

Figure 12

Concepts Review

1. The graph of $r = 3 + 2 \cos \theta$ is a _____ .

2. The graph $r = 2 + 2 \cos \theta$ is a _____ .

3. The graph of $r = 4 \sin n\theta$ is a _____ .

4. The graph of $r = \theta/3$ is a _____ .

Problem Set 10.4

In Problems 1–32, sketch the graph of the given polar equation. Identify the polar coordinates of several key points on the graph and describe what happens to r as θ goes from its minimum value to its maximum value.

1. $\theta^2 - 1 = 0$

2. $(r - 2)\left(\theta + \dfrac{\pi}{4}\right) = 0$

3. $r \sin \theta + 6 = 0$

4. $r = -2 \sec \theta$

5. $r = 6 \sin \theta$

6. $r = 4 \cos \theta$

7. $r = \dfrac{4}{1 - \cos \theta}$

8. $r = \dfrac{2}{1 + \sin \theta}$

9. $r = 5 - 5 \sin \theta$ (cardioid)

10. $r = 4 - 4 \cos \theta$ (cardioid)

11. $r = 3 - 3 \cos \theta$ (cardioid)

12. $r = 4 + 4 \sin \theta$ (cardioid)

13. $r = 2 - 4 \cos \theta$ (limaçon)

14. $r = 4 - 2 \cos \theta$ (limaçon)

15. $r = 4 - 3 \sin \theta$ (limaçon)

16. $r = 3 - 4 \sin \theta$ (limaçon)

17. $r^2 = 9 \sin 2\theta$ (lemniscate)

18. $r^2 = 4 \cos 2\theta$ (lemniscate)

19. $r^2 = -16 \cos 2\theta$ (lemniscate)

20. $r^2 = -4 \sin 2\theta$ (lemniscate)

21. $r = 5 \cos 3\theta$ (three-leaved rose)

22. $r = 3 \sin 3\theta$ (three-leaved rose)

23. $r = 6 \sin 2\theta$ (four-leaved rose)

24. $r = 4 \cos 2\theta$ (four-leaved rose)

25. $r = 7 \cos 5\theta$ (five-leaved rose)

26. $r = 3 \sin 5\theta$ (five-leaved rose)

27. $r = \frac{1}{2}\theta, \theta \geq 0$ (spiral of Archimedes)

28. $r = 2\theta, \theta \geq 0$ (spiral of Archimedes)

29. $r = e^{\theta}, \theta \geq 0$ (logarithmic spiral)

30. $r = e^{\theta/2}, \theta \geq 0$ (logarithmic spiral)

31. $r = \dfrac{2}{\theta}, \theta > 0$ (reciprocal spiral)

32. $r = -\dfrac{1}{\theta}, \theta > 0$ (reciprocal spiral)

In Problems 33–38, sketch the given curves and find their points of intersection.

33. $r = 6, r = 4 + 4 \cos \theta$

34. $r = 1 - \cos \theta, r = 1 + \cos \theta$

35. $r = 3\sqrt{3} \cos \theta, r = 3 \sin \theta$

36. $r = 5, r = \dfrac{5}{1 - 2 \cos \theta}$

37. $r = 6 \sin \theta, r = \dfrac{6}{1 + 2 \sin \theta}$

38. $r^2 = 4 \cos 2\theta, r = 2\sqrt{2} \sin \theta$

39. Let a and b be fixed positive numbers and suppose AOP is a line segment with A on the line $x = a$ and $|AP| = b$. Find both the polar equation and the rectangular equation for the set of points P (called a *conchoid*) and sketch its graph.

40. Let F and F' be fixed points with polar coordinates $(a, 0)$ and $(-a, 0)$, respectively. Show that the set of points P satisfying $|PF| \, |PF'| = a^2$ is a lemniscate by finding its polar equation.

41. A line segment L of length $2a$ has its two endpoints on the x- and y-axis, respectively. The point P is on L and is such that OP is perpendicular to L. Show that the set of points P satisfying this condition is a four-leaved rose by finding it polar equation.

Answers to Concepts Review: 1. Limaçon 2. Cardioid 3. Rose 4. Spiral

10.5 Calculus in Polar Coordinates

The two most basic problems in calculus are the determinations of the slope of a tangent line and the area of a curved region. Here we consider both problems, but in the context of polar coordinates.

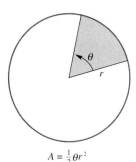

$A = \frac{1}{2}\theta r^2$

Figure 1

In Cartesian coordinates, the functional building block in area problems is the rectangle. In polar coordinates, it is the sector of a circle (a pie-shaped region like that in Figure 1). From the fact that the area of a circle in πr^2, we infer that the area of a sector with central angle θ radians is $(\theta/2\pi)\pi r^2$; that is,

$$\text{Area of a sector:} \quad A = \frac{1}{2}\theta r^2$$

Area in Polar Coodinates   To begin, let $r = f(\theta)$ determine a curve in the plane, where f is a continuous, nonnegative function for $\alpha \le \theta \le \beta$ and $\beta - \alpha \le 2\pi$. The curves $r = f(\theta)$, $\theta = \alpha$, and $\theta = \beta$ bound a region R (the one shown at the left in Figure 2), whose area $A(R)$ we wish to determine.

Partition the interval $[\alpha, \beta]$ into n equal subintervals by means of numbers $\alpha = \theta_0 < \theta_1 < \theta_2 < \cdots < \theta_n = \beta$, thereby slicing R into n smaller pie-shaped regions $R_1, R_2, \ldots, R_n$, as shown in the right half of Figure 2. Clearly $A(R) = A(R_1) + A(R_2) + \cdots + A(R_n)$.

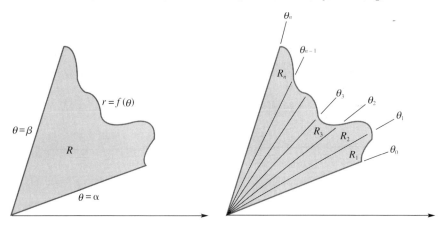

Figure 2

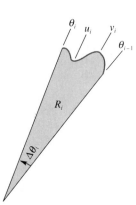

Figure 3

We approximate the area $A(R_i)$ of the ith slice; in fact, we do it in two ways. On the ith interval $[\theta_{i-1}, \theta_i]$, f achieves its minimum value and maximum value — for instance, at u_i and v_i, respectively (Figure 3). The area $A(R_i)$ must lie between the area of a sector with $r = f(u_i)$ and one with $r = f(v_i)$. Thus, if $\Delta\theta_i = \theta_i - \theta_{i-1}$,

$$\frac{1}{2}[f(u_i)]^2 \Delta\theta_i \le A(R_i) \le \frac{1}{2}[f(v_i)]^2 \Delta\theta_i$$

and so

$$\sum_{i=1}^{n} \frac{1}{2}[f(u_i)]^2 \Delta\theta_i \le \sum_{i=1}^{n} A(R_i) \le \sum_{i=1}^{n} \frac{1}{2}[f(v_i)]^2 \Delta\theta_i$$

The first and third members of this inequality are Riemann sums for the same integral, namely, $\int_{\alpha}^{\beta} \frac{1}{2}[f(\theta)]^2 d\theta$. When we let $n \to \infty$, we obtain the area formula

$$A = \frac{1}{2}\int_{\alpha}^{\beta} [f(\theta)]^2 \, d\theta$$

This formula can, of course, be memorized. We prefer that you remember how it was derived. In fact, you will note that the three familiar words, *slice, approximate, integrate,* are also the key to area problems in polar coordinates. We illustrate what we mean.

EXAMPLE 1: Find the area of the region inside the limaçon $r = 2 + \cos \theta$.

SOLUTION: The graph is sketched in Figure 4; note that θ varies from 0 to 2π. We slice, approximate, integrate.

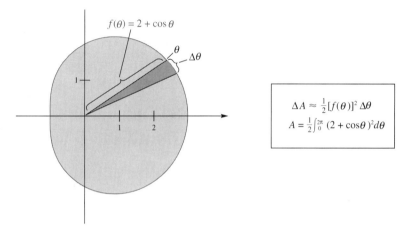

$$\Delta A \approx \tfrac{1}{2}[f(\theta)]^2 \, \Delta\theta$$
$$A = \tfrac{1}{2}\int_0^{2\pi} (2 + \cos\theta)^2 d\theta$$

Figure 4

By symmetry, we can double the integral from 0 to π. Thus,

$$A = \int_0^{\pi} (2 + \cos\theta)^2 \, d\theta = \int_0^{\pi} (4 + 4\cos\theta + \cos^2\theta) \, d\theta$$

$$= \int_0^{\pi} 4 \, d\theta + 4\int_0^{\pi} \cos\theta \, d\theta + \frac{1}{2}\int_0^{\pi}(1 + \cos 2\theta) \, d\theta$$

$$= \int_0^{\pi} \frac{9}{2} \, d\theta + 4\int_0^{\pi} \cos\theta \, d\theta + \frac{1}{4}\int_0^{\pi} \cos 2\theta \cdot 2 \, d\theta$$

$$= \left[\frac{9}{2}\theta\right]_0^{\pi} + [4\sin\theta]_0^{\pi} + \left[\frac{1}{4}\sin 2\theta\right]_0^{\pi}$$

$$= \frac{9\pi}{2} \qquad \blacktriangleleft$$

EXAMPLE 2: Find the area of one leaf of the four-leaved rose $r = 4 \sin 2\theta$.

SOLUTION: The complete rose was sketched in Example 3 of the previous section. Here we show only the first-quadrant leaf (Figure 5). This leaf is 4 units long and averages about 1.5 units in width, giving 6 as an estimate for its area. The exact area A is given by

$$A = \frac{1}{2}\int_0^{\frac{\pi}{2}} 16\sin^2 2\theta\, d\theta = 8\int_0^{\frac{\pi}{2}} \frac{1 - \cos 4\theta}{2}\, d\theta$$

$$= 4\int_0^{\frac{\pi}{2}} d\theta - \int_0^{\frac{\pi}{2}} \cos 4\theta \cdot 4\, d\theta$$

$$= \left[4\theta \right]_0^{\frac{\pi}{2}} - \left[\sin 4\theta \right]_0^{\frac{\pi}{2}} = 2\pi$$

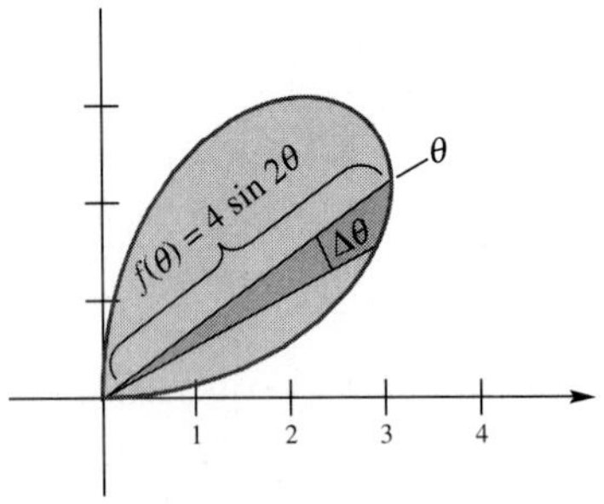

$$\Delta A \approx \frac{1}{2}[f(\theta)]^2 \Delta\theta$$
$$A = \frac{1}{2}\int_0^{\pi/2}(4 + \sin 2\theta)^2 d\theta$$

Figure 5   ◄

EXAMPLE 3: Find the area of the region outside the cardioid $r = 1 + \cos\theta$ and inside the circle $r = \sqrt{3}\sin\theta$.

SOLUTION: The graphs of the two curves are sketched in Figure 6. We will need the θ-coordinates of the points of intersection. Let's try solving the two equations simultaneously.

$$1 + \cos\theta = \sqrt{3}\sin\theta$$

$$1 + 2\cos\theta + \cos^2\theta = 3\sin^2\theta$$

$$1 + 2\cos\theta + \cos^2\theta = 3(1 - \cos^2\theta)$$

$$4\cos^2\theta + 2\cos\theta - 2 = 0$$

$$2\cos^2\theta + \cos\theta - 1 = 0$$

$$(2\cos\theta - 1)(\cos\theta + 1) = 0$$

$$\cos\theta = \frac{1}{2}, -1 \qquad \theta = \frac{\pi}{3}, \pi$$

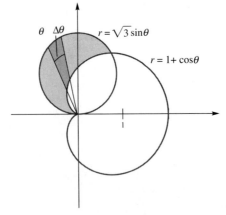

$$\Delta A \approx \frac{1}{2}[3\sin^2\theta - (1 + \cos\theta)^2]\,\Delta\theta$$
$$A = \frac{1}{2}\int_{\pi/3}^{\pi}[3\sin^2\theta - (1 + \cos\theta)^2]d\theta$$

Figure 6

Alternatively, you could have solved the equation $1 + \cos\theta = \sqrt{3}\sin\theta$ numerically, using the techniques of Section 2.5, to get the approximations $\theta \approx 1.0472$, $\theta \approx 3.1416$. (Try this.) Then

$$A = \frac{1}{2}\int_{\frac{\pi}{3}}^{\pi}[3\sin^2\theta - (1 + \cos\theta)^2]d\theta$$

$$= \frac{1}{2}\int_{\frac{\pi}{3}}^{\pi}[3\sin^2\theta - 1 - 2\cos\theta - \cos^2\theta]d\theta$$

$$= \frac{1}{2}\int_{\frac{\pi}{3}}^{\pi}\left[\frac{3}{2}(1 - \cos 2\theta) - 1 - 2\cos\theta - \frac{1}{2}(1 + \cos 2\theta)\right]d\theta$$

$$= \frac{1}{2}\int_{\frac{\pi}{3}}^{\pi}[-2\cos\theta - 2\cos 2\theta]d\theta$$

$$= \frac{1}{2}\left[-2\sin\theta - \sin 2\theta\right]_{\frac{\pi}{3}}^{\pi}$$

$$= \frac{1}{2}\left[2\frac{\sqrt{3}}{2} + \frac{\sqrt{3}}{2}\right] = \frac{3\sqrt{3}}{4} \approx 1.299$$

You should check this integral numerically to verify that this is the correct result.   ◄

Tangents in Polar Coordinates

In Cartesian coordinates, the slope m of the tangent line to a curve is given by $m = dy/dx$. We quickly reject $dr/d\theta$ as the corresponding slope formula in polar coordinates. Rather, if $r = f(\theta)$ determines the curve, we write

$$y = r\sin\theta = f(\theta)\sin\theta$$

$$x = r\cos\theta = f(\theta)\cos\theta$$

Thus,

$$\frac{dy}{dx} = \lim_{\Delta x \to 0}\frac{\Delta y}{\Delta x} = \lim_{\Delta\theta \to 0}\frac{\Delta y/\Delta\theta}{\Delta x/\Delta\theta} = \frac{dy/d\theta}{dx/d\theta}$$

Now, using

$$\frac{dx}{d\theta} = \frac{d}{d\theta}(f(\theta)\cos\theta) = f'(\theta)\cos\theta - f(\theta)\sin\theta$$

and

$$\frac{dy}{d\theta} = \frac{d}{d\theta}(f(\theta)\sin\theta) = f'(\theta)\sin\theta + f(\theta)\cos\theta$$

we get

$$m = \frac{f(\theta)\cos\theta + f'(\theta)\sin\theta}{-f(\theta)\sin\theta + f'(\theta)\cos\theta}$$

The formula just derived simplifies when the graph of $r = f(\theta)$ passes through the pole. Suppose, for example, that for some angle α, $r = f(\alpha) = 0$ and $f'(\alpha) \neq 0$. Then (at the pole) our formula for m is

$$m = \frac{f'(\alpha)\sin\alpha}{f'(\alpha)\cos\alpha} = \tan\alpha$$

Because the line $\theta = \alpha$ also has slope $\tan \alpha$, we conclude that this line is tangent to the curve at the pole. We infer the useful fact that *tangent lines at the pole can be found by solving the equation $f(\theta) = 0$.* We illustrate below.

EXAMPLE 4: Consider the polar equation $r = 4 \sin 3\theta$.
  (a) Find the slope of the tangent line at $\theta = \pi/6$ and $\theta = \pi/4$.
  (b) Find the tangent lines at the pole.
  (c) Sketch the graph.
  (d) Find the area of one leaf.

SOLUTION:

(a) $m = \dfrac{f(\theta)\cos \theta + f'(\theta)\sin \theta}{-f(\theta)\sin \theta + f'(\theta)\cos \theta} = \dfrac{4 \sin 3\theta \cos \theta + 12 \cos 3\theta \sin \theta}{-4 \sin 3\theta \sin \theta + 12 \cos 3\theta \cos \theta}$

At $\theta = \pi/6$,

$$m = \frac{4 \cdot 1 \cdot \dfrac{\sqrt{3}}{2} + 12 \cdot 0 \cdot \dfrac{1}{2}}{-4 \cdot 1 \cdot \dfrac{1}{2} + 12 \cdot 0 \cdot \dfrac{\sqrt{3}}{2}} = -\sqrt{3}$$

At $\theta = \pi/4$,

$$m = \frac{4 \cdot \dfrac{\sqrt{2}}{2} \cdot \dfrac{\sqrt{2}}{2} - 12 \cdot \dfrac{\sqrt{2}}{2} \cdot \dfrac{\sqrt{2}}{2}}{-4 \cdot \dfrac{\sqrt{2}}{2} \cdot \dfrac{\sqrt{2}}{2} - 12 \cdot \dfrac{\sqrt{2}}{2} \cdot \dfrac{\sqrt{2}}{2}} = \frac{2 - 6}{-2 - 6} = \frac{1}{2}$$

(b) We set $f(\theta) = 4 \sin 3\theta = 0$ and solve. This yields $\theta = 0$, $\theta = \pi/3$, $\theta = 2\pi/3$, $\theta = \pi$, $\theta = 4\pi/3$, and $\theta = 5\pi/3$.

(c) We use a graphing calculator or computer algebra system to generate a polar graph on the interval $0 \le \theta \le 2\pi$. See Figure 7.

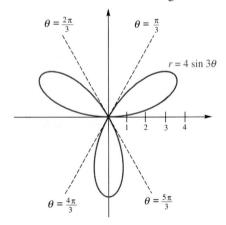

Figure 7

(d)
$$A = \frac{1}{2} \int_0^{\frac{\pi}{3}} (4 \sin 3\theta)^2 \, d\theta = 8 \int_0^{\frac{\pi}{3}} \sin^2 3\theta \, d\theta$$

$$= 4 \int_0^{\frac{\pi}{3}} (1 - \cos 6\theta) \, d\theta = 4 \int_0^{\frac{\pi}{3}} d\theta - \frac{4}{6} \int_0^{\frac{\pi}{3}} \cos 6\theta \cdot 6 \, d\theta$$

$$= \left[4\theta - \frac{2}{3} \sin 6\theta \right]_0^{\frac{\pi}{3}} = \frac{4\pi}{3}$$

Concepts Review

1. The formula for the area A of a sector of a circle of radius r and angle θ (in radians) is $A =$ _____ .

2. The formula in Question 1 leads to the formula for the area A of the region bounded by the curve $r = f(\theta)$ between $\theta = \alpha$ and $\theta = \beta$, namely, $A =$ _____ .

3. From the formula of Question 2, we conclude that the area A of the region inside the cardioid $r = 2 + 2 \cos \theta$ can be expressed as $A =$ _____ .

4. The tangent lines to the polar curve $r = f(\theta)$ at the pole can be found by solving the equation _____ .

Problem Set 10.5

In the problems below, you may choose either an exact or a numerical method to find the necessary definite integrals, whichever you think is appropriate.

In each of Problems 1–10, sketch the graph of the given equation and find the area of the region bounded by it.

1. $r = a, a > 0$

2. $r = 2a \sin \theta, a > 0$

3. $r = 3 + \cos \theta$

4. $r = 4 + 3 \cos \theta$

5. $r = 4 - 4 \cos \theta$

6. $r = 7 - 7 \sin \theta$

7. $r = a(1 + \sin \theta), a > 0$

8. $r^2 = 5 \cos 2\theta$

9. $r^2 = 4 \sin 2\theta$

10. $r^2 = a \sin 2\theta, a > 0$

11. Sketch the limaçon $r = 2 - 4 \cos \theta$, and find the area of the region inside its small loop.

12. Sketch the limaçon $r = 3 - 6 \sin \theta$, and find the area of the region inside its small loop.

13. Sketch the limaçon $r = 2 - 4 \sin \theta$, and find the area of the region inside its large loop.

14. Sketch one leaf of the four-leaved rose $r = 3 \cos 2\theta$, and find the area of the region enclosed by it.

15. Sketch the three-leaved rose $r = 4 \cos 3\theta$, and find the area of the total region enclosed by it.

16. Sketch the three-leaved rose $r = 2 \sin 3\theta$, and find the area of the region bounded by it.

17. Find the area of the region between the two concentric circles $r = 7$ and $r = 10$.

18. Sketch the region that is inside the circle $r = 3 \sin \theta$ and outside the cardioid $r = 1 + \sin \theta$, and find its area.

19. Sketch the region that is outside the circle $r = 2$ and inside the lemniscate $r^2 = 8 \cos 2\theta$, and find its area.

20. Sketch the limaçon $r = 3 - 6 \sin \theta$ and find the area of the region that is inside its large loop and outside its small loop.

21. Sketch the region in the first quadrant that is inside the cardioid $r = 3 + 3 \cos \theta$ and outside the cardioid $r = 3 + 3 \sin \theta$, and find its area.

22. Sketch the region in the second quadrant that is inside the cardioid $r = 2 + 2 \sin \theta$ and outside the cardioid $r = 2 + 2 \cos \theta$, and find its area.

23. Find the slope of the tangent line to each of the following curves at $\theta = \pi/3$.

(a) $r = 2 \cos \theta$ (b) $r = 1 + \sin \theta$

(c) $r = \sin 2\theta$ (d) $r = 4 - 3 \cos \theta$

24. Find all points on the cardioid $r = a(1 + \cos \theta)$ where the tangent line is (a) horizontal, and (b) vertical.

25. Find all points on the limaçon $r = 1 - 2 \sin \theta$ where the tangent line is horizontal.

26. Let $r = f(\theta)$, where f is continuous on the closed interval $[\alpha, \beta]$. Derive the following formula for the length L of the corresponding polar curve from $\theta = \alpha$ to $\theta = \beta$.

$$L = \int_{\alpha}^{\beta} \sqrt{[f(\theta)]^2 + [f'(\theta)]^2} \, d\theta$$

27. Use the formula of Problem 26 to find the perimeter of the cardioid $r = a(1 + \cos \theta)$.

28. Find the length of the logarithmic spiral $r = e^{\frac{\theta}{2}}$ from $\theta = 0$ to $\theta = 2\pi$.

29. Find the total length of the limaçons $r = 2 + \cos t$ and $r = 2 + 4 \cos t$.

30. Find the total area and the total length of the three-leaved rose $r = 4 \sin 3t$.

31. Find the total area and the total length of the lemniscate $r^2 = 8 \cos 2t$.

32. Draw the curve $r = 4 \sin(3t/2)$, $0 \le t \le 4\pi$, and then find its total length.

33.   Find the total area of the rose $r = a \cos n\theta$, where n is a positive integer.

34.   Sketch the graph of the *strophoid* $r = \sec \theta - 2 \cos \theta$ and find the area of its loop.

35.   Consider the two circles $r = 2a \sin \theta$ and $r = 2b \cos \theta$ with a and b positive.

(a) Find the area of the region inside both circles.

(b) Show that the two circles intersect at right angles.

36.   Assume that a planet of mass m is revolving around the sun (located at the pole) with constant angular momentum $mr^2 \, d\theta/dt$. Deduce Kepler's Second Law: The line from the sun to the planet sweeps out equal areas in equal time.

37.   (First Old Goat Problem)   A goat is tethered to the edge of a circular pond of radius a by a rope of length $ka(0 < k < 2)$. Use the method of this section to find its grazing area (the shaded area in Figure 8).

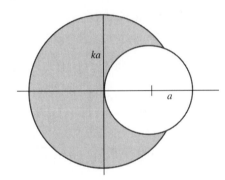

Figure 8

38.   (Second Old Goat Problem)   Do Problem 37 again but assume that the pond has a fence around it so that in forming the wedge A, the rope wraps around the fence (Figure 9). *Hint*: If you are exceedingly ambitious, try the method of this section. Better, note that in the wedge A, $\Delta A \approx \left(\frac{1}{2}\right)|PT|^2 \Delta\phi$, which leads to a Riemann sum for an integral. The final answer is $a^2(\pi k^2/2 + k^3/3)$, a result needed in Problem 39.

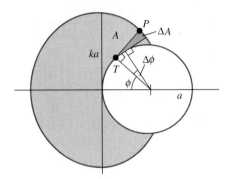

Figure 9

39.   (Third Old Goat Problem)   An untethered goat grazes inside a yard enclosed by a circular fence of radius a; another grazes outside the fence, tethered as in Problem 38. Find the length of the rope if the two goats have the same grazing area.

Answers to Concepts Review: 1. $\frac{1}{2}r^2\theta$ 2. $\frac{1}{2}\int_{\alpha}^{\beta}[f(\theta)]^2 \, d\theta$

3. $\frac{1}{2}\int_{0}^{2\pi}(2 + 2\cos\theta)^2 \, d\theta$ 4. $f(\theta) = 0$

LAB 23:   *ORBITS OF THE PLANETS*

Mathematical and Physical Background

Fix two points F_1 and F_2. For any other point P let d_1 and d_2 be the distances to F_1 and F_2. Now consider the set of all points P for which the sum $d_1 + d_2$ is constant. This set of points forms an ellipse. F_1 and F_2 are called foci; each is a focus.

Johann Kepler (1571–1630) formulated three laws of planetary motion. They were:

(1) The orbit of each planet is an ellipse with the sun at one focus.

(2) The line connecting the planet to the sun sweeps out area at a constant rate.

(3) The square of the period (time) of revolution of a planet is proportional to the cube of the major semiaxis of its elliptical orbit.

These laws can be used to solve problems of planetary motion.

The orbits of the planets around the sun can be expressed very easily in polar coordinates. In polar coordinates, an ellipse with one focus at the origin can be described by the equation

$$r = \frac{a(1 - e^2)}{1 + e \cos \theta} \tag{1}$$

where a and e are parameters, and where $0 < e < 1$. The parameter a is called the major semiaxis, and e the eccentricity. Thus, the above equation, when graphed in polar coordinates, gives the orbit of a planet with the sun at the origin.

The formulas for area and arc length in polar coordinates are

$$\text{area} = \int_a^b \frac{1}{2} [f(\theta)]^2 \, d\theta \tag{2}$$

and

$$\text{arc length} = \int_a^b \sqrt{[f(\theta)]^2 + [f'(\theta)]^2} d\theta \tag{3}$$

where a and b are the θ-boundaries of the region.

Lab Introduction

A year can be divided into four seasons, but what do we really mean by a season? Asked another way, what does it mean for Earth to go one-quarter of the way around the sun? We could measure the one-quarter mark by time, angular distance, distance traveled (arc length), or by distance to the sun. Are each of these the same, or are they different? Our goal is to answer this question, as well as to develop an understanding for the parameters a and e in equation (1) above.

First we will graph various possible orbits, and then find the values of a and e that correspond to Earth's orbit. To do this, the only data we need is the closest distance from Earth to the sun, and the farthest distance. Then, with Earth starting at its farthest point from the sun, we will find the points that correspond to each of the one-quarter marks described above; here we will need Kepler's second law and equations (2) and (3). We can use equation (2) to get the time it takes to travel from one point on the orbit to another, because the second law means that area is proportional to time. Kepler's first law is contained in equation (1) in that it assumes elliptical orbits with the sun at the focus (origin).

Data

The closest that Earth gets to the sun is 91.45 million miles. The farthest is 94.56 million miles.

Experiment

(1) Use a computer or calculator to graph equation (1) above for $-200 \leq x \leq 200$, $-200 \leq y \leq 200$, and $0 \leq \theta \leq 2\pi$ for various values of a and e. Start with $a = 100$ and $e = 0.5$. Next, try various a-values between 50 and 150. Now go back to $a = 100$ and vary e between 0 and 1 (include 0 and 1 also). Sketch the orbits as a varies on one graph, and sketch the orbits as e varies on another. Estimate from the graph where the closest and

farthest points from the origin (the sun) occur for each of the sketches, and write down the corresponding θ-values for each sketch.

(2) Using the data from above, plot the data points on the orbit sketches you did in (1) above. Now find the approximate values of a and e for Earth. To do this, you could use (1) to write two equations, one for when Earth is closest to the sun and one for when Earth is farthest from the sun, in the unknowns a and e, and solve them simultaneously. Or just adjust the parameters a and e until you fit the data.

Now sketch Earth's orbit using the values for a and e that you found, and label the closest and farthest points in polar coordinates. Label the farthest point as A and the closest as B; let the corresponding θ-values by θ_A and θ_B.

(3) Use the values of a and e you found in (2) to find how long it takes Earth to travel from its farthest point from the sun (point A, $\theta = \theta_A$) to a point one-quarter of the way around measured by angle—that is, $\theta = \theta_A + 0.5\pi$. First find the total area swept out in one year; thus, integrate from $\theta = 0$ to $\theta = 2\pi$ using formula (2) above. This area represents 365.26 days (one year). Now integrate from point $\theta = \theta_A$ to the point $\theta = \theta_A + 0.5\pi$ using the same formula, and divide by the total area to get the time in years (because one orbit = one year).

(4) Find the position of Earth when its distance from the sun is halfway between the closest and the farthest points. Thus, you want to find the θ-value for which $r = \dfrac{91.45 + 94.56}{2}$, using equation (1). First estimate this θ-value from your sketches, and then find a more accurate value by solving equation (1) (numerically if necessary). Now find how long in years it takes for Earth to get from point A to this point, as you did in part (3).

(5) Again, suppose that Earth starts out at point A, $\theta = \theta_A$. Find out where it will be $\frac{1}{4}$ year later. Kepler's second law says that one-fourth of this area will be swept out in one-fourth of a year. Thus, you now want to find the value of b for which

$$\int_{\theta_A}^{b} \frac{1}{2}[f(\theta)]^2 \, d\theta = \frac{1}{4} \text{ total area}$$

This b is the θ-coordinate of the planet after one-quarter year; from this you can get the r-coordinate. Use numerical integration and trial-and-error.

(6) Find out where Earth will be when it completes $\frac{1}{4}$ of its orbit as measured by total distance traveled (arc length), starting at the farthest point from the sun ($\theta = \theta_A$). Thus, you will repeat the process in part (5) using the arc length formula (3) instead of the area formula (2). First find the total distance around Earth's orbit by integrating from 0 to 2π, then integrate from θ_A to b for various b's until you get one-fourth of the total length. The b that you get is the θ that gives the one-fourth point; then you can get the corresponding r-value. Finally, find how long (in years) it takes to get from point A to this point.

Discussion

1. Describe how a and e affect the shape of the orbits. Refer to the sketches of orbits you made in (1) above. Show exactly what a represents on one sketch of an orbit; look at your data from (1) above. Can you find a connection between the number a and the definition of an ellipse given at the beginning of this lab?

2. Describe your strategy for finding a and e in part (2) above.

3. Make a hand sketch of Earth's orbit exaggerating the shape of the ellipse and on it label each of the points you found in parts (3)–(6) of the Experiment section, that is, the one-quarter points as measured by distance to the sun, angle, time, and distance traveled. Also make a table with the r-, θ-, and time (in both years and days) values for each of the four points. Are any of the four the same? Discuss your results.

4. Based on the information above, try to determine whether Earth is moving fastest when it is closer to or farther from the sun.

10.6 | Plane Curves: Parametric Representation

Not simple, not closed

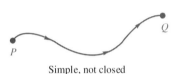

Simple, not closed

Closed, not simple

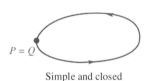

Simple and closed

Figure 1

We gave the general definition of a plane curve in Section 6.4 in connection with our derivation of the arc length formula. A **plane curve** is determined by a pair of parametric equations

$$x = f(t), \qquad y = g(t), \qquad t \text{ in } I$$

with f and g continuous on the interval I. Usually, I is a closed interval $[a, b]$. Think of t, called the **parameter**, as measuring time. As t advances from a to b, the point (x, y) traces out the curve in the xy-plane. The points $P = (x(a), y(a))$ and $Q = (x(b), y(b))$ are the initial and final **endpoints**. If the endpoints coincide, the curve is **closed**. If distinct values of t yield distinct points in the plane (except possibly for $t = a$ and $t = b$), we say the curve is a **simple** curve (Figure 1).

Eliminating the Parameter To recognize a curve given by parametric equations, it may be desirable to eliminate the parameter. Sometimes this can be accomplished by solving one equation for t and substituting in the other (Example 1). Often we can make use of a familiar identity, as in Example 2.

EXAMPLE 1: Eliminate the parameter in

$$x = t^2 + 2t, \qquad y = t - 3, \qquad -2 \le t \le 3$$

Then identify the corresponding curve and sketch its graph.

SOLUTION: From the second equation, $t = y + 3$. Substituting this expression for t in the first equation gives

$$x = (y + 3)^2 + 2(y + 3) = y^2 + 8y + 15$$

or

$$x + 1 = (y + 4)^2$$

This we recognize as a parabola with vertex at $(-1, -4)$ and opening to the right.

In graphing the given equation, we must be careful to display only that part of the parabola corresponding to $-2 \le t \le 3$. A table of values and the graph are shown in Figure 2. The arrow indicates the direction of increasing t. ◄

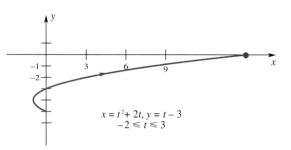

t	x	y
–2	0	–5
–1	–1	–4
0	0	–3
1	3	–2
2	8	–1
3	15	0

$x = t^2 + 2t, y = t - 3$
$-2 \leq t \leq 3$

Figure 2

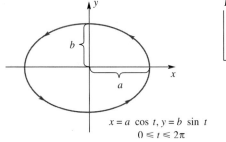

$x = a \cos t, y = b \sin t$
$0 \leq t \leq 2\pi$

Figure 3

EXAMPLE 2: Show that

$$x = a \cos t, \qquad y = b \sin t, \qquad 0 \leq t \leq 2\pi$$

represents the ellipse shown in Figure 3.

SOLUTION: We solve the equations for $\cos t$ and $\sin t$, then square, and add.

$$\left(\frac{x}{a}\right)^2 + \left(\frac{y}{b}\right)^2 = \cos^2 t + \sin^2 t = 1$$

$$\frac{x^2}{a^2} + \frac{y^2}{b^2} = 1$$

A quick check of a few values for t convinces us that we do get the complete ellipse. In particular, $t = 0$ and $t = 2\pi$ give the same point, namely, $(a, 0)$.

If $a = b$, we get the circle $x^2 + y^2 = a^2$.   ◄

Technology and Parametric Graphs

Graphing calculators and computer algebra systems can be used to generate graphs whose equations are given parametrically. As with rectangular graphs and polar graphs, the calculator or computer generates an internal table of values first, then connects the points with short straight line segments.

Different pairs of parametric equations may have the same graph, as the next two examples illustrate.

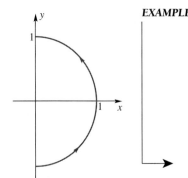

Figure 4

EXAMPLE 3: Show that each of the following pairs of parametric equations have the same graph—namely, the semicircle shown in Figure 4.

(a) $x = \sqrt{1 - t^2}, y = t, -1 \leq t \leq 1$

(b) $x = \cos t, y = \sin t, -\dfrac{\pi}{2} \leq t \leq \dfrac{\pi}{2}$

(c) $x = \dfrac{1 - t^2}{1 + t^2}, y = \dfrac{2t}{1 + t^2}, -1 \leq t \leq 1$

SOLUTION: In each case, we discover that

$$x^2 + y^2 = 1$$

It is then just a matter of checking a few values of t to make sure that the given intervals for t yield the same section of the circle. We can also check each set of equations by graphing them with a calculator in parametric mode, or with a computer algebra system using the parametric plot command; the result is indeed the semicircle, as you should check.   ◄

EXAMPLE 4: Show that each of the following pairs of parametric equations yields one branch of a hyperbola.

(a) $x = a \sec t, \ y = b \tan t, \ -\dfrac{\pi}{2} < t < \dfrac{\pi}{2}$

(b) $x = a \cosh t, \ y = b \sinh t, \ -\infty < t < \infty$

Assume in both cases that $a > 0$ and $b > 0$. Then graph each pair of equations using a calculator or computer for $a = 3$ and $b = 2$, and label a few points on each graph with the corresponding t-value.

SOLUTION:

(a) In the first case,

$$\left(\frac{x}{a}\right)^2 - \left(\frac{y}{b}\right)^2 = \sec^2 t - \tan^2 t = 1$$

(b) In the second case,

$$\left(\frac{x}{a}\right)^2 - \left(\frac{y}{b}\right)^2 = \cosh^2 t - \sinh^2 t$$

$$= \left(\frac{e^t + e^{-t}}{2}\right)^2 - \left(\frac{e^t - e^{-t}}{2}\right)^2 = 1$$

Checking a few t-values shows that, in both cases, we obtain the branch of the hyperbola $x^2/a^2 - y^2/b^2 = 1$ shown in Figure 5.

We cannot graph these equations on a computer or calculator as they are, because they involve the unknown constants a and b. However, by assigning values to these constants and then graphing, we obtain another check on our work. In Figure 6 we show a brief table of values and computer-generated parametric plots of $x = 3 \sec t, \ y = 2 \tan t$ for $-1.57 < t < 1.57$ and $-1.58 < t < 1.58$, and in Figure 7 we show a table and graph for $x = 3 \cosh t$, $y = 2 \sinh t$ with $-1 < t < 1$.

First look at Figure 6. Because $\frac{\pi}{2} \approx 1.571$, $-1.57 < t < 1.57$ represents a slightly smaller interval than $-\frac{\pi}{2} < t < \frac{\pi}{2}$ and $-1.58 < t < 1.58$ a slightly larger one. When using the interval $-1.58 < t < 1.58$, the asymptotes of the hyperbola seem to appear on the graph. This is similar to what we saw and discussed in Section 2.1 concerning graphs of functions with vertical asymptotes. In this

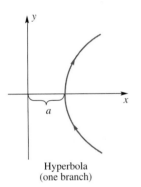

Hyperbola
(one branch)

Figure 5

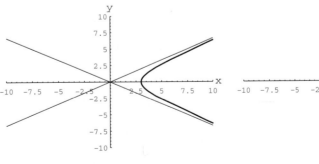

t	x	y
-1	5.55	-3.11
0	3.00	0.00
1	5.55	3.11

$x = 3 \sec t, \ y = 2 \tan t, \ -1.58 < t < 1.58$

$x = 3 \sec t, \ y = 2 \tan t, \ -1.57 < t < 1.57$

Figure 6

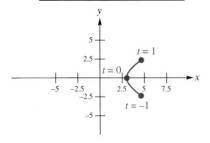

t	x	y
-1	4.63	-2.35
0	3.00	0.00
1	4.63	2.35

$x = 3 \cosh t,\ y = 2 \sinh t,\ -1 < t < 1$

Figure 7

case, points are being plotted in opposite corners of the graph window as the t-values cross over $-\frac{\pi}{2}$ and $\frac{\pi}{2}$ (where the secant and tangent functions are infinite). Thus, the graph that uses the smaller interval $-1.57 < t < 1.57$ is closer to what we expect the graph of $x = 3 \sec t,\ y = 2 \tan t$ for $-\frac{\pi}{2} < t < \frac{\pi}{2}$ to look like. ◄

What if we used an "exact" value for $\frac{\pi}{2}$? Graphing calculators have a π key and computer algebra systems have a symbol (such as Pi) to represent this special constant. The problem is that when the machine goes to generate a graph by plotting points, it must round the value of π either up a little bit or down a little bit. Thus, your graph may sometimes appear to have the asymptotes drawn in, and sometimes not. This is true of both graphing calculators and computer algebra systems. Thus, you must employ your knowledge of the functions involved, as well as some common sense about machine graphs, to decide which parts of the graph are "real" and which are not.

In order to graph $x = 3 \cosh t,\ y = 2 \sinh t,\ -\infty < t < \infty$ we have an even bigger problem. There is no way to "approximate" infinity; we must choose finite values for the endpoints of the t-interval. In Figure 7 we choose the endpoints -1 and 1; as a result we get only a finite portion of the entire curve. This is typical of parametric graphs. In order to give the appearance of generating the entire curve, you must choose the endpoints large enough that the curve extends beyond the graph window you have chosen.

The Cycloid   A cycloid is the curve traced by a point P on the rim of a wheel as the wheel rolls along a straight line without slipping (Figure 8). The Cartesian equation of a cycloid is quite complicated, but simple parametric equations are readily found, as shown in the next example.

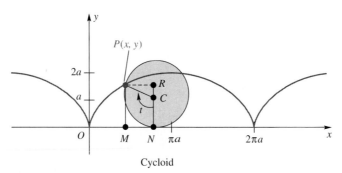

Cycloid

Figure 8

EXAMPLE 5: Find parametric equations for the cycloid.

SOLUTION: Let the wheel roll along the x-axis with P initially at the origin. Denote the center of the wheel by C and let a be its radius. Choose for a parameter the radian measure t of the clockwise angle through which the line segment CP has turned from its vertical position when P was at the origin. All of this is shown in Figure 8.

Because $|ON| = \text{arc } PN = at,$

$$x = |OM| = |ON| - |MN| = at - a \sin t = a(t - \sin t)$$

and

$$y = |MP| = |NR| = |NC| + |CR| = a - a \cos t = a(1 - \cos t)$$

Thus, the parametric equations for the cycloid are

$$x = a(t - \sin t), \qquad y = a(1 - \cos t) \qquad \blacktriangleleft$$

The cycloid has a number of interesting applications, especially in mechanics. It is the "curve of fastest descent." If a particle, acted on only by gravity, is allowed to slide down some curve from a point A to a lower point B not on the same vertical line, it completes its journey in the shortest time when the curve is an inverted cycloid (Figure 9). Of course, the shortest distance is along the straight line segment AB, but the least time is used when the path is along a cycloid; this is because the acceleration when it is released depends on the steepness of descent, and along a cycloid it builds up velocity much more quickly than it does along a straight line.

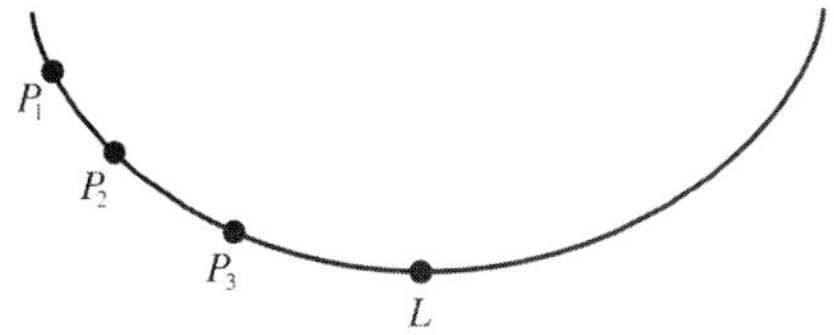

Another interesting property is this: If L is the lowest point on an arch of an inverted cycloid, the time it takes a particle P to slide down the cycloid to L is the same no matter where P starts from on the inverted arch; thus, if several particles, P_1, P_2, and P_3, in different positions on the cycloid (Figure 10) start to slide at the same instant, all will reach the low point L at the same time.

In 1673, the Dutch astronomer Christian Huygens published a description of an ideal pendulum clock. Because the bob swings between cycloidal "cheeks," the path of the bob is a cycloid (Figure 11). This means that the period of the swing is independent of the amplitude, and so the period does not change as the clock's spring unwinds.

A surprising fact is that the three results just mentioned all date from the seventeenth century. To demonstrate them is not an easy task, as you may discover by looking at any book on the history of calculus.

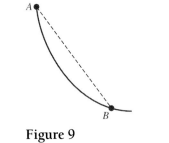

Figure 9

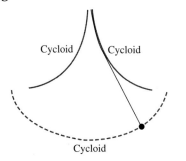

Figure 10

Cycloid   Cycloid

Cycloid

Figure 11

Calculus for Curves Defined Parametrically

Can we find the slope of the tangent line to a curve given parametrically without first eliminating the parameter? The answer is yes, according to the following theorem.

Theorem A

Let f and g be continuously differentiable with $f'(t) \neq 0$ on $\alpha \leq t \leq \beta$. Then the parametric equations

$$x = f(t), \qquad y = g(t)$$

define y as a differentiable function of x and

$$\frac{dy}{dx} = \frac{dy/dt}{dx/dt}$$

Proof   Because $f'(t) \neq 0$ for $\alpha \leq t \leq \beta$, f is strictly monotonic and so has a differentiable inverse f^{-1} (see the Inverse Function Theorem (Theorem 7.1B)). Define F by $F = g \circ f^{-1}$, so that

$$y = g(t) = g(f^{-1}(x)) = F(x) = F(f(t))$$

Then by the Chain Rule,

$$\frac{dy}{dt} = F'(f(t)) \cdot f'(t) = \frac{dy}{dx} \cdot \frac{dx}{dt}$$

Because $dx/dt \neq 0$, we have

$$\frac{dy}{dx} = \frac{dy}{dt} \div \frac{dx}{dt}$$

◀

EXAMPLE 6: Find the first two derivatives dy/dx and d^2y/dx^2 for the function determined by

$$x = 5 \cos t, \qquad y = 4 \sin t, \qquad 0 < t < 3$$

and evaluate them at $t = \pi/6$ (see Example 2).

SOLUTION: Let y' denote dy/dx. Then

$$\frac{dy}{dx} = \frac{dy}{dt} \div \frac{dx}{dt} = \frac{4 \cos t}{-5 \sin t} = -\frac{4}{5} \cot t$$

$$\frac{d^2y}{dx^2} = \frac{dy'}{dx} = \frac{dy'}{dt} \div \frac{dx}{dt} = \frac{\frac{4}{5} \csc^2 t}{-5 \sin t} = -\frac{4}{25} \csc^3 t$$

At $t = \pi/6$,

$$\frac{dy}{dx} = \frac{-4\sqrt{3}}{5}, \qquad \frac{d^2y}{dx^2} = \frac{-4}{25}(8) = \frac{-32}{25}$$

The first value is the slope of the tangent line to the ellipse $x^2/25 + y^2/16 = 1$ at the point $(5\sqrt{3}/2, 2)$. You can check that this is so by implicit differentiation. ◀

Sometimes a definite integral involves two variables, such as x and y, in the integrand and differential, and y may be defined as a function of x by equations that give x and y in terms of a parameter such as t. In such cases, it is often convenient to evaluate the definite integral by expressing the integrand and the differential in terms of t and dt and adjusting the limits of integration before integrating with respect to t.

EXAMPLE 7: Evaluate (a) $\int_1^3 y \, dx$ and (b) $\int_1^3 xy^2 \, dx$, using $x = 2t - 1$ and $y = t^2 + 2$.

SOLUTION: From $x = 2t - 1$ we have $dx = 2 \, dt$; when $x = 1$, $t = 1$; and when $x = 3$, $t = 2$.

(a) $\int_1^3 y \, dx = \int_1^2 (t^2 + 2)2 \, dt = 2\left[\frac{t^3}{3} + 2t\right]_1^2 = \frac{26}{3}$

(b) $\int_1^3 xy^2 \, dx = \int_1^2 (2t - 1)(t^2 + 2)^2 \, 2 \, dt$

$$= 2\int_1^2 (2t^5 - t^4 + 8t^3 - 4t^2 + 8t - 4) \, dt = 86\tfrac{14}{15}$$ ◀

EXAMPLE 8: Find the area A under one arch of a cycloid (Figure 12). Also, find the length L of this arch.

SOLUTION: From Example 5, we know that we may represent one arch of the cycloid by

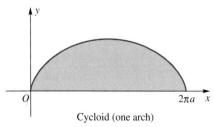

Cycloid (one arch)

Figure 12

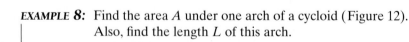

$$x = a(t - \sin t), \qquad y = a(1 - \cos t), \qquad 0 \le t \le 2\pi$$

Thus,

$$A = \int_0^{2\pi a} y \, dx = \int_0^{2\pi} a(1 - \cos t) d[a(t - \sin t)]$$

$$= a^2 \int_0^{2\pi} (1 - \cos t)(1 - \cos t) \, dt$$

$$= a^2 \int_0^{2\pi} (1 - 2\cos t + \cos^2 t) \, dt$$

$$= a^2 \int_0^{2\pi} (1 - 2\cos t + \tfrac{1}{2} + \tfrac{1}{2}\cos 2t) \, dt$$

$$= a^2 [\tfrac{3}{2} t - 2\sin t + \tfrac{1}{4}\sin 2t]_0^{2\pi} = 3\pi a^2$$

To calculate L, we recall the arc-length formula from Section 6.4.

$$L = \int_\alpha^\beta \sqrt{\left(\frac{dx}{dt}\right)^2 + \left(\frac{dy}{dt}\right)^2} \, dt$$

which in our case reduces to

$$L = \int_0^{2\pi} \sqrt{a^2(1 - \cos t)^2 + a^2(\sin^2 t)} \, dt$$

$$= a \int_0^{2\pi} \sqrt{2(1 - \cos t)} \, dt$$

$$= a \int_0^{2\pi} \sqrt{4\sin^2 \frac{t}{2}} \, dt$$

$$= 2a \int_0^{2\pi} \sin \frac{t}{2} \, dt$$

$$= \left[-4a \cos \frac{t}{2} \right]_0^{2\pi} = 8a$$

◄

Two Fleas on a Trike

Two fleas are arguing about who will get the longest ride when Angie pedals her tricycle home from the park. A will ride between the treads of the front tire; B will ride between the treads of one of the rear tires. Settle the argument by showing that their paths will have equal lengths. Example 8 should help.

Concepts Review

1. A circle is a prime example of a curve that is both _____ and _____; a figure eight is an example of a closed curve that is not _____ .

2. We call two equations $x = f(t)$ and $y = g(t)$ a _____ representation of a curve, and t is called a _____ .

3. The path of a point on the rim of a rolling circle is called a _____ .

4. The formula for dy/dx, given the representation $x = f(t)$ and $y = g(t)$, is $dy/dx =$ _____ .

Problem Set 10.6

In each of Problems 1–12, a parametric representation of a curve is given.

(a) Sketch the curve by assigning a few values to the parameter, and then graphing with a calculator or computer. Identify the t-values of some of the points on the sketch.

(b) Which of the following applies to the curve: simple or closed?

(c) Obtain the Cartesian equation of the curve by eliminating the parameter. (See Examples 1 and 4 in Section 1.)

1. $x = 2t, y = 3t; -\infty < t < \infty$

2. $x = 4t - 1, y = 2t; 0 \le t \le 3$

3. $x = t - 4, y = \sqrt{t}; 0 \le t \le 4$

4. $x = t, y = \dfrac{1}{t}; t > 0$

5. $x = t^2, y = t^3; -1 \le t \le 2$

6. $x = t^2 - 1, y = t^3 - t; -3 \le t \le 3$

7. $x = t^3 - 4t, y = t^2 - 4; -3 \le t \le 3$

8. $x = 3\sqrt{t - 3}, y = 2\sqrt{4 - t}; 3 \le t \le 4$

9. $x = 3 \sin t, y = 5 \cos t; 0 \le t \le 2\pi$

10. $x = 3 \sin \theta - 1, y = 2 \cos \theta + 2; 0 \le \theta \le \pi$

11. $x = 4 \sin^4 t, y = 4 \cos^4 t; 0 \le t \le \frac{1}{2}\pi$

12. $x = 2 \cos \theta, y = 2 \cos \frac{1}{2}\theta; -\infty < \theta < \infty$

In Problems 13–18, find dy/dx and d^2y/dx^2 without eliminating the parameter.

13. $x = 3t^2, y = 2t^3; t \ne 0$

14. $x = 6t^2, y = t^3; t \ne 0$

15. $x = 2t - \dfrac{3}{t}, y = 2t + \dfrac{3}{t}; t \ne 0$

16. $x = 1 - \cos t, y = 2 + 3 \sin t; t \ne n\pi$

17. $x = 3 \tan t - 1, y = 5 \sec t + 2; t \ne \dfrac{(2n + 1)\pi}{2}$

18. $x = \dfrac{2}{1 + t^2}, y = \dfrac{2}{t(1 + t^2)}; t \ne 0$

In Problems 19–22, find the equation of the tangent to the given curve at the given point without eliminating the parameter. Make a sketch and identify the t-values of some of the points on the sketch.

19. $x = t^2, y = t^3; t = 2$

20. $x = 3t, y = 8t^3; t = -\frac{1}{2}$

21. $x = 2 \sec t, y = 2 \tan t; t = -\dfrac{\pi}{6}$

22. $x = 2e^t, y = \frac{1}{3}e^{-t}; t = 0$

In Problems 23–24, evaluate the integrals.

23. $\displaystyle\int_0^1 (x^2 - 4y)\, dx$, where $x = t + 1, y = t^3 + 4$.

24. $\displaystyle\int_1^{\sqrt{3}} xy\, dy$, where $x = \sec t, y = \tan t$.

25. Find the area of the region between the curve $x = e^{2t}, y = e^{-t}$, and the x-axis from $t = 0$ to $t = \ln 5$. Make a sketch.

26. Find the area of the region bounded by the curve

$$x = t + \frac{1}{t}, \qquad y = t - \frac{1}{t}$$

and the line $3x - 10 = 0$, without eliminating the parameter. Make a sketch.

27. Modify the text discussion of the cycloid (and its accompanying diagram) to handle the case where the point P is $b < a$ units from the center of the wheel. Show that the corresponding parametric equations are

$$x = at - b \sin t \qquad y = a - b \cos t$$

Sketch the graph of these equations (called a **curtate cycloid**) when $a = 8$ and $b = 4$, and identify the t-values of some of the points on the sketch.

28. Follow the instructions of Problem 27 for the case $b > a$ (a flanged wheel, as on a train) showing that you get the same parametric equations. Sketch the graph of these equations (called a **prolate cycloid**) when $a = 6$ and $b = 8$ and identify the t-values of some of the points on the sketch.

29. The path of a projectile fired from level ground with a speed of v_0 feet per second at an angle α with the ground, is given by the parametric equations

$$x = (v_0\cos \alpha)t, \qquad y = -16t^2 + (v_0\sin \alpha)t$$

(a) Show that the path is a parabola.

(b) Find the time of flight.

(c) Show that the range (horizontal distance traveled) is $(v_0^2/32)\sin 2\alpha$.

(d) For a given v_0, what value of α gives the largest possible range?

30. Each value of the parameter t in the equations of a cycloid,

$$x = a(t - \sin t), \qquad y = a(1 - \cos t),$$

determines a unique point $P(x, y)$ on the cycloid and also determines a unique position of the rolling circle that generates the cycloid. Let $P_1(x_1, y_1)$ be the point on the cycloid determined by the value t_1 of the parameter t. Prove that the tangent to the cycloid at P_1 passes through the highest point on the rolling circle when the circle is in the position determined by t_1. Through what point on the circle does the normal pass?

31. Let a circle of radius b roll, without slipping, inside a fixed circle of radius a, $a > b$. A point P on the circumference of the rolling circle traces out a curve called a **hypocycloid**. Find parametric equations of the hypocycloid. *Hint:* Place the origin O of Cartesian coordinates at the center of the fixed, larger circle and let the point $A(a, 0)$ be one position of the tracing point P. Denote by B the moving point of tangency of the two circles and let t, the radian measure of the angle AOB, be the parameter (see Figure 13).

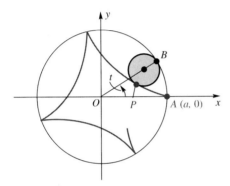

Figure 13

32. Show that if $b = a/4$ in Problem 31, the parametric equations of the hypocycloid may be simplified to

$$x = a \cos^3 t, \qquad y = a \sin^3 t$$

This is called a **hypocycloid of four cusps**. Sketch it carefully and show that its Cartesian equation is $x^{\frac{2}{3}} + y^{\frac{2}{3}} = a^{\frac{2}{3}}$.

33. The curve traced by a point on the circumference of a circle of radius b as it rolls without slipping on the outside of a fixed circle of radius a is called an **epicycloid**. Show that it has parametric equations

$$x = (a + b) \cos t - b \cos \frac{a + b}{b} t$$

$$y = (a + b) \sin t - b \sin \frac{a + b}{b} t$$

(See the hint in Problem 31.)

34. If $b = a$, the equations in Problem 33 are

$$x = 2a \cos t - a \cos 2t$$

$$y = 2a \sin t - a \sin 2t$$

Show that this special epicycloid is the cardioid $r = 2a(1 - \cos \theta)$, where the pole of the polar coordinate system is the point $(a, 0)$ in the Cartesian system and the polar axis has the direction of the positive x-axis. *Hint:* Find a Cartesian equation of the epicycloid by eliminating the parameter t between the equations. Then show that the equations connecting the Cartesian and polar systems are

$$x = r \cos \theta + a \qquad y = r \sin \theta$$

and use these equations to transform the Cartesian equation into $r = 2a(1 - \cos \theta)$.

35. Consider a circle of radius a centered at $(0, a)$, as in Figure 14. Let a line OA intersect the line $y = 2a$ at A and the circle C at B. Finally, let P be the point of intersection of a horizontal line through B and a vertical line through A. As θ, the angle OA makes with the positive x-axis, varies, P traces out a curve, the **witch of Agnesi**.

(a) Find parametric equations for the witch using θ as a parameter.

(b) Find the Cartesian equation of the witch.

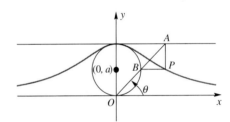

Figure 14

36. If $b = a/3$ in Problem 31, we obtain a hypocycloid of three cusps, called a **deltoid**, with parametric equations

$$x = \left(\frac{a}{3}\right)(2 \cos t + \cos 2t), \qquad y = \left(\frac{a}{3}\right)(2 \sin t - \sin 2t)$$

Find the length of the deltoid.

37. Find the length of the epicycloid of Problem 33 assuming a/b is an integer and $0 \le t \le 2\pi$.

38. Consider the ellipse $x^2/a^2 + y^2/b^2 = 1$.

(a) Show that its perimeter is

$$P = 4a \int_0^{\pi/2} \sqrt{1 - e^2 \cos^2 t}\, dt,$$

where e is the eccentricity.

(b) The integral in part a is called an *elliptic integral*. It has been studied at great length, and it is known that the integrand does not have an elementary antiderivative, so we must turn to approximate methods to evaluate P. Do so when $a = 1$ and $e = \frac{1}{4}$. (Your answer should be near 2π. Why?)

39.   Draw the graph of each of the following for the interval $0 \leq t \leq 4\pi$ and identify the t-values of some of the points on the sketch.

(a) $x = 2(t - \sin t), y = 2(1 - \cos t)$ (a cycloid)

(b) $x = 2t - \sin t, y = 2 - \cos t$ (a curtate cycloid; see Problem 27)

(c) $x = t - 2 \sin t, y = 1 - 2 \cos t$ (a prolate cycloid; see Problem 28)

40.   Draw the graph of the hypocycloid (see Problem 31)

$$x = (a - b) \cos t + b \cos \frac{a - b}{b} t$$

$$y = (a - b) \sin t - b \sin \frac{a - b}{b} t$$

for appropriate values of t in each of the following cases.

(a) $a = 4, b = 1$ (b) $a = 3, b = 1$

(c) $a = 5, b = 2$ (d) $a = 7, b = 4$

Experiment with other positive integer values of a and b and then make conjectures about the length of the t-inter-

val required for the curve to return to its starting point and about the number of cusps. What can you say if a/b is irrational?

41.   Draw the graph of the epicycloid (see Problem 33)

$$x = (a + b) \cos t - b \cos \frac{a + b}{b} t$$

$$y = (a + b) \sin t - b \sin \frac{a + b}{b} t$$

for various values of a and b. What conjectures can you make?

42.   Draw the Folium of Descartes $x = 3t/(t^3 + 1)$, $y = 3t^2/(t^3 + 1)$. Then tell for what values of t this graph is in each of the four quadrants.

Answers to Concepts Review: 1. Simple; closed; simple 2. Parametric, parameter 3. Cycloid 4.$(dy/dt)/(dx/dt)$

$\boxed{10.7}$ CHAPTER REVIEW

Concepts Test

Respond with true or false to each of the following assertions, and provide an explanation of of your response.

1.   The graph of $y = ax^2 + bx + c$ is a parabola for all choices of a, b, and c.

2.   The vertex of a parabola is midway between the focus and the directrix.

3.   A vertex of an ellipse is closer to a directrix than to a focus.

4.   The point on a parabola closest to its focus is the vertex.

5.   The hyperbolas $x^2/a^2 - y^2/b^2 = 1$ and $y^2/b^2 - x^2/a^2 = 1$ have the same asymptotes.

6.   The circumference C of the ellipse $x^2/a^2 + y^2/b^2 = 1$ with $b < a$ satisfies $2\pi b < C < 2\pi a$.

7.   The smaller the eccentricity e of an ellipse, the more nearly circular the ellipse is.

8.   The ellipse $6x^2 + 4y^2 = 24$ has its foci on the x-axis.

9.   The equation $x^2 + y^2 = 0$ represents a hyperbola.

10.   The equation $(y^2 - 4x + 1)^2 = 0$ represents a parabola.

11.   If $k \neq 0$, $x^2/a^2 - y^2/b^2 = k$ is the equation of a hyperbola.

12.   If $k \neq 0$, $x^2/a^2 + y^2/b^2 = k$ is the equation of an ellipse.

13.   The distance between the foci of the graph of $x^2/a^2 + y^2/b^2 = 1$ is $2\sqrt{a^2 - b^2}$.

14.   The graph of $x^2/9 + y^2/8 = -2$ does not intersect the x-axis.

15.   Light emanating from a point between a focus and the nearest vertex of an elliptical mirror will be reflected beyond the other focus.

16.   An ellipse that is drawn using a string of length 8 units attached to foci 2 units apart will have minor diameter of length $\sqrt{60}$ units.

17.   The graph of $2x^2 + y^2 + Cx + Dy + F = 0$ cannot be a single point.

18.   The graph of the polar equation $r = 4 \cos(\theta - \pi/3)$ is a circle.

19.   Every point in the plane has infinitely many sets of polar coordinates.

20.   All points of intersection of the graphs of the polar equations $r = f(\theta)$ and $r = g(\theta)$ can be found by solving these two equations simultaneously.

21. The graph of $r = 4\cos 3\theta$ is a rose of 3 leaves whose area is less than half that of the circle $r = 4$.

22. The area inside the graph of $r = \cos\theta$ is given by $A = \dfrac{1}{2}\displaystyle\int_0^{2\pi}\cos^2\theta\,d\theta$.

23. The graph of $r^2 = \cos 2\theta$ is the same as the graph of $r = \sqrt{\cos 2\theta}$.

24. The graph of $r = 4$ will always appear as a circle when plotted on a graphing calculator or computer algebra system.

25. The parametric representation of a curve is unique.

26. The graph of $x = 2t^3$, $y = t^3$ is a straight line.

27. If $x = f(t)$ and $y = g(t)$, then we can find a function h such that $y = h(x)$.

28. The curve with parametric representation $x = \ln t$ and $y = t^2 - 1$ passes through the origin.

29. If $x = f(t)$ and $y = g(t)$ and if both f'' and g'' exist, then $d^2y/dx^2 = g''(t)/f''(t)$ wherever $f''(t) \neq 0$.

30. A curve may have more than one tangent line at a point on the curve.

Sample Test Problems

1. From the numbered list, pick the correct responses to put in the blanks that follow.

(1) No graph. (2) A single point.
(3) A single line. (4) Two parallel lines.
(5) Two intersecting lines (6) A circle.
(7) A parabola. (8) An ellipse.
(9) A hyperbola. (10) None of the above.

_____ (a) $x^2 - 4y^2 = 0$
_____ (b) $x^2 - 4y^2 = 0.001$
_____ (c) $x^2 - 4 = 0$
_____ (d) $x^2 - 4x + 4 = 0$
_____ (e) $x^2 + 4y^2 = 0$
_____ (f) $x^2 + 4y^2 = x$
_____ (g) $x^2 + 4y^2 = -x$
_____ (h) $x^2 + 4y^2 = -1$
_____ (i) $(x^2 + 4y - 1)^2 = 0$
_____ (j) $3x^2 + 4y^2 = -x^2 + 1$

In each of Problems 2–10, name the conic that has the given equation. Find its vertices and foci and sketch its graph.

2. $y^2 - 6x = 0$
3. $9x^2 + 4y^2 - 36 = 0$
4. $25x^2 - 36y^2 + 900 = 0$
5. $x^2 - 9y = 0$
6. $x^2 - 4y^2 - 16 = 0$
7. $9x^2 + 25y^2 - 225 = 0$
8. $9x^2 + 9y^2 - 225 = 0$
9. $r = \dfrac{5}{2 + 2\sin\theta}$
10. $r(2 + \cos\theta) = 3$

In each of Problems 11–18, find the Cartesian equation of the conic with the given properties.

11. Vertices $(\pm 4, 0)$ and eccentricity $\frac{1}{2}$.
12. Eccentricity 1, focus $(0, -3)$, and vertex $(0, 0)$.
13. Eccentricity 1, vertex $(0, 0)$, symmetric with respect to the x-axis, and passing through the point $(-1, 3)$.
14. Eccentricity $\frac{5}{3}$ and vertices $(0, \pm 3)$.
15. Vertices $(\pm 2, 0)$ and asymptotes $x \pm 2y = 0$.
16. Parabola with focus $(3, 2)$ and vertex $(3, 3)$.
17. Ellipse with center $(1, 2)$, focus $(4, 2)$, and major diameter 10.
18. Hyperbola with vertices $(2, 0)$, $(2, 6)$, and eccentricity $\frac{10}{3}$.

In Problems 19–22, use the process of completing the square to reduce the given equation to a standard form. Then name the corresponding curve and sketch its graph.

19. $4x^2 + 4y^2 - 24x + 36y + 81 = 0$
20. $4x^2 + 9y^2 - 24x + 36y + 36 = 0$
21. $x^2 - 8x + 6y + 28 = 0$
22. $3x^2 - 10y^2 + 36x - 20y + 68 = 0$

In Problems 23–34, analyze the given polar equation and sketch its graph. Label several important points on each graph with their polar coordinates.

23. $r = 6\cos\theta$
24. $r = \dfrac{5}{\sin\theta}$
25. $r = \cos 2\theta$
26. $r = \dfrac{3}{\cos\theta}$
27. $r = 4$
28. $r = 5 - 5\cos\theta$
29. $r = 4 - 3\cos\theta$
30. $r = 2 - 3\cos\theta$
31. $\theta = \frac{2}{3}\pi$
32. $r = 4\sin 3\theta$
33. $r^2 = 16\sin 2\theta$
34. $r = -\theta,\ \theta \geq 0$

35. Find the slope of the tangent to the graph of

$$r = 3 + 3\cos\theta$$

at the point on the graph where $\theta = \frac{1}{6}\pi$.

36. Sketch the graphs of

$$r = 5\sin\theta \quad \text{and} \quad r = 2 + \sin\theta$$

and find their points of intersection.

37. Find the area of the region bounded by the graph of

$$r = 5 - 5\cos\theta$$

38. Find the area of the region that is outside the limaçon $r = 2 + \sin\theta$ and inside the circle $r = 5\cos\theta$. Use either exact or approximate methods, and give reasons for your choice of method.

39. A racing car on the elliptical race track $x^2/400 + y^2/100 = 1$ went out of control at the point $(16, 6)$ and thereafter continued on the tangent line until it hit a tree at $(14, k)$. Determine k.

In Problems 40–43, a parametric representation of a curve is given. Eliminate the parameter to obtain the corresponding Cartesian equation. Sketch the given curve, and check using the parametric graphing capability of a computer or calculator.

40. $x = 6t + 2, y = 2t; -\infty < t < \infty$

41. $x = 4t^2, y = 4t; -1 \le t \le 2$

42. $x = 4\sin t - 2, y = 3\cos t + 1; 0 \le t \le 2\pi$

43. $x = 2\sec t, y = \tan t; -\dfrac{\pi}{2} < t < \dfrac{\pi}{2}$

In Problems 44 and 45, find the equation of the tangent line at $t = 0$.

44. $x = 2t^3 - 4t + 7, y = t + \ln(t + 1)$

45. $x = 3e^{-t}, y = \frac{1}{2}e^t$

46. Find the length of the curve

$$x = \cos t + t\sin t$$

$$y = \sin t - t\cos t$$

from 0 to 2π. Make a sketch.

Appendix:
Data for the Labs

Lab 3—The Damped Harmonic Oscillator   Three trials were performed for each data point: the second return of the pendulum, the fourth return, and the sixth return. Time is in seconds and displacements in inches.

Second return		Fourth return		Sixth return	
time	displacement	time	displacement	time	displacement
4.63	9.25	9.55	8.875	13.92	8.25
4.49	9.125	9.27	8.875	13.81	8.25
4.65	9.25	9.54	9.00	14.00	8.375

Lab 8—The Draining Can   Three trials were performed; on each trial the times at which the water level reached the 9-ounce, 6-ounce, 3-ounce, and 0-ounce levels were recorded. At time 0 the level was 12 ounces. Time is in seconds.

ounces	time (first trial)	time (second trial)	time (third trial)
12	0	0	0
9	6.52	6.95	6.91
6	14.91	14.99	14.93
3	25.67	25.70	25.31
0	41.61	39.49	41.25

Lab 12—Falling Objects   Three trials were performed for both the 6-foot drop height and the 3-foot drop height. In each case the time in seconds when the balloon hit the ground was recorded.

six-foot drop	three-foot drop
2.20	1.29
2.22	1.23
2.29	1.17

Lab 16—Probability and Improper Integrals   The following are the heights of 14 women chosen at random. All heights are in inches.

66, 61, 65, 71, 63, 65.5, 64.5, 62.5, 66, 64, 67, 69, 59, 60

Lab 18—Bouncing Balls and Infinite Series   For each of three drop heights, the rebound height was recorded by watching the top of the ball against a background of parallel lines corresponding to different heights. Eventually a mark was found that the ball reached on nearly every bounce, and that height was recorded as the rebound height. (This worked better than the usual procedure of taking several trials and recording the results individually.) Also, for each drop height, the time until the ball stopped was recorded; for this, three trials were taken for each drop height. We used a lacrosse ball 2.5 inches in diameter. Heights are in inches and times in seconds. *Note:* You should change the units below to feet (or meters) to be consistent with the units of the acceleration of gravity.

	60-inch drop	30-inch drop	15-inch drop
rebound height:	40.5	20.5	10.5
time to stop:	7.18, 7.07, 7.14	5.58, 5.67, 5.59	4.16, 4.26, 4.07

Index

DERIVATIVES

$D_x\, x^r = rx^{r-1}$ $\qquad D_x\, |x| = \dfrac{|x|}{x}$

$D_x \sin x = \cos x$ $\qquad D_x \cos x = -\sin x$

$D_x \tan x = \sec^2 x$ $\qquad D_x \cot x = -\csc^2 x$

$D_x \sec x = \sec x \tan x$ $\qquad D_x \csc x = -\csc x \cot x$

$D_x \sinh x = \cosh x$ $\qquad D_x \coth x = -\operatorname{csch}^2 x$

$D_x \cosh x = \sinh x$ $\qquad D_x \operatorname{sech} x = -\operatorname{sech} x \tanh x$

$D_x \tanh x = \operatorname{sech}^2 x$ $\qquad D_x \operatorname{csch} x = -\operatorname{csch} x \coth x$

$D_x \ln x = \dfrac{1}{x}$ $\qquad D_x \log_a x = \dfrac{1}{x \ln a}$

$D_x e^x = e^x$ $\qquad D_x a^x = a^x \ln a$

$D_x \sin^{-1} x = \dfrac{1}{\sqrt{1-x^2}}$ $\qquad D_x \cos^{-1} x = \dfrac{-1}{\sqrt{1-x^2}}$

$D_x \tan^{-1} x = \dfrac{1}{1+x^2}$ $\qquad D_x \sec^{-1} x = \dfrac{1}{|x|\sqrt{x^2-1}}$

INTEGRALS

1. $\int u\, dv = uv - \int v\, du$

2. $\int u^n\, du = \dfrac{1}{n+1} u^{n+1} + C, \quad n \neq -1$

3. $\int \dfrac{1}{u}\, du = \ln|u| + C$

4. $\int e^u\, du = e^u + C$

5. $\int a^u\, du = \dfrac{a^u}{\ln a} + C$

6. $\int \sin u\, du = -\cos u + C$

7. $\int \cos u\, du = \sin u + C$

8. $\int \sec^2 u\, du = \tan u + C$

9. $\int \csc^2 u\, du = -\cot u + C$

10. $\int \sec u \tan u\, du = \sec u + C$

11. $\int \csc u \cot u\, du = -\csc u + C$

12. $\int \tan u\, du = \ln|\sec u| + C$

13. $\int \cot u\, du = \ln|\sin u| + C$

14. $\int \sec u\, du = \ln|\sec u + \tan u| + C$

15. $\int \csc u\, du = \ln|\csc u - \cot u| + C$

16. $\int \dfrac{1}{\sqrt{a^2-u^2}}\, du = \sin^{-1}\dfrac{u}{a} + C$

17. $\int \dfrac{1}{a^2+u^2}\, du = \dfrac{1}{a}\tan^{-1}\dfrac{u}{a} + C$

18. $\int \dfrac{1}{a^2-u^2}\, du = \dfrac{1}{2a}\ln\left|\dfrac{u+a}{u-a}\right| + C$

19. $\int \dfrac{1}{u\sqrt{u^2-a^2}}\, du = \dfrac{1}{a}\sec^{-1}\left|\dfrac{u}{a}\right| + C$

GEOMETRY

Triangles

Pythagorean Theorem

$a^2 + b^2 = c^2$

Angles $\alpha + \beta + \gamma = 180°$

Area $A = \frac{1}{2}bh$

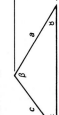

Right triangle

Any triangle

Circles

Circumference $\quad C = 2\pi r$

Area $\quad A = \pi r^2$

Cylinders

Surface area $\quad S = 2\pi r^2 + 2\pi r h$

Volume $\quad V = \pi r^2 h$

Cones

Surface area $\quad S = \pi r^2 + \pi r \sqrt{r^2 + h^2}$

Volume $\quad V = \frac{1}{3}\pi r^2 h$

Spheres

Surface area $\quad S = 4\pi r^2$

Volume $\quad V = \frac{4}{3}\pi r^3$

Conversions

1 inch = 2.54 centimeters

1 liter = 1000 cubic centimeters

1 kilogram = 2.20 pounds

1 kilometer = .62 miles

1 liter = 1.057 quarts

1 pound = 453.6 grams

π radians = 180 degrees

TRIGONOMETRY

Basic Identities

$$\tan t = \frac{\sin t}{\cos t} \qquad \cot t = \frac{\cos t}{\sin t} \qquad \csc t = \frac{1}{\sin t}$$

$$\sec t = \frac{1}{\cos t} \qquad \sin^2 t + \cos^2 t = 1$$

$$1 + \tan^2 t = \sec^2 t \qquad 1 + \cot^2 t = \csc^2 t$$

Cofunction Identities

$$\sin\left(\frac{\pi}{2} - t\right) = \cos t \qquad \cos\left(\frac{\pi}{2} - t\right) = \sin t \qquad \tan\left(\frac{\pi}{2} - t\right) = \cot t$$

Odd-even Identities

$$\sin(-t) = -\sin t \qquad \cos(-t) = \cos t \qquad \tan(-t) = -\tan t$$

Addition Formulas

$$\sin(s + t) = \sin s \cos t + \cos s \sin t \qquad \sin(s - t) = \sin s \cos t - \cos s \sin t$$

$$\cos(s + t) = \cos s \cos t - \sin s \sin t \qquad \cos(s - t) = \cos s \cos t + \sin s \sin t$$

$$\tan(s + t) = \frac{\tan s + \tan t}{1 - \tan s \tan t} \qquad \tan(s - t) = \frac{\tan s - \tan t}{1 + \tan s \tan t}$$

Double Angle Formulas

$$\sin 2t = 2 \sin t \cos t$$

$$\cos 2t = \cos^2 t - \sin^2 t = 1 - 2 \sin^2 t = 2 \cos^2 t - 1$$

$$\tan 2t = \frac{2 \tan t}{1 - \tan^2 t}$$

Half Angle Formulas

$$\sin\frac{t}{2} = \pm\sqrt{\frac{1 - \cos t}{2}} \qquad \cos\frac{t}{2} = \pm\sqrt{\frac{1 + \cos t}{2}} \qquad \tan\frac{t}{2} = \frac{1 - \cos t}{\sin t}$$

Product Formulas

$$2 \sin s \cos t = \sin(s + t) + \sin(s - t)$$

$$2 \cos s \cos t = \cos(s + t) + \cos(s - t)$$

$$2 \sin s \sin t = \cos(s - t) - \cos(s + t)$$

Factoring Formulas

$$\sin s + \sin t = 2 \cos\frac{s - t}{2} \sin\frac{s + t}{2}$$

$$\sin s - \sin t = 2 \cos\frac{s + t}{2} \sin\frac{s - t}{2}$$

$$\cos s + \cos t = 2 \cos\frac{s - t}{2} \cos\frac{s + t}{2}$$

$$\cos s - \cos t = -2 \sin\frac{s + t}{2} \sin\frac{s - t}{2}$$

Laws of Sines and Cosines

$$\frac{\sin \alpha}{a} = \frac{\sin \beta}{b} = \frac{\sin \gamma}{c}$$

$$a^2 = b^2 + c^2 - 2bc \cos \alpha$$

Graphs

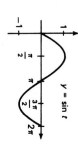

$$\sin t = \sin \theta = y = \frac{b}{r}$$

$$\cos t = \cos \theta = x = \frac{a}{r}$$

$$\tan t = \tan \theta = \frac{y}{x} = \frac{b}{a} \qquad \cot t = \cot \theta = \frac{x}{y} = \frac{a}{b}$$

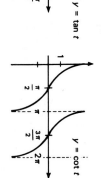

$y = \sin t$

$y = \cos t$

$y = \tan t$

$y = \cot t$

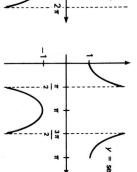

$y = \csc t$

$y = \sec t$

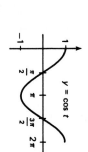

Inverse Trigonometric Functions

$$y = \sin^{-1} x \Longleftrightarrow x = \sin y, \; -\pi/2 \le y \le \pi/2$$

$$y = \cos^{-1} x \Longleftrightarrow x = \cos y, \; 0 \le y \le \pi$$

$$y = \tan^{-1} x \Longleftrightarrow x = \tan y, \; -\pi/2 < y < \pi/2$$

$$y = \sec^{-1} x \Longleftrightarrow x = \sec y, \; 0 \le y \le \pi, \, y \ne \pi/2$$

$$\sec^{-1} x = \cos^{-1}(1/x)$$

Hyperbolic Functions

$$\sinh x = \frac{1}{2}(e^x - e^{-x}) \qquad \cosh x = \frac{1}{2}(e^x + e^{-x})$$

$$\tanh x = \frac{\sinh x}{\cosh x} \qquad \coth x = \frac{\cosh x}{\sinh x}$$

$$\operatorname{sech} x = \frac{1}{\cosh x} \qquad \operatorname{csch} x = \frac{1}{\sinh x}$$

Series

$$\frac{1}{1 - x} = 1 + x + x^2 + x^3 + \cdots - 1 < x < 1$$

$$e^x = 1 + x + \frac{x^2}{2!} + \frac{x^3}{3!} + \cdots$$

$$\ln(1 + x) = x - \frac{x^2}{2} + \frac{x^3}{3} - \frac{x^4}{4} + \cdots, \; -1 < x \le 1$$

$$\sin x = x - \frac{x^3}{3!} + \frac{x^5}{5!} - \frac{x^7}{7!} + \cdots$$

$$\cos x = 1 - \frac{x^2}{2!} + \frac{x^4}{4!} - \frac{x^6}{6!} + \cdots$$

$$\sinh x = x + \frac{x^3}{3!} + \frac{x^5}{5!} + \frac{x^7}{7!} + \cdots$$

$$\cosh x = 1 + \frac{x^2}{2!} + \frac{x^4}{4!} + \frac{x^6}{6!} + \cdots$$

$$\tan^{-1} x = x - \frac{x^3}{3} + \frac{x^5}{5} - \frac{x^7}{7} + \cdots - 1 \le x \le 1$$

$$(1 + x)^p = 1 + \binom{p}{1}x + \binom{p}{2}x^2 + \binom{p}{3}x^3 + \cdots - 1 < x < 1$$

$$\binom{p}{k} = \frac{p(p - 1)(p - 2)\cdots(p - k + 1)}{k!}$$

Table of Integrals

1 $\displaystyle\int u\, dv = uv - \int v\, du$ **2** $\displaystyle\int u^n\, du = \frac{1}{n+1}\, u^{n+1} + C$ if $n \ne -1$ **3** $\displaystyle\int \frac{du}{u} = \ln|u| + C$ **4** $\displaystyle\int e^u\, du = e^u$

5 $\displaystyle\int a^u\, du = \frac{a^u}{\ln a} + C$ **6** $\displaystyle\int \sin u\, du = -\cos u + C$ **7** $\displaystyle\int \cos u\, du = \sin u + C$

8 $\displaystyle\int \sec^2 u\, du = \tan u + C$ **9** $\displaystyle\int \csc^2 u\, du = -\cot u + C$ **10** $\displaystyle\int \sec u \tan u\, du = \sec u + C$

11 $\displaystyle\int \csc u \cot u\, du = -\csc u + C$ **12** $\displaystyle\int \tan u\, du = \ln|\sec u| + C$ **13** $\displaystyle\int \cot u\, du = \ln|\sin u| + C$

14 $\displaystyle\int \sec u\, du = \ln|\sec u + \tan u| + C$ **15** $\displaystyle\int \csc u\, du = \ln|\csc u - \cot u| + C$ **16** $\displaystyle\int \frac{du}{\sqrt{a^2 - u^2}} = \sin^{-1} \frac{u}{a} + C$

17 $\displaystyle\int \frac{du}{a^2 + u^2} = \frac{1}{a} \tan^{-1} \frac{u}{a} + C$ **18** $\displaystyle\int \frac{du}{a^2 - u^2} = \frac{1}{2a} \ln\left|\frac{u+a}{u-a}\right| + C$ **19** $\displaystyle\int \frac{du}{u\sqrt{u^2 - a^2}} = \frac{1}{a} \sec^{-1}\left|\frac{u}{a}\right| + C$

TRIGONOMETRIC FORMS

20 $\displaystyle\int \sin^2 u\, du = \frac{1}{2}u - \frac{1}{4}\sin 2u + C$ **21** $\displaystyle\int \cos^2 u\, du = \frac{1}{2}u + \frac{1}{4}\sin 2u + C$ **22** $\displaystyle\int \tan^2 u\, du = \tan u - u + C$

23 $\displaystyle\int \cot^2 u\, du = -\cot u - u + C$ **24** $\displaystyle\int \sin^3 u\, du = -\frac{1}{3}(2 + \sin^2 u)\cos u + C$

25 $\displaystyle\int \cos^3 u\, du = \frac{1}{3}(2 + \cos^2 u)\sin u + C$ **26** $\displaystyle\int \tan^3 u\, du = \frac{1}{2}\tan^2 u + \ln|\cos u| + C$

27 $\displaystyle\int \cot^3 u\, du = -\frac{1}{2}\cot^2 u - \ln|\sin u| + C$ **28** $\displaystyle\int \sec^3 u\, du = \frac{1}{2}\sec u \tan u + \frac{1}{2}\ln|\sec u + \tan u| + C$

29 $\displaystyle\int \csc^3 u\, du = -\frac{1}{2}\csc u \cot u + \frac{1}{2}\ln|\csc u - \cot u| + C$

30 $\displaystyle\int \sin au \sin bu\, du = \frac{\sin(a-b)u}{2(a-b)} - \frac{\sin(a+b)u}{2(a+b)} + C$ if $a^2 \ne b^2$

31 $\displaystyle\int \cos au \cos bu\, du = \frac{\sin(a-b)u}{2(a-b)} + \frac{\sin(a+b)u}{2(a+b)} + C$ if $a^2 \ne b^2$

32 $\displaystyle\int \sin au \cos bu\, du = -\frac{\cos(a-b)u}{2(a-b)} - \frac{\cos(a+b)u}{2(a+b)} + C$ if $a^2 \ne b^2$

33 $\displaystyle\int \sin^n u\, du = -\frac{1}{n}\sin^{n-1} u \cos u + \frac{n-1}{n}\int \sin^{n-2} u\, du$ **34** $\displaystyle\int \cos^n u\, du = \frac{1}{n}\cos^{n-1} u \sin u + \frac{n-1}{n}\int \cos^{n-2} u\, du$

35 $\displaystyle\int \tan^n u\, du = \frac{1}{n-1}\tan^{n-1} u - \int \tan^{n-2} u\, du$ if $n \ne 1$ **36** $\displaystyle\int \cot^n u\, du = \frac{-1}{n-1}\cot^{n-1} u - \int \cot^{n-2} u\, du$ if $n \ne 1$

37 $\displaystyle\int \sec^n u\, du = \frac{1}{n-1}\sec^{n-2} u \tan u + \frac{n-2}{n-1}\int \sec^{n-2} u\, du$ if $n \ne 1$

38 $\displaystyle\int \csc^n u\, du = \frac{-1}{n-1}\csc^{n-2} u \cot u + \frac{n-2}{n-1}\int \csc^{n-2} u\, du$ if $n \ne 1$

39a $\displaystyle\int \sin^n u \cos^m u\, du = -\frac{\sin^{n-1} u \cos^{m+1} u}{n+m} + \frac{n-1}{n+m}\int \sin^{n-2} u \cos^m u\, du$ if $n \ne -m$

39b $\displaystyle\int \sin^n u \cos^m u\, du = \frac{\sin^{n+1} u \cos^{m-1} u}{n+m} + \frac{m-1}{n+m}\int \sin^n u \cos^{m-2} u\, du$ if $m \ne -n$

40 $\displaystyle\int u \sin u\, du = \sin u - u \cos u + C$ **41** $\displaystyle\int u \cos u\, du = \cos u + u \sin u + C$

42 $\displaystyle\int u^n \sin u\, du = -u^n \cos u + n \int u^{n-1} \cos u\, du$ **43** $\displaystyle\int u^n \cos u\, du = u^n \sin u - n \int u^{n-1} \sin u\, du$

$$\frac{u}{2}\sqrt{u^2 \pm a^2} \pm \frac{a^2}{2}\ln|u + \sqrt{u^2 \pm a^2}| + C \qquad 45 \int \frac{du}{\sqrt{u^2 \pm a^2}} = \ln|u + \sqrt{u^2 \pm a^2}| + C$$

$$u = \sqrt{u^2 + a^2} - a\ln\left(\frac{a + \sqrt{u^2 + a^2}}{u}\right) + C \qquad 47 \int \frac{\sqrt{u^2 - a^2}}{u}\,du = \sqrt{u^2 - a^2} - a\sec^{-1}\frac{u}{a} + C$$

$$\overline{a^2}\,du = \frac{u}{8}(2u^2 \pm a^2)\sqrt{u^2 \pm a^2} - \frac{a^4}{8}\ln|u + \sqrt{u^2 \pm a^2}| + C$$

$$\frac{u}{\pm a^2} = \frac{u}{2}\sqrt{u^2 \pm a^2} \mp \frac{a^2}{2}\ln|u + \sqrt{u^2 \pm a^2}| + C \qquad 50 \int \frac{du}{u^2\sqrt{u^2 \pm a^2}} = \mp \frac{\sqrt{u^2 \pm a^2}}{a^2 u} + C$$

$$\frac{\pm a^2}{u^2}\,du = -\frac{\sqrt{u^2 \pm a^2}}{u} + \ln|u + \sqrt{u^2 \pm a^2}| + C \qquad 52 \int \frac{du}{(u^2 \pm a^2)^{3/2}} = \frac{\pm u}{a^2\sqrt{u^2 \pm a^2}} + C$$

$$(u^2 \pm a^2)^{3/2}\,du = \frac{u}{8}(2u^2 \pm 5a^2)\sqrt{u^2 \pm a^2} + \frac{3a^4}{8}\ln|u + \sqrt{u^2 \pm a^2}| + C$$

ORMS INVOLVING $\sqrt{a^2 - u^2}$

$$54 \int \sqrt{a^2 - u^2}\,du = \frac{u}{2}\sqrt{a^2 - u^2} + \frac{a^2}{2}\sin^{-1}\frac{u}{a} + C \qquad 55 \int \frac{\sqrt{a^2 - u^2}}{u}\,du = \sqrt{a^2 - u^2} - a\ln\left|\frac{a + \sqrt{a^2 - u^2}}{u}\right| + C$$

$$56 \int \frac{u^2\,du}{\sqrt{a^2 - u^2}} = -\frac{u}{2}\sqrt{a^2 - u^2} + \frac{a^2}{2}\sin^{-1}\frac{u}{a} + C$$

$$57 \int u^2\sqrt{a^2 - u^2}\,du = \frac{u}{8}(2u^2 - a^2)\sqrt{a^2 - u^2} + \frac{a^4}{8}\sin^{-1}\frac{u}{a} + C$$

$$58 \int \frac{du}{u^2\sqrt{a^2 - u^2}} = -\frac{\sqrt{a^2 - u^2}}{a^2 u} + C \qquad 59 \int \frac{\sqrt{a^2 - u^2}}{u^2}\,du = -\frac{\sqrt{a^2 - u^2}}{u} - \sin^{-1}\frac{u}{a} + C$$

$$60 \int \frac{du}{u\sqrt{a^2 - u^2}} = -\frac{1}{a}\ln\left|\frac{a + \sqrt{a^2 - u^2}}{u}\right| + C \qquad 61 \int \frac{du}{(a^2 - u^2)^{3/2}} = \frac{u}{a^2\sqrt{a^2 - u^2}} + C$$

$$62 \int (a^2 - u^2)^{3/2}\,du = \frac{u}{8}(5a^2 - 2u^2)\sqrt{a^2 - u^2} + \frac{3a^4}{8}\sin^{-1}\frac{u}{a} + C$$

EXPONENTIAL AND LOGARITHMIC FORMS

$$63 \int u e^u\,du = (u - 1)e^u + C \qquad 64 \int u^n e^u\,du = u^n e^u - n\int u^{n-1}e^u\,du$$

$$65 \int \ln u\,du = u\ln u - u + C \qquad 66 \int u^n \ln u\,du = \frac{u^{n+1}}{n+1}\ln u - \frac{u^{n+1}}{(n+1)^2} + C$$

$$67 \int e^{au}\sin bu\,du = \frac{e^{au}}{a^2 + b^2}(a\sin bu - b\cos bu) + C \qquad 68 \int e^{au}\cos bu\,du = \frac{e^{au}}{a^2 + b^2}(a\cos bu + b\sin bu) + C$$

INVERSE TRIGONOMETRIC FORMS

$$69 \int \sin^{-1} u\,du = u\sin^{-1} u + \sqrt{1 - u^2} + C \qquad 70 \int \tan^{-1} u\,du = u\tan^{-1} u - \frac{1}{2}\ln(1 + u^2) + C$$

$$71 \int \sec^{-1} u\,du = u\sec^{-1} u - \ln|u + \sqrt{u^2 - 1}| + C \qquad 72 \int u\sin^{-1} u\,du = \frac{1}{4}(2u^2 - 1)\sin^{-1} u + \frac{u}{4}\sqrt{1 - u^2} + C$$

$$73 \int u\tan^{-1} u\,du = \frac{1}{2}(u^2 + 1)\tan^{-1} u - \frac{u}{2} + C \qquad 74 \int u\sec^{-1} u\,du = \frac{u^2}{2}\sec^{-1} u - \frac{1}{2}\sqrt{u^2 - 1} + C$$

$$75 \int u^n \sin^{-1} u\,du = \frac{u^{n+1}}{n+1}\sin^{-1} u - \frac{1}{n+1}\int \frac{u^{n+1}}{\sqrt{1 - u^2}}\,du + C \quad \text{if } n \neq -1$$

$$76 \int u^n \tan^{-1} u\,du = \frac{u^{n+1}}{n+1}\tan^{-1} u - \frac{1}{n+1}\int \frac{u^{n+1}}{1 + u^2}\,du + C \quad \text{if } n \neq -1$$

$$77 \int u^n \sec^{-1} u\,du = \frac{u^{n+1}}{n+1}\sec^{-1} u - \frac{1}{n+1}\int \frac{u^n}{\sqrt{u^2 - 1}}\,du + C \quad \text{if } n \neq -1$$